AF324927

SELECTED
PAPERS OF
WANG YUAN

SELECTED PAPERS OF
WANG YUAN

editor

Wang Yuan
Chinese Academy of Sciences, China

World Scientific

NEW JERSEY · LONDON · SINGAPORE · BEIJING · SHANGHAI · HONG KONG · TAIPEI · CHENNAI

Published by

World Scientific Publishing Co. Pte. Ltd.

5 Toh Tuck Link, Singapore 596224

USA office: 27 Warren Street, Suite 401-402, Hackensack, NJ 07601

UK office: 57 Shelton Street, Covent Garden, London WC2H 9HE

British Library Cataloguing-in-Publication Data
A catalogue record for this book is available from the British Library.

The editor and publisher would like to thank the publishers of the following periodicals for their assistance and permission to reproduce the articles found in this volume:

Acta Arith.
Acta Math. Sin.
Acta Math. Sin. (New Series)
Ann. Pol. Math.
Chin. Ann. Math.
J. Number Theory
Kexue Tongbao
Sci. Rec. (New Series)
Sci. Sin.
Sci. Sin. (Series A)
Shuxue Jinzhan

While every effort has been made to contact the publishers of reprinted papers prior to publication, we have not been successful in some cases. Where we could not contact the publishers, we have acknowledged the source of the material. Proper credit will be accorded to these publishers in future editions of this work after permission is granted.

SELECTED PAPERS OF WANG YUAN

ISBN 981-256-197-8

Printed in Singapore by B & JO Enterprise

PREFACE

Earlier in 1998, Professors C. D. Pan and L. Yang, and many of my friends and colleagues urged and encouraged me to publish a volume of my selected papers since most of these papers were published in China and hard to find elsewhere. The Hunan Normal Publisher and in particular, Ms S. H. Meng published a Chinese edition which appeared in 1999.

Now World Scientific Publishing Company and Dr K. K. Phua has invited me to publish an English version of my selected papers. This is really my honour and I accepted the offer. Ms R. T. Tan and Ms E. H. Chionh helped me with the editing work.

I have been working in Chinese Academy of Sciences since 1952 when I graduated from the Department of Mathematics, Zhe Jiang University. My teacher, Professor L. K. Hua, led me to the field of Number Theory. We also cooperated for a long time on the applications of number theory to numerical analysis. My other longtime collaborator is Professor K. T. Fang. Our joint work is in experimental designs. In order for the readers to understand my life and works, the volume contains a related paper of Professors W. L. Li and X. D. Yuan.

The readers can also see that my works are related and influenced by N. C. Ankeny, N. S. Bahvalov, A. Baker, V. Brun, A. A. Buchstab, D. A. Burgess, J. R. Chen, T. Cochrane, H. Davenport, P. Erdös, K. T. Fang, G. H. Hardy, L. K. Hua, N. M. Korobov, P. Kuhn, Yu. V. Linnik, J. E. Littlewood, T. Mitsui, C. D. Pan, W. M. Schmidt, A. Selberg, C. L. Siegel and H. Weyl.

Finally, I would like to take this opportunity to express my sincere thanks to those colleagues and institutions for their kind help.

Wang Yuan

CONTENTS

2. Numerical Analysis and Statistics

WANG YUAN: A BRIEF OUTLINE OF HIS LIFE AND WORKS

LI WENLIN AND YUAN XIANGDONG

Institute of Mathematics, Academia Sinica

Wang Yuan was born on 29 April 1930 in Lan-Xi county of Zhe-Jiang province. His father Wang Mao-Qing was the county magistrate. Their family moved to Si-Chuan province at the beginning of Anti-Japanese War on 1937. They lived in Yue-Lai Town, Jiang-Bei county of Chong-Qing municipality. At that time, Wang Yuan's family was quite poor and he was graduated from a rural primary school.

In 1942, Wang Yuan enrolled in National Second Middle School in He-Chuan county. At that time, his father was appointed to the position of Chief Secretary of the general office, Academia Sinica. Wang Yuan returned with his family to Nan-Jing and transferred to the middle school attached to Social Education College in 1946. The name of school was changed to Nan-Jing Sixth Middle School next year.

In 1948, Wang Yuan enrolled in the Department of Mathematics, Ying-Shi University, when he graduated from middle school. In 1949, Wang Yuan was transferred to the Department of Mathematics, Zhe-Jiang University when Ying-Shi University was merged into Zhe-Jiang University. Professors Chen Jian-Gong and Su Bu-Qing, the famous chinese mathematicians, had worked in Zhe-Jiang University since 1930; in particular, they organized the seminar for senior graduate students which is a good way to nurture their ability to work independently. Wang Yuan was very interested in mathematics in the good circumstances of Zhe-Jiang University. In his fourth grade, Wang Yuan gave a series of lectures in the seminars according to the famous book of A. E. Ingham, *The Distribution of Prime Numbers*, Camb. Tracts **30** (1932). He was deeply impressed by the beauty of Analytic Number Theory.

Wang Yuan graduated with excellent grade from Zhe-Jiang University in 1952, and he was assigned by the government to work at the Institute of Mathematics, Academia Sinica on the recommendation of professors Chen Jian-Gong and Su Bu-Qing. He then joined the Number Theory Section and worked with professor Hua Loo-Keng on Analytic Number Theory one year later.

Wang Yuan showed his mathematical ability soon after he entered the Number Theory Section. In 1954, Polish mathematician K. Kuratowski visited China and gave some reprints of W. Sierpinski and A. Schinzel on the distribution of arithmetic functions to Hua Loo-Keng. Wang Yuan made some improvements on their results very soon and his results with those of Schinzel were published in Poland in two joint papers with the same title: *A Note on some Properties of the Functions $\varphi(n)$, $\sigma(n)$ and $\theta(n)$*. Then Wang Yuan began to study C. Goldbach conjecture, and he established (2, 3) in 1957, *i.e.* every large even integer is a sum of a product of at most 2 primes and a product of at most 3 primes.

In 1958, Hua Loo-Keng and Wang Yuan cooperated to study the applications of number theory to numerical analysis, in particular, the numerical integration of multiple integrals. They proposed a method for numerical evaluation of multiple integrals based on classical algebraic number theory and diophantine approximations. Their cooperation on this field was fruitful, and had been prolonged for more than 20 years.

In 1966, Wang Yuan's research work was stopped, since the so-called cultural revolution started in China. He was subject to harassing interrogation and unjust criticisms. His mathematical research was resumed in 1972. He has studied the diophantine equations and inequalities in algebraic number fields, and the number-theoretic methods in statistics wich Fang Kai-Tai ever since. He has also studied the history of modern chinese mathematics since 1985.

Wang Yuan wrote a lot of monographs and books since 1978: *Applications of Number Theory to Numerical Analysis* (1978, with *Hua Loo-Keng*), *Goldbach conjecture* (1984), *Popularizing Mathematical Methods in the People's Republic of China* (1989, with *Hua Loo-Keng*), *Diophantine Equations and Inequalities in Algebraic Number Fields* (1991), *Number Theoretic Methods in Statistics* (1994, with *Fang Kai-Tai*). *Hua Loo-Keng* (1995) and *Calculus* (1997, with *Fong Yuen*).

Wang Yuan was promoted to the research member (professor) of the Institute of Mathematics, Academia Sinica in 1978 and elected as the member (academician) of Academia Sinica in 1980 owing to his achievements. He was given the honorary degree by Hong Kong Baptist University in 1998. He was awarded the first grade National Prize of Natural Sciences (with *Chen Jing-Run and Pan Cheng-Dong*), the prize of Chen Jia-Geng Physical Sciences (with *Hua Loo-Keng*) the He Liang He Li Mathematics Prize, the Hua Loo-Keng Mathematics Prize and the Wu Da You Prize for promoting sciences books.

Wang Yuan was appointed the position of director of the Institute of Mathematics, Academia Sinica in 1984, and elected the president of Chinese Mathematical Society in 1988, holding these positions from 1984 to 1987 and 1988 to 1992 respectively. He was the member of Chinese People's Political Consultative Conference from 1986 to 2002.

Mathematical Work

1. Number Theory

1) *Sieve Method and Goldbach Conjecture*

In a letter to L. Euler in 1742, Goldbach proposed two conjectures on the representations of integers as sum of primes. These conjectures may be stated as follows:

(A) *Every even integer > 5 is the sum of two odd primes.*
(B) *Every odd integer > 8 is the sum of three odd primes.*

Evidently (B) can be derived from (A). These two conjectures are still unsolved till now. However, there are great achievements on these conjectures by the applications of circle method and sieve method since 1920s.

The historical origin of sieve method may be traced back to the *Sieve of Eratosthenes*. V. Brun gave it an important improvement and applied his method successfully to the conjecture (A). He proved $(9, 9)$, *i.e.* every large even integer is the sum of two numbers each being a product of at most 9 primes. Similarly, we may define (a, b). There are some other improvements on sieve methods after Brun's work, for example, A. Selberg's sieve in 1947, A. A. Buchstab's identities for sieving functions in 1937 and P. Kuhn's weighted sieve in 1954. Brun's result was also improved to $(4, 4)$ (Buchstab, 1940) and (a, b) $(a + b \leq 6)$ (Kuhn, 1954) up to 1954.

By the combination of the above methods, Wang Yuan $[1, 7, 11]^*$ proved in 1956 and 1957 the following results:

$$(3, 4) \ (1956), \quad (2, 3) \ (1957). \tag{1}$$

The idea of sieve method may be sketched as follows: Consider the set $P_\alpha = \{m : 1 \leq m < n, \ p | m(n - m) \Rightarrow p > n^{1/\alpha}\}$, where $\alpha \geq 2, n$ is even and p denotes prime number. If we can prove that $P_{l+1} > 0$ when l is a positive integer and n is sufficiently large, then we obtain (l, l). Consider further $Q_\alpha = \{q : 1 < q < n, p | (n - q) \Rightarrow p > n^{1/\alpha}\}$, where q denotes prime number. Then $(1, l)$ follows from $Q_{l+1} > 0$.

T. Estermann was the first who proved in 1932 that every large even integer is the sum of a prime and a product of at most 6 primes under the assumption of Generalised Riemann Hypothesis (GRH). We denote this result by $(1, 6)_R$.

Wang Yuan $[2, 5, 16]$ improved Estermann's result to

$$(1, 4)_R \ (1956), \quad (1, 3)_R \ (1957). \tag{2}$$

By the use of Brun's method, theory of prime number distribution and Yu. V. Linnik's large sieve, A. Renyi proved $(1, c)$ in 1948, where c is an absolute constant. The main part of Renyi's proof of his $(1, c)$ is the following mean

---

*The number within the square brackets refers to Sec. I of the list of publications in this volume (pp. 481–485).

value theorem: There exists a constant $\delta > 0$ such that

$$(M_\delta) \qquad \sum_{k \leq x^\delta} \max_{(l,k)=1} \left| \pi(x; k, l) - \frac{1}{\varphi(k)} \int_2^x \frac{dt}{\ln t} \right| = O\left(\frac{x}{(\ln x)^{c_1}} \right)$$

where c_1 is a constant ≥ 6, $\varphi(k)$ the Euler function and $\pi(x; k, l) = \sum_{\substack{p \leq x \\ p \equiv l \,(\mathrm{mod}\, k)}} 1$ in which primes over all primes $\leq x$. If the range of k in (M_δ) can be extended to $x^{1/2-\varepsilon}$, where ε is any positive number, then (M_δ) can be used instead of (GRH) in the proof of $(1,3)_R$. M. B. Barban (1961) and Pan Cheng-Dong (1962) established independently (M_δ) with $\delta = \frac{1}{6}-\varepsilon$ and $\delta = \frac{1}{3}-\delta$, respectively. Moreover, Pan Cheng-Dong derived $(1,5)$ from his $(M_\delta)(\delta = \frac{1}{3} - \varepsilon)$. In 1963, they proved independently (M_s) with $\delta = \frac{3}{8} - \varepsilon$ and derived $(1,4)$.

Wang Yuan [16] pointed out that $(1,4)$ can be derived from Pan Cheng-Dong's $(M_\delta)(\delta = \frac{1}{3} - \varepsilon)$.

In 1965, E. Bombieri and A. I. Vinogradov proved that the range of k in (M_δ) can be extended to $k \leq x^{1/2}/(\ln x)^A$ and $x^{1/2-\varepsilon}$ respectively, where A is a constant > 0. Consequently, we have $(1,3)$ by the argument of Wang Yuan's proof of $(1,3)_R$. Finally in 1966, Chen Jing-Run improved the argument of $(1,3)_R$, and established the so-called Chen's theorem $(1,2)$. In 1975, Ding Xia-Xi, Pan Cheng-Dong and Wang Yuan [39, 40] gave a simplified proof of Chen's theorem.

Wang Yuan [6, 12] also applied the method for treating the conjecture (A) to the problems of distribution of almost primes in a short interval, and in the sequence $\{F(x) : x = 1, 2, \ldots\}$, where $F(x)$ is an integral valued polynomial. The so-called almost prime is an integer whose number of prime factors does not exceed a fixed number. We denote by P_k the almost prime having at most k prime factors. Wang Yuan improved the previous results, in particular, he first treated the distribution of P_2 in a short interval. He proved that if x is sufficiently large, there is a P_2 such that

$$x < P_2 \leq x + x^{\frac{10}{17}}. \tag{3}$$

This leads to a series of works on the improvements of (3) later.

2) *Circle Method and Goldbach Conjecture*

The circle method has its genesis in a paper of G. H. Hardy and S. Ramanujan in 1918 concerned with the partition function and the problem of representing numbers as the sums of squares. More general, in a series of papers, Hardy and J. E. Littlewood developed systematically a new analytic method, the circle method, and applied it to Goldbach conjecture and G. Waring's problem. Under the assumption of (GRH), they established in essence the conjecture (B), *i.e.* every large odd integer is a sum of three prime numbers. Using his ingenious method on the estimation of trigonometrical sum with prime variable, I. M. Vinogradov removed the unproved (GRH) in the Hardy and Littlewood's proof of conjecture (B).

In 1951, Yu. V. Linnik applied circle method to treat conjecture (A) in short intervals. His investigation involved the density hypothesis of $\zeta(s)$: Let $0 \leq \nu \leq \frac{1}{2}, T > 0$, and $N(T, \nu)$ be the number of zeros of $\zeta(s)$ in the rectangle

$$\frac{1}{2} + \nu \leq \sigma \leq 1, \quad |t| \leq T.$$

The estimation

$$(DH) \quad N(T, \nu) \geq CT^{1-2\nu} \ln(T + 2), \quad 0 < \nu \leq \frac{1}{2}$$

is called the density hypothesis which is a consequence of (RH). It is still an open problem now. Under the assumption of (DH), Linnik proved in 1951 that for any given integer n and positive number ε, there exist prime numbers p_1 and p_2 such that

$$|n - p_1 - p_2| \leq c(\varepsilon)(\ln n)^{7+\varepsilon},$$

where $c(\varepsilon)$ is a constant depending on ε, but not always with the same value in different occurrences.

In 1977, Wang Yuan [43] pointed out that there is a gap in the proof of Linnik's result, and he gave also a correct proof. More precise, under a weaker assumption

$$(DH)' \quad N(T, \nu) \leq c(\varepsilon)T^{1-2\nu} \ln(T + 2) \quad 0 < \nu \leq \frac{12}{37} + \varepsilon, \varepsilon > 0,$$

he proved that for any given integer n, there are primes p_1, p_2 such that

$$|n - p_1 - p_2| \leq c(\varepsilon)(\ln n)^{\frac{148}{13}+\varepsilon}. \tag{4}$$

Wang Yuan and Shan Zun [58] also proved the following result: Let q and n be two positive integers such that $q \leq \frac{n}{c(\ln n)^2}$ and n is even if q is even. Then under the assumption of GRH, the equation

$$n = p_1 + p_2 + hq$$

always has a solution in prime numbers p_1, p_2 and integer h satisfying $0 \leq h \leq c(\ln n)^2$. This gives an improvement of a result due to Linnik. In his original result, the range of h is $0 \leq h \leq c(\varepsilon)(\ln n)^{6+\varepsilon}(\varepsilon > 0)$.

3) Least Primitive Root Modulo a Prime p

The so-called primitive root modulo p is a generator of the reduced residue system $\{1, 2, \ldots, p - 1\}$ modulo p. Let $g(p)$ be the least positive primitive root modulo p. I. M. Vinogradov first proved in 1930 that

$$g(p) < 2^m p^{1/2} \ln p,$$

where $m = \omega(p - 1)$ denotes the number of distinct prime factors of $p - 1$, and then he pushed his own result to $g(p) < 2^m p^{1/2} \ln \ln p$. Later, Hua Loo-Keng, P. Erdös, and Erdös and H. N. Shapiro proved

$$g(p) < 2^{m+1}p^{1/2}, \quad g(p) = O(p^{1/2}\ln^{17}p) \quad \text{and} \quad g(p) = O(m^c p^{1/2})$$

respectively, where c and the constant in O are absolute constants. As for the lower estimation of $g(p)$. P. Turan established that $g(p) = \Omega(\ln p)$. However, under the assumption of (GRH), N. C. Ankeny proved that $g(p) = O(2^m\ln^2 p\ln^2(2^m\ln^2 p))$.

Using D. A. Burgess' method, Wang Yuan [13, 14] (1959) and Burgess (1962) proved independently that

$$g(p) \le c(\varepsilon)p^{\frac{1}{4}+\varepsilon}. \tag{5}$$

Wang Yuan [13, 14] also improved the Ankeny's result to

$$g(p) = O(m^6\ln^2 p) \tag{6}$$

under the same assumption (GRH).

By the similar method, Wang Yuan [21] improved the previous records on the bounds of least positive nth non-residue modulo p, and the least solution of J. Pell's equation.

4) *Distribution of Arithmetic Functions*

By the use of Brun's method, Wang Yuan and A. Schinzel [3, 8] proved in 1956 the following result: For any given set of h non-negative integers $\mathbf{a} = (a_1, \ldots, a_h)$ and $\varepsilon > 0$, there exists a positive integer n such that

$$\left|\frac{\varphi(n+i)}{\varphi(n+i+1)} - a_i\right| < \varepsilon, \quad 1 \le i \le h. \tag{7}$$

More precise, there exist constant $c_0(\mathbf{a}, \varepsilon)$ and $X_0(\mathbf{a}, \varepsilon)$ such that in any interval $1 \le n \le X$, the number of n satisfying (7) is greater than $c_0 X/(\ln X)^{h+1}$ whenever $X > X_0$.

Previously Schinzel can only prove the existence of n such that (7) holds but he cannot give a bound for the member of integers in a given intervals with property (7).

By the combination of Brun's method and the large sieve of Linnik and Renyi, Wang Yuan [9] proved further that for any given non-negative integral vector $\mathbf{a}$ and $\varepsilon > 0$, there is a prime p such that

$$\left|\frac{\varphi(p+\nu)}{\varphi(p+\nu+1)} - a_\nu\right| < \varepsilon, \quad 1 \le \nu \le h. \tag{8}$$

More precise, there exists constants $c_1(\mathbf{a}, \varepsilon)$ and $X_1(\mathbf{a}, \varepsilon)$ such that in any interval $1 \le n \le X$, the number of primes satisfying (8) is not less than $c_1 X/(\ln X)^{h+2}\ln\ln X$ whenever $X > X_1$.

Similar results hold also if Euler function is replaced by some other arithmetic functions.

5) *Small Solutions of System of Diophantine Equations and Inequalities*

Given a vector with complex components $\mathbf{x} = (x_1, \ldots, x_s)$, let $|\mathbf{x}| = \max_{1 \geq i \geq s} |x_i|$. Given a form F, let $|F|$ be the maximum absolute value of its coefficients. With every form F of degree k, a form $\hat{F}(\mathbf{x}_1, \ldots, \mathbf{x}_k)$ is associated which is linear in each vector $\mathbf{x}_i (1 \leq i \leq k)$ and such that $F(\mathbf{x}) = \hat{F}(\mathbf{x}, \ldots, \mathbf{x})$.

W. M. Schmidt proved in 1980 the following result: Given integers $h \geq 1, m \geq 1$ and odd numbers $k_1, \ldots, k_h$ and given a positive number E, however large, there is a constant $c_0 = c_0(k_1, \ldots, k_h; m, E)$ as follows. If M is ≥ 1 and if $F_1, \ldots, F_h$ are forms with real coefficients of respective degree $k_1, \ldots, k_h$ in $\mathbf{x} = (x_1, \ldots, x_s)$, where $s \geq c_0$ then there are m linearly independent points $\mathbf{x}(1), \ldots, \mathbf{x}(m)$ in $\mathbb{Z}^m$ with

$$|\mathbf{x}(i)| \leq M, \quad 1 \leq i \leq m$$

and

$$|\hat{F}_j(\mathbf{x}(i_1), \ldots, \mathbf{x}(i_{k_j}))| \ll M^{-E}|F_j|, \quad 1 \leq j \leq h, \ 1 \leq i_1, \ldots, i_{k_j} \leq m.$$

We can derive the following result from the above theorem: Given $h \geq 1, m \geq 1$ and odd numbers $k_1, \ldots, k_h$, and $\varepsilon > 0$, however small, there is a constant $c_1 = c_1(k_1, \ldots, k_h; m, \varepsilon)$ such that if $G_1, \ldots, G_h$ are forms of respective degrees $k_1, \ldots, k_h$ in $\mathbf{x} = (x_1, \ldots, x_s)$ with integral coefficients, where $x \geq c_1$, then $G_1, \ldots, G_h$ vanish on an m-dimensional subspace which is spanned by integer points $\mathbf{x}(1), \ldots, \mathbf{x}(m)$ having

$$|\mathbf{x}(i)| \ll G^\varepsilon, \quad 1 \leq i \leq m, \ G = \max(1, |G_1|, \ldots, G_h 1).$$

Taking a form $F = x_1^2 + \cdots + x_s^2$, we see immediately the restriction that the degrees of forms which are all odd cannot be removed in Schmidt's results. Wang Yuan solve the problem by applying the diophantine inequalities for forms in purely complex algebraic number fields so that there is no restriction on the degree of forms. Let K be a purely complex algebraic number field of degree $2r$ and J be its ring of integers. For $\xi \in K$, we denote by $\xi^{(i)} (1 \leq i \leq 2r)$ the conjugates of ξ and $\|\xi\| = \max_{1 \leq i \leq 2r} |\xi^{(i)}|$. Let $\|\boldsymbol{\lambda}\| = \max_{1 \leq j \leq s} \|\lambda_j\|$ for a vector $\boldsymbol{\lambda} = (\lambda_1, \ldots, \lambda_s)$.

In 1988, Wang Yuan [57, 59–61, 64] proved the following result: Given positive integers h, m and $k_1, \ldots, k_h$, and given a positive number E, however large, there is a constant $c_2 = c_2(k_1, \ldots, k_h; m, r, E)$ as follows: If $M \geq 1$ and $F_1, \ldots, F_h$ are forms with complex coefficients of respective degrees $k_1, \ldots, k_h$ in $\boldsymbol{\lambda} = (\lambda_1, \ldots, \lambda_s)$, where $s \geq c_2$, then there are m linearly independent points $\boldsymbol{\lambda}(1), \ldots, \boldsymbol{\lambda}(m)$ in J^s with

$$\|\boldsymbol{\lambda}(i)\| \leq M$$

and

$$|\hat{F}_j(\boldsymbol{\lambda}(i_1), \ldots, \boldsymbol{\lambda}(i_{k_j}))| \ll M^{-E}|F_j|, \quad 1 \leq i \leq h, \ 1 \leq i_1, \ldots, i_{k_j} \leq m.$$

If the coefficients of forms $F_1, \ldots, F_h$ are integers in J, we obtain similar result on the small solutions of a system of diophantine equations for forms.

For the problem of small solutions of a quadratic congruence modulo p, D. R. Heath-Brown proved in 1985 the following result: Let $Q(\mathbf{x}) = Q(x_1, \ldots, x_4)$ be a quadratic form with integral coefficients and p be a prime number. Then the congruence $Q(\mathbf{x}) \equiv 0 \pmod{p}$ has a solution $\mathbf{x}$ satisfying

$$0 < \max_{1 \leq i \leq 4} |x_i| \ll \sqrt{p} \ln p.$$

Todd Cochrane improved the right-hand side of the above inequality to $\sqrt{p}$ which is of best possible result.

Wang Yuan [66, 77] generalised in 1989 and 1993 the above two results to any finite fields. Wang Yuan [76] also studied the bound for small solutions of a system of congruences in number fields.

2. Numerical Analysis and Statistics

6) *Number Theoretic Methods in Numerical Analysis*

The theoretic basis of number-theoretic methods in numerical analysis is the theory of uniform distribution of H. Weyl. The main idea is to find a set of points in space with lower discrepancy. We call this set of points the quasi-random numbers which may be used instead of random numbers in Monte Carlo method for statistical simulation. A satisfactory example of number theoretic method is to apply it to the problem of numerical evaluation of multiple integrals in high dimensional space:

Suppose that $f(\mathbf{x}) = f(x_1, \ldots, x_s)$ is a function defined on the unit cube $G_s = \{\mathbf{x} : 0 \leq x_i \leq 1, \ 1 \leq i \leq s\}$. We want to evaluate the approximate value of the definite integral

$$I = \int_{G_s} f(\mathbf{x})d\mathbf{x}, \quad d\mathbf{x} = dx_1 \cdots dx_s.$$

Without loss of generality, we may assume that $f(\mathbf{x})$ is a periodic function on G_s and each variable of $f(\mathbf{x})$ has periodic 1. Suppose further that $f(\mathbf{x})$ has an absolutely convergent Fourier expansion

$$f(\mathbf{x}) = \sum_{-\infty}^{\infty} C(\mathbf{m})e^{2\pi i(\mathbf{m},\mathbf{x})}$$

$$= \sum_{m_1=-\infty}^{\infty} \cdots \sum_{m_s=-\infty}^{\infty} C(m_1, \ldots, m_s)e^{2\pi i(m_1 x_1 + \cdots + m_s x_s)}$$

where

$$|C(\mathbf{m})| \leq \frac{C}{\|\mathbf{m}\|^{\alpha}}, \quad \|\mathbf{m}\| = \bar{m}_1 \cdots \bar{m}_s, \quad \bar{m} = \max(1, |m|)$$

in which $\alpha > 1$ and $C > 0$ are absolute constants. The class of these functions is denoted by $E_s^\alpha(C)$.

N. M. Korobov (1959) and E. Hlawka (1962) proved independently that for any given prime number p, there exists an integral vector $\mathbf{a} = (a_1, \ldots, a_s)$ such that

$$\sup_{f \in E_s^\alpha(C)} \left| \int_{G_s} f(\mathbf{x}) d\mathbf{x} - \frac{1}{p} \sum_{k=-1}^{p} f\left(\frac{k\mathbf{a}}{p}\right) \right| = O\left(\frac{C(\ln p)^{\alpha s}}{p^\alpha}\right),$$

where the constant in O depends only on α's. The error term of this formula is much better than $O(n^{-\alpha/s})$ of Catesian product formula of classical one-dimensional quadrature formula and also the error $O(\frac{1}{\sqrt{n}})$ in the sense of probability in Monte Carlo method, where n denotes the number of points required in the quadrature techniques. Since the amount of elementary operations for obtaining $\mathbf{a}(p)$ is $O(p^2)$, it is important to find out a direct method to determine a vector $\mathbf{a}(N)$ depending on the number of points N.

Let F_n $(n = 0, 1, \ldots)$ be the L. P. Fibonacci sequence, that is, the sequence of integers defined by the recurrent formula $F_0 = F_1 = 1$, $F_{n+2} = F_{n+1} + F_n$ $(n \geq 0)$. N. S. Bahvalov (1959) and Hua Loo-Keng and Wang Yuan (1960) [15] established independently the following

$$\sup_{f \in E_s^2(C)} \left| \int_0^1 \int_0^1 f(x, y) dx \, dy - \frac{1}{F_n} \sum_{k=1}^{F_n} f\left(\frac{k}{F_n}, \frac{F_{n-1}k}{F_n}\right) \right| = O\left(\frac{C \ln 3F_n}{F_n^\alpha}\right). \quad (9)$$

Hua Loo-Keng and Wang Yuan proved also that the right-hand side of (9) is of best possible result. The amount of elementary operations for obtaining F_n is only $O(\ln 3F_n)$.

In a series of papers $[23, 24, 29, 32, 34, 37, 38]$ in 1964 to 1974, Hua Loo-Keng and Wang Yuan generalised their method to the case $s > 2$. Their method may be sketched as follows: Starting from a set of independent units of a totally real algebraic number field K of degree s, we may construct a set of units η_k $(k = 1, 2, \ldots)$ of K satisfying

$$\eta_k = \eta_k^{(1)} > k, \quad \eta_k^{(i)} \ll \eta_k^{\frac{1}{s-1}}, \qquad 2 \leq i \leq s.$$

Suppose that ω_j $(1 \leq j \leq s)$ is a basis of K, where ω_j $(1 \leq j \leq s-1)$ are irrational numbers. Then

$$\sum_{i=1}^{s} \eta_k^{(i)} = n_k \quad \text{and} \quad \sum_{i=1}^{s} \omega_j^{(i)} \eta_k^{(i)} = h_{jk} \quad 1 \leq j \leq s-1$$

are rational integers. Therefore we have the simultaneously rational approximations

$$\frac{h_{jk}}{n_k} = \omega_j + O\left(n_k^{-1-\frac{1}{s-1}}\right), \quad 1 \leq j \leq s-1.$$

Set $\mathbf{h}_k = (1, h_{1k}, \ldots, h_{s-1\,\varepsilon})$. This is the integral vector corresponding to n_k. We have the quadrature formula

$$\sup_{f \in E_s^\alpha(C)} \left| \int_{G_s} f(\mathbf{x})d\mathbf{x} - \frac{1}{n_k} \sum_{m=1}^{n_k} f\left(\frac{m\mathbf{h}_k}{n_k}\right) \right| = O\left(n_k^{-\frac{\alpha}{2} - \frac{\alpha}{2(s-1)} + \varepsilon}\right) \qquad (10)$$

where the constant in O depends on ε. The amount of elementary operations for obtaining $\mathbf{h}_k$ is only $O(\ln n_k)$. In practical manipulation, Hua Loo-Keng and Wang Yuan suggested the use of cyclotomic field $\mathbb{Q}(\cos \frac{2\pi}{m})(m \geq 5)$. They also gave another generalisation of (9) based on the C. Jacobi–O. Perron algorithm.

Let $\mathbf{r}_i = (\mathbf{r}_{1i}, \ldots, \mathbf{r}_{ti})$, $1 \leq i \leq s$ be real vectors, $\mathbf{q} = (q_1, \ldots, q_t)$ integral vector and $(\mathbf{x}, \mathbf{y}) = \sum_{i=1}^t x_i y_i$ the inner product of $\mathbf{x}$ and $\mathbf{y}$. Wang Yuan [52, 54, 55] proved in 1982 the following result: There exist $\mathbf{r}_1, \ldots, \mathbf{r}_s$ such that

$$\sup_{f \in E_s^2(C)} \left| \int_{G_s} f(\mathbf{x})d\mathbf{x} - \frac{1}{n^t} \sum_{\substack{q_k = -n \\ 1 \leq k \leq t}}^{n} \prod_{i=1}^{t} \left(1 - \frac{|q_i|}{n}\right) f((\mathbf{r}_1, \mathbf{q}), \ldots, (\mathbf{r}_s, \mathbf{q}_t)) \right|$$
$$= O(Cn^{-2t+\varepsilon}). \qquad (11)$$

There is also formula for the general $E_s^\alpha(C)(\alpha > 1)$, but the weights are more complicated. When $t = 1$, the results were obtained by Bahvalov, Haselgrove, Hua and Wang, Korobov and Niederreiter. The generalisation in rational form was obtained by I. H. Sloan later (see I. H. Sloan and S. Joe, *Lattice Method for Multiple Integration*, Oxford, 1994).

7) *Number Theoretic Methods in Statistics*

The problems in statistics often need the use of small sample of quasi-random numbers. In 1981, Wang Yuan and Fang Kai-Tai first found a set of small samples of quasi-random numbers and gave applications to the problems of experimental design: Suppose that there are s factors and each factor has q levels in an experiment. Then there are q^s distinct combinations among levels of s factors. Each combination may correspond to a lattice point in G_s. The main idea of number theoretic method is to find out a set U of $O(q)$ points among these q^s points which have lower discrepancy. Then we arrange the experiments on each point of U and use the regression analysis to obtain a better combination. This method is called uniform design, where the number of experiments is far less than the number of experiments $O(q^2)$ required in orthogonal array. As a result of the applications of uniform design in Chinese industrial departments, the precision of results is similar to that obtained by orthogonal design.

To apply number theoretic methods to the problems of statistics, we shall also need sets of points which are uniformly scattered (i.e. set with lower discrepancy) on some domains other than G_s. Wang Yuan and Fang Kai-Tai [67, 68] proposed in 1990 an algorithm to treat this problem: Starting from a set of n points which is uniformly scattered on G_s, we may derive a set of n points with lower discrepancy on any one of the following domains: $A_s = \{\mathbf{x} : 0 \le x_1 \le x_2 \le \cdots \le x_s \le 1\}$, $B_s = \{\mathbf{x} : x_1^2 + \cdots + x_s^2 \le 1\}$, $S_{s-1} = \{\mathbf{x} : x_1^2 + \cdots + x_s^2 = 1\}$ and $T_{s-1} = \{\mathbf{x} : x_1 t + \cdots + x_s = 1,\ x_i \ge 0,\ 1 \le i \le s\}$, and also a set of approximately n points which is uniformly scattered on the domain

$$T_{s-1}(\mathbf{a}, \mathbf{b}) = \{\mathbf{x} : x_1 + \cdots + x_s = 1, a_i \le x_i \le b_i, 1 \le i \le s\},$$

where $a_i \ge 0, 1 \le i \le s, \sum_{i=1}^{s} a_i < 1$ and $\sum_{i=1}^{s} b_i > 1$. A set on $T_{s-1}(\mathbf{a}, \mathbf{b})$ is, in fact, corresponding to an experimental design with mixture (see [79]).

The evaluation of probability and moment is reduced to the numerical evaluation of multiple integral, so we may try to use number theoretic method. Moreover, the number theoretic method can also be applied to the problems of optimization, of finding out a set of representative points for a multivariate distribution, and of statistical inference in statistical sciences.

3. Miscellaneous

8) Orthogonal Latin Squares

Let $(1, 2, \ldots, s)$ be arranged in an $s \times s$ square in such a way that every number occurs exactly once in every row and once in every column. Such a square is called a Latin square of order s. Two Latin squares are called orthogonal, if one of the squares is superposed on the other, every number of the first square occurs with every number of the second square once and only once. We denote by $N(s)$ the maximal number of mutually orthogonal Latin squares of order s. Euler proposed a conjecture that $N(s) = 0$ if $s > 6$ and $s \equiv 2 \pmod 4$. R. C. Bose, S. S. Shrikhande and E. T. Parker made great contribution on Euler conjecture. They proved in 1960 that $N(s) \ge 2$ when $s > 6$. Using their method and Brun's method, S. Chowla, Erdös and E. G. Straus proved that there exists a constant s_0 such that $N(s) > \frac{1}{3} s^{\frac{1}{91}}$ whenever $s > s_0$. Wang Yuan [22, 31] gave in 1964 the following improvement: There exists a number s_1 such that if $s > s_1$, we have

$$N(s) > s^{\frac{1}{26}}. \tag{12}$$

9) History of Modern Chinese Mathematics

Wang Yuan wrote a book "Hua Loo Keng". Besides the life and mathematical works of the great modern chinese mathematician Hua Loo-Keng, this book also describes some pieces of the process in which the modern mathematics was drawn into China from America, West Europe and Japan and its developments around Hua Loo-Keng as a centre.

For the references of this paper, we refer to the list of Publications by Wang Yuan.

References

[1] Li Wen-Lin and Wang Yuan, *The Biography of Modern Chinese Scientists*, Vol. 1, ed. Lu Jia-Xi (Science Press, 1991), pp. 81–88.

[2] Yuan Xiang-Dong, *Wang Yuan and Number Theory, The Eminent Contemporary Chinese Scientists in Mathematics and Information Sciences*, ed. Lu Jia-Xi (Hei Long-jiang Normal Publishing House, 1994), pp. 13–25.

ON THE REPRESENTATION OF LARGE EVEN INTEGER AS A SUM OF A PRODUCT OF AT MOST 3 PRIMES AND A PRODUCT OF AT MOST 4 PRIMES*

WANG YUAN

Institute of Mathematics, Academia Sinica

Received 8 July 1955

0. Introduction

V. Brun [1] first proved in 1920 the following result:

Every large even integer is the sum of two integers each being a product of at most 9 primes. We denote this theorem by $(9, 9)$, and we may define (a, b) similarly.

Brun's method and his result were improved by several mathematicians, namely:

$(7, 7)$ (Rademacher, 1924) [2]
$(6, 6)$ (Estermann, 1932) [3]
$(5, 7), (4, 9), (3, 15), (2, 366)$ (Ricci, 1937) [11]
$(5, 5)$ (Buchstab, 1938) [4]
$(4, 4)$ (Buchstab, 1940) [5]

Professor Hua Loo Keng pointed out that $(4, 4)$ may be possibly improved by the combination of the methods of Selberg [6], Brun and Buchstab. The purpose of this paper is to prove $(3, 4)$, i.e. the following:

Theorem 1. *Every large even integer can be represented as a sum of a product of at most 3 primes and a product of at most 4 primes.*

Theorem 2. *There are infinitely many integers n such that n has at most 3 prime factors and $n + 2$ has at most 4 prime factors.*

In this paper, we use $p, p', p'', \ldots, p_1, p_2, \ldots$ to denote prime numbers.

It seems possible to use the present method to prove $(3, 3)$ but it needs some complicated numerical calculations.

---

*Acta Mathematica Sinica, **6**:3 (1956) 500–513.

1. Some Computations

Lemma 1. *If $x \geq 1$ and $N \geq 1$, then*

$$\sum_{\substack{n \leq N \\ (n,x)=1}} \frac{|\mu(n)|2^{\Omega(n)}}{n} = \frac{1}{2} \prod_{p|x} \frac{p}{p+2} \prod_p \left(1 - \frac{1}{p}\right)^2 \left(1 + \frac{2}{p}\right) \log^2 N$$

$$+ O(\log 2N \cdot \log\log 3xN) + O((\log\log 3x)^2),$$

where $\mu(n)$ denotes the Möbius function and $\Omega(n)$ the number of distinct prime factors of n.

We refer [7] for the proof.

Lemma 2. *Let $z \geq 1$, $g(1) = 1$, $g(2) = 1/2$, $g(p) = 2/p(p > 2)$ and $g(n) = \prod_{p|n} g(p)$ for square free number n. Then*

$$\sum_{\substack{n \leq z \\ 2 \nmid n}} |\mu(n)|g(n) \prod_{p|n}(1 - g(p))^{-1} = \frac{1}{8} \prod_{p>2} \frac{(p-1)^2}{p(p-2)} \log^2 z + O(\log 2z \cdot \log\log 3z).$$

Proof. Set $\psi(q) = \prod_{p|q}(p - 2)$. Then

$$\sum_{\substack{2 \leq z \\ 2 \nmid n}} |\mu(n)|g(n) \prod_{p|n}(1 - g(p))^{-1}$$

$$= \sum_{\substack{2 \nmid n \\ n \leq z}} |\mu(n)| \frac{2^{\Omega(n)}}{n} \prod_{p|n} \frac{p}{p-2}$$

$$= \sum_{\substack{n \leq z \\ 2 \nmid n}} |\mu(n)| \frac{2^{\Omega(n)}}{n} \prod_{p|n} \left(1 + \frac{2}{p-2}\right)$$

$$= \sum_{\substack{n \leq z \\ 2 \nmid n}} |\mu(n)| \frac{2^{\Omega(n)}}{n} \sum_{r|n} \frac{2^{\Omega(r)}}{\psi(r)}$$

$$= \sum_{\substack{r \leq z \\ 2 \nmid r}} |\mu(r)| \frac{2^{2\Omega(r)}}{\psi(r)r} \sum_{\substack{s \leq z/r \\ (s,2r)=1}} \frac{|\mu(s)|2^{\Omega(s)}}{s}$$

$$= \sum_{\substack{r \leq z \\ 2 \nmid r}} |\mu(r)| \frac{2^{2\Omega(r)}}{\psi(r)r} \left\{ \frac{1}{2} \prod_p \frac{(p-1)^2(p+2)}{p^3} \prod_{p|2r} \frac{p}{p+2} \cdot \log^2 \frac{z}{r} + O(\log 2z \cdot \log\log 3z) \right\}$$

$$= \frac{1}{4} \prod_p \frac{(p-1)^2(p+2)}{p^3} \log^2 z \cdot \sum_{\substack{r \leq z \\ 2 \nmid r}} \frac{4^{\Omega(r)}|\mu(r)|}{\prod_{p|r}(p^2-4)}$$

$$+ O\left(\log 2z \cdot \sum_{\substack{r \leq z \\ 2 \nmid r}} \frac{4^{\Omega(r)}|\mu(r)| \log r}{\prod_{p|r}(p^2-4)} \right) + O(\log 2z \cdot \log\log 3z)$$

$$= \frac{1}{8} \prod_{p>2} \frac{(p-1)^2}{p(p-2)} \cdot \log^2 z + O(\log 2z \cdot \log\log 3z).$$

The lemma is proved.

Lemma 3. *Let $z \geq 1$ and*

$$f(n) = \sum_{d|n} \frac{\mu(d)}{g\left(\frac{n}{d}\right)} = \frac{1}{g(n)} \prod_{p|n}(1 - g(p)),$$

for square free number n. Then

$$\sum_{n \leq z} \frac{|\mu(n)|}{f(n)} = \frac{1}{4} \prod_{p>2} \frac{(p-1)^2}{p(p-2)} \log^2 z + O(\log 2z \cdot \log\log 3z).$$

Proof.

$$\sum_{n \leq z} \frac{|\mu(n)|}{f(n)} = \sum_{\substack{n \leq z \\ 2 \nmid n}} \frac{|\mu(n)|}{f(n)} + \sum_{\substack{n \leq z \\ 2 | n}} \frac{|\mu(n)|}{f(n)}$$

$$= \sum_{\substack{n \leq z \\ 2 \nmid n}} |\mu(n)|g(n) \prod_{p|n}(1 - g(p))^{-1} + \frac{1}{f(2)} \sum_{\substack{n \leq z/2 \\ 2 \nmid n}} |\mu(n)|g(n) \prod_{p|n}(1 - g(p))^{-1}$$

$$= \frac{1}{8} \prod_{p>2} \frac{(p-1)^2}{p(p-2)} \log^2 z + \frac{1}{8f(2)} \prod_{p>2} \frac{(p-1)^2}{p(p-2)} \log^2 \frac{z}{2} + O(\log 2z \cdot \log\log 3z)$$

$$= \frac{1}{4} \prod_{p>2} \frac{(p-1)^2}{p(p-2)} \log^2 z + O(\log 2z \cdot \log\log 3z).$$

Lemma 4. *Let α, β be two numbers such that $2 < \alpha < \beta$. Then*

$$\sum_{x^{1/\beta} < p \leq x^{1/\alpha}} \frac{1}{p \log^2 \frac{x}{p}} = \frac{1}{\log^2 x} \left\{ \log \frac{\beta - 1}{\alpha - 1} + \frac{1}{\alpha - 1} - \frac{1}{\beta - 1} \right\} + O\left(\frac{1}{\log^3 x} \right).$$

See [4] and [8].

2. Theorem A

Let

$$(w) \quad a = 0 \text{ or } 1, \qquad 0 \le a_i, b_i < p_i, \qquad a_i \ne b_i \quad (1 \le i \le r)$$

be a set of numbers, where $3 = p_1 < p_2 < \cdots < p_r \le \xi$ are all the odd prime numbers $\le \xi$. Let $P_w(x, \xi)$ be the number of integers satisfying the following conditions:

$$n \le x, \quad n \equiv a \,(\mathrm{mod}\,2), \quad n \not\equiv a_i \,(\mathrm{mod}\,p_i), \quad n \not\equiv b_i \,(\mathrm{mod}\,p_i) \,(1 \le i \le r). \qquad (1)$$

It follows by Sun Zi theorem (Chinese Remainder Theorems) that the systems of congruences

$$\begin{cases} y \equiv 1 + a \,(\mathrm{mod}\,2), \\ y \equiv a_i \quad (\mathrm{mod}\,p_i) \quad (1 \le i \le r); \end{cases} \qquad \begin{cases} y \equiv 1 + a \,(\mathrm{mod}\,2), \\ y \equiv b_i \quad (\mathrm{mod}\,p_i) \quad (1 \le i \le r), \end{cases}$$

have unique solutions a^*, b^* respectively in the interval $0 \le y < 2p_1 \cdots p_r$.

Now we proceed to show that the number of integers n satisfying (1) is equal to the number satisfying the following:

$$n \le x, \quad (n - a^*)(n - b^*) \not\equiv 0 \,(\mathrm{mod}\,p_i) \,(1 \le i \le r), \quad (n - a^*)(n - b^*) \not\equiv 0 \,(\mathrm{mod}\,2). \qquad (2)$$

In fact, if n satisfies (1), then

$$(n - a^*)(n - b^*) \equiv (n - a_i)(n - b_i) \not\equiv 0 \,(\mathrm{mod}\,p_i) \,(1 \le i \le r),$$
$$(n - a^*)(n - b^*) \equiv (a - 1 - a)^2 \equiv 1 \,(\mathrm{mod}\,2).$$

On the other hand, if n satisfies (2), then

$$(n - a_i)(n - b_i) \equiv (n - a^*)(n - b^*) \not\equiv 0 \,(\mathrm{mod}\,p_i) \,(1 \le i \le r),$$

i.e.

$$n \not\equiv a_i \,(\mathrm{mod}\,p_i), \quad n \not\equiv b_i \,(\mathrm{mod}\,p_i) \,(1 \le i \le r).$$

We also have

$$(n - 1 - a)^2 \equiv (n - a^*)(n - b^*) \not\equiv 0 \,(\mathrm{mod}\,2),$$

and thus $n \not\equiv 1 + a \,(\mathrm{mod}\,2)$, i.e. $n \equiv a \,(\mathrm{mod}\,2)$. Therefore n also satisfies (1).

Theorem A. *Let $c > 0$ and $P = \prod_{p \le \xi} p$. Then*

$$P_w(x, \xi) \le \frac{x}{\sum\limits_{\substack{1 \le k \le \xi^c \\ k|p}} \frac{\mu^2(k)}{f(k)}} + O\left(\sum_{\substack{1 \le k_1,\, k_2 \le \xi^c \\ k_1|P \\ k_2|P}} |\lambda_{k_1} \lambda_{k_2}| 2^{\Omega(k_1)} 2^{\Omega(k_2)} \right),$$

holds uniformly for any given (w), where $g(1) = 1$, $g(2) = 1/2$, $g(p) = 2/p(p > 2)$, $g(n) = \prod_{p|n} g(p)$ and $f(n) = \sum_{d|n} \mu(d) / g\left(\frac{n}{d}\right)$ for square free number n,

$$\lambda_n = \frac{\mu(n)}{g(n)f(n)} \sum_{\substack{1 \le m \le \xi^c/n \\ (n,m)=1 \\ m|P}} \frac{\frac{\mu^2(m)}{f(m)}}{\sum\limits_{\substack{1 \le l \le \xi^c \\ l|P}} \frac{\mu^2(l)}{f(l)}}.$$

Proof. Let $k|P$. Then

$$\sum_{\substack{k|(n-a^*)(n-b^*) \\ n \le x}} 1 = 2^{\Omega(k)-\Omega(k,2)} \left[\frac{x}{k}\right] + O(2^{\Omega(k)})$$

$$= \frac{2^{\Omega(k)-\Omega((k,2))}}{k} x + O(2^{\Omega(k)})$$

$$= g(k)x + O(2^{\Omega(k)}).$$

Since the number of integers satisfying (1) is equal to the number of integers satisfying (2), $\lambda_1 = 1$, and $\lambda_d = 0$ $(d > \xi^c)$, we have

$$P_w(x, \xi) = \sum_{\substack{n \le x \\ ((n-a^*)(n-b^*),P)=1}} 1 = \sum_{n \le x} \sum_{d|((n-a^*)(n-b^*),P)} \mu(d)$$

$$\le \sum_{n \le x} \left(\sum_{d|((n-a^*)(n-b^*),P)} \lambda_d \right)^2 = \sum_{\substack{d_1|P \\ d_1 \le \xi^c}} \sum_{\substack{d_2|P \\ d_2 \le \xi^c}} \lambda_{d_1} \lambda_{d_2} \sum_{\substack{\frac{d_1 d_2}{(d_1,d_2)} |(n-a^*)(n-b^*) \\ n \le x}} 1$$

$$= x \sum_{\substack{d_1|P \\ d_1 \le \xi^c}} \sum_{\substack{d_2|P \\ d_2 \le \xi^c}} \lambda_{d_1} \lambda_{d_2} g\left(\frac{d_1 d_2}{(d_1, d_2)} \right)$$

$$+ O\left(\sum_{\substack{d_1|P \\ d_1 \le \xi^c}} \sum_{\substack{d_2|P \\ d_2 \le \xi^c}} |\lambda_{d_1} \lambda_{d_2}| 2^{\Omega(d_1)+\Omega(d_2)} \right) = xQ + R.$$

When n is square free, we have

$$\frac{1}{g(n)} = \sum_{\tau|n} \frac{1}{g(\tau)} \sum_{d|n/\tau} \mu(d) = \sum_{d\tau|n} \frac{\mu(d)}{g(\tau)} = \sum_{k|n} \sum_{d|k} \frac{\mu(d)}{g\left(\frac{k}{d}\right)} = \sum_{k|n} f(k)$$

and

$$Q = \sum_{\substack{1 \le d_1 \le \xi^c \\ d_1 | P}} \sum_{\substack{1 \le d_2 \le \xi^c \\ d_2 | P}} \lambda_{d_1} \lambda_{d_2} \frac{g(d_1)g(d_2)}{g((d_1,d_2))}$$

$$= \sum_{\substack{1 \le d_1 \le \xi^c \\ d_1 | P}} \sum_{\substack{1 \le d_2 \le \xi^c \\ d_2 | P}} \lambda_{d_1} \lambda_{d_2} g(d_1) g(d_2) \sum_{d | (d_1,d_2)} f(d)$$

$$= \sum_{\substack{1 \le d \le \xi^c \\ d | P}} f(d) \left(\sum_{\substack{1 \le k \le \xi^c \\ d | k | P}} \lambda_k g(k) \right)^2 .$$

Let

$$S = \sum_{\substack{1 \le m \le \xi^c \\ m | p}} \frac{\mu^2(m)}{f(m)},$$

then

$$\lambda_k g(k) = \frac{1}{S} \sum_{\substack{1 \le m \le \xi^c/k \\ (m,k)=1 \\ m | p}} \frac{\mu(k)\mu^2(m)}{f(k)f(m)} = \frac{1}{S} \sum_{\substack{1 \le m \le \xi^c/k \\ (m,k)=1 \\ m | p}} \frac{\mu(mk)\mu(m)}{f(mk)},$$

$$\sum_{\substack{d | k | P \\ 1 \le k \le \xi^c}} \lambda_k g(k) = \frac{1}{S} \sum_{\substack{d | k | P \\ 1 \le k \le \xi^c}} \sum_{\substack{1 \le m \le \xi^c/k \\ m | P}} \frac{\mu(mk)\mu(m)}{f(mk)}$$

$$= \frac{1}{S} \sum_{\substack{1 \le r \le \xi^c \\ d | r | P}} \frac{\mu(r)}{f(r)} \sum_{d | k | r} \mu\left(\frac{r}{k}\right)$$

$$= \frac{1}{S} \frac{\mu(d)}{f(d)}.$$

Therefore

$$Q = \frac{1}{S}.$$

The theorem follows.

3. Applications of Theorem A

First, we estimate the second term of the inequality in Theorem A. Let $\xi > 3$ and $d(k) = \sum_{\tau | k} 1$. Then

$$2^{\Omega(k)} \le d(k) \quad \text{and} \quad \sum_{k \le \xi} d(k) = O(\xi \log \xi).$$

Hence

$$R = O\left(\sum_{\substack{k_1 \le \xi^c \\ k_1 | P}} \sum_{\substack{k_2 \le \xi^c \\ k_2 | P}} |\lambda_{k_1} \lambda_{k_2}| 2^{\Omega(k_1)} \cdot 2^{\Omega(k_2)} \right)$$

$$= O\left(\left(\sum_{\substack{1 \le k \le \xi^c \\ k | P}} |\lambda_k| 2^{\Omega(k)} \right)^2 \right) = O\left(\left(\sum_{\substack{1 \le k \le \xi^c \\ k | P}} \frac{|\mu(k)|}{|f(k)g(k)|} 2^{\Omega(k)} \right)^2 \right)$$

$$= O\left(\left(\sum_{1 \le k \le \xi^c} \frac{|\mu(k)| 2^{\Omega(k)}}{\prod_{2 < p \le \xi^c} \left(1 - \frac{2}{p} \right)} \right)^2 \right) = O\left(\log^4 \xi \cdot \left(\sum_{1 \le k \le \xi^c} d(k) \right)^2 \right)$$

$$= O(\xi^{2c} \log^6 \xi). \tag{3}$$

If l is an integer ≥ 1 and $l < c \le l+1$, then

$$\sum_{\substack{1 \le n \le \xi^c \\ n | p}} \frac{|\mu(n)|}{f(n)} = \sum_{1 \le n \le \xi^c} \frac{|\mu(n)|}{f(n)} - \sum_{\xi < p \le \xi^c} \sum_{\substack{1 \le n \le \xi^c \\ p | n}} \frac{|\mu(n)|}{f(n)}$$

$$+ \sum_{\substack{\xi < p_1 < p_2 \\ p_1 p_2 \le \xi^c}} \sum_{\substack{1 \le n \le \xi^c \\ p_1 p_2 | n}} \frac{|\mu(n)|}{f(n)} - \cdots + \cdots + (-)^l \sum_{\substack{\xi < p_1 < \cdots < p_l \\ p_1 p_2 \cdots p_l \le \xi^c}} \sum_{\substack{1 \le n \le \xi^c \\ p_1 \cdots p_l | n}} \frac{|\mu(n)|}{f(n)}$$

$$= \sum_{1 \le n \le \xi^c} \frac{|\mu(n)|}{f(n)} - \sum_{\xi < p \le \xi^c} \frac{1}{f(p)} \sum_{\substack{1 \le n \le \xi^c/p \\ (p,n)=1}} \frac{|\mu(n)|}{f(n)}$$

$$+ \sum_{\substack{\xi < p_1 < p_2 \\ p_1 p_2 \le \xi^c}} \frac{1}{f(p_1)f(p_2)} \sum_{\substack{1 \le n \le \xi^c/p_1 p_2 \\ (p_1 p_2, n)=1}} \frac{|\mu(n)|}{f(n)}$$

$$- \cdots + \cdots + (-1)^l \sum_{\substack{\xi < p_1 < \cdots < p_l \\ p_1 \cdots p_l \le \xi^c}} \frac{1}{f(p_1) \cdots f(p_l)} \sum_{\substack{1 \le n \le \xi^c/p_1 \cdots p_l \\ (n, p_1 \cdots p_l)=1}} \frac{|\mu(n)|}{f(n)}. \tag{4}$$

(i) Suppose $1 < c \le 2$. Since

$$\sum_{\xi < p \le \xi^c} \frac{1}{f(p)} - \sum_{\xi < p \le \xi^c} \frac{2}{p} = \sum_{\xi < p \le \xi^c} \left(\frac{2}{p-2} - \frac{2}{p} \right) = O\left(\sum_{n > \xi} \frac{1}{n^2} \right) = O\left(\frac{1}{\xi} \right),$$

it follows from (4) and Lemma 3 that

$$\sum_{\substack{1\le n\le \xi^c \\ n|P}} \frac{|\mu(n)|}{f(n)} = \sum_{1\le n\le \xi^c} \frac{|\mu(n)|}{f(n)} - \sum_{\xi<p\le \xi^c} \frac{2}{p} \sum_{1\le n\le \xi^c/p} \frac{|\mu(n)|}{f(n)} + O\left(\frac{\log^2\xi}{\xi}\right)$$

$$= \frac{1}{4}\prod_{p>2}\frac{(p-1)^2}{p(p-2)}\log^2\xi^c - \sum_{\xi<p\le \xi^c}\frac{2}{p}\cdot\frac{1}{4}\prod_{p>2}\frac{(p-1)^2}{p(p-2)}\log^2\frac{\xi^c}{p}$$

$$+ O(\log\xi\log\log\xi)$$

$$= \frac{1}{4}\prod_{p>2}\frac{(p-1)^2}{p(p-2)}\{(2c-1)^2 - 2c^2\log c\}\log^2\xi + O(\log\xi\log\log\xi),$$

where the following well-known formulas are used:

$$\sum_{p\le x}\frac{1}{p} = \log\log x + c_1 + O\left(\frac{1}{\log x}\right), \quad c_1 \text{ is a constant,}$$

$$\sum_{p\le x}\frac{\log p}{p} = \log x + O(1),$$

$$\sum_{p\le x}\frac{\log^2 p}{p} = \frac{1}{2}\log^2 x + O(\log x).$$

Take $\xi = x^{1/2c}/\log^5 x$, and $2c = d$. Then it follows by Theorem A and (3) that

$$P_w(x, x^{1/d}) \le P_w\left(x, \frac{x^{1/d}}{\log^5 x}\right)$$

$$\le 2e^{-2\gamma}\prod_{p>2}\left(1 - \frac{1}{(p-1)^2}\right)\cdot\Lambda(d)\frac{x}{\log^2 x} + O\left(\frac{x\log\log x}{\log^3 x}\right), \qquad (5)$$

where γ is Euler constant and

$$\Lambda(d) = 2e^{2\gamma}\left[\frac{d^2}{(d-1)^2 - 2\left(\frac{d}{2}\right)^2\log\frac{d}{2}}\right], \quad 2\le d\le 4. \qquad (6)$$

(ii) Suppose $2 < c \le 3$. Since

$$\sum_{\xi<p\le \xi^c}\frac{1}{f(p)}\sum_{1\le n\le \xi^c/p}\frac{|\mu(n)|}{f(n)} - \sum_{\xi<p\le \xi^c}\frac{1}{f(p)}\sum_{\substack{1\le n\le \xi^c/p \\ (p,n)=1}}\frac{|\mu(n)|}{f(n)}$$

$$= O\left(\sum_{\xi<p\le \xi^c}\frac{1}{f(p)}\sum_{\substack{1\le n\le \xi^c/p \\ p|n}}\frac{|\mu(n)|}{f(n)}\right) = O\left(\sum_{\xi<p\le \xi^c}\frac{1}{f(p)^2}\log^2\xi\right) = O\left(\frac{\log^2\xi}{\xi}\right)$$

and

$$\sum_{\substack{\xi<p<p'\\ pp'\leq\xi^c}} \frac{1}{f(p)f(p')} - \sum_{\substack{\xi<p<p'\\ pp'\leq\xi^c}} \frac{4}{pp'} = \sum_{\substack{\xi<p<p'\\ pp'\leq\xi^c}} \left(\frac{2}{p-2}\cdot\frac{2}{p'-2} - \frac{2}{p-2}\cdot\frac{2}{p'}\right)$$

$$+ \sum_{\substack{\xi<p<p'\\ pp'\leq\xi^c}} \left(\frac{2}{p-2}\cdot\frac{2}{p'} - \frac{4}{pp'}\right)$$

$$= O\left(\sum_{\xi<p\leq\xi^c}\frac{1}{p}\sum_{p'>\xi}\frac{1}{p'^2}\right) + O\left(\sum_{\xi<p'\leq\xi^c}\frac{1}{p'}\sum_{p>\xi}\frac{1}{p^2}\right)$$

$$= O\left(\frac{1}{\xi}\right),$$

we have

$$\sum_{\substack{n\leq\xi^c\\ n|P}} \frac{|\mu(n)|}{f(n)} = \sum_{n\leq\xi^c}\frac{|\mu(n)|}{f(n)} - \sum_{\xi<p\leq\xi^c}\frac{2}{p}\sum_{1\leq n\leq\xi^c/p}\frac{|\mu(n)|}{f(n)}$$

$$+ \sum_{\substack{\xi<p<p'\\ pp'\leq\xi^c}}\frac{4}{pp'}\sum_{1\leq n\leq\xi^c/pp'}\frac{|\mu(n)|}{f(n)} + O\left(\frac{\log^2\xi}{\xi}\right)$$

$$= \frac{1}{4}\prod_{p>2}\frac{(p-1)^2}{p(p-2)}\left[(2c-1)^2 - 2c^2\log c\right.$$

$$\left. + \left(\sum_{\substack{\xi<p<p'\\ pp'\leq\xi^c}}\frac{4}{pp'}\log^2\frac{\xi^c}{pp'}\right)\frac{1}{\log^2\xi}\right]\log^2\xi + O(\log\xi\log\log\xi).$$

By (4) and Lemma 3. Take $\xi = x^{1/2c}/\log^5 x$ and $d = 2c$. Then we have by Theorem A and (3) that (5) holds also for $4 < d \leq 6$, where

$$\Lambda(d) = 2e^{2\gamma}\left(\frac{d^2}{(d-1)^2 - 2\left(\frac{d}{2}\right)^2\log\frac{d}{2} + \delta\left(\frac{d}{2}\right)}\right), \quad 4\leq d\leq 6, \tag{7}$$

and

$$\delta(c) = 4\sum_{\substack{\xi<p<p'\\ pp'\leq\xi^c}}\frac{1}{pp'}\log^2\frac{\xi^c}{pp'}\bigg/\log^2\xi, \quad c\geq 2. \tag{8}$$

Similarly we have

$$\Lambda(d) = 2e^{2\gamma}\left(\frac{d^2}{(d-1)^2 - 2\left(\frac{d}{2}\right)^2\log\frac{d}{2} + \delta\left(\frac{d}{2}\right) - \kappa\left(\frac{d}{2}\right)}\right), \quad 6\leq d\leq 8, \tag{9}$$

where

$$\kappa(c) = 8 \sum_{\substack{pp'p'' \le \xi^c \\ \xi < p < p' < p''}} \frac{1}{pp'p''} \log^2 \frac{\xi^c}{pp'p''} \Big/ \log^2 \xi, \quad c \ge 3. \tag{10}$$

The case of $d > 8$ may also be treated.

4. Two Iteration Theorems

Theorem B$_1$. *If $\Lambda_k(\alpha)$ and $\lambda_i(\alpha)$ $(2 \le \alpha \le 15)$ are two increasing functions with at most finite number of discontinuities of first kind such that*

$$P_w(x, x^{1/\alpha}) > \lambda_i(\alpha)\frac{cx}{\log^2 x} + O\left(\frac{x}{\log^3 x} \log\log x\right), \quad 2 \le \alpha \le 15, \tag{11}$$

and

$$P_w(x, x^{1/\alpha}) < \Lambda_k(\alpha)\frac{cx}{\log^2 x} + O\left(\frac{x}{\log^3 x} \log\log x\right), \quad 2 \le \alpha \le 15 \tag{12}$$

holds uniformly for (w), where $c = 2e^{-2\gamma} \prod_{p>2} \left(1 - \frac{1}{(p-1)^2}\right)$ and γ denotes Euler constant, then

$$\psi(\alpha) = \begin{cases} 0, & 2 \le \alpha \le \tau, \\ \lambda_i(\beta) - 2 \int_{\alpha-1}^{\beta-1} \Lambda_k(z)\frac{z+1}{z}dz, & 2 \le \tau \le \alpha \le \beta \le 15, \end{cases}$$

and

$$w(\alpha) = \Lambda_k(\beta) - 2 \int_{\alpha-1}^{\beta-1} \lambda_i(z)\frac{z+1}{z^2}dz, \quad 2 \le \tau \le \alpha \le \beta \le 15$$

satisfies (11) and (12) respectively. We usually write $\psi(\alpha) = \lambda_{i+1}(\alpha)$ and $w(\alpha) = \Lambda_{k+1}(\alpha)$.

We refer to [4] for the proof.

Let

$$(w) \quad a = 0 \text{ or } 1, \quad 0 \le a_i < p_i \ (1 \le i \le r), \quad 0 \le b_j < p_j \ (1 \le j \le s),$$
$$a_v \ne b_v \quad (v = 1, 2, \ldots, \min(r, s)),$$

be a set of integers, where $\xi = p_1 < \cdots < p_r \le y$ are all the odd prime numbers $\le z$. Let $P_w(x, y, z)$ be the number of integers satisfying

$$n \le x, \quad n \equiv a \,(\mathrm{mod}\, 2), \quad n \not\equiv a_i \,(\mathrm{mod}\, p_i)\,(1 \le i \le r),$$
$$n \not\equiv b_i \,(\mathrm{mod}\, p_i)\,(1 \le i \le s). \tag{13}$$

In particular, we have $P_w(x, y, z) = P_w(x, y)$, for $y = z$.

Theorem B$_2$. *If the inequality*

$$P_w(x, x^{1/\alpha}) < \Lambda(\alpha)\frac{cx}{\log^2 x} + O\left(\frac{x}{\log^3 x}\log\log x\right), \quad 2 \le \alpha \le 15$$

holds uniformly for (w), where $\Lambda(d)$ is an increasing function which has at most finite number of discontinuities of the first kind, then for any given three positive numbers α, β, γ with $2 < \gamma \le \beta \le \alpha$, we have

$$P_w(x, x^{1/\gamma}, x^{1/\alpha}) > P_w(x, x^{1/\beta}, x^{1/\alpha}) - \Lambda\left(\frac{(\beta-1)\alpha}{\beta}\right) c\int_{\gamma-1}^{\beta-1} \frac{z+1}{z^2}dz \cdot \frac{x}{\log^2 x}$$
$$+ O\left(\frac{x}{\log^3 x}\log\log x\right).$$

Proof. If $p_m \ge p_r$, then the difference between $P_w(x, p_m, p_r)$ and $P_w(x, p_{m+1}, p_r)$ is equal to the number of integers satisfying

$$n \le x, \quad n \equiv a\,(\text{mod}\,2), \quad n \not\equiv b_i\,(\text{mod}\,p_i)\,(1 \le i \le r),$$
$$n \not\equiv a_j\,(\text{mod}\,p_j)\,(1 \le j \le m), \quad n \equiv a_{m+1}\,(\text{mod}\,p_{m+1}). \tag{14}$$

Denote by a_i^*, b_i^* and $\tilde{a}_m$ the respective solutions of the congruences

$$\begin{cases} p_{m+1}y + a_{m+1} \equiv a_i\,(\text{mod}\,p_i) & (0 \le y < p_i), \\ p_{m+1}y + a_{m+1} \equiv b_i\,(\text{mod}\,p_i) & (0 \le y \le p_i), \\ p_{m+1}y + a_{m+1} \equiv a\,(\text{mod}\,2) & (0 \le y \le 1). \end{cases}$$

Then $a_i^* \ne b_i^*$ $(1 \le i \le r)$ and therefore the number of integers satisfying (14) is equal to the number of integers satisfying

$$n \le \frac{x - a_{m+1}}{p_{m+1}}, \quad n \equiv \tilde{a}_m\,(\text{mod}\,2), \quad n \not\equiv a_i^*\,(\text{mod}\,p_i)\,(1 \le i \le m),$$
$$n \not\equiv b_i^*\,(\text{mod}\,p_i)\,(i \le r). \tag{15}$$

Let

$$(w_m)\quad 0 \le \tilde{a}_m \le 1, \quad 0 \le a_i^* < p_i\,(i \le m), \quad 0 \le b_j^* < p_j\,(j \le r).$$

Then the number of integers satisfying (15) is $P_{w_m}(x - a_{m+1}/p_{m+1}, p_m, p_r)$. Hence

$$P_w(x, p_m, p_r) - P_w(x, p_{m+1}, p_r) = P_{w_m}\left(\frac{x - a_{m+1}}{p_{m+1}}, p_m, p_r\right)$$
$$\le P_{w_m}\left(\frac{x}{p_{m+1}}, p_m, p_r\right) \le P_{w_m}\left(\frac{x}{p_{m+1}}, p_r\right). \tag{16}$$

Now let us arrange the prime numbers between $x^{1/\beta}$ and $x^{1/\gamma}$ as follows:

$$p_t \le x^{1/\beta} < p_{t+1} < \cdots < p_s \le x^{1/\gamma} < p_{s+1}.$$

We have

$$P_w(x, x^{1/\beta}, x^{1/\alpha}) = P_w(x, p_t, x^{1/\alpha}), \quad P_w(x, x^{1/\gamma}, x^{1/\alpha}) = P_w(x, p_s, x^{1/\alpha}).$$

Using (16) successively, we obtain

$$P_w(x, x^{1/\beta}, x^{1/\alpha}) \leq P_w(x, x^{1/\gamma}, x^{1/\alpha}) + \sum_{x^{1/\beta} < p_{i+1} \leq x^{1/\gamma}} P_{w_i}\left(\frac{x}{p_{i+1}}, x^{1/\alpha}\right)$$

$$= P_w(x, x^{1/\gamma}, x^{1/\alpha}) + \sum_{x^{1/\beta} < p_{i+1} \leq x^{1/\gamma}} P_{w_i}\left(\frac{x}{p_{i+1}}, \left(\frac{x}{p_{i+1}}\right)^{\frac{\log x^{1/\alpha}}{\log \frac{x}{p_{i+1}}}}\right)$$

$$\leq P_w(x, x^{1/\gamma}, x^{1/\alpha}) + \sum_{x^{1/\beta} < p_{i+1} \leq x^{1/\gamma}} P_{w_i}\left(\frac{x}{p_{i+1}}, \left(\frac{x}{p_{i+1}}\right)^{\frac{1}{(\beta-1)\alpha}{\beta}}\right)$$

$$\leq P_w(x, x^{1/\gamma}, x^{1/\alpha}) + \sum_{x^{1/\beta} < p_{i+1} \leq x^{1/\gamma}} \Lambda\left(\frac{(\beta-1)\alpha}{\beta}\right) \frac{cx}{p_{i+1} \log^2 \frac{x}{p_{i+1}}}$$

$$+ O\left(\sum_{x^{1/\beta} < p_{i+1} \leq x^{1/\gamma}} \frac{x}{p_{i+1} \log^3 \frac{x}{p_{i+1}}} \log\log \frac{x}{p_{i+1}}\right).$$

Therefore

$$P_w(x, x^{1/\beta}, x^{1/\alpha}) \leq P_w(x, x^{1/\gamma}, x^{1/\alpha}) + \Lambda\left(\frac{(\beta-1)\alpha}{\beta}\right)\left(\log\frac{\beta-1}{\gamma-1} + \frac{1}{\gamma-1} - \frac{1}{\beta-1}\right)$$

$$\times \frac{cx}{\log^2 x} + O\left(\frac{x}{\log^3 x} \log\log x\right)$$

$$= P_w(x, x^{1/\gamma}, x^{1/\alpha}) + \Lambda\left(\frac{(\beta-1)\alpha}{\beta}\right) \int_{\gamma-1}^{\beta-1} \frac{z+1}{z^2} dz \cdot \frac{cx}{\log^2 x}$$

$$+ O\left(\frac{x}{\log^3 x} \log\log x\right),$$

by Lemma 4, the theorem is proved.

5. Proofs of Theorems

Let $\lambda(\alpha)$ and $\Lambda(\alpha)$ be two increasing functions with at most finite number of discontinuities such that

$$\lambda(\alpha)\frac{cx}{\log^2 x} + O\left(\frac{x}{\log^3 x} \log\log x\right)$$

$$< P_w(x, x^{1/\alpha}) < \Lambda(\alpha)\frac{cx}{\log^2 x} + O\left(\frac{x}{\log^3 x} \log\log x\right), \quad 2 \leq \alpha \leq 15,$$

holds uniformly on (w). We denote these functions by $\lambda_0(\alpha)$, $\Lambda_0(\alpha)$, $\lambda_1(\alpha)$, $\Lambda_1(\alpha), \ldots$.

By (6), (7), (9) and Theorem B_1, we have

α	10	9	8	7	6	5	4
$\lambda_0(\alpha)$	99.98181*	79.78469	60.88817	43.51554	26.70925	9.18109	0
$\Lambda_0(\alpha)$	100.02073*	82.7207	68.52511	54.39352	43.0082	34.89666	29.39023

Divide the interval $\alpha - 1 < x \leq \beta - 1$ into n subintervals $u_i < x \leq u_{i+1}$ ($i = 0, 1, \ldots, n - 1$), where $u_0 = \alpha - 1$ and $u_n = \beta - 1$. Since $\lambda(\alpha)$ and $\Lambda(\alpha)$ are increasing functions, we have

$$\int_{\alpha-1}^{\beta-1} \lambda(z) \frac{z+1}{z^2} dz \geq \sum_{s=0}^{n-1} \lambda(u_s) \int_{u_s}^{u_{s+1}} \frac{z+1}{z^2} dz,$$

and

$$\int_{\alpha-1}^{\beta-1} \Lambda(z) \frac{z+1}{z^2} dz \leq \sum_{s=0}^{n-1} \Lambda(u_{s+1}) \int_{u_s}^{u_{s+1}} \frac{z+1}{z^2} dz.$$

Take $u_{s+1} - u_s = 0.02$. Starting from $\lambda_0(\alpha)$ and $\Lambda_0(\alpha)$, we obtain the following table by Theorem B_1 with several iterations

α	10	9	$\cdots$	6	5
$\lambda_i(\alpha)$	99.98181*	80.71187		29.28627	11.75811
$\Lambda_i(\alpha)$	100.02073*	81.36441		43.0082	34.89666

Therefore by Theorem B_2,

$$P_w(x, x^{1/4}, x^{1/5}) > P_w(x, x^{\frac{1}{4.2}}, x^{1/5}) - \Lambda \left(\frac{3.2 \times 5}{4.2} \right) \int_3^{3.2} \frac{z+1}{z} dz \cdot \frac{cx}{\log^2 x}$$

$$+ O\left(\frac{x}{\log^3 x} \log \log x \right)$$

$$> P_w(x, x^{1/5}, x^{1/5}) - \left\{ \Lambda \left(\frac{3.2 \times 5}{4.2} \right) \int_3^{3.2} \frac{z+1}{z^2} dz \right.$$

$$+ \Lambda \left(\frac{3.4 \times 5}{4.4} \right) \int_{3.2}^{3.4} \frac{z+1}{z^2} dz + \Lambda \left(\frac{3.6 \times 5}{4.6} \right) \int_{3.4}^{3.6} \frac{z+1}{z^2} dz$$

$$+ \Lambda \left(\frac{3.8 \times 5}{4.8} \right) \int_{3.6}^{3.8} \frac{z+1}{z^2} dz + \Lambda(4) \int_{3.8}^{4} \frac{z+1}{z^2} dz \left. \right\} \frac{cx}{\log^2 x}$$

$$+ O\left(\frac{x}{\log^3 x} \log \log x \right)$$

$$> (11.75811 - 10.75728) \frac{cx}{\log^2 x} + O\left(\frac{x}{\log^3 x} \log \log x \right)$$

$$= 1.00083 \frac{cx}{\log^2 x} + O\left(\frac{x}{\log^3 x} \log \log x \right). \tag{17}$$

---

*$\Lambda(10) = 100.02073$ and $\lambda(10) = 99.98181$ are taken from Buchstab [5].

(a) Suppose that x is an even number. Let $a = 1, a_i = 0, b_i = x \pmod{p_i}$ $(i = 1, 2, \ldots)$. Then it follows by (17) that there exists a constant x_0 such that

$$P_w(x, x^{1/4}, x^{1/5}) = \sum_{\substack{n \leq x \\ p|n \Rightarrow p > x^{1/4} \\ p|(x-n) \Rightarrow p > x^{1/5}}} 1 > \frac{cx}{\log^2 x}, \quad x > x_0.$$

This shows that there exists an integer $n \leq x$ such that the prime divisors of n and $x = n$ are greater than $x^{1/4}$ and $x^{1/5}$ respectively, i.e. the respective numbers of prime factors of n and $x - n$ are at most 3 and 4. Theorem 1 is proved.

(b) Take $a = 1, a_i = 0, b_i = p_i - 2$ $(i = 1, 2, \ldots)$. Then (17) becomes

$$P_w(x, x^{1/4}, x^{1/5}) = \sum_{\substack{n \leq x \\ p|n \Rightarrow p > x^{1/4} \\ p|(n+2) \Rightarrow p > x^{1/5}}} 1 > \frac{cx}{\log^2 x}, \quad x > x_0.$$

Theorem 2 follows.

A. Selberg announced that some results might be possibly obtained by his method, for example, $(2, 3)$ in [9] and $(3, 3)$ in [10]. However, the proofs of these results did not appeared in the literature till now.

The present method can also be used to prove the following results which will be published in other papers.

Let $F(x)$ be an irreducible polynomial of degree k without any fixed prime divisor and $\pi(N; F(x))$ be the number of integers x in the interval $1 \leq x \leq N$ such that $F(x)$ are primes. Then we have the following results:

(a) There exist infinitely many x such that $F(x)$ is a product of at most $[2.1k]$ primes.

(b)

$$\pi(N; F(x)) \leq 2e^{\gamma} \mu_F \frac{N}{\log N} + o\left(\frac{N}{\log N}\right),$$

where γ is Euler constant, μ_F a constant depending on $F(x)$ and the constant implicit in "o" depending on $F(x)$ only.

Under the assumption of Generalised Riemann hypothesis (GRH), i.e. the assumption that the real parts of zeros of all Dirichlet's L-functions are $\leq 1/2$, we have the following two results:

(c) Every large even integer is a sum of a prime and a product of at most 4 primes.

(d) $(p, p + 2)$ is called a pair of twin primes, if p and $p + 2$ are all prime numbers. Let $Z_2(N)$ be the number of primes such that $p \leq N$ and $(p, p + 2)$ is twin primes. Then

$$Z_2(N) \le (8 + \varepsilon) \prod_{p>2} \left(1 - \frac{1}{(p-1)^2}\right) \frac{N}{\log^2 N} + O\left(\frac{N}{\log^3 N}\right),$$

where ε is any pre-assigned positive number and the constant implicit in "O" depends on ε only.

References

[1] V. Brun, de cribe d'Eratosthéne et le théorème de Goldbach, Videnskabs—selskabt: Kristiania Skrifter I, *Mate.—Naturvidenskapelig Klasse.* **3** (1920) 1–36.

[2] H. Rademacher, Beitrage zur Viggo Brunschen Methode in der Zahlen—Theorie, *Abh. Math. Sem. Hamburgischen Univ.* **3** (1924) 12–30.

[3] T. Estermann, Eine neue Darstellung und neue Anwendungen der Viggo Brunschen Methode, *J. Reine Angew. Math.* **168** (1932) 106–116.

[4] A. A. Buchstab, New improvements on Eratosthenes sieve method, *Math. Sb.* **4** (1938) 375–387.

[5] A. A. Buchstab, On representation of even number as a sum of two integers with bounded number of divisors, *Dokl. Akad. Nauk SSSR* **29** (1940) 544–548.

[6] A. Selberg, On an elementary method in the theory of primes, *Norske Vid. Selsk. Forhdl.* 19 No. **18** (1947) 64–67.

[7] H. N. Shapiro and J. Warga, On representation of large integers as sum of primes, part I. *Comm. Pure Appl. Math.* **3** (1950) 153–176.

[8] A. A. Buchstab, Asymptotic estimation of a general number theoretic function, *Math. Sb.* **2** (1937) 1239–1245.

[9] A. Selberg, The general sieve method and its place in prime number theory, *Proc. of the Int. Congr. Math.* **1** (1950) 286–292.

[10] A. Selberg, On elementary methods in prime number theory and their limitations, *Den 11ᴸᵗᵉ Skandinaviske Mat.* (1952) 13–22.

[11] G. Ricci, Su la congettura di Goldbach e la constante di schnirelmann, *Ann. della R. Scuola Norm. Sup. Pisa* (2) **6** (1937) 70–115.

ON SOME PROPERTIES OF INTEGRAL VALUED POLYNOMIALS*

WANG YUAN

Institute of Mathematics, Academia Sinica

Let $F(x)$ be an irreducible integral valued polynomial of degree k without any fixed prime divisor. Let $\pi(N; F(x))$ be the number of primes[a] represented by $F(x)$ for $1 \leq x \leq N$. Nagell [1] first proved that

$$\pi(N; F(x)) = o(N), \tag{1}$$

where the constant in "o" depends only on $F(x)$. Using Bruns's method, Heilbronn [2] established that

$$\pi(N; F(x)) = O\left(\frac{N}{\log N}\right), \tag{2}$$

where the constant in "O" depends only on $F(x)$, and then Ricci [3] gave the following improvement:

$$\varlimsup_{N \to \infty} \frac{\pi(N; F(x)) \log N}{N} < \frac{29}{4} \mu F, \tag{3}$$

where μ_F is a constant depending on $F(x)$ and its definition will be given later.

In this paper we shall use Selberg's method to prove the following:

Theorem 1. *We have*

$$\pi(N; F(x)) \leq 2e^{\gamma} \mu_F \frac{N}{\log N} + o\left(\frac{N}{\log N}\right), \tag{4}$$

where γ denotes Euler constant and the constant in "o" depends on $F(x)$ only.

Here $2e^{\gamma} < 3.564 < \frac{29}{4}$. It deserves mentioning that there requires complicated numerical calculations in the proof of Ricci's (3), but it does not need the numerical computations in the proof of Theorem 1.

---

*Shuxue Jinzhan, **3**:3 (1957) 416–423.

[a]In this paper, we use $p, p_1, p_2, \ldots$ to denote prime numbers.

To prove the theorem, we need the following three lemmas.

Let $h_m(k!\,m)$ be the number of solutions of the congruence

$$F(x) \equiv 0 \ (\mathrm{mod}\ m) \ (0 \le x < k!\,m)$$

and

$$\omega(m) = \frac{h_m(k!\,m)^{\mathrm{b}}}{k!}. \tag{5}$$

Lemma 1. $\omega(q)$ *is a multiplicative function of* q, *i.e.* $\omega(q_1)\omega(q_2) = \omega(q_1 q_2)$.

Proof. (i) Suppose that $q = q_1 q_2$, $(q_1, q_2) = 1$ and the prime divisor of q are all $> k$. Then $q | F(x)$ and $q | k!\,F(x)$ are equivalent. Since $k!\,F(x)$ is a polynomial of integral coefficients, $\omega(q)$ is also the number of solutions of the congruence

$$k!\,F(x) \equiv 0 \ (\mathrm{mod}\ q) \ (0 \le x < q),$$

and therefore

$$\omega(q) = \omega(q_1)w(q_2).$$

(ii) Suppose that $q = q_1 q_2$, where the prime divisors of q_1 and q_2 are all $> k$ and $\le k$ respectively. Let $x = b_m$ $(m = 1, 2, \ldots, h_{q_2}(k!\,q_2!))$ be the solutions of the congruence

$$F(x) \equiv 0 \ (\mathrm{mod}\ q_2) \ (0 \le x < k!\,q_2). \tag{6}$$

By (i), we may denote by $x = a_l$ $(l = 1, \ldots, \omega(q_1))$ the solutions of the congruence

$$F(x) \equiv 0 \ (\mathrm{mod}\ q_1) \ (0 \le x < q_1). \tag{7}$$

If follows by Sun Zi theorem that the system of congruences

$$\left. \begin{array}{l} x \equiv a_1 \ (\mathrm{mod}\ q_1) \\[2mm] x \equiv b_m \ (\mathrm{mod}\ k!\,q_2) \end{array} \right\} \tag{8}$$

has a unique solution

$$x \equiv c_n \ (\mathrm{mod}\ k!\,q_1 q_2) \ (0 \le c_n < k!\,q_1 q_2). \tag{9}$$

Evidently

$$F(c_n) \equiv o \ (\mathrm{mod}\ q_1 q_2).$$

On the other hand, we have a unique pain (a_l, b_m) satisfying (6)–(8) for a given c_n and consequently

$$\omega(q_1)\omega(q_2) = \frac{h_{q_2}(k!\,q_2)\omega(q_1)}{k!} = \frac{h_{q_1 q_2}(k!\,q_1 q_2)}{k!} = \omega(q_1 q_2).$$

---

[b]Since $F(x)$ has no fixed prime divisor, we have $\omega(p) < p$.

(iii) Suppose that the prime divisors of q are all $\leq k$. Let $q = \prod_{i=1}^{t} p_i^{r_i}$ be the standard decomposition of q and $s_i = \sum_{l=1}^{\infty} \left[\frac{k}{p_i^l}\right]$. It is evident that

$$F(x) \equiv F(x + p_i^{r_i+s_i}) \,(\mathrm{mod}\, p_i^{r_i}).\tag{10}$$

In fact, the integral polynomial of degree k can be expressed by $F(x) = a_0\binom{x}{k} + a_1\binom{x}{k-1} + \cdots + a_k$, where a_i's are integers and $\binom{x}{k} = \frac{x(x+1)\cdots(x+k-1)}{k!}$.

Let $k_q(q\prod_{i=1}^{t} p_i^{s_i})$ and $k_{p^{r_i}}(p_i^{r_i+s_i})\,(1 \leq i \leq t)$ be the respective numbers of solutions of the congruences

$$F(x) \equiv 0 \,(\mathrm{mod}\, q) \quad \left(0 \leq x < q\prod_{i=1}^{t} p_i^{s_i}\right)$$

and

$$F(x) \equiv 0 \,(\mathrm{mod}\, p_i^{r_i}) \qquad (0 \leq x < p_i^{r_i+s_i}) \quad (1 \leq i \leq t).$$

Then it follows from Sun Zi theorem that

$$h_q\left(q\prod_{i=1}^{t} p_i^{s_i}\right) = \prod_{i=1}^{t} h_{p_i^{r_i}}(p_i^{r_i+s_i}).\tag{11}$$

And we have

$$\frac{k!}{\prod_{i=1}^{t} p_i^{s_i}}\, h_q\left(q\prod_{i=1}^{t} p_i^{s_i}\right) = h_q(k!\,q)$$

and

$$\frac{k!}{p_i^{s_i}}\, h_{p_i^{r_i}}(p_i^{r_i+s_i}) = h_{p_i^{r_i}}(k!\,p_i^{r_i}) \quad (1 \leq i \leq t)$$

by (10). Substituting into (11), we obtain

$$\omega(q) = \frac{h_q(k!\,q)}{k!} = \frac{h_q\left(q\prod_{i=1}^{t} p_i^{s_i}\right)}{\prod_{i=1}^{t} n_i^{s_i}} = \frac{\prod_{i=1}^{t} h_{p_i^{r_i}}(p_i^{r_i+s_i})}{\prod_{j=1}^{t} n_j^{s_j}}$$

$$= \prod_{i=1}^{t}\left(\frac{h_{p_i^{r_i}}(k!\,p_i^{r_i})}{k!}\right) = \omega(p_1^{r_1})\omega(p_2^{r_2})\cdots\omega(p_t^{r_t}).$$

The lemma follows by the combination of (i)–(iii).

Lemma 2. (Nagell [4])

$$\prod_{p\leq x}\left(1 - \frac{\omega(p)}{p}\right) = \frac{\mu_F}{\log x} + O\left(\frac{1}{\log^2 x}\right),$$

where the constant in "O" depends on $F(x)$ only.

Lemma 3. *If n is a square free number and $\omega(n) \neq 0$, we set $f(n) = \prod_{p|n} f(p)$, where $f(p) = \frac{p}{\omega(p)} - 1$. Then*

$$\sum_{\substack{n \leq z \\ \omega(n) \neq 0}} \frac{\mu^2(n)}{f(n)} = \prod_p \frac{\left(1 - \frac{1}{p}\right)}{\left(1 - \frac{\omega(p)}{p}\right)} \log z + o(\log z)$$

where the constant in "o" depends on $F(x)$ only.

Proof. Since $\omega(p) \leq k$ and $\omega(n) \leq k^{\Omega(n)} = O(n^\varepsilon)$ for any square free number n, where $\Omega(n)$ denotes the number of prime factors of n, ε is any preassigned positive number and the constant in "O" depends on ε and k only. The series $\sum_{\substack{n=1 \\ \omega(n) \neq 0}}^{\infty} \frac{\mu^2(n)}{f(n)n^s}$ is absolutely convergent if $s > 0$ and

$$\sum_{\substack{n=1 \\ \omega(n) \neq 0}}^{\infty} \frac{\mu^2(n)}{f(n)n^s} = \prod_{\omega(p) \neq 0} \left(1 + \frac{1}{f(p)p^s}\right) = \prod_{\omega(p) \neq 0} \left(1 + \frac{\omega(p)}{(p - \omega(p))p^s}\right)$$

$$= \prod_p \left(1 + \frac{\omega(p)}{(p - \omega(p))p^s}\right)\left(1 - \frac{1}{p^{s+1}}\right) \cdot \zeta(1 + s)$$

$$\sim \prod_p \frac{\left(1 - \frac{1}{p}\right)}{\left(1 - \frac{\omega(p)}{p}\right)} \cdot \frac{1}{s} \quad (\text{as } s \to 0^+). \tag{12}$$

By Lemma 2 and Merten's theorem, we have

$$\prod_p \frac{\left(1 - \frac{\omega(p)}{p}\right)}{\left(1 - \frac{1}{p}\right)} = \lim_{\xi \to \infty} \prod_{p \leq \xi} \frac{\left(1 - \frac{\omega(p)}{p}\right)}{\left(1 - \frac{1}{p}\right)} = \lim_{\xi \to \infty} \frac{\frac{\mu_F}{\log \xi} + O\left(\frac{1}{\log^2 \xi}\right)}{\frac{e^{-r}}{\log \xi} + O\left(\frac{1}{\log^2 \xi}\right)} = e^r \mu_F,$$

i.e.

$$\mu_F = e^{-r} \prod_p \frac{\left(1 - \frac{\omega(p)}{p}\right)}{\left(1 - \frac{1}{p}\right)}. \tag{13}$$

On the other hand, let $\alpha(t) = \sum_{\substack{n \leq z \\ \omega(n) \neq 0}} \frac{\mu^2(n)}{f(n)}$. Then

$$\sum_{\substack{n=1 \\ \omega(n) \neq 0}}^{\infty} \frac{\mu^2(n)}{f(n)n^2} = \int_1^{\infty} z^{-s} d\alpha(z) = \int_0^{\infty} e^{-st} d\alpha(e^t), \quad s > 0$$

and therefore it follows by (12) and a Tauberian theorem[c] that

$$a(e^t) \sim \prod_p \frac{\left(1 - \frac{1}{p}\right)}{\left(1 - \frac{\omega(p)}{p}\right)} \cdot t,$$

i.e.

$$\alpha(z) = \sum_{\substack{n \leq z \\ \omega(n) \neq 0}} \frac{\mu^2(n)}{f(n)} = \prod_p \frac{\left(1 - \frac{1}{p}\right)}{\left(1 - \frac{\omega(p)}{p}\right)} \log z + o(\log z).$$

The lemma is proved.

Proof (Theorem 1). Let $x > \xi > k$ and $P = \prod_{p \leq \xi} p$. Let $g(n) = \frac{\omega(n)}{n}$, $f(n) = \sum_{d|n} \frac{\mu(d)}{g(\frac{n}{d})}$ and

$$\lambda_n = \frac{\mu(n)}{g(n)f(n)} \sum_{\substack{1 \leq m \leq \xi/n \\ (m,n)=1 \\ m|P \\ \omega(m) \neq 0}} \frac{\mu^2(m)}{f(m)} \Big/ \sum_{\substack{1 \leq l \leq \xi \\ l|P \\ \omega(l) \neq 0}} \frac{\mu^2(l)}{f(l)}$$

for square free number n and $\omega(n) \neq 0$. Suppose $m|P$. Then

$$\sum_{\substack{m|F(n) \\ n \leq x}} 1 = \sum_{\substack{F(n) \equiv 0 (\mathrm{mod}\, m) \\ 1 \leq n \leq [\frac{x}{k!\, m}]k!\, m}} 1 + \theta_m h_m(k!\, m) \quad (0 \leq \theta_m \leq 1)$$

$$= \left[\frac{x}{k!\, m}\right] h_m(k!\, m) + \theta_m h_m(k!\, m) = \frac{x\omega(m)}{m} + \bar{\theta}_m k!\, \omega(m) \quad (|\bar{\theta}_m| \leq 1). \quad (14)$$

Since $\lambda_1 = 1$, we have

$$P(x,\xi) = \sum_{\substack{1 \leq n \leq x \\ (F(n),P)=1}} 1 = \sum_{n \leq x} \sum_{d|(F(n),P)} \mu(d) \leq \sum_{n \leq x} \left(\sum_{\substack{d|(F(n),P) \\ \omega(d) \neq 0}} \lambda_d \right)^2$$

$$= \sum_{\substack{d_1 \leq \xi \\ d_1|P \\ \omega(d_1) \neq 0}} \sum_{\substack{d_2 \leq \xi \\ d_2|P \\ \omega(d_2) \neq 0}}' \lambda_{d_1} \lambda_{d_2} \sum_{\substack{\frac{d_1 d_2}{(d_1,d)}|F(n) \\ n \leq x}} 1$$

$$\leq x \sum_{\substack{d_1 \leq \xi \\ d_1|P \\ \omega(d_1) \neq 0}} \sum_{\substack{d_2 \leq \xi \\ d_2|P \\ \omega(d_2) \neq 0}} \lambda_{d_1} \lambda_{d_2} \frac{g(d_1)g(d_2)}{g((d_1,d_2))}$$

$$+ 2 \cdot k! \sum_{\substack{d_1 \leq \xi \\ d_1|P \\ \omega(d_1) \neq 0}} \sum_{\substack{d_2 \leq \xi \\ d_2|P \\ \omega(d_2) \neq 0}} |\lambda_{d_1} \lambda_{d_2}| \omega\left(\frac{d_1 d_2}{(d_1,d_2)}\right) = xQ + R \quad (15)$$

---

[c]The following theorem is used: Let $\alpha(t)$ be an increasing function such that $f(s) = \int_0^\infty e^{-st} d\alpha(t)$ if $s > 0$ and that $f(s) \sim \frac{1}{s^\beta}$ $(\beta > 0)$ as $s \to 0^+$. Then $\alpha(t) \sim \frac{t^\beta}{\Gamma(\beta+1)}$ (as $t \to \infty$). See [5].

by (14) and Lemma 1. For square free number n with $\omega(n) \neq 0$, we obtain

$$\frac{1}{g(n)} = \sum_{\tau \mid n} \frac{1}{g(\tau)} \sum_{d \mid \frac{n}{\tau}} \mu(d) = \sum_{s \mid n} \sum_{d \mid s} \frac{\mu(d)}{g\left(\frac{s}{d}\right)} = \sum_{s \mid n} f(s)$$

and therefore

$$Q = \sum_{\substack{d_1 \leq \xi \\ d_1 \mid P \\ \omega(d_1) \neq 0}} \sum_{\substack{d_2 \leq \xi \\ d_2 \mid P \\ \omega(d_2) \neq 0}} \lambda_{d_1} \lambda_{d_2} g(d_1) g(d_2) \sum_{d \mid (d_1, d_2)} f(d)$$

$$= \sum_{\substack{d \leq \xi \\ d \mid P \\ \omega(d) \neq 0}} f(d) \left(\sum_{\substack{d \mid n \mid P \\ n \leq \xi \\ \omega(n) \neq 0}} \lambda_n g(n) \right)^2 .$$

Let

$$S = \sum_{\substack{1 \leq n \leq \xi \\ n \mid P \\ \omega(n) \neq 0}} \frac{\mu^2(n)}{f(n)}.$$

Then by Lemma 1, we have

$$\lambda_n g(n) = \frac{1}{S} \sum_{\substack{1 \leq m \leq \xi/n \\ (m,n)=1 \\ m \mid P \\ \omega(m) \neq 0}} \frac{\mu(n)\mu^2(m)}{f(n)f(m)}$$

$$= \frac{1}{S} \sum_{\substack{1 \leq m \leq \xi/n \\ (m,n)=1 \\ m \mid P \\ \omega(m) \neq 0}} \frac{\mu(mn)\mu(m)}{f(mn)},$$

$$\sum_{\substack{d \mid n \mid P \\ n \leq \xi \\ \omega(n) \neq 0}} \lambda_n g(n) = \frac{1}{S} \sum_{\substack{d \mid n \mid P \\ n \leq \xi \\ \omega(n) \neq 0}} \sum_{\substack{1 \leq m \leq \xi/n \\ (m,n)=1 \\ m \mid P \\ \omega(m) \neq 0}} \frac{\mu(mn)\mu(m)}{f(mn)}$$

$$= \frac{1}{S} \sum_{\substack{1 \leq r \leq \xi \\ d \mid r \mid P \\ \omega(r) \neq 0}} \frac{\mu(r)}{f(r)} \sum_{d \mid n \mid r} \mu\left(\frac{r}{n}\right) = \frac{\mu(d)}{Sf(d)},$$

where n is square free, $\omega(n) \neq 0$, $g(n) = \prod_{p \mid n} g(p)$ and $f(n) = \frac{1}{g(n)} \prod_{p \mid n} (1 - g(p))$, and consequently

$$Q = \frac{1}{S}. \tag{16}$$

Now we proceed to estimate the error term R: Since $\omega(q) = \frac{h_q(k!\,q)}{k!} \geq \frac{1}{k!}$ if $\omega(q) \neq 0$, we have by Lemmas 1 and 2 that

$$R \leq 2 \cdot (k!)^2 \left(\sum_{\substack{d \leq \xi \\ \omega(d) \neq 0 \\ d \mid P}} |\lambda_d| \omega(d) \right)^2 \leq 2 \cdot k!^2 \left(\sum_{\substack{d \leq \xi \\ \omega(d) \neq 0 \\ d \mid P}} \frac{|\mu(d)|\omega(d)}{\prod_{p \mid d} \left(1 - \frac{\omega(p)}{p}\right)} \right)$$

$$\leq 2 \cdot k!^2 \left(\sum_{\substack{d \leq \xi \\ \omega(d) \neq 0 \\ d \mid P}} d|\mu(d)| \prod_{p \mid d} \frac{w(p)}{p - \omega(p)} \right)^2 \leq 2 \cdot k!^2 \xi^2 \prod_{p \leq \xi} \left(1 + \frac{\omega(p)}{p - \omega(p)}\right)^2$$

$$= 2 \cdot k!^2 \xi^2 \prod_{p \leq \xi} \left(1 - \frac{\omega(p)}{p}\right)^{-2} = O(\xi^2 \log^2 \xi), \tag{17}$$

where the constant in "O" depends on $F(x)$ only. Therefore by (15)–(17) and Lemma 3, we obtain

$$P(x, \xi) \leq \frac{x}{\displaystyle\sum_{\substack{1 \leq n \leq \xi \\ \omega(n) \neq 0}} \frac{\mu^2(n)}{f(n)}} + O(\xi^2 \log^2 \xi)$$

$$= \prod_p \frac{\left(1 - \frac{\omega(p)}{p}\right)}{\left(1 - \frac{1}{p}\right)} \cdot \frac{x}{\log \xi} + o\left(\frac{x}{\log \xi}\right) + O(\xi^2 \log^2 \xi). \tag{18}$$

We may assume that the leading coefficient of $F(x)$ is positive. Then $F(n) > N^{1/2}$, if $n > N^{2/3}$ and N is sufficiently large. Let Σ_1 be the number of n such that $1 \leq n \leq N$ and $F(x)$ is a prime number $\leq N^{1/2}$ and let Σ_2 denote the number of remaining primes represented by $F(n)$ for $1 \leq n \leq N$. Then

$$\Sigma_1 \leq N^{\frac{2}{3}}, \quad \Sigma_2 \leq P(N, N^{\frac{1}{2}}).$$

Since

$$\pi(N; F(x)) = \Sigma_1 + \Sigma_2,$$

we have

$$\pi(N; F(x)) \leq P\left(N; \frac{N^{\frac{1}{2}}}{\log^2 N}\right) + O(N^{\frac{2}{3}}) \leq 2e^r \mu_F \frac{N}{\log N} + o\left(\frac{N}{\log N}\right)$$

by (13) and (18). The theorem is proved.

The above method is also applicable if $F(x)$ is a reducible polynomial. We have the following:

Theorem 2. *Let $F(x)$ be an integral valued polynomial of degree k without any fixed prime divisor and let*

$$F(x) = F_1(x) \cdots F_f(x),$$

where $F_\nu(x)\,(\nu = 1, \ldots, f)$ are irreducible polynomials and are distinct from one another. Denote by $C_f(N; F(x))$ the number of x such that $1 \le x \le N$ and the prime divisors of x are all $> N^{1/2}$. Let

$$\varlimsup_{N \to \infty} C_f(N; F(x)) \frac{\log^f N}{N} = \kappa. \tag{19}$$

Then

$$\kappa \le f!\, 2^f e^{fr} \rho_F, \tag{20}$$

where $\rho_F = e^{-fr} \prod_p \dfrac{\left(1 - \frac{\omega(p)}{p}\right)}{\left(1 - \frac{1}{p}\right)^f}$, $\omega(m) = \dfrac{h_m(m \cdot k!)}{k!}$ and $h_m(k!\, m)$ is the number of solutions of the congruence $F(x) \equiv 0 \pmod{m}\,(0 \le x < k!\, m)$.

The proof of Theorem 2 is the same as that of Theorem 1, where Lemma 2 should be replaced by

Lemma 4. (Nagell [4]) *There exists a constant ρ_F depending on $F(x)$ such that*

$$\prod_{p \le N} \left(1 - \frac{\omega(p)}{p}\right) \sim \frac{\rho_F}{\log^f N}\,(as\ N \to \infty).$$

Take $F(x) = x(x + u_1) \cdots (x + u_{m-1})$, where $0, u_1, \ldots, u_{m-1}$ are non-negative integers such that they are distinct from one another and that if U_p denotes the number of different residues of $\{0, u_1, \ldots, u_{m-1}\}$ modulo p, then $U_p < p$ for any prime p. We derive immediately from Theorem 2 the following:

Theorem 3. (Tchudnovski [6]) *Let $\{0, u_1, \ldots, u_{m-1}\}$ be a set of integers satisfying the above conditions and $Z_{u_1, \ldots, u_{m-1}}(N)$ be the number of x such that $1 \le x \le N$ and that $x, x + u_1, \ldots, x + u_{m-1}$ are all prime numbers. Then*

$$\varlimsup_{N \to \infty} Z_{u_1, \ldots, u_{m-1}}(N) \frac{\log^m N}{N} \le 2^m \cdot m! \prod_p \frac{1 - \frac{U_p}{p}}{\left(1 - \frac{1}{p}\right)^m}. \tag{21}$$

Take $F(x) = (x^2 + 1)(x^2 + 3)$. Then we derive from Theorem 2 the following:

Theorem 4. *Let $Z((x^2 + 1)(x^2 + 3))$ be the number of integers x such that $x^2 + 1$ and $x^2 + 3$ are all prime numbers and $1 \leq x \leq N$. Then*

$$Z((x^2 + 1)(x^2 + 3), N) \leq 24 \prod_{p \geq 5} \left(1 + \frac{2p - a_p p - 1}{(p - 1)^2} \right) \frac{N}{\log^2 N}(1 + o(1)),$$

where

$$a_p = \begin{cases} 0, & \text{if } -1 \text{ and } -3 \text{ are all quadratic non-residues} \bmod p, \\ 2, & \text{if one of } -1 \text{ and } 3 \text{ is a quadratic non-residue} \bmod p. \\ 4 & \text{if } -1 \text{ and } -3 \text{ are all quadratic residues} \bmod p. \end{cases}$$

References

[1] T. Nagell, Zur Arithmetik der Polynome, *Abh. Math. Sem. Hamb. Univ.* **1** (1922) 179–194.

[2] H. Heilbronn, Über die Verteilung der Primzahlen in Polynomen, *Math. Ann.* **104** (1931) 794–799.

[3] G. Ricci, Su la congettura di Goldbach e la constante di Schnirelmann, *Ann. della R. Scu. Normale Sup. di Pisa* (2) **6** (1973) 70–115.

[4] T. Nagell, Généralisation dún théorème de Tchbycheff, *J. Math.* (8) **4** (1921) 343–356.

[5] D. V. Widder, *The Laplace Transform* (Princeton Univ. Press, 1946), p. 192.

[6] E. V. Chulanovski, Some estimations connected with Seiberg's new method, *Dokl. Akad. Nauk SSSR* **43** (1948) 491–494.

SCIENCE RECORD

Vol. 1, No. 1, 1957

MATHEMATICS

ON SIEVE METHODS AND SOME OF THE RELATED PROBLEMS*

Wang Yuan (王　元)

Institute of Mathematics, Academia Sinica

(Communicated by Prof. Hua, L. K., Member of Academia Sinica)

First of all, let us state the grand Riemann hypothesis as follows:

(R) The real parts of all zeros of all the Dirichlet's L-functions $L(s, \chi)$ are $\leqslant \frac{1}{2}$.

From (R) we may derive the following

(R^*) Let $(k, l) = 1$. Then for any given $\varepsilon > 0$, we have

$$\pi(x; k, l) = \sum_{\substack{p \leqslant x \\ p \equiv l \,(\mathrm{mod}\ k)}} 1^{**} = \frac{\mathrm{li}\,x}{\varphi(k)} + O\left(x^{\frac{1}{2} + \varepsilon}\right),$$

where $\mathrm{li}\,x = \displaystyle\int_2^x \frac{dt}{\log t}$ and the constant implied by the symbol "O" depends on ε only.

Under the truth of (R^*), T. Estermann[1] first proved in 1932 that every large even integer can be represented as a sum of a prime and a product of at most 6 primes. 6 has been reduced to 4 by the author[2] and А. И. Виноградов[3] independently in 1955. In this paper, the author improves the result from 4 to 3 by the method used in the previous paper[2] with the help of Kuhn's[4] method. Thus, we obtain

Theorem 1. Under the truth of (R^*), every sufficiently large even integer is a sum of a prime and a product of at most 3 primes.

Let x be even integer and ξ be a real number not exceeding x. Let a, q be integers satisfying $(a, q) = 1$ and $1 \leqslant q \leqslant x$. Let p_i be the i-th prime number. Let $P(x, a, q, \xi)$ be the number of integers $x - p$ satisfying the following conditions:

(1) $2 < p < x$, $p \equiv a \,(\mathrm{mod}\ q)$, $x - p \not\equiv 0 \,(\mathrm{mod}\ p_i)$, $1 \leqslant i \leqslant r$, where $p_r \leqslant \xi < p_{r+1}$.

---

* Received Nov. 29, 1956.

** Hereafter p denotes prime numbers; the same holds for p'.

10

I et $p_s \leqslant x^{\frac{1}{5}} < p_{s+1}$ and $p_t \leqslant x^{\frac{3}{10}} < p_{t+1}$. Let $\mathfrak{M}$ denote the set of integers $x - p$ satisfying the following conditions:

(2) $2 < p < x$, $x - p \not\equiv 0 \pmod{p_i}$, $1 \leqslant i \leqslant s$, $x - p \not\equiv 0 \pmod{p_{s+j}^2}$,

$1 \leqslant j \leqslant t - s$. We denote the number of elements of $\mathfrak{M}$ by $M(x, x^{\frac{1}{5}}, x^{\frac{3}{10}})$.

The proof of Theorem 1 depends on the following two lemmas.

Lemma 1. $M(x, x^{\frac{1}{5}}, x^{\frac{3}{10}}) = P(x, 1, 1, x^{\frac{1}{5}}) + O(x^{\frac{4}{5}})$.

Proof:

$$P(x,1,1,x^{\frac{1}{5}}) - M(x,x^{\frac{1}{5}},x^{\frac{3}{10}}) \leqslant \sum_{j \leqslant t-s} \sum_{\substack{2 < p < x \\ P_{s+j}^2 \mid (x-p)}} 1 = \sum_{j \leqslant t-s} \pi(x; p_{s+j}^2, x) \leqslant$$

$$\leqslant \sum_{x^{\frac{1}{5}} < n \leqslant x^{\frac{3}{10}}} (1 + \frac{x}{n^2}) = O(x^{\frac{4}{5}}).$$

Lemma 2. The number of elements in $\mathfrak{M}$ with at least 3 prime divisors in the interval $x^{\frac{1}{5}} < p' \leqslant x^{\frac{3}{10}}$ does not exceed

$$\frac{1}{3} \left(\sum_{\substack{j \leqslant t-s \\ p_{s+j} \nmid x}} P(x, x, p_{s+j}, x^{\frac{1}{5}}) + 4 \right).$$

Proof: Let $1 \leqslant j \leqslant t - s$. Let Γ_j be the subset of $\mathfrak{M}$ whose elements are divided by p_{s+j}. Indeed, Γ_j is the set of integers $x - p$ satisfying the following conditions:

(3) $2 < p < x$, $x - p \not\equiv 0 \pmod{p_i}$, $1 \leqslant i \leqslant s$, $x - p \not\equiv 0 \pmod{p_{s+k}^2}$, $1 \leqslant k \leqslant t - s$, $x - p \equiv 0 \pmod{p_{s+j}}$.

If $p_{s+j} \mid x$, then the number of elements of Γ_j is equal 1 or 0, otherwise, the number of elements of Γ_j does not exceed $P(x, x, p_{s+j}, x^{\frac{1}{5}})$.

If $x - p$ belongs to $\mathfrak{M}$ and has at least 3 prime divisors in the interval $x^{\frac{1}{5}} < p' \leqslant x^{\frac{3}{10}}$, then $x-p$ belongs to at least 3 different sets Γ_j. Hence we have the lemma.

Proof of Theorem 1: Under the truth of (R^*), we may estimate with the aid of the methods of Brun, Бухштаб and Selberg the upper bounds and lower bounds of $P(x, a, q, \xi)$ for some given a, q and ξ (Cf. [2]). For fixed ε sufficiently small in (R^*), then for $x > x_0(\varepsilon)$, the number of those elements in $\mathfrak{M}$ which has at most 2 prime divisors in the interval $x^{\frac{1}{5}} < p \leqslant x^{\frac{3}{10}}$ is greater or equal to

11

$$M\left(x, x^{\frac{1}{5}}, x^{\frac{3}{10}}\right) - \frac{1}{3}\left(\sum_{\substack{j\leq t-s \\ p_{s+j}\nmid x}} P\left(x, x, p_{s+j}, x^{\frac{1}{5}}\right) + 4\right) =$$

$$= P\left(x, 1, 1, x^{\frac{1}{5}}\right) - \frac{1}{3}\left(\sum_{\substack{j\leqslant t-s \\ p_{s+j}\nmid x}} P\left(x, x, p_{s+j}, x^{\frac{1}{5}}\right) + 4\right) + O(x^{\frac{4}{5}}) >$$

$$> 0.4\, e^{-\gamma} \prod_{\substack{p\,\mid\,x \\ p>2}} \frac{p-1}{p-2} \prod_{p>2} \left(1 - \frac{1}{(p-1)^2}\right) \frac{\text{li } x}{\log x} + O\left(\frac{x\,(\log\log x)^2}{\log^3 x}\right) > 1,$$

where γ is Euler's constant.

It means that for $x > x_0$, there exists $p > 2$ such that $x - p$ has no prime divisor $\leqslant x^{\frac{1}{5}}$ and at most two prime factors in the interval $x^{\frac{1}{5}} < p' \leqslant x^{\frac{3}{10}}$. That is, $x - p$ is a product of at most 3 primes. Thus we have the theorem.

By the same way, we have

Theorem 2. Under the truth of (R^*), there are infinitely many primes p such that $p + 2$ is a product of at most 3 primes.

The method used here has also many other important applications. For example, without any hypothesis, we have

1°. Every sufficiently large even integer is a sum of two integers c_1 and c_2 with the property that $c_1\, c_2$ is a product of at most 5 primes.

2°. Every sufficiently large even integer is a sum of two integers, each of which has at most 3 prime factors.[*]

We take the method used here just for the sake of avoidance of the advanced analysis. If the method suggested by Selberg[5] (Cf. also [3]) is used to estimate directly the lower bounds of the sums of the type $P(x,a,q,\xi)$ and more complicated numerical calculations have been taken, some of the above results can further be improved. For example, it is possible to prove that every large even integer can be written as a sum of two integers, of which one contains at most 2 and the other 3 prime factors.

---

[*] This result has already been obtained by А. И. Виноградов[3], but in his proof the theory of Riemann ζ-function was used.

12

References

[1] Estermann, T., 1932. Eine neue Dastellung und Anwendungen der Viggo Brunschen Methode, *J. Reine Angew. Math.*, **168**, 106 — 116.

[2] Wang, Y., 1956. On the Representation of Large Even Integer as a Sum of a Prime and a Product of at most 4 Primes, *Acta Math. Sinica*, **4**, 565—582.

[3] Виноградов, А. И., 1956. О связях решета Эратосфена с ζ-функцей Римана, Вестник Ленинградского Университета, **13**, 3, 142 — 146.

[4] Kuhn, P., 1942. Zur Viggo Brun'schen Siebmethode I, Norske Vid. Selsk. Forh. Trondhjem, **39**, 145 — 148.

[5] Selberg, A., 1952. On Elementary Methods in Prime Number Theory and Their Limitations, Den 11 - te Scand. Mat. Kongr. Trondheim, 13 — 22.

SCIENCE RECORD

New Ser. Vol. I, No. 3, 1957

MATHEMATICS

ON SIEVE METHODS AND SOME OF THEIR APPLICATIONS*

Wang Yuan （王　元）

Institute of Mathematics, Academia Sinica

(*Communicated by Prof. Hua, L. K., Member of Academia Sinica*)

In 1919, V. Brun[1] first proved the following result:

For sufficiently large x, there exists, between $x - x^{\frac{1}{2}}$ and x, an integer of at most 11 primes factors.

Brun's method and his results were improved by several Mathematicians. The best result before my work was due to P. Kuhn[2]: he obtained 4 instead of 11. Recently, Kuhn[3] proves a more general result. We state it as follows:

For sufficiently large x, there exists always an integer n between $x - x^{\frac{1}{v}}$ and x, such that n has at most $k = v + w$ prime factors, where v is a positive integer and w is the least integer satisfying the inequality

$$\log (6v - w) \leqslant 0.968 (w + 1). \tag{1}$$

For examples: $v = 2, k = 4$; $v = 3, k = 5$; $v = 100, k = 106$.

In this paper, the author has improved the inequality (1) by the methods used in previous papers (Cf.[4], [5], [6]). Namely, we have the following

Theorem 1. Let v be a positive integer. Let w be the least integer satisfying the inequality

$$\frac{5.64527}{4.8396} + \frac{3.65}{4.8396} \log \frac{5v - w}{w + 5} \leqslant w + 1. \tag{2}$$

Then for sufficiently large x, there is always an integer between $x - x^{\frac{1}{v}}$ and x which has at most $k = w + v$ prime factors.

For examples: $v = 2, k = 3$; $v = 3, k = 4$; $v = 100, k = 104$.

For some special given values of v (in general, v not an integer), the more sharper results have been obtained. For examples, we have

---

* Received March 19, 1957.

2

Theorem 2. For sufficiently large x, there exists always a number

(i) in interval $x - x^{\frac{10}{17}} < n \leqslant x$ with at most 2 prime factors;

(ii) in interval $x - x^{\frac{20}{49}} < n \leqslant x$ with at most 3 prime factors;

(iii) in interval $x - x^{\frac{1}{5}} < n \leqslant x$ with at most 6 prime factors and

(iv) in interval $x - x^{\frac{1}{100+\frac{1}{3}}} < n \leqslant x$ with at most 103 prime factors.

Let p_i be the i-th prime. Let $A > x > \xi$. Let $P(x, A, \xi)$ be the number of integers satisfying the following conditions:

$$A - x < n \leqslant A, \quad n \not\equiv 0 \,(\text{mod } p_i), \quad 1 \leqslant i \leqslant r, \tag{3}$$

where $p_r \leqslant \xi < p_{r+1}$.

Let $\beta > a > 1$ be two given positive numbers and $v > 1$. Let $\mathfrak{M}$ denote the set of integers n satisfying the following conditions:

$$x^v - x < n \leqslant x^v, \quad n \not\equiv 0 \,(\text{mod } p_i), \quad 1 \leqslant i \leqslant s, \quad n \not\equiv 0 \,(\text{mod } \bar{p}^2_{s+j}), \tag{4}$$
$$1 \leqslant j \leqslant t - s,$$

where $p_s \leqslant x^{\frac{1}{\beta}} < p_{s+1}$ and $p_t \leqslant x^{\frac{1}{\alpha}} < p_{t+1}$. We denote the number of

elements of $\mathfrak{M}$ by $M(x, x^v, x^{\frac{1}{\beta}}, x^{\frac{1}{\alpha}})$.

Lemma 1. $M(x, x^v, x^{\frac{1}{\beta}}, x^{\frac{1}{\alpha}}) = P(x, x^v, x^{\frac{1}{\beta}}) + O(x^{1-\frac{1}{\beta}}) + O(x^{\frac{1}{\alpha}})$.

Proof.

$$P(x, x^v, x^{\frac{1}{\beta}}) - M(x, x^v, x^{\frac{1}{\beta}}, x^{\frac{1}{\alpha}})$$

$$\leqslant \sum_{x^{\frac{1}{\beta}} < p \leqslant x^{\frac{1}{\alpha}}} \sum_{\substack{x^v - x < n \leqslant x^v \\ n \equiv 0 \,(\text{mod } p^2)}} 1 = O\left(\sum_{x^{\frac{1}{\beta}} < n \leqslant x^{\frac{1}{\alpha}}} \frac{x}{p^2} + 1 \right) = O(x^{1-\frac{1}{\beta}}) + O(x^{\frac{1}{\alpha}}).$$

Lemma 2. In $\mathfrak{M}$, the number of elements with at least m prime

divisors in the interval $x^{\frac{1}{\beta}} < p \leqslant x^{\frac{1}{\alpha}}$ is not exceeding

$$\frac{1}{m} \sum_{1 \leqslant j \leqslant t-s} P\left(\frac{x}{p_{s+j}}, \frac{x^v}{p_{s+j}}, x^{\frac{1}{\beta}} \right).$$

Proof. For $1 \leqslant j \leqslant t - s$, let Γ_j be the subset of $\mathfrak{M}$ whose elements are divided by p_{s+j}. Indeed, Γ_j is the set of integers n satisfying the following conditions:

$$x^v - x < n \leqslant x^v, \quad n \not\equiv 0 \,(\text{mod } p_i), \quad 1 \leqslant i \leqslant s, \quad n \not\equiv 0 \,(\text{mod } p^2_{s+k}),$$
$$1 \leqslant k \leqslant t - s, \quad n \equiv 0 \,(\text{mod } p_{s+j}). \tag{5}$$

3

Evidently, the number of elements of Γ_j is not exceeding the number of integers satisfying the following conditions:

$$\frac{x^v}{p_{s+j}} - \frac{x}{p_{s+j}} < m \leqslant \frac{x^v}{p_{s+j}}, \qquad m \not\equiv 0 \,(\mathrm{mod}\ p_i),\ 1 \leqslant i \leqslant s, \tag{6}$$

i. e. $P\left(\dfrac{x}{p_{s+j}},\ \dfrac{x^v}{p_{s+j}},\ x^{\frac{1}{\beta}}\right)$.

If n belongs to $\mathfrak{M}$ and n has at least m prime divisors in the interval $x^{\frac{1}{\beta}} < p \leqslant x^{\frac{1}{\alpha}}$, then n belongs to at least m different sets Γ_j. Hence we have the lemma.

Lemma 3. Let $C > 10$ be a given number. Then there exist two non-negative and non-decreasing functions $\lambda(z)$ and $\Lambda(z)$ $(0 < z \leqslant C.)$ with the properties that each has at most finite discontinuities, such that the following inequality holds uniformly in $A > x$

$$\lambda(z)\frac{e^{-\gamma}x}{\log x} + O\left(\frac{x}{\log^{\frac{3}{2}} x}\right) \leqslant P(x, A, x^{\frac{1}{z}}) \leqslant \Lambda(z)\frac{e^{-\gamma}x}{\log x} + O\left(\frac{x}{\log^{\frac{3}{2}} x}\right), \tag{7}$$

where γ is Euler's constant and the constants implied by the symbol O are absolute constants.

The lemma can be followed immediately by Brun's method. By means of the methods of Brun, Бухштаб and Selberg (Cf. [4], [5]), we have the following tables:

α	...	6.5	...	5	...	4	...	3	...	$0 < \alpha \leqslant 2$	
$\Lambda(\alpha)$	...	...	...	5.197	...	4.42	...	3.848	...	3.564	(8)
$\lambda(\alpha)$	...	6.4524	...	4.8396	...	3.6432	...	2.1371	...	0	

Remark: If we use Бухштаб's method and by more complicated numerical calculations, then the values in the above table can be improved further.

Fundamental lemma. Let $\lambda(\alpha)$ and $\Lambda(\alpha)$ $(0 < \alpha \leqslant C)$ be two functions with the properties as stated in lemma 3. Let w be an integer and $\beta > a > 1$ be two positive numbers. If

$$\lambda(\beta) - \frac{1}{w+1}\int_{a-1}^{\beta-1} \Lambda\left(\frac{\beta z}{z+1}\right)\frac{dz}{z} > 0 \tag{9}$$

and if m is the least integer satisfying the inequality $\dfrac{w}{\beta} + \dfrac{m+1}{a} \geqslant v$, then for sufficiently large x, there is always an integer n between $x^v - x$ and x^v such that n has at most $k = w + m$ prime factors.

4

Proof. By lemma 1, 2 and 3, for sufficiently large x, the number of elements in $\mathfrak{M}$ with at most w prime divisors in the interval $x^{\frac{1}{\beta}} < p \leqslant x^{\frac{1}{\alpha}}$ is not less than

$$M\left(x, x^v, x^{\frac{1}{\beta}}, x^{\frac{1}{\alpha}}\right) - \frac{1}{w+1} \sum_{1 \leqslant j \leqslant t-s} P\left(\frac{x}{p_{s+j}}, \frac{x^v}{p_{s+j}}, x^{\frac{1}{\beta}}\right) =$$

$$= P\left(x, x^v, x^{\frac{1}{\beta}}\right) - \frac{1}{w+1} \sum_{1 \leqslant j \leqslant t-s} P\left(\frac{x}{p_{s+j}}, \frac{x^v}{p_{s+j}}, x^{\frac{1}{\beta}}\right) + O\left(x^{1-\frac{1}{\beta}}\right) +$$

$$+ O\left(x^{\frac{1}{\alpha}}\right) \geqslant \left(\lambda\left(\beta\right) - \frac{1}{w+1} \int_{\alpha-1}^{\beta-1} \Lambda\left(\frac{\beta z}{z+1}\right) \frac{dz}{z}\right) \frac{e^{-\gamma} x}{\log x} + O\left(\frac{x}{\log^{3/2} x}\right) > 1.$$

This means that for sufficiently large x, there exists in the interval $x^v - x < n \leqslant x^v$ an integer n having not prime divisor less than or equal to $x^{\frac{1}{\beta}}$ and having at most w prime divisors in the interval $x^{\frac{1}{\beta}} < p \leqslant x^{\frac{1}{\alpha}}$. That is, n is a product of at most $k = w + m$ primes. Thus we have the lemma.

(i) Let v be a positive integer. Let $\beta = 5$ and $\alpha = \dfrac{5(v+1)}{5v-w}$, where w is the least integer satisfying (2). Then we have

$$\lambda(5) - \frac{1}{w+1} \int_{\frac{5(v+1)}{5v-w}-1}^{4} \Lambda\left(\frac{5z}{z+1}\right) \frac{dz}{z} > 0$$

by table (8). Hence Theorem 1 follows from the fundamental lemma.

(ii) By table (8), we have

$$\lambda(5) - \frac{1}{2} \int_{\frac{1}{3}}^{4} \Lambda\left(\frac{5z}{z+1}\right) \frac{dz}{z} > 0, \qquad \lambda(5) - \frac{1}{3} \int_{\frac{10}{119}}^{4} \Lambda\left(\frac{5z}{z+1}\right) \frac{dz}{z} > 0$$

and

$$\lambda(5) - \frac{1}{5} \int_{\frac{1}{180}}^{4} \Lambda\left(\frac{5z}{z+1}\right) \frac{dz}{z} > 0.$$

Hence by the fundamental lemma, we have Theorem 2.

The method used here has also many other applications. For examples, we have

Theorem 3. Let $F(x)$ be an irreducible integral valued polynomial of degree k without any fixed prime divisor. Let

$$n = \begin{cases} k+1, & \text{if } 1 \leqslant k \leqslant 5, \\ k+w, & \text{if } k > 5, \end{cases}$$

where w is the least integer satisfying

$$w + 1 \geqslant \frac{5.64527}{4.8396} + \frac{3.65}{4.8396} \log \frac{5k-w}{w+5} \cdot \cdot$$

5

Then there are infinitely many integers x such that $F(x)$ is a product of at most n primes.

For examples: there are infinitely many integers x, such that $x^3 + 2$ has at most 4 prime factors.

REFERENCES

[1] Brun, V., 1920, Le crible d'Eratosthène et le théorème de Goldbach, *Vid.-Selsk. Skr. 1. M. N. KL.* **3**, 1-36.

[2] Kuhn, P., 1942, Zur Viggo Brun'schen Siebmethode, I *Norske Vid. Selsk. Forh. Trondhjem*, **39**, 145-148.

[3] Kuhn, P., 1953, Neue Abschätzungen auf Grund der Viggo Brunschen Siebmethode, *Tolfte Skandinaviska Matematikerkongressen*, Lund, 160-168.

[4] Wang, Y., 1956, On the Representation of Large Even Integer as a Sum of a Product of at Most 3 Primes and a Product of at Most 4 Primes, *Acta Mathematica Sinica*, 6:3, 500-513.

[5] Wang, Y., 1956, On the Representation of Large Even Integer as a Sum of a Prime and a Product of at Most 4 Primes, *Acta Mathematica Sinica*, 6:4, 565-582.

[6] Wang, Y., 1957, On Sieve Methods and Some of the Related Problems, *Science Record*, Academia Sinica, New Series, 1:1, 9-12.

SCIENCE RECORD

New Ser. Vol. I, No. 5, 1957

MATHEMATICS

ON THE REPRESENTATION OF LARGE EVEN NUMBER AS A SUM OF TWO ALMOST-PRIMES* †

WANG YUAN (王　元)

Institute of Mathematics, Academia Sinica
(*Communicated by Prof. Hua, L. K., Member of Academia Sinica*)

For the sake of briefness, we write the following proposition by (a, b).

Every sufficiently large even integer can be represented as a sum of two integers > 1, of which one contains at most a and other at most b prime factors.

The aim of the present note is to prove $(3, 3)$ and (a, b) $(a + b \leqslant 5)$ by the method used in previous papers[1, 2]. These results improve the $(3, 4)$[3] of the auther in 1955. Moreover, using Бухштаб's[4] method with more complicated numerical calculations, we have $(2, 3)$. Recently, we have found some mistakes in numerical calculations in the proof of $(3, 3)$ of А. И. Виноградов[5]. We shall state it at the end of this note.

In this paper, p denotes prime number and p_i denotes i-th odd prime.

Let x be an even integer and ξ be a real number. Let
$$(w) \quad a; \ a_i, \ b_i \ (1 \leqslant i \leqslant r)$$
be a sequence of integers satisfying the following conditions:

(1)　$a = 0$ or 1;　$0 \leqslant a_i, b_i < p_i$, if p_i/x,　then $a_i = b_i$, otherwise $a_i \neq b_i$ $(1 \leqslant i \leqslant r)$,

where $p_r \leqslant \xi < p_{r+1}$. Let $P_w(x, \xi)$ be the number of integers n satisfying the following conditions:

(2)　$1 \leqslant n \leqslant x$,　$n \equiv a \pmod 2$,　$n \not\equiv a_i \pmod{p_i}$,　$n \not\equiv b_i \pmod{p_i}$ $(1 \leqslant i \leqslant r)$.

Let $v > u > 1$ be two given positive numbers. Let $\mathfrak{N}$ denote the set of integers $n\,(x - n)$ satisfying the following conditions:

(3)　$1 \leqslant n \leqslant x$, $n\,(x - n) \not\equiv 0 \pmod 2$, $n\,(x - n) \not\equiv 0 \pmod{p_i}$ $(1 \leqslant i \leqslant s)$,

where　$p_s \leqslant x^{\frac{1}{v}} < p_{s+1}$. Let $\mathfrak{M}$ denote the set of integers $n\,(x - n)$

---

*　Received June 29, 1957.

†　An integer is called almost prime, if the number of its prime factors is not exceeding a fixed constant.

satisfying (3) and the following conditions:

(4) $\quad n\,(x-n) \not\equiv o \;(\mathrm{mod}\; p_{s+j}^2)\quad 1 \leqslant j \leqslant t-s,$

where $\quad p_t \leqslant x^{\frac{1}{u}} < p_{t+1}.$ We denote the number[1] of elements of $\mathfrak{R}$ and $\mathfrak{M}$ by $N\,(x, x^{\frac{1}{v}})$ and $M\,(x, x^{\frac{1}{v}}, x^{\frac{1}{u}})$ respectively.

Lemma 1. $\quad M\,(x, x^{\frac{1}{v}}, x^{\frac{1}{u}}) = N\,(x, x^{\frac{1}{v}}) + O\,(x^{1-\frac{1}{v}}) + O\,(x^{\frac{1}{u}}).$

Proof. $\quad N\,(x, x^{\frac{1}{v}}) - M\,(x, x^{\frac{1}{v}}, x^{\frac{1}{u}}) \leqslant \sum_{x^{\frac{1}{v}} < p \leqslant x^{\frac{1}{u}}} \; \sum_{\substack{1 \leqslant n \leqslant x \\ n\,(x-n) \equiv 0 \;(\mathrm{mod}\; p^2)}} 1 =$

$$= \sum_{x^{\frac{1}{v}} < p \leqslant x^{\frac{1}{u}}} S_p.$$

(i) $\quad p \,|\, x:\; S_p \leqslant \sum_{\substack{1 \leqslant n \leqslant x \\ n \equiv o \;(\mathrm{mod}\; p)}} 1 = \left[\frac{x}{p}\right] + 1 = O\,(x^{1-\frac{1}{v}}).$

(ii) $\quad p \nmid x:\; S_p = \sum_{\substack{1 \leqslant n \leqslant x \\ n\,(x-n) \equiv o \;(\mathrm{mod}\; p^2)}} 1 \leqslant 2\left[\frac{x}{p^2}\right] + 2.$

Hence, we have

$$\sum_{x^{\frac{1}{v}} < p \leqslant x^{\frac{1}{u}}} S_p = O\left(\sum_{\substack{p \,|\, x \\ p > x^{\frac{1}{v}}}} x^{1-\frac{1}{v}}\right) + O\left(\sum_{m > x^{\frac{1}{v}}} \frac{x}{m^2}\right) + O\left(\sum_{m \leqslant x^{\frac{1}{u}}} 1\right) =$$

$$= O\,(x^{1-\frac{1}{v}}) + O\,(x^{\frac{1}{u}}).$$

Thus we have the lemma.

Lemma 2. There exist sequences of integers (w_j) $(1 \leqslant j \leqslant t-s)$ satisfying (1), such that the number of elements in $\mathfrak{M}$ with at least l prime divisors in the interval $x^{\frac{1}{v}} < p \leqslant x^{\frac{1}{u}}$ is not exceeding

$$\frac{2}{l} \sum_{\substack{1 \leqslant j \leqslant t-s \\ p_{s+j} \nmid x}} P_{w_j}\left(\frac{x}{p_{s+j}}, x^{\frac{1}{v}}\right) + O\left(x^{1-\frac{1}{v}}\right).$$

Proof. For $1 \leqslant j \leqslant t-s$, let Γ_j be the subset of $\mathfrak{M}$ whose elements are divided by p_{s+j}. Indeed, Γ_j is the set of integers $n\,(x-n)$ satisfying (3), (4) and the following condition:

(5) $\quad n(x-n) \equiv 0 \;(\mathrm{mod}\; p_{s+j}).$

(i) $\quad p_{s+j} \,|\, x:$ Evidently, the number of elements of Γ_j is not exceeding

---

1) If $n = x-n'$ and $n\,(x-n)\,(= n'\,(x-n'))$ belongs to $\mathfrak{R}$ or $\mathfrak{M}$, then we agree that $n\,(x-n)$ be computed twice.

$$\sum_{\substack{1 \leqslant n \leqslant x \\ n \equiv o \,(\mathrm{mod}\, p_{s+j})}} 1 = O\left(x^{1-\frac{1}{v}}\right).$$

(ii) $p_{s+j}\nmid x$: From condition (5), we deduce that $n \equiv o \,(\mathrm{mod}\, p_{s+j})$ or $n \equiv x \,(\mathrm{mod}\, p_{s+j})$ Let

(w_j) $a = 1$; if $p_i \,/\, x$, then $a_i = b_i = 0$, otherwise $a_i \equiv 0$,

$\qquad b_i\, p_{s+j} \equiv x \,(\mathrm{mod}\, p_i)\ (1 \leqslant i \leqslant s)$.

Evidently, the number of elements of Γ_j is not exceeding

$$2 P_{w_j}\left(\frac{x}{p_{s+j}},\ x^{\frac{1}{v}}\right).$$

If $n\,(x-n)$ belongs to $\mathfrak{M}$ and $n\,(x-n)$ has at least l prime divisors in the interval $x^{\frac{1}{v}} < p \leqslant x^{\frac{1}{u}}$, then $n\,(x-\mathrm{n})$ belongs to at least l different sets Γ_j. Hence we have the lemma.

Lemma 3. Let $C > 1$ be a given number. Then there exist two non-decreasing and non-negative functions $\lambda\,(a)$ and $\Lambda\,(a)$ $(o < a \leqslant C)$ with the properties that each of $\lambda\,(a)$ and $\Lambda\,(a)$ has at most finite discontinuities, such that the following inequality holds uniformly in (w) and z

$$(6)\quad \lambda\,(z)\,\frac{c_x\,x}{\log^2 x} + O\left(\frac{c_x\,x}{\log^2 x \log\log x}\right) \leqslant P_w\left(x,\,x^{\frac{1}{z}}\right) \leqslant \Lambda\,(z)\,\frac{c_x\,x}{\log^2 x} +$$

$$+ O\left(\frac{c_x\,x}{\log^2 x \log\log x}\right)(0 < z \leqslant C),$$

where $\displaystyle c_x = 2\,e^{2\gamma} \prod_{p>2}\left(1-\frac{1}{(p\text{-}1)^2}\right) \prod_{\substack{p\,|\,x \\ p>2}}\frac{p-1}{p-2}$ and γ is Euler's constant.

The lemma follows immediately by Brun's method.

Fundamental Theorem. Let m be a non-negative integer and $C > v > u > 1$ be three positive numbers. Let $\lambda\,(a)$ and $\Lambda\,(a)$ $(0 < a \leqslant C)$ be two functions with the properties as stated in lemma 3. If

$$(7)\quad \lambda\,(v) - \frac{1}{m+1}\int_{u-1}^{v-1} \Lambda\left(\frac{vz}{z+1}\right)\frac{z+1}{z^2}\,dz > 0,$$

then for all sufficiently large x, there is always an integer n in the interval $1 < n < x-1$, such that $n\,(x-n)$ has not prime divisor less than or equal to $x^{\frac{1}{v}}$ and has at most m prime divisors in the interval $x^{\frac{1}{v}} < p \leqslant x^{\frac{1}{u}}$.

Proof. Take

$(\overline{\omega})$ $a = 1$; if p_i/x, then $a_i = b_i = 0$, otherwise $a_i = 0$, $b_i \equiv x \,(\mathrm{mod}\, p_i)$, $i > 1, 2, \cdots$.

Then we have $N(x, x^{\frac{1}{v}}) = P_{\bar{w}}(x, x^{\frac{1}{v}})$. By lemma 1, 2 and 3, for sufficiently large x, the number of elements in $\mathfrak{M}$ with at most m prime divisors in the interval $x^{\frac{1}{v}} < p \leqslant x^{\frac{1}{u}}$ is not less than

$$M(x, x^{\frac{1}{v}}, x^{\frac{1}{u}}) - \frac{1}{m+1} \sum_{\substack{1 \leqslant j \leqslant t-s \\ p_{s+j} \nmid x}} P_{w_j}\left(\frac{x}{p_{s+j}}, x^{\frac{1}{v}}\right) + O(x^{1-\frac{1}{v}})$$

$$= P_{\bar{w}}(x, x^{\frac{1}{v}}) - \frac{1}{m+1} \sum_{\substack{1 \leqslant j \leqslant t-s \\ p_{s+j} \nmid x}} P_{w_j}\left(\frac{x}{p_{s+j}}, x^{\frac{1}{v}}\right) + O(x^{\frac{1}{u}}) + O(x^{1-\frac{1}{v}})$$

$$\geqslant \left(\lambda(\beta) - \frac{1}{m+1} \int_{u-1}^{v-1} \Lambda\left(\frac{vz}{z+1}\right) \frac{z+1}{z^2} \, dz \right) \frac{c_x \, x}{\log^2 x} + $$

$$+ O\left(\frac{c_x \, x}{\log^2 x \log \log x}\right) > 3.$$

This means that for sufficiently large x, there exists in the interval $1 < n < x-1$ an integer n having not prime divisor $\leqslant x^{\frac{1}{v}}$ and having at most m prime divisors in the interval $x^{\frac{1}{v}} < p \leqslant x^{\frac{1}{u}}$. Thus we have the Theorem.

By Brun – Бухштаб – Selberg's method $(Cf [3])$, we have the following table:

(8)

α	$\cdots$	8	$\cdots$	6	$\cdots$	5	$\cdots$	4	$\cdots$
$\Lambda(\alpha)$	$\cdots$	68.52511	$\cdots$	43.0082	$\cdots$	34.89666	$\cdots$	29.39023	$\cdots$
$\lambda(\alpha)$	$\cdots$	60.88817	$\cdots$	26.70925	$\cdots$	9.18109	$\cdots$	0	$\cdots$

From table (8), we deduce that

$$\lambda(6) - \frac{2}{3} \int_2^5 \Lambda\left(\frac{6z}{z+1}\right) \frac{z+1}{z^2} \, dz > 0.33829$$

and

$$\lambda(8) - \frac{1}{2} \int_1^7 \Lambda\left(\frac{8z}{z+1}\right) \frac{z+1}{z^2} \, dz > 0.56125.$$

Hence, $(3, 3)$ and (a, b) $(a+b \leqslant 5)$ follow from the Fundamental theorem.

By Бухштаб's method[4] with more complicated numerical calculations, we have improved the values in table (8) and obtain the following

(9)

α	$\cdots$	8	$\cdots$	7	$\cdots$	6	$\cdots$	5
$\Lambda(\alpha)$	$\cdots$	64.403149	$\cdots$	50.529826	$\cdots$	41.01897	$\cdots$	34.89666
$\lambda(\alpha)$	$\cdots$	63.59931	$\cdots$	47.471252	$\cdots$	31.004145	$\cdots$	13.61559

From table (9), we have

$$\lambda\,(8) - \frac{2}{3}\int_{\frac{9}{7}}^{7}\Lambda\left(\frac{8\,z}{z+1}\right)\frac{z+1}{z^2}\,dz > 0.43$$

Hence, (2, 3) follows from the Fundamental Theorem.

Finally, we indicate some mistakes in the proof of (3, 3) of A. И. Виноградов*. Here, we use his notations and omit the explanations. He has obtained that

(10) $3.2\,I - 2\,I_1 < 0.3167.$

Hence for sufficiently large z, we have

$$(11)\qquad \int_{1\,1}^{\lambda}\varepsilon\,(u)\,(2u+1)\,d\,u \leqslant \frac{e^{2w-1.1}}{\pi}(3\cdot2\,I - 2\,I_1) + O\left(\frac{1}{\sqrt{\log z}}\right) < 0.5.$$

On the other hand, we know that $\varepsilon\,(u)$ is a non-negative and non-increasing function and $1 - \varepsilon\,(2) = (4.5 - 4\log 2)\,e^{-2r}$. Hence, it follows that

$$(12)\quad \int_{1.1}^{\lambda}\varepsilon\,(u)\,(2u+1)\,d\,u \geqslant \int_{1.1}^{2}\varepsilon\,(u)\,(2u+1)\,d\,u \geqslant \varepsilon\,(2)\int_{1\,1}^{2}(2u+1)\,du > 1.5,$$

which is a contradiction with (11). Moreover, one can prove

$$(13)\qquad (4\cdot1)^2 - 4\int_{1.55}^{\lambda}\frac{\varepsilon\,(u)}{1-\varepsilon(u)}\,(2u+1)\,d\,u < 16\cdot81 -$$

$$- 4\int_{1.55}^{3}\frac{\varepsilon\,(u)}{1-\varepsilon(u)}\,(2u+1)\,d\,u < -1.$$

Corresponding results of twin-primes pro' ems have also been obtained.

REFERENCES

[1] Wang, Y., 1957, On Sieve Methods and Some of the Related Problems, *Science Record*, Academia Sinica, New Ser. I, 1, 9—12.

[2] Wang, Y., 1957, On Sieve Methods and Their Some Applications, *Science Record*, Academia Sinica, New Ser. I, 3. 1—6.

[3] Wang, Y., 1956. On the Representation of Large Even Integer as a Sum of a Product of at most 3 Primes and a Product of at most 4 Primes, *Acta Math. Sinica*, 3, 500—513.

[4] Бухштаб, А. А., 1940. О Разложении Чётных Чисел на Сумму Двух Слагаемых с Ограниченным Числом Множителей, *ДАН СССР;* 29, 544 548.

[5] Виноградов, А. И., 1957. Применение ς (s) к Решету Эратосфена, *матем, сб;* 41: 1 pp. 49—80.

---

* Added 6. Sept. 1957. I have been informed that this mistake has also been discovered and corrected by himself (Cf. Матем. сб. 41 (83): 3,   415, 1957).

ANNALES
POLONICI MATHEMATICI
IV. 2. (1958)

A note on some properties of the functions
$\varphi(n)$, $\sigma(n)$ and $\theta(n)$

by A. Schinzel (Warszawa) and Y. Wang (Peking)

§ 1. Introduction. A. Schinzel has proved in [4] that for every sequence a of h positive numbers $a_1, a_2, \ldots, a_h$ and $\varepsilon > 0$ there exist natural numbers n and n' such that

$$\left| \frac{\varphi(n+i)}{\varphi(n+i-1)} - a_i \right| < \varepsilon, \qquad \left| \frac{\sigma(n'+i)}{\sigma(n'+i-1)} - a_i \right| < \varepsilon \qquad (i = 1, 2, \ldots, h)(^1).$$

Professor Hua Loo-Keng has pointed out that by Brun's method we can prove the existence of positive constants $c = c(a, \varepsilon)$ and $X_0 = X_0(a, \varepsilon)$ such that the number of numbers n satisfying the first of these inequalities in the interval $1 \leqslant n \leqslant X$ is greater than

$$cX/\log^{h+1}X \qquad \text{for} \qquad X > X_0.$$

In the present paper we give the proof of this theorem, of an analogous theorem on the function $\sigma(n)$ and of a theorem on the function $\theta(n)(^2)$ which is weaker but gives a positive solution of the problem put forward in paper [2] of A. Schinzel and comprises the theorem from paper [3] of A. Schinzel.

The question whether a theorem analogous to the theorems on functions φ and σ is true for the function θ remains open.

§ 2. An auxiliary theorem. Let

$$A_0 = h!\, q_1 \ldots q_s q_{01} \ldots q_{0t_0}, \qquad A_i = q_{i1} \ldots q_{it_i} \qquad (1 \leqslant i \leqslant h)$$

be positive integers, where $q_1, q_2, \ldots, q_s$ are all the prime numbers in the interval $0 < x \leqslant 10(h+1)$ and q_{ij} $(0 \leqslant i \leqslant h, 1 \leqslant j \leqslant t_i)$ are primes greater than $10(h+1)$ such that $A_0, A_1, \ldots, A_h > 1$ are relatively prime in pairs.

(^{1}) $\varphi(n)$ denotes the Euler function, $\sigma(n)$ — the sum of divisors of number n.

(^{2}) $\theta(n)$ denotes the number of divisors of n.

If $Z > A_0 A_1^2 \ldots A_h^2$, then let us denote by $N_Z(X)$ the number of integral solutions $(x_0, \ldots, x_h)$ of the system of equations

$$(1) \qquad A_0 x_0 + i = i A_i x_i \qquad (1 \leqslant i \leqslant h)$$

satisfying the conditions

$$(2) \quad 1^\circ \ 1 \leqslant x_0 \leqslant X, \quad \text{and} \quad 2^\circ \ \text{if } p|x_i, \text{ then } p > Z \ (0 \leqslant i \leqslant h),$$

where p denotes a prime.

THEOREM 1. *There exist positive constants* c_1, *depending on* h *only, and* c_2, X_1, *depending on* $A_i's$ *only, such that*

$$N_{X^{c_1}}(X) > c_2 X / \log^{h+1} X \qquad (X > X_1).$$

Proof. If $Z > A_0 A_1^2 \ldots A_h^2$, λ is a given integer in the interval $0 \leqslant \lambda < A_0 A_1^2 \ldots A_h^2$ and $p_1 < p_2 < \ldots < p_r$ are all prime numbers not dividing $A_0 A_1 \ldots A_h$ and not exceeding Z and if $a_{ji} \ (1 \leqslant j \leqslant r, \ 0 \leqslant i \leqslant h)$ are given integers satisfying the conditions $0 \leqslant a_{ji} < p_j \ (1 \leqslant j \leqslant r, \ 0 \leqslant i \leqslant h)$ and $a_{ji_1} \neq a_{ji_2}$ for $i_1 \neq i_2 \ (1 \leqslant j \leqslant r)$, then we can define $M_Z(X)$ as the number of x satisfying the conditions

$$(3) \qquad 1 \leqslant x \leqslant X, \quad x \equiv \lambda \pmod{A_0 A_1^2 \ldots A_h^2}, \quad x \equiv a_{ji} \pmod{p_j}$$

$$(1 \leqslant j \leqslant r, \ 0 \leqslant i \leqslant h).$$

It is evident, that theorem 1 is a consequence of the following two lemmas:

LEMMA 1. *There exist* λ *and* $a_{ji}'s$ *such that*

$$N_Z(X) \geqslant M_Z(X).$$

LEMMA 2. *There exist positive constants* c_1, *depending on* h *only, and* c_2, X_1, *depending on* $A_i's$ *only, such that*

$$M_{X^{c_1}}(X) \geqslant c_2 X / \log^{h+1} X \qquad (X \geqslant X_1)$$

for any given λ *and* $a_{ji}'s$.

Proof of lemma 1. First we shall define λ and $a_{ji}'s$ as follows: By lemma 2 [4] there exists m such that

$$A_i | m+i, \quad \left(A_i, \frac{m+i}{A_i} \right) = 1, \quad A_0^2 A_1^2 \ldots A_h^2 > m+i > 0 \qquad (0 \leqslant i \leqslant h).$$

Let $\lambda = m/A_0$. The solution of the following congruence

$$(4) \qquad A_0 y + i \equiv 0 \pmod{p_j} \qquad (0 \leqslant y < p_j)$$

will be denoted by $a_{ji} \ (1 \leqslant j \leqslant r, \ 0 \leqslant i \leqslant h)$. It is evident that $a_{j0} = 0$ $(1 \leqslant j \leqslant r)$. If $i_1 \neq i_2$ and $a_{ji_1} = a_{ji_2}$, then from (4) we have $p_j | i_1 - i_2$. We obtain a contradiction since $0 < |i_1 - i_2| < h$ and $p_j > 10(h+1)$.

We take x satisfying (3) with these λ and $a'_{ji}s$ and define $x = x_0$. From (1)-(3) of [4]

$$(5) \qquad A_0 x_0 + i = i A_i x_i \qquad (1 \leqslant i \leqslant h),$$

where

$$(6) \qquad (x_i, A_0 A_1 \ldots A_h) = 1 \qquad (0 \leqslant i \leqslant h).$$

From (4), since $x \not\equiv a_{ji} \pmod{p_j}$ $(1 \leqslant j \leqslant r, 0 \leqslant i \leqslant h)$, we have

$$(7) \qquad \big(A_0 x_0 (A_0 x_0 + 1) \ldots (A_0 x_0 + h), p_1 \ldots p_r \big) = 1.$$

From (5), (6), (7) we find that if $p | x_i$, then $p > Z$ $(0 \leqslant i \leqslant h)$. Thus we have proved that there exist λ and $a'_{ji}s$ such that from any x satisfying conditions (3) we can construct a solution of (1) satisfying (2) and different x correspond to the different solutions of (1). Thus we have proved lemma 1. In § 6 a proof of lemma 2 is given. This proof is obtained by a modification of a method elaborated by H. Rademacher [1] in the case $h = 1$. We shall precede it by lemmas and estimations in § 3-§ 5.

§ 3. Some lemmas. Let $A = A_0 A_1^2 \ldots A_h^2$ and write

$$(8) \qquad M_Z(X) = P(\lambda, a, A, X; p_1, \ldots, p_r)$$

for given λ and a_{ji} $(1 \leqslant j \leqslant r, 0 \leqslant i \leqslant h)$. In particular, let $P(\lambda, A, X)$ denote the number of x satisfying the conditions $1 \leqslant x \leqslant X$ and $x \equiv \lambda \pmod{A}$.

LEMMA 3. *There exist integers λ_i $(0 \leqslant i \leqslant h)$ satisfying $0 \leqslant \lambda_i \leqslant A p_r$ $(0 \leqslant i \leqslant h)$ and $\lambda_{i_1} \neq \lambda_{i_2}$ $(i_1 \neq i_2)$ such that*

$$(9) \qquad P(\lambda, a, A, X; p_1, \ldots, p_r) = P(\lambda, a, A, X; p_1, \ldots, p_{r-1}) -$$

$$- \sum_{i=0}^{h} P(\lambda_i, a, A p_r, X; p_1, \ldots, p_{r-1}).$$

Proof. By definition, $P(\lambda, a, A, X; p_1, \ldots, p_r)$ is equal to the difference between the number of x satisfying the conditions

$$(10) \qquad 1 \leqslant x \leqslant X, \qquad x \equiv \lambda \pmod{A}, \qquad x \not\equiv a_{ji} \pmod{p_j}$$

$$(1 \leqslant j \leqslant r-1, 0 \leqslant i \leqslant h)$$

and the number of x satisfying (10) and one of the following congruences:

$$(11) \qquad x \equiv a_{ri} \pmod{p_r} \qquad (0 \leqslant i \leqslant h).$$

A. Schinzel and Y. Wang

Since $(A, p_r) = 1$, each of the following systems of congruences is solvable and has a unique solution in the interval $0 \leqslant x < Ap_r$

$$\begin{cases} x \equiv \lambda \,(\mathrm{mod}\, A), \\ x \equiv a_{r0}\,(\mathrm{mod}\, p_r), \end{cases} \quad \begin{cases} x \equiv \lambda\,(\mathrm{mod}\, A), \\ x \equiv a_{r1}\,(\mathrm{mod}\, p_r), \end{cases} \cdots \quad \begin{cases} x \equiv \lambda\,(\mathrm{mod}\, A), \\ x \equiv a_{rh}\,(\mathrm{mod}\, p_r). \end{cases}$$

Denote these solutions by $\lambda_0, \lambda_1, \ldots, \lambda_h$ respectively. Since

$$a_{ri_1} \not\equiv a_{ri_2}\,(\mathrm{mod}\, p_r) \quad (i_1 \neq i_2)$$

we have $\lambda_{i_1} \neq \lambda_{i_2}$ $(i_1 \neq i_2)$. Hence $P(\lambda, a, A, X; p_1, \ldots, p_r)$ is equal to the difference between the number x satisfying (10) and the number of x satisfying one of the following conditions (as $i = 0, 1, \ldots,$):

$$(12) \quad 1 \leqslant x \leqslant X, \quad x \equiv \lambda_i\,(\mathrm{mod}\, Ap_r), \quad x \equiv a_{ji}\,(\mathrm{mod}\, p_j)$$

$$(1 \leqslant j \leqslant r-1, 0 \leqslant i \leqslant h).$$

This proves lemma 3.

Briefly we write

$$(13) \quad \begin{aligned} P(\lambda, a, D, X; p_1, \ldots, p_r) &= P(D, X; p_1, \ldots, p_r), \\ P(\lambda, D, X) &= P(D, X). \end{aligned}$$

Hence, it follows from (9) that, with the usual convention in notation, we have

$$(14) \quad P(A, X; p_1, \ldots, p_r)$$

$$= P(A, X; p_1, \ldots, p_{r-1}) - (h+1)P(Ap_r, X; p_1, \ldots, p_{r-1}).$$

Using lemma 3 r times successively we get

LEMMA 4.

$$P(A, X; p_1, \ldots, p_r) = P(A, X) - (h+1) \sum_{a=1}^{r} P(Ap_a, X; p_1, \ldots, p_{a-1}).$$

Let

$$(15) \quad r = r_0 \geqslant r_1 \geqslant \ldots \geqslant r_t = 1$$

be any given sequence of t positive integers.

LEMMA 5.

$$P(A, X; p_1, \ldots, p_r) \geqslant P(A, X) - (h+1) \sum_{a=1}^{r} P(Ap_a, X) +$$

$$+ (h+1)^2 \sum_{a=1}^{r} \sum_{\substack{a_1 \leqslant r_1 \\ a_1 < a}} P(Ap_a p_{a_1}, X; p_1, \ldots, p_{a_1-1}).$$

Proof. By lemma 4, we have

$$P(A, X; p_1, \ldots, p_r) = P(A, X) - (h+1) \sum_{a=1}^{r} P(A p_a, X) +$$

$$+ (h+1)^2 \sum_{a=1}^{r} \sum_{a_1 < a} P(A p_a p_{a_1}, X; p_1, \ldots, p_{a_1-1}).$$

This proves lemma 5.

Using lemma 5 t times successively and observing that $r_t = 1$, we get

LEMMA 6.

$$P(A, X; p_1, \ldots, p_r) \geqslant P(A, X) - (h+1) \sum_{a=1}^{r} P(A p_a, X) +$$

$$+ (h+1)^2 \sum_{\substack{a=1}}^{r} \sum_{\substack{a_1 \leqslant r_1 \\ a_1 < a}} P(A p_a p_{a_1}, X) - (h+1)^3 \sum_{a=1}^{r} \sum_{\substack{a_1 \leqslant r_1 \\ a_1 < a}} \sum_{\substack{\beta_1 \leqslant r_1 \\ \beta_1 < a_1}} P(A p_a p_{a_1} p_{\beta_1}, X) + \ldots +$$

$$+ (h+1)^{2t} \sum_{a=1}^{r} \sum_{\substack{a_1 \leqslant r_1 \\ a_1 < a}} \sum_{\substack{\beta_1 \leqslant r_1 \\ \beta_1 < a_1}} \cdots \sum_{\substack{a_{t-1} \leqslant r_{t-1} \\ a_{t-1} < \beta_{t-2}}} \sum_{\substack{\beta_{t-1} \leqslant r_{t-1} \\ \beta_{t-1} < a_{t-1}}} \sum_{\substack{a_t \leqslant r_t \\ a_t < \beta_{t-1}}} P(A p_a p_{a_1} p_{\beta_1} \cdots p_{a_t}, X).$$

By the definition of $P(\lambda, D, X)$ we have

$$P(\lambda, D, X) = [X/D] + \theta_\lambda, \qquad \theta_\lambda = 0 \text{ or } 1.$$

Hence, for all λ, we have $|P(\lambda, D, X) - X/D| \leqslant 1$ and from lemma 6 we get

LEMMA 7. *For any given λ and $a'_{ji}s$*

$$M_Z(X) = P(A, X; p_1, \ldots, p_r) \geqslant X \frac{E}{A} - R$$

where

$$E = 1 - (h+1) \sum_{a=1}^{r} \frac{1}{p_a} + (h+1)^2 \sum_{a=1}^{r} \sum_{\substack{a_1 \leqslant r_1 \\ a_1 < a}} \frac{1}{p_a p_{a_1}} -$$

$$- (h+1)^3 \sum_{a=1}^{r} \sum_{\substack{a_1 \leqslant r_1 \\ a_1 < a}} \sum_{\substack{\beta_1 \leqslant r_1 \\ \beta_1 < a_1}} \frac{1}{p_a p_{a_1} p_\beta} + \ldots +$$

$$+ (h+1)^{2t} \sum_{a=1}^{r} \sum_{\substack{a_1 \leqslant r_1 \\ a_1 < a}} \sum_{\substack{\beta_1 \leqslant r_1 \\ \beta_1 < a_1}} \cdots \sum_{\substack{a_{t-1} \leqslant r_{t-1} \\ a_{t-1} < \beta_{t-2}}} \sum_{\substack{\beta_{t-1} \leqslant r_{t-1} \\ \beta_{t-1} < a_{t-1}}} \sum_{\substack{a_t \leqslant r_t \\ a_t < \beta_{t-1}}} \frac{1}{p_a p_{a_1} p_{\beta_1} \cdots p_{a_t}}$$

and

$$R \leqslant 1 + (h+1) \sum_{a=1}^{r} 1 + (h+1)^2 \sum_{a=1}^{r} \sum_{\substack{a_1 \leqslant r_1 \\ a_1 \leqslant a}} 1 + (h+1)^3 \sum_{a=1}^{r} \sum_{\substack{a_1 \leqslant r_1 \\ a_1 < a}} \sum_{\substack{\beta_1 \leqslant r_1 \\ \beta_1 < a_1}} 1 + \cdots +$$

$$+ (h+1)^{2t} \sum_{a=1}^{r} \sum_{\substack{a_1 \leqslant r_1 \\ a_1 < a}} \sum_{\substack{\beta_1 \leqslant r_1 \\ \beta_1 < a_1}} \cdots \sum_{\substack{a_{t-1} \leqslant r_{t-1} \\ \beta_{t-1} < a_{t-1}}} \sum_{\substack{\beta_{t-1} \leqslant r_{t-1} \\ \beta_{t-1} < a_{t-1}}} \sum_{\substack{a_t \leqslant r_t \\ a_t < \beta_{t-1}}} 1$$

$$\leqslant [1 + (h+1)r] \prod_{n=1}^{t} [1 + (h+1)r_n]^2 .$$

§ 4. Estimation of R. In this section and the next we shall always assume that $c_3, c_4, \ldots$ are positive constants depending on h only. Let r_1 denote the least positive integer for which

$$\pi_1 = \prod_{r_1 < s \leqslant r_0} \left(1 - \frac{h+1}{p_s}\right) \geqslant \frac{1}{1,3} .$$

Similarly, we define r_n as the least positive integer for which

$$(16) \qquad \pi_n = \prod_{r_n < s \leqslant r_{n-1}} \left(1 - \frac{h+1}{p_s}\right) \geqslant \frac{1}{1,3} \qquad (1 \leqslant n \leqslant t-1),$$

$$(17) \qquad \pi_t = \prod_{r_t \leqslant s \leqslant r_{t-1}} \left(1 - \frac{h+1}{p_s}\right) \geqslant \frac{1}{1,3} .$$

(Since $1 - (h+1)/p_s > 1 - (h+1)/10(h+1) = 9/10 > 1/1,3$, we finally have $r_t = 1$.) From the definition of r_n, we have

$$\frac{9}{10}\,\pi_n = \left(1 - \frac{h+1}{10(h+1)}\right) \pi_n < \left(1 - \frac{h+1}{p_{r_n}}\right) \pi_n < \frac{1}{1,3} < \frac{4}{5} \qquad (1 \leqslant n \leqslant t-1)$$

or

$$(18) \qquad \pi_n < \tfrac{8}{9} \qquad (1 \leqslant n \leqslant t-1).$$

Hence

$$\log[1 + (h+1)r_n]^2 \leqslant c_3 \log h r_n \leqslant c_4 \log p_{r_n} < c_5 \prod_{s=1}^{r_n} (1 - 1/p_s)^{-1} \prod_{p \mid A} (1 - 1/p)^{-1}$$

$$< c_5 \prod_{j=1}^{n} \prod_{r_j < s \leqslant r_{j-1}} (1 - (h+1)/p_s) \prod_{s=1}^{r} (1 - (h+1)/p_s)^{-1} \prod_{p \mid A} (1 - (h+1)/p)^{-1}$$

$$< c_5 \prod_{p \leqslant Z} (1 - (h+1)/p)^{-1} (8/9)^n \qquad (0 \leqslant n \leqslant t-1).$$

Then by lemma 7

$$\log R \leqslant \log\left\{[1+(h+1)r_0]\prod_{n=1}^{t-1}[1+(h+1)r_n]^2(h+2)^2\right\}$$

$$< c_6\prod_{p\leqslant Z}(1-1/p)^{-1}\sum_{n=0}^{\infty}(8/9)^n = 9c_6\prod_{p\leqslant Z}(1-1/p)^{-1} < c_7\log Z \ (^3)$$

i. e.,

$$(19)\qquad\qquad R \leqslant \exp(c_7\log Z) = Z^{c_7}.$$

§ 5. Estimation of E. Let $S_n^{(l)}$ $(1 \leqslant n \leqslant t-1)$ be the l-th elementary symmetric function of

$$\{(h+1)/(p_{r_n-1}), \ldots, (h+1)/(p_{r_{n-1}})\}$$

and let $S_t^{(l)}$ be the l-th elementary symmetric function of

$$\{(h+1)/p_{r_t}, \ldots, (h+1)/p_{r_{t-1}}\}.$$

Put

$$(20)\qquad E = E_t, \qquad E_n = E_n^{(0)} - E_n^{(1)} + \ldots - E_n^{(2n-1)} + E_n^{(2n)}$$

where $E_n^{(0)} = 1$ $(1 \leqslant n \leqslant t)$, $E_n^{(\nu)}$ $(1 \leqslant n \leqslant t-1)$ denotes the absolute value of the sum of terms of E with exactly ν prime factors the indices of those prime factors being greater than r_n, and $E_t^{(\nu)}$ denotes the absolute value of the sum of all terms of E with exactly ν prime factors. We have

$$(21)\qquad E_n^{(\nu)} = E_{n-1}^{(\nu)} + E_{n-1}^{(\nu-1)}S_n^{(1)} + \ldots + E_{n-1}^{(1)}S_n^{(\nu-1)} + S_n^{(\nu)}$$

$$(2 \leqslant n \leqslant t, \ 1 \leqslant \nu \leqslant 2n-1).$$

It is clear that $E_n^{(\nu)} = 0$ if $n \leqslant t-1$, $\nu \geqslant 2n$ and

$$(22)\qquad\begin{aligned}E_t^{(2t)} &\leqslant E_{t-1}^{(2t)} + E_{t-1}^{(2t-1)}S_t^{(1)} + \ldots + E_{t-1}^{(1)}S_t^{(2t-1)} + S_t^{(2t)},\\ E_t^{(\nu)} &= 0 \quad (\nu > 2t).\end{aligned}$$

$(^3)$ The fact that $\prod_{p\leqslant x}(1-1/p)^{-1} \leqslant a\log x$, for $x \geqslant 2$, where a is an absolute positive constant, implies that

$$\prod_{h+1<p\leqslant x}\left(1-\frac{h+1}{p}\right)^{-1} \leqslant a(h)\log^{h+1}x,$$

for $x > h+1$, where $a(h)$ is a positive constant depending on h only.

                    A. Schinzel and Y. Wang

From (16) and (17) we have

$$S_n^{(1)} = \sum_{r_n < s \leqslant r_{n-1}} \frac{h+1}{p_s} < -\log \pi_n < \log 1{,}3 < 0{,}3 \quad (1 \leqslant n \leqslant t-1),$$

$$(23)$$

$$S_t^{(1)} = \sum_{r_t \leqslant s \leqslant r_{t-1}} \frac{h+1}{p_s} < -\log \pi_t < \log 1{,}3 < 0{,}3.$$

If $\nu > 1$ and $1 \leqslant n \leqslant t-1$, we get

$$\nu S_n^{(\nu)} = \sum_{p_s' q' = q} 1 \sum_q \frac{(h+1)^\nu}{q} = \sum_{\substack{p_s', q' \\ (p_s', q') = 1}} \frac{(h+1)^\nu}{p_s' q'}$$

$$\leqslant \sum_{p_s'} \frac{h+1}{p_s'} \sum_{q'} \frac{(h+1)^{\nu-1}}{q'} = S_n^{(1)} S_n^{(\nu-1)},$$

where p_s' runs over all prime numbers such that $p_{r_n} < p_s' \leqslant p_{r_{n-1}}$, q runs over all products $p_1' \ldots p_\nu'$ of ν different prime numbers, q' runs over all products $p_1' \ldots p_{\nu-1}'$ of $\nu-1$ different prime numbers.

Similarly, we have $\nu S_t^{(\nu)} \leqslant S_t^{(1)} S_t^{(\nu-1)}$ $(\nu > 1)$. Hence, by (23), we have

$$(24) \qquad S_n^{(\nu)} < S_n^{(\nu-1)} \quad (\nu > 1), \qquad S_n^{(\nu)} \leqslant \frac{(S_n^{(1)})^\nu}{\nu!} < \frac{(0{,}3)^\nu}{\nu!}$$

$$(1 \leqslant n \leqslant t, \ \nu = 1, 2, \ldots).$$

From (24) we immediately get

$$(25) \qquad \sum_{j \geqslant 2n-i} (-1)^j S_n^{(j)} \leqslant S_n^{(2n-i)} \quad (1 \leqslant n \leqslant t).$$

By (20) and (21), if $2 \leqslant n \leqslant t-1$, we have

$$E_n = \sum_{\nu=0}^{2n-1} (-1)^\nu E_n^{(\nu)} = \sum_{\nu=0}^{2n-1} (-1)^\nu \sum_{i+j=\nu} E_{n-1}^{(i)} S_n^{(j)} = \sum_{i+j<2n} (-1)^{i+j} S_n^{(j)} E_{n-1}^{(i)}$$

$$= \sum_{i<2n-2} (-1)^i E_{n-1}^{(i)} \sum_{j<2n-i} (-1)^j S_n^{(j)}$$

$$= \sum_{i<2n-2} (-1)^i E_{n-1}^{(i)} \Big[\pi_n - \sum_{j \geqslant 2n-i} (-1)^j S_n^{(j)} \Big]$$

$$= \pi_n E_{n-1} - \sum_{i<2n-2} (-1)^i E_{n-1}^{(i)} \sum_{j \geqslant 2n-i} (-1)^j S_n^{(j)}.$$

Similarly we have

$$E_t = \sum_{r=0}^{2t} (-1)^r E_t^{(r)} = \pi_t E_{t-1} - \sum_{i<2t-2} (-1)^i E_{t-1}^{(i)} \sum_{j>2t-i} (-1)^j S_t^{(j)} + E_t^{(2t)}.$$

Hence, from (24) and (25), we have

$$(26) \qquad E_n \geqslant \pi_n E_{n-1} - \Phi_n \qquad (2 \leqslant n \leqslant t),$$

where

$$(27) \qquad \Phi_n = \sum_{i \leqslant 2n-3} E_{n-1}^{(i)} S_n^{(2n-i)} \qquad (2 \leqslant n \leqslant t).$$

Hence, we get

$$(28) \qquad E = E_t \geqslant \pi_t E_{t-1} - \Phi_t \geqslant \pi_t(E_{t-2}\pi_{t-1} - \Phi_{t-1}) - \Phi_t$$

$$\geqslant \pi_2 \dots \pi_t \left\{ E_1 - \frac{\Phi_2}{\pi_2} - \frac{\Phi_3}{\pi_2 \pi_3} - \dots - \frac{\Phi_t}{\pi_2 \pi_3 \dots \pi_t} \right\}$$

$$> \prod_{s=1}^{r} \left(1 - \frac{h+1}{p_s}\right) \left\{ 1 - S_1^{(1)} - \frac{\Phi_2}{\pi_2} - \frac{\Phi_3}{\pi_2 \pi_3} - \dots - \frac{\Phi_t}{\pi_2 \pi_3 \dots \pi_t} \right\}.$$

From (21), (22), (23), (24), we have for all $u \geqslant 0$, $t \geqslant 2$

$$2^u(0,3)^{-u} E_2^{(u)} \leqslant 2^u(0,3)^{-u} \sum_{i+j=u} E_1^{(i)} S_2^{(j)} \leqslant \sum_{i+j=u} E_1^{(i)} \frac{2^u(0,3)^{-i}}{j!}$$

$$\leqslant \frac{2^u}{u!} + \frac{(0,3)2^u(0,3)^{-1}}{(u-1)!} = \frac{2^u}{u!} + \frac{2^u}{(u-1)!} \leqslant 6.$$

By induction we get

$$(29) \qquad 2^v(0,3)^{-v} E_n^{(v)} \leqslant 6e^{2n-4} \qquad (t \geqslant n \geqslant 2) \qquad \text{for all } v.$$

We have shown that (29) is true for $n = 2$. Now suppose it is true for $n \geqslant 2$; then

$$2^v(0,3)^{-v} E_{n+1}^{(v)} \leqslant 2^v(0,3)^{-v} \sum_{\mu=0}^{v} E_n^{(\mu)} S_{n+1}^{(v-\mu)}$$

$$< 2^v(0,3)^{-v} \sum_{\mu=0}^{v} 6e^{2n-4} 2^{-\mu}(0,3)^{\mu} \frac{(0,3)^{v-\mu}}{(v-\mu)!}$$

$$= 6e^{2n-4} \sum_{\mu=0}^{v} \frac{2^{v-\mu}}{(v-\mu)!} < 6e^{2n-2}.$$

It completes the proof.

If $t \geqslant n \geqslant 3$, we get

$$(30) \qquad \Phi_n = \sum_{r=0}^{2n-3} E_{n-1}^{(r)} S_n^{(2n-r)} \leqslant 6e^{2n-6} \sum_{r=0}^{2n-3} 2^{-r}(0,3)^r \frac{(0,3)^{2n-r}}{(2n-r)!}$$

$$= 6e^{2n-6} \left(\frac{0,3}{2}\right)^{2n} \sum_{r=0}^{2n-3} \frac{2^{2n-r}}{(2n-r)!} < 6e^{2n-6}(0,2)^{2n}(e^2-5),$$

and we also easily get

$$(31) \qquad \Phi_2 = S_2^{(4)} + S_1^{(1)} S_2^{(3)} \leqslant \frac{(0,3)^4}{4!} + \frac{(0,3)^4}{3!} < 0,0018.$$

From (16), (17), (28), (30) and (31) we get

$$(32) \qquad E > \prod_{s=1}^{r} \left(1 - \frac{h+1}{p_s}\right) \times$$

$$\times \left\{1 - 0,3 - 1,3 \cdot 0,0018 - 6(e^2-5) \sum_{r=3}^{\infty} (0,2)^{2r} e^{2r-6} (1,3)^{r-1}\right\}$$

$$> 0,5 \prod_{s=1}^{r} \left(1 - \frac{h+1}{p_s}\right) \geqslant 0,5 \prod_{p \leqslant Z} \left(1 - \frac{h+1}{p}\right) > \frac{c_8}{\log^{h+1} Z} \, .$$

§ 6. Proof of lemma 2. From lemma 7, (19) and (32) we have

$$M_Z(X) > \frac{c_8}{A} \cdot \frac{X}{\log^{h+1} Z} - Z^{c_7}$$

for any given λ and $a'_{ji}s$. We take $c_1 = 1/(c_7+1)$ and $Z = X^{c_1}$. It is obvious that there exist c_2 and X_1, depending on $A'_i s$ only, such that

$$M_{X^{c_1}}(X) > \frac{c_2 X}{\log^{h+1} X} \quad \text{for} \quad X > X_1, \quad \text{q. e. d.}$$

§ 7. Theorems on the functions φ and σ.

THEOREM 2. *For any given sequence a of h non-negative numbers $a_1, a_2, \ldots, a_h$ and $\varepsilon > 0$, there exists a positive integer n such that*

$$(33) \qquad \left|\frac{\varphi(n+i)}{\varphi(n+i-1)} - a_i\right| < \varepsilon \qquad (1 \leqslant i \leqslant h).$$

There exist positive constants $c = c(a, \varepsilon)$ and $X_0 = X_0(a, \varepsilon)$ such that in any interval $1 \leqslant n \leqslant X$ the number of n satisfying (33) is greater than $cX/\log^{h+1} X$ whenever $X > X_0$.

Proof. To begin with, by similar arguments as in the proofs of lemma 3a and of theorem 1 [4], we can choose $A_0, A_1, \ldots, A_h$, depending on $a_i's$ and ε only and satisfying the same conditions as in theorem 1, such that

$$(34) \quad \left|\frac{\varphi(A_1)/A_1}{\varphi(A_0)/A_0} - a_1\right| < \frac{\varepsilon}{2} \quad \text{and} \quad \left|\frac{\varphi(iA_i)/iA_i}{\varphi[(i-1)A_{i-1}]/(i-1)A_{i-1}} - a_i\right| < \frac{\varepsilon}{2}$$
$$(2 \leqslant i \leqslant h).$$

For those $A_i's$ we assume that $(x_0, \ldots, x_h)$ is a solution of (1) satisfying (2) with $Z = X^{c_1}$. If we take $A_0 x_0 = n$, then $iA_i x_i = n+i$ $(1 \leqslant i \leqslant h)$. Since $(x_i, A_i) = 1$ $(0 \leqslant i \leqslant h)$, we have

$$\frac{\varphi(n+1)}{\varphi(n)} = \frac{\varphi(A_1 x_1)}{\varphi(A_0 x_0)} = \frac{\big(\varphi(A_1)/A_1\big)\big(\varphi(x_1)/x_1\big)A_1 x_1}{\big(\varphi(A_0)/A_0\big)\big(\varphi(x_0)/x_0\big)A_0 x_0}$$

$$(35) \qquad = \frac{\varphi(A_1)/A_1}{\varphi(A_0)/A_0} \cdot \frac{\varphi(x_1)/x_1}{\varphi(x_0)/x_0} \cdot \frac{n+1}{n} ,$$

$$\frac{\varphi(n+i)}{\varphi(n+i-1)} = \frac{\varphi(iA_i)/iA_i}{\varphi[(i-1)A_{i-1}]/(i-1)A_{i-1}} \cdot \frac{\varphi(x_i)/x_i}{\varphi(x_{i-1})/x_{i-1}} \cdot \frac{n+i}{n+i-1} .$$

On account of $x_i \leqslant c_9 X$ $(0 \leqslant i \leqslant h)$ we can choose $X_2(A) = X_2(a, \varepsilon) > X_1$ such that the number of prime divisors (identical or different) of each x_i $(0 \leqslant i \leqslant h)$ does not exceed $c_{10} = [1/c_1]+2$ for $X > X_2$. Hence

$$1 \geqslant \frac{\varphi(x_i)}{x_i} = \prod_{p|x_i}\left(1 - \frac{1}{p}\right) \geqslant \left(1 - \frac{1}{X^{c_1}}\right)^{c_{10}} \to 1 \quad (\text{as } X \to \infty).$$

From (34) and (35) we can choose $X_3(a, \varepsilon) > X_2$ such that if $n = A_0 x_0 > X_3$, then

$$\left|\frac{\varphi(n+i)}{\varphi(n+i-1)} - a_i\right| < \varepsilon \quad (1 \leqslant i \leqslant h).$$

Thus we have proved that from every solution of (1) satisfying (2) with $Z = X^{c_1}$ and such that $A_0 x_0 > X_3$ we can define a positive integer $n = A_0 x_0$ satisfying (33), and that evidently different solutions correspond to different n. It is clear that the number of solutions of (1) satisfying (2) with $Z = X^{c_1}$ and such that $A_0 x_0 \leqslant X_3$ is less than X_3^{h+1}.

Hence, by theorem 1, there exist positive constants X_0 and c, depending on $a_i's$ and ε only, such that, if $X > X_0$ in any interval $1 \leqslant n \leqslant X$, the number of n satisfying (33) is greater than $cX/\log^{h+1}X$, q. e. d.

THEOREM 3. *In theorem 2 the function* φ *can be replaced by* σ *(clearly the constants* c *and* X_0 *must be changed).*

The proof of theorem 3 is analogous to the preceding one but based on the proofs of lemma 3b and theorem 2 [4].

A. Schinzel and Y. Wang

§ 8. Theorems on the function θ. We now prove

THEOREM 4. *For any given positive integer h, there exists a constant $b = b(h)$ such that for any given sequence a of h integers $a_1, a_2, \ldots, a_h > 1$ there exists a positive integer n such that*

$$(36) \qquad \theta(i)a_i < \theta(n+i) < b\theta(i)a_i \qquad (1 \leqslant i \leqslant h).$$

There exist positive constants $c' = c'(a)$ and $X_0' = X_0'(a)$ such that in any interval $1 \leqslant n \leqslant X$ the number of n satisfying (36) is greater than $c'X/\log^{h+1}X$, whenever $X > X_0'$.

Proof. We can choose $A_0, A_1, \ldots, A_h$, depending on $a_i's$ only and satisfying the same conditions as in theorem 1, such that $\theta(A_i) = a_i$ $(1 \leqslant i \leqslant h)$. For those $A_i's$ we assume that $(x_0, \ldots, x_h)$ is a solution of (1) satisfying (2) with $Z = X^{c_1}$. If we take $A_0 x_0 = n$, then $iA_i x_i = n+i$ $(1 \leqslant i \leqslant h)$. Since $(x_i, iA_i) = 1$ $(1 \leqslant i \leqslant h)$, we have

$$\theta(n+i) = \theta(iA_i)\theta(x_i) = \theta(i)a_i\theta(x_i) \qquad (1 \leqslant i \leqslant h).$$

As in the proof of theorem 2, we can choose $X_2'(a)$ such that the number of prime divisors (identical or different) of each x_i $(1 \leqslant i \leqslant h)$ does not exceed c_{10} for $X > X_2'$. Hence for $X > X_2'$

$$\theta(i)a_i < \theta(n+i) < \theta(i)a_i 2^{c_{10}} = b\theta(i)a_i \qquad (1 \leqslant i \leqslant h).$$

Further proof is analogous to the proof of theorem 2.

From theorem 4 we can directly obtain

THEOREM 5. *For any given sequence of h numbers $a_1, \ldots, a_h$, where $a_i = 0$ or $+\infty$ $(1 \leqslant i \leqslant h)$ there exists an infinite sequence of natural numbers $n_1, n_2, \ldots$ such that*

$$\lim_{k \to \infty} \frac{\theta(n_k+i)}{\theta(n_k+i-1)} = a_i \qquad (1 \leqslant i \leqslant h).$$

In the course of publication theorems 3 and 5 were proved independently by Shao Pin-Tsung (to appear in Shou Hsueh Chin-Chan (Progress of Mathematics) and Pei-Ta-Hsueh Pao (Transactions of Peking University)) (cf. [5]).

References

[1] H. Rademacher, *Beiträge zur Viggo Brunschen Methode in der Zahlentheorie*, Abh. Math. Sem. Hamburgischen Univ. 3 (1923), p. 12-30.

[2] A. Schinzel, *Quelques théorèmes sur les fonctions $\varphi(n)$ et $\sigma(n)$*, Bull. Acad. Polon. Sci., Cl. III, 2 (1954), p. 467-469.

[3] — *Sur une propriété du nombre de diviseurs*, Publ. Math. 3, Debrecen 1954, p. 261-262.

[4] — *On functions $\varphi(n)$ and $\sigma(n)$*, Bull. Acad. Polon. Sci., Cl. III, 3 (1955), p. 415-419.

[5] S h a o P i n - T s u n g, *On the distribution of the values of a class of arithmetical functions*, Bull. Acad. Polon. Sci., Cl. III, 4 (1956), p. 569-572.

Reçu par la Rédaction le 1. 3. 1956

Note added in the proof by A. Schinzel. Theorem 1 easily results from the following theorem of G. Ricci (see G. Ricci, *Su la congettura di Goldbach e la constante di Schnirelman*. Annali della R. Scuola Normale Superiore di Pisa 6 (2) 1937, p. 83):

Let $a_1 x + b_1$, $a_2 x + b_2$, ... $a_f x + b_f$, where $(a_i, b_i) = 1$ $(i = 1, 2, ..., f)$ are f different arithmetical progressions. let D be a fixed divisor of the polynomial

$$(a_1 x + b_1)(a_2 x + b_2) \ldots (a_f x + b_f)$$

and put $a_1 x + b_1 = d_1 P_1$, $a_2 x + b_2 = d_2 P_2, \ldots, a_f x + b_f = d_f P_f$, where $d_1 d_2 \ldots d_f = D$.

The number of natural numbers $x \leqslant \xi$ such that all integers $P_1, P_2, \ldots P_f$ have no prime factors $\leqslant \xi^{1/(1+2\tau(f))}$ is of the same order of magnitude as $\xi/\log^f \xi$.

In fact, in virtue of lemma 2 [4] there exists a natural number m such that

$$A_i | m + i. \quad (A_i, (m+i)/A_i) = 1 \quad (0 \leqslant i \leqslant h)$$

and in virtue of the formulas (1)-(3) of [4]

$$(A_0^2 A_1^2 \ldots A_h^2, m) = A_0. \quad (A_0^2 A_1^2 \ldots A_h^2, m+i) = iA_i \quad (1 \leqslant i \leqslant h).$$

Put

$$a_1 = A_0 A_1^2 \ldots A_h^2, \quad b_1 = m/A_0,$$

$$a_i = A_0^2 A_1^2 \ldots A_h^2/(i-1)A_{i-1} \quad (1 < i \leqslant h+1).$$

$$b_i = (m+i-1)/(i-1)A_{i-1}$$

We therefore get $(a_i, b_i) = 1$ $(1 \leqslant i \leqslant h+1)$ and $(A_0, b_1 b_2 \ldots b_{h+1}) = 1$ whence also $((h+1)!, b_1 b_2 \ldots b_{h+1}) = 1$. From the last equality it follows that the polynomial

$$(a_1 x + b_1)(a_2 x + b_2) \ldots (a_{h+1} x + b_{h+1})$$

has no fixed divisor > 1, since such divisor D divides $(h+1)!$.

Putting in the above mentioned theorem of Ricci $d_i = 1$, $a_i x + b_i = x_{i-1}$ $(1 \leqslant i \leqslant h)$ we find that the number of natural numbers $x \leqslant \xi$ such that all the numbers x_i $(0 \leqslant i \leqslant h)$ have no prime factors $\leqslant \xi^{1/(1+2\tau(h+1))}$ is of the same order of magnitude as $\xi/\log^{h+1} \xi$. Put

$$\xi = \frac{A_0 x - m}{A_0^2 A_1^2 \ldots A_h^2}, \quad z = \xi^{1/(1+2\tau(h+1))}.$$

As the number x_i satisfy the system of equations (1) and for $x \leqslant \xi$ the conditions (2), we get the inequality

$$N_{x c_1}(X) > \frac{c_2 X}{\log^{h+1} X} \quad (X > X_1)$$

where $c_2 > 0$ and X_1 are constants depending only on A_i and c_1 is an arbitrary constant $< \dfrac{1}{1 + 2\tau(h+1)}$, depending therefore only on h.

ANNALES
POLONICI MATHEMATICI
XIX (1967)

Corrigendum to "A note on some properties of the functions $\varphi(n)$, $\sigma(n)$ and $\theta(n)$" by A. Schinzel and Y. Wang

(these Annales 4 (1958))

The deduction of formula (19) from (18) given on pp. 206-207 is incorrect and should be replaced by the following text:

Hence

$$\prod_{r_j < s \leqslant r_{j-1}} (1-1/p_s) < \pi_j^{9/10(h+1)} < c_3 < 1 \qquad (1 \leqslant j \leqslant t-1),$$

$$\log[1+(h+1)r_n]^2 < c_4 \log p_{r_n} < c_5 \prod_{s=1}^{r_n} (1-1/p_s)^{-1} \prod_{p|A} (1-1/p)^{-1}$$

$$= c_5 \prod_{j=1}^{n} \prod_{r_j < s \leqslant r_{j-1}} (1-1/p_s) \prod_{s=1}^{r} (1-1/p_s)^{-1} \prod_{p|A} (1-1/p)^{-1}$$

$$< c_5 \prod_{p \leqslant Z} (1-1/p)^{-1} c_3^{n}.$$

Then by lemma 7

$$\log R \leqslant \log \left\{ [1+(h+1)r_0] \prod_{n=1}^{t-1} [1+(h+1)r_n]^2 (h+2)^2 \right\}$$

$$< c_6 \prod_{p \leqslant Z} (1-1/p)^{-1} \sum_{n=0}^{\omega} c_3^{n} < c_7 \log Z \ldots$$

Reçu par la Rédaction le 25. 4. 1966

A NOTE ON SOME PROPERTIES OF THE ARITHMETICAL FUNCTIONS $\varphi(n), \sigma(n)$ AND $d(n)^*$

WANG YUAN

Institute of Mathematics, Academia Sinica

1. Introduction

Let $f(n)$ be an arithmetic function and consider the problem of distribution of the ratios $f(n+1)/f(n)$ $(n = 1, 2, \ldots)$. Somayajulu [1], Sierpiński [2] and Schinzel [3] treated the functions by their elementary methods. Hua Loo-keng pointed out the possibility of applying Brun's method to treat this problem, and then the author and Schinzel [4], and Shou Pin-tsung [5] obtained more precise results in this way, for example, they proved the following result:

Let $\varphi(n)$ be Euler function. Then for any given k non-negative real numbers $a_1, \ldots, a_k$, there exists a sequence of integers $\{n_j\}$ such that

$$\lim_{j \to \infty} \frac{\varphi(n_j + \nu + 1)}{\varphi(n_j + \nu)} = a_\nu, \quad 1 \le \nu \le k.$$

The aim of the present paper is to treat this problem by Linnik–Renyi's method, and to improve the above result: under the same assumption, we shall prove that there exists a sequence of prime numbers $\{p_j\}$ such that

$$\lim_{j \to \infty} \frac{\varphi(p_j + \nu + 1)}{\varphi(p_j + \nu)} = a_\nu, \quad 1 \le \nu \le k.$$

The proof of the above result depends on the following:

Fundamental lemma. *Let k be a positive integer and*

$$m_0 = (k+1)!^2 q_{01} \cdots q_{0t_0}, \quad m_i = q_{i1} \cdots q_{it_i}, \quad 1 \le i \le k \tag{1}$$

be integers which are coprime from one another, where $q_{\mu\nu}$ $(0 \le \mu \le k, 1 \le \nu \le t_\mu)$ are prime numbers $> k + 1$. When $x > Z > (m_0 m_1 \cdots m_k)^2$, let $N_Z(x)$ denote the

---

*Acta Mathematica Sinica, **8**:1, 1958, 1–11

number of integral solutions $(p, x_0, x_1, \ldots, x_k)$ *of the system of equations*

$$\begin{cases} p + 1 = m_0 x_0, \\ p + \nu + 1 = v m_\nu x_\nu, \quad 1 \le \nu \le k \end{cases} \tag{2}$$

satisfying the conditions

$$1 < p \le x, \quad \text{and if } p' | x_\nu, \text{ then } p' > Z \quad 0 \le \nu \le k, \tag{3}$$

where p and p' denote prime numbers.[a] Then there exist two positive constants c_1 and X_1, and a positive constant α depending only on $m'_i s$ such that

$$N_{x^\alpha}(x) > \frac{c_1 x}{\log^{k+2} x \log \log x}, \quad x > X_1.$$

2. The Proof of Fundamental Lemma

Let $M = (m_0, m_1, \ldots, m_k)^2$ and λ be an integer such that $1 \le \lambda \le M$ and $(\lambda, M) = 1$. Let $p_1 < p_2 < \cdots < p_r \le Z$ be all primes not exceeding Z and not dividing M and a_{ij} $(1 \le i \le r, 1 \le j \le k + 1)$ be positive integers satisfying the conditions: $1 \le a_{ij} < p_i$ and $a_{ij_1} \ne a_{ij_2}$ for $j_1 \ne j_2$. When $x > Z > M$, we use $M_Z(x)$ to denote the number of prime numbers satisfying

$$1 < p \le x, \quad p \equiv \lambda \pmod{M}, \quad p \not\equiv a_{ij} \pmod{p_i}, \quad 1 \le i \le r, \quad 1 \le j \le k + 1. \tag{4}$$

Lemma 1. *There exist λ and a_{ij}'s such that*

$$N_Z(x) > M_Z(x).$$

Proof. It follows by Sun-Zi theorem that the system of congruences

$$y + \nu + 1 \equiv m_\nu \pmod{m_\nu^2}, \quad 0 \le \nu \le k \tag{5}$$

has a unique solution λ in the interval $1 \le y \le M$.

Since $m_\nu | (\lambda + \nu + 1)$, we have $(m_\nu, \lambda) = 1$ by the definition of m_ν. Hence

$$(\lambda, M) = 1, \quad \left(m_\nu, \frac{\lambda + \nu + 1}{m_\nu} \right) = 1, \quad 0 \le \nu \le k. \tag{6}$$

Let

$$a_{ij} = p_i - j, \quad 1 \le i \le r, \quad 1 \le j \le k + 1. \tag{7}$$

---

[a] Hereafter we use $p, p', p_1, p_2, \ldots, p'_1, p'_2, \ldots$ to denote prime numbers.

Take p satisfying (4) for the λ and a_{ij}'s. Then

$$\begin{cases} p + 1 = m_0 x_0 \\ p + \nu + 1 = m_0 x_0 + \nu = \nu m_\nu x_\nu, \quad 1 \le \nu \le k \end{cases}$$

by (5), and

$$(x_0, m_0) = \left(\frac{p+1}{m_0}, m_0 \right) = \left(\frac{\lambda+1}{m_0}, m_0 \right) = 1,$$

$$(x_0, m_i) = (m_0 x_0, m_i) = (p + 1, m_i) = (p + i + 1 - i, m_i) = (-i, m_i)$$
$$= 1, \quad 1 \le i \le k,$$

$$(x_i, m_j) = (i m_i x_i, m_j) = (p + i + 1, m_j) = (p + j + 1 + i - j, m_j)$$
$$= (i - j, m_j) = 1, \quad i \ne 0, \ i \ne j, \ j \ne 0,$$

$$(x_i, m_i) = (i x_i, m_i) = \left(\frac{i m_i x_i}{m_i}, m_i \right) = \left(\frac{p + i + 1}{m_i}, m_i \right) = 1, \quad i \ne 0,$$

$$(i x_i, m_0) = (i m_i x_i, m_0) = (p + i + 1, m_0) = (i, m_0) = i, \quad i \ne 0;$$

by (6). Since $i^2 \mid m_0$, we have $(x_i, m_0) = 1$ $(i \ne 0)$. Therefore

$$(x_0 x_1 \cdots x_k, \ m_0 m_1 \cdots m_k) = 1 \tag{8}$$

and

$$((p + 1)(p + 2) \cdots (p + k + 1), \ p_1 \cdots p_r) = 1 \tag{9}$$

by (7). It follows by (8) and (9) that for such p, we have a solution $(p, x_0, x_1, \ldots, x_k)$ satisfying the requirement in Fundamental lemma. The distinct primes correspond evidently to distinct solutions. The lemma is proved.

Lemma 2. *There exist a positive constant β depending on k and two positive constants c_2 and X_2 depending on k and M only such that*

$$M_{x^\beta}(x) > \frac{c_2 x}{\log^{k+2} x \log \log x}, \quad x > X_2.$$

Fundamental lemma is clearly a consequence of Lemmas 1 and 2. For the proof of Lemma 2 the reader may refer to Renyi [6] for $k = 0$. There is no essential difficulty to extend his proof to the case $k > 0$, and we still give a sketch of the proof for completeness.

3. Brun's Sieve Method

We suppose that M, λ and a_{ij}'s are integers satisfying the conditions stated in Sec. 2. Set

$$a_p = e^{-p\frac{\log x}{x}},$$

$$P(x,Q) = \sum_{\substack{p \leq x \\ p \equiv l (\mathrm{mod}\, Q)}} a_p = \frac{x}{\varphi(Q)\log x} + R_Q(x), \quad (l,Q)=1,$$

$$\tilde{M}_Z(x) = \sum_{\substack{p \leq x \\ p \equiv \lambda (\mathrm{mod}\, M) \\ p \not\equiv a_{ij} (\mathrm{mod}\, p_i)\ (1 \leq i \leq r, 1 \leq j \leq k+1)}} a_p.$$

Lemma 3. *Let $r = r_0 \geq r_1 \geq \cdots \geq r_n \geq 1$ be any given set of integers. Then*

$$\tilde{M}_Z(x) \geq \frac{x\bar{E}}{\varphi(M)\log x} - \bar{R},$$

where

$$\bar{E} = 1 - (k+1)\sum_{\alpha \leq r}\frac{1}{\varphi(p_\alpha)} + (k+1)^2\sum_{\substack{\alpha \leq r\ \beta \leq r_1 \\ \alpha > \beta}}\frac{1}{\varphi(p_\alpha)\varphi(p_\beta)} - \cdots$$

$$- (k+1)^{2n+1}\overbrace{\sum_{\substack{\alpha \leq r\ \beta \leq r_1\ \gamma \leq r_1\ \delta \leq r_2 \\ \alpha > \beta > \gamma > \delta > \cdots > \nu}}\sum\sum\sum \cdots \sum_{\nu \leq r_n}}^{2n+1}\frac{1}{\varphi(p_\alpha)\varphi(p_\beta)\cdots\varphi(p_\nu)},$$

$$\bar{R} = |R_M(x)| + (k+1)\sum_{\alpha \leq r}|R_{Mp_\alpha}(x)| + (k+1)^2\sum_{\substack{\alpha \leq r\ \beta \leq r_1 \\ \alpha > \beta}}|R_{Mp_\alpha p_\beta}(x)| + \cdots$$

$$+ (k+1)^{2n+1}\overbrace{\sum_{\substack{\alpha \leq r\ \beta \leq r_1\ \gamma \leq r_1\ \delta \leq r_2 \\ \alpha > \beta > \cdots > \nu}}\sum\sum\sum \cdots \sum_{\nu \leq r_n}}^{2n+1}|R_{Mp_\alpha \cdots p_\nu}(x)|.$$

The estimation of $\bar{E}$: Take $h = (1.25)^{\frac{1}{k+1}}$ and $h_0 = \sqrt[4]{e}$. Then there exists a constant $\delta_0 \geq M$ such that for $\delta > \delta_0$, we have

$$\sum_{\delta < p \leq \delta^h}\frac{k+1}{\varphi(p)} < \log h_0 = \tau, \qquad \prod_{\delta < p \leq \delta^h}\left(1 - \frac{k+1}{\varphi(p)}\right)^{-1} < h_0.$$

Let p_{r_j} $(0 \leq j \leq t)$ be the largest primes $\leq Z^{\frac{1}{h_j}}$, where t satisfies $Z^{\frac{1}{h_t}} > \delta_0 \geq Z^{\frac{1}{h_{t+1}}}$. Take $n = t + r_t$ and $r_s = r_t (t \leq s \leq n)$. Then

$$\bar{E} > \left(1 - \sum_{n=1}^{\infty} \frac{(h_0 \tau^2 n^2)^n}{(2n)!}\right) \prod_{\substack{p \leq Z \\ p \nmid M}} \left(1 - \frac{k+1}{\varphi(p)}\right)$$

$$> 0.75 \prod_{\substack{p \leq Z \\ p \nmid M}} \left(1 - \frac{k+1}{\varphi(p)}\right) > \frac{c_3}{\log^{k+1} Z},$$

where c_3 is a positive constant depending on k and M only (see Wang Yuan [7]).

4. Several Lemmas

We use χ_D to denote the character modulo D. Let $D = p_1^{a_1} \cdots p_l^{a_l}$ be the standard decomposition of D. Then $\chi_D = \chi_{p_1^{a_1}} \cdots \chi_{p_l^{a_l}}$. If $\chi_{p_i^{a_i}}$ is primitive, then we say that χ_D is primitive with respect to $p_i^{a_i}$. χ_D is primitive in the usual sense if it is primitive with respect to all $p_i^{a_i}$ $(1 \leq i \leq l)$.

Lemma A. *Let q, A be two integers such that $A > c_4$, where c_4 is an absolute constant, and p be a prime number satisfying $A \leq p < 2A$ and $(p, q) = 1$. If*

$$e^{(\log x)^{2/5}} \leq Aq \leq \frac{\sqrt{x}}{2}, \quad k_1 = \frac{\log q}{\log \frac{p}{2}} + 1 \geq \frac{\log q}{\log A} + 1 = \tilde{k},$$

and $k_1 < \log^3 A$, then except $O(p^{3/4})$ primes in the interval $A \leq p < 2A$, the estimation

$$\left| \sum_{p \leq x} \chi(p) \log p \cdot e^{-\frac{p \log x}{x}} \right| \leq c_5 x^{1 - \frac{\delta_2}{k_1 + 1}} \log x,$$

holds for any $\chi(n) \bmod pq$ which is primitive with respect to p, where c_5 is an absolute constant and $\delta_2 = (4 \times 10^4 \, k_3)^{-1}$ in which k_3 is defined in Linnik [8].

We refer Linnik [8]. Renyi [6] and Littlewood [9] for the proof.

Lemma B. *Suppose that $(l, Q) = 1$ and $Q \leq e^{\sqrt{\log x}}$, then*

$$P(x, Q) = \begin{cases} \frac{x}{\varphi(Q) \log x} + O\left(x e^{-c_6 \sqrt{\log x}}\right), & if \ Q_1 \nmid Q, \\ \frac{x}{\varphi(Q) \log x} + O\left(x e^{-c_6 \sqrt{\log x}}\right) + O\left(\frac{x^{1 - c(\varepsilon)/Q_1^\varepsilon}}{\varphi(Q)}\right), & if \ Q_1 \mid Q, \end{cases}$$

where ε is any pre-assigned positive number, c_6 and the constant in "O" are absolute constants, and Q_1 is the modulus of "exceptional" character $\tilde{\chi}$.

See Titchmarsh [10], Page [11] and Siegel [12].

Since $a_p \leq \log x$, we have

Lemma C. *The estimation*

$$P(x, Q) < \frac{2x \log x}{\varphi(Q)}$$

holds uniformly for $1 \leq Q < \sqrt{x}$.

We denote by E the set of positive integers $Q = p_1' \cdots p_u' M$ satisfying

$$p_1' > \cdots > p_u', p_i' + M, p_i' \leq Z^{\frac{1}{h\left[\frac{i}{2}\right]}} \quad 1 \leq i \leq 2t+1, \quad p_i' \leq p_{r_t}, \quad i > 2t+1.$$

If $Q \in E$, we write $Q = p_1' q_1, q_1 = p_2' q_2, \ldots, q_{u-1} = p_u' q_u, q_u = M$ and call integers $q_1, \ldots, q_u$ the diagonal factors of Q. It is evident that all diagonal factors of Q belong to E if $Q \in E$.

Set $K = \prod_{\substack{p \leq p_{r_t} \\ p \nmid M}} p$. We omit the proofs of the following lemmas, since they can be easily derived from the definition of E. And we use $c_7, c_8, \ldots$ to denote positive constants depending only on k and M if there is no special explanation.

Lemma 4. *Let $\nu(Q)$ denote the number of prime divisors of Q. Then $\nu(Q) < 10(k+1) \log \log x$ if $x > c_7$.*

Lemma 5. *The number of elements of E is $\leq K Z^f$, where $f = 1 + \frac{2}{h-1}$.*

Lemma 6. *Suppose that $Q \in E$, $Q = p_1' q_1$, $p_1' < q_1^{1/\nu}$ and $p_1' > MK$. Then the number of such Q is not exceeding $Z^{g\nu/h^{\nu/2}}$, where g is a positive constant depending on k only.*

Lemma 7. *Let $\{p^*\}$ be a set of integers with the property. For any given number A, there are at most $A^{3/4}$ elements of the set which are contained in the interval $A \leq p^i < 2A$. Then*

$$\sum_{p^* > B} \frac{1}{p^*} \leq \left(\frac{1}{1 - 2^{-1/4}}\right) B^{-1/4}.$$

Lemma 8.

$$\sum_{n \leq x} \frac{1}{\varphi(n)} = \prod_p \left(1 + \frac{1}{p(p-1)}\right) \log x + O(1).$$

Lemma 9. *If $n \geq 2$, then $\varphi(n) > c_8 n / \log n$.*

5. The Proof of Lemma 2

Set

$$K_{\chi_Q}(x) = \sum_{p \le x} \chi_Q(p) a_p.$$

Then for $(l, Q) = 1$, we have

$$P(x, Q) = \frac{1}{\varphi(Q)} \sum_{(\chi_Q)} \bar{\chi}_Q(l) K_{\chi_Q}(x). \tag{10}$$

It follows from Lemma 4 and the definition of E that

$$\bar{R} \le (k+1)^{2n+1} \sum_{Q \in E} |R_Q(x)| \le e^{(2n+1)\log(k+3)} \sum_{Q \in E} |R_Q(x)|$$

$$\le e^{10(k+1)\log(k+3)\log\log x} \sum_{Q \in E} |R_Q(x)|. \tag{11}$$

If $Q = p_1' q_1 \in E$ and $Q > e^{(\log x)^{2/5}}$, then

$$p_1' > Q^{\frac{1}{\nu(Q)}} > 2e^{(\log x)^{1/3}}, \quad x > c_9$$

by Lemma 4. Since $q_1 < p_1'^{\nu(Q)}$, we have

$$k_1 = \frac{\log q_1}{\log \frac{p_1'}{2}} + 1 = \frac{\log q_1}{\log p_1'}\left(1 + O\left(\frac{1}{\log p_1'}\right)\right) < 11(k+1)\log\log x, \quad x > c_{10}. \tag{12}$$

Take $B = 2e^{(\log x)^{1/3}}$ and $A = 2^n B (n = 1, 2, \ldots)$. For a given q_1, we use Lemma A to the interval $A \le p < 2A$ and introduce the following two conditions:

(i) If $Q > e^{(\log x)^{2/5}}$, then we say that condition I is satisfied.

(ii) If $Q \in E, Q = p_1' q_1$ and p_1' is not the exception with respect to q_1 in the sense of Lemma A, then condition II is said to be satisfied.

If the above two conditions are all satisfied, then it follows from Lemma A and (10) that

$$P(x, Q) = \frac{1}{\varphi(p_1')} P(x, q_1) + O\left(x^{1 - \frac{\delta_2}{k_1 + 1}} \log x\right).$$

If q_1 still satisfies these two conditions, then the process may be continued and so on until one of the conditions I or II is not satisfied. Suppose that such processes are performed S times. Then

$$P(x, Q) = \frac{1}{\varphi(p_1' \cdots p_S')} P(x, q_S) + O\left(\sum_{l=1}^{S} \frac{x^{1 - \frac{\delta_2}{k_1 + 1}}}{\varphi\left(\frac{Q}{q_{l-1}}\right)} \log x\right),$$

where $q_0 = Q$ and $k_l = \log q_l / \log \frac{p_l'}{2} + 1$.

If condition I is not satisfied, i.e. $q_s < e^{(\log x)^{2/5}}$, then we have by Lemma B that

$$P(x, q_s) = \begin{cases} \frac{x}{\varphi(q_s)\log x} + O(xe^{-c_6\sqrt{\log x}}), & \text{if } Q_1 \nmid q_s \\ \frac{x}{\varphi(q_s)\log x} + O(xe^{-c_6\sqrt{\log x}}) + O\left(\frac{x^{1-c(\varepsilon)/Q_1^\varepsilon}}{\varphi(q_s)}\right), & \text{if } Q_1 \mid q_s. \end{cases}$$

If condition II is not satisfied, then

$$P(x, q_s) = \frac{x}{\varphi(q_s)\log x} + O\left(\frac{x\log x}{\varphi(q_s)}\right)$$

by Lemma C.

In conclusion we have the following four types of error terms:

$$\text{(I)} \ \ \frac{x\log x}{\varphi(Q)}, \quad \text{(II)} \ \ \frac{xe^{-c_6\sqrt{\log x}}}{\varphi\left(\frac{Q}{q_s}\right)}, \quad \text{(III)} \ \ \frac{x^{1-c(\varepsilon)/Q_1^\varepsilon}}{\varphi(Q)}, \quad \text{(IV)} \ \ \frac{x^{1-\frac{\delta_2}{k_l+1}}}{\varphi\left(\frac{Q}{q_{l-1}}\right)} \log x.$$

We denote by $R_{\mathrm{I}}, R_{\mathrm{II}}, R_{\mathrm{III}}$ and R_{IV} for those terms in $\sum_{Q\in E}|R_Q(x)|$ that belong to types I, II, III and IV respectively.

(i) By Lemmas 4, 7 and 8, we have

$$R_{\mathrm{I}} < c_{11}\, x\log^3 x / e^{1/4(\log x)^{1/3}}, \quad x > c_{12}.$$

(ii) By Lemma 8, we obtain

$$R_{\mathrm{II}} < c_{13}\, x\log x \cdot e^{(\log x)^{2/5} - c_6\sqrt{\log x}}, \quad x > c_{14}.$$

(iii) Take $\varepsilon = \frac{1}{18(k+1)\log(k+3)}$. Then it follows from Lemmas 8 and 9 that

$$R_{\mathrm{III}} < c_{15}\, x\log^3 x \cdot e^{-18(k+1)\log(k+3)\log\log x}, \quad x > c_{16}.$$

(iv) We use $R_{\mathrm{IV}}^{(1)}$ to denote the part in R_{IV} with $k_l \leq 2$. Take

$$Z = x^{1/N} \quad \text{and} \quad N \geq \frac{4f}{\delta_2},$$

where N will be determined later. Then by Lemma 5, we have

$$R_{\mathrm{IV}}^{(1)} < c_{17}Z^f \cdot x^{1-\frac{\delta_2}{3}}\log x = c_{17}x^{1-\frac{\delta_2}{12}}\log x, \quad x > c_{18}.$$

(v) Let $R_{\text{IV}}^{(2)}$ be the part in R_{IV} with $k_l > 2$. Then k_l satisfies $2\nu \le k_l < 2\nu + 2$, $\nu = 1, 2, \ldots, \left[\frac{11(k+1)\log\log x}{2}\right]$ by (12). Since $p_l' < q_l^{1/\nu}$ and $p_l' > 2e^{(\log x)^{1/3}} > MK$ $(x > c_{19})$, we have

$$R_{\text{IV}}^{(2)} < c_{20} \sum_{\nu=1}^{\left[\frac{11}{2}(k+1)\log\log x\right]} x^{1-\frac{\delta_2}{2\nu+4}} Z^{\frac{g\nu}{h^{\nu/2}}} \log^2 x, \quad x > c_{21}$$

by Lemma 6. Take

$$N = \text{Max}\left(\frac{4f}{\delta_2}, \frac{4dg}{\delta_2}\right), \quad d = \underset{\nu=1,2,\ldots}{\text{Max}} \frac{\nu(\nu+2)}{h^{\nu/2}}^{\text{b}}. \tag{13}$$

We have

$$R_{\text{IV}}^{(2)} < c_{22}\, x \log^3 x \cdot e^{-\frac{\delta_2 \log x}{22(k+1)\log\log x - \delta}}, \quad x > c_{23}.$$

By the combination of (i), (ii), (iii), (iv), (v) and (11), we have

$$\bar{R} \le c_{24}\, xe^{-8(k+1)\log(k+3)\log\log x} \log^3 x < c_{24}\frac{x}{\log^{5(k+1)} x}, \quad x > c_{25}. \tag{14}$$

Hence it follows from Lemma 3:

Lemma 10. *There exist c_{26} and X_3 depending on k and M such that*

$$\tilde{M}_{x^{\frac{1}{N}}}(x) > \frac{c_{26}\, x}{\log^{k+2} x}, \quad x > X_3,$$

where N depends only on k.

The Proof of Lemma 2.   Set $\beta = \frac{1}{N}$. Then

$$\tilde{M}_{x^\beta}(x) \le \log x \cdot e^{-2\frac{\log x}{x}} M_{x^\beta}\left(\frac{(k+3)x\log\log x}{\log x}\right) + \log x \cdot e^{-(k+3)\log\log x}\pi(x)$$

$$\le \log x \cdot M_{x^\beta}\left(\frac{(k+3)x\log\log x}{\log x}\right) + O\left(\frac{x}{\log^{k+3} x}\right)$$

and

$$M_{x^\beta}\left(\frac{(k+3)x\log\log x}{\log x}\right) \ge \frac{c_{27}x}{\log^{k+3} x}, \quad x > c_{28}$$

---

[b] Since $h = (1.25)^{\frac{1}{k+1}} > 1$, we have $\lim_{y\to\infty} \frac{y(y+2)}{h^{y/2}} = 0$. Hence the existence of d is proved and it depends on k only.

by Lemma 10. Let $\frac{(k+3)x\log\log x}{\log x} = y$. Then

$$M_{y^\beta}(y) > \frac{c_2\, y}{\log^{k+2} y \log\log y}, \quad y > X_2.$$

The lemma follows.

6. Applications of Fundamental Lemma

Lemma 11.   *Let $\sigma_0 = 1$ and $\sigma_\nu = \nu$ $(1 \le \nu \le k)$. For any given k non-negative real numbers $a_1, \ldots, a_k$ and $\varepsilon > 0$, there exist positive integers $m_0, \ldots, m_k$ depending only on a_i's and ε such that*

$$\left| \frac{\frac{\varphi(\sigma_\nu m_\nu)}{\sigma_\nu m_\nu}}{\frac{\varphi(\sigma_{\nu-1} m_{\nu-1})}{\sigma_{\nu-1} m_{\nu-1}}} - a_\nu \right| < \frac{\varepsilon}{2}, \quad 1 \le \nu \le k. \tag{15}$$

Proof.   Let $\rho_\nu = \frac{\varphi(\nu)}{\nu}$ $(1 \le \nu \le k)$ and $\rho_0 = \varphi((k+1)!^2)/(k+1)!^2$. Then there are rational numbers $b_\nu/d_\nu > 0$ $(1 \le \nu \le k)$ such that

$$\left| \frac{b_\nu}{d_\nu} - a_\nu \frac{\rho_{\nu-1}}{\rho_\nu} \right| < \frac{\varepsilon}{3} \cdot \frac{\rho_{\nu-1}}{\rho_\nu}, \quad 1 \le \nu \le k,$$

i.e.

$$\left| \frac{b_1 \cdots b_{\nu-1} b_\nu d_{\nu+1} \cdots d_k}{b_1 \cdots b_{\nu-1} d_\nu d_{\nu+1} \cdots d_k} - a_\nu \cdot \frac{\rho_{\nu-1}}{\rho_\nu} \right| < \frac{\varepsilon}{3} \cdot \frac{\rho_{\nu-1}}{\rho_\nu}, \quad 1 \le \nu \le k.$$

Let

$$\frac{b_1 \cdots b_{\nu-1} d_\nu d_{\nu+1} \cdots d_k}{b_1 \cdots b_k d_1 \cdots d_k} = \eta_{\nu-1}, \quad 1 \le \nu \le k+1.$$

Then

$$\left| \frac{\eta_\nu}{\eta_{\nu-1}} - a_\nu \frac{\rho_{\nu-1}}{\rho_\nu} \right| < \frac{\varepsilon}{3} \cdot \frac{\rho_{\nu-1}}{\rho_\nu}, \quad 1 \le \nu \le k.$$

For any given $\varepsilon' > 0$, we may choose positive integers $m_0', m_1, \ldots, m_k$ which are coprime from one another such that their prime divisors are all $> k+1$ and

$$\left| \eta_\nu - \frac{\varphi(m_\nu)}{m_\nu} \right| < \varepsilon', \quad 1 \le \nu \le k, \quad \left| \frac{\varphi(m_0')}{m_0'} - \eta_0 \right| < \varepsilon',$$

since $0 < \eta_\nu \leq 1 \, (0 \leq \nu \leq k)$ and $\prod_p \left(1 - \frac{1}{p}\right) = 0$. Hence we may take $\varepsilon' = \varepsilon'(a_i\text{'s}, \varepsilon, \rho)$ sufficiently small such that

$$\left| \frac{\frac{\varphi(m_\nu)}{m_\nu}}{\frac{\varphi(m_{\nu-1})}{m_{\nu-1}}} - a_\nu \frac{\rho_{\nu-1}}{\rho_\nu} \right| < \frac{\varepsilon}{2} \cdot \frac{\rho_{\nu-1}}{\rho_\nu}, \quad 2 \leq \nu \leq k, \qquad \left| \frac{\frac{\varphi(m_1)}{m_1}}{\frac{\varphi(m_0')}{m_0'}} - a_1 \frac{\rho_0}{\rho_1} \right| < \frac{\varepsilon}{2} \cdot \frac{\rho_0}{\rho_1}.$$

Set $m_0 = (k+1)!^2 m_0'$. Then we have the lemma.

Lemma 12. *This lemma is obtained by changing the function $\varphi(n)$ by $\sigma(n)$ in Lemma 11.*

Theorem 1. *For any given k non-negative real numbers $a_1, \ldots, a_k$ and $\varepsilon > 0$, there exists a prime number p such that*

$$\left| \frac{\varphi(p+\nu+1)}{\varphi(p+\nu)} - a_\nu \right| < \varepsilon, \quad 1 \leq \nu \leq k. \tag{16}$$

Moreover, there are two positive constants c_{29} and X_4 depending on a_i's and ε only such that the number of primes satisfying (16) in any interval $1 < p \leq x$ is not less than $c_{29} \frac{x}{\log^{k+2} x \log \log x}$ whenever $x > X_4$.

Proof. First we choose using Lemma 11 the positive integers $m_0, m_1, \ldots, m_k$ depending on a_i's and ε only and satisfying the requirement of Fundamental lemma such that (15) hold. Then let $(p, x_0, \ldots, x_k)$ be a solution of (2) satisfying (3) and $Z = x^\alpha$. Since $(x_i, m_0, \ldots, m_k) = 1$, we have

$$\frac{\varphi(p+\nu+1)}{\varphi(p+\nu)} = \frac{\varphi(\sigma_\nu m_\nu x_\nu)}{\varphi(\sigma_{\nu-1} m_{\nu-1} x_{\nu-1})} = \frac{\frac{\varphi(\sigma_\nu m_\nu)}{\sigma_\nu m_\nu}}{\frac{\varphi(\sigma_{\nu-1} m_{\nu-1})}{\sigma_{\nu-1} m_{\nu-1}}} \cdot \frac{\frac{\varphi(x_\nu)}{x_\nu}}{\frac{\varphi(x_{\nu-1})}{x_{\nu-1}}} \cdot \frac{p+\nu+1}{p+\nu}. \tag{17}$$

Since the prime divisors of x_ν are all $> x^\alpha$ and $x_\nu \leq x$, the number of prime divisors of x_ν is at most $\left[\frac{1}{\alpha}\right]$. Hence

$$1 \geq \frac{\varphi(x_\nu)}{x_\nu} = \prod_{p' \mid x_\nu} \left(1 - \frac{1}{p'}\right) \geq \left(1 - \frac{1}{x^\alpha}\right)^{\frac{1}{\alpha}} \tag{18}$$

and consequently, it derives from (15), (17) and (18) that there exists c_{30} such that (16) holds whenever $x > p > c_{30}(\alpha, \varepsilon)$. This proves that we may obtain a prime p satisfying (16) from a solution $(p, x_0, \ldots, x_k)$ of (2) satisfying (3), $Z = x^\alpha$ and $p > c_{30}$. Since the number of solutions of (2) with $p \leq c_{30}$ is at most c_{30}, Theorem 1 follows immediately from Fundamental lemma.

Theorem 2 is obtained by replacing the function $\varphi(n)$ by $\sigma(n)$ in Theorem 1. Of course, the constants c_{29} and X_4 should be changed respectively by c_{31} and X_5.

Theorem 3. *Let k be a positive integer. Then there exists a constant γ depending on k only such that for any given $k+1$ positive integers $a_0, a_1, \ldots, a_k$, there is a prime number p satisfying*

$$a_\nu \leq d(p + \nu + 1) \leq \gamma a_\nu, \quad 0 \leq \nu \leq k. \tag{19}$$

Moreover, there are two constants c_{32} and X_6 depending on a_ν's only such that the number of primes satisfying (19) *in any interval $1 < p \leq x$ is not less than $\frac{c_{32} x}{\log^{k+2} x \log \log x}$ whenever $x > X_6$.*

Proof. Assume that $a_0, \ldots, a_k$ satisfy

$$2^{\alpha_\nu} \leq a_\nu < 2^{\alpha_\nu + 1}, \quad 0 \leq \nu \leq k.$$

We take $t_\nu = \alpha_\nu + 1$ in (1). For a set of m_i's, let $(p, x_0, \ldots, x_k)$ be a solution of (2) satisfying (3) and $Z = x^\alpha$. Since the prime divisors of x_ν are all $> x^\alpha$, x_ν is a product of at most $\left[\frac{1}{\alpha}\right]$ primes. Set

$$\gamma = 2^{\frac{1}{\alpha}+1}(k+1)!^2. \tag{20}$$

Then

$$d(p + \nu + 1) = d(\sigma_\nu m_\nu x_\nu) \geq d(m_\nu)d(x_\nu) \geq 2^{t_\nu} > a_\nu$$
$$d(p + \nu + 1) = d(\sigma_\nu m_\nu x_\nu) \leq d(\sigma_0)d(m_\nu)d(x_\nu)$$
$$< (k+1)!^2 2^{t_\nu} 2^{\left[\frac{1}{\alpha}\right]} < \gamma a_\nu, \quad 0 \leq \nu \leq k.$$

The theorem follows.

From Theorem 3, we derive the following

Theorem 4. *For any given $a_1, \ldots, a_k$, where a_i equals to 0 or $+\infty$ $(1 \leq i \leq k)$, there exists a sequence of prime numbers $\{p_j\}$ such that*

$$\lim_{j \to \infty} \frac{d(p_j + \nu + 1)}{d(p_j + \nu)} = a_\nu, \quad 1 \leq \nu \leq k.$$

Finally the author would like to ask a question: Is Theorem 1 still true if $\varphi(n)$ is replaced by the divisor function $d(n)$?

References

[1] B. S. K. R. Somayajulu, The Euler's totient function $\varphi(n)$, *Math. Stud.* **18** (1950) 31.

[2] A. Schinzel and W. Sierpiński, Sur quelques propiétés des fonctions $\varphi(n)$ et $\sigma(n)$, *Bull. Acad. Polon. Sci. Cl. III* **2** (1954) 463.

[3] A. Schinzel, Quelques théorèmes sur les fonctions $\varphi(n)$ et $\sigma(n)$, *Bull. Acad. Polon. Sci. Cl. III* **2** (1954) 467; Sur une propriété du nombre de diviseurs, *Pub. Math.* **3** (1954) 261; On Functions $\varphi(n)$ and $\sigma(n)$, *Bull. Acad. Polon. Sci. Cl. III* **8** (1955) 415.

[4] A. Schinzel and Y. Wang, On some properties of the functions, *Polon. Akad. Nauk* **3** (1956) 201; A note on some properties of the functions $\varphi(n)$, $\sigma(n)$ and $\theta(n)$, *Ann. Polon. Math.* (in press).

[5] Shuo Pin-tsung, On distribution of values of a certain class of arithmetical functions, *Bull. Beijing Univ.* **3** (1956) 261.

[6] A. Renyi, On representation of even integer as a sum of a prime number and an almost prime number, *Dokl. Akad. Nauk SSSR* **12** (1948) 57.

[7] Wang Yuan, On representation of large even integer as a sum of a prime and a product of at most 4 primes, *Acta Math. Sinica* **6**:4 (1956) 565.

[8] Yu. V. Linnik, On Dirichlet's L-series and a sum of prime numbers, *Math. Sb.* **15** (1944) 3.

[9] J. E. Littlewood, On the Class number of corpus $P(\sqrt{-k})$., *Proc. Lond. Math. Soc.* **27** (1928) 315.

[10] E. C. Titchmarsh, A divisor problem, *Rend. del circolo Mat. di Palermo* **54** (1930) 414.

[11] A. Page, On number of primes in the arithmetic progression, *Proc. Lond. Math. Soc.* (2) **39** (1935) 116.

[12] C. L. Siegel, Uber die classenzahle quadratischen Zahlenkoper, *Acta Arith.* **I** (1936) 83.

SCIENTIA SINICA

Vol. VIII, No. 4, 1959

MATHEMATICS

ON SIEVE METHODS AND SOME OF THEIR APPLICATIONS*

WANG YUAN (王 元)

(*Institute of Mathematics, Academia Sinica*)

§1. INTRODUCTION

For the sake of brevity, we denote the following proposition by (a, b).

Every sufficiently large even integer can be represented as a sum of two integers > 1, of which one contains at most a and the other at most b prime factors.

In 1919, v. Brun[1] first gave an essential improvement of Eratosthenes' sieve method and proved (9, 9). Brun's method and his results were improved and extended by several mathematicians, for examples:

(7,7)   (H. Rademacher[2]),

(6,6)   (T. Estermann[3]),

(5,7), (4,9), (3,15), (2,366) (G. Ricci[4]),

(5,5)   (A. A. Бухштаб[5]),

(4,4)   (A. A. Бухштаб[6])[1]),

(a, b),   where $a + b \leqslant 6$ (P. Kuhn)[9-11].

In 1947, A. Selberg[12-14] published his new improvements of Eratosthenes' sieve method. By the combination of the methods of Brun, Бухштаб and Selberg, we have (3, 4)[15]. By the use of Selberg's method and some results in the theory of the Riemann ζ-function, A. И. Виноградов[16,17] obtained (3, 3).

The aim of the present paper is to prove (3, 3) and $\cdot$ (a, b) $(a + b \leqslant 5)$. Moreover, using Бухштаб's method with more complicated numerical calculations, we have (2, 3)[18-20], that is:

---

*First published in Chinese in *Acta Mathematica Sinica*, Vol. 8, No. 3, pp. 413—429, 1958·

1) Cf. also В. А. Тартаковский[7,8].

Theorem 1. Every sufficiently large even integer can be written as a sum of two positive numbers > 1, of which one contains at most 2 and the other at most 3 prime factors.

We have also the corresponding results for the problem of twin-primes and the problem of representation of large odd number as a sum of two almost-primes[1]. Namely, we have the following

Theorem 2. For any given even number k, there are infinitely many integers n, such that each of n and $n + k$ has at most 3 prime factors and $n(n + k)$ is a product of not more than 5 primes.

Theorem 3. Every sufficiently large odd integer can be represented as $2N + 1 = 2P + Q$ $(P > 1,\ Q > 1)$, where the number of prime factors of P and also of Q is not more than 3 and PQ is a product of at most 5 primes.

In this paper, $p,\ p',\ p'',\ \cdots,\ p_1,\ p_2,\ \cdots$ denote primes.

$$\S\,2.\quad \text{Computations}$$

Lemma 1. Let $\Omega(n)$ be the number of different prime factors of n, that is $\Omega(n) = \sum\limits_{p \mid n} 1$. If $x \geqslant 1$ and $z \geqslant 1$, then

$$\sum_{\substack{n \leqslant z \\ (n,x)=1}} \frac{|\mu(n)|\,2^{\Omega(n)}}{n} = \frac{1}{2} \prod_{p \mid x} \frac{p}{p+2} \prod_{p} \left(1 - \frac{1}{p}\right)^2 \left(1 + \frac{2}{p}\right) \cdot \log^2 z$$
$$+ O(\log 2z \cdot \log\log 3zx) + O((\log\log 3x)^2),$$

where $\mu(n)$ denotes the Möbius function[21].

Lemma 2. If $2 \mid x$ and $z \geqslant 1$, then

$$\sum_{\substack{n \leqslant z \\ (n,x)=1}} \frac{|\mu(n)|\,2^{\Omega(n)}}{n} \prod_{p \mid n} \left(1 + \frac{2}{p-2}\right) = \frac{1}{8} \prod_{p>2} \frac{(p-1)^2}{p(p-2)} \prod_{\substack{p \mid x \\ p>2}} \frac{p-2}{p} \log^2 z$$
$$+ O(\log xz \log\log 3xz).$$

Proof. Let $\Psi(r) = \prod\limits_{p \mid r} (p - 2)$. Then by Lemma 1, we have

$$\sum_{\substack{n \leqslant z \\ (n,x)=1}} \frac{|\mu(n)|\,2^{\Omega(n)}}{n} \prod_{p \mid n} \left(1 + \frac{2}{p-2}\right) = \sum_{\substack{n \leqslant z \\ (n,x)=1}} \frac{|\mu(n)|\,2^{\Omega(n)}}{n} \sum_{r \mid n} \frac{2^{\Omega(r)}}{\Psi(r)}$$

$$= \sum_{\substack{r \leqslant z \\ (r,x)=1}} \frac{|\mu(r)|\,2^{2\Omega(r)}}{r\,\Psi(r)} \sum_{\substack{s \leqslant z/r \\ (s,rx)=1}} \frac{|\mu(s)|\,2^{\Omega(s)}}{s}$$

---

1) An integer is called almost-prime, if the number of its prime factors does not exceed a fixed constant.

$$
= \sum_{\substack{r \leqslant z \\ (r,x)=1}} \frac{|\mu(r)| 2^{2\Omega(r)}}{r \Psi(r)} \left\{ \frac{1}{2} \prod_p \frac{(p-1)^2 (p+2)}{p^3} \prod_{p|rx} \frac{p}{p+2} \log^2 \frac{z}{r} \right.
$$

$$
\left. + O(\log zx \log\log 3 zx) \right\}
$$

$$
= \frac{1}{2} \prod_p \frac{(p-1)^2 (p+2)}{p^3} \prod_{p|x} \frac{p}{p+2} \log^2 z \cdot \sum_{\substack{r \leqslant z \\ (r,x)=1}} \frac{|\mu(r)| 4^{\Omega(r)}}{\prod_{p|r} (p^2-4)}
$$

$$
+ O(\log zx \log\log 3 zx)
$$

$$
= \frac{1}{8} \prod_{p>2} \frac{(p-1)^2}{p(p-2)} \prod_{\substack{p|x \\ p>2}} \frac{p-2}{p} \log^2 z + O(\log zx \log\log 3 zx) .
$$

Thus we have the lemma.

Lemma 3. For any given $\eta > 0$, there exists a positive constant $x_0 = x_0(\eta)$ such that the inequality

$$
2^{\Omega(x)} \leqslant d(x) \leqslant 2^{(1+\eta) \frac{\log x}{\log\log x}}
$$

holds for $x > x_0$, where $d(x)$ denotes the number of divisors of x[22]

Lemma 4. Let $\triangle(x) = e^{\frac{\log 3x}{\log\log 9x}}$, where $x \geqslant 1$. Then there exists $c_1 > 0$ such that

$$
\prod_{p|x} \left(1 - \frac{1}{p}\right)^{-1} - \left\{ 1 + \sum_{\substack{p|x \\ p \leqslant \triangle(x)}} \frac{1}{p} \left(1 - \frac{1}{p}\right)^{-1} + \sum_{\substack{pp'|x \\ pp' \leqslant \triangle(x) \\ p \neq p'}} \frac{1}{pp'} \left(1 - \frac{1}{p}\right)^{-1} \left(1 - \frac{1}{p'}\right)^{-1} \right.
$$

$$
\left. + \cdots \right\} = O(e^{-c_1 \frac{\log 3x}{\log\log 9x}}) .
$$

Proof. Let p_i be the ith prime number. Then by Mertens theorem we have

$$
\prod_{p|x} \left(1 - \frac{1}{p}\right)^{-1} = O\left(\prod_{i \leqslant \Omega(x)} \left(1 - \frac{1}{p_i}\right)^{-1} \right) = O\left(\prod_{p \leqslant c \log 2x} \left(1 - \frac{1}{p}\right)^{-1} \right)
$$

$$
= O(\log\log 3x) .
$$

Hence, it follows by Lemma 3 that

$$
\prod_{p|x} \left(1 - \frac{1}{p}\right)^{-1} - \left\{ 1 + \sum_{\substack{p|x \\ p \leqslant \triangle(x)}} \frac{1}{p} \left(1 - \frac{1}{p}\right)^{-1} + \sum_{\substack{pp'|x \\ pp' \leqslant \triangle(x) \\ p \neq p'}} \frac{1}{pp'} \left(1 - \frac{1}{p}\right)^{-1} \left(1 - \frac{1}{p'}\right)^{-1} \right.
$$

$$
\left. + \cdots \right\} = \sum_{\substack{p|x \\ p > \triangle(x)}} \frac{1}{p} \left(1 - \frac{1}{p}\right)^{-1} + \sum_{\substack{pp'|x \\ pp' > \triangle(x) \\ p \neq p'}} \frac{1}{pp'} \left(1 - \frac{1}{p}\right)^{-1} \left(1 - \frac{1}{p'}\right)^{-1} + \cdots
$$

360

$$\leqslant \frac{1}{\Delta(x)} \prod_{p \mid x} \left(1 - \frac{1}{p}\right)^{-1} \left\{1 + \sum_{p \mid x} 1 + \sum_{\substack{pp' \mid x \\ p \neq p'}} 1 + \cdots\right\}$$

$$= \frac{2^{\Omega(x)}}{\Delta(x) \prod_{p \mid x} \left(1 - \frac{1}{p}\right)}$$

$$= O\left(e^{-c_1 \frac{\log 3x}{\log \log 9x}}\right).$$

Thus we have

Lemma 5. Let $q = O(1)$ be a given integer and y be an integer. Let $g(1) = 1$, $g(p) = \frac{1}{p}$ $(p \mid y)$, $g(p) = \frac{2}{p}$ $(p + y)$, $g(n) = \prod_{p \mid n} g(p)$ and $f(n) = \frac{1}{g(n)} \prod_{p \mid n} (1 - g(p))$. If $z \geqslant 1$, then

$$\sum_{\substack{n \leqslant z \\ (n,q)=1}} \frac{|\mu(n)|}{f(n)} = \frac{1}{4} \prod_{p \mid q} \frac{p-1}{p} \prod_{p > 2} \frac{(p-1)^2}{p(p-2)} \prod_{\substack{p \mid qy \\ p > 2}} \frac{p-2}{p-1} \log^2 z$$

$$+ O\left(\prod_{\substack{p \mid qy \\ p > 2}} \frac{p-2}{p-1} \cdot \frac{\log^2 2\, yz}{\log \log 3\, yz}\right).$$

Proof.   (i) If $z \geqslant \Delta(y) \geqslant \log 2z$, then $\log \log 3z = O(\log 2y)$ and $\log \log 3y = O(\log \log 3z)$.   By Lemmas 2 and 4, we have

$$\sum_{\substack{n \leqslant z \\ (n,q)=1}} \frac{|\mu(n)|}{f(n)} = \sum_{\substack{n \leqslant z \\ (n,qy)=1}} \frac{|\mu(n)| 2^{\Omega(n)}}{n} \prod_{p \mid n} \left(1 - \frac{2}{p}\right)^{-1}$$

$$+ \sum_{\substack{p' \mid y \\ p' + q}} \frac{1}{p'} \left(1 - \frac{1}{p'}\right)^{-1} \sum_{\substack{n \leqslant z/p' \\ (n,qy)=1}} \frac{|\mu(n)| 2^{\Omega(n)}}{n} \prod_{p \mid n} \left(1 - \frac{2}{p}\right)^{-1}$$

$$+ \sum_{\substack{p'p'' \mid y \\ (p'p'',q)=1 \\ p' \neq p''}} \frac{1}{p'\, p''} \left(1 - \frac{1}{p'}\right)^{-1} \left(1 - \frac{1}{p''}\right)^{-1} \sum_{\substack{n \leqslant z/p'p'' \\ (n,qy)=1}} \frac{|\mu(n)| 2^{\Omega(n)}}{n} \prod_{p \mid n} \left(1 - \frac{2}{p}\right)^{-1} + \cdots$$

$$\geqslant \left\{1 + \sum_{\substack{p' \mid y \\ p' \leqslant \Delta(y) \\ p' + q}} \frac{1}{p'} \left(1 - \frac{1}{p'}\right)^{-1} + \sum_{\substack{p'p'' \mid y \\ (p'p'',q)=1 \\ p' \neq p'' \\ p'p'' \leqslant \Delta(y)}} \frac{1}{p'\, p''} \left(1 - \frac{1}{p'}\right)^{-1} \left(1 - \frac{1}{p''}\right)^{-1} + \cdots\right\}$$

$$\sum_{\substack{n \leqslant z/\Delta(y) \\ (n,qy)=1}} \frac{|\mu(n)| 2^{\Omega(n)}}{n} \prod_{p \mid n} \left(1 + \frac{2}{p-2}\right)$$

$$= \left(\prod_{\substack{p \mid y \\ p + q}} \left(1 - \frac{1}{p}\right)^{-1} + O\left(e^{-c_1 \frac{\log 3y}{\log \log 9y}}\right)\right) \left(\frac{1}{8} \prod_{p > 2} \frac{(p-1)^2}{p(p-2)} \prod_{\substack{p \mid qy \\ p > 2}} \frac{p-2}{p} \log^2 \frac{z}{\Delta(y)}\right.$$

$$\left. + O\left(\log 2\, zy \log \log 3\, zy\right)\right)$$

$$= \frac{1}{4} \prod_{p|q} \frac{p-1}{p} \prod_{p>2} \frac{(p-1)^2}{p(p-2)} \prod_{\substack{p|qy \\ p>2}} \frac{p-2}{p-1} \log^2 z + O\left(\prod_{\substack{p|qy \\ p>2}} \frac{p-2}{p-1} \cdot \frac{\log 2z \, \log 2y}{\log\log 3y} \right)$$

$$+ O\left(\prod_{\substack{p|qy \\ p>2}} \frac{p-2}{p-1} \cdot \frac{\log^2 2y}{(\log\log 3y)^2} \right) + O\left(e^{-c_1 \frac{\log 3y}{\log\log 9y}} \log^2 2\,yz \right)$$

$$+ O\left(\log 2\,zy \cdot (\log\log 3\,zy)^2 \right)$$

$$= \frac{1}{4} \prod_{p|q} \frac{p-1}{p} \prod_{p>2} \frac{(p-1)^2}{p(p-2)} \prod_{\substack{p-qy \\ p>2}} \frac{p-2}{p-1} \log{}^2 z + O\left(\prod_{\substack{p|qy \\ p>2}} \frac{p-2}{p-1} \frac{\log^2 2zy}{\log\log 3zy} \right).$$

On the other hand, we have

$$\sum_{\substack{n\leqslant z \\ (n,q)=1}} \frac{|\mu(n)|}{f(n)} \leqslant \left\{ 1 + \sum_{\substack{p'|y \\ p'+q}} \frac{1}{p'} \left(1 - \frac{1}{p'} \right)^{-1} \right.$$

$$\left. + \sum_{\substack{p'p''|y \\ (p'p'',q)=1 \\ p'\neq p''}} \frac{1}{p'\,p''} \left(1 - \frac{1}{p'} \right)^{-1} \left(1 - \frac{1}{p''} \right)^{-1} + \cdots \right\} \sum_{\substack{n\leqslant z \\ (n,qy)=1}} \frac{|\mu(n)|2^{\Omega(n)}}{n} \prod_{p|n} \left(1 - \frac{2}{p} \right)^{-1} \right\}$$

$$= \prod_{\substack{p|y \\ p+q}} \left(1 - \frac{1}{p} \right)^{-1} \left(\frac{1}{8} \prod_{p>2} \frac{(p-1)^2}{p(p-2)} \prod_{\substack{p|qy \\ p>2}} \frac{p-2}{p} \log^2 z + O(\log 2\,zy \cdot \log\log 3\,zy) \right)$$

$$= \frac{1}{4} \prod_{p|q} \frac{p-1}{p} \prod_{p>2} \frac{(p-1)^2}{p(p-2)} \prod_{\substack{p|qy \\ p>2}} \frac{p-2}{p-1} \log^2 z + O\left(\prod_{\substack{p|qy \\ p>2}} \frac{p-2}{p-1} \cdot \frac{\log^2 2\,zy}{\log\log 3\,zy} \right).$$

This proves the Lemma.

(ii) If $\Delta(y) < \log 2z$, then $y < e^{c(\log\log 3z)^2}$ $(c > 0)$. Hence, we have

$$\sum_{\substack{n\leqslant z \\ n,q)=1}} \frac{|\mu(n)|}{f(n)} \geqslant \left\{ 1 + \sum_{\substack{p'|y \\ p'+q}} \frac{1}{p'} \left(1 - \frac{1}{p'} \right)^{-1} + \sum_{\substack{p'p''|y \\ (p'p'',q)=1 \\ p'\neq p''}} \frac{1}{p'\,p''} \left(1 - \frac{1}{p'} \right)^{-1} \left(1 - \frac{1}{p''} \right)^{-1} \right.$$

$$\left. + \cdots \right\} \sum_{\substack{n\leqslant z/e^{c(\log\log 3z)^2} \\ (n,qy)=1}} \frac{|\mu(n)|2^{\Omega(n)}}{n} \prod_{p|n} \left(1 - \frac{2}{p} \right)^{-1}$$

$$= \frac{1}{4} \prod_{p|q} \frac{p-1}{p} \prod_{p>2} \frac{(p-1)^2}{p(p-2)} \prod_{\substack{p|qy \\ p>2}} \frac{p-2}{p-1} \log^2 z + O\left(\prod_{\substack{p|qy \\ p>2}} \frac{p-2}{p-1} \frac{\log^2 2\,zy}{\log\log 3\,zy} \right)$$

and

$$\sum_{\substack{n\leqslant z \\ (n,q)=1}} \frac{|\mu(n)|}{f(n)} \leqslant \left\{ 1 + \sum_{\substack{p'|y \\ p'+q}} \frac{1}{p'} \left(1 - \frac{1}{p'} \right)^{-1} + \sum_{\substack{p'p''|y \\ (p'p'',q)=1 \\ p'\neq p''}} \frac{1}{p'\,p''} \left(1 - \frac{1}{p'} \right)^{-1} \left(1 - \frac{1}{p''} \right)^{-1} \right.$$

$$\left. + \cdots \right\} \sum_{\substack{n\leqslant z \\ (n,qy)=1}} \frac{|\mu(n)|\,2^{\Omega(n)}}{n} \prod_{p|n} \left(1 - \frac{2}{p} \right)^{-1}$$

362

$$= \frac{1}{4} \prod_{p|q} \frac{p-1}{p} \prod_{p>2} \frac{(p-1)^2}{p(p-2)} \prod_{\substack{p|qy \\ p>2}} \frac{p-2}{p-1} \log^2 z + O\left(\prod_{\substack{p|qy \\ p>2}} \frac{p-2}{p-1} \cdot \frac{\log^2 2 zy}{\log\log 3 zy} \right).$$

Thus the Lemma follows.

(iii) If $\Delta(y) > z$, then by Lemma 2, we have

$$\sum_{\substack{n \leqslant z \\ (n,q)=1}} \frac{|\mu(n)|}{f(n)} = O\left(\sum_{\substack{n \leqslant z \\ (n,q)=1}} \frac{|\mu(n)| 2^{\Omega(n)}}{n} \prod_{p|n} \left(1 - \frac{2}{p}\right)^{-1} \right)$$

$$= O(\log^2 2 z) = O\left(\frac{\log^2 2y}{(\log\log 3y)^2} \right) = O\left(\prod_{\substack{p|qy \\ p>2}} \frac{p-2}{p-1} \cdot \frac{\log^2 2 zy}{\log\log 3 zy} \right)$$

and

$$\frac{1}{4} \prod_{p|q} \frac{p-1}{p} \prod_{p>2} \frac{(p-1)^2}{p(p-2)} \prod_{\substack{p|qy \\ p>2}} \frac{p-2}{p-1} \log^2 z = O(\log^2 2z)$$

$$= O\left(\prod_{\substack{p|qy \\ p>2}} \frac{p-2}{p-1} \cdot \frac{\log^2 2 zy}{\log\log 3 zy} \right).$$

Hence,

$$\sum_{\substack{n \leqslant z \\ (n,q)=1}} \frac{|\mu(n)|}{f(n)} = \frac{1}{4} \prod_{p|q} \frac{p-1}{p} \prod_{p>2} \frac{(p-1)^2}{p(p-2)} \prod_{\substack{p|qy \\ p>2}} \frac{p-2}{p-1} \log^2 z$$

$$+ O\left(\prod_{\substack{p|qy \\ p>2}} \frac{p-2}{p-1} \cdot \frac{\log^2 2 zy}{\log\log 3 zy} \right).$$

Thus the Lemma follows by (i) (ii) and (iii).

Lemma 6. Let $\beta > \alpha > 1$ be two real numbers. Then for $x \geqslant 2$, we have

$$\sum_{x^{1/\beta} < p \leqslant x^{1/\alpha}} \frac{1}{p \log^2 \frac{x}{p}} = \frac{1}{\log^2 x} \left\{ \log\frac{\beta-1}{\alpha-1} + \frac{1}{\alpha-1} - \frac{1}{\beta-1} \right\} + O\left(\frac{1}{\log^3 x} \right)^{[23]}.$$

§3. Theorem A

Let $2 \leqslant y \leqslant x$ be two integers. Let

$$a, q; \quad a_i, b_i \quad (1 \leqslant i \leqslant r) \tag{ω}$$

be a sequence of integers satisfying the following conditions:

$$2|q, \quad q = O(1); \quad \text{if} \quad p_i|y,$$

$$\text{then } a_i \equiv b_i \pmod{p_i}, \text{ otherwise } a_i \not\equiv b_i \pmod{p_i} \quad (1 \leqslant i \leqslant r), \tag{1}$$

where $2 < p_1 < \cdots < p_r \leqslant \xi$ are all the primes not dividing q and not exceeding ξ and where ξ is a real number and $\xi > q$.

Let $P_\omega(x, q, \xi)$ be the number of integers n satisfying the following conditions:

$$1 \leqslant n \leqslant x, \quad n \equiv a \pmod{q}, \quad n \not\equiv a_i \pmod{p_i}, \quad n \not\equiv b_i \pmod{p_i}$$
$$(1 \leqslant i \leqslant r). \qquad (2)$$

It follows by the Chinese Remainder Theorem that each of the systems of congruences

$$y \equiv a_i \pmod{p_i} \qquad (1 \leqslant i \leqslant r)$$

and

$$y \equiv b_i \pmod{p_i} \qquad (1 \leqslant i \leqslant r)$$

has a unique solution in the interval $1 \leqslant y \leqslant p_1 p_2 \cdots p_r$. Denote these solutions by a^* and b^* respectively. It is evident that the conditions (2) are equivalent to the following conditions:

$$1 \leqslant n \leqslant x, \quad n \equiv a \pmod{q}, \quad (n - a^*)(n - b^*) \not\equiv 0 \pmod{p_i}$$
$$(1 \leqslant i \leqslant r). \qquad (3)$$

Theorem A. Let $c > 0$ and $P = \prod_{i=1}^{r} p_i = \prod_{\substack{p \leqslant \xi \\ p + q}} p$. Then the estimation

$$P_\omega'(x, q, \xi) \leqslant \frac{x}{q \sum_{\substack{m \leqslant \xi^c \\ m \mid P}} \dfrac{|\mu(m)|}{f(m)}} + O(\xi^{2c} \log^6 \xi)$$

holds uniformly in (ω), where

$$g(1) = 1, \quad g(p) = \frac{1}{p} \; (p \mid y), \quad g(p) = \frac{2}{p} \, (p + y),$$

$$g(n) = \prod_{p \mid n} g(p) \quad \text{and} \quad f(n) = \frac{1}{g(n)} \prod_{p \mid n} (1 - g(p)).$$

Proof. If $k \mid P$, then $(k, q) = 1$. It follows by the Chinese Remainder Theorem that the number of solutions of the following system of congruences

$$\left. \begin{array}{l} (n - a^*)(n - b^*) \equiv 0 \pmod{k} \\[4pt] n \equiv a \pmod{q} \end{array} \right\}$$

in the interval $1 \leqslant n \leqslant kq$ is $2^{\Omega(k) - \Omega((k, y))}$. Hence

$$\sum_{\substack{k\mid(n-a^*)(n-b^*)\\n\equiv a(\mathrm{mod}\ q)\\1\leqslant n\leqslant x}} 1 = 2^{\Omega(k)-\Omega((k,q))}\left[\frac{x}{kq}\right] + O(2^{\Omega(k)})$$

$$= g(k)\,\frac{x}{q} + O(2^{\Omega(k)})\ .$$

Let

$$\lambda_k = \frac{\mu(k)}{g(k)\,f(k)} \sum_{\substack{1\leqslant n<\xi^c/k\\(n,k)=1\\n\mid P}} \frac{|\mu(n)|}{f(n)} \Big/ \sum_{\substack{1\leqslant l\leqslant\xi^c\\l\mid P}} \frac{|\mu(l)|}{f(l)}\ ,$$

where $k\mid P$.

Since $\lambda_1 = 0$ and $\lambda_d = 0$ for $d > \xi^c$, therefore

$$P_\omega(x,q,\xi) = \sum_{\substack{1\leqslant n\leqslant x\\((n-a^*)(n-b^*),P)=1\\n\equiv a\ (\mathrm{mod}\ q)}} 1$$

$$\leqslant \sum_{\substack{1\leqslant n\leqslant x\\n\equiv a\ (\mathrm{mod}\ q)}} \left(\sum_{d\mid((n-a^*)(n-b^*),P)} \lambda_d\right)^2$$

$$=. \sum_{\substack{d_1\leqslant\xi^c\\d_1\mid P}} \sum_{\substack{d_2\leqslant\xi^c\\d_2\mid P}} \lambda_{d_1}\lambda_{d_2} \sum_{\substack{1\leqslant n\leqslant x\\n\equiv a\ (\mathrm{mod}\ q)\\\{d_1,d_2\}\mid(n-a^*)(n-b^*)}} 1$$

$$= \frac{x}{q} \sum_{\substack{d_1\leqslant\xi^c\\d_1\mid P}} \sum_{\substack{d_2\leqslant\xi^c\\d_2\mid P}} \lambda_{d_1}\lambda_{d_2}\, g(\{d_1,d_2\})$$

$$+ O\left(\sum_{\substack{d_1\leqslant\xi^c\\d_1\mid P}} \sum_{\substack{d_2\leqslant\xi^c\\d_2\mid P}} |\lambda_{d_1}\lambda_{d_2}|\, 2^{\Omega(d_1)+\Omega(d_2)}\right)$$

$$= \frac{x}{q}\,Q + R\ ,$$

where $\{d_1,d_2\}$ denotes the least common multiple of d_1 and d_2. If n is a square-free number, then

$$\frac{1}{g(n)} = \sum_{\tau\mid n} \frac{1}{g(\tau)} \sum_{d\mid\tau\mid n} \mu(d) = \sum_{d\tau\mid n} \frac{\mu(d)}{g(\tau)} = \sum_{k\mid n} \sum_{d\mid k} \frac{\mu(d)}{g\left(\frac{k}{d}\right)}$$

$$= \sum_{k\mid n} \frac{1}{g(k)} \prod_{p\mid k} (1-g(p)) = \sum_{k\mid n} f(k)\ .$$

Hence,

$$Q = \sum_{\substack{1\leqslant d_1<\xi^c\\d_1\mid P}} \sum_{\substack{1\leqslant d_2<\xi^c\\d_2\mid P}} \lambda_{d_1}\lambda_{d_2}\, \frac{g(d_1)g(d_2)}{g(\{d_1,d_2\})}$$

$$= \sum_{\substack{1 \leqslant d_1 < \xi^c \\ d_1 | P}} \sum_{\substack{1 \leqslant d_2 < \xi^c \\ d_2 | P}} \lambda_{d_1} \lambda_{d_2} g(d_1) g(d_2) \sum_{d | (d_1, d_2)} f(d)$$

$$= \sum_{\substack{1 \leqslant d \leqslant \xi^c \\ d | P}} f(d) \left(\sum_{\substack{1 \leqslant k \leqslant \xi^c \\ d | k | P}} \lambda_k g(k) \right)^2 .$$

Let

$$S = \sum_{\substack{1 \leqslant m \leqslant \xi^c \\ m | P}} \frac{|\mu(m)|}{f(m)} .$$

Then, we have for $k | P$

$$\lambda_k g(k) = \frac{1}{S} \sum_{\substack{1 \leqslant m \leqslant \xi^c / k \\ (m, k) = 1 \\ m | P}} \frac{\mu(k) \mu^2(m)}{f(k) f(m)} = \frac{1}{S} \sum_{\substack{1 \leqslant m \leqslant \xi^c / k \\ (m, k) = 1 \\ m | P}} \frac{\mu(mk) \mu(m)}{f(mk)} ,$$

and

$$\sum_{\substack{d | k | P \\ 1 \leqslant k \leqslant \xi^c}} \lambda_k g(k) = \frac{1}{S} \sum_{\substack{d | k | P \\ 1 \leqslant k \leqslant \xi^c}} \sum_{\substack{1 \leqslant m \leqslant \xi^c / k \\ m | P}} \frac{\mu(mk) \mu(m)}{f(mk)}$$

$$= \frac{1}{S} \sum_{\substack{1 \leqslant r \leqslant \xi^c \\ d | r | P}} \frac{\mu(r)}{f(r)} \sum_{d | k | r} \mu\left(\frac{r}{k}\right)$$

$$= \frac{1}{S} \frac{\mu(d)}{f(d)} .$$

Hence,

$$Q = \frac{1}{S} .$$

By the Mertens theorem, we have

$$R = O\left(\left(\sum_{\substack{k < \xi^c \\ k | P}} |\lambda_k| \, 2^{\Omega(k)} \right)^2 \right) = O\left(\left(\sum_{\substack{k < \xi^c \\ k | P}} \frac{|\mu(k)|}{f(k) \, g(k)} \, 2^{\Omega(k)} \right)^2 \right)$$

$$= O\left(\left(\sum_{\substack{1 < k < \xi^c \\ k | P}} \frac{|\mu(k)| \, 2^{\Omega(k)}}{\prod_{2 < p < \xi^c} \left(1 - \frac{2}{p} \right)} \right)^2 \right) = O\left(\left(\sum_{k < \xi^c} d(k) \log^2 \xi \right)^2 \right)$$

$$= O\left(\xi^{2c} \log^6 \xi \right) .$$

Thus we have the theorem.

366

§4. Applications of Theorem A

Let l be a positive integer and $l < c \leqslant l + 1$. Then by Eratosthenes' sieve method we have

$$
\sum_{\substack{1 \leqslant n \leqslant \xi^c \\ n \mid P}} \frac{|\mu(n)|}{f(n)} = \sum_{\substack{1 \leqslant n \leqslant \xi^c \\ (n,q)=1}} \frac{|\mu(n)|}{f(n)} - \sum_{\xi < p \leqslant \xi^c} \sum_{\substack{1 \leqslant n \leqslant \xi^c \\ (n,q)=1 \\ p \mid n}} \frac{|\mu(n)|}{f(n)}
$$

$$
+ \sum_{\substack{\xi < p_1 < p_2 \\ p_1 p_2 \leqslant \xi^c}} \sum_{\substack{1 \leqslant n \leqslant \xi^c \\ p_1 p_2 \mid n \\ (n,q)=1}} \frac{|\mu(n)|}{f(n)} - + \cdots + (-1)^l \sum_{\substack{\xi < p_1 < \cdots < p_l \\ p_1 \cdots p_l \leqslant \xi^c}} \sum_{\substack{1 \leqslant n \leqslant \xi^c \\ p_1 \cdots p_l \mid n \\ (n,q)=1}} \frac{|\mu(n)|}{f(n)}
$$

$$
= \sum_{\substack{1 \leqslant n \leqslant \xi^c \\ (n,q)=1}} \frac{|\mu(n)|}{f(n)} - \sum_{\xi < p \leqslant \xi^c} \frac{1}{f(p)} \sum_{\substack{n \leqslant \xi^c/p \\ (n,pq)=1}} \frac{|\mu(n)|}{f(n)} \tag{4}
$$

$$
+ \sum_{\substack{\xi < p_1 < p_2 \\ p_1 p_2 \leqslant \xi^c}} \frac{1}{f(p_1) f(p_2)} \sum_{\substack{1 \leqslant n \leqslant \xi^c/p_1 p_2 \\ (n, p_1 p_2 q)=1}} \frac{|\mu(n)|}{f(n)} - + \cdots
$$

$$
+ (-1)^l \sum_{\substack{\xi < p_1 < \cdots < p_l \\ p_1 p_2 \cdots p_l \leqslant \xi^c}} \frac{1}{f(p_1) \cdots f(p_l)} \sum_{\substack{1 \leqslant n \leqslant \xi^c/p_1 \cdots p_l \\ (n, p_1 \cdots p_l q)=1}} \frac{|\mu(n)|}{f(n)} .
$$

$1°$. If $1 \leqslant c \leqslant 2$, then for sufficiently large x, we take

$$
\xi = \frac{x^{\frac{1}{2c}}}{\log^5 x} \; (> q) \text{ and write } 2c = d. \text{ Since}
$$

$$
\sum_{\xi < p \leqslant \xi^c} \frac{1}{f(p)} - \sum_{\xi < p \leqslant \xi^c} \frac{2}{p} = O\Big(\sum_{\substack{\xi < p \leqslant \xi^c \\ p+y}} \frac{1}{p^2} \Big) + O\Big(\sum_{\substack{p \mid y \\ p > \xi}} \frac{1}{p} \Big) = O\Big(\frac{1}{\xi} \Big)
$$

$$
= O\Big(\frac{\log^5 x}{x^{1/d}} \Big),
$$

thus by (4) and Lemma 5, we have

$$
\sum_{\substack{m \leqslant \xi^c \\ m \mid P}} \frac{|\mu(m)|}{f(m)} = \sum_{\substack{m \leqslant \xi^c \\ (m,q)=1}} \frac{|\mu(m)|}{f(m)} - \sum_{\xi < p \leqslant \xi^c} \frac{2}{p} \sum_{\substack{m \leqslant \xi^c/p \\ (m,q)=1}} \frac{|\mu(m)|}{f(m)} + O\Big(\frac{\log^7 x}{x^{1/d}} \Big)
$$

$$
= \left\{ (d-1)^2 - 2\left(\frac{d}{2}\right)^2 \log \frac{d}{2} \right\} x \cdot \frac{1}{4} \prod_{p \mid q} \frac{p-1}{p} \prod_{p > 2} \frac{(p-1)^2}{p(p-2)}
$$

$$
\times \prod_{\substack{p \mid qy \\ p > 2}} \frac{p-2}{p-1} \log^2 \xi + O\Big(\prod_{\substack{p \mid qy \\ p > 2}} \frac{p-2}{p-1} \cdot \frac{\log^2 x}{\log \log x} \Big).
$$

Therefore, it follows by Theorem A that

$$P_\omega(x, q, x^{1/d}) \leqslant P_\omega\left(x, q, \frac{x^{1/d}}{\log^5 x}\right)$$

$$\leqslant \Lambda(d)\, c_{qy}\, \frac{x}{\log^2 x} + O\left(\prod_{\substack{p|qy \\ p>2}} \frac{p-2}{p-1} \cdot \frac{\log^2 x}{\log\log x}\right), \tag{5}$$

$$\Lambda(d) = 2e^{2\gamma}\left[\frac{d^2}{(d-1)^2 - 2\left(\frac{d}{2}\right)^2 \log\frac{d}{2}}\right] \quad (2 \leqslant d \leqslant 4), \tag{6}$$

$$c_{qy} = \frac{2e^{-2\gamma}}{q} \prod_{p>2}\left(1 - \frac{1}{(p-1)^2}\right) \prod_{p|q} \frac{p}{p-1} \prod_{\substack{p|qy \\ p>2}} \frac{p-1}{p-2}, \tag{7}$$

where the constant implied by the symbol "O" is an absolute constant, and γ denotes Euler's constant.

2°. If $2 \leqslant c \leqslant 3$, then we take also $\xi = \dfrac{x^{\frac{1}{2c}}}{\log^5 x}$ and write $2c = d$. Since

$$\sum_{\xi < p \leqslant \xi^c} \frac{1}{f(p)} \sum_{\substack{1 \leqslant n \leqslant \xi^c/p \\ (n,q)=1}} \frac{|\mu(n)|}{f(n)} - \sum_{\xi < p \leqslant \xi^c} \frac{1}{f(p)} \sum_{\substack{n \leqslant \xi^c/p \\ (n,qp)=1}} \frac{|\mu(n)|}{f(n)}$$

$$= O\left(\sideset{}{'}\sum_{\xi < p \leqslant \xi^c} \frac{1}{f(p)^2} \sum_{\substack{n \leqslant \xi^c/p^2 \\ (n,q)=1}} \frac{|\mu(n)|}{f(n)}\right) = O\left(\frac{\log^2 \xi}{\xi}\right) = O\left(\frac{\log^7 x}{x^{1/d}}\right),$$

and

$$\sum_{\substack{\xi < p < p' \\ pp' \leqslant \xi^c}} \frac{1}{f(p)\,f(p')} - \sum_{\substack{\xi < p < p' \\ pp' \leqslant \xi^c}} \frac{4}{pp'} = \sum_{\substack{\xi < p < p' \\ pp' \leqslant \xi^c}} \left(\frac{2}{p-2} \cdot \frac{2}{p'-2} - \frac{2}{p-2} \cdot \frac{2}{p'}\right)$$

$$+ \sum_{\substack{\xi < p < p' \\ pp' \leqslant \xi^c}} \left(\frac{2}{p-2} \cdot \frac{2}{p'} - \frac{4}{pp'}\right) + O\left(\frac{1}{\xi}\right) = O\left(\sum_{\xi < p \leqslant \xi^c} \frac{1}{p} \sum_{\xi' > \xi} \frac{1}{p'^2}\right)$$

$$+ O\left(\frac{1}{\xi}\right) = O\left(\frac{1}{\xi}\right) = O\left(\frac{\log^5 x}{x^{1/d}}\right),$$

therefore by (4) and Lemma 5, we have

$$\sum_{\substack{n \leqslant \xi^c \\ n|p}} \frac{|\mu(n)|}{f(n)} = \sum_{\substack{n \leqslant \xi^c \\ (n,q)=1}} \frac{|\mu(n)|}{f(n)} - \sum_{\xi < p \leqslant \xi^c} \frac{2}{p} \sum_{\substack{n \leqslant \xi^c/p \\ (n,q)=1}} \frac{|\mu(n)|}{f(n)}$$

358

$$+ \sum_{\substack{\xi<p<p' \\ pp'\leqslant\xi^c}} \frac{4}{pp'} \sum_{\substack{n\leqslant\xi^c/pp' \\ (n,q)=1}} \frac{|\mu(n)|}{f(n)} + O\left(\frac{\log^7 x}{x^{1/4}}\right)$$

$$= \left\{(d-1)^2 - 2\left(\frac{d}{2}\right)^2 \log\frac{d}{2} + \frac{4}{\log^2\xi} \sum_{\substack{\xi<p<p' \\ pp'\leqslant\xi^c}} \frac{1}{pp'} \log^2\frac{\xi^c}{pp'}\right\} \times$$

$$\times \frac{1}{4} \prod_{p>2} \frac{(p-1)^2}{p(p-2)} \prod_{p|q} \frac{p-1}{p} \prod_{\substack{p|qy \\ p>2}} \frac{p-2}{p-1} \log^2\xi + O\left(\prod_{\substack{p|qy \\ p>2}} \frac{p-2}{p-1} \cdot \frac{\log^2 x}{\log\log x}\right).$$

Hence it follows by Theorem A that for any given $\varepsilon > 0$, there exists a positive constant $x_0 = x_0(\varepsilon)$ such that (5) holds for $x > x_0$ and

$$\Lambda(d) = 2e^{2\gamma}\left(\frac{d^2}{(d-1)^2 - 2\left(\frac{d}{2}\right)^2 \log\frac{d}{2} + \delta\left(\frac{d}{2}\right) - \varepsilon}\right) \quad (4\leqslant d\leqslant 6), \quad (8)$$

where

$$\delta(\alpha) = \lim_{\xi\to\infty} \frac{4}{\log^2\xi} \sum_{\substack{\xi<p<p' \\ pp'\leqslant\xi^\alpha}} \frac{1}{pp'} \log^2\frac{\xi^\alpha}{pp'} \quad (\alpha\geqslant 2). \quad (9)$$

Eq. (5) also holds for $x > x_1(\varepsilon)$, $6 \leqslant d \leqslant 8$ and

$$\Lambda(d) = 2e^{2\gamma}\left(\frac{d^2}{(d-1)^2 - 2\left(\frac{d}{2}\right)^2 \log\frac{d}{2} + \delta\left(\frac{d}{2}\right) - K\left(\frac{d}{2}\right) - \varepsilon}\right)$$
$$(6\leqslant d\leqslant 8), \quad (10)$$

where

$$K(\alpha) = \overline{\lim_{\xi\to\infty}} \frac{8}{\log^2\xi} \sum_{\substack{pp'p''\leqslant\xi \\ \xi<p<p'<p''}} \frac{1}{pp'p''} \log^2\frac{\xi^\alpha}{pp'p''} \quad (\alpha\geqslant 3). \quad (11)$$

In the same way we have the expressions of $\Lambda(d)$ for $d > 8$. Let us put $\Lambda(d) = \Lambda(2)$ for $0 < d \leqslant 2$, then it is evident that (5) holds for $0 < d \leqslant 2$.

We may estimate $\delta(\alpha)$ and $K(\alpha)$ by Mertens' theorem. For example,

$$\lim_{\xi\to\infty} \sum_{\substack{\xi<p<p' \\ \xi^{2.1}<pp'\leqslant\xi^{2.2}}} \frac{1}{pp'} \geqslant \lim_{\xi\to\infty}\left(\sum_{\xi<p\leqslant\xi^{1.01}} \frac{1}{p} \sum_{\xi^{1.1}<p'\leqslant\xi^{1.19}} \frac{1}{p'} + \right.$$

$$+ \sum_{\substack{\xi^{1.01}<p\leqslant\xi^{1.02}}} \frac{1}{p} \sum_{\substack{\xi^{1.09}<p'\leqslant\xi^{1.18}}} \frac{1}{p'} + \cdots + \sum_{\substack{\xi^{1.08}<p\leqslant\xi^{1.09}}} \frac{1}{p} \sum_{\substack{\xi^{1.09}<p'\leqslant\xi^{1.11}}} \frac{1}{p'}$$

$$+ \sum_{\substack{\xi^{1.09}<p\leqslant\xi^{1.095}}} \frac{1}{p} \sum_{\substack{\xi^{1.095}<p'\leqslant\xi^{1.105}}} \frac{1}{p'} \Big)$$

$$= \log 1.01 \log \frac{1.19}{1.1} + \log \frac{1.02}{1.01} \log \frac{1.18}{1.09} + \cdots + \log \frac{1.09}{1.08} \log \frac{1.11}{1.09}$$

$$+ \log \frac{1.095}{1.09} \log \frac{1.105}{1.095} > 0.00561 \; .$$

We have also the estimations of $\sum_{\substack{\xi<p<p' \\ pp'\leqslant\xi^{2.1}}} \dfrac{1}{pp'}$ and $\sum_{\substack{\xi<p<p' \\ \xi^{2.i}<pp'\leqslant\xi^{2.i+0.1}}} \dfrac{1}{pp'}$ $(2 \leqslant i \leqslant 9)$. It follows that

$$\delta(3) \geqslant 4 \lim_{\xi\to\infty} \Big(0.81 \sum_{\substack{\xi<p<p' \\ pp'\leqslant\xi^{2.1}}} \frac{1}{pp'} + 0.64 \sum_{\substack{\xi<p<p' \\ \xi^{2.1}<pp'\leqslant\xi^{2.2}}} \frac{1}{pp'} + \cdots$$

$$+ 0.01 \sum_{\substack{\xi^{2.8}<pp'\leqslant\xi^{2.9} \\ \xi<p<p'}} \frac{1}{pp'} \Big) > 0.087202 \; .$$

§5. Theorem B

Theorem B_1. Let $1 < \alpha < \beta \leqslant 15$ be two given numbers. If there exists a non-negative and non-decreasing function $\Lambda(d)$ $(0 < d \leqslant 14)$ with the property that $\Lambda(d)$ has at most a finite number of discontinuities, such that the following inequality holds uniformly in (ω) and d:

$$P_\omega(x, q, x^{1/d}) < \Lambda(z) \frac{c_{qy}\,x}{\log^2 x} + O\Big(\prod_{\substack{p\mid qy \\ p>2}} \frac{p-1}{p-2} \cdot \frac{x}{\log^2 x \log\log x} \Big),$$

then

$$\sum_{\substack{x^{1/\beta}<p\leqslant x^{1/\alpha} \\ p\nmid y}} P_{\omega_p}\Big(\frac{x}{p}, q, x^{1/\beta}\Big) \leqslant \Big(\int_{\alpha-1}^{\beta-1} \Lambda\Big(\frac{\beta z}{z+1}\Big) \frac{z+1}{z^2} \, dz \Big) \frac{c_{qy}\,x}{\log^2 x}$$

$$+ O\Big(\frac{c_{qy}\,x}{\log^2 x \log\log x} \Big)$$

holds uniformly in (ω_p).

370

Proof. Let $n = [\sqrt{\log x}]$ and $u_l = \alpha + \dfrac{\beta - 1}{n} l - 1 \quad (0 \leqslant l \leqslant n)$.
Then by Lemma 6

$$T_l = \sum_{\substack{x^{\frac{1}{u_{l+1}+1}} < p \leqslant x^{\frac{1}{u_l+1}} \\ p+y}} P\omega_p\left(\frac{x}{p}, q, x^{1/\beta}\right) = \sum_{\substack{x^{\frac{1}{u_{l+1}+1}} < p \leqslant x^{\frac{1}{u_l+1}} \\ p+y}} P\omega_p\left(\frac{x}{p}, q, \left(\frac{x}{p}\right)^{\frac{\log x}{\beta \log x/p}}\right)$$

$$\leqslant \sum_{\substack{x^{\frac{1}{u_{l+1}+1}} < p \leqslant x^{\frac{1}{u_l+1}}}} \left\{ \Lambda\left(\frac{\beta \log x/p}{\log x}\right) \frac{c_{qy} x}{p \log^2 x/p} + O\left(\frac{c_{qy} x}{p \log^2 x/p \log \log x}\right) \right\}$$

$$\leqslant \Lambda\left(\frac{\beta u_{l+1}}{u_{l+1} + 1}\right) \frac{c_{qy} x}{\log^2 x} \left(\log \frac{u_{l+1}}{u_l} + \frac{u_{l+1} - u_l}{u_{l+1} u_l} \right) + O\left(\frac{c_{qy} x}{\log^3 x}\right)$$

$$+ O\left(\frac{c_{qy} x}{\log^2 x \log \log x}\left(\log \frac{u_{l+1}}{u_l} + \frac{u_{l+1} - u_l}{u_{l+1} u_l} \right)\right).$$

Since $\dfrac{x}{1 + x}$ is an increasing function of x for $x \geqslant 1$, hence·

$$\sum_{l=0}^{n-1} \Lambda\left(\frac{\beta u_{l+1}}{u_{l+1} + 1}\right) \left\{ \log \frac{u_{l+1}}{u_l} + \frac{u_{l+1} - u_l}{u_{l+1} u_l} \right\} - \int_{\alpha-1}^{\beta-1} \Lambda\left(\frac{\beta z}{z + 1}\right) \frac{z + 1}{z^2} dz$$

$$= \sum_{l=0}^{n-1} \int_{u_l}^{u_{l+1}} \left(\Lambda\left(\frac{\beta u_{l+1}}{u_{l+1} + 1}\right) - \Lambda\left(\frac{\beta z}{z + 1}\right) \right) \frac{z + 1}{z^2} dz$$

$$\leqslant \sum_{l=0}^{n-1} \left(\Lambda\left(\frac{\beta u_{l+1}}{u_{l+1} + 1}\right) - \Lambda\left(\frac{\beta u_l}{u_l + 1}\right) \right) \max_{0 \leqslant l \leqslant n-1} \int_{u_l}^{u_{l+1}} \frac{z + 1}{z^2} dz$$

$$= o\left(\frac{1}{n}\right).$$

It follows that

$$\sum_{\substack{x^{1/\beta} < p \leqslant x^{1/\alpha} \\ p+y}} P\omega_p\left(\frac{x}{p}, q, x^{1/\beta}\right) = \sum_{l=0}^{n-1} T_l$$

$$\leqslant \left(\int_{\alpha-1}^{\beta-1} \Lambda\left(\frac{\beta z}{z + 1}\right) \frac{z + 1}{z^2} dz \right) \frac{c_{qy} x}{\log^2 x}$$

$$+ O\left(\frac{c_{qy} x}{\log^2 x \log \log x}\right).$$

Thus we have the theorem.

Theorem B$_2$. Let $2 \leqslant \alpha < \beta \leqslant 15$ be two given numbers. If there exist two non-negative and non-decreasing functions $\lambda(z)$ and

$\Lambda(z)$ $(0 < z \leqslant 15)$ with the property that each has at most a finite number of discontinuities, such that

$$P_\omega(x, q, x^{1/z}) \geqslant \lambda(z)\, \frac{c_{qy}\, x}{\log^2 x} + O\!\left(\frac{c_{qy}\, x}{\log^2 x \log\log x}\right) \quad (0 < z \leqslant 15) \qquad (12)$$

and

$$P_\omega(x, q, x^{1/z}) \leqslant \Lambda(z)\, \frac{c_{qy}\, x}{\log^2 x} + O\!\left(\frac{c_{qy}\, x}{\log^2 x \log\log x}\right) \quad (0 < z \leqslant 15), \qquad (13)$$

where the constants implied by the symbol "O" are absolute constants, then

$$\lambda_1(\alpha) = \begin{cases} 0, & \text{if } 2 < \alpha \leqslant \tau; \\[2mm] \lambda(\beta) - 2\displaystyle\int_{\alpha-1}^{\beta-1} \Lambda(z)\, \frac{z+1}{z^2}\, dz, & \text{if } 2 \leqslant \alpha \leqslant \beta \leqslant 15 \end{cases} \qquad (14)$$

and

$$\Lambda_1(\alpha) = \Lambda(\beta) - 2\int_{\alpha-1}^{\beta-1} \lambda(z)\, \frac{z+1}{z^2}\, dz, \quad 2 \leqslant \alpha \leqslant \beta \leqslant 15 \qquad (15)$$

have the properties of the functions $\lambda(z)$ and $\Lambda(z)$ respectively.

Proof. Let $n = [\sqrt{\log x}\,]$ and $u_l = \alpha + \dfrac{\beta-\alpha}{n}\, l$ $(0 \leqslant l \leqslant n)$. The difference of $P_\omega(x, q, p_s)$ and $P_\omega(x, q, p_{r+1})$ is the number of integers satisfying the following conditions

$$1 \leqslant n \leqslant x, \quad n \equiv a \,(\mathrm{mod}\, q), \quad n \not\equiv a_i \,(\mathrm{mod}\, p_i), \quad n \not\equiv b_i \,(\mathrm{mod}\, p_i)$$

$$(1 \leqslant i \leqslant r), \quad n \equiv c_{r+1} \,(\mathrm{mod}\, p_{r+1})\ (c_{r+1} = a_{r+1} \text{ or } b_{r+1}). \qquad (16)$$

Let

$$a^{(r)}, q;\ a_i^{(r)}, b_i^{(r)} \qquad (1 \leqslant i \leqslant r) \qquad\qquad (\omega_r)$$

and

$$\tilde{a}^{(r)}, q;\ \tilde{a}_i^{(r)}, \tilde{b}_i^{(r)} \qquad (1 \leqslant i \leqslant r), \qquad\qquad (\tilde{\omega}_r)$$

where $a^{(r)}$, $\tilde{a}^{(r)}$, $a_i^{(r)}$, $b_i^{(r)}$, $\tilde{a}_i^{(r)}$ and $\tilde{b}_i^{(r)}$ denote the solutions of the congruences

$$mp_{r+1} + a_{r+1} \equiv a \,(\mathrm{mod}\, q),$$
$$mp_{r+1} + b_{r+1} \equiv a \,(\mathrm{mod}\, q),$$
$$mp_{r+1} + a_{r+1} \equiv a_i \,(\mathrm{mod}\, p_i),$$
$$mp_{r+1} + a_{r+1} \equiv b_i \,(\mathrm{mod}\, p_i),$$
$$mp_{r+1} + b_{r+1} \equiv a_i \,(\mathrm{mod}\, p_i),$$

and

$$mp_{r+1} + b_{r+1} \equiv b_i \,(\mathrm{mod}\, p_i)$$

372

respectively. Hence

$$P_\omega(x,q,p_r) - P_\omega(x,q,p_{r+1}) \leqslant \begin{cases} P_{\omega_r}\left(\dfrac{x}{p_{r+1}}, q, p_r\right) + P_{\tilde\omega_r}\left(\dfrac{x}{p_{r+1}}, q, p_r\right), & \text{if } p_{r+1} \nmid y; \\[3ex] P_{\omega_r}\left(\dfrac{x}{p_{r+1}}, q, p_r\right), & \text{if } p_{r+1} \mid y. \end{cases}$$

If the primes in the interval $x^{\frac{1}{u_{l+1}}} < p < x^{\frac{1}{u_l}}$ are

$$p_t \leqslant x^{1/u_{l+1}} < p_{t+1} < \cdots < p_s \leqslant x^{1/u_l} < p_{s+1},$$

then

$$P_\omega(x, q, x^{1/u_{l+1}}) - P_\omega(x, q, x^{1/u_l}) \geqslant \sum_{i=t+1}^{s} P_{\omega_{i-1}}\left(\frac{x}{p_i}, q, p_{i-1}\right)$$

$$+ \sum_{i=t+1}^{s} P_{\tilde\omega_{i-1}}\left(\frac{x}{p_i}, q, p_{i-1}\right) + O\left(x^{1-\frac{1}{u_{l+1}}}\right)$$

$$\geqslant \sum_{x^{\frac{1}{u_{l+1}}} < p \leqslant x^{\frac{1}{u_l}}} P_{\omega_p}\left(\frac{x}{p}, q, \left(\frac{x}{p}\right)^{\frac{\log p}{\log x/p}}\right) + \sum_{x^{\frac{1}{u_{l+1}}} < p \leqslant x^{\frac{1}{u_l}}} P_{\tilde\omega_p}\left(\frac{x}{p}, q, \left(\frac{x}{p}\right)^{\frac{\log p}{\log x/p}}\right)$$

$$+ O\left(x^{1-\frac{1}{15}}\right)$$

$$\geqslant 2\lambda(u_l - 1)\, x \sum_{x^{\frac{1}{u_{l+1}}} < p \leqslant x^{\frac{1}{u_l}}} \frac{c_{qy}}{p \log^2 x/p} + O\left(\frac{c_{qy}\, x}{\log\log x} \sum_{x^{\frac{1}{u_{l+1}}} < p \leqslant x^{\frac{1}{u_l}}} \frac{1}{p \log^2 x/p}\right)$$

$$+ O\left(x^{1-\frac{14}{15}}\right)$$

$$\geqslant 2\lambda(u_l - 1)\frac{x\, c_{qy}}{\log^2 x}\left(\log\frac{u_{l+1} - 1}{u_{l-1}} + \frac{1}{u_l - 1} - \frac{1}{u_{l+1} - 1}\right)$$

$$+ O\left(\frac{x\, c_{qy}}{\log^3 x}\right) + O\left(\frac{c_{qy}\, x}{\log^2 x \log\log x}\left(\log\frac{u_{l+1} - 1}{u_l - 1} + \frac{1}{u_l - 1} - \frac{1}{u_{l+1} - 1}\right)\right).$$

Hence

$$P_\omega(x, q, x^{1/\beta}) - P_\omega(x, q, x^{1/\alpha}) = \sum_{l=0}^{n-1} \left(P_\omega(x, q, x^{\frac{1}{u_{l+1}}}) - P_\omega(x, q, x^{\frac{1}{u_l}})\right)$$

$$\geqslant 2\left(\int_{\alpha-1}^{\beta-1} \lambda(z)\, \frac{z+1}{z^2}\, dz\right) \frac{x\, c_{qy}}{\log^2 x} + O\left(\frac{x\, c_{qy}}{\log^{2.5} x}\right)$$

$$+ O\left(\frac{c_{qy}\, x}{\log^2 x \log\log x}\right).$$

This proves (15). In the same way we have (14).

§6. Theorem C

Let $2 \leqslant y \leqslant x$ be two given integers. Let

$$a, q; \quad a_i, b_i \qquad (i = 1, 2, \cdots) \tag{ω}$$

be a sequence of integers satisfying the following conditions:

$$2 \mid q, \quad q = O(1); \quad \text{if} \quad p_i \mid y, \quad \text{then} \quad a_i \equiv b_i \pmod{p_i},$$
$$\text{otherwise} \quad a_i \not\equiv b_i \pmod{p_i} \qquad (i = 1, 2, \cdots), \tag{17}$$

where $2 < p_1 < p_2 < \cdots$ are all the primes not dividing q.

Let $15 \geqslant v > u > 1$ be two given numbers. Let $\mathfrak{M}$ denote the set of integers n satisfying the following:

$$1 \leqslant n \leqslant x, \quad n \equiv a \pmod{q}, \quad n \not\equiv a_i \pmod{p_i}, \quad n \not\equiv b_i \pmod{p_i} \quad (1 \leqslant i \leqslant s),$$
$$n \not\equiv a_{s+j} \pmod{p_{s+j}^2}, \quad n \not\equiv b_{s+j} \pmod{p_{s+j}^2} \quad (1 \leqslant j \leqslant t - s), \tag{18}$$

where $p_s \leqslant x^{1/v} < p_{s+1}$ and $p_t \leqslant x^{1/u} < p_{t+1}$. The number of · elements of $\mathfrak{M}$ is denoted by $M(x, x^{1/v}, x^{1/u})$.

The purpose of this section is to prove the following

Theorem C. The number of elements n of $\mathfrak{M}$ satisfying at most m of the following congruences

$$n \equiv c_{s+j} \pmod{p_{s+j}} \quad (1 \leqslant j \leqslant t - s, c_{s+j} = a_{s+j} \text{ or } b_{s+j}) \tag{19}$$

is not less than

$$P_\omega(x, q, x^{1/v}) - \left(\frac{2}{m+1} \int_{u-1}^{v-1} \Lambda \left(\frac{vz}{z+1} \right)^{z+1} \frac{dz}{z^2} \right) \frac{c_{qy}\, x}{\log^2 x} + O\left(\frac{c_{qy}\, x}{\log^2 x \log \log x} \right).$$

The proof of the Theorem depends on the following two Lemmas.
Lemma 7.

$$M(x, x^{1/v}, x^{1/u}) = P_\omega(x, q, x^{1/v}) + O(x^{1/u}) + O(x^{1-1/v}).$$

Proof.

$$P_\omega(x, q, x^{1/v}) - M(x, x^{1/v}, x^{1/u})$$

$$\leqslant \sum_{j < t-s} \left\{ \sum_{\substack{n < x \\ n \equiv a_{s+j} (\mathrm{mod}\ p_{s+j}^2)}} 1 + \sum_{\substack{n < x \\ n \equiv b_{s+j} (\mathrm{mod}\ p_{s+j}^2)}} 1 \right\}$$

$$\leqslant \sum_{j \leqslant t-s} \left(2 \left[\frac{x}{p_{s+j}^2} \right] + 2 \right)$$

$$= O \left(\sum_{n > x^{1/v}} \frac{1}{n^2} \right) + O \left(\sum_{n \leqslant x^{1/u}} 1 \right) = O\left(x^{1-1/v}\right) + O\left(x^{1/u}\right) .$$

Thus, the lemma follows.

Lemma 8. There exist two sequences of integers (ω_j) and $(\tilde{\omega}_j)$ $(1 \leqslant j \leqslant t - s)$ such that the number of elements of $\mathfrak{M}$ satisfying at least l congruences of (19) is at most

$$\frac{1}{l} \left\{ \sum_{\substack{j \leqslant t-s \\ p_{s+j} + y}} P_\omega \left(\frac{x}{p_{s+j}}, q, x^{1/v} \right) + P_{\tilde{\omega}_j} \left(\frac{x}{p_{s+j}}, q, x^{1/v} \right) \right\} + O\left(x^{1-1/v}\right) .$$

Proof. For $1 \leqslant j \leqslant t-s$, let Γ_j be the subset of $\mathfrak{M}$ whose elements satisfy the congruence

$$n \equiv c_{s+j} \pmod{p_{s+j}} \qquad (c_{s+j} = a_{s+j} \text{ or } b_{s+j}). \tag{20}$$

(i) If $p_{s+j} | y$, then $a_{s+j} \equiv b_{s+j} \pmod{p_{s+j}}$. For $1 \leqslant i \leqslant s$, we denote the solutions of the congruences

$$\begin{cases} a_{s+j} + mp_{s+j} \equiv a \pmod{q}, & (1 \leqslant i \leqslant q), \\ a_{s+j} + mp_{s+j} \equiv a_i \pmod{p_i}, & (1 \leqslant m \leqslant p_i), \\ a_{s+j} + mp_{s+j} \equiv b_i \pmod{p_i}, & (1 \leqslant m \leqslant p_i), \end{cases}$$

by $a^{(j)}$, $a_i^{(j)}$, $b_i^{(j)}$ respectively. It is evident that $a_i^{(j)} \equiv b_i^{(j)} \pmod{p_i}$ for $p_i | y$, otherwise $a_i^{(j)} \not\equiv b_i^{(j)} \pmod{p_i}$. Let

$$a^{(j)}, \ q; \ a_i^{(j)}, \ b_i^{(j)}. \tag{ω_j}$$

Then the number of elements of Γ_j is at most $P_{\omega_j} \left(\dfrac{x}{p_{s+j}}, q, x^{1/v} \right)$.

(ii) If $p_{s+j} + y$, then $a_{s+j} \not\equiv b_{s+j} \pmod{p_{s+j}}$. For $1 \leqslant i \leqslant s$. We denote the solutions of the congruences

$$\begin{cases} b_{s+j} + mp_{s+j} \equiv a \pmod{q} & (1 \leqslant m \leqslant q), \\ b_{s+j} + mp_{s+j} \equiv a_i \pmod{p_i} & (1 \leqslant m \leqslant p_i), \\ b_{s+j} + mp_{s+j} \equiv b_i \pmod{p_i} & (1 \leqslant m \leqslant p_i) \end{cases}$$

by $\tilde{a}^{(j)}$, $\tilde{a}_i^{(j)}$, $\tilde{b}_i^{(j)}$ respectively. Let

$$\tilde{a}^{(j)}, \ q; \ \tilde{a}_i^{(j)}, \ \tilde{b}_i^{(j)}. \tag{$\tilde{\omega}_j$}$$

Then the number of elements of Γ_j is not more than

$$P_{\omega_j}\left(\frac{x}{p_{s+j}}, q, x^{1/v}\right) + P_{\tilde{\omega}_j}\left(\frac{x}{p_{s+j}}, q, x^{1/v}\right).$$

Since $p_{s+j} > x^{1/v}$, therefore

$$P_\omega\left(\frac{x}{p_{s+j}}, q, x^{1/v}\right) = O\left(\sum_{n \leqslant \frac{x}{p_{s+j}}} 1\right) = O\left(x^{1-1/v}\right).$$

If the element of $\mathfrak{M}$ satisfies at least l congruences of (20), then it belongs to at least l different sets Γ_j. Hence the number of elements of $\mathfrak{M}$ satisfying at least l congruences of (20) does not exceed

$$\frac{1}{l}\left\{\sum_{\substack{j\leqslant t-s \\ p_{s+j}+y}}\left(P_{\omega_j}\left(\frac{x}{p_{s+j}}, q, x^{1/v}\right) + P_{\tilde{\omega}_j}\left(\frac{x}{p_{s+j}}, q, x^{1/v}\right)\right)\right\}$$

$$+ \frac{1}{l}\sum_{\substack{j\leqslant t-s \\ p_{s+j}|y}} P_{\omega_j}\left(\frac{x}{p_{s+j}}, q, x^{1/v}\right)$$

$$= \frac{1}{l}\sum_{\substack{j\leqslant t-s \\ p_{s+j}+y}}\left(P_{\omega_j}\left(\frac{x}{p_{s+j}}, q, x^{1/v}\right) + P_{\tilde{\omega}_j}\left(\frac{x}{p_{s+j}}, q, x^{1/v}\right)\right) + O\left(x^{1-1/v}\right).$$

Thus, we have the Lemma.

The proof of Theorem C. It follows by Lemmas 7 and 8 and Theorem B_1 that the number of elements of $\mathfrak{M}$ satisfying at most m congruences of (19) is not less than

$$M(x, x^{1/v}, x^{1/u}) - \frac{1}{m+1}\sum_{\substack{j\leqslant t-s \\ p_{s+j}+y}}\left(P_{\omega_j}\left(\frac{x}{p_{s+j}}, q, x^{1/v}\right)\right.$$

$$\left.+ P_{\tilde{\omega}_j}\left(\frac{x}{p_{s+j}}, q, x^{1/v}\right)\right) + O\left(x^{1-1/v}\right)$$

$$= P_\omega\left(\frac{x}{p_{s+j}}, q, x^{1/v}\right) - \frac{1}{m+1}\sum_{\substack{j\leqslant t-s \\ p_{s+j}+y}}\left(P_{\omega_j}\left(\frac{x}{p_{s+j}}, q, x^{1/v}\right)\right.$$

$$\left.+ P_{\tilde{\omega}_j}\left(\frac{x}{p_{s+j}}, q, x^{1/v}\right)\right) + O\left(x^{1-1/v}\right) + O\left(x^{1/u}\right)$$

$$\geqslant P_\omega(x, q, x^{1/v}) - \left(\frac{2}{m+1}\int_{u-1}^{v-1} \Lambda\left(\frac{vz}{z+1}\right)\frac{z+1}{z^2}\, dz\right)\frac{c_{qy}\, x}{\log^2 x}$$

$$+ O\left(\frac{c_{qy}\, x}{\log^2 x \log\log x}\right).$$

376

Thus, we have the Theorem.

§7. The Proof of the Main Results

Let $\lambda(\alpha)$ and $\Lambda(\alpha)$ $(0 < \alpha \leqslant 10)$ be two non-negative and non-decreasing functions with the property that each has at most finite discontinuities such that

$$\lambda(\alpha)\,\frac{c_{qy}\,x}{\log^2 x} + O\left(\frac{c_{qy}\,x}{\log^2 x \,\log\log x}\right) \leqslant P_\omega(x,\,q,\,x^{1/v})$$

$$\leqslant \Lambda(\alpha)\,\frac{c_{qy}\,x}{\log^2 x} + O\left(\frac{c_{qy}\,x}{\log^2 x \,\log\log x}\right) \qquad (20)$$

holds uniformly in (ω) and α.

Denote these functions by $\lambda_0(\alpha)$, $\Lambda_0(\alpha)$; $\lambda_1(\alpha)$, $\Lambda_1(\alpha)$; $\cdots$.

Let $x_i = 3.5 + 0.01\,i$ $(0 \leqslant i \leqslant 210)$ and $x_{210+i} = 5.6 + 0.1\,i$ $(1 \leqslant i \leqslant 34)$. For sufficiently small ε, we have $\Lambda(x_i)$ $(0 \leqslant i \leqslant 232)$ by (6), (8) and (10). By the use of Бухштаб's methods and his estimations of $\Lambda(10) = 100.02073$ and $\lambda(10) = 99.98181$[5,6], we have $\Lambda(x_i)$ $(233 \leqslant i \leqslant 244)$ and $\lambda(x_i)$ $(0 \leqslant i \leqslant 244)$. Define

$$\begin{cases} \Lambda_0(x) = \Lambda(x_{i+1}), & \text{if } x_i < x \leqslant x_{i+1}; \\ \lambda_0(x) = \lambda(x_i), & \text{if } x_i \leqslant x < x_{i+1}. \end{cases}$$

Now we list the values of $\lambda_0(x)$ and $\Lambda_0(x)$ for integers:

α	10	9	8	7	6	5	4
$\lambda_0(\alpha)$	99.98181	79.78469	60.88817	43.51554	26.70925	9.18109	0
$\Lambda_0(\alpha)$	100.02073	82.7207	68.52511	54.39352	43.0082	34.89666	29.39023

$$(21)$$

Divide the interval $(\alpha-1) < x \leqslant (\beta-1)$ into u sub-intervals $u_i < x \leqslant u_{i+1}$. Since $\lambda(\alpha)$ and $\Lambda(\alpha)$ are increasing functions, hence

$$\int_{a-1}^{\beta-1} \lambda(z)\,\frac{z+1}{z^2}\,dz \geqslant \sum_{s=0}^{n-1} \lambda(u_s) \int_{u_s}^{u_{s+1}} \frac{z+1}{z^2}\,dz,$$

and

$$\int_{a-1}^{\beta-1} \Lambda(z)\,\frac{z+1}{z^2}\,dz \leqslant \sum_{s=0}^{n-1} \Lambda(u_{s+1}) \int_{u_s}^{u_{s+1}} \frac{z+1}{z^2}\,dz.$$

Take $u_{i+1} - u_i = 0.01$. It follows by Theorem B_2 that

α	10	9	8	7	6	$\cdots$	$0<\alpha<5.53$
$\lambda_{11}(\alpha)$	99.98181	80.892035	63.59931	47.471252	31.004145	$\cdots$	$-$
$\Lambda_{12}(\alpha)$	100.02073	81.11841	64.403149	50.529826	41.01897	$\cdots$	$\Lambda_{12}(x) = \Lambda_0(\alpha)$

$$(22)$$

I. Let $x = y$ be an even integer. Let

$$a = 1, \quad q = 2; \qquad a_i = 0, \quad b_i = x \quad (i = 1, 2, \cdots). \qquad (\omega_1)$$

Let p_i be the ith odd prime.

(i) By (21), we have

$$P_{\omega_1}(x, 2, x^{1/8}) - \left(\frac{1}{2} \int_1^7 \Lambda_0\left(\frac{8z}{z+1}\right) \frac{z+1}{z^2} \cdot dz \right) \frac{c_{2x}\, x}{\log^2 x}$$

$$+ O\left(\frac{c_{2x}\, x}{\log^2 x \log\log x} \right)$$

$$> \left\{ 60.88817 - \frac{1}{2} \left[\Lambda_0(7) \int_4^7 \frac{z+1}{z^2}\, dz + \Lambda_0(6.4) \int_3^4 \frac{z+1}{z^2}\, dz \right.\right.$$

$$\left.\left. + \Lambda_0(6) \int_2^3 \frac{z+1}{z^2}\, dz + \Lambda_0(5.4) \int_{1.23}^2 \frac{z+1}{z^2}\, dz + \Lambda_0(4.5) \int_1^{1.23} \frac{z+1}{z^2}\, dz \right]^{1)} \right\} \frac{c_{2x}\, x}{\log^2 x}$$

$$+ O\left(\frac{c_{2x}\, x}{\log^2 x \log\log x} \right)$$

$$> 0.56125 \frac{c_{2x}\, x}{\log^2 x} + O\left(\frac{c_{2x}\, x}{\log^2 x \log\log x} \right) > 3 .$$

for $x > x_0$. Hence, It follows by Theorem C that for $x > x_0$, there exists an integer n such that $1 < n < (x-1)$ and $n(x-n)$ have no prime divisors $\leqslant x^{1/8}$ and have at most 3 prime divisors in the interval $x^{1/8} < p \leqslant x^{1/2}$. Hence, $n(x-n)$ is a product of not more than 5 primes. Since $x = n + (x-n)$, thus we obtain (a, b), where $a + b \leqslant 5$.

(ii) By (21), we have

$$P_{\omega_1}(x, 2, x^{1/6}) - \left(\frac{2}{3} \int_2^5 \Lambda_0\left(\frac{6z}{z+1}\right) \frac{z+1}{z^2}\, dz \right) \frac{c_{2x}\, x}{\log^2 x} + O\left(\frac{c_{2x}\, x}{\log^2 x \log\log x} \right)$$

$$> \left\{ 26.70925 - \frac{2}{3} \left[\Lambda_0(5) \int_{4.9/1.1}^5 \frac{z+1}{z^2}\, dz + \Lambda_0(4.9) \int_1^{4.9/1.1} \frac{z+1}{z^2}\, dz \right.\right.$$

$$\left. + \Lambda_0(4.8) \int_{4.6/1.4}^4 \frac{z+1}{z^2}\, dz + \Lambda_0(4.6) \int_{4.4/1.6}^{4.6/1.4} \frac{z+1}{z^2}\, dz + \Lambda_0(4.4) \int_{4.2/1.8}^{4.4/1.6} \frac{z+1}{z^2}\, dz \right.$$

$$\left.\left. + \Lambda_0(4.2) \int_2^{4.2/1.8} \frac{z+1}{z^2}\, dz \right] \right\} \frac{c_{2x}\, x}{\log^2 x} + O\left(\frac{c_{2x}\, x}{\log^2 x \log\log x} \right) > 3$$

---

1) Here, we use the following simple fact: If $g(x)$ is a non-negative function and $f(x)$ is a non-negative and non-decreasing function in the interval $a \leqslant x \leqslant b$, then

$$\int_a^b g(x)f(x)dx < f(b) \int_{c-\delta}^b g(x)\, dx + f(c) \int_a^{c-\delta} g(x)dx$$

holds for any $a < (c - \delta)$ and $c < b$.

378

for $x > x_0$. Hence, we have (3.3).

(iii) By (22), we have

$$P_{\omega_1}(x, 2, x^{1/8}) - \left(\frac{2}{3}\int_{9/7}^{7} \Lambda_{12}\left(\frac{8z}{z+1}\right)\frac{z+1}{z^2}\,dz\right)\frac{c_{2x}\,x}{\log^2 x} + O\left(\frac{c_{2x}\,x}{\log^2 x \log\log x}\right)$$

$$> \left(\lambda_{11}(8) - \frac{128}{3}\sum_{y=0}^{124}\frac{\Lambda_{12}(4.5 + 0.02y)}{(4.5 + 0.02y)^2}\int_{\substack{4.5+0.02y\\3.5-0.02y}}^{\substack{4.5+0.02y+0.02\\3.5-0.02y-0.02}}\frac{dz}{z+1}\right)\frac{c_{2x}\,x^{1)}}{\log^2 x}$$

$$+ O\left(\frac{c_{2x}\,x}{\log^2 x \log\log x}\right)$$

$$> 0.43\,\frac{c_{2x}\,x}{\log^2 x} + O\left(\frac{c_{2x}\,x}{\log^2 x \log\log x}\right) > 3$$

for $x > x_0$. Hence, we have Theorem 1.

II. Let x be an integer and $y = k$ be a fixed even number. Let

$$a = 1, \quad q = 2; \qquad a_i = 0, \quad b_i = -k \quad (i = 1, 2, \cdots). \qquad (\omega_2)$$

Let p_i be the ith odd prime. By (22) we have

$$P_{\omega_2}(x, 2, x^{1/8}) - \left(\frac{2}{3}\int_{9/7}^{7}\Lambda_{12}\left(\frac{8z}{z+1}\right)\frac{z+1}{z^2}\,dz\right)\frac{c_{2k}\,x}{\log^2 x}$$

$$+ O\left(\frac{x}{\log^2 x \log\log x}\right) > 0.43\,\frac{c_{2k}\,x}{\log^2 x} + O\left(\frac{x}{\log^2 x \log\log x}\right)$$

$$> 0.4\,\frac{c_{2k}\,x}{\log^2 x}$$

for $x > x_0$. Hence, it follows by Theorem C that for $x > x_0$, there exist not less than $0.4\,\dfrac{C_{2k}x}{\log^2 x}$ integers n in the interval $1 \leqslant n \leqslant x$ such that $n(n + k)$ has no prime divisors $\leqslant x^{1/8}$ and has at most 2 primes in the interval $x^{1/8} < p \leqslant x^{7/16}$. Thus we have Theorem 2.

III. Take $x = y$ be an odd integer. Take

$$a = x - 2, \quad q = 4; \quad a_i = 0, \quad b_i = x \quad (i = 1, 2, \cdots). \qquad (\omega_3)$$

---

1) It follows by Бухштаб's[6] result that $\dfrac{\Lambda_{12}(z)}{z^2}$ is a decreasing function in $5.53 \leqslant z \leqslant 10$. Since $\Lambda_{12}(z) = \Lambda_0(z)$ for $0 < z < 5.53$, hence $\dfrac{\Lambda_{12}(z)}{z^2} \leqslant \dfrac{\Lambda_{12}(4.5 + 0.02y)}{(4.5 + 0.02y)^2}$ for $0 \leqslant y \leqslant 51$ and $4.5 + 0.02y \leqslant z \leqslant 4.5 + 0.02y + 0.02$.

Let p_i be the ith odd prime. By (22), we have .

$$P_\omega(x, 4, x^{1/8}) - \left(\frac{2}{3} \int_{9/7}^{7} \Lambda_{12}\left(\frac{8z}{z+1}\right) \frac{z+1}{z^2} \, dz\right) \frac{c_{4x} \, x}{\log^2 x}$$

$$+ O\left(\frac{c_{4x} \, x}{\log^2 x \log \log x}\right) > 3$$

for $x > x_0$. Hence, it follows by Theorem C that for $x > x_0$, there exists an integer n such that $1 < n < x - 1$, $2 \mid n(x - n)$, and $\frac{n(x-n)}{2}$ has no prime factors $\leq x^{1/8}$ and has at most 2 prime factors in the interval $x^{1/8} < p \leq x^{7/16}$. Thus, we have Theorem 3.

§ 8. Other Applications

I. First of all, let us state the grand Riemann hypothesis as follows:

The real parts of all zeros of all Dirichlet's L-functions $L(s, x)$ are $\leq \dfrac{1}{2}$. (R)

From (R) we may derive the following[24]:

Let $(k, l) = 1$. Then

$$\pi(x; k, l) = \sum_{\substack{p \leq x \\ p \equiv l \,(\text{mod } k)}} 1 = \frac{\mathrm{li} \, x}{\varphi(k)} + O\left(x^{1/2} \log x\right), \qquad (R^*)$$

where $\mathrm{li} \, x = \displaystyle\int_2^x \frac{dt}{\log t}$ and the constant implied by the symbol "O" is an absolute constant.

Assuming the truth of (R^*), we have[18,25] the following:

1°. Every large even integer can be represented as a sum of a prime and a product of at most 3 primes.

2°. There are infinitely many primes p such that $p + 2$ is a product of not more than 3 primes.

3°. Every sufficiently large odd integer can be represented as $2N + 1 = p + 2Q$, where p is a prime number and the number of prime factors of Q is at most 3.

4°. Let $Z_2(N)$ be the number of twin primes $(p, p + 2)$ not exceeding N. Then

$$Z_2(N) \leq 8 \prod_{p > 2} \left(1 - \frac{1}{(p - 1)^2}\right) \frac{N}{\log^2 N} + O\left(\frac{N}{\log^2 N \log \log N}\right).$$

380

II. If $F(x)$ denotes an irreducible–integral valued polynomial of degree k without any fixed prime divisor, then we have[26,27] the following:

5°. Let $\pi(N; F(x))$ be the number of primes represented by $F(x)$ as $x = 1, 2, \cdots, N$. Then

$$\pi(N; F(x)) \leqslant 2e^\gamma \mu_F \frac{N}{\log N} + O\left(\frac{N}{\log N}\right),$$

where the constant μ_F and the constant implied by the symbol "O" depend on $F(x)$ only and γ denotes Euler's constant.

6°. Let

$$n = \begin{cases} k + 1, & \text{if } 1 \leqslant k \leqslant 5; \\ k + w, & \text{if } k > 5, \end{cases}$$

where w is the least integer satisfying

$$w + 1 \geqslant \frac{5.64527}{4.8396} + \frac{3.65}{4.8396} \log \frac{5k - w}{w + 5}.$$

Then there are infinitely many integers x such that $F(x)$ is a product of at most n primes.

III. For the problem of the distribution of consecutive almost, primes, we have[27]

7°. For sufficiently large x, there exists always a number

(i) in the interval $x - x^{10/17} < n \leqslant x$ with at most 2 prime factors;

(ii) in the interval $x - x^{20/49} < n \leqslant x$ with at most 3 prime factors and

(iii) in the interval $x - x^{1/5} < n \leqslant x$ with at most 6 prime factors.

8°. Let v be a positive integer. Let w be the least integer satisfying the inequality

$$\frac{5.64527}{4.8396} + \frac{3.65}{4.8396} \log \frac{5v - w}{w + 5} \leqslant w + 1.$$

Then for sufficiently large x, there is always an integer between $x - x^{1/v}$ and x which has at most $k = w + v$ prime factors.

REFERENCES

[1] Brun, V. *Vid.-Selsk. Skr.* 1. M. N. KL. 3, 1—36.
[2] Rademacher, H. 1924 *Abh. Math. Hamb. Univ.* **3**, 12—30.

[3] Estermann, T. 1932 *J. reine Angew. Math.* **168,** 106—116.

[4] Ricci, G. 1937 *Annali della R. Scuola Normale Superiore di Pisa* (2) 6, 70—115.

[5] Бухштаб, А. А. 1938 *Матем. сб* , **4**(46), 375—387.

[6] —————————. 1940 *ДАН СССР,* **29,** 544—548.

[7] Тартаковский, В. А. 1939 *ДАН СССР,* **23,** 122—126.

[8] —————————. 1939 *ДАН СССР,* **23,** 127—130.

[9] Kuhn, P. 1942 *Norske Vid. Selsk. Forh. Trondhjem* **39,** 145—148.

[10] —————. 1953 Tolfte Skandinaviska Matematikerkongressen, Lund. 160—168.

[11] —————. 1954 (Sept.) Abstract from the *Proc. Inter. Math. Congress,* Amsterdam.

[12] Selberg, A. 1947 *Norske Vid. Selsk. Forhdl.* **19**(18), 64—67.

[13] —————. 1950 *Proc. Inter. Congress Math.* **1,** 286—292.

[14] —————. 1952 Den 11-te Skandinaviska Matematikerkongress, 13—22.

[15] Wang, Y. 1956 *Acta Math. Sinica,* **3,** 500—513.

[16] Виноградов, А. И. 1957 *Матем. со.* **41**(1), 49—80.

[17] —————————. 1957 *Матем. сб.* **41**(3), 415—416.

[18] Wang, Y. 1957 *Science Record* (New Ser.), **1,** (1), 9—12.

[19] —————. 1957 *Science Record* (New Ser.), **1,** (5), 15—20.

[20] —————. 1958 *Acta Math. Sinica,* (3), 413—429.

[21] Shapiro, H. N. & Warga, J. 1950 *Comm. Pure Appl. Math;* (3) 153—176.

[22] Hua, L. K. 1955 *An introduction to the theory of numbers* (in Chinese).

[23] Бухштаб, А. А. 1937 *Матем. сб.* (2), 1239—1245.

[24] Чудаков, Н. Г. 1948 *ДАН СССР, серия матем.* (12), 31—46.

[25] Wang, Y. 1956 *Acta Math. Sinica,* (4), 565—582.

[26] —————. 1957 *Prog. Math. Academia,* (3) 416—426.

[27] —————. 1957 *Science Record* (New Ser.), **1** (3), 1—6.

SCIENTIA SINICA

Vol. XI, No. 12, 1962

MATHEMATICS

ON SIEVE METHODS AND SOME OF THEIR APPLICATIONS*

WANG YUAN (王 元)

(*Institute of Mathematics, Academia Sinica*)

1. STATEMENT OF RESULTS

In this paper, we shall give the detailed proofs of the following three theorems (cf. [1]).

Theorem 1. *Let $F(x)$ be an irreducible integer valued polynomial of degree k without any fixed prime divisor. Let*

$$n = \begin{cases} k + 1, & \text{if } 1 \leqslant k \leqslant 5; \\ k + w, & \text{if } k > 5, \end{cases} \tag{1}$$

where w is the least integer satisfying

$$w + 1 \geqslant \frac{5.64527}{4.8396} + \frac{3.65}{4.8396} \log \frac{5k - w}{w + 5}. \tag{2}$$

Then there are infinitely many integers x such that $F(x)$ will be a product of at most n primes.

Theorem 2. *Let k be a positive integer. Let n be an integer satisfying (1) and (2). Then for sufficiently large x, there is always an integer between x and $x + x^{1/k}$ which has at most n prime factors.*

Theorem 3. *For sufficiently large x, there exists always a number*

(i) *in interval $x < m \leqslant x + x^{\frac{10}{17}}$ with at most 2 prime factors;*

(ii) *in interval $x < m \leqslant x + x^{\frac{20}{49}}$ with at most 3 prime factors; and*

(iii) *in interval $x < m \leqslant x + x^{\frac{1}{100+\frac{2}{9}}}$ with at most 103 prime factors.*

---

* First published in Chinese in *Acta Mathematica Sinica*, Vol. IX, No. 2, pp. 87—100, 1959.

1608

No complicated numerical computations are needed in the present paper. As to the history of the above problems, we refer to Brun[2], Rademacher[3], Ricci[4] [5], Kuhn[6] [7] and Линник[8].

In this paper, $p, p', p'', \cdots; p_1, p_2, \cdots$ denote primes.

2. Several Lemmas

Let $F(x)$ be an irreducible integer valued polynomial of degree k without any fixed prime divisor. Let $\omega(p)$ denote the number of solutions of the congruence

$$F(x) \equiv 0 \pmod{p}, \quad (1 \leqslant x \leqslant p),$$

where $p > k$. Let $\omega(n) = \prod_{p \mid n} \omega(p)$, where n is a square free number and its prime divisor is always greater than k.

Lemma 1. *According to Nagell[9]*

$$R(x) = \sum_{k < p \leqslant x} \frac{\omega(p)}{p} \log p = \log x + O(1),$$

$$S(x) = \sum_{k < p \leqslant x} \frac{\omega(p)}{p} = \log \log x + \rho_F + O\left(\frac{1}{\log x}\right),$$

$$T(x) = \prod_{k < p \leqslant x} \left(1 - \frac{\omega(p)}{p}\right) = \frac{\mu_F}{\log x} + O\left(\frac{1}{\log^2 x}\right),$$

where ρ_F, μ_F and the constants implied by the symbol "O" depend only on $F(x)$.

Lemma 2. *If n is a square free number and its prime divisor is always greater than k and if $\omega(n) \neq 0$ and $f(n) = \prod_{p \mid n} f(p)$, where*

$$f(p) = \frac{p}{\omega(p)} - 1, \quad then$$

$$\sum_{\substack{n \leqslant z \\ p \mid n \Rightarrow p > k \\ \omega(n) \neq 0}} \frac{\mu^2(n)}{f(n)} = \alpha_F \log z + o(\log z),$$

where $\alpha_F = \prod\limits_{p \leqslant k} \dfrac{p-1}{p} \prod\limits_{p>k} \dfrac{\left(1-\dfrac{1}{p}\right)^{1)}}{\left(1-\dfrac{\omega(p)}{p}\right)}$ *and the constant implied by the*

symbol "*O*" *depends only on* $F(x)$.

Proof. Since $\omega(p) \leqslant k$ and n is a square free number, therefore $\omega(n) \leqslant k^{\Omega(n)} = O(n^{\varepsilon})$, where $\Omega(n)$ denotes the number of prime divisors of n, ε is any given positive number and the constant implied by the symbol "O" depends only on ε and $F(x)$. Since the series $\sum\limits_{\substack{n=1 \\ p\,|\,n \Rightarrow p>k \\ \omega(n)\neq 0}}^{\infty} \dfrac{\mu^2(n)}{f(n)n^s}$ is absolutely convergent for $s>0$, hence we have

$$\sum_{\substack{n=1 \\ p\,|\,n \Rightarrow p>k \\ \omega(n)\neq 0}}^{\infty} \frac{\mu^2(n)}{f(n)n^s} = \prod_{\substack{p>k \\ \omega(p)\neq 0}} \left(1 + \frac{1}{f(p)p^s}\right) =$$

$$= \prod_{p>k} \left[\left(1 + \frac{\omega(p)}{(p-\omega(p))p^s}\right)\left(1 - \frac{1}{p^{s+1}}\right)\right] \prod_{p\leqslant k} \left(1 - \frac{1}{p^{s+1}}\right) \cdot \zeta(1+s) \sim$$

$$\sim \prod_{p>k} \frac{\left(1 - \dfrac{1}{p}\right)}{\left(1 - \dfrac{\omega(p)}{p}\right)} \prod_{p\leqslant k} \frac{p-1}{p} \cdot \frac{1}{s} \ (\text{as } s \to 0^+).$$

On the other hand, let $\alpha(z) = \sum\limits_{\substack{n\leqslant z \\ p\,|\,n \Rightarrow p>k \\ \omega(n)\neq 0}} \dfrac{\mu^2(n)}{f(n)}$. Then

$$\sum_{\substack{n=1 \\ p\,|\,n \Rightarrow p>k \\ \omega(n)\neq 0}}^{\infty} \frac{\mu^2(n)}{f(n)n^s} = \int_1^{\infty} z^{-s}\, d\alpha(z) = \int_0^{\infty} e^{-st}\, d\alpha(e^t)$$

for $s>0$. Hence $\alpha(z) = \alpha_F \log z(1 + O(1))$. (Cf. [10]).

---

1) It follows by Mertens' Theorem that $\prod\limits_{p>k} \dfrac{\left(1-\dfrac{1}{p}\right)}{\left(1-\dfrac{\omega(p)}{p}\right)} = \lim\limits_{\xi\to\infty} \dfrac{\prod\limits_{k<p\leqslant\xi}\left(1-\dfrac{1}{p}\right)}{\prod\limits_{k<p\leqslant\xi}\left(1-\dfrac{\omega(p)}{p}\right)} =$

$= \dfrac{e^{-\gamma}}{\mu_F \prod\limits_{p\leqslant k}\left(1-\dfrac{1}{p}\right)}$. Hence $\alpha_F = (e^{\gamma}\mu_F)^{-1}$, where γ denotes Euler's constant.

1610

Lemma 3. *Let* $1 < \alpha < \beta$ *be two given numbers. Then*

$$\sum_{k < x^{\frac{1}{\beta}} < p \leqslant x^{\frac{1}{\alpha}}} \frac{\omega(p)}{p \log \dfrac{x}{p}} = \log \frac{\beta - 1}{\alpha - 1} \cdot \frac{1}{\log x} + O\left(\frac{1}{\log^2 x}\right),$$

where the constant implied by the symbol "O" depends only on $F(x)$.

Proof. It follows from Lemma 1 that

$$R(n) = \log n + r_n, \quad r_n = O(1).$$

Hence

$$\sum_{k < x^{\frac{1}{\beta}} < p \leqslant x^{\frac{1}{\alpha}}} \frac{\omega(p)}{p \log \dfrac{x}{p}} = \sum_{k < x^{\frac{1}{\beta}} < n \leqslant x^{\frac{1}{\alpha}}} \frac{R(n) - R(n-1)}{\log n \log \dfrac{x}{n}} =$$

$$= \sum_{k < x^{\frac{1}{\beta}} < n \leqslant x^{\frac{1}{\alpha}}} \frac{\log n - \log(n-1)}{\log n \log \dfrac{x}{n}} + \sum_{k < x^{\frac{1}{\beta}} < n \leqslant x^{\frac{1}{\alpha}}} \frac{r_n - r_{n-1}}{\log n \log \dfrac{x}{n}} = \Sigma_1 + \Sigma_2;$$

$$\Sigma_1 = \sum_{k < x^{\frac{1}{\beta}} < n \leqslant x^{\frac{1}{\alpha}}} \frac{1}{n \log n \log \dfrac{x}{n}} + O\left(x^{-\frac{1}{\beta}}\right) =$$

$$= \int_{x^{\frac{1}{\beta}}}^{x^{\frac{1}{\alpha}}} \frac{dt}{t \log t \log \dfrac{x}{t}} + O\left(x^{-\frac{1}{\beta}}\right) \left(\log t = \frac{\log x}{z + 1}\right) =$$

$$= \frac{1}{\log x} \int_{\alpha-1}^{\beta-1} \frac{dz}{z} + O\left(x^{-\frac{1}{\beta}}\right) = \frac{1}{\log x} \log \frac{\beta - 1}{\alpha - 1} + O\left(x^{-\frac{1}{\beta}}\right),$$

$$\Sigma_2 = \sum_{k < x^{\frac{1}{\beta}} < n \leqslant x^{\frac{1}{\alpha}}} r_n \left(\frac{1}{\log n \log \dfrac{x}{n}} - \frac{1}{\log(n+1) \log \dfrac{x}{n+1}}\right) + O\left(\frac{1}{\log^2 x}\right) =$$

$$= O\left(\frac{1}{\log x} \sum_{n > x^{\frac{1}{\beta}}} \frac{1}{n \log^2 n}\right) + O\left(\frac{1}{\log^2 x}\right) = O\left(\frac{1}{\log^2 x}\right).$$

Thus we have the Lemma.

3. Theorem A

Let $s_i = \sum_{l=1}^{\infty} \left[\dfrac{k}{q_i^l}\right]$ $(1 \leqslant i \leqslant m)$, where $q_1, \cdots, q_m$ are all primes not exceeding k. Let $K = \prod_{i=1}^{m} q_i^{s_i+1}$ and

$$a, K; \quad 1 \leqslant a_{i_1}, \cdots, a_{i\omega(p_i)} \leqslant p_i^2 \quad (i = 1, 2, \cdots), \qquad (\omega)$$

be a sequence of integers satisfying $a_{ij_1} \not\equiv a_{ij_2} \pmod{p_i} \ (j_1 \neq j_2)$, where $p_1 < p_2 < \cdots$ are all the primes greater than k and $\omega(p) \neq 0$. (For the definition of $\omega(p)$, cf. §2). Let $P_\omega(x, \xi)$ be the number of integers such that

$$1 \leqslant n \leqslant x, \ n \equiv a \pmod{K}, \ n \not\equiv a_{ij} \pmod{p_i} (1 \leqslant i \leqslant r, 1 \leqslant j \leqslant \omega(p_i)), \quad (3)$$

where p_r is the greatest prime not exceeding ξ.

Since $\omega(p) \leqslant k$, let us denote $a_{i\omega(p_i)+1} = a_{i\omega(p_i)+2} = \cdots = a_{ik} = a_{i\omega(p_i)}$ $(i = 1, 2, \cdots)$. Then it follows from Chinese Remainder Theorem that for $1 \leqslant j \leqslant k$, the system of congruences

$$n \equiv a_{ij} \pmod{p_i} \quad (1 \leqslant i \leqslant r)$$

has an unique solution in the interval $1 \leqslant n \leqslant p_1 \cdots p_r$. Denote this solution by a_j. Hence $P_\omega(x, \xi)$ is equal to the number of integers satisfying the following conditions

$$1 \leqslant n \leqslant x, \ n \equiv a \pmod{K}, \ (n - a_1) \cdots (n - a_k) \not\equiv 0 \pmod{p_i} \qquad (4)$$

$$(1 \leqslant i \leqslant r).$$

Theorem A. *Let $c > 0$, $P = \prod_{i=1}^{r} p_i$. Then the estimation*

$$P_\omega(x, \xi) \leqslant \frac{x}{K \displaystyle\sum_{\substack{m \leqslant \xi^c \\ m \mid P}} \frac{\mu^2(m)}{f(m)}} + O(\xi^{2c} \log^{6k} \xi)$$

holds uniformly in (ω), where $g(1) = 1$; $g(p) = \dfrac{\omega(p)}{p}$ for $\omega(p) \neq 0$ and $p > k$; and $g(n) = \prod_{p \mid n} g(p)$, $f(n) = \dfrac{1}{g(n)} \prod_{p \mid n} (1 - g(p))$ for n being a square free number and its prime divisor all greater than k, with $\omega(n) = \prod_{p \mid n} \omega(p) \neq 0$.

Proof. Let

$$\lambda_n = \frac{\mu(n)}{g(n)f(n)S} \sum_{\substack{m \leqslant \xi^c/n \\ (m, n)=1 \\ m \mid P}} \frac{\mu^2(m)}{f(m)},$$

1612

where $n \mid P$ and $S = \sum_{\substack{n \leqslant \xi^c \\ n \mid P}} \frac{\mu^2(n)}{f(n)}$. Then

$$P_\omega(x, \xi) = \sum_{\substack{n \leqslant x \\ n \equiv a \,(\mathrm{mod}\, K) \\ ((n-a_1)\cdots(n-a_k),\, P)=1}} 1 \leqslant \sum_{\substack{n \leqslant x \\ n \equiv a \,(\mathrm{mod}\, K)}} \left(\sum_{\substack{d \mid (n-a_1)\cdots(n-a_k) \\ d \mid P}} \lambda_d \right)^2 =$$

$$= \sum_{\substack{d_1 \leqslant \xi^c \\ d_1 \mid P}} \sum_{\substack{d_2 \leqslant \xi^c \\ d_2 \mid P}} \lambda_{d_1} \lambda_{d_2} \sum_{\substack{n \leqslant x \\ n \equiv a \,(\mathrm{mod}\, K) \\ (d_1,\, d_2) \mid (n-a_1)\cdots(n-a_k)}} 1 =$$

$$= \frac{x}{K} \sum_{\substack{d_1 \leqslant \xi^c \\ d_1 \mid P}} \sum_{\substack{d_2 \leqslant \xi^c \\ d_2 \mid P}} \lambda_{d_1} \lambda_{d_2} \frac{\omega(\{d_1, d_2\})}{\{d_1, d_2\}} +$$

$$+ O\left(\sum_{\substack{d_1 \leqslant \xi^c \\ d_1 \mid P}} \sum_{\substack{d_2 \leqslant \xi^c \\ d_2 \mid P}} |\lambda_{d_1} \lambda_{d_2}| \, k^{\Omega(\{d_1,\, d_2\})} \right) =$$

$$= \frac{x}{K} Q + R,$$

where $\{d_1, d_2\}$ denotes the least common multiple of d_1 and d_2.

$$Q = \sum_{\substack{d_1 \leqslant \xi^c \\ d_1 \mid P}} \sum_{\substack{d_2 \leqslant \xi^c \\ d_2 \mid P}} \lambda_{d_1} \lambda_{d_2} g(d_1) g(d_2) \sum_{d \mid (d_1,\, d_2)} f(d) =$$

$$= \sum_{\substack{d \leqslant \xi^c \\ d \mid P}} f(d) \left\{ \sum_{\substack{n \leqslant \xi^c \\ d \mid n \mid P}} \lambda_n g(n) \right\}^2.$$

Since

$$\lambda_n g(n) = \frac{1}{S} \sum_{\substack{m \leqslant \xi^c/n \\ m \mid P \\ (m,\, n)=1}} \frac{\mu(n)\mu^2(m)}{f(n)f(m)}$$

for $n \mid P$, hence

$$\sum_{\substack{n \leqslant \xi^c \\ d \mid n \mid P}} \lambda_n g(n) = \frac{1}{S} \sum_{\substack{n \leqslant \xi^c \\ d \mid n \mid P}} \sum_{\substack{m \leqslant \xi^c/n \\ (m,\, n)=1 \\ m \mid P}} \frac{\mu(nm)\mu(m)}{f(nm)} =$$

$$= \frac{1}{S} \sum_{\substack{r \leqslant \xi^c \\ r \mid P}} \frac{\mu(r)}{f(r)} \sum_{d \mid n \mid r} \mu\left(\frac{r}{n}\right) = \frac{1}{S} \cdot \frac{\mu(d)}{f(d)}.$$

Then

$$Q = \frac{1}{S}.$$

$$R = O\left(\left(\sum_{\substack{d \leqslant \xi^c \\ d|P}} |\lambda_d| k^{\Omega(d)}\right)^2\right) = O\left(\left(\sum_{\substack{d \leqslant \xi^c \\ d|P}} \frac{|\mu(d)| k^{\Omega(d)}}{\prod_{k < p \leqslant \xi^c} \left(1 - \frac{k}{p}\right)}\right)^2\right) =$$

$$= O\left(\log^{2k}\xi \cdot \left(\sum_{d \leqslant \xi^c} k^{\Omega(d)}\right)^2\right) = O\left(\log^{2k}\xi \cdot \left(\sum_{n \leqslant \xi^c} d(n)^{\frac{\log k}{\log 2}}\right)^2\right) =$$

$$= O\left(\xi^{2c}\log^{6k}\xi\right),$$

where $d(n)$ denotes the number of divisors of n. Thus we have the Theorem.

4. The Estimation of the Upper Bound of $P_\omega(x, \xi)$

Let $\xi > K$. Let $l < c \leqslant l+1$, where l is a positive integer. Then

$$\sum_{\substack{m \leqslant \xi^c \\ m|P}} \frac{\mu^2(m)}{f(m)} = \sum_{\substack{m \leqslant \xi^c \\ (m,K)=1 \\ \omega(m) \neq 0}} \frac{\mu^2(m)}{f(m)} - \sum_{\substack{\xi < p \leqslant \xi^c \\ \omega(p) \neq 0}} \frac{1}{f(p)} \sum_{\substack{m \leqslant \xi^c/p \\ (m, Kp)=1 \\ \omega(m) \neq 0}} \frac{\mu^2(m)}{f(m)} + \cdots +$$

$$+ (-1)^l \sum_{\substack{\xi < p' < \cdots < p^{(l)} \\ p' \cdots p^{(l)} \leqslant \xi^c \\ \omega(p' \cdots p^{(l)}) \neq 0}} \frac{1}{f(p') \cdots f(p^{(l)})} \sum_{\substack{m \leqslant \xi^c/p' \cdots p^{(l)} \\ (m, Kp' \cdots p^{(l)})=1 \\ \omega(m) \neq 0}} \frac{\mu^2(m)}{f(m)}.$$

$1°$.   Take $\xi = \dfrac{x^{\frac{1}{2c}}}{\log^{5k}x}$ for $1 \leqslant c \leqslant 2$. Denote $2c = d$. Since

$$\sum_{\substack{\xi < p \leqslant \xi^c \\ \omega(p) \neq 0}} \frac{1}{f(p)} - \sum_{\substack{\xi < p \leqslant \xi^c \\ \omega(p) \neq 0}} \frac{\omega(p)}{p} = \sum_{\substack{\xi < p \leqslant \xi^c \\ \omega(p) \neq 0}} \frac{\omega^2(p)}{p(p - \omega(p))} = O\left(\frac{1}{\xi}\right),$$

therefore from Lemma 2, we have

$$\sum_{\substack{m \leqslant \xi^c \\ m|P}} \frac{|\mu(m)|}{f(m)} = \sum_{\substack{m \leqslant \xi^c \\ (m,K)=1 \\ \omega(m) \neq 0}} \frac{|\mu(m)|}{f(m)} - \sum_{\substack{\xi < p \leqslant \xi^c \\ \omega(p) \neq 0}} \frac{\omega(p)}{p} \sum_{\substack{m \leqslant \xi^c/p \\ (m,K)=1 \\ \omega(m) \neq 0}} \frac{|\mu(m)|}{f(m)} + O\left(\frac{\log \xi}{\xi}\right) =$$

$$= \alpha_F \log \xi^c - \sum_{\xi < p \leqslant \xi^c} \frac{\omega(p)}{p} \alpha_F \log \frac{\xi^c}{p} + O\left(\frac{\log \xi}{\xi}\right) + o(\log \xi) =$$

$$= (2c - 1 - c \log c)\alpha_F \log \xi + o(\log \xi).$$

1614

Hence it follows from Theorem A that

$$P_\omega(x, x^{\frac{1}{d}}) \leqslant P_\omega\left(x, \frac{x^{\frac{1}{d}}}{\log^{5k} x}\right) \leqslant \frac{x}{\alpha_F K\left(d - 1 - \frac{d}{2}\log\frac{d}{2}\right)\log \xi} +$$

$$+ o\left(\frac{x}{\log x}\right) = \Lambda(d)\frac{\mu_F x}{K \log x} + o\left(\frac{x}{\log x}\right), \tag{5}$$

where the constants implied by the symbol "o" depend only on $F(x)$ and

$$\Lambda(d) = \frac{d}{d - 1 - \frac{d}{2}\log\frac{d}{2}}\, e^\gamma \quad (2 \leqslant d \leqslant 4). \tag{6}$$

$2°$. Take $\xi = \dfrac{x^{\frac{1}{2c}}}{\log^{5k} x}$ for $2 < c \leqslant 3$. Denote $2c = d$. It follows from Theorem A that for any given $\varepsilon > 0$, there exists a positive constant x_0 depending only on ε and $F(x)$ such that (5) holds for $x > x_0$ and

$$\Lambda(d) = \frac{d}{d - 1 - \frac{d}{2}\log\frac{d}{2} + \delta\left(\frac{d}{2}\right) - \varepsilon}\, e^\gamma \quad (4 \leqslant d \leqslant 6), \tag{7}$$

where

$$\delta(c) = \lim_{\xi \to \infty} \sum_{\substack{\xi < p \leqslant p' \\ pp' \leqslant \xi^c \\ \omega(pp') \neq 0}} \frac{\omega(p)\omega(p')}{pp'}\log\frac{\xi^c}{pp'} \Big/ \log \xi \quad (c > 2). \tag{8}$$

By the same way, we have the expressions of $\Lambda(d)$ for $d > 6$.

5. The Estimation of the Lower Bound of $P_\omega(x, \xi)$

For simplicity, we denote

$$P_\omega(d; p_1, \cdots, p_s) = \sum_{\substack{1 \leqslant n \leqslant x \\ n \equiv a \,(\mathrm{mod}\, d) \\ n \not\equiv a_{ij}(\mathrm{mod}\, p_i)(1 \leqslant i \leqslant s,\, 1 \leqslant j \leqslant \omega(p_i))}} 1, \quad P_\omega(d) = \sum_{\substack{1 \leqslant n \leqslant x \\ n \equiv a\,(\mathrm{mod}\, d)}} 1,$$

where $a_{ij_1} \not\equiv a_{ij_2} \,(\mathrm{mod}\, p_i)$ $(j_1 \neq j_2)$, $k < p_1 < \cdots < p_s$ and $(d, p_1 \cdots p_s) = 1$. In particular, $P_\omega(K; p_1, \cdots, p_r) = P_\omega(x, \xi)$.

The aim of the present section is to prove the following theorem.

Theorem B. *The estimation*

$$P_\omega(x, x^{\frac{1}{6.5}}) > 6.4524 \, \frac{\mu_F \, x}{K \log x} + O\left(\frac{x}{\log^2 x}\right)$$

holds uniformly in (ω), where the constant implied by the symbol "O" depends only on $F(x)$.

Before the proof, let us state the well-known lemma.

Lemma 4. *Let $r \geqslant r_1 \geqslant \cdots \geqslant r_n \geqslant 1$ be a given set of integers. Then*

$$P_\omega(x, \xi) > \frac{x}{K} E - R,$$

where

$$E = 1 - \sum_{\alpha \leqslant r} \frac{\omega(p_\alpha)}{p_\alpha} + \sum_{\substack{\alpha \leqslant r \\ \alpha > \beta}} \sum_{\beta \leqslant r_1} \frac{\omega(p_\alpha)\omega(p_\beta)}{p_\alpha p_\beta} - \cdots + \cdots -$$

$$- \overbrace{\sum_{\alpha \leqslant r} \sum_{\beta \leqslant r_1} \sum_{\substack{\gamma \leqslant r_1 \\ \alpha > \beta > \cdots > \mu}} \sum_{\delta \leqslant r_2} \cdots \sum_{\mu \leqslant r_n}}^{2n+1} \frac{\omega(p_\alpha)\omega(p_\beta)\cdots\omega(p_\mu)}{p_\alpha p_\beta \cdots p_\mu}$$

and

$$R = (1 + kr)(1 + kr_1)^2 \cdots (1 + kr_n)^2.$$

The proof of Theorem B. Let $\varepsilon > 0$ be a given sufficiently small number. Let $h = \dfrac{55}{35} + \varepsilon$. Then there exists $\delta_0(>k)$ depending only on ε and $F(x)$ such that

$$\sum_{\delta < p \leqslant \delta^h} \frac{\omega(p)}{p} < \log(h + \varepsilon) < 0.452 = \tau,$$

$$\prod_{\delta < p \leqslant \delta^h} \left(1 - \frac{\omega(p)}{p}\right)^{-1} < h + \varepsilon < 1.572 = \lambda,$$

for $\delta > \delta_0$.

Let $p_r = p_{r_1}$ be the greatest prime not exceeding $x^{\frac{1}{6.5}}$. If $2 \leqslant m \leqslant t + 1$, then p_{r_m} denotes the greatest prime not exceeding $x^{\frac{1}{6.5\,h^{m-1}}}$, where $p_{r_{t+1}}$ has the property that $p_{r_{t+1}}^{\frac{1}{h}} < \delta_0 \leqslant p_{r_{t+1}}$. Let n be an integer such

1616

that $2n > 2t + r_{t+1}$. Let $p_{r_m} = p_{r_{t+1}} (t+1 \leqslant m \leqslant n)$. Then we have (cf. [11])

$$P_\omega(x, \xi) > \left(1 - \sum_{n=0}^{\infty} \frac{\lambda^{n+1}[(n+1)\tau]^{2n+4}}{(2n+4)!}\right) \frac{x}{K} \prod_{i=1}^{r} \left(1 - \frac{\omega(p_i)}{p_i}\right) +$$

$$+ O\left(x^{\frac{3}{6.5} + \frac{2}{6.5\left(\frac{20}{35}+\varepsilon\right)}}\right) >$$

$$> (1 - 0.0073193) \frac{x}{K} \prod_{k < p \leqslant x^{\frac{1}{6.5}}} \left(1 - \frac{\omega(p)}{p}\right) + O\left(\frac{x}{\log^2 x}\right) >$$

$$> \frac{6.4524 \mu_F x}{K \log x} + O\left(\frac{x}{\log^2 x}\right).$$

Thus we have the theorem.

6. Two Theorems

Let $\lambda(\alpha)$ and $\Lambda(\alpha)$ $(0 < \alpha \leqslant 6.5)$ be two non-negative and non-decreasing functions such that

$$\lambda(\alpha) \frac{\mu_F x}{K \log x} + o\left(\frac{x}{\log x}\right) \leqslant P_\omega(x, x^{\frac{1}{a}}) \leqslant \Lambda(\alpha) \frac{\mu_F x}{K \log x} + o\left(\frac{x}{\log x}\right) \quad (9)$$

holds uniformly in (ω) and α, where the constants implied by the symbol "o" depend only on $F(x)$.

For the existence of $\lambda(\alpha)$ and $\Lambda(\alpha)$, see §§4—5.

Theorem C$_1$. *Let $\lambda(\alpha)$ and $\Lambda(\alpha)$ $(0 < \alpha \leqslant 6.5)$ be two functions satisfying the inequality (9), then the functions*

$$\lambda_1(\alpha) = \begin{cases} \lambda(\beta) - \displaystyle\int_{a-1}^{\beta-1} \Lambda(z)\, \frac{dz}{z}, & \tau \leqslant \alpha \leqslant \beta \leqslant 6.5; \\ 0, & \alpha \leqslant \tau \end{cases}$$

and

$$\Lambda_1(\alpha) = \Lambda(\beta) - \int_{a-1}^{\beta-1} \lambda(z)\, \frac{dz}{z}, \quad \alpha \leqslant \beta \leqslant 6.5$$

have the same properties as those of the functions $\lambda(\alpha)$ and $\Lambda(\alpha)$ respectively.

Proof. Take x such that $x^{\frac{1}{6.5}} > k$. Now, we estimate the lower bound of $P_\omega(x, x^{\frac{1}{a}})$. Arrange the primes between $x^{\frac{1}{\beta}}$ and $x^{\frac{1}{a}}$ as follows:

$$p_t \leqslant x^{\frac{1}{\beta}} < p_{t+1} < \cdots < p_r \leqslant x^{\frac{1}{a}} < p_{r+1}.$$

Then $P_\omega(x, x^{\frac{1}{\beta}}) = P_\omega(x, p_t)$ and $P_\omega(x, x^{\frac{1}{a}}) = P_\omega(x, p_r)$.

If $x^{\frac{1}{\beta}} < p_{s+1} \leqslant x^{\frac{1}{a}}$, then the difference between $P_\omega(x, p_s)$ and $P_\omega(x, p_{s+1})$ is equal to the number of integers n satisfying one of the following conditions:

$$1 \leqslant n \leqslant x, \ n \equiv a \pmod{K}, \ n \not\equiv a_{ij} \pmod{p_i} \ (1 \leqslant i \leqslant s, \ 1 \leqslant j \leqslant \omega(p_i)),$$
$$n \equiv a_{s+1j} \pmod{p_{s+1}} \quad (1 \leqslant j \leqslant \omega(p_{s+1})). \tag{10}$$

For $1 \leqslant i \leqslant s$, $1 \leqslant l \leqslant \omega(p_i)$, $1 \leqslant j \leqslant \omega(p_{s+1})$, each of the following congruences

$$mp_{s+1} + a_{s+1j} \equiv a \pmod{K} \quad (1 \leqslant m \leqslant K)$$

and

$$mp_{s+1} + a_{s+1j} \equiv a_{il} \pmod{p_i} \quad (1 \leqslant m \leqslant p_i)$$

has an unique solution. Denote these solutions by $\tilde{a}_{sj}$ and $\tilde{a}_{sjil}$ respectively. Let

$$\tilde{a}_{sj}, K; \ \tilde{a}_{sjil}(1 \leqslant i \leqslant s, \ 1 \leqslant j \leqslant \omega(p_{s+1}), \ 1 \leqslant l \leqslant \omega(p_i)),$$
$$a_{s+1l}(1 \leqslant l \leqslant \omega(p_{s+1})). \tag{ω_{sj}}$$

Then it is evident that the conditions (10) are equivalent to the following conditions:

$$1 \leqslant m \leqslant \frac{x - a_{s+1j}}{p_{s+1}}, \quad m \equiv \tilde{a}_{sj} \pmod{K}, \quad m \not\equiv \tilde{a}_{sjil} \pmod{p_i}$$
$$(1 \leqslant i \leqslant s, \ 1 \leqslant j \leqslant \omega(p_{s+1}), \ 1 \leqslant l \leqslant \omega(p_i)). \tag{11}$$

Hence

$$P_\omega(x, p_{s+1}) = P_\omega(x, p_s) - \sum_{j=1}^{\omega(p_{s+1})} P_{\omega_{sj}}\left(\frac{x - a_{s+1j}}{p_{s+1}}, p_s\right) =$$

$$= P_\omega(x, p_s) - \sum_{j=1}^{\omega(p_{s+1})} P_{\omega_{sj}}\left(\frac{x}{p_{s+1}}, p_s\right) + O(1),$$

$$P_\omega(x, x^{\frac{1}{a}}) = P_\omega(x, x^{\frac{1}{\beta}}) - \sum_{v=t}^{r-1} \sum_{j=1}^{\omega(p_{v+1})} P_{\omega_{vj}}\left(\frac{x}{p_{v+1}}, p_v\right) + O\left(x^{\frac{1}{a}}\right).$$

1618

Let $n = [\sqrt{\log x}]$, $u_l = \alpha + \dfrac{\beta - \alpha}{n} l$ $(0 \leqslant l \leqslant n)$. Then it follows by Lemma 3 that

$$
T_l = \sum_{\substack{\frac{1}{x^{u_{l+1}}} < p_{\nu+1} \leqslant x^{\frac{1}{u_l}}}} \sum_{j=1}^{\omega(p_{\nu+1})} P_{\omega_{\nu j}}\left(\frac{x}{p_{\nu+1}}, p_\nu\right) \leqslant
$$

$$
\leqslant \sum_{\substack{\frac{1}{x^{u_{l+1}}} < p_{\nu+1} \leqslant x^{\frac{1}{u_l}}}} \sum_{j=1}^{\omega(p_{\nu+1})} P_{\omega_{\nu j}}\left(\frac{x}{p_{\nu+1}}, \left(\frac{x}{p_{\nu+1}}\right)^{\frac{\log p_{\nu+1}}{\log \frac{x}{p_{\nu+1}}}}\right) + O\left(x^{1-\frac{1}{\beta}}\right) \leqslant
$$

$$
\leqslant \sum_{\substack{\frac{1}{x^{u_{l+1}}} < p \leqslant x^{\frac{1}{u_l}}}} \Lambda(u_{l+1} - 1) \frac{x \omega(p) \mu_F}{K p \log \frac{x}{p}} + o\left(x \sum_{\substack{\frac{1}{x^{u_{l+1}}} < p \leqslant x^{\frac{1}{u_l}}}} \frac{\omega(p)}{p \log \frac{x}{p}}\right) \leqslant
$$

$$
\leqslant \Lambda(u_{l+1} - 1) \frac{\mu_F x}{K \log x} \log \frac{u_{l+1} - 1}{u_l - 1} + o\left(\frac{x}{\log x} \log \frac{u_{l+1} - 1}{u_l - 1}\right) + O\left(\frac{x}{\log^2 x}\right).
$$

Since

$$
\sum_{l=0}^{n-1} \Lambda(u_{l+1} - 1) \log \frac{u_{l+1} - 1}{u_l - 1} - \int_{\alpha-1}^{\beta-1} \Lambda(u) \frac{du}{u} =
$$

$$
= \sum_{l=0}^{n-1} \int_{u_l-1}^{u_{l+1}-1} \left(\Lambda(u_{l+1} - 1) - \Lambda(u)\right) \frac{du}{u} \leqslant
$$

$$
\leqslant \sum_{l=0}^{n-1} \{\Lambda(u_{l+1} - 1) - \Lambda(u_l - 1)\} \log \frac{u_{l+1} - 1}{u_l - 1} \leqslant
$$

$$
\leqslant \max_{0 \leqslant l \leqslant n-1} \log \frac{u_{l+1} - 1}{u_l - 1} \sum_{\nu=0}^{n-1} \left(\Lambda(u_{\nu+1} - 1) - \Lambda(u_\nu - 1)\right) =
$$

$$
= O\left(\frac{1}{\sqrt{\log x}}\right),
$$

therefore

$$
P_\omega\left(x, x^{\frac{1}{\alpha}}\right) \geqslant P_\omega\left(x, x^{\frac{1}{\beta}}\right) - \sum_{l=0}^{n-1} T_l + O\left(x^{\frac{1}{\alpha}}\right) \geqslant
$$

$$
\geqslant \lambda(\beta) \frac{\mu_F x}{K \log x} + o\left(\frac{x}{\log x}\right) - \sum_{l=0}^{n-1} \Lambda(u_{l+1} - 1) \log \frac{u_{l+1} - 1}{u_l - 1} .
$$

$$\cdot \ \frac{\mu_F x}{K \log x} + o\left(\frac{x}{\log x} \sum_{l=0}^{n-1} \log \frac{u_{l+1} - 1}{u_l - 1}\right) + O\left(\frac{nx}{\log^2 x}\right) \geqslant$$

$$\geqslant \left(\lambda(\beta) - \int_{a-1}^{\beta-1} \Lambda(u)\, \frac{du}{u}\right) \frac{\mu_F x}{\log x} + o\left(\frac{x}{\log x}\right).$$

Define

$$\lambda_1(\alpha) = \begin{cases} \lambda(\beta) - \displaystyle\int_{a-1}^{\beta-1} \Lambda(u)\, \frac{du}{u}, & \text{if } \lambda(\beta) - \displaystyle\int_{a-1}^{\beta-1} \Lambda(u)\, \frac{du}{u} \geqslant 0; \\[2em] 0, & \text{if } \lambda(\beta) - \displaystyle\int_{a-1}^{\beta-1} \Lambda(u)\, \frac{du}{u} < 0. \end{cases}$$

Hence $\lambda_1(\alpha)$ satisfies the left hand of the inequality (9). The other hand of the inequality (9) may be proved similarly.

Theorem C_2. *Let $1 < a < \beta \leqslant 6.5$ be two given numbers. Then the inequality*

$$\sum_{\substack{\frac{1}{\beta} \\ x < p \leqslant x^{\frac{1}{a}}}} \sum_{l=1}^{\omega(p)} P_{\omega_{pl}}\left(\frac{x}{p},\ x^{\frac{1}{\beta}}\right) \leqslant \left(\int_{a-1}^{\beta-1} \Lambda\left(\frac{\beta z}{z+1}\right) \frac{dz}{z}\right) \frac{\mu_F x}{K \log x} + o\left(\frac{x}{\log x}\right)$$

holds uniformly in (ω_{pl}), where $\Lambda(z)$ satisfies (9) and the constant implied by the symbol "o" depends only on $F(x)$.

Proof. Let $n = [\sqrt{\log x}]$, $u_l = \alpha + \dfrac{\beta - \alpha}{n} l - 1$ $(0 \leqslant l \leqslant n)$. Then

$$T_l = \sum_{\substack{\frac{1}{u_{l+1}+1} \\ x < p \leqslant x^{\frac{1}{u_l+1}}}} \sum_{l=1}^{\omega(p)} P_{\omega_{pl}}\left(\frac{x}{p},\ x^{\frac{1}{\beta}}\right) =$$

$$= \sum_{\substack{\frac{1}{u_{l+1}+1} \\ x < p \leqslant x^{\frac{1}{u_l+1}}}} \sum_{l=1}^{\omega(p)} P_{\omega_{pl}}\left(\frac{x}{p},\ \left(\frac{x}{p}\right)^{\frac{\log x}{\beta \log \frac{x}{p}}}\right) \leqslant$$

$$\leqslant \Lambda\left(\frac{\beta u_{l+1}}{u_{l+1}+1}\right) \sum_{\substack{\frac{1}{u_{l+1}+1} \\ x < p \leqslant x^{\frac{1}{u_l+1}}}} \left[\frac{\mu_F x \omega(p)}{K p \log \frac{x}{p}} + o\left(\frac{x}{p \log \frac{x}{p}}\right)\right] \leqslant$$

$$\leqslant \Lambda\left(\frac{\beta u_{l+1}}{u_{l+1}+1}\right) \frac{\mu_F x}{K \log x} \log \frac{u_{l+1}}{u_l} + O\left(\frac{x}{\log^2 x}\right) + o\left(\frac{x}{\log x} \log \frac{u_{l+1}}{u_l}\right).$$

1620

Hence

$$\sum_{x^{\frac{1}{\beta}}<p\leqslant x^{\frac{1}{\alpha}}} \sum_{l=1}^{\omega(p)} P_{\omega_p l}\left(\frac{x}{p},\, x^{\frac{1}{\beta}}\right) = \sum_{l=0}^{n-1} T_l =$$

$$= \frac{\mu_F x}{K\log x} \sum_{l=0}^{n-1} \Lambda\left(\frac{\beta u_{l+1}}{u_{l+1}+1}\right)\log\frac{u_{l+1}}{u_l} + O\left(\frac{nx}{\log^2 x}\right) + o\left(\frac{x}{\log x}\sum_{l=0}^{n-1}\log\frac{u_{l+1}}{u_l}\right) \leqslant$$

$$\leqslant \left(\int_{\alpha-1}^{\beta-1}\Lambda\left(\frac{\beta z}{z+1}\right)\frac{dz}{z}\right)\frac{\mu_F x}{K\log x} + o\left(\frac{x}{\log x}\right).$$

Thus we have the Theorem.

7. The Set $\mathfrak{M}$ and Some of Its Properties

Let $1<u<v\leqslant 6.5$ be two given numbers. Let $\mathfrak{M}$ denote the set of integers satisfying the following conditions

$$1\leqslant n\leqslant x,\ n\equiv a(\mathrm{mod}\,K),\ n\not\equiv a_{ij}(\mathrm{mod}\,p_i)\ (1\leqslant i\leqslant s,\ 1\leqslant j\leqslant\omega(p_i)),$$

$$n\not\equiv a_{s+ij}(\mathrm{mod}\,p_{s+ij}^2)\ (1\leqslant i\leqslant t-s,\ 1\leqslant j\leqslant\omega(p_{s+i})) \tag{12}$$

where $p_s\leqslant x^{\frac{1}{v}}<p_{s+1},\ p_t\leqslant x^{\frac{1}{u}}<p_{t+1}$. The number of elements of $\mathfrak{M}$ is denoted by $M(x, x^{\frac{1}{v}}, x^{\frac{1}{u}})$.

The aim of the present section is to prove the following

Theorem D. *Let $\lambda(\alpha)$ and $\Lambda(\alpha)$ $(0<\alpha\leqslant 6.5)$ be two functions satisfying (9). Then the number of elements of $\mathfrak{M}$ satisfying at most m of the following congruences*

$$n\equiv a_{s+ij}(\mathrm{mod}\,p_{s+i})\ \ (1\leqslant j\leqslant\omega(p_{s+i})) \tag{13}_i$$

is not less than

$$\left(\lambda(v)-\frac{1}{m+1}\int_{u-1}^{v-1}\Lambda\left(\frac{vz}{z+1}\right)\frac{dz}{z}\right)\frac{\mu_F x}{K\log x} + o\left(\frac{x}{\log x}\right),$$

where the constant implied by the symbol "o" depends only on $F(x)$.

The proof of the Theorem depends on the following two Lemmas.

Lemma 5.

$$P_\omega(x, x^{\frac{1}{v}}) = M(x, x^{\frac{1}{v}}, x^{\frac{1}{u}}) + O(x^{1-\frac{1}{v}}) + O(x^{\frac{1}{u}}).$$

Proof.

$$P_\omega(x, x^{\frac{1}{v}}) - M(x, x^{\frac{1}{v}}, x^{\frac{1}{u}}) \leqslant \sum_{1 \leqslant i \leqslant t-s} \sum_{j=1}^{\omega(p_{s+i})} \sum_{\substack{n \leqslant x \\ n \equiv a_{s+ij} \,(\mathrm{mod}\ p_{s+i}^2)}} 1 =$$

$$= O\left(\sum_{n > x^{\frac{1}{v}}} \frac{x}{n^2}\right) + O\left(\sum_{n \leqslant x^{\frac{1}{u}}} 1\right) = O(x^{1-\frac{1}{v}}) + O(x^{\frac{1}{u}}).$$

Lemma 6. *There exist sequences of integers (ω_{il}) such that the number of elements of $\mathfrak{M}$ satisfying at least m congruences of (13) is at most*

$$\frac{1}{m} \sum_{i=1}^{t-s} \sum_{l=1}^{\omega(p_{s+i})} P_{\omega_{il}}\left(\frac{x}{p_{s+i}}, x^{\frac{1}{v}}\right).$$

Proof. For $1 \leqslant i \leqslant t-s$, let Γ_i be the subset of $\mathfrak{M}$ whose elements satisfy the congruences $(13)_i$. Denote the solutions of the congruences

$$p_{s+i}m + a_{s+il} \equiv a\,(\mathrm{mod}\ K) \quad (1 \leqslant m \leqslant K)$$

and

$$p_{s+i}m + a_{s+il} \equiv a_{uv}\,(\mathrm{mod}\ p_u) \quad (1 \leqslant m \leqslant p_u)$$

by $\tilde{a}_{il}(1 \leqslant i \leqslant t-s, 1 \leqslant l \leqslant \omega(p_{s+i}))$ and $\tilde{a}_{iuvl}(1 \leqslant i \leqslant t-s, 1 \leqslant u \leqslant s, 1 \leqslant v \leqslant \omega(p_u))$ respectively. Suppose we have

$$\tilde{a}_{il}, K; \ \tilde{a}_{iuvl} \quad (1 \leqslant u \leqslant s, 1 \leqslant v \leqslant \omega(p_u)). \tag{ω_{il}}$$

Evidently

$$\tilde{a}_{iuv_1l} \not\equiv \tilde{a}_{iuv_2l}\,(\mathrm{mod}\ p_u) \quad (1 \leqslant u \leqslant s)$$

for $a_{uv_1} \not\equiv a_{uv_2}\,(\mathrm{mod}\ p_u)$. Hence the number of elements of Γ_i is at most

$$\sum_{l=1}^{\omega(p_{s+i})} P_{\omega_{il}}\left(\frac{x}{p_{s+i}}, x^{\frac{1}{v}}\right).$$

If the element of $\mathfrak{M}$ satisfies at least m congruences of $(13)_i$, then it belongs to at least m different sets Γ_i. Hence the number of elements of $\mathfrak{M}$ satisfying at least m congruences of $(13)_i$ does not exceed

$$\frac{1}{m} \sum_{i=1}^{t-s} \sum_{l=1}^{\omega(p_{s+i})} P_{\omega_{il}}\left(\frac{x}{p_{s+i}}, x^{\frac{1}{v}}\right).$$

The proof of Theorem D. It follows by Lemmas 5 and 6 and Theorem C_2 that the number of elements of $\mathfrak{M}$ satisfying at most m congruences of $(13)_i$ is not less than

1622

$$M(x, x^{\frac{1}{v}}, x^{\frac{1}{u}}) - \frac{1}{m+1} \sum_{i=1}^{t-s} \sum_{l=1}^{\omega(p_{s+i})} P_{\omega_{il}} \left(\frac{x}{p_{s+i}}, x^{\frac{1}{v}}\right) =$$

$$= P_{\omega}(x, x^{\frac{1}{v}}) - \frac{1}{m+1} \sum_{i=1}^{t-s} \sum_{l=1}^{\omega(p_{s+i})} P_{\omega_{il}} \left(\frac{x}{p_{s+i}}, x^{\frac{1}{v}}\right) + O\left(x^{1-\frac{1}{v}}\right) + O\left(x^{\frac{1}{u}}\right) =$$

$$= \left(\lambda(v) - \frac{1}{m+1} \int_{u-1}^{v-1} \Lambda\left(\frac{vz}{z+1}\right) \frac{dz}{z}\right) \frac{\mu_F x}{K \log x} + o\left(\frac{x}{\log x}\right).$$

Thus we have the Theorem.

8. THE PROOF OF THE MAIN THEOREMS

Let $\lambda(\alpha)$ and $\Lambda(\alpha)$ $(0 < \alpha \leqslant 6.5)$ be two non-negative and non-decreasing functions with the properties that each has at most finite discontinuities such that

$$\lambda(\alpha) \frac{\mu_F x}{K \log x} + o\left(\frac{x}{\log x}\right) \leqslant P_{\omega}(x, x^{\frac{1}{a}}) \leqslant \Lambda(\alpha) \frac{\mu_F x}{K \log x} + o\left(\frac{x}{\log x}\right)$$

holds uniformly in (ω) and α, where the constants implied by the symbol "o" depend only on $F(x)$. For the existence of $\lambda(\alpha)$ and $\Lambda(\alpha)$ see §§ 3—5.

By the method given in § 3, we have the following table:

d	5.5	5.2	5	4.8	4.4	4.2
$\Lambda(d)$	5.65	5.372	5.197	5.036	4.7085	4.56

d	4	3.8	3.6	3.4	3.2	3	2.8	2.6	2.5	$0 < d \leqslant 2$
$\Lambda(d)$	4.42	4.288	4.161	4.049	3.941	3.848	3.76	3.6834	3.65	3.564

$$\Bigg\} . \quad (14)$$

Then it follows from Theorem B and Theorem C_1 that

$$\lambda(5) > 6.4524 - \int_4^{5.5} \Lambda(z) \frac{dz}{z} >$$

$$> 6.4524 - \Lambda(5.5) \int_{5.2}^{5.5} \frac{dz}{z} - \cdots - \Lambda(4.2) \int_4^{4.2} \frac{dz}{z} > 4.8396.$$

(i) For $k > 5$, let $a_{1i}, \cdots, a_{\omega(p_i)i}$ denote the solutions of the congruences

$$F(y) \equiv 0 \pmod{p_i^2} \quad (0 < y \leqslant p_i^2)^{1)}$$

---

1) For the sake of convenience, suppose the congruences $F(x) \equiv 0 \pmod{p}$ and $F'(x) \equiv 0 \pmod{p}$ have no common solution.

1623

respectively, where $k < p_1 < p_2 < \cdots$ are all the primes greater than k. Let $b_i^{1)}$ be the integer satisfying

$$F(y) \not\equiv 0 \,(\mathrm{mod}\, q_i), \quad (0 < y \leqslant q_i^{s_i+1}),$$

where $s_i = \sum_{l=1}^{\infty} \left[\dfrac{k}{q_i^l}\right]$ and $q_1 < \cdots < q_m \leqslant k$ are all the primes not exceeding k. Denote the solution of the system of congruences

$$y \equiv b_i \,(\mathrm{mod}\, q_i^{s_i+1}), \quad (1 \leqslant i \leqslant m)$$

by a. Let

$$a, K; \ a_{ij} \ (i = 1, 2, \cdots; 1 \leqslant j \leqslant \omega(p_i)). \qquad (\bar{\omega})$$

Then $\mathfrak{M}$ is the set of integers satisfying the following conditions

$$1 \leqslant n \leqslant x, \ F(n) \not\equiv 0 \,(\mathrm{mod}\, q_i) \ (1 \leqslant i \leqslant m),$$

$$F(n) \not\equiv 0 \,(\mathrm{mod}\, p_i) \ (1 \leqslant i \leqslant s), \quad F(n) \not\equiv 0 \,(\mathrm{mod}\, p_i^2) \ (s < i \leqslant t). \qquad (15)$$

If w is the least integer satisfying (2), then from the table (14), we have

$$\lambda(5) - \frac{1}{w+1} \int_{\frac{5(k+1)}{5k-w}-1}^{4} \Lambda\left(\frac{5z}{z+1}\right) \frac{dz}{z} \geqslant$$

$$\geqslant 4.8396 - \frac{1}{w+1} \int_{\frac{2}{3}}^{4} \Lambda\left(\frac{5z}{z+1}\right) \frac{dz}{z} -$$

$$- \Lambda(2) \int_{\frac{5(k+1)}{5k-w}-1}^{\frac{2}{3}} \frac{dz}{z} > 4.8396 - \frac{1}{w+1} \left\{ \Lambda(4) \int_{3.16}^{4} \frac{dz}{z} + \right.$$

$$+ \Lambda(3.8) \int_{2.57}^{3.16} \frac{dz}{z} + \Lambda(3.6) \int_{2.12}^{2.57} \frac{dz}{z} + \Lambda(3.4) \int_{1.77}^{2.12} \frac{dz}{z} +$$

$$+ \Lambda(3.2) \int_{1.5}^{1.77} \frac{dz}{z} + \Lambda(3) \int_{1.27}^{1.5} \frac{dz}{z} + \Lambda(2.8) \int_{1.08}^{1.27} \frac{dz}{z} +$$

$$\left. + \Lambda(2.6) \int_{1}^{1.08} \frac{dz}{z} + \Lambda(2.5) \int_{\frac{2}{3}}^{1} \frac{dz}{z} \right\} - \frac{3.564}{w+1} \int_{\frac{5(k+1)}{5k-w}-1}^{\frac{2}{3}} \frac{dz}{z} >$$

$$> 4.8396 - \frac{5.64527}{w+1} - \frac{3.65}{w+1} \log \frac{5k-w}{w+5} = 2\delta > 0.$$

For x sufficiently large, $\dfrac{2\delta \mu_F x}{K \log x} + o\left(\dfrac{x}{\log x}\right) > \dfrac{\delta \mu_F x}{\log x}$. Then it follows from Theorem D that for x sufficiently large, there are not less than $\dfrac{\delta \mu_F x}{\log x}$ integers y in the interval $1 \leqslant y \leqslant x$ such that $F(y)$ has no prime

---

1) The existence of b_i is due to the fact that $F(y)$ has no fixed prime divisor.

1624

divisor $\leqslant x^{\frac{1}{5}}$ and has at most w prime divisors in the interval $x^{\frac{1}{5}} < y \leqslant x^{\frac{5k-w}{5(k+1)}}$. Hence $F(y)$ is a product of at most $n = k + w$ primes. Thus we have Theorem 1 for the case $k > 5$. In particular, take $F(y) = x^k + y$ $(1 \leqslant y \leqslant x)$. We obtain Theorem 2.

(ii)   From the table (14), we have

$$\lambda(5) - \frac{1}{2} \int_{\frac{1}{3}}^{4} \Lambda\left(\frac{5z}{z+1}\right) \frac{dz}{z} > 0, \quad \lambda(5) - \frac{1}{3} \int_{\frac{10}{119}}^{4} \Lambda\left(\frac{5z}{z+1}\right) \frac{dz}{z} > 0$$

and

$$\lambda(5) - \frac{1}{5} \int_{\frac{1}{180}}^{4} \Lambda\left(\frac{5z}{z+1}\right) \frac{dz}{z} > 0.$$

Hence it follows from Theorem D that we have Theorem 3 and also Theorems 1 and 2 for the case $k \leqslant 5$.

Remark. Making use of Theorem 2, but with more complicated numerical calculations (cf. [12]), we may refine table (14).

References

[1]   Wang Yuan, 1957 On sieve methods and some of their applications, *Science Record*, **1**:3, 1—5.

[2]   Brun, V. 1920 Le crible d'Eratosthène et le théorème de Goldbach, *Vid.- Selsk. Skr. 1. M.N.KL.*, **3**, 1—36.

[3]   Rademacher, H. 1924 Beitrage zur Viggo Brunschen Methode in der Zahlen-theorie, *Abh. Math. Hamb. Univ.*, **3**, 12—30.

[4]   Ricci, G. 1933 Ricerche aritmetiche sui polinomi, *Rend. Circ. Mat. Palermo*, **57**, 433—475.

[5]   Ricci, G. 1937 Su la congettura di Goldbach e la constante di Schnirelmann, *Annali della R. Scuola Normale Superiore di Pisa*, (2) **6**, 70—115.

[6]   Kuhn, P. 1942 Zur Viggo Brunschen Siebmethode I, *Norske Vid. Selsk. Forh. Trondheim*, **39**, 145—148.

[7]   Kuhn, P. 1953 Neue Abschatzungen auf Grund der Viggo Brunschen Siebmethode, *Tolfte skandinaviska Matematiker-kongressen*, 160—168.

[8]   Линник, Ю. В. 1950 Замечание о произведении трёх простых чисел, *ДАН СССР*, **72**, 9—10.

[9]   Nagell, T. 1921 Généralisation d'un théorème de Tchbycheff, *Journ. de Math.* (8) **4**, 343—356.

[10]   Widder, D. V. *The Laplace Transform*, Princeton Press.

[11]   Wang Yuan, 1956 On the representation of large even integer as a sum of a prime and a product of at most 4 primes, *Acta Math. Sinica*, **6**:4, 565—582.

[12]   Wang Yuan, 1959 On sieve method and some of their applications (I), *Acta Math. Sinica*, **8**:3, 413—429.

SCIENTIA SINICA

Vol. X, No. 1, 1961

MATHEMATICS

ON THE LEAST PRIMITIVE ROOT OF A PRIME*

Wang Yuan (王　元)

(Institute of Mathematics, Academia Sinica)

I. Introduction

Let p be a prime number[1] and $g(p)$ denote the least positive primitive root of p. Let $\omega(n)$ denote the number of different prime factors of n. Виноградов[1] first proved that

$$g(p) < 2^m p^{1/2} \log p, \tag{1}$$

where $m = \omega(p-1)$. The historical records of the estimations of $g(p)$ are as follows: $g(p) < 2^m p^{1/2} \log\log p$ of Виноградов[2]; $g(p) < 2^{m+1}\sqrt{p}$ of Hua[3]; $g(p) = O(p^{1/2} \log^{17} p)$ of Erdös[4] and $g(p) = O(m^c p^{1/2})$ of Erdös and Shapiro[5], where c is an absolute constant.

On the assumption of the grand Riemann hypothesis, Ankeny[7] proved that

$$g(p) = O(2^m \log^2 p \log^2 (2^m \log^2 p)). \tag{2}$$

As to the lower estimation of $g(p)$, Turán[6] has proved that

$$g(p) = \Omega(\log p). \tag{3}$$

In this paper, we give in detail the proofs of the following two results (cf. [8]).

Theorem 1. *For any given positive number* ε, *we have*

$$g(p) = O(p^{1/4+\varepsilon}), \tag{4}$$

where the constant implied by the symbol "O" depends on ε *only.*

Here the main order $1/2$ is now replaced by $1/4 + \varepsilon$.

Theorem 2. *On the assumption of the grand Riemann hypothesis, we have*

$$g(p) = O(m^6 \log^2 p). \tag{5}$$

---

* First published in Chinese in *Acta Mathematica Sinica*, Vol. IX, No. 4, 1959.

1) In this paper, p, q, p_1, p_2, $\cdots$ denote primes.

2

Hence, the upper bound on (5) cannot be improved beyond $O(m^6 \log p)$.

I am indebted to Professor Hua Loo-keng for his valuable suggestions and assistance.

II. CHARACTER SUM (I)

The aim of this section is to prove the following

Theorem A. *Let* $\delta < \dfrac{1}{6}$ *be a positive number. Then there exists a positive constant* $P(\delta)$, *such that*

$$\left| \sum_{n=N+1}^{N+H} \chi(n) \right| < \frac{H}{p^\eta} \tag{6}$$

holds for all $p > P(\delta)$, $H > p^{1/4+\delta}$ *and any integer* N, *where* $\eta = \dfrac{\delta^2}{6}$ *and* χ *is a nonprincipal character modulo* p.

The proof of Theorem A depends on the following three Lemmas.

Lemma 1.[9,10] *Let* $[p]$ *denote the finite field of* p *elements. Let* $f_1(x), \cdots, f_n(x)$ *be different normalized*[1] *polynomials, each irreducible over* $[p]$. *Let* $k_1, \cdots, k_n$ *denote the degrees of these polynomials, and let* $K = k_1 + \cdots + k_n$. *Let* $\chi_1, \cdots, \chi_n$ *be nonprincipal characters of* $[p]$ *with the convention* $\chi(0) = 0$. *Then*

$$\left| \sum_x \chi_1(f_1(x)) \cdots \chi_n(f_n(x)) \right| \leqslant (k-1)\sqrt{p},$$

where the summation is over a complete set of residues (mod p).

Lemma 2. *Let* r *be a positive integer, and* h *be a positive integer satisfying* $1 < h < p$. *Let*

$$S_h(x) = \sum_{m=1}^{h} \chi(x + m).$$

Then

$$\sum_x |S_h(x)|^{2r} < (2r)^{7r} p h^r + 2r\sqrt{p}\, h^{2r},$$

where χ *is a nonprincipal character modulo* p.

---

1) A normalized polynomial is one in which the leading coefficient is 1.

Proof. If χ is a character of order $d(>1)$, then

$$\sum_x |S_h(x)|^{2r} = \sum_{m_1=1}^{h} \cdots \sum_{m_r=1}^{h} \sum_{n_1=1}^{h} \cdots \sum_{n_r=1}^{h} \sum_x \chi((x+m_1)\cdots(x+m_r))$$

$$\bar{\chi}((x+n_1)\cdots(x+n_r)).$$

Divide the set of integers $\{m_1, \cdots, m_r, n_1, \cdots, n_r\}$ $(1\leqslant m_i \leqslant h, 1\leqslant n_i \leqslant h, 1\leqslant i\leqslant r)$ into two classes σ_1 and σ_2. The class σ_1 contains those which satisfy $n_{i_1}=m_{i_1}, \cdots, n_{i_s}=m_{i_s}; m_{i_t}\neq n_{i_u}$ $(s+1\leqslant t, u\leqslant r)$, and each of $\{n_{i_{s+1}}, \cdots, n_{i_r}\}$ and $\{m_{i_{s+1}}, \cdots, m_{i_r}\}$ consists of at most $\dfrac{r-s}{d}$ distinct integers, each occurring dl times $(l\geqslant 1)$, where $(i_1, \cdots, i_r)$ and $(j_1, \cdots, j_r)$ are permutations of $(1, 2, \cdots, r)$. The other sets are in σ_2. Thus

$$\sum_x |S_h(x)|^{2r} = \sum_{\sigma_1} \sum_x \chi((x+m_1)\cdots(x+m_r))\bar{\chi}((x+n_1)\cdots(x+n_r)) +$$

$$+ \sum_{\sigma_2} \sum_x \chi((x+m_1)\cdots(x+m_r))\bar{\chi}((x+n_1)\cdots(x+n_r)) =$$

$$= \sum_1 + \sum_2.$$

Since the number of elements of σ_1 does not exceed

$$\sum_{s=1}^{r} rC_s^2 h^s \left(\sum_{t=1}^{[\frac{r-s}{d}]} {}_{[\frac{r-s}{d}]}C_t h^t \sum_{\substack{l_1+\cdots+l_s=[\frac{r-s}{d}] \\ l_i>0}} 1 \right)^2 <$$

$$< r^{6r} \sum_{s=1}^{r} h^s \left(\sum_{t=1}^{[\frac{r-s}{d}]} h^t \right)^2 = r^{6r} \sum_{s=1}^{r} h^s \left(\frac{h^{[\frac{r-s}{d}]+1} - h}{h-1} \right)^2 <$$

$$< r^{6r} \sum_{s=1}^{r} h^s \left(\frac{h^{[\frac{r-s}{d}]+1}}{\frac{h}{2}} \right)^2 =$$

$$= 4r^{6r} \sum_{s=1}^{r} h^s h^{2[\frac{r-s}{d}]} < (2r)^{7r} h^r,$$

therefore

$$\left| \sum_1 \right| < (2r)^{7r} p h^r.$$

4

Also, we have

$$\left|\sum_2\right| = \left|\sum_{\sigma_2}\sum_x \chi((x+m_1)^{l_1}\cdots(x+m_j)^{l_j}) \cdot \right.$$
$$\left. \bar\chi((x+n_1)^{f_1}\cdots(x+n_k)^{f_k}) \right|,$$

where $1 \leqslant l_s < d$, $1 \leqslant f_t < d$, $m_s \neq n_t$ $(1 \leqslant s \leqslant j,\ 1 \leqslant t \leqslant k,\ 1 \leqslant j,\ k \leqslant r)$, $m_s \neq m_t$ $(s \neq t)$ and $n_f \neq n_g$ $(f \neq g)$. Let $\chi^{l_s} = \chi_s$ $(1 \leqslant s \leqslant j)$ and $\bar\chi^{f_t} = \chi_{j+t}$ $(1 \leqslant t \leqslant k)$. Then $\chi_1, \cdots, \chi_{j+k}$ are nonprincipal characters. It follows by Lemma 1 that

$$\left|\sum_2\right| = \left|\sum_{\sigma_2}\sum_x \chi_1(x+m_1)\cdots\chi_j(x+m_j)\chi_{j+1}(x+n_1)\cdots\chi_{j+k}(x+n_k)\right| \leqslant$$
$$\leqslant h^{2r}(2r-1)\sqrt{p} < 2rh^{2r}\sqrt{p}.$$

Thus we have the Lemma.

For any of the integers $H > 0$, $q > 0$, t and N, we define the interval $I(q,t)$ as consisting of all integers z satisfying

$$\frac{N+tp}{q} < z \leqslant \frac{N+H+tp}{q}. \tag{7}$$

Lemma 3.[11] *Let q run through a set of distinct positive integers, Q in number, all satisfying*

$$q_1 < q < q_2$$

and all being relatively prime in pairs. Suppose that

$$2Hq_2 < p.$$

Then (for given p, N, H) it is possible to associate with each q a set $T(q)$ of integers t, with $0 \leqslant t < q$, their number being $q - Q$, in such a way that the intervals $I(q,t)$, for all q and all t in $T(q)$, are disjoint.

Proof of Theorem A. It follows by Pólya's Theorem[12] that we may assume

$$p^{1/4+\delta} < H < p^{1/2+\delta}. \tag{8}$$

Suppose that

$$\left|\sum_{n=N+1}^{N+H}\chi(n)\right| \geqslant \frac{H}{p^\eta}$$

5

for some N and H satisfying (8); we deduce a contradiction if $p > P(\delta)$.

For $q < p$, we have

$$\sum_{n=N+1}^{N+H} \chi(n) = \sum_{t=0}^{q-1} \sum_{\substack{n=N+1 \\ n \equiv -tp \,(\mathrm{mod}\, q)}}^{N+H} \chi(n) = \sum_{t=0}^{q-1} \sum_{z \in I(q,\,t)} \chi(z).$$

Hence

$$\sum_{t=0}^{q-1} \left| \sum_{z \in I(q,\,t)} \chi(z) \right| \geqslant \frac{H}{p^\eta}. \tag{9}$$

Now we apply Lemma 3, taking the set of q to consist of all the primes in the interval

$$p^{1/4} - \frac{p^{1/4}}{p^\eta} < q < p^{1/4}. \tag{10}$$

It follows by Ingham's Theorem[13] that the number of q is

$$Q = \pi(p^{1/4}) - \pi\left(p^{1/4} - \frac{p^{1/4}}{p^\eta}\right) = \frac{4p^{1/4-\eta}}{\log p}\,(1 + o(1)). \tag{11}$$

Hence

$$\sum_q \sum_{t \in T(q)} \left| \sum_{z \in I(q,\,t)} \chi(z) \right| \geqslant \frac{HQ}{p^\eta} - \sum_q 2HQq^{-1} \geqslant$$

$$\geqslant \frac{HQ}{p^\eta} - 2HQ^2 \cdot 2p^{-1/4} > \frac{HQ}{2p^\eta} \qquad (p > P_1(\delta)),$$

where the summation is over all primes in the interval of (10). We rewrite the result as

$$\sum_I \left| \sum_{z \in I} \chi(z) \right| > \frac{HQ}{2p^\eta}. \tag{12}$$

Since the number of $I(q, t)$ is at most $p^{1/4}Q$ and

$$\sum_{n \in I} \sum_{m=1}^{h} \chi(n + m) = h \sum_{z \in I} \chi(z) + \sum_{m=1}^{h} \psi_m,$$

where $|\psi_m| \leqslant 2m$, therefore

$$\sum_I \sum_{n \in I} |S_h(n)| > \frac{HQh}{2p^\eta} - 2p^{1/4}Qh^2.$$

Let

$$h = \left[\frac{Hp^{-1/4}}{8p^\eta}\right].$$

114

6

Then

$$\sum_I \sum_{n \in I} |S_h(n)| > \frac{HQh}{4p^\eta}.$$

Since the number of integers in $I(q, t)$ is at most $3 p^{-1/4} H$, therefore it follows by Hölder's inequality that

$$\sum_I \sum_{n \in I} |S_h(n)|^{2r} > \left(\frac{1}{4p^\eta} HQh\right)^{2r} (p^{1/4}Q \cdot 3p^{-1/4}H)^{1-2r}.$$

Since the intervals I are disjoint, then

$$\sum_x |S_h(x)|^{2r} > \left(\frac{1}{12p^\eta}\right)^{2r} HQh^{2r}.$$

Combining this with the result of Lemma 2, we obtain

$$\left(\frac{1}{12p^\eta}\right)^{2r} HQh^{2r} < p(2r)^{7r} h^r + 2r\sqrt{p}\, h^{2r}.$$

Take $r = \left[\dfrac{2}{\delta}\right] + 1$. Then

$$h^r \geqslant \left(\frac{p^\delta}{9p^\eta}\right)^r > p^{\frac{\delta r}{2}} > p$$

and

$$\left(\frac{1}{12p^\eta}\right)^{2r} HQh^{2r} < 3r\sqrt{p}\, h^{2r} \tag{13}$$

for $p > P_2(\delta) > P_1$. On the other hand, it follows by (8), (11) and (14) that (13) is impossible for $p > P_3(\delta) > P_2$, in virtue of

$$\delta - 2r\eta - 2\eta = \delta - 2\left(\left[\frac{2}{\delta}\right] + 2\right)\frac{\delta^7}{6} \geqslant \delta - \frac{4}{6}\delta - \delta^2 \geqslant \frac{\delta}{6}. \tag{14}$$

Thus we get a contradiction and this proves the theorem.

III. Character Sum (II)

The purpose of this section is to prove the following

Theorem B. *Let χ be a nonprincipal character modulo p. Then, on the assumption of the grand Riemann hypothesis, we have*

$$\sum_{n=1}^{\infty} \Lambda(n)\chi(n) e^{-\frac{n}{x}} = O(x^{1/2} \log p), \tag{15}$$

where $\Lambda(n)$ denotes the von Mangoldt function.

Proof. We need to mention the following well-known facts:

(i) There exists T_m such that

$$m < T_m < m + 1$$

and

$$\frac{L'}{L}(\sigma + iT_m, \chi) = O(\log^2 p(|T_m| + 1)) \quad (1/3 \leqslant \sigma \leqslant 2).$$

(ii) If $1/3 \leqslant \sigma \leqslant 2$ and $|t| \geqslant 1$, then

$$\Gamma(s) = O(e^{-\frac{\pi}{2}|t|} |t|^{\sigma - 1/2}).$$

(iii) On the assumption of the grand Riemann hypothesis, we have

$$\frac{L'}{L}(1/3 + it, \chi) = O(\log p(|t| + 1)).$$

Hence, we have

$$\int_{2 + iT_m}^{2 + i\infty} \Gamma(s) x^s \frac{L'}{L}(s, \chi)\, ds = O\left(x^2 \int_{T_m}^{\infty} e^{-\frac{\pi}{2}t} t^2\, dt\right) = o(1) \quad (\text{as } m \to \infty),$$

$$\int_{2 - i\infty}^{2 - iT_m} \Gamma(s) x^s \frac{L'}{L}(s, \chi)\, ds = o(1) \quad (\text{as } m \to \infty),$$

and

$$\int_{1/3}^{2} \Gamma(\sigma \pm iT_m) \frac{L'}{L}(\sigma \pm iT_m, \chi) x^s\, ds =$$

$$= O(x^2 \log^2 p(|T_m| + 1) e^{-\frac{\pi}{2}|T_m|} |T_m|^{3/2}) = o(1) \quad (\text{as } m \to \infty).$$

It follows by Mellin's transform and Cauchy's theorem that

$$\sum_{n=1}^{\infty} \Lambda(n)\chi(n) e^{-\frac{n}{x}} = -\frac{1}{2\pi i} \int_{2 - i\infty}^{2 + i\infty} \Gamma(s) x^s \frac{L'}{L}(s, \chi)\, ds =$$

$$= \sum_{\rho} x^\rho \Gamma(\rho) + \frac{1}{2\pi i} \int_{1/3 - i\infty}^{1/3 + i\infty} x^s \Gamma(s) \frac{L'}{L}(s, \chi)\, ds =$$

$$= \sum_{\rho} x^\rho \Gamma(\rho) + O(x^{1/3} \log p),$$

where ρ runs over all the nontrivial zeros of $L(s, \chi)$. Since the

8

number of zeros of $L(s, \chi)$ satisfying $n \leqslant \gamma = \mathscr{I}(\rho) \leqslant n + 1$ is $O(\log p(|n| + 1))$, hence

$$\sum_{\rho} x^{\rho} \Gamma(\rho) = O\left(x^{1/2} \sum_{n=-\infty}^{\infty} \sum_{n \leqslant \gamma \leqslant n+1} |\Gamma(1/2 + i\gamma)|\right) =$$

$$= O\left(x^{1/2} \sum_{n=0}^{\infty} e^{-\frac{\pi}{2}n} \log p(n + 1)\right) = O(x^{1/2} \log p).$$

Thus we have the theorem.

IV.   Sieve Method of Brun

If Q is a positive integer and $(Q, p_1 \cdots p_s) = 1$, then let us denote by $P(Q; p_1, \cdots, p_s)$ the sum of nonnegative numbers a_n, where n satisfies the following conditions:

$$n \geqslant 1, \text{ind } n \not\equiv 0 \ (\text{mod } p_i) \quad (1 \leqslant i \leqslant s), \text{ ind } n \equiv 0 \ (\text{mod } Q)^{1)}. \qquad (16)$$

Especially $P(T)$ denotes the sum of a_n, where n satisfies

$$n \geqslant 1, \text{ind } n \equiv 0 (\text{mod } T). \qquad (17)$$

It is evident that $P(Q; p_1, \cdots, p_{s+1})$ is equal to the difference between $P(Q; p_1, \cdots, p_s)$ and the sum of a_n, where n satisfies

$$n \geqslant 1, \text{ind } n \not\equiv 0 \ (\text{mod } p_i) \ (1 \leqslant i \leqslant s), \text{ind } n \equiv 0 \ (\text{mod } Q),$$
$$\text{ind } n \equiv 0 \ (\text{mod } p_{s+1}). \qquad (18)$$

Since the conditions

$$\text{ind } n \equiv 0 \ (\text{mod } Q)$$

and

$$\text{ind } n \equiv 0 \ (\text{mod } p_{s+1})$$

are equivalent to the condition

$$\text{ind } n \equiv 0 \ (\text{mod } Q p_{s+1}),$$

therefore we have

$$P(Q; p_1, \cdots, p_{s+1}) = P(Q; p_1, \cdots, p_s) - P(Q p_{s+1}; p_1, \cdots, p_s).$$

---

1) If g is a primitive root of p, then for any integer n in the interval $1 \leqslant n < p$ there exists a positive number a such that

$$g^a \equiv n (\text{mod } p).$$

We write $a = \text{ind}_g n$ or briefly $a = \text{ind } n$.

Using this formula successively for r times, we get

$$P(Q; p_1, \cdots, p_r) = P(Q) - \sum_{a=1}^{r} P(Qp_a; p_1, \cdots, p_{a-1}). \qquad (19)$$

Let

$$r = r_0 \geqslant r_1 \geqslant \cdots \geqslant r_t$$

be any given sequence of t positive integers. Then we have by (19) that

$$P(Q; p_1, \cdots, p_r) \geqslant P(Q) - \sum_{a \leqslant r} P(Qp_a) +$$

$$+ \sum_{\substack{a \leqslant r \ a_1 \leqslant r_1 \\ a_1 < a}} \sum P(Qp_ap_{a_1}; p_1, \cdots, p_{a_1-1}). \qquad (20)$$

Using (20) successively for t times and observing that

$$P(Qp_ap_{a_1}, \cdots, p_{a_t}p_{\beta_t}; p_1, \cdots, p_{\beta_t-1}) \leqslant P(Qp_a, \cdots, p_{\beta_t}),$$

we get

$$P(Q; p_1, \cdots, p_r) \geqslant P(Q) - \sum_{a \leqslant r} P(Qp_a) + \sum_{\substack{a \leqslant r \ a_1 \leqslant r_1 \\ a_1 < a}} \sum P(Qp_ap_{a_1}) -$$

$$- + \cdots - \overbrace{\sum_{a \leqslant r} \sum_{a_1 \leqslant r_1} \sum_{\beta_1 \leqslant r_1} \sum_{\substack{a_2 \leqslant r_2 \\ a > a_1 > \cdots > \beta_t}} \cdots \sum_{\beta_t \leqslant r_t}}^{2n+1} P(Qp_ap_{a_1}, \cdots, p_{\beta_t}). \qquad (21)$$

If there exist M and N such that

$$P(T) = \frac{M}{T} + O(N), \qquad (22)$$

where the constant implied by the symbol "O" is an absolute constant, then we have by (21) and (22) that

$$P(Q; p_1, \cdots, p_r) \geqslant \frac{EM}{Q} - KR|N| \quad (K > 0),$$

where

$$E = 1 - \sum_{a \leqslant r} \frac{1}{p_a} + \sum_{\substack{a \leqslant r \ a_1 \leqslant r_1 \\ a > a_1}} \sum \frac{1}{p_a p_{a_1}} - + \cdots -$$

$$- \sum_{a \leqslant r} \sum_{a_1 \leqslant r_1} \sum_{\beta_1 \leqslant r_1} \sum_{\substack{a_2 \leqslant r_2 \\ a > a_1 > \cdots > \beta_t}} \cdots \sum_{\beta_t \leqslant r_t} \frac{1}{p_a p_{a_1} \cdots p_{\beta_t}}$$

10

and

$$R \leqslant 1 + \sum_{a \leqslant r} 1 + \sum_{\substack{a \leqslant r \ a_1 \leqslant r_1 \\ a > a_1}} \sum 1 + \cdots + \sum_{a \leqslant r} \sum_{a_1 \leqslant r_1} \sum_{\substack{\beta_1 \leqslant r_1 \ a_2 \leqslant r_2 \\ a > a_1 > \cdots > \beta_t}} \sum \cdots \sum_{\beta_t \leqslant r_t} 1 \leqslant$$

$$\leqslant r \prod_{i=1}^{t} (1 + r_i)^2.$$

By the suitable choice of $r, r_1, \cdots, r_t$[14], we have

Lemma 4. *If* (22) *holds, then there exists a positive constant* σ *such that*

$$P(Q_i; p_1, \cdots, p_r) \geqslant \frac{\sigma M}{Q} \prod_{i=1}^{r} \left(1 - \frac{1}{p_i}\right) + O(p_r^{2.99}|N|).$$

V. Proof of Theorem 1

Let $x = [p^{1/4+\epsilon}]$. Then we have by Theorem A that

$$\sum_{n=1}^{x} \chi(n) = O\left(\frac{x}{p^\zeta}\right) \quad \left(\zeta = \frac{\epsilon^2}{6}\right), \tag{23}$$

where χ is a nonprincipal character (mod p).

If $q \mid p-1$, then let us denote by χ_q the character of order q for modulo p. We have by (23) that

$$\sum_{\substack{n \leqslant x \\ \text{ind } n \equiv 0 (\bmod q)}} 1 = \frac{1}{q} \sum_{n \leqslant x} \sum_{\chi_q} \chi_q(n) - \frac{x}{q} + O\left(\frac{x}{p^\zeta}\right). \tag{24}$$

Let

$$P = \prod_{q \mid p-1} q, \quad P_1 = \prod_{\substack{q \mid p-1 \\ q \leqslant \log^2 p}} q.$$

Next, let $N(x)$ denote the number of primitive roots not exceeding x. Then

$$N(x) = \sum_{\substack{n \leqslant x \\ (\text{ind } n, P)=1}} 1 \geqslant \sum_{\substack{n \leqslant x \\ (\text{ind } n, P_1)=1}} 1 - \sum_{\substack{q \mid p-1 \\ q > \log^2 p}} \sum_{\substack{n \leqslant x \\ \text{ind } n \equiv 0 (\bmod q)}} 1 = \Sigma_1 - \Sigma_2.$$

Hence, by (24) we have

$$\Sigma_2 = O\left(\sum_{\substack{q\,|\,p-1 \\ q > \log^2 p}} \frac{x}{q}\right) + O\left(\sum_{\substack{q\,|\,p-1 \\ q > \log^2 p}} \frac{x}{p^\zeta}\right) = O\left(\frac{x}{\log p}\right).$$

Let

$$a_n = \begin{cases} 1, & \text{if } n \leqslant x; \\ 0, & \text{if } n > x. \end{cases}$$

Then, it follows by Lemma 4 and (24) that

$$\Sigma_1 > \sigma x \prod_{\substack{q\,|\,p-1 \\ q \leqslant \log^2 p}} \left(1 - \frac{1}{q}\right) + O\left(\log^6 p \cdot \frac{x}{p^\zeta}\right) >$$

$$> \frac{\alpha x}{\log \log p} \quad (\alpha > 0,\ p > P_1(\varepsilon)).$$

Hence

$$N(x) > \beta \frac{x}{\log \log p} \quad (\beta > 0,\ p > P_2(\varepsilon) > P_1).$$

This proves Theorem 1.

VI. Proof of Theorem 2

If $d\,|\,p-1$, then let us denote by n_d the residue modulo p, which satisfies

$$n_d^d \equiv 1 \pmod{p}.$$

Then

$$G_d(x) = \sum_{n_d > 1} \Lambda(n_d) e^{-\frac{n_d}{x}} = \frac{1}{d} \sum_{n=1}^{\infty} \Lambda(n) e^{-\frac{n}{x}} \sum_{\chi_d} \chi_d(n) =$$

$$= \frac{1}{d} \sum_{\substack{n=1 \\ p\,\nmid\,n}}^{\infty} \Lambda(n) e^{-\frac{n}{x}} + \frac{1}{d} \sum_{\chi_d}' \sum_{n=1}^{\infty} \Lambda(n) \chi_d(n) e^{-\frac{n}{x}},$$

where $\sum_{\chi_d}'$ denotes the omission of the principal character. Hence, we have by Theorem B that

$$G_d(x) = \frac{1}{d} \sum_{n=1}^{\infty} \Lambda(n) e^{-\frac{n}{x}} + O\left(x^{1/2} \log p\right). \tag{25}$$

120

12

By partial summations and Чебышев's Theorem[1], we have

$$\sum_{n>c_3x} \Lambda(n)e^{-\frac{n}{x}} \leqslant c_2 e^{-c_3}(c_3 + 1)x \quad (c_3 > 1),\tag{26}$$

$$\sum_{n\leqslant c_0x} \Lambda(n)e^{-\frac{n}{x}} \leqslant c_2 c_3 x, \quad (c_3 > 1),\tag{27}$$

and

$$\sum_{n\leqslant c_3x} \Lambda(n)e^{-\frac{n}{x}} \geqslant e^{-1}c_1 x \quad (c_3 > 1).\tag{28}$$

Let

$$P = \prod_{q|p-1} q, \quad P_1 = \prod_{\substack{q|p-1 \\ q\leqslant c_4 m \log 4m}} q \, (c_4 \geqslant 2);$$

and

$$N(x) = \sum_{n} \Lambda(n)e^{-\frac{n}{x}},$$

where the summation is over all the positive primitive roots modulo p. Hence, it follows by Theorem B that

$$N(x) = \sum_{\substack{n=1 \\ (\text{ind } n,\, P)=1}}^{\infty} \Lambda(n)e^{-\frac{n}{x}} \geqslant \sum_{\substack{n=1 \\ (\text{ind } n, P_1)=1}}^{\infty} \Lambda(n)e^{-\frac{n}{x}} - \sum_{\substack{q|p-1 \\ q>c_4 m \log 4m}} G_q(x) =$$

$$= M(x) - \sum_{\substack{q|p-1 \\ q>c_4 m \log 4m}} \frac{1}{q}H(x) + O(mx^{1/2}\log p),\tag{29}$$

where

$$M(x) = \sum_{\substack{n=1 \\ (\text{ind } n,\, P_1)=1}}^{\infty} \Lambda(n)e^{-\frac{n}{x}}, \quad H(x) = \sum_{\substack{n=1 \\ p\nmid n}}^{\infty} \Lambda(n)e^{-\frac{n}{x}}.\tag{30}$$

Let

$$a_n = \begin{cases} \Lambda(n)e^{-\frac{n}{x}}, & \text{if } p\nmid n_j; \\ 0, & \text{if } p|n. \end{cases}$$

1) There exist two positive constants c_1 and c_2 such that $c_1 x \leqslant \sum_{n\leqslant x} \Lambda(n) = \psi(x) \leqslant c_2 x$ for $x \geqslant 2$.

13

Then, we have by (25) and Lemma 4 that

$$M(x) \geqslant \sigma \prod_{q \mid p_1} \left(1 - \frac{1}{q}\right) H(x) + O\left((c_4 m)^{2.991} x^{1/2} \log p\right) \quad (\sigma > 0)$$

$$\geqslant \frac{\alpha H(x)}{\log c_4 m} + O\left((c_4 m)^{2.991} x^{1/2} \log p\right) \quad (\alpha > 0), \tag{31}$$

where α (and β and γ used in the sequel) is an absolute constant. On taking c_3 sufficiently large, we have by (26), (27), and (28) that

$$H(x) \leqslant \sum_{n \leqslant c_3 x} \Lambda(n) e^{-\frac{n}{x}} + \sum_{n > c_3 x} \Lambda(n) e^{-\frac{n}{x}} \leqslant 2 c_2 c_3 x$$

and

$$H(x) \geqslant \sum_{n \leqslant c_3 x} \Lambda(n) e^{-\frac{n}{x}} \geqslant e^{-1} c_1 x.$$

Take $c_4 = c_4(c_3)$. Then we have by (29) and (31) that

$$N(x) \geqslant \frac{\alpha e^{-1} c_1 x}{\log c_4 \cdot \log 4m} - \frac{2 c_2 c_3 x}{c_4 \log 4m} + O\left((c_4 m)^{2.991} x^{1/2} \log p\right) \geqslant$$

$$\geqslant \frac{\beta x}{\log c_4 \cdot \log 4m} + O\left((c_4 m)^{2.991} x^{1/2} \log p\right) \quad (\beta > 0). \tag{32}$$

Write

$$N(x) = \sum_{n \leqslant c_5 x \log 4m} \Lambda(n) e^{-\frac{n}{x}} + \sum_{n > c_5 x \log 4m} \Lambda(n) e^{-\frac{n}{x}} = \sum\nolimits_1 + \sum\nolimits_2.$$

Then, by (26) we have

$$\sum\nolimits_2 \leqslant c_2 e^{-c_5 \log 4m} (c_5 \log 4m + 1) x.$$

Take $c_5 = c_5(c_4)$. Then it follows by (32) that

$$\sum\nolimits_1 \geqslant \frac{\gamma x}{\log c_4 \cdot \log 4m} + O\left((c_4 m)^{2.991} x^{1/2} \log p\right) \quad (\gamma > 0).$$

Take $c_6 = c_6(c_5)$ and

$$x = c_6 m^{5.99} \log^2 p. \tag{33}$$

14

Then we have

$$\Sigma_1 > 0.$$

That is,

$$g(p) \leqslant c_5\, x\, \log 4m.$$

Hence, we have Theorem 2.

REFERENCES

[1] Landau, E. 1927 *Vorlesungen über Zahlentheorie* **2**, 178 Leipzig.

[2] Виноградов, И. М. 1930 ДАН СССР, 7—11.

[3] Hua, L. K. 1942 *Bull. Amer. Math. Soc.*, **48**, 726—730.

[4] Erdös, P. 1945 *Bull. Amer. Math. Soc.*, **51**, 131—132.

[5] Erdös, P. and Shapiro, H. N. 1957 *Pacific Jour. Math.*, **7**, 861—865.

[6] Túran, P. 1950 *Math. Lapok*, 243—266.

[7] Ankeny, N. C. 1952 *Ann. Math.*, **55**, 65—72.

[8] Wang Yuan 1959 *Sci. Record*, New Ser. **III**, 5, 174—179.

[9] Weil, A. 1948 *Pub. Inst. Math. Strasbourg* (N. S; Nr. 2), 1—85.

[10] Davenport, H. 1939 *Acta Math.*, **71**, 99—122.

[11] Burgess, D. A. 1957 *Mathematica*, London, **4**, 106—112.

[12] Pólya, G. 1918 *Nachr. Wiss. Gött.*, 21—29.

[13] Ingham, A. E. 1937 *Quart. Jour. Math.* Oxford, **8**, 255—266.

[14] Ricci, G. 1937 *Ann. del. R. Scu. Nor. Sup. di Pisa*, **6**, 70—115.

SCIENTIA SINICA
Vol. XI, No. 8, 1962

MATHEMATICS

ON THE REPRESENTATION OF LARGE INTEGER AS A SUM OF A PRIME AND AN ALMOST PRIME*

WANG YUAN (王　元)

(*Institute of Mathematics, Academia Sinica*)

§ 1

In this paper, we shall give the detailed proofs of certain results obtained upon assuming the truth of the grand Riemann hypothesis. (Cf. [1], [2]). First of all, let us state the grand Riemann hypothesis as follows:

(R) *The real parts of all zeros of all Dirichlet's L — functions $L(s, \chi)$ are $\leqslant 1/2$.*

From (R) we derive the following[3]

(R*) *Let $(l, k) = 1$. Then*

$$\pi(x; k, l) = \sum_{\substack{p \leqslant x \\ p \equiv l \,(\mathrm{mod}\, k)}} 1 = \frac{\mathrm{li}\, x}{\varphi(k)} + O(x^{1/2} \log x),$$

where $\mathrm{li}\, x = \displaystyle\int_2^x \frac{dt}{\log t}.$

Now we state the results as follows:

Theorem 1. *Under the truth of (R*), every sufficiently large even integer is a sum of a prime and a product of at most 3 primes.*

Theorem 2. *Under the truth of (R*), there exist infinitely many primes p such that $p + 2k$ is a product of at most 3 primes, where k is a given positive integer.*

Theorem 3. *Under the truth of (R*), every sufficiently large odd integer N can be represented as $N = p + 2P$, where p is a prime number and P is an almost prime of not more than 3 prime divisors.*

---

* This paper has been published previously in Chinese in 'Acta Math. Sinica, Vol. X, No. 2, pp. 168—181, 1960, but the Appendix is added during translation.

1034

Theorem 4. *Let $Z_k(x)$ be the number of prime pairs of the form $(p, p+2k)$ not exceeding x. Then*

$$Z_k(x) \leqslant 8 \prod_{\substack{p \mid 2k \\ p > 2}} \frac{p-1}{p-2} \prod_{p > 2} \left(1 - \frac{1}{(p-1)^2}\right) \frac{x}{\log^2 x} + O\left(\frac{x}{\log^3 x} \log \log x\right).$$

Theorems $1, 2, 3$ improve the results which were obtained independently and simultaneously by the author[4] and А. И. Виноградов[5]. Our original results were obtained by replacing 3 by 4 in these Theorems.

It is well known that if $\pi(x; k, l)$ is represented by

$$P(x; k, l) = \sum_{\substack{p \leqslant x \\ p \equiv l \,(\mathrm{mod}\, k)}} e^{-\frac{p \log x}{x}} \log p,$$

then Theorems 1, 2, 3 may be derived from the following weaker hypothesis (R^{**})

(R^{**}) *Let χ be the character* mod D. *Then $L(s, \chi)$ has no zeros in the domain*

$$|t| \leqslant \log^3 D, \quad \sigma > \frac{1}{2} \quad (s = \sigma + it).$$

In this paper, $p, p', p'', \cdots; p_1, p_2, \cdots$ denote primes.

§ 2

Lemma 1. *If $x \geqslant 1$ and $z \geqslant 1$, then*

$$\sum_{\substack{1 \leqslant n \leqslant z \\ (n, x) = 1}} \frac{\mu^2(n)}{\varphi(n)} = \frac{\varphi(x)}{x} \log z + O(\log \log 3x).$$

(Cf. [4]).

Lemma 2. *Let $f(k) = \varphi(k) \prod_{p \mid k} \frac{p-2}{p-1}$. If $1 \leqslant z \leqslant x$, $1 \leqslant y \leqslant x$, then*

$$\sum_{\substack{1 \leqslant k \leqslant z \\ (k, 2y) = 1}} \frac{\mu^2(k)}{f(k)} = \frac{1}{2} \prod_{\substack{p \mid y \\ p > 2}} \frac{p-2}{p-1} \prod_{p > 2} \left(1 + \frac{1}{p(p-2)}\right) \log z + O(\log \log 3x).$$

Proof. Let $\psi(q) = \prod_{p \mid q} \frac{p-2}{p-1}$. Then

$$\sum_{\substack{1 \leqslant k \leqslant z \\ (k, 2y) = 1}} \frac{\mu^2(k)}{f(k)} = \sum_{\substack{1 \leqslant k \leqslant z \\ (k, 2y) = 1}} \frac{\mu^2(k)}{\varphi(k)} \prod_{p \mid k} \left(1 + \frac{1}{p-2}\right) = \sum_{\substack{1 \leqslant k \leqslant z \\ (k, 2y) = 1}} \frac{\mu^2(k)}{\varphi(k)} \sum_{q \mid k} \frac{1}{\psi(q)} =$$

$$= \sum_{\substack{q < z \\ (q,2y)=1}} \frac{\mu^2(q)}{\varphi(q)\psi(q)} \sum_{\substack{t \leqslant z/q \\ (t,2qy)=1}} \frac{\mu^2(t)}{\varphi(t)} =$$

$$= \sum_{\substack{q \leqslant z \\ (q,2y)=1}} \frac{\mu^2(q)}{\varphi(q)\psi(q)} \left[\frac{\varphi(2qy)}{2qy} \log \frac{z}{q} + O(\log\log 6qy) \right] =$$

$$= \frac{\varphi(2y)}{2y} \sum_{\substack{q \leqslant z \\ (q,2y)=1}} \frac{\mu^2(q)}{q\psi(q)} \log z + O(\log\log 3x) =$$

$$= \frac{\varphi(2y)}{2y} \prod_{p \nmid 2y} \left(1 + \frac{1}{p(p-2)} \right) \log z + O(\log\log 3x) =$$

$$= \frac{1}{2} \prod_{\substack{p|y \\ p>2}} \frac{p-2}{p-1} \prod_{p>2} \left(1 + \frac{1}{p(p-2)} \right) \log z + O(\log\log 3x).$$

Thus we have the Lemma.

$$\S\,3$$

Let $2 \leqslant y \leqslant x$ be two given integers. Let

$$(\omega) \qquad\qquad a, q; a_i \quad (1 \leqslant i \leqslant r)$$

be a sequence of integers satisfying the following conditions:

$$(1) \qquad q \leqslant x, (a,q)=1; \text{ if } p_i|y, \text{ then } a_i \equiv 0(\mathrm{mod}\ p_i), \text{ otherwise}$$
$$a_i \not\equiv 0(\mathrm{mod}\ p_i) \quad (1 \leqslant i \leqslant r),$$

where $2 < p_1 < \cdots < p_r \leqslant \xi$ are all primes not exceeding ξ and not dividing q.

Let $P_\omega(x, q, \xi)$ be the number of primes p satisfying the following conditions:

$$(2) \qquad p \leqslant x, p \equiv a(\mathrm{mod}\ q), p \not\equiv a_i(\mathrm{mod}\ p_i) \quad (1 \leqslant i \leqslant r).$$

It follows from the Chinese Remainder Theorem that the system of congruences

$$y \equiv a_i(\mathrm{mod}\ p_i) \quad (1 \leqslant i \leqslant r)$$

has a unique solution in the interval $1 \leqslant y \leqslant p_1 \cdots p_r$. Denote this solution by a^*. Hence $P_\omega(x, q, \xi)$ is equal to the number of primes satisfying the following conditions:

$$(3) \qquad p \leqslant x, p \equiv a(\mathrm{mod}\ q), p \not\equiv a^*(\mathrm{mod}\ p_i) \quad (1 \leqslant i \leqslant r).$$

1036

Theorem A. *Let* $c > 0$; $P = \prod_{i=1}^{r} p_i$. *Then under the truth of* (R^*), *the estimation*

$$P_\omega(x, q, \xi) \leqslant \frac{\operatorname{li} x}{\varphi(q) \sum\limits_{\substack{1 \leqslant k \leqslant \xi^c \\ k \mid P \\ (k, y) = 1}} \frac{\mu^2(k)}{f(k)}} + O(x^{1/2} \log x \cdot \xi^{2c} \log^2 \xi)$$

holds uniformly in (ω), *where* $f(k) = \varphi(k) \prod\limits_{p \mid k} \dfrac{p - 2}{p - 1}$.

Proof. Denote $g(k) = \varphi(k)^{-1}$. If $(k, y) = 1$ and $k \mid P$, then it follows from (R^*) that

$$\sum_{\substack{p \leqslant x \\ p \equiv a \pmod q \\ p \equiv a^* \pmod k}} 1 = \frac{\operatorname{li} x}{\varphi(q)\varphi(k)} + O(x^{1/2} \log x).$$

Let

$$\lambda_d = \frac{\mu(d)}{f(d) g(d)} \sum_{\substack{1 \leqslant k \leqslant \xi^c/d \\ (k, d) = 1 \\ k \mid P \\ (k, y) = 1}} \frac{\mu^2(k)}{f(k)} \Bigg/ \sum_{\substack{1 \leqslant l \leqslant \xi^c \\ l \mid P \\ (l, y) = 1}} \frac{\mu^2(l)}{f(l)},$$

where $d \mid P$. Then

$$P_\omega(x, q, \xi) = \sum_{\substack{p \leqslant x \\ p \equiv a \pmod q \\ (p - a^*, P) = 1}} 1 \leqslant \sum_{\substack{p \leqslant x \\ p \equiv a \pmod q}} \left(\sum_{\substack{d \mid (p - a^*, P) \\ (d, y) = 1}} \lambda_d \right)^2 =$$

$$= \sum_{\substack{d_1 \leqslant \xi^c \\ d_1 \mid P \\ (d_1, y) = 1}} \sum_{\substack{d_2 \leqslant \xi^c \\ d_2 \mid P \\ (d_2, y) = 1}} \lambda_{d_1} \lambda_{d_2} \sum_{\substack{p \leqslant x \\ p \equiv a \pmod q \\ p \equiv a^* \left(\operatorname{mod} \frac{d_1 d_2}{(d_1, d_2)} \right)}} 1 =$$

$$= \frac{\operatorname{li} x}{\varphi(q)} \sum_{\substack{d_1 \leqslant \xi^c \\ d_1 \mid P \\ (d_1, y) = 1}} \sum_{\substack{d_2 \leqslant \xi^c \\ d_2 \mid P \\ (d_2, y) = 1}} \lambda_{d_1} \lambda_{d_2} \frac{g(d_1) g(d_2)}{g((d_1, d_2))} +$$

$$+ O\left(x^{1/2} \log x \left(\sum_{\substack{d \leqslant \xi^c \\ d \mid P}} |\lambda_d| \right)^2 \right) = \frac{\operatorname{li} x}{\varphi(q)} Q + R.$$

Let

$$S = \sum_{\substack{l \leqslant \xi^c \\ l \mid P \\ (l, y) = 1}} \frac{\mu^2(l)}{f(l)}.$$

Then

$$\lambda_k g(k) = \frac{1}{S} \sum_{\substack{m \leqslant \xi^c / k \\ (m, k)=1 \\ m \mid P \\ (m, y)=1}} \frac{\mu(k)}{f(k)} \cdot \frac{\mu^2(m)}{f(m)} = \frac{1}{S} \sum_{\substack{m \leqslant \xi^c / k \\ (m, y)=(m, k)=1 \\ m \mid P}} \frac{\mu(mk)\mu(m)}{f(mk)}.$$

For $(d, y) = 1$, we have

$$\sum_{\substack{d \mid k \mid P \\ k \leqslant \xi^c \\ (k, y)=1}} \lambda_k g(k) = \frac{1}{S} \sum_{\substack{d \mid k \mid P \\ k \leqslant \xi^c \\ (k, y)=1}} \sum_{\substack{m \leqslant \xi^c / k \\ (m, y)=(m, k)=1 \\ m \mid P}} \frac{\mu(mk)\mu(m)}{f(mk)} =$$

$$= \frac{1}{S} \sum_{\substack{r \leqslant \xi^c \\ r \mid P \\ (r, y)=1}} \frac{\mu(r)}{f(r)} \sum_{d \mid k \mid r} \mu\left(\frac{r}{k}\right) = \frac{1}{S} \cdot \frac{\mu(d)}{f(d)}.$$

Hence

$$Q = \sum_{\substack{d_1 \leqslant \xi^c \\ d_1 \mid P \\ (d_1, y)=1}} \sum_{\substack{d_2 \leqslant \xi^c \\ d_2 \mid P \\ (d_2, y)=1}} \lambda_{d_1} \lambda_{d_2} g(d_1) g(d_2) \sum_{d \mid (d_1, d_2)} f(d) =$$

$$= \sum_{\substack{d \leqslant \xi^c \\ d \mid P \\ (d, y)=1}} f(d) \left(\sum_{\substack{k \leqslant \xi^c \\ d \mid k \mid P \\ (k, y)=1}} \lambda_k g(k) \right)^2 = \frac{1}{S}.$$

By Merten's Theorem, we have

$$|\lambda_d| \leqslant \frac{|\mu(d)|}{|f(d)g(d)|} \leqslant \prod_{p \mid d} \frac{p-1}{p-2} = O(\log \xi),$$

for $d \mid P$ and $d \leqslant \xi^c$. Hence

$$R = O(x^{1/2} \log x \cdot \xi^{2c} \log^2 \xi).$$

Thus we have the Theorem.

$$\S\,4$$

Let $\xi > 2$. Let $l < c \leqslant l + 1$, where l is a positive integer. Then

$$\sum_{\substack{1 \leqslant n \leqslant \xi^c \\ n \mid P \\ (n, y)=1}} \frac{\mu^2(n)}{f(n)} = \sum_{\substack{1 \leqslant n \leqslant \xi^c \\ (n, 2qy)=1}} \frac{\mu^2(n)}{f(n)} - \sum_{\substack{\xi < p \leqslant \xi^c}} \sum_{\substack{1 \leqslant n \leqslant \xi^c \\ (n, 2qy)=1 \\ p \mid n}} \frac{\mu^2(n)}{f(n)} + \sum_{\substack{\xi < p < p' \\ pp' \leqslant \xi^c}} \sum_{\substack{n \leqslant \xi^c \\ pp' \mid n \\ (n, 2qy)=1}} \frac{\mu^2(n)}{f(n)} +$$

1038

$$+ \cdots + (-1)^l \sum_{\substack{\xi < p' < \cdots < p^{(l)} \\ p' \cdots p^{(l)} \leqslant \xi^c \\ (n, 2qy)=1}} \sum_{\substack{n \leqslant \xi^c \\ p' \cdots p^{(l)} \mid n}} \frac{|\mu(n)|}{f(n)} =$$

$$= \sum_{\substack{1 \leqslant n \leqslant \xi^c \\ (n, 2qy)=1}} \frac{\mu^2(n)}{f(n)} - \sum_{\substack{\xi < p \leqslant \xi^c \\ (p, qy)=1}} \frac{1}{f(p)} \sum_{\substack{n \leqslant \xi^c/p \\ (n, 2pyq)=1}} \frac{\mu^2(n)}{f(n)} +$$

$$(4) \qquad + \cdots + (-1)^l \sum_{\substack{\xi < p' < \cdots < p^{(l)} \\ p' \cdots p^{(l)} \leqslant \xi^c \\ (p' \cdots p^{(l)}, qy)=1}} \frac{1}{f(p') \cdots f(p^{(l)})} \sum_{\substack{n \leqslant \frac{\xi^c}{p' \cdots p^{(l)}} \\ (n, 2p' \cdots p^{(l)} qy)=1}} \frac{\mu^2(n)}{f(n)}.$$

1°. If $3 \leqslant u \leqslant 6$ and $x^{1/u} < q \leqslant c_0 x^{1/u}$, then we take $\xi = \dfrac{x^{1/u}}{\log^{12} x}$ and $c = \dfrac{u-2}{4} < 1$. It follows from Lemma 2 and (4) that

$$\sum_{\substack{n \leqslant \xi^c \\ n \mid P \\ (n, y)=1}} \frac{\mu^2(n)}{f(n)} = \sum_{\substack{n \leqslant \xi^c \\ (n, 2qy)=1}} \frac{\mu^2(n)}{f(n)} = \frac{1}{2} \prod_{\substack{p \mid qy \\ p > 2}} \frac{p-2}{p-1} \prod_{p > 2} \left(1 + \frac{1}{p(p-2)}\right) \log \xi^c +$$

$$+ O(\log \log 3x) =$$

$$= \frac{u-2}{8u} \prod_{\substack{p \mid qy \\ p > 2}} \frac{p-2}{p-1} \prod_{p > 2} \left(1 + \frac{1}{p(p-2)}\right) \log x + O(\log \log 3x).$$

Hence by Theorem A we have

$$(5) \qquad P_\omega(x, q, x^{1/u}) \leqslant P_\omega\left(x, q, \frac{x^{1/u}}{\log^{12} x}\right) \leqslant$$

$$\leqslant \Lambda(u) \frac{c_{qy} x}{\varphi(q) \log^2 x} + O\left(\frac{c_{qy} x}{\varphi(q)} \cdot \frac{\log \log x}{\log^3 x}\right),$$

where

$$(6) \qquad \Lambda(u) = \frac{8u}{u-2} e^\gamma,$$

$$c_{qy} = e^{-\gamma} \prod_{\substack{p \mid qy \\ p > 2}} \frac{p-1}{p-2} \prod_{p > 2} \left(1 - \frac{1}{(p-1)^2}\right), \quad \gamma \text{ denotes Euler's constant.}$$

2°. If $6 \leqslant u \leqslant 13$ and $x^{1/u} < q \leqslant c_0 x^{1/u}$, then we take $\xi = \dfrac{x^{1/u}}{\log^{12} x}$ and $c = \dfrac{u-2}{4} < 3$. Since

$$\sum_{\substack{\xi < p \leqslant \xi^c \\ p \nmid qy}} \frac{1}{f(p)} - \sum_{\xi < p \leqslant \xi^c} \frac{1}{p} = O\left(\sum_{p > \xi} \frac{1}{p^2}\right) + O\left(\sum_{\substack{p > \xi \\ p \mid qy}} \frac{1}{p}\right) = O\left(\frac{1}{\xi}\right),$$

$$\sum_{\substack{\xi<p\leqslant\xi^c\\p\nmid qy}}\frac{1}{f(p)}\sum_{\substack{n\leqslant\xi^c/p\\(n,\,2qy)=1}}\frac{\mu^2(n)}{f(n)}-\sum_{\substack{\xi<p\leqslant\xi^c\\p\nmid qy}}\frac{1}{f(p)}\sum_{\substack{n\leqslant\xi^c/p\\(n,\,2pqy)=1}}\frac{\mu^2(n)}{f(n)}=$$

$$=O\left(\sum_{\substack{\xi<p\leqslant\xi^c\\p\nmid qy}}\frac{1}{f(p)}\sum_{\substack{n\leqslant\xi^c/p\\(n,\,2qy)=1\\p\mid n}}\frac{\mu^2(n)}{f(n)}\right)=$$

$$=O\left(\sum_{\substack{\xi<p\leqslant\xi^c\\p\nmid qy}}\frac{1}{f(p)^2}\sum_{\substack{n\leqslant\xi^c/p^2\\(n,\,2qy)=1}}\frac{\mu^2(n)}{f(n)}\right)=O\left(\frac{\log x}{\xi}\right),$$

therefore

$$\sum_{\substack{\xi<p\leqslant\xi^c\\(n,\,2qy)=1}}\frac{1}{p}\sum_{\substack{n\leqslant\xi^c/p\\(n,\,2qy)=1}}\frac{\mu^2(n)}{f(n)}-\sideset{}{'}\sum_{\substack{\xi<p\leqslant\xi^c\\p\nmid qy}}\frac{1}{f(p)}\sum_{\substack{n\leqslant\xi^c/p\\(n,\,2pqy)=1}}\frac{\mu^2(n)}{f(n)}=O\left(\frac{\log x}{\xi}\right).$$

From Lemma 2 and (4) we have

$$\sum_{\substack{n\leqslant\xi^c\\n\mid P\\(n,\,y)=1}}\frac{\mu^2(n)}{f(n)}\geqslant\sum_{\substack{n\leqslant\xi^c\\(n,\,2qy)=1}}\frac{\mu^2(n)}{f(n)}-\sum_{\xi<p\leqslant\xi^c}\frac{1}{p}\sum_{\substack{n\leqslant\xi^c/p\\(n,\,2qy)=1}}\frac{\mu^2(n)}{f(n)}+O\left(\frac{\log x}{\xi}\right)=$$

$$=\frac{1}{2}\left(2c-1-c\log c\right)\prod_{\substack{p\mid qy\\p>2}}\frac{p-2}{p-1}\prod_{p>2}\frac{(p-1)^2}{p(p-2)}\log\xi+O(\log\log x).$$

Hence it follows from Theorem A that (5) holds for

$$(7)\qquad \Lambda(u)=\frac{2u}{\dfrac{u-2}{2}-1-\dfrac{u-2}{4}\log\dfrac{u-2}{4}}\,e^{\gamma}\quad(6\leqslant u\leqslant 13).$$

Let U denote the root of the equation

$$\frac{u}{4}-2-\frac{1}{2}\log\frac{u-2}{4}=0.$$

Then

$$\frac{d\Lambda(u)}{du}=\Lambda'(u)\begin{cases}>0,&\text{if }13\geqslant u>U;\\<0,&\text{if }U>u\geqslant3.\end{cases}$$

Hence $\Lambda(u)$ is decreasing in the interval $(3,U)$ and increasing in the interval $(U,13)$. By numerical calculations, we have

$$7.35<U<7.4.$$

1040

$$\S 5$$

Let $v > 4$.

$1°$. If $q = x^{1/u}$ and $2 < u \leqslant \dfrac{1}{\dfrac{1}{2} - \dfrac{2}{v}}$, then we take $c = \dfrac{v}{2}\left(\dfrac{1}{2} - \dfrac{1}{u}\right) \leqslant 1$

and $\xi = \dfrac{x^{1/v}}{\log^{3/c} x}$. Since

$$\sum_{\substack{n \leqslant \xi^c \\ n \mid P \\ (n,\, y)=1}} \frac{\mu^2(n)}{f(n)} = \sum_{\substack{n \leqslant \xi^c \\ (n,\, 2qy)=1}} \frac{\mu^2(n)}{f(n)} = \frac{u-2}{8u} \prod_{\substack{p \mid qy \\ p>2}} \frac{p-2}{p-1} \prod_{p>2} \left(1 + \frac{1}{p(p-2)}\right) \log x +$$

$$+ O(\log \log x),$$

therefore from Theorem A we have

$$P_\omega(x, q, x^{1/v}) \leqslant P_\omega(x, q, \xi) \leqslant$$

$$\leqslant \frac{8u}{u-2} \prod_{\substack{p \mid qy \\ p>2}} \frac{p-1}{p-2} \prod_{p>2} \left(1 - \frac{1}{(p-1)^2}\right) \frac{x}{\varphi(q) \log^2 x} +$$

$$(8) \qquad + O\left(\frac{x c_{qy} \log \log x}{\log^3 x}\right) = \Lambda_1(u) \frac{c_{qy} x}{\varphi(q) \log^2 x} + O\left(\frac{x c_{qy} \log \log x}{\varphi(q) \log^3 x}\right),$$

where

$$(9) \qquad \Lambda_1(u) = \frac{8u}{u-2} e^\gamma.$$

$2°$. If $q = x^{1/u}$ and

$$\frac{1}{\dfrac{1}{2} - \dfrac{2}{v}} < u < \begin{cases} \dfrac{1}{\dfrac{1}{2} - \dfrac{4}{v}} & (v > 8); \\[2ex] \infty & (v \leqslant 8), \end{cases}$$

then we take $\xi = \dfrac{x^{1/u}}{\log^3 x}$ and $c = \dfrac{v}{2}\left(\dfrac{1}{2} - \dfrac{1}{u}\right)$. Since

$$\sum_{\substack{n \leqslant \xi^c \\ n \mid P \\ (n,\, y)=1}} \frac{\mu^2(n)}{f(n)} = \sum_{\substack{n \leqslant \xi^c \\ (n,\, 2qy)=1}} \frac{\mu^2(n)}{f(n)} - \sum_{\xi < p \leqslant \xi^c} \frac{1}{p} \sum_{\substack{n \leqslant \xi^c/p \\ (n,\, 2qy)=1}} \frac{\mu^2(n)}{f(n)} + O\left(\frac{\log x}{\xi}\right) =$$

$$= \frac{1}{2v}\left[v\left(\frac{1}{2} - \frac{1}{u}\right) - 1 - \frac{v}{2}\left(\frac{1}{2} - \frac{1}{u}\right) \log \frac{v}{2}\left(\frac{1}{2} - \frac{1}{u}\right)\right] \times$$

$$\times \prod_{\substack{p \mid qy \\ p>2}} \frac{p-2}{p-1} \prod_{p>2} \frac{(p-1)^2}{p(p-2)} \log x + O(\log \log x),$$

therefore it follows from Theorem A that (8) also holds with

$$(10) \qquad \Lambda_1(u) = \cfrac{2ve^{\gamma}}{v\left(\dfrac{1}{2} - \dfrac{1}{u}\right) - 1 - \dfrac{v}{2}\left(\dfrac{1}{2} - \dfrac{1}{u}\right)\log\dfrac{v}{2}\left(\dfrac{1}{2} - \dfrac{1}{u}\right)}.$$

Since

$$\Lambda_1'(u) = \frac{d}{du}\Lambda_1(u) = \begin{cases} \dfrac{-16}{(u-2)^2} < 0 \quad (\text{if } 0 < c \leqslant 1), \\[3mm] \dfrac{-\dfrac{v^2}{u^2}(1 - \log c)}{(2c - 1 - c\log c)^2} < 0 \quad (\text{if } 1 < c \leqslant 2) \end{cases}$$

for given $4 < v < 8$, hence $\Lambda_1(u)$ is a decreasing function for $u \leqslant$

$$\frac{1}{\dfrac{1}{2} - \dfrac{4}{v}}.$$

§ 6

Theorem B. *Under the truth of* (R^*), *the estimation*

$$P_\omega(x, q, x^{1/13}) > 25.8096e^{-\gamma} \prod_{\substack{p>2 \\ p \mid qy}} \frac{p-1}{p-2} \cdot$$

$$\prod_{p>2}\left(1 - \frac{1}{(p-1)^2}\right)\frac{x}{\varphi(q)\log^2 x} + O\left(\frac{xc_{qy}}{\log^3 x}\right)$$

holds uniformly in (ω), *where* q *is a given integer.*

Lemma 3. *Let* $r \geqslant r_1 \geqslant \cdots \geqslant r_n \geqslant 1$ *be a given set of integers. Then under the truth of* (R^*), *the estimation*

$$P_\omega(x, q, p_r) > \frac{E\,\mathrm{li}\,x}{\varphi(q)} - |R|$$

holds uniformly in (ω), *where*

$$E = 1 - \sum_{\substack{a \leqslant r \\ (p_a, y)=1}} \frac{1}{\varphi(p_a)} + \sum_{a \leqslant r}\sum_{\substack{\beta \leqslant r_1 \\ (p_a p_\beta, y)=1 \\ a > \beta}} \frac{1}{\varphi(p_a)\varphi(p_\beta)} - \cdots -$$

$$- \overbrace{\sum_{a \leqslant r}\sum_{\substack{\beta \leqslant r_1 \\ a > \beta > \cdots > \mu \\ (p_a p_\beta \cdots p_\mu, y)=1}}\sum_{\gamma \leqslant r_1}\sum_{\delta \leqslant r_2} \cdots \sum_{\mu \leqslant r_n}}^{2n+1} \frac{1}{\varphi(p_a)\cdots\varphi(p_\mu)},$$

$$R = O((1 + r)(1 + r_1)^2 \cdots (1 + r_n)^2 x^{1/2}\log x).$$

1042

Proof. Let $P_\omega(q; p_1, \cdots, p_r) = P_\omega(x, q, p_r)$. Especially, we have $P_\omega(q) = \pi(x, q, a)$. The difference between $P_\omega(q; p_1, \cdots, p_{r-1})$ and $P_\omega(q; p_1, \cdots, p_r)$ is equal to the number of primes p satisfying the following conditions:

$$p \leqslant x,\ p \equiv a(\mathrm{mod}\ q),\ p \not\equiv a_i(\mathrm{mod}\ p_i)(1 \leqslant i \leqslant r-1),\ p \equiv a_r(\mathrm{mod}\ p_r).$$

It follows from the Chinese Remainder Theorem that the system of congruences

$$\begin{cases} y \equiv a_r(\mathrm{mod}\ p_r), \\ y \equiv a(\mathrm{mod}\ q) \end{cases}$$

has a unique solution a^* in the interval $1 \leqslant a^* \leqslant qp_r$. If $p_r \nmid y$, then $(a^*, qp_r) = 1$; otherwise $(a^*, qp_r) > 1$. For the sake of brevity, we write all the (ω_r) as (ω). Hence

$$P_\omega(q; p_1, \cdots, p_{r-1}) - P_\omega(x; p_1, \cdots, p_r) =$$
$$= \begin{cases} = P_\omega(qp_r; p_1, \cdots, p_{r-1}), \text{ if } p_r \nmid y; \\ \leqslant 1, \text{ if } p_r \mid y, \end{cases}$$

$$P_\omega(q; p_1, \cdots, p_r) = P_\omega(q) - \sum_{\substack{a=1 \\ p_a \nmid y}}^{r} P_\omega(qp_a; p_1, \cdots, p_{a-1}) - \theta r,$$

$$0 \leqslant \theta \leqslant 1.$$

Using this formula r times and with the restriction $\beta \leqslant r_1$, we have

$$P_\omega(q; p_1, \cdots, p_r) \geqslant P_\omega(q) - \sum_{\substack{a=1 \\ p_a \nmid y}}^{r} P_\omega(qp_a) +$$

$$+ \sum_{\substack{a \leqslant r\\ \beta \leqslant r_1 \\ a > \beta \\ (p_a p_\beta,\ y)=1}} \sum P_\omega(qp_a p_\beta; p_1, \cdots, p_{\beta-1}) - \theta(r + rr_1),\ 0 \leqslant \theta \leqslant 1.$$

Let $r \geqslant r_1 \geqslant \cdots \geqslant r_n \geqslant 1$ be a given sequence of integers. Since

$$P_\omega(qp_a \cdots p_\mu; p_1, \cdots, p_{\mu-1}) \leqslant P_\omega(qp_a \cdots p_\mu),$$

therefore we have

$$P_\omega(q; p_1, \cdots, p_r) \geqslant P_\omega(q) - \sum_{\substack{a \leqslant r \\ (p_a,\ y)=1}} P_\omega(qp_a) + \sum_{\substack{a \leqslant r \\ \beta \leqslant r_1 \\ a > \beta \\ (p_a p_\beta,\ y)=1}} \sum P_\omega(qp_a p_\beta) -$$

$$- \cdots - \overbrace{\sum_{\substack{a \leqslant r}} \sum_{\substack{\beta \leqslant r_1 \\ a > \beta > \cdots > \mu \\ (p_a p_\beta \cdots p_\mu,\ y)=1}} \cdots \sum_{\substack{\mu \leqslant r_n}}}^{2n+1} P_\omega(qp_a \cdots p_\mu) - (1+r)(1+r_1)^2 \cdots (1+r_n)^2.$$

Since we assume the truth of (R^*), hence

$$P_\omega(q; p_1, \cdots, p_r) \geq \frac{E \operatorname{li} x}{\varphi(q)} - |R|.$$

Thus we have the Lemma.

The proof of Theorem B. Let ε be a given sufficiently small positive number. Let $h = \frac{55}{35} + \varepsilon$. Then there exists δ_0 such that

$$\begin{cases} \displaystyle\sum_{\substack{\delta < p \leq \delta^h \\ p \nmid qy}} \frac{1}{\varphi(p)} < \log(h + \varepsilon) < 0.452 = \tau, \\[2em] \displaystyle\prod_{\substack{\delta < p \leq \delta^h \\ p \nmid qy}} \left(1 - \frac{1}{\varphi(p)}\right)^{-1} < h + \varepsilon < 1.572 = \lambda \end{cases}$$

for $\delta > \delta_0$.

Let $p_r = p_{r_1}$ be the greatest prime not exceeding $x^{1/13}$. If $2 \leq k \leq t+1$, then we denote the greatest prime not exceeding $x^{\frac{1}{13 h^{k-1}}}$ by p_{r_k}, where $p_{r_{t+1}}$ is the least prime with the property that $p_{r_{t+1}}^{1/h} < \delta_0 \leq p_{r_{t+1}}$. Let n be an integer such that $2n > 2t + r_{t+1}$. Let $r_k = r_{t+1}(t+1 \leq k \leq n)$. Then we have (cf. [4])

$$P_\omega(x, q, x^{1/13}) > \frac{E \operatorname{li} x}{\varphi(q)} - |R|,$$

where

$$E > (1 - 0.0073193) \prod_{\substack{2 < p \leq x^{1/13} \\ p \nmid qy}} \left(1 - \frac{1}{\varphi(p)}\right) >$$

$$> 25.8096 e^{-\gamma} \prod_{\substack{p \mid qy \\ p > 2}} \frac{p-1}{p-2} \prod_{p > 2} \left(1 - \frac{1}{(p-1)^2}\right) \frac{1}{\log x} + O\left(\frac{c_{qy}}{\log^2 x}\right),$$

$$R = O(x^{\frac{1}{2} + \frac{3}{13} + \frac{2}{13h} + \frac{2}{13h^2} + \cdots} \log x) = O(x^{\frac{1}{2} + \frac{3}{13} + \frac{2}{13(h-1)}} \log x) = O\left(\frac{x}{\log^3 x}\right).$$

Thus we have the Theorem.

§ 7

Theorem C_1. *Let α, β be two positive numbers satisfying $8 > \beta > 4$ and $\beta \geq \alpha > 2$. Then*

1044

$$\sum_{\substack{x^{1/\beta}<p\leqslant x^{1/\alpha}\\p\nmid qy}} P_\omega(x,pq,x^{1/\beta}) \leqslant \left(\frac{c_{qy}}{\varphi(q)}\int_\alpha^\beta \frac{\Lambda_1(u)}{u}\,du\right)\frac{x}{\log^2 x} + O\left(\frac{c_{qy}x}{\log^3 x}\log\log x\right),$$

where q is a given positive integer.

Proof. Let $n=[\log x]$, $u_l = \alpha + \dfrac{\beta-\alpha}{n}l\,(0\leqslant l\leqslant n)$. Then

$$\sum_{\substack{x^{1/\beta}<p\leqslant x^{1/\alpha}\\p\nmid qy}} P_\omega(x,pq,x^{1/\beta}) = \sum_{l=0}^{n-1}\;\sum_{\substack{x^{\frac{1}{u_{l+1}}}<p\leqslant x^{\frac{1}{u_l}}\\p\nmid qy}} P_\omega(x,x^{\frac{1}{\log x}{\log pq}},x^{1/\beta}) = \sum_{l=0}^{n-1} T_l,$$

$$T_l = \sum_{\substack{x^{\frac{1}{u_{l+1}}}<p\leqslant x^{\frac{1}{u_l}}\\p\nmid qy}} P_\omega(x,x^{\frac{1}{\log x}{\log pq}},x^{1/\beta}) \leqslant$$

$$\leqslant \sum_{\substack{x^{\frac{1}{u_{l+1}}}<p\leqslant x^{\frac{1}{u_l}}\\p\nmid qy}} \Lambda_1\left(\frac{\log x}{\log qp}\right)\frac{c_{qy}x}{\varphi(p)\varphi(q)\log^2 x} + O\left(\frac{c_{qy}x}{\log^3 x}\log\log x\log\frac{u_{l+1}}{u_l}\right).$$

Since $\Lambda_1\left(u+O\left(\dfrac{1}{\log x}\right)\right)=\Lambda_1(u)+O\left(\dfrac{1}{\log x}\right)$ and $\Lambda_1(u)$ is a decreasing function, therefore

$$T_l \leqslant \Lambda_1(u_l)\frac{c_{qy}x}{\varphi(q)\log^2 x}\log\frac{u_{l+1}}{u_l} + O\left(\frac{c_{qy}x}{\log^3 x}\log\log x\log\frac{u_{l+1}}{u_l}\right).$$

Since

$$\sum_{l=0}^{n-1}\Lambda_1(u_l)\log\frac{u_{l+1}}{u_l} - \int_\alpha^\beta\frac{\Lambda_1(u)}{u}\,du \leqslant \sum_{l=0}^{n-1}(\Lambda_1(u_{l+1})-\Lambda_1(u_l))\max_{0\leqslant l\leqslant n-1}\log\frac{u_{l+1}}{u_l} =$$

$$= O\left(\frac{1}{n}\right) = O\left(\frac{1}{\log x}\right),$$

hence

$$\sum_{l=0}^{n-1} T_l \leqslant \left(\frac{c_{qy}}{\varphi(q)}\int_\alpha^\beta\frac{\Lambda_1(u)}{u}\,du\right)\frac{x}{\log^2 x} + O\left(\frac{c_{qy}x}{\log^3 x}\log\log x\right).$$

This proves the Theorem.

Theorem C_2. *Let $3 \leqslant \alpha < \beta \leqslant 13$ be two given numbers. Then*

$$P_\omega(x, q, x^{1/\alpha}) \geqslant P_\omega(x, q, x^{1/\beta}) - \frac{c_{qy}x}{\varphi(q)\log^2 x}\int_\alpha^\beta \frac{\Lambda(u)}{u}\,du +$$

$$+ O\left(\frac{c_{qy}x}{\log^3 x}\log\log x\right),$$

where q is a given positive integer.

Proof. It is evident that we may assume $\alpha < U < \beta$. We estimate the differences $P_\omega(x, q, x^{1/\beta}) - P_\omega(x, q, x^{1/U})$ and $P_\omega(x, q, x^{1/U}) - P_\omega(x, q, x^{1/\alpha})$.

The difference between $P_\omega(x, q, p_m)$ and $P_\omega(x, q, p_{m+1})$ is equal to the number of primes satisfying the following conditions:

$$p \leqslant x,\; p \equiv a(\mathrm{mod}\, q),\; p \not\equiv a_i(\mathrm{mod}\, p_i)(i \leqslant m),\; p \equiv a_{m+1}(\mathrm{mod}\, p_{m+1}).$$

If $p_{m+1} \nmid y$, then by the definition we know that it is equal to $P_\omega(x, qp_{m+1}, p_m)$; otherwise it is equal to 0 or 1. Hence

$$P_\omega(x, q, p_m) - P_\omega(x, q, p_{m+1})\begin{cases} = P_\omega(x, qp_{m+1}, p_m), & \text{if } p_{m+1} \nmid y; \\ \leqslant 1, & \text{if } p_{m+1} \mid y. \end{cases}$$

Arrange the primes between $x^{1/U}$ and $x^{1/\alpha}$ as follows:

$$p_t \leqslant x^{1/U} < p_{t+1} < \cdots < p_s \leqslant x^{1/\alpha} < p_{s+1}.$$

Then

$$P_\omega(x, q, x^{1/U}) = P_\omega(x, q, x^{1/\alpha}) + \sum_{\substack{t \leqslant i < s \\ p_{i+1} \nmid qy}} P_\omega(x, p_{i+1}, q, p_i) + O(1).$$

Let $n = [\log x]$. $u_m = \alpha + \dfrac{U - \alpha}{n}m\,(0 \leqslant m \leqslant n)$. And put

$$T = \sum_{\substack{t \leqslant i < s \\ p_{i+1} \nmid qy}} P_\omega(x, qp_{i+1}, p_i) = \sum_{m=0}^{n-1} T_m.$$

Since $p_i < qp_{i+1} < 4qp_i$ and $\Lambda(u)$ is decreasing in the interval (α, U), hence

$$T_m = \sum_{\substack{x^{\frac{1}{u_{m+1}}} < p_{i+1} \leqslant x^{\frac{1}{u_m}} \\ p_{i+1} \nmid qy}} P_\omega(x, qp_{i+1}, p_i) = \sum_{\substack{x^{\frac{1}{u_{m+1}}} < p_{i+1} \leqslant x^{\frac{1}{u_m}} \\ p_{i+1} \nmid qy}} P_\omega\left(x, qp_{i+1}, x^{\frac{\log p_i}{\log x}}\right) \leqslant$$

$$\leqslant \Lambda(u_m)\frac{c_{qy}x}{\varphi(q)\log^2 x}\log\frac{u_{m+1}}{u_m} + O\left(\frac{c_{qy}x}{\log^3 x}\log\log x\log\frac{u_{m+1}}{u_m}\right).$$

1046

Since

$$\sum_{m=0}^{n-1} \Lambda(u_m) \log \frac{u_{m+1}}{u_m} - \int_a^U \frac{\Lambda(u)}{u}\, du = O\left(\frac{1}{\log x}\right),$$

therefore

$$T \leqslant \left(\frac{c_{qy}}{\varphi(q)} \int_a^U \frac{\Lambda(u)}{u}\, du\right)\frac{x}{\log^2 x} + O\left(\frac{c_{qy}x}{\log^3 x} \log \log x\right).$$

Hence

$$P_\omega(x, q, x^{1/U}) \leqslant P_\omega(x, q, x^{1/a}) + \left(\frac{c_{qy}}{\varphi(q)} \int_a^U \frac{\Lambda(u)}{u}\, du\right)\frac{x}{\log^2 x} +$$

$$+ O\left(\frac{c_{qy}x}{\log^3 x} \log \log x\right).$$

Similarly, we have

$$P_\omega(x, q, x^{1/\beta}) \leqslant P_\omega(x, q, x^{1/U}) + \left(\frac{c_{qy}}{\varphi(q)} \int_U^\beta \frac{\Lambda(u)}{u}\, du\right)\frac{x}{\log^2 x} +$$

$$+ O\left(\frac{c_{qy}x}{\log^3 x} \log \log x\right).$$

Thus we have the Theorem.

$$\S\,8$$

Let $4 < v < 8$, $2 < u \leqslant v$ be two given numbers. Let $\mathfrak{M}$ denote the set of primes satisfying the following conditions:

(11)
$$p \leqslant x,\ p \equiv a (\mathrm{mod}\ q),\ p \not\equiv a_i (\mathrm{mod}\ p_i)(i \leqslant s),$$
$$p \not\equiv a_{s+j} (\mathrm{mod}\ p_{s+j}^2)(j < t - s),$$

where $p_s \leqslant x^{1/v} < p_{s+1}$, $p_t \leqslant x^{1/u} < p_{t+1}$ and q is a given positive integer.

The number of elements of $\mathfrak{M}$ is denoted by $M_\omega(x, x^{1/v}, x^{1/u})$.

Lemma 4. *There exist sequences of integers (ω_i) such that the number of elements of $\mathfrak{M}$ satisfying at least l congruences of*

(12)
$$p \equiv a_{s+j} (\mathrm{mod}\ p_{s+j})(1 \leqslant j \leqslant t - s)$$

is at most

$$\frac{1}{l} \sum_{\substack{j < t-s \\ p_{s+j} \nmid y}} P_{\omega_j}(x, q p_{s+j}, x^{1/v}).$$

Proof. Let Γ_j be the subset of $\mathfrak{M}$ whose elements satisfy the congruence

$$p \equiv a_{s+j}(\mathrm{mod}\ p_{s+j}).$$

Now, we estimate the number of elements of Γ_j. If $p_{s+j}|y$, then the number elements of Γ_j is equal to 0 or 1. Assume $p_{s+j}\nmid y$. Denote the solution of the system of congruences

$$\begin{cases} n \equiv a_{s+j}(\mathrm{mod}\ p_{s+j}), \\ n \equiv a(\mathrm{mod}\ q) \end{cases}$$

by $\tilde{a}_{s+j}$. Let

$$(\omega_j) \qquad\qquad \tilde{a}_{s+j},\ qp_{s+j};\ a_i(1 \leqslant i \leqslant t).$$

Then the number of elements of Γ_j is not more than $P_{\omega_j}(x, qp_{s+j}, x^{1/v})$.

If the element of $\mathfrak{M}$ satisfies at least l congruences of (12), then it belongs to at least l different sets Γ_j. Hence the number of elements of $\mathfrak{M}$ satisfying at least l congruences of (12) does not exceed

$$\frac{1}{l} \sum_{\substack{j \leqslant t-s \\ p_{s+j}\nmid y}} P_{\omega_j}(x,\ qp_{s+j},\ x^{1/v}).$$

Thus we have the Lemma.

Theorem D. *The number of elements of $\mathfrak{M}$ satisfying at most m congruences of (12) is not less than*

$$P_{\omega}(x,\ q,\ x^{1/v}) - \frac{1}{m+1}\left(\int_u^v \Lambda_1(t)\ \frac{dz}{z}\right)\frac{c_{qy}x}{\varphi(q)\log^2 x} + O\left(\frac{c_{qy}x}{\log^3 x}\log\log x\right).$$

Proof. Since

$$P_{\omega}(x,\ q,\ x^{1/v}) - M_{\omega}(x,\ x^{1/v},\ x^{1/u}) = \sum_{j \leqslant t-s}\ \sum_{\substack{p\equiv a_{s+j}(\mathrm{mod}\ p_{s+j}^2) \\ p\leqslant x}} 1 \leqslant$$

$$\leqslant \sum_{1\leqslant j\leqslant t-s}\left(\frac{x}{p_{s+j}^2} + 1\right) = O(x^{1/u}) + O(x^{1-\frac{1}{v}}),$$

therefore it follows from Lemma 4 and Theorem C_1 that the number of elements of $\mathfrak{M}$ satisfying at most m congruences of (12) is not less than

$$M_{\omega}(x,\ x^{1/v},\ x^{1/u}) - \frac{1}{m+1} \sum_{\substack{j\leqslant t-s \\ p_{s+j}\nmid y}} P_{\omega_j}(x,\ qp_{s+j},\ x^{1/v}) + O(1) \geqslant$$

1048

$$\geq P_{\omega}(x, q, x^{1/v}) - \frac{1}{m+1}\left(\int_{u}^{v} \Lambda_1(z)\,\frac{dz}{z}\right)\frac{c_{qy}x}{\varphi(q)\log^2 x} + O\left(\frac{c_{qy}x}{\log^3 x}\log\log x\right).$$

Thus we have the Theorem.

$$\S 9$$

It follows from Theorem B and Theorem C_2 that

$$P_{\omega}(x, q, x^{1/6}) > \left(25.8096 - \int_{6}^{13}\frac{\Lambda(u)}{u}\,du\right)\frac{c_{qy}x}{\varphi(q)\log^2 x} +$$

$$+ O\left(\frac{c_{qy}x}{\log^3 x}\log\log x\right) > 8.4\,\frac{c_{qy}x}{\varphi(q)\log^2 x} + O\left(\frac{c_{qy}x}{\log^3 x}\log\log x\right).$$

(i) Let $x = y$ be an even integer. Let

$$(\omega_1) \qquad\qquad a = 1, q = 2; a_i = x(i = 1, 2, \cdots).$$

From (9) we know that there exists a positive constant x_1 such that

$$P_{\omega_1}(x, 2, x^{1/6}) - \frac{1}{3}\left(\int_{3}^{6}\frac{\Lambda_1(u)}{u}\,du\right)\frac{c_{2x}x}{\log^2 x} + O\left(\frac{c_{2x}x}{\log^3 x}\log\log x\right) >$$

$$> (8.4 - 6.588)\frac{c_{2x}x}{\log^2 x} + O\left(\frac{c_{2x}x}{\log^3 x}\log\log x\right) > \frac{c_{2x}x}{\log^2 x} > 1$$

for $x > x_1$. Hence it follows from Theorem D that for $x > x_1$ there exists a prime p such that $1 < p < x-1$ and $x - p$ has no prime divisors $\leq x^{1/6}$ and has at most 2 prime divisors in the interval $x^{1/6} < p' \leq x^{1/3}$. Hence $x - p$ is a product of at most 3 primes. Since $x = p + x - p$, we obtain Theorem 1.

(ii) Let $x = y$ be an odd integer. Let

$$(\omega_2) \qquad\qquad a = x - 2, q = 4; a_i = x(i = 1, 2, \cdots).$$

From (9) we know that there exists a positive constant x_2 such that

$$P_{\omega_2}(x, 4, x^{1/6}) - \frac{1}{3}\left(\int_{3}^{6}\frac{\Lambda_1(u)}{u}\,du\right)\frac{c_{4x}x}{2\log^2 x} + O\left(\frac{c_{4x}x}{\log^3 x}\log\log x\right) > \frac{c_{4x}x}{2\log^2 x} > 2$$

for $x > x_2$. Hence it follows from Theorem D that for $x > x_2$ there exists a prime number p such that $p < x - 3$ and $\dfrac{x-p}{2}$ has no prime divisors $\leq x^{1/6}$ and has at most 2 prime divisors in the intervals $x^{1/6} < p' \leq x^{1/3}$. Hence $\dfrac{x-p}{2}$ is a product of at most 3 primes. Thus we have Theorem 3.

(iii) Let k be a given integer. Let

$$(\omega_3) \qquad\qquad a = 1, q = 2; a_i = -2k \quad (i = 1, 2, \cdots).$$

It follows from (9) that there exists a positive constant x_3 such that

$$P_{\omega_3}(x, 2, x^{1/6}) - \frac{1}{3}\left(\int_3^6 \frac{\Lambda_1(u)}{u}\,du\right)\frac{c_4 k x}{\log^2 x} + O\left(\frac{x}{\log^3 x}\log\log x\right) > \frac{c_4 k x}{\log^2 x}$$

for $x > x_3$. Hence from (9) we know that for $x > x_3$ there exist not less than $\frac{c_4 k x}{\log^2 x}$ prime numbers p in the interval $1 < p \leqslant x$ such that $p + 2k$ is a product of at most 3 primes. Thus we have Theorem 2.

From Lemma 2 and Theorem A with $c = 1$, we have

$$P_{\omega_3}\left(x, 2, \frac{x^{1/4}}{\log^3 x}\right) \leqslant 8 \prod_{\substack{p|2k \\ p>2}} \frac{p-1}{p-2} \prod_{p>2}\left(1 - \frac{1}{(p-1)^2}\right)\frac{x}{\log^2 x} + O\left(\frac{x}{\log^3 x}\log\log x\right).$$

Since

$$Z_k(x) = \sum_{\substack{p \leqslant \frac{x^{1/4}}{\log^3 x} \\ p+2k=p'}} 1 + \sum_{\substack{\frac{x^{1/4}}{\log^3 x} < p \leqslant x \\ p+2k=p'}} 1 \leqslant O(x^{1/4}) + P_{\omega_3}\left(x, 2, \frac{x^{1/4}}{\log^3 x}\right) \leqslant$$

$$\leqslant 8 \prod_{\substack{p|2k \\ p>2}} \frac{p-1}{p-2} \prod_{p>2}\left(1 - \frac{1}{(p-1)^2}\right)\frac{x}{\log^2 x} + O\left(\frac{x}{\log^3 x}\log\log x\right),$$

we have Theorem 4.

REFERENCES

[1] Wang Yuan 1957 On Sieve Methods and Some of the Related Problems, *Science Record*, Vol. I, No. 1, 9—11.

[2] Wang Yuan 1959 On Sieve Methods and Some of Their Applications, *Scientia Sinica*, **8**, 375—381.

[3] Чудаков, Н. Г. 1948 О конечной разности для функций $\Psi(x, k, l)$, *ИАН СССР, серия матем.*, **12**, 31—46.

[4] Wang Yuan 1956 On the representation of large integer as a sum of a prime and a product of at most 4 primes, *Acta Mathematica Sinica*, **6** (4), 565—582.

[5] Виноградов, А. И. 1957 Применение $\zeta(s)$ к решету Эратосфена, *Мат. Сб*; **41**, 49—80.

1050

APPENDIX

$1°$.   We state the generalized weak Riemann hypothesis as follows:

(R_δ) *The real parts of all zeros of all Dirichlet's L-functions $L(s, \chi)$ are $\leqslant \delta^{-1}$, where $1 < \delta \leqslant 2$.*

Especially, (R_2) is the well-known grand Riemann hypothesis.

For the sake of brevity, we denote the following proposition by $(1, A)$:

Every sufficiently large even integer is a sum of a prime and an almost prime of at most A prime divisors.

Here, we state the refined result:

Theorem I. $(1, 3)$ *may be derived from (R_{δ_1}), where $\delta_1 \geqslant \dfrac{2.475}{1.475}$ and $(1, 4)$ is the consequence of (R_{δ_2}), where $\delta_2 \geqslant \dfrac{3.237}{2.237}$.*

All the following results are obtained under the truth of (R_δ).

$2°$.   Let $\eta = \dfrac{\delta}{\delta - 1}$.   Let $x^{1/u} \leqslant q \leqslant c_0 x^{1/u}$, where c_0 is a given constant.   Then the estimation

$$(1) \qquad P_\omega(x, q, x^{1/u}) \leqslant \Lambda(u) \frac{c_{qy} x}{\varphi(q) \log^2 x} + O\left(\frac{c_{qy} x \log \log x}{\varphi(q) \log^3 x}\right)$$

holds uniformly in (ω), where

$$(2) \qquad \Lambda(u) = \frac{4\eta u}{u - \eta} e^\gamma \quad (\text{if } \eta < u \leqslant 3\eta)$$

and

$$(3) \qquad \Lambda(u) = \frac{2u e^\gamma}{\dfrac{u - \eta}{\eta} - 1 - \dfrac{u - \eta}{2\eta} \log \dfrac{u - \eta}{2\eta}} \quad (\text{if } 3\eta < u \leqslant 7\eta).$$

$3°$.   Let $v > 2\eta$ be a given number and $q = x^{1/u}$.   Then

$$(4) \qquad P_\omega(x, q, x^{1/v}) \leqslant \Lambda_1(u) \frac{c_{qy} x}{\varphi(q) \log^2 x} + O\left(\frac{x c_{qy} \log \log x}{\varphi(q) \log^3 x}\right)$$

holds uniformly in (ω), where

$$(5) \qquad \Lambda_1(u) = \frac{4\eta u}{u - \eta} e^\gamma$$

for $\eta < u \leqslant \dfrac{1}{\dfrac{1}{\eta} - \dfrac{2}{v}}$, and

$$(6) \qquad \Lambda_1(u) = \frac{2ve^{\gamma}}{v\left(\dfrac{1}{\eta} - \dfrac{1}{u}\right) - 1 - \dfrac{v}{2}\left(\dfrac{1}{\eta} - \dfrac{1}{u}\right)\log\dfrac{v}{2}\left(\dfrac{1}{\eta} - \dfrac{1}{u}\right)}$$

for

$$\frac{1}{\dfrac{1}{\eta} - \dfrac{2}{v}} < v < \begin{cases} \dfrac{1}{\dfrac{1}{\eta} - \dfrac{4}{v}} & (\text{if } v > 4\eta); \\[2em] \infty & (\text{if } v \leqslant 4\eta). \end{cases}$$

$4°.$ Let q be a given integer. Then

$$(7) \qquad P_\omega(x, q, x^{1/6.5\eta}) > \lambda(6.5\eta)\frac{c_{qy}x}{\varphi(q)\log^2 x} + O\left(\frac{xc_{qy}}{\log^3 x}\right),$$

where

$$(8) \qquad \lambda(6.5\eta) \geqslant 2\eta \times 6.453306.$$

The proof is similar to that of Theorem B with the essential difference that here we put $\tau = 0.452$ and $\lambda = 1.5715$ so as to obtain the more exact estimation

$$\sum_{k=0}^{\infty}\frac{\lambda^{k+1}[(k+1)\tau]^{2k+4}}{(2k+4)!} < \sum_{k=0}^{6}\frac{\lambda^{k+1}[(k+1)\tau]^{2k+4}}{(2k+4)!} +$$

$$+ \frac{\lambda^8(8\tau)^{18}}{18!}\sum_{k=0}^{\infty}\left(\frac{\lambda\tau^2 e^2}{4} \times \frac{6561}{6080}\right)^k < 0.007183682.$$

$5°.$ Let $\eta < \alpha < \beta \leqslant 6.5\eta$ be two given numbers. Let q denote a given positive integer. Then

$$P_\omega(x, q, x^{1/\alpha}) \geqslant P_\omega(x, q, x^{1/\beta}) - \frac{c_{qy}x}{\varphi(q)\log^2 x}\int_{\alpha}^{\beta}\frac{\Lambda(u)}{u}\,du +$$

$$(9) \qquad + O\left(\frac{c_{qy}x}{\log^3 x}\log\log x\right).$$

$6°.$ Let u, v be two given numbers satisfying $2\eta < v < 10\eta$ and $\eta < u < v$. Let $\mathfrak{M}$ denote the set of primes p satisfying the following conditions:

$$p \leqslant x, \, p \equiv a(\text{mod } q), \, p \not\equiv a_i(\text{mod } p_i)(i \leqslant s),$$

$$(10) \qquad p \not\equiv a_{s+j}(\text{mod } p_{s+j}^2)(j \leqslant t - s),$$

1052

where $p_s \leqslant x^{1/v} < p_{s+1}$, $p_t \leqslant x^{1/u} < p_{t+1}$ and q is a given integer. Then the number of elements of $\mathfrak{M}$ satisfying at most m congruences of

$$(11) \qquad p \equiv a_{s+i} \pmod{p_{s+i}}(1 \leqslant i \leqslant t - s)$$

is not less than

$$(12) \quad P_\omega(x, q, x^{1/v}) - \frac{1}{m+1}\left(\int_u^v \Lambda_1(z)\,\frac{dz}{z}\right)\frac{c_{qy}x}{\varphi(q)\log^2 x} + O\left(\frac{c_{qy}x}{\log^3 x}\log\log x\right).$$

$7°$. Computation of integrals.

(i) $\quad \Delta_1 = \displaystyle\int_{5\eta}^{6.5\eta} \frac{\Lambda(u)}{u}\,du = 2e^\gamma \int_{5\eta}^{6.5\eta} \frac{du}{\dfrac{u-\eta}{\eta}-1-\dfrac{u-\eta}{2\eta}\log\dfrac{u-\eta}{2\eta}}$

$\qquad = 2\eta e^\gamma \displaystyle\int_4^{5.5} \frac{dw}{w-1-\dfrac{w}{2}\log\dfrac{w}{2}} = 2\eta e^\gamma \int_4^{5.5} f(w)\,dw$

$\qquad < 2\eta e^\gamma\left(\displaystyle\sum_{i=0}^{71} f(4+0.02i) + \sum_{j=0}^{2} f(5.46+0.02j)\right)$

$\qquad < 2\eta e^\gamma \times 0.89050652.$

(ii) Let $v = 5\eta$ in $3°$. Then

$$\Delta_2 = \int_{\frac{5\eta}{3}}^{5\eta} \frac{\Lambda_1(u)}{u}\,du = \int_{\frac{5\eta}{3}}^{5\eta} \frac{2ve^\gamma}{v\left(\dfrac{1}{\eta}-\dfrac{1}{u}\right)-1-\dfrac{v}{2}\left(\dfrac{1}{\eta}-\dfrac{1}{u}\right)\log\dfrac{v}{2}\left(\dfrac{1}{\eta}-\dfrac{1}{u}\right)}\cdot\frac{du}{u}$$

$$= 20\eta e^\gamma\int_1^2 \frac{dz}{(5-2z)(2z-1-z\log z)} = 20\eta e^\gamma\int_1^2 g(z)\,dz$$

$$< 20\eta e^\gamma\left(\sum_{i=0}^{7} g(1+0.02i) + \sum_{j=0}^{41} g(1.18+0.02j)\right)$$

$$< 20\eta e^\gamma \times 0.3972371.$$

$8°$. The proof of Theorem I. Let $x = y$ be an even integer. Let

$$(\omega) \qquad a = 1,\ q = 2;\ a_i = x \quad (i = 1, 2, \cdots).$$

(i) Let $\eta = 2.475$. Then from $2° — 7°$, we know that there exists a constant x_1 such that

$$P_\omega(x, 2, x^{1/5\eta}) - \frac{1}{2}\left(\int_{\frac{15\eta}{5\eta-1}}^{5\eta}\frac{\Lambda_1(u)}{u}\,du\right)\frac{c_{2x}x}{\log^2 x} + O\left(\frac{c_{2x}x}{\log^3 x}\log\log x\right) >$$

$$> P_\omega(x, 2, x^{1/6 \cdot 5\eta}) - \left(\int_{5\eta}^{6 \cdot 5\eta} \frac{\Lambda(u)}{u}\, du + \frac{1}{2} \int_{\frac{15\eta}{5\eta - 1}}^{5\eta} \frac{\Lambda_1(u)}{u}\, du \right) \frac{c_{2x} x}{\log^2 x} +$$

$$+ O\left(\frac{c_{2x} x}{\log^2 x} \log \log x \right) > 2\eta [6.453306 - e^\gamma (0.89050652 +$$

$$+ 1.9861855 + 0.738218)] \frac{c_{2x} x}{\log^2 x} + O\left(\frac{c_{2x} x}{\log^3 x} \log \log x \right) >$$

$$> 0.05 \frac{c_{2x} x}{\log^2 x} + O\left(\frac{c_{2x} x}{\log^3 x} \log \log x \right) > 1,$$

for $x > x_1$. Hence it follows from $6°$ that for $x > x_1$ there exists a prime p such that $p < x - 1$ and $x - p$ has no prime divisors $\leqslant x^{1/5\eta}$ and has at most one prime divisor in the interval $x^{1/5\eta} < p' \leqslant x^{\frac{5\eta - 1}{15\eta}}$. Hence $x - p$ is a product of at most 3 primes. Thus we have $(1, 3)$.

(ii)   Let $\eta = 3.237$. Then from $2° — 7°$, we know that there exists a constant x_2 such that

$$P_\omega(x, 2, x^{1/5\eta}) - \frac{1}{2} \left(\int_{\frac{20\eta}{5\eta - 1}}^{5\eta} \frac{\Lambda_1(u)}{u}\, du \right) \frac{c_{2x} x}{\log^2 x} + O\left(\frac{c_{2x} x}{\log^3 x} \log \log x \right) >$$

$$> 0.01 \frac{c_{2x} x}{\log^2 x} + O\left(\frac{c_{2x} x}{\log^3 x} \log \log x \right) > 1$$

for $x > x_2$. Hence it follows from $6°$ that for $x > x_2$ there exists a prime number p such that $p < x - 1$ and $x - p$ has no prime divisor $\leqslant x^{1/5\eta}$ and has at most one prime divisor in the interval $x^{1/5\eta} < p' \leqslant x^{\frac{5\eta - 1}{20\eta}}$. Hence $x - p$ is a product of at most 4 primes. Thus we have $(1, 4)$.

$9°$.   It is well known that in the proof of Theorem I, the hypothesis (R_δ) may be replaced by

$$(\tilde{R}_\delta) \qquad \sum_{D < x^{1/\eta}} \mu^2(D) \operatorname*{Max}_{\substack{l \,(\mathrm{mod}\, D) \\ (l, D) = 1}} \left| \pi(x, D, l) - \frac{\mathrm{li}\, x}{\varphi(D)} \right| = O\left(\frac{x}{\log^A x} \right),$$

where A is any given positive constant and the constant implied by the symbol "O" depends only on δ and A.

Similarly, (R_δ) may also be replaced by

$$(\tilde{\tilde{R}}_\delta) \qquad \sum_{D < x^{1/\eta}} \mu^2(D) \operatorname*{Max}_{\substack{l \,(\mathrm{mod}\, D) \\ (l, D) = 1}} \left| P(x, D, l) - \frac{x}{\varphi(D) \log x} \right| = O\left(\frac{x}{\log^A x} \right),$$

1054

where $P(x, D, l) = \sum_{\substack{p \leqslant x \\ p \equiv l \ (\mathrm{mod}\ D)}} \log p \cdot e^{-\frac{\log x}{x} p}$ (cf. [1], [2]).

Барбан[3, 4] first proved $(\widetilde{R}_{1.2})$. Later, Pan Chin Tong[2] obtained $(\widetilde{\widetilde{R}}_{1.5})$ independently, from which he deduced $(1, 5)$. From Theorem I, it can be easily seen that $(\widetilde{\widetilde{R}}_{1.5})$ implies $(1, 4)$. In other words, we have proved

Theorem II. *Every sufficiently large even integer is a sum of a prime and a product of at most 4 primes.*

Remark. $(1, 4)$ has also been proved by Pan and Барбан independently, but their proofs are more complicated than that given here, in fact, their proofs are based on $(\widetilde{R}_{1.6})$ and $(\widetilde{\widetilde{R}}_{1.6})$ respectively (cf. [5]). I am grateful to Messrs. Pan and Барбан for their kindly informing me of their results.

References

[1] Реньи, А. 1948 О представлении четных чисел в виде суммы простого и почти простого числа, *ИАН СССР*, **2**, 57—78.

[2] Pan Chin Tong, 1962 On the representation of large even integer as a sum of a prime and an almost prime, *Acta Math. Sinica*, Vol. 12, No. 1, 95—106.

[3] Барбан, М. Б. 1961 Арифметические функции на редких множествах, *Доклады Академии Наук УзССР*, **8**, 9—11.

[4] Барбан, М. Б. 1961 Новые применении большого решета Ю. В. Линника, *Труды Института Математики, им. В. И. Романовского*, вып. 22.

[5] Линник, Ю. В. 1960 Асимптотическая формула в аддитивной проблеме Гарди Литтльвуда, *ИАН СССР*, Том 24, № 5, 629—706.

ON THE ESTIMATION OF CHARACTER SUM AND ITS APPLICATIONS*

WANG YUAN

Institute of Mathematics, Academia Sinica

Received 25 September 1962

1. Introduction

By the use of Weil's [1] famous result on analogue of Riemann hypothesis for algebraic function fields over a finite field, Burgess [1] first proposed a method for the estimation of character sum which gave an improvement of the well-known Polya's inequality [3] for the case of real primitive character modulo p (prime number). Later, the author [4] and Burgess [5,6] himself gave some further generalisations, modifications and applications on Burgess's estimation. His method may be stated as follows:

Theorem A. [6]   *Let χ be a primitive character modulo k. Let η be a given positive number and r be an integer ≥ 1. If k is a square free number or $r = 2$, then the estimation*

$$\left| \sum_{n=N+1}^{N+H} \chi(n) \right| \leq c_1(r,\eta) H^{1-\frac{1}{r+1}} k^{\frac{1}{4r}+\eta},$$

holds for any pair of integers N and H ($H > 0$).

We have the following two consequences:

Corollary 1. [6]   *Suppose that $\chi(n) = \left(\frac{f}{n}\right)$ is the real primitive character modulo f, where $\left(\frac{f}{n}\right)$ denotes the Kronecker symbol. Then*

$$\left| \sum_{n=1}^{H} \left(\frac{f}{n}\right) \right| \leq c_2(r,\eta) H^{1-\frac{1}{r+1}} f^{\frac{1}{4r}+\eta},$$

---

*Shuxue Jinzhan **1** (1964) 78–83.

Corollary 2. [4] *Suppose that χ is a non-principal character modulo p (prime number). Then*

$$\left| \sum_{n=1}^{H} \chi(n) \right| < c_3(\eta) \frac{H}{p^{\eta^2/6}},$$

holds for $H > p^{\frac{1}{4}+\eta}$.

The aim of present paper is to apply these estimations to the problems for estimations of the least solutions of Pell's equations and the nth non-residue modulo p. The latter problem will also be treated under the assumption of generalised Riemann hypothesis (GRH).

Let d be a positive integer not a perfect square, $d \equiv 0$ or $1 \pmod 4$. If the integer point (x_0, y_0) $(x_0 > 0, y_0 > 0)$ is a solution of the Pell's equation

$$x^2 - dy^2 = 4,$$

such that $x_0 + \sqrt{d}y_0$ attains its minimum, then it is called to be a least solution. Let

$$\varepsilon = \frac{x_0 + \sqrt{d}y_0}{2}.$$

Theorem 1. *For any $\delta > 0$, there exists $c_4(\delta)$ such that*

$$\ln \varepsilon < \left(\frac{1}{4} + \delta \right) \sqrt{d} \ln d,$$

holds for $d > c_4(\delta)$.

This gives an improvement of a result due to Hua Loo-keng [7].

Let $n \geq 2$ and $n|(p-1)$. If the congruence

$$x^n \equiv c \pmod p, \quad 1 \leq x \leq p$$

has no solution, then c is called an nth non-residue modulo p. Otherwise it is called an nth residue modulo p. Denote by $N(p,n)$ the least positive nth non-residue modulo p.

Theorem 2. *Let δ be a given positive number. Then for a sufficiently large prime p, we have*

$$\text{(i) } N(p,n) \leq p^{4e^{-\frac{1}{n-1}} + \delta}, \quad n \geq 2,$$

$$\text{(ii) } N(p,n) \leq p^{\frac{1}{12}}, \quad n \geq 21$$

and

$$\text{(iii) } N(p,n) \leq p^{\frac{\ln \ln n + 2}{4 \ln n}}, \quad n > e^{33}.$$

(i), (ii) and (iii) are the respective improvements of results due to Vinogradov [8] and Buchstab [9].

2. Proof of Theorem 1

Write

$$\sigma(a) = \sum_{n=1}^{a} \left(\frac{d}{n}\right) \quad \text{and} \quad K(d) = \sum_{n=1}^{\infty} \left(\frac{d}{n}\right)\frac{1}{n}.$$

Lemma 1. *Let r be an integer ≥ 4 and $\tau = \frac{1}{r}$. Then*

$$|\sigma(a)| \leq c_5(\tau)ad^{-\tau^2/6}$$

holds for $a \geq d^{\frac{1}{4}+\tau}$.

Proof. Let $d = fm^2$, where f is the fundamental discriminate. Then

$$\sigma(a) = \sum_{n=1}^{a} \left(\frac{d}{n}\right) = \sum_{\substack{n=1 \\ (n,m)=1}}^{a} \left(\frac{f}{n}\right) = \sum_{n=1}^{a} \left(\frac{f}{n}\right) \sum_{k|(n,m)} \mu(k)$$

$$= \sum_{k|m} \mu(k)\left(\frac{f}{k}\right) \sum_{n\leq a/k} \left(\frac{f}{n}\right)$$

and

$$|\sigma(a)| \leq \sum_{k|m} \left| \sum_{n\leq a/k} \left(\frac{f}{n}\right) \right|.$$

Therefore by Corollary 1 we obtain

$$|\sigma(a)| \leq \sum_{k|m} c_6(\tau)\left(\frac{a}{k}\right)^{1-\frac{1}{r+1}} f^{\frac{1}{4r}+\frac{1}{4r(r+1)}} \leq c_5(\tau)a^{1-\frac{1}{r+1}}d^{\frac{1}{4r}+\frac{1}{2r(r+1)}}$$

$$\leq c_5(\tau)ad^{-\frac{1}{4(r+1)}-\frac{1}{r(r+1)}+\frac{1}{4r}+\frac{1}{2r(r+1)}} = c_5(\tau)ad^{-\frac{1}{4r(r+1)}} < c_5(\tau)ad^{-\tau^2/6}.$$

The lemma is proved.

Lemma 2. *For any given δ with $0 < \delta < \frac{1}{2}$, there exists $c_7(\delta)$ such that*

$$K(d) < \left(\frac{1}{4} + \delta\right)\ln d,$$

holds for $d > c_7(\delta)$.

Proof.

$$K(d) = \sum_{n=1}^{\infty} \frac{\sigma(n) - \sigma(n-1)}{n} = \sum_{n=1}^{\infty} \frac{\sigma(n)}{n(n+1)} = \Sigma_1 + \Sigma_2 + \Sigma_3.$$

Let $\tau = \frac{1}{r} \leq \frac{\delta}{2} < 2\tau$, where r is an integer. Then

$$|\Sigma_1| \leq \left| \sum_{1 \leq n < d^{\frac{1}{4}+\tau}} \frac{\sigma(n)}{n(n+1)} \right| \leq \sum_{1 \leq n < d^{\frac{1}{4}+\tau}} \frac{1}{n+1}$$

$$< \int_1^{d^{\frac{1}{4}+\tau}} \frac{dt}{t} + \frac{1}{d^{\frac{1}{4}+\tau}} \leq \left(\frac{1}{4} + \frac{\delta}{2} \right) \ln d + \frac{1}{d^{\frac{1}{4}+\frac{\delta}{4}}}.$$

By Lemma 1 we derive

$$|\Sigma_2| = \left| \sum_{d^{\frac{1}{4}+\tau} \leq n < d} \frac{\sigma(n)}{n(n+1)} \right| \leq c_8(\delta) d^{-\frac{\delta^2}{96}} \sum_{1 \leq n < d} \frac{1}{n+1} < c_8(\delta) d^{-\frac{\delta^2}{96}} \ln d.$$

Finally it follows from Polya's theorem that

$$|\sigma(a)| \leq \sum_{k|m} \sqrt{f} \ln f < \sqrt{d} \ln d$$

and therefore

$$|\Sigma_3| \leq \sqrt{d} \ln d \cdot \sum_{n \geq d} \frac{1}{n(n+1)} = \frac{\ln d}{\sqrt{d}}.$$

The lemma follows.

Theorem 1 is the consequence of Lemma 2 (see [10]).

3. Proof of Theorem 2

Lemma 3. [9]   *Let $\Psi(x, y)$ be the number of integers in the interval $1 \leq n \leq x$ such that the prime divisors of n are all $\leq y$. Then*

$$\Psi(x, x^{\frac{1}{a}}) = \rho(\alpha)x + O\left(\frac{x}{\sqrt{\ln x}} \right),$$

where the constant in "O" depends only on α, and

$$\rho(\alpha) = \begin{cases} 1, & \text{if } 0 < \alpha \leq 1; \\[2mm] 1 - \displaystyle\int_1^\alpha \frac{dz_1}{z_1} + \int_2^\alpha \int_1^{z_1-1} \frac{dz_1 dz_1}{z_1 z_2} - \cdots \\[4mm] \quad + (-1)^{[\alpha]} \displaystyle\int_{[\alpha]}^\alpha \int_{[\alpha-1]}^{z_1-1} \cdots \int_1^{z[\alpha-1]^{-1}} \frac{dz_1 \cdots dz_{[\alpha]}}{z_1 \cdots z_{[\alpha]}}, & \text{if } \alpha > 1. \end{cases}$$

Therefore $\rho(\alpha)$ is a continuous and non-increasing function of α such that

$$\rho(\alpha) > e^{-\alpha\left(\ln \alpha + \ln \ln \alpha + \frac{6 \ln \ln \alpha}{\ln \alpha}\right)}.$$

Lemma 4. *Let R be the number of nth residue modulo p in the interval $1 \leq c \leq H$. If $n|(p-1), n \geq 2$ and $H > p^{\frac{1}{4}+\eta}(\eta > 0)$, then*

$$R = \frac{H}{n} + S,$$

where

$$|S| < c_9(\eta)Hp^{-\eta^2/6}.$$

Proof. Let $\chi(x) = e^{2\pi i \operatorname{Ind} x/n}$. Then by Corollary 2, it yields that

$$R = \sum_{x=1}^H \frac{1}{n} \sum_{a=1}^n e^{2\pi i a \operatorname{Ind} x/n} = \frac{H}{n} + \frac{1}{n} \sum_{a=1}^{n-1} \sum_{x=1}^H \chi(x)^a = \frac{H}{n} + S,$$

where

$$|S| < c_9(\eta)Hp^{-\eta^2/6}.$$

The lemma is proved.

Lemma 5. *Suppose that $n \geq 2$, $n|(p-1)$ and $\rho(\alpha) > \frac{1}{n} + \delta(\delta > 0)$. Then there exists $c_{10}(\alpha, \delta)$ such that $N(p,n) \leq p^{\frac{1}{4\alpha}}$, whenever $p > c_{10}(\alpha, \delta)$.*

Proof. Take $H = [p^{\frac{1}{4}+\eta}] + 1 \ (\eta > 0)$. If $N(p,n) > p^{\frac{1}{4\alpha}}$, then the integer in $1 < n \leq H$ such that it contains only the prime divisors $\leq p^{\frac{1}{4\alpha}}$ is an nth residue modulo p. If $\eta = \eta(\delta)$ is sufficiently small, we have

$$R \geq \Psi\left(H, p^{\frac{1}{4\alpha}}\right) \geq \rho((1+5\eta)\alpha)H + O\left(\frac{H}{\sqrt{\ln p}}\right)$$
$$> \left(\frac{1}{n} + \frac{\delta}{2}\right)H + O\left(\frac{H}{\sqrt{\ln p}}\right)$$

and

$$\frac{H}{n} + O\!\left(Hp^{-\eta^2/6}\right) \geq \left(\frac{1}{n} + \frac{\delta}{2}\right)H + O\!\left(\frac{H}{\sqrt{\ln p}}\right),$$

by Lemma 4, where the constant in "O" depends only on α and δ. This is impossible if p is large and therefore the lemma follows.

Now we proceed to prove Theorem 2 as follows:

(i) Take $\alpha = e^{\frac{n-1}{n} - \tau}\,(\frac{1}{2} > \tau > 0)$. Then

$$\rho(\alpha) \geq 1 - \ln\alpha = \frac{1}{n} + \tau.$$

Hence there exists $c_{11}(n,\tau)$ such that

$$N(p,n) \leq p^{4e^{\frac{1}{n-1} + \delta}},$$

whenever $p > c_{11}(n,\tau)$ and $n \geq 2$. Here $\delta > 0$ and $\lim_{\tau \to 0} \delta = 0$.

(ii) Take $\alpha = 3$. Then

$$\rho(3) = 1 - \ln 3 + \int_2^3 \int_1^{z_1 - 1} \frac{dz_1 dz_2}{z_1 z_2} > 0.4804 > \frac{1}{21}.$$

Hence there exists c_{12} such that

$$N(p,n) \leq p^{\frac{1}{12}}$$

holds for $p > c_{12}$ and $n \geq 21$.

(iii) Take $\alpha = \frac{\ln n}{\ln\ln n + 2}$. Then we can choose $\delta = \delta(n) > 0$ for $n > e^{33}$ such that

$$\rho(\alpha) > \frac{1}{n} + \delta.$$

Hence there is a $c_{13}(n)$ such that

$$N(p,n) \leq p^{\frac{\ln\ln n + 2}{4\ln n}}$$

holds for $p > c_{13}(n)$ and $n > e^{33}$.

4. Conditional Result

Lemma 6. $[4,11]$ *Under the assumption of GRH, we have*

$$\sum_{n=1}^{\infty} \Lambda(n)\chi(n)e^{-\frac{n}{x}} = \begin{cases} x + O(x^{\frac{1}{2}} \ln p), & \text{if } \chi = \chi_0 \text{ is principal character modulo } p, \\ O(x^{\frac{1}{2}} \ln p), & \text{otherwise.} \end{cases}$$

Hereafter the constant in "O" is an absolute constant.

Theorem 3. *Under the assumption of GRH, we have*

$$N(p, n) = O(\ln^2 p), \quad n \geq 2.$$

Proof. Since there is a c_{14} such that

$$\Psi(x) = \sum_{n \leq x} \Lambda(n) \leq c_{14} x, \quad x \geq 2.$$

We derive from summation by parts that

$$\sum_{n > c_{15} x} \Lambda(n) e^{-\frac{n}{x}} \leq c_{14} e^{-c_{15}} (c_{15} + 1) x$$

holds for a constant c_{15}. Consider now

$$R(x) = \sum_{r_n \leq c_{15} x} \chi_0(r_n) \Lambda(r_n) e^{-\frac{r_n}{x}},$$

where $\sum_{r_n \leq c_{15} x}$ denotes a sum where r_n runs over all positive nth non-residues modulo p. Hence

$$R(x) \geq \sum_{r_n \geq 1} \chi_0(r_n) \Lambda(r_n) e^{-\frac{r_n}{n}} - \sum_{m > c_{15} x} \Lambda(m) e^{-\frac{m}{x}}$$

$$\geq \sum_{m=1}^{\infty} \chi_0(m) \Lambda(m) e^{-\frac{m}{x}} \left(1 - \frac{1}{n} \sum_{a=1}^{n} e^{2\pi i a \operatorname{Ind} m/n} \right) - c_{14} e^{-c_{15}} (c_{15} + 1) x$$

$$= \left(1 - \frac{1}{n} \right) \sum_{m=1}^{\infty} \chi_0(m) \Lambda(m) e^{-\frac{m}{x}} - \frac{1}{n} \sum_{a=1}^{n-1} \sum_{m=1}^{\infty} \Lambda(m) \chi_0(m) e^{2\pi i a \operatorname{Ind} m/n} e^{-\frac{m}{x}}$$

$$- c_{14} e^{-c_{15}} (c_{15} + 1) x$$

$$\geq \left(1 - \frac{1}{n} - c_{14} e^{-c_{15}} (c_{15} + 1) \right) x + O(x^{\frac{1}{2}} \ln p).$$

Take $x = c_{16} \log^2 p$. Then

$$R(x) > 0,$$

whenever c_{15} and c_{16} are sufficiently large, i.e.

$$N(p, n) \leq c_{15} c_{16} \ln^2 p.$$

The theorem is proved.

References

[1] A. Weil, Sur les courbes algébriques et les variétés qui s'en déduisent, *Publ. Inst. Math. Strasbourg* (N. S. Nr. 2) (1948) 1–85.
[2] D. A. Burgess, The distribution of quadratic residues and non-residues, *Mathematica Lond.* **4** (1957) 106–112.

[3] P. Pólya, Über die Verteilung der Quadratischen Reste und Nichtreste, *Nachr. Ges. Wiss. Göttingen* (1918) 21–29.

[4] Wang Yuan, On the least primitive root of a prime, *Acta. Math. Sin.* **4** (1959) 432–441.

[5] D. A. Burgess, On character sums and primitive roots, *Proc. Lond. Math. Soc.* **XII** (1962) 179–192.

[6] D. A. Burgess, On character sums and *L*-series, *Proc. Lond. Math. Soc.* **XII** (1962) 193–206.

[7] Hua Loo-keng, On the least solution of Pell's equation, *Bull. Amer. Math. Soc.* **48** (1942) 731–735.

[8] I. M. Vinogradov, On the bound of least n-th non-residue, *Dokl. Akad. Nauk* **20** (1926) 47–58.

[9] A. A. Buchstab, On estimations of the numbers in an arithmetic progression whose prime divisors being less than a given order, *Dokl. Akad. Nauk SSSR* **67** (1947) 5–8.

[10] Hua Loo-keng, *Introduction to Number Theory* (Science Press, 1957).

[11] N. C. Ankeny, The least quadratic non-residue, *Ann. Math.* **55** (1952) 65–72.

ON THE MAXIMAL NUMBER OF PAIRWISE ORTHOGONAL LATIN SQUARE OF ORDER s (APPLICATION OF SIEVE METHOD)*

WANG YUAN

Institute of Mathematics, Academia Sinica

Received 25 September 1964

1. Introduction

In this paper, we shall give in detail the proofs of results announced in [1] with a slight modifications.

Let a set of s distinct symbols, for example, $1, 2, \ldots, s$ be arranged in an $s \times s$ square in such a way that every symbol occurs exactly once in every row and once in every column. Such a square is called a Latin square of order s. Two Latin squares are called orthogonal if, when one of the squares is superposed on the other, every symbol of the first square occurs with every symbol of the second square once and only once.

We denote by $N(s)$ the maximal number of mutually orthogonal Latin squares of order s. Euler proposed a famous conjecture on $N(s)$:

$$N(s) = 0,$$

whenever $s \geq 10$ and $s \equiv 2 \pmod 4$.

Bose, Shrikhande and Parker [2] made a great contribution on Euler's conjecture. They proved that

$$N(s) \geq 2 \quad (s > 6).$$

By the combination of their method and Brun's sieve method, Chowla, Erdös and Straus [3] established that there exists a constant s_0 such that

$$N(s) > \frac{1}{3} s^{\frac{1}{91}} \quad (s > s_0).$$

---

*Acta Mathematica Sinica, **16**:3 (1966) 400–410.

The author improved $\frac{1}{3}s^{\frac{1}{91}}$ to $s^{\frac{1}{28}}$ in [1]. In this paper we give the following

Theorem 1.   *There exists a constant s_1 such that*

$$N(s) > s^{\frac{1}{26}}, \quad s > s_1.$$

In Sec. 2, we shall generalise the method of Chowla, Erdös and Straus, and give the relation between $N(s)$ and sieve method. Theorem 2 shows that the estimation of $N(s)$ depends on the estimations of $P_\omega(x, q; \xi, \eta)$ $(\xi < \eta)$ and $P_\omega(x, q; \xi)$ (see Sec. 2 for the definitions). The upper and lower estimations of $P_\omega(x, q; \xi)$ may be obtained directly by the use of sieve method, and then the upper and lower estimations of $P_\omega(x, q; \xi, \eta)$ are derived by the simple inequalities

$$P_\omega(x, q; \eta) \leq P_\omega(x, q; \xi, \eta) \leq P_\omega(x, q; \xi), \quad \xi \leq \eta. \tag{1}$$

In this paper, we obtain a more precise estimation for $P_\omega(x, q; \xi)$ by the combination of the sieve methods of Brun, Buchstab and Selberg instead of the simple application of Brun's method (see Theorem 4). The combination of these methods first appeared in author's works [4–7] on Goldbach conjecture. In this way, we can establish that if s is sufficiently large, then

$$N(s) > s^{\frac{1}{34}}.$$

To obtain more precise result, we need to give a more accurate estimation on $P_\omega(x, q; \xi, \eta)$ $(\xi < \eta)$. Hence we introduce in Sec. 3 a recurrence formula (i.e. Theorem 3) such that we can obtain more precise upper and lower estimations for $P_\omega(x, q; \xi, \eta)$ from the results of $P_\omega(x, q; \xi)$ (see Theorems 5 and 6), and consequently we have Theorem 1.

In this paper, we use $c_1, c_2, \ldots$ to denote absolute constants and $p, p', p_1, p_2, \ldots$ prime numbers. The constants in "O" and "o" are absolute constants.

2. $N(s)$ and Sieve Methods

Let x be a positive integer and

$$(\omega) \qquad a, q; a_i, b_i, \quad i = 1, 2, \ldots$$

be a set of integers satisfying

$$2 \mid q, \qquad q = O(x^{c_1}) \ \ (0 \leq c_1 < 1), \ \ q_1 \cdots q_t = O(1),$$
$$a_i \not\equiv b_i \, (\mathrm{mod}\, p_i) \quad (i = 1, 2, \ldots),$$

where $2 = q_1 < q_2 < \cdots < q_t$ are all prime divisors of q and $p_1 < p_2 < \cdots$ the prime numbers not dividing q. Let $P_\omega(x, q; \xi, \eta)$ $(\xi \leq \eta)$ be the number of integers n

satisfying the following conditions

$$n \not\equiv b_i \,(\mathrm{mod}\, p_i) \quad (p_i \leq \eta). \tag{2}$$

In particular

$$P_\omega(x, q; \xi, \xi) = P_\omega(x, q; \xi), \tag{3}$$

$$P_\omega(x, 2; \xi, \eta) = P_\omega(x; \xi, \eta). \tag{4}$$

Theorem 2. *Suppose that*

$$P_\omega\left(x, q; \left(\frac{2x}{q}\right)^{\frac{1}{a}}\right) > c_2 \frac{c(q)x}{q \ln^2 \frac{x}{q}} + o\left(\frac{x}{q \ln^2 x}\right) \tag{5}$$

and

$$P_\omega\left(x; x^{\frac{1}{c}}, x^{\frac{1}{b}}\right) > c_3 \frac{c(2)x}{\ln^2 x} + o\left(\frac{x}{\ln^2 x}\right) \tag{6}$$

hold uniformly in (ω), where $q = O(x^{\frac{1}{a+1}})$, a, b, c are constants satisfying $a > 2$, $b > 2$, $(a+1)b \geq c \geq b$, and

$$c(q) = 4e^{-2\gamma} \prod_{\substack{p|q \\ p > 2}} \frac{p}{p-2} \prod_{p>2} \left(1 - \frac{1}{(p-1)^2}\right) = O(1), \tag{7}$$

in which γ denotes Euler constant. Then for any positive number ε there exists a constant $s_0 = s_0(\varepsilon)$ such that

$$N(s) > s^{\frac{1}{(a+b)(b+2)} - \varepsilon}. \tag{8}$$

whenever $s > s_0$.

To prove Theorem 2, we need the following two theorems:

Theorem A. [2] *If $1 \leq k \leq N(m) + 1$ and $1 < u < m$, then*

$$N(km + u) \geq \min\{N(k), N(k+1), 1 + N(m), 1 + N(u)\} - 1.$$

Theorem B. [8] (i) $N(uv) \geq \min\{N(u), N(v)\}$,
  (ii) $N(p^r) = p^r - 1.$

Proof of Theorem 2. (1) Suppose $2/s$. Take k such that

$$1 \le k \le \left(\frac{s}{2}\right)^{\frac{1}{b+2}}, \quad k \equiv -1 \left(\bmod \, 2^{\left[\frac{1}{(a+1)(b+2)} \log_2 \frac{s}{2}\right]}\right),$$

$$k \not\equiv 0, \quad -1 \, (\bmod \, p) \left(3 \le p \le \left(\frac{s}{2}\right)^{\frac{1}{(a+1)(b+2)}}\right). \tag{9}$$

Then it follows from (5) that the number of k satisfying (9) is not less than

$$c_4 \frac{s^{\frac{a}{s(a+1)(b+2)}}}{\ln^2 s} > 1,$$

if s is sufficiently large, and therefore from Theorem B, we have

$$N(k+1) > 2^{\left[\frac{1}{(a+1)(b+2)} \log_2 \frac{s}{2}\right]} - 1 > s^{\frac{1}{(a+1)(b+2)} - \varepsilon} + 1, \tag{10}$$

$$N(k) > \left(\frac{s}{2}\right)^{\frac{1}{(a+1)(b+2)}} - 1 > s^{\frac{1}{(a+1)(b+2)} - \varepsilon} + 1, \tag{11}$$

if s is large. Set $s = s_1 + s_2 k, 0 < s_1 \le k$ and $u = s_1 + u_1 k$, where u_1 satisfies the following conditions:

$$1 \le u_1 \le \left(\frac{s}{2}\right)^{\frac{b}{b+2}}, \tag{12}$$

$$u_1 \equiv s_1 + 1 \, (\bmod \, 2), \tag{13}$$

$$u_1 \not\equiv -s_1 \bar{k}_p \, (\bmod \, p) \left(3 \le p \le \left(\frac{s}{2}\right)^{\frac{1}{(a+1)(b+2)}}\right), \tag{14}$$

$$u_1 \not\equiv s_2 \, (\bmod \, p) \left(3 \le p \le \left(\frac{s}{2}\right)^{\frac{1}{b+2}}\right), \tag{15}$$

in which $\bar{k}_p$ denotes the solution of $ky \equiv 1 \, (\bmod \, p)$. The existence and uniqueness of $\bar{k}_p \, (\bmod \, p)$ for $3 \le p \le \left(\frac{s}{2}\right)^{\frac{1}{(a+1)(b+2)}}$ are followed by (9).

Suppose $3 \le p \le \left(\frac{s}{2}\right)^{\frac{1}{(a+1)(b+2)}}$. If $p|s$, then $s_2 \equiv -s_1 \bar{k}_p \, (\bmod \, p)$. Otherwise $s_2 \not\equiv -s_1 \bar{k}_p \, (\bmod \, p)$. Hence we may choose a set of integers (ω) such that the number of integers u_1 satisfying (12)–(15) is not less than

$$P_\omega \left(\left(\frac{s}{2}\right)^{\frac{b}{b+2}}; \left(\frac{s}{2}\right)^{\frac{1}{(a+1)(b+2)}}, \left(\frac{s}{2}\right)^{\frac{1}{b+2}}\right).$$

It is clearly

$$P_\omega \left(\left(\frac{s}{2}\right)^{\frac{b}{b+2}}; \left(\frac{s}{2}\right)^{\frac{b}{c(b+2)}}, \left(\frac{s}{2}\right)^{\frac{1}{b+2}}\right)$$

$$\le P_\omega \left(\left(\frac{s}{2}\right)^{\frac{b}{b+2}}; \left(\frac{s}{2}\right)^{\frac{1}{(a+1)(b+2)}}, \left(\frac{s}{2}\right)^{\frac{1}{b+2}}\right),$$

whenever $b \leq c \leq (a+1)b$. Hence we derive by (6) that the number of u_1 satisfying (12)–(15) is not less than

$$c_s \frac{s^{\frac{b}{b+2}}}{\ln^2 s} > 1,$$

if s is sufficiently large.

We know $2 \nmid u$ by (13) and $p \nmid u$ by (14) if $3 \leq p \leq \left(\frac{s}{2}\right)^{\frac{1}{(a+1)(b+2)}}$. Therefore it yields by Theorem B that

$$N(u) > \left(\frac{s}{2}\right)^{\frac{1}{(a+1)(b+2)}} - 1 > s^{\frac{1}{(a+1)(b+2)} - \varepsilon} \tag{16}$$

holds if s is large. Let $m = \frac{s-u}{k} = s_2 - u$. Then

$$m > s^{\frac{b+1}{b+2}} - \left[\left(\frac{s}{2}\right)^{\frac{b}{b+2}} + 1\right] > \left(\frac{s}{2}\right)^{\frac{1}{b+2}}\left[\left(\frac{s}{2}\right)^{\frac{b}{b+2}} + 1\right] \geq u > 1, \tag{17}$$

if s is large. Since $2 \nmid u$, $2 \nmid k$ and $2 \mid s$, we have $2 \nmid m$. It follows from (15) that the prime divisors of m are all $> k$, and therefore by Theorem B, we obtain

$$N(m) \geq k > N(k). \tag{18}$$

From (10), (11), (16)–(18) and Theorem A, we derive that there exists a constant $s_0 = s_0(\varepsilon)$ such that

$$N(s) > s^{\frac{1}{(a+1)(b+2)} - \varepsilon}, \quad s > s_0.$$

2) Suppose $2 \nmid s$. Take k satisfying (9)–(11). Let $s = s_1 + s_2(k+1), 0 < s_1 < (k+1)^\delta$ and $u = s_1 + u_1(k+1)$, where u_1 satisfies the following:

$$1 \leq u_1 \leq \left(\frac{s}{2}\right)^{\frac{b}{b+2}}, \tag{19}$$

$$u_1 \equiv s_2 + 1 \pmod 2, \tag{20}$$

$$u_1 \not\equiv -s_1 \overline{(k+1)}_p \pmod p \left(3 \leq p \leq \left(\frac{s}{2}\right)^{\frac{1}{(a+1)(b+2)}}\right), \tag{21}$$

$$u_1 \not\equiv s_2 \pmod p \left(3 \leq p \leq \left(\frac{s}{2}\right)^{\frac{1}{b+2}}\right). \tag{22}$$

Since $2 \mid (k+1)$ and $2 \nmid s$, we have $2 \nmid s$ and so $2 \nmid u$. We know also that the prime divisors of u are all $> \left(\frac{s}{2}\right)^{\frac{1}{(a+1)(b+2)}}$ by (21). Let $m = \frac{s-u}{k+1} = s_2 - u_1$. Then by (20) and (22), we know that the prime divisors of m are all $> k$. On the other hand,

$$m > u > 1$$

if s is sufficiently large. The theorem follows from Theorem A.

3. A Recurrent Formula

The aim of this section is to prove the following

Theorem 3. *Let α, β be two numbers satisfying $2 \le \alpha < \beta \le 15$. If there exist two non-negative functions $\lambda(u, v)$ and $\Lambda(u, v)$ $(0 < v \le u \le 15)$ with the properties that if one of the two variables u or v is fixed, these functions are increasing with only finite number of discontinuities in another variable such that*

$$P_\omega(x; x^{\frac{1}{\beta}}, x^{\frac{1}{\alpha}}) \ge \lambda(\beta, \alpha) \frac{cx}{\ln^2 x} + O\left(\frac{x}{\ln^{2.5} x}\right) \tag{23}$$

and

$$P_\omega(x; x^{\frac{1}{\beta}}, x^{\frac{1}{\alpha}}) \le \Lambda(\beta, \alpha) \frac{cx}{\ln^2 x} + O\left(\frac{x}{\ln^{2.5} x}\right) \tag{24}$$

hold, where

$$c = 2e^{-2\gamma} \prod_{p>2} \left(1 - \frac{1}{(p-1)^2}\right), \tag{25}$$

in which γ is Euler constant, then the functions

$$\lambda_1(\beta, \alpha) = \max\left(0, \lambda(\beta, \delta) - \int_{a-1}^{\delta-1} \Lambda\left(\frac{\beta z}{z+1}, z\right) \frac{z+1}{z^2} dz\right) \tag{26}$$

and

$$\Lambda_1(\beta, \alpha) = \Lambda(\beta, \delta) - \int_{a-1}^{\delta-1} \lambda\left(\frac{\beta z}{z+1}, z\right) \frac{z+1}{z^2} dz \tag{27}$$

have the same properties as $\lambda(\beta, \alpha)$ and $\Lambda(\beta, \alpha)$, where δ is a number with $\alpha \le \delta \le \beta$.

To prove Theorem 3, we need the following two lemmas.

Lemma 1. *Suppose $\xi \le \zeta \le \eta$. There exists a sequence of sets of integers (ω_{n_1})'s such that*

$$P_\omega(x; \xi, \eta) = P_\omega(x; \xi, \zeta) - \sum_{\zeta < p_{n_1} \le \eta} P_{\omega_{n_1}}\left(\frac{x - b_{n_1}}{p_{n_1}}; \xi, p_{n_1-1}\right). \tag{28}$$

Proof. By Eratosthenes sieve we obtain

$$P_\omega(x; \xi, \eta) = P_\omega(x; \xi, \zeta) - \sum_{\zeta < p_{n_1} \le \eta} P_{\tilde{\omega}_{p_{n_1}}}(x; \xi, \zeta)$$

$$+ \sum_{\zeta < p_{n_1} \le \eta} \sum_{\zeta < p_{n_2} < p_{n_1}} P_{\tilde{\omega}_{p_{n_1} p_{n_2}}}(x; \xi, \zeta)$$

$$- \sum_{\zeta < p_{n_1} \le \eta} \sum_{\zeta < p_{n_2} < p_{n_1}} \sum_{\zeta < p_{n_3} < p_{n_2}} p_{\tilde{\omega}_{p_{n_1} p_{n_2} p_{n_3}}}(x; \xi, \zeta) + \cdots, \tag{29}$$

where $p_{\tilde{\omega}_{p_{n_1} \cdots p_{n_t}}}(x; \xi, \zeta)$ denotes the number of integers satisfying

$$0 \le n \le x, n \equiv a \,(\mathrm{mod}\, 2), \quad n \equiv b_{n_1} \,(\mathrm{mod}\, p_{n_1}), \ldots, \quad n \equiv b_{n_t} \,(\mathrm{mod}\, p_{n_t}),$$

$$n \not\equiv a_i \,(\mathrm{mod}\, p_i), \quad n \not\equiv b_i \,(\mathrm{mod}\, p_i) \ (p_i \le \zeta). \tag{30}$$

We may assume that $0 \le a_i, \ b_i < p_i$. Let $\tilde{a}_{n_1}, a_{in_1}$ and b_{in_1} be the respective solutions of congruences

$$mp_{n_1} + b_{n_1} \equiv a \,(\mathrm{mod}\, 2),$$

$$mp_{n_1} + b_{n_1} \equiv a_i \,(\mathrm{mod}\, p_i)$$

and

$$mp_{n_1} + b_{n_1} \equiv b_i \,(\mathrm{mod}\, p_i)$$

whenever $p_i \ne p_{n_1}$. Let

$$(\omega_{n_1}) \qquad \tilde{a}_{n_1}; a_{in_1}, b_{in_1} \qquad i \ne n_1$$

and $p_{\tilde{\omega}_{n_1 p_{n_2} \cdots p_{n_t}}}(x; \xi, \zeta)$ be the number of integers satisfying

$$0 \le n \le x, n \equiv \tilde{a}_{n_1} \,(\mathrm{mod}\, 2), \quad n \equiv b_{n_2 n_1} \,(\mathrm{mod}\, p_{n_2}), \ldots, \quad n \equiv b_{n_t n_1} \,(\mathrm{mod}\, p_{n_t}),$$

$$n \not\equiv a_{in_1} \,(\mathrm{mod}\, p_i) \ (p \le \xi), \quad n \not\equiv b_{in_1} \,(\mathrm{mod}\, p_i) \ (p_i \le \zeta). \tag{31}$$

Then

$$P_{\tilde{\omega}_{p_{n_1}\cdots p_{n_t}}}(x;\xi,\zeta) = P_{\tilde{\omega}_{n_1,p_{n_2}\cdots p_{n_t}}}\left(\frac{x-b_{n_1}}{p_{n_1}};\xi,\zeta\right) \tag{32}$$

and consequently

$$P_{\tilde{\omega}_{p_{n_1}}}(x;\xi,\zeta) - \sum_{\zeta<p_{n_2}<p_{n_1}} P_{\tilde{\omega}_{p_{n_1}p_{n_2}}}(x;\xi,\zeta)$$

$$+ \sum_{\zeta<p_{n_2}<p_{n_1}}\sum_{\zeta<p_{n_3}<p_{n_2}} P_{\tilde{\omega}_{p_{n_1}p_{n_2}p_{n_3}}}(x;\xi,\zeta) - \cdots$$

$$= P_{\omega_{n_1}}\left(\frac{x-b_{n_1}}{p_{n_1}};\xi,\zeta\right) - \sum_{\zeta<p_{n_2}<p_{n_1}} P_{\tilde{\omega}_{n_1,p_{n_2}}}\left(\frac{x-b_{n_1}}{p_{n_1}};\xi,\zeta\right)$$

$$+ \sum_{\zeta<p_{n_2}<p_{n_1}}\sum_{\zeta<p_{n_3}<p_{n_2}} P_{\tilde{\omega}_{n_1,p_{n_2}p_{n_3}}}\left(\frac{x-b_{n_1}}{p_{n_1}};\xi,\zeta\right) - \cdots$$

$$= P_{\omega_{n_1}}\left(\frac{x-b_{n_1}}{p_{n_1}};\xi,p_{n_1}-1\right). \tag{33}$$

Substituting (33) into (29), we have the lemma.

Lemma 2. [9] *Let* α,β *be two numbers such that* $\beta>\alpha>1$. *Then*

$$\sum_{x^{\frac{1}{\beta}}<p\le x^{\frac{1}{\alpha}}} \frac{1}{p\ln^2\frac{x}{p}} = \left(\int_{\alpha-1}^{\beta-1}\frac{z+1}{z^2}dz\right)\frac{1}{\ln^2 x}+O\left(\frac{1}{\ln^3 x}\right) \quad x\ge 2. \tag{34}$$

Proof of Theorem 3. Set $\xi=x^{\frac{1}{\beta}},\zeta=x^{\frac{1}{\delta}}$ and $\eta=x^{\frac{1}{\alpha}}$ in Lemma 1. Let $n=[\sqrt{\ln x}]$ and $u_l=\alpha+\frac{\beta-\alpha}{n}l-1$ $(0\le l\le n)$. Then

$$T_l = \sum_{x^{\frac{1}{u_{l+1}+1}}<p_{n_1}\le x^{\frac{1}{u_l+1}}} P_{\omega_{n_1}}\left(\frac{x-b_{n_1}}{p_{n_1}};x^{\frac{1}{\beta}},p_{n_1}-1\right)$$

$$= \sum_{x^{\frac{1}{u_{l+1}+1}}<p\le x^{\frac{1}{u_l+1}}} P_{\omega_{n_1}}\left(\frac{x}{p};x^{\frac{1}{\beta}},p\right)+O(x^{\frac{1}{\alpha}})+O(x^{1-\frac{1}{\delta}})$$

$$= \tilde{T}_l + O(x^{\frac{1}{\alpha}})+O(x^{1-\frac{1}{\delta}}), \tag{35}$$

$$\tilde{T}_l = \sum_{x^{\frac{1}{u_{l+1}+1}} < p \le x^{\frac{1}{u_l+1}}} P_{\omega_{n_1}}\left(\frac{x}{p}; \left(\frac{x}{p}\right)^{\frac{\ln x}{\beta \ln \frac{x}{p}}}, \left(\frac{x}{p}\right)^{\frac{\ln p}{\ln \frac{x}{p}}}\right)$$

$$\le \sum_{x^{\frac{1}{u_{l+1}+1}} < p \le x^{\frac{1}{u_l+1}}} \Lambda\left(\frac{\beta \ln \frac{x}{p}}{\ln x}, \frac{\ln \frac{x}{p}}{\ln p}\right) \frac{cx}{p \ln^2 \frac{x}{p}} + O\left(\frac{x}{\ln^{2.5} x} \int_{ul}^{u_{l+1}} \frac{z+1}{z^2} dz\right)$$

$$\le \Lambda\left(\frac{\beta u_{l+1}}{u_{l+1}+1}, u_{l+1}\right) \sum_{x^{\frac{1}{u_{l+1}+1}} < p \le x^{\frac{1}{u_l+1}}} \frac{cx}{p \ln^2 \frac{x}{p}} + O\left(\frac{x}{\ln^3 x}\right)$$

$$\le \left[\Lambda\left(\frac{\beta u_{l+1}}{u_{l+1}+1}, u_{l+1}\right) \int_{ul}^{u_{l+1}} \frac{z+1}{z^2} dz\right] \frac{cx}{\ln^2 x} + O\left(\frac{x}{\ln^3 x}\right). \tag{36}$$

Since

$$\sum_{l=0}^{n-1} \Lambda\left(\frac{\beta u_{l+1}}{u_{l+1}+1}, u_{l+1}\right) \int_{ul}^{u_{l+1}} \frac{z+1}{z^2} dz - \int_{\alpha-1}^{\delta-1} \Lambda\left(\frac{\beta z}{z+1}, z\right) \frac{z+1}{z^2} dz$$

$$= \sum_{l=0}^{n-1} \int_{ul}^{u_{l+1}} \left[\Lambda\left(\frac{\beta u_{l+1}}{u_{l+1}+1}, u_{l+1}\right) - \Lambda\left(\frac{\beta z}{z+1}, z\right)\right] \frac{z+1}{z^2} dz$$

$$\le \sum_{l=0}^{n-1} \left[\Lambda\left(\frac{\beta u_{l+1}}{u_{l+1}+1}, u_{l+1}\right) - \Lambda\left(\frac{\beta u_l}{u_l+1}, u_l\right)\right] \max_{0 \le l \le n} \int_{ul}^{u_{l+1}} \frac{z+1}{z^2} dz$$

$$= O\left(\frac{1}{n}\right) = O\left(\frac{1}{\sqrt{\ln x}}\right), \tag{37}$$

we have

$$\sum_{x^{\frac{1}{\delta}} < p_{n_1} \le x^{\frac{1}{\alpha}}} P_{\omega_{n_1}}\left(\frac{x - b_{n_1}}{p_{n_1}}; x^{\frac{1}{\beta}}, p_{n_1-1}\right)$$

$$= \sum_{l=0}^{n-1} T_l$$

$$= \sum_{l=0}^{n-1} \tilde{T}_l + O(x^{\frac{1}{\alpha}} \sqrt{\ln x}) + O(x^{1-\frac{1}{\delta}} \sqrt{\ln x})$$

$$\le \left(\sum_{l=0}^{n-1} \Lambda\left(\frac{\beta u_{l+1}}{u_{l+1}+1}, u_{l+1}\right) \int_{ul}^{u_{l+1}} \frac{z+1}{z^2} dz\right) \frac{cx}{\ln^2 x} + O\left(\frac{x}{\ln^{2.5} x}\right)$$

$$= \left(\int_{\alpha-1}^{\delta-1} \Lambda\left(\frac{\beta z}{z+1}, z\right) \frac{z+1}{z^2} dz\right) \frac{cx}{\ln^2 x} + O\left(\frac{x}{\ln^{2.5} x}\right), \tag{38}$$

and therefore

$$P_\omega(x; x^{\frac{1}{\beta}}, x^{\frac{1}{\alpha}}) \ge \max\left(0, \lambda(\beta, \delta) - \int_{\alpha-1}^{\delta-1} \Lambda\left(\frac{\beta z}{z+1}, z\right) \frac{z+1}{z^2} dz\right) \frac{cx}{\ln^2 x} + O\left(\frac{x}{\ln^{2.5} x}\right)$$

by Lemma 1. On the other hand, since

$$\tilde{T}_l \geq \left[\lambda\left(\frac{\beta u_l}{u_{l+1}}, u_l\right) \int_{u_l}^{u_{l+1}} \frac{z+1}{z^2}\, dz \right] \frac{cx}{\ln^2 x} + O\left(\frac{x}{\ln^3 x}\right)$$

and

$$\int_{\alpha-1}^{\delta-1} \lambda\left(\frac{\beta z}{z+1}, z\right) \frac{z+1}{z^2}\, dz - \sum_{l=0}^{n-1} \lambda\left(\frac{\beta u_l}{u_l+1}, u_1\right) \int_{u_l}^{u_{l+1}} \frac{z+1}{z^2}\, dz = O\left(\frac{1}{\sqrt{\ln x}}\right)$$

we have

$$\sum_{x^{\frac{1}{\delta}} < p_{n_1} \leq x^{\frac{1}{\alpha}}} P_{\omega n_1}\left(\frac{x - b_{n_1}}{p_{n_1}}; x^{\frac{1}{\beta}}, p_{n_1-1}\right)$$

$$\geq \left(\int_{\alpha-1}^{\delta-1} \lambda\left(\frac{\beta z}{z+1}, z\right) \frac{z+1}{z^2}\, dz \right) \frac{cx}{\ln^2 x} + O\left(\frac{x}{\ln^{2.5} x}\right)$$

and therefore

$$P_\omega(x; x^{\frac{1}{\beta}}, x^{\frac{1}{\alpha}}) \leq \left(\Lambda(\beta, \delta) - \int_{\alpha-1}^{\delta-1} \lambda\left(\frac{\beta z}{z+1}, z\right) \frac{z+1}{z^2}\, dz \right) \frac{cx}{\ln^2 x} + O\left(\frac{x}{\ln^{2.5} x}\right),$$

by Lemma 1. (27) follows and the theorem is proved.

Remark. Since

$$P_\omega(x; \eta, \eta) \leq P_\omega(x; \xi, \eta) \leq P_\omega(x; \xi, \xi), \quad \xi \leq \eta,$$

the functions

$$\lambda(u, v) = \lambda(v, v) = \lambda(v)$$

and

$$\Lambda(u, v) = \Lambda(u, u) = \Lambda(u)$$

clearly satisfied the requirements in Theorem 3 (see [4–7]).

4. Estimation of $P_\omega(x, q; \xi)$

We use the notations in Secs. 2 and 3. The aim of this section is to prove the following:

Theorem 4. *The inequality*

$$P_\omega\left(x, q; \left(\frac{2x}{q}\right)^{\frac{1}{4.31}}\right) > 0.05 \frac{c(q)x}{x \ln^2 \frac{x}{q}} + O\left(\frac{x \ln \ln x}{q \ln^3 x}\right), \tag{39}$$

holds uniformly in (ω), where $q = O(x^{\frac{1}{3}})$.

Proof. 1) It follows by Selberg's method that the estimation

$$P_\omega\left(x, q; \left(\frac{2x}{q}\right)^{\frac{1}{d}}\right) \leq \Lambda(d)\frac{c(q)x}{q\ln^2\frac{x}{q}} + O\left(\frac{x\ln\ln x}{q\ln^3 x}\right),\tag{40}$$

holds uniformly in (ω) and $q = O(x^{\frac{2}{3}})$, where

$$\Lambda(d) = \begin{cases} 2e^{2\gamma}\left[\dfrac{d^2}{(d-1)^2 - \frac{d^2}{2}\ln\frac{d}{2}}\right], & 2 \leq d \leq 4, \\[3ex] 2e^{2\gamma}\left[\dfrac{d^2}{(d-1)^2 - \frac{d^2}{2}\ln\frac{d}{2} + \delta\left(\frac{d}{2}\right) - \eta}\right], & 4 \leq d \leq 6, \end{cases}\tag{41}$$

in which η is a preassigned positive number and

$$\delta(d) = \lim_{\xi\to\infty}\frac{4}{\ln^2\xi}\sum_{\substack{\xi < p < p' \\ pp' \leq \xi^d}}\frac{1}{pp'}\ln^2\frac{\xi^d}{pp'}, \quad d \geq 2.\tag{42}$$

$\Lambda(d)$ has a similar expression for $d > 6$, and we may take $\Lambda(d) = \Lambda(2)$ for $0 < d \leq 2$.

We refer to Wang Yuan [4–7] for the proof of these formulas but we should take $\xi = \left(\frac{2x}{q}\right)^{\frac{1}{d}}/\log^5 x$ here.

2) By Brun's method, we know that the estimation

$$P_\omega\left(x, q; \left(\frac{2x}{q}\right)^{\frac{1}{15}}\right) > 224.9997\frac{c(q)x}{q\ln^2\frac{x}{q}} + O\left(\frac{x}{q\ln^3 x}\right)\tag{43}$$

holds in (ω) and $q = O(x^{\frac{2}{3}})$.

We refer to Buchstab [10, 11] for the proof, but we should make the following changes: p_r is the largest prime number $\leq(\frac{2x}{q})^{\frac{1}{15}}, p_{r_3} = p_{r_2} = p_{r_1} = p_r$, and $p_{r_k}(4 \leq k \leq t+1)$ are the respective largest primes $\leq(\frac{2x}{q})^{\frac{1}{15h^{k-3}}}$, where $h = \frac{5}{4} + \delta$ and δ is any preassigned positive number.

We may prove similarly

$$P_\omega\left(x, q; \left(\frac{2x}{q}\right)^{\frac{1}{14}}\right) < 196.0022\frac{c(q)x}{q\ln^2\frac{x}{q}} + O\left(\frac{x}{q\ln^3 x}\right),$$

$$P_\omega\left(x, q; \left(\frac{2x}{q}\right)^{\frac{1}{10}}\right) > 98\frac{c(q)}{q\ln^2\frac{x}{q}} + O\left(\frac{x}{q\ln^3 x}\right),$$

and

$$P_\omega\left(x, q; \left(\frac{2x}{q}\right)^{\frac{1}{10}}\right) < 101.6\frac{c(q)x}{q\ln^2\frac{x}{q}} + O\left(\frac{x}{q\ln^3 x}\right).$$

3) Suppose that α, β are two real numbers satisfying $2 \leq \alpha < \beta \leq 15$ and that there are two non-negative and non-decreasing functions $\lambda(z)$ and $\Lambda(z)$ with the following properties: They have at most finite number of discontinuities and the formulas

$$P_\omega\left(x, q; \left(\frac{2x}{q}\right)^{\frac{1}{z}}\right) \geq \lambda(z)\frac{c(q)x}{q\ln^2\frac{x}{q}} + O\left(\frac{x\ln\ln x}{q\ln^3 x}\right) \tag{44}$$

and

$$P_\omega\left(x, q; \left(\frac{2x}{q}\right)^{\frac{1}{z}}\right) \leq \Lambda(z)\frac{c(q)x}{q\ln^2\frac{x}{q}} + O\left(\frac{x\ln\ln x}{q\ln^3 x}\right) \tag{45}$$

holds uniformly in (ω). Then

$$\lambda_1(\alpha) = \max\left(0, \lambda(\beta) - 2\int_{\alpha-1}^{\beta-1} \Lambda(z)\frac{z+1}{z^2}dz\right) \tag{46}$$

and

$$\Lambda_1(\alpha) = \Lambda(\beta) - 2\int_{\alpha-1}^{\beta-1} \lambda(z)\frac{z+1}{z^2}dz \tag{47}$$

have the same properties as $\lambda(z)$ and $\Lambda(z)$ respectively.

We refer to Buchstab [10, 11] for the proof. Note that $x^{\frac{1}{2}} \leq \frac{x}{p}$ if $\left(\frac{2x}{q}\right)^{\frac{1}{\beta}} \leq p \leq \left(\frac{2x}{q}\right)^{\frac{1}{\alpha}}$, and so $q = O\left(x^{\frac{1}{3}}\right) = O\left(\left(\frac{x}{p}\right)^{\frac{2}{3}}\right)$. The existence of the functions $\lambda(z)$ and $\Lambda(z)$ satisfying (44) and (45) follows from 1) and 2).

4) From 1)–3), we have the following table:

d	10	9	8	7	$\cdots$
$\lambda_0(d)$	99.98181	79.78469	60.88817	43.51554	$\cdots$
$\Lambda_0(d)$	100.02073	82.7207	68.52511	54.39352	$\cdots$

$$\tag{48}$$

(see [7, 10, 11]). Take $u_{t+1} - u_t = 0.01$. Then by 3) and the formulas

$$\int_{\alpha-1}^{\beta-1} \lambda(z)\frac{z+1}{z^2}dz \geq \sum_{t=0}^{n-1} \lambda(u_t) \int_{u_t}^{u_{t+1}} \frac{z+1}{z^2}dz \tag{49}$$

and

$$\int_{\alpha-1}^{\beta-1} \Lambda(z)\frac{z+1}{z^2}dz \leq \sum_{t=0}^{n-1} \Lambda(u_{t+1}) \int_{u_t}^{u_{t+1}} \frac{z+1}{z^2}dz \tag{50}$$

for iterations, where $\frac{\beta-\alpha}{n} = 0.01$, we obtain

d	10	9	8	$\cdots$	4.31	$\cdots$
$\lambda_{11}(d)$	99.98181	80.892035	63.59931	$\cdots$	0.05	$\cdots$
$\Lambda_{12}(d)$	100.02073	81.11841	64.403149	$\cdots$	30.821	$\cdots$

$$\tag{51}$$

The theorem is proved.

5. Estimation of $P_\omega(x; \xi, \eta)$

Theorem 5. *The inequality*

$$P_\omega(x; x^{\frac{1}{8}}, x^{\frac{1}{3.21}}) > 0.01 \frac{cx}{\ln^2 x} + O\left(\frac{x}{\ln^{2.5} x}\right) \tag{52}$$

holds uniformly in (ω).

Proof. Take $\beta = \delta = 8$ and $\alpha = 3.24$ in Theorem 3 and set

$$\Lambda(u, v) = \Lambda_{12}(u) \quad \text{and} \quad \lambda(u, v) = \Lambda_{11}(v).$$

Then

$$\lambda_{11}(8) - \int_{2.24}^{7} \Lambda_{12}\left(\frac{8z}{z+1}\right) \frac{z+1}{z^2} dz > 0.01. \tag{53}$$

The theorem follows.

Theorem 6. *The estimation*

$$P_\omega(x; x^{\frac{1}{10}}, x^{\frac{1}{2.84}}) > 0.29 \frac{cx}{\ln^2 x} + O\left(\frac{x}{\ln^{2.5} x}\right) \tag{54}$$

holds uniformly in (ω).

Proof. Let

$$\lambda_0(u, v) = \lambda_{11}(v) \quad \text{and} \quad \Lambda_0(u, v) = \Lambda_{12}(u).$$

Let $u_{t+1} - u_t = 0.01$ and $u_{t+1} \le \beta$. Then by Theorem 3 and the formulas

$$\lambda_{i+1}(\beta, u_t) = \lambda_{i+1}(\beta, u_{t+1}) - \int_{u_t-1}^{u_{t+1}-1} \Lambda_i\left(\frac{\beta z}{z+1}, z\right) \frac{z+1}{z^2} dz$$

$$\ge \lambda_{i+1}(\beta, u_{t+1}) - \Lambda_i\left(\frac{\beta(u_{t+1}-1)}{u_{t+1}}, u_{t+1}-1\right) \int_{u_t-1}^{u_{t+1}-1} \frac{z+1}{z^2} dz \tag{55}$$

and

$$\Lambda_{i+1}(\beta, u_t) = \Lambda_{i+1}(\beta, u_{t+1}) - \int_{u_t-1}^{u_{t+1}-1} \lambda_i\left(\frac{\beta z}{z+1}, z\right) \frac{z+1}{z^2} dz$$

$$\le \Lambda_{i+1}(\beta, u_{t+1}) - \lambda_i\left(\frac{\beta(u_t-1)}{u_t}, u_t-1\right) \int_{u_t-1}^{u_{t+1}-1} \frac{z+1}{z^2} dz \tag{56}$$

we have

(u, v)	$(8.05, 8)$	$(8.05, 7)$	$(8.05, 6)$	$\cdots$
$\lambda_1(u, v)$	64.011998	55.559551	46.582488	$\cdots$

(u, v)	$(7.05, 7)$	$(6.9, 6)$	$(6.76, 5)$	$\cdots$
$\Lambda_1(u, v)$	50.833685	43.869985	39.994537	$\cdots$

$$\tag{57}$$

and

$$
\begin{array}{c|c|c|c|c|c|c}
(u,v) & (10,9) & (10,8) & (10,7) & \cdots & (10,2.84) & \cdots \\
\hline
\lambda_2(u,v) & 89.912569 & 79.816805 & 69.635682 & \cdots & 0.29 & \cdots
\end{array}
$$

$$
\begin{array}{c|c|c|c|c}
(u,v) & (8.89,8) & (8.75,7) & (8.58,6) & \cdots \\
\hline
\Lambda_2(u,v) & 71.344245 & 61.677931 & 52.667582 & \cdots
\end{array}
\tag{58}
$$

The theorem is proved.

6. Proof of Main Theorem

Take $\varepsilon = \frac{1}{5.31 \times 5.24} - \frac{1}{28}$. Then by Theorems $2, 4$ and 5, it follows that

$$
N(s) > s^{\frac{1}{5.31 \times 5.24} - \varepsilon} = s^{\frac{1}{28}}, \tag{59}
$$

if s is sufficiently large. This is the main result announced in [1]. Then take $\varepsilon = \frac{1}{5.31 \times 4.84} - \frac{1}{26}$. By Theorems 2, 4 and 6, we obtain

$$
N(s) > s^{\frac{1}{5.31 \times 4.84} - \varepsilon} = s^{\frac{1}{26}}, \tag{60}
$$

if s is sufficiently large. Theorem 1 follows.

Remark. If $x^{\frac{1}{10}}$ is replaced by $x^{\frac{1}{12}}$ or $x^{\frac{1}{13}}$ in Theorem 6, then it is possible to establish $N(s) > s^{\frac{1}{25}}$ by Theorem 3 with more complicated numerical calculations.

Remark added on 7 April 1966. We know recently that Kenneth Rogers has proved

$$
N(s) > s^{\frac{1}{42}}, \quad s > s_0
$$

(see *Pacific J. Math.* **14** (1964) 1395–1397).

References

[1] Wang Yuan, A note on the maximal number of pairwise orthogonal Latin squares of a given order, *Sci. Sinica* **8** (1964) 841–843.

[2] R. C. Bose, S. S. Shrikhande and E. T. Parker, Further results on the construction of mutually orthogonal Latin squares and the falsity of Euler's conjecture, *Can. J. Math.* **7** (1960) 189–203.

[3] S. Chowla, P. Erdös and E. G. Straus, On the maximal number of pairwise orthogonal Latin squares of a given order, *Can. J. Math.* **7** (1960) 204–208.

[4] Wang Yuan, On the respresentation of large even integer as a sum of a product of at most 3 primes and a product of atmost 4 primes, *Acta Math. Sin.* **6**:3 (1956) 500–513.

[5] Wang Yuan, On the representation of large even integer as a sum of a prime and a product of at most 4 primes, *Acta Math. Sin.* **6**:4 (1956) 565–582.

[6] Wang Yuan, On the representation of large even number as a sum of two almost primes, *Sci. Rec. (New Series)* **1**:5 (1957) 15–19.

[7] Wang Yuan, On sieve methods and some of their applications, *Sci. Sinica* **8** (1959) 357–381.

[8] H. F. MacNeish, Euler squares, *Ann. Math.* **22** (1922) 221–227.

[9] A. A. Buchstab, Asymptotic estimation of a general number theoretic function, *Math. Sb.* **2** (1937) 1239–1245.

[10] A. A. Buchstab, New improvements on Eratosthenes sieve method, *Math. Sb.* **4** (1938) 375–387.

[11] A. A. Buchstab, On representation of even number as a sum of two integers with bounded number of divisors, *Dokl. Akad. Nauk SSSR* **29** (1940) 544–548.

Vol. XVIII No. 5   **SCIENTIA SINICA**   Sept. - Oct. 1975

Science Articles

ON THE REPRESENTATION OF EVERY LARGE EVEN INTEGER AS A SUM OF A PRIME AND AN ALMOST PRIME

PAN CHENG-DONG （潘承洞）

(*Department of Mathematics, Shandong University*)

DING XIA-XI （丁夏畦）   WANG YUÁN （王　元）

(*Institute of Mathematics, Academia Sinica*)

Received Oct. 21, 1974.

ABSTRACT

In this paper, we give a modified proof of Chen's theorem "every sufficiently large even integer is a sum of a prime and a product of at most 2 primes".

1. INTRODUCTION

For brevity, we denote the following proposition by $(1, a)$:

Every sufficiently large even integer is a sum of a prime and an almost prime of at most 2 prime factors.

The proposition was studied by T. Estermann[4], A. Renyi[18], Wang Yuán[9,10], М. Б. Барбан[11,12], Pan Cheng-dong[6,17], Б. В. Левин[16], А. А. Бухштаб[13,14], А. И. Виноградов[15], H. E. Richert[8] and Chen Jing-run[2,3] successively by means of sieve method and large sieve method. The best result is due to Chen. He proved

Theorem 1 $(1, 2)$

Chen gave the previous method an important improvement. Especially, he introduced and estimated the

$$\Omega = \sum_{\substack{(p_{1,2}) \\ p_3 \leqslant x/p_1 p_2 \\ x-p = p_1 p_2 p_3}} 1, \tag{1.1}$$

where p, p_1, p_2, p_3 are primes and $(p_{1,2})$ denotes the condition $x^{\frac{1}{10}} < p_1 \leqslant x^{\frac{1}{3}} \leqslant p_2 \leqslant \left(\dfrac{x}{p_1}\right)^{\frac{1}{2}}$.

The aim of the present paper is to give a modified proof of $(1, 2)$. First, we shall show that the estimation of Ω may be derived from the following mean value theorem. Let $2 \leqslant y \leqslant x$. Let

$$\pi(y, a, q, l) = \sum_{\substack{n \leqslant y/a \\ a_n \equiv l \,(\mathrm{mod}\, q)}} a_n$$

where

$$a_n = \begin{cases} 1, & \text{for prime } n, \\ 0, & \text{otherwise.} \end{cases}$$

Theorem 2. *For any given positive constant A and positive number $\varepsilon(<1)$, the estimation*

$$I = \sum_{q \leqslant x^{1/2}\log^{-B}x} \max_{y \leqslant x} \max_{(l,q)=1} \left| \sum_{\substack{A_1 < a \leqslant A_2 \\ (a,q)=1}} f(a)\left(\pi(y,a,q,l) - \frac{\mathrm{li}\,\dfrac{y}{a}}{\varphi(q)} \right) \right|$$
$$= O\left(\frac{x}{\log^A x} \right) \tag{1.2}$$

holds for $\log^{2B}y < A_1 \leqslant A_2 < y^{1-\varepsilon}$, where $|f(a)| \leqslant 1$, $B = A + 7$ and the constant implied by the symbol "O" depends only on ε and A.

Next, we use the comparatively elementary sieve method given in the previous work of one of the authors (Cf. [10]) to replace the Richert's method.

We have by the same way that

Theorem 3. *There exist infinitely many primes p such that $p + 2k$ is a product of at most 2 primes, where k is a given positive integer.*

Theorem 4. *Every sufficiently large odd integer N can be represented as $N = p + 2P^{(2)}$, where p is a prime and $P^{(2)}$ is an almost prime of not more than 2 prime divisors.*

The other famous problems concerning the distribution of almost primes can be treated similarly.

II.   The Proof of Theorem 2

To prove Theorem 2, we shall need

Theorem A (large sieve). *Let $1 < P < Q$. Let M and N be positive integers and let $b'_n s$ be any complex numbers. Then*

$$\sum_{P < q \leqslant Q} \frac{1}{\varphi(q)} \sum_{\chi_q}^{*} \left| \sum_{n=M+1}^{M+N} b_n \chi(n) \right|^2 \ll \left(Q + \frac{N}{P} \right) \sum_{n=M+1}^{M+N} |b_n|^2, \tag{2.1}$$

where $\displaystyle\sum_{\chi_q}^{*}$ denotes that the sum is over the primitive characters mod q.

For brevity, we omit the index q of χ_q.

Cf. P. X. Gallagher [5] and Chen Jing-Run [3].

The Proof of Theorem 2.

1)   Denote

$$D_1 = \log^B x, \quad D = x^{\frac{1}{2}}\log^{-B}x. \tag{2.2}$$

Then we have

$$\pi(y, a, q, l) = \sum_{\substack{n \leqslant y/a \\ a_n \equiv l (\mathrm{mod}\, q)}} a_n = \frac{1}{\varphi(q)} \sum_{\chi_q} \bar{\chi}(l)\chi(a) \sum_{n \leqslant y/a} a_n \chi(n), \qquad (2.3)$$

where $(a, q) = (l, q) = 1$ and $\sum\limits_{\chi_q}$ denotes a sum in which χ_q runs over all characters mod q.   Hence

$$\pi(y, a, q, l) - \frac{1}{\varphi(q)} \sum_{\substack{n \leqslant y/a \\ (n,q)=1}} a_n = \frac{1}{\varphi(q)} \sum_{\chi_q \neq \chi_0} \bar{\chi}(l)\chi(a) \sum_{n \leqslant y/a} a_n \chi(n)$$

$$= \frac{1}{\varphi(q)} \sum_{q_1 | q} \sum_{\chi_{q_1}}^{*} \bar{\chi}(l)\chi(a) \sum_{\substack{n \leqslant y/a \\ (n, q/q_1)=1}} a_n \chi(n). \qquad (2.4)$$

It follows from prime number theorem that

$$\sum_{\substack{n \leqslant y/a \\ (n,q)=1}} a_n = \pi\left(\frac{y}{a}\right) + O\left(\sum_{p|q} 1\right) = \mathrm{li}\, \frac{y}{a} + O\left(\frac{y}{a} e^{-\varepsilon\sqrt{\log y}}\right) + O(q^\varepsilon). \qquad (2.5)$$

Therefore we have from (2.3), (2.4) and (2.5) that

$$I \leqslant \sum_{q_1 \leqslant D} \frac{1}{\varphi(q_1)} \sum_{q_2 \leqslant D} \frac{1}{\varphi(q_2)} \max_{y \leqslant x} \sum_{\chi_{q_1}}^{*} \left| \sum_{\substack{A_1 < a \leqslant A_2 \\ (a,q_2)=1}} f(a)\chi(a) \sum_{\substack{n \leqslant y/a \\ (n,q_2)=1}} a_n \chi(n) \right|$$

$$+ O\left(\frac{x}{\log^A x}\right) \leqslant \max_{y \leqslant x} \max_{m \leqslant D} \log x \cdot I_{y,m} + O\left(\frac{x}{\log^A x}\right), \qquad (2.6)$$

where

$$I_{y,m} = \sum_{q \leqslant D} \frac{1}{\varphi(q)} \sum_{\chi_q}^{*} \left| \sum_{A_1 < a \leqslant A_2} g(a)\chi(a) \sum_{n \leqslant y/a} d_n \chi(n) \right|, \qquad (2.7)$$

in which

$$\begin{cases} g(n) = f(n), \quad d_n = a_n, \ \text{for} \ (n, m) = 1, \\ g(n) = d_n = 0, \quad \text{otherwise}. \end{cases} \qquad (2.8)$$

    2)   Let

$$I_{y,m} = I_{y,m}^{(1)} + I_{y,m}^{(2)}, \qquad (2.9)$$

where

$$I_{y,m}^{(1)} = \sum_{q \leqslant D_1} \frac{1}{\varphi(q)} \sum_{\chi_q}^{*} \left| \sum_{A_1 < a \leqslant A_2} g(a)\chi(a) \sum_{n \leqslant y/a} d_n \chi(n) \right|, \qquad (2.10)$$

$$I_{y,m}^{(2)} = \sum_{D_1 < q \leqslant D} \frac{1}{\varphi(q)} \sum_{\chi_q}^{*} \left| \sum_{A_1 < a \leqslant A_2} g(a)\chi(a) \sum_{n \leqslant y/a} d_n \chi(n) \right|. \qquad (2.11)$$

Then it follows by Siegel-Walfisz theorem (Cf. K. Prachar [7]) that there exists a positive number $\varepsilon_1 = \varepsilon_1(\varepsilon)$ such that

$$I_{y,m}^{(1)} \leqslant \sum_{q \leqslant D_1} \frac{1}{\varphi(q)} \sum_{\chi_q}^{*} \left| \sum_{A_1 < a \leqslant A_2} g(a)\chi(a) \sum_{n \leqslant y/a} a_n \chi(n) \right|$$

$$+ \sum_{q \leqslant D_1} \frac{1}{\varphi(q)} \sum_{\chi_q}^{*} \sum_{A_1 < a \leqslant A_2} \sum_{\substack{n \leqslant y/a \\ (n,m) > 1}} a_n$$

$$= O(D_1 x e^{-\varepsilon_1 \sqrt{\log x}} \log x) + O(D_1 A_2 m^{\varepsilon})$$

$$= O\left(\frac{x}{\log^{A+1} x} \right). \tag{2.12}$$

3) Obviously, we have

$$I_{y,m}^{(2)} \leqslant \sum_{j=0}^{J} \sum_{k=0}^{K} I_{y,m}^{(2)}(j, k), \tag{2.13}$$

where $2^J D_1 < D \leqslant 2^{J+1} D_1$, $2^K A_1 < A_2 \leqslant 2^{K+1} A_1$ and

$$I_{y,m}^{(2)}(j, k) = \sum_{2^j D_1 < q \leqslant 2^{j+1} D_1} \frac{1}{\varphi(q)} \sum_{\chi_q}^{*} \left| \sum_{2^k A_1 < a \leqslant 2^{k+1} A_1} g(a)\chi(a) \sum_{n \leqslant y/a} d_n \chi(n) \right|, \tag{2.14}$$

for $k < K$ and the $2^{k+1} A_1$ in (2.14) should be replaced by A_2 when $k = K$. Let

$$f(s, \chi) = \sum_{n=1}^{\infty} \frac{d_n \chi(n)}{n^s}, \quad \left(\sigma = 1 + \frac{1}{\log x} \right). \tag{2.15}$$

Take

$$T = e^{2(\log x)^{\iota}}. \tag{2.16}$$

Then from Perron's formula (Cf. K. Prachar |7|), we have

$$\sum_{n \leqslant y/a} d_n \chi(n) = \frac{1}{2\pi i} \int_{\sigma - iT}^{\sigma + iT} \frac{f(s, \chi)}{s} \left(\frac{y}{a} \right)^s ds + O\left(\frac{y}{T} \right) + \theta(a), \tag{2.17}$$

where

$$\theta(a) = \begin{cases} O(1), & \text{for } a \mid y, \\ 0, & \text{for } a \nmid y. \end{cases} \tag{2.18}$$

Suppose that $H < T$. Denote

$$f_1(s, \chi) = \sum_{n \leqslant H} \frac{d_n \chi(n)}{n^s}, \quad f_2(s, \chi) = \sum_{H < n \leqslant T} \frac{d_n \chi(n)}{n^s}. \tag{2.19}$$

Then we have immediately that

$$f(s, \chi) = f_1(s, \chi) + f_2(s, \chi) + O(x^{-2}). \tag{2.20}$$

Let

$$g_k(s, \chi) = \sum_{2^k A_1 < a \leqslant 2^{k+1} A_1} \frac{g(a)\chi(a)}{a^s} \tag{2.21}$$

and

$$P_{j,k}^{(l)}(s) = \sum_{2^j D_1 < q \leqslant 2^{j+1} D_1} \frac{1}{\varphi(q)} \sum_{\chi_q}^{*} |g_k(s, \chi) f_l(s, \chi)|, \quad (l = 1, 2). \tag{2.22}$$

Since

$$\int_{\sigma-iT}^{\sigma+iT} g_k(s,\chi)f_1(s,\chi)\frac{y^s}{s}ds - \int_{\frac{1}{2}-iT}^{\frac{1}{2}+iT} g_k(s,\chi)f_1(s,\chi)\frac{y^s}{s}ds$$

$$= O\left(\frac{y}{T}\left(\sum_{n\leqslant H}n^{-\frac{1}{2}}\right)\left(\sum_{a\leqslant A_2}a^{-\frac{1}{2}}\right)\right) = O\left(\frac{y^{\frac{3}{2}}\sqrt{H}}{T}\right) = O(x^{-2}), \qquad (2.23)$$

therefore from (2.14)—(2.23), we have

$$I^{(2)}_{y,m}(j,k) \leqslant \frac{1}{2\pi}\int_{\frac{1}{2}-iT}^{\frac{1}{2}+iT} P^{(1)}_{j,k}(s)\frac{y^{\frac{1}{2}}}{|s|}|ds| + \frac{e}{2\pi}\int_{\sigma-iT}^{\sigma+iT} P^{(2)}_{j,k}(s)\frac{y}{|s|}|ds|$$

$$+ O\left(\frac{x}{\log^{A+3}x}\right). \qquad (2.24)$$

4) By Schwarz inequality, we have

$$P^{(l)}_{j,k}(s) \leqslant \left(\sum_{2^j D_1 < q \leqslant 2^{j+1} D_1}\frac{1}{\varphi(q)}\sum_{\chi_q}{}^{*}|g_k(s,\chi)|^2\right)^{\frac{1}{2}}$$

$$\times \left(\sum_{2^j D_1 < q \leqslant 2^{j+1} D_1}\frac{1}{\varphi(q)}\sum_{\chi_q}{}^{*}|f_l(s,\chi)|^2\right)^{\frac{1}{2}} \quad (l=1,2). \qquad (2.25)$$

Take

$$H = (2^j D_1)^2. \qquad (2.26)$$

Then we have from Theorem A that

$$P^{(1)}_{j,k}(s) \ll \left(\left(2^j D_1 + \frac{2^k A_1}{2^j D_1}\right)\sum_{a\leqslant A_2}\frac{1}{a}\right)^{\frac{1}{2}}\left(\left(2^j D_1 + \frac{H}{2^j D_1}\right)\sum_{n\leqslant H}\frac{1}{n}\right)^{\frac{1}{2}}$$

$$\ll (H + 2^k A_1)^{\frac{1}{2}}\log x \ll x^{\frac{1}{2}}\log^{-B+1}x \qquad (2.27)$$

for $s = \frac{1}{2} + it(-T\leqslant t\leqslant T)$.

Let $2^R H < T \leqslant 2^{R+1}H$ and

$$f^{(r)}_2(s,\chi) = \begin{cases} \displaystyle\sum_{2^r H < n\leqslant 2^{r+1}H}\frac{d_n\chi(n)}{n^s}, & \text{for } 0\leqslant r < R, \\[3ex] \displaystyle\sum_{2^R H < n\leqslant T}\frac{d_n\chi(n)}{n^s}, & \text{for } r = R. \end{cases}$$

Then $R \ll \log^2 x$ and

$$f_2(s,\chi) = \sum_{r=0}^{R} f^{(r)}_2(s,\chi).$$

Hence, it follows by Theorem A that

$$P^{(2)}_{j,k}(s) \ll \max_{0\leqslant r\leqslant R}\left(\left(2^j D_1 + \frac{2^k A_1}{2^j D_1}\right)\sum_{a>2^k A_1}\frac{1}{a^2}\right)^{\frac{1}{2}}\left(\left(2^j D_1 + \frac{2^r H}{2^j D_1}\right)\sum_{n>2^r H}\frac{1}{n^2}\right)^{\frac{1}{2}}\log^2 x$$

$$\ll \left(\frac{2^j D_1}{A_1} + \frac{1}{2^j D_1}\right)^{\frac{1}{2}} \left(\frac{2^j D_1}{H} + \frac{1}{2^j D_1}\right)^{\frac{1}{2}} \log^2 x$$

$$\ll \left(\frac{1}{A_1} + \frac{1}{H}\right)^{\frac{1}{2}} \log^2 x \ll \log^{-B} y \cdot \log^2 x \tag{2.28}$$

for $s = \sigma + it(-T \leqslant t \leqslant T)$.

Substituting (2.27) and (2.28) into (2.24), we have

$$\max_{y \leqslant x} I_{y,m}^{(2)}(j, k) \ll x \log^{-B+4} x \ll x \log^{-A-3} x. \tag{2.29}$$

Hence the theorem follows from (2.6), (2.12), (2.13) and (2.29).

We deduce from Theorem 2 immediately (Cf. Pan Cheng-dong [6])

Corollary. *Under the conditions of Theorem 2, we have*

$$J = \sum_{q \leqslant x^{1/2} \log^{-B} x} 3^{\nu(q)} |\mu(q)| \max_{y \leqslant x} \max_{(l,q)=1} \left| \sum_{\substack{A_1 < a \leqslant A_2 \\ (a,q)=1}} f(a) \left(\pi(y, a, q, l) - \frac{\mathrm{li}\,\frac{y}{a}}{\varphi(q)} \right) \right|$$

$$= O\left(\frac{x}{\log^A x}\right), \tag{2.30}$$

where $\nu(q)$ *denotes the number of prime divisors of* q *and* $B = 2A + 24$.

Remark. Starting from the function

$$\psi_k(y, a, q, l) = \sum_{\substack{n \leqslant y/a \\ a_n \equiv l \pmod q}} \Lambda(n) \log^k \frac{y}{a_n},$$

we may prove the similar result too.

III. The Sieve Methods

1. *Mean Value Theorem of Bombieri-Виноградов*

Let $\eta = \frac{1}{2} - \varepsilon$, where $\varepsilon \left(< \frac{1}{4}\right)$ is any given positive number. Let q be a positive integer and $\xi > 0$. Let $\Omega(x, q, \xi)$ denote the set of all integers with the form $k = qm$, where $m \leqslant \frac{x^\eta}{q}$ and the largest prime divisor of m is not exceeding ξ. Further let

$$R(x, q, \xi) = \sum_{k \in \Omega(x,q,\xi)} 3^{\nu(k)} |\mu(k)| \max_{y \leqslant x} \max_{(l,k)=1} \left| \pi(y, k, l) - \frac{\mathrm{li}\,y}{\varphi(k)} \right|, \tag{3.1}$$

where $\pi(y, k, l) = \pi(y, 1, k, l)$.

Theorem B. *For any positive constant* A, *we have*

$$R(x, 1, x^\eta) = O\left(\frac{x}{\log^A x}\right), \tag{3.2}$$

where the constant implied by the symbol "O" depends only on ε and A

Cf. E. Bombieri [1] and А. И. Виноградов [15].

2. *Brun's Method*

Let $2 \leqslant y \leqslant x$ be two integers. Let

$$a, q; d_i, \quad (1 \leqslant i \leqslant r) \tag{ω}$$

be a sequence of integers satisfying

$$q < x^{\eta-\varepsilon}, \quad (a, q) = 1, \quad \alpha_i \not\equiv 0 (\bmod p_i), \quad (1 \leqslant i \leqslant r), \tag{3.3}$$

where $2 < p_1 < \cdots < p_r \leqslant \xi$ are all primes not exceeding ξ and not dividing qy. Further let $P_\omega(x, q, \xi)$ denote the number of primes p satisfying

$$p \leqslant x, \quad p \equiv a(\bmod q), \quad p \not\equiv \alpha_i(\bmod p_i), \quad (1 \leqslant i \leqslant r). \tag{3.4}$$

Denote

$$C_{qy} = e^{-\gamma} \prod_{\substack{p \mid qy \\ p > 2}} \frac{p-1}{p-2} \prod_{p > 2} \left(1 - \frac{1}{(p-1)^2}\right), \tag{3.5}$$

where γ is the Euler constant.

Theorem C. *Suppose that C is a positive constant. Then there exist two non-negative and non-decreasing functions $\lambda(\alpha)$ and $\Lambda(\alpha)$ $(0 < \alpha \leqslant C)$, each of which has only finite discontinuities, such that*

$$\lambda(\alpha)\frac{C_{qy}\operatorname{li} x}{\varphi(q)\log\frac{x^\eta}{q}}\left(1 + O\left(\frac{\log\log 3x}{\log x}\right)\right) + O\left(\log^2 x \cdot R\left(x, q, \left(\frac{x^\eta}{q}\right)^{\frac{1}{\alpha}}\right)\right)$$

$$< P_\omega\left(x, q, \left(\frac{x^\eta}{q}\right)^{\frac{1}{\alpha}}\right) < \Lambda(\alpha)\frac{C_{qy}\operatorname{li} x}{\varphi(q)\log\frac{x^\eta}{q}}\left(1 + O\left(\frac{\log\log 3x}{\log x}\right)\right)$$

$$+ O\left(\log^2 x \cdot R\left(x, q, \left(\frac{x^\eta}{q}\right)^{\frac{1}{\alpha}}\right)\right) \tag{3.6}$$

holds uniformly in α and (ω).

In fact, we may take

$$\lambda(\alpha) = \begin{cases} 2\alpha\left(1 - \displaystyle\sum_{k=0}^{\infty} \frac{\lambda^{k+1}((k+1)\tau)^{2k+4}}{(2k+4)!}\right), & \text{for } \alpha \geqslant 7, \\[2mm] 0, & \text{for } \alpha < 7, \end{cases} \tag{3.7}$$

and

$$\Lambda(\alpha) = \begin{cases} 2\alpha\left(1 + \displaystyle\sum_{k=0}^{\infty} \frac{\lambda^{k+1}((k+1)\tau)^{2k+5}}{(2k+5)!}\right), & \text{for } \alpha \geqslant 8, \\[2mm] \Lambda(8), & \text{for } \alpha \leqslant 8, \end{cases} \tag{3.8}$$

where $\lambda = 1.5 + \varepsilon$ and $\tau = \log 1.5 + \varepsilon$, in which ε is any given positive number.

3. Selberg's Upper Bound Method

Let $c > 0, P = \prod_{i=1}^{r} p_i$ and $\xi^{2c} \leqslant \dfrac{x^{\eta}}{q}$. Let

$$\lambda_d = \frac{\mu(d)\varphi(d)}{f(d)} \sum_{\substack{1 \leqslant k \leqslant \xi^c/d \\ k \mid P}} \frac{\mu^2(k)}{f(k)} \bigg/ \sum_{\substack{1 \leqslant l \leqslant \xi^c \\ l \mid P}} \frac{\mu^2(l)}{f(l)} \tag{3.9}$$

for $d \mid P$, where $f(k) = \varphi(k) \prod_{p \mid k} \dfrac{p-2}{p-1}$. Further let

$$Q = Q(x, q, \xi) = \sum_{\substack{d_1 \leqslant \xi^c \\ d_1 \mid P}} \sum_{\substack{d_2 \leqslant \xi^c \\ d_2 \mid P}} \lambda_{d_1} \lambda_{d_2} \frac{\text{li } x}{\varphi(q)\varphi([d_1, d_2])}, \tag{3.10}$$

where $[d_1, d_2]$ denotes the least common multiple of d_1 and d_2.

Theorem D. *The estimation*

$$P_\omega(x, q, \xi) \leqslant Q(x, q, \xi) + O(\log^2 x \cdot R(x, q, \xi)) \tag{3.11}$$

and

$$Q\left(x, q, \left(\frac{x^{\eta}}{q}\right)^{\frac{1}{\alpha}}\right) \leqslant \Lambda(\alpha) \frac{C_{q\psi} \text{ li } x}{\varphi(q) \log \dfrac{x^{\eta}}{q}} \left(1 + O\left(\frac{\log \log 3x}{\log x}\right)\right) \tag{3.12}$$

hold uniformly in $\alpha(0 \leqslant \alpha \leqslant 6)$ and (ω), where

$$\Lambda(\alpha) = \begin{cases} 4e^{\gamma}, & \text{for } 0 < \alpha \leqslant 2, \\[2mm] \dfrac{2\alpha e^{\gamma}}{\alpha - 1 - \dfrac{\alpha}{2} \log \dfrac{\alpha}{2}}, & \text{for } 2 \leqslant \alpha \leqslant 4, \\[4mm] \dfrac{2\alpha e^{\gamma}}{\alpha - 1 - \dfrac{\alpha}{2} \log \dfrac{\alpha}{2} + \delta(\alpha)}, & \text{for } 4 \leqslant \alpha \leqslant 6, \end{cases} \tag{3.13}$$

in which

$$\delta(\alpha) = \int_1^{\frac{\alpha}{4}} \int_s^{\frac{\alpha}{2}-s} \frac{\left(\dfrac{\alpha}{2} - s - t\right)}{st} \, dt \, ds, \quad (\alpha \geqslant 4). \tag{3.14}$$

4. Бухштаб Method.

Theorem E. *Let $\lambda(\alpha)$ and $\Lambda(\alpha)$ be two functions with the properties as those stated in Theorem C. Then the functions defined by*

$$\lambda_1(\alpha) = \begin{cases} \max\left(\lambda(\alpha), \lambda(\beta) - \displaystyle\int_{\alpha-1}^{\beta-1} \frac{\Lambda(z)}{z} \, dz\right), & \text{for } 1 + \varepsilon \leqslant \alpha \leqslant \beta \leqslant C, \\[4mm] \lambda(\alpha), & \text{for } 0 < \alpha \leqslant 1 + \varepsilon, \end{cases} \tag{3.15}$$

and

$$\Lambda_1(\alpha) = \begin{cases} \min\left(\Lambda(\alpha), \ \Lambda(\beta) - \int_{\alpha-1}^{\beta-1} \frac{\lambda(z)}{z}\, dz\right), & \text{for } 1 + \varepsilon < \alpha \leqslant \beta \leqslant C, \\ \Lambda(\alpha), & \text{for } 0 < \alpha \leqslant 1 + \varepsilon \end{cases} \tag{3.16}$$

have the same properties as those of the functions $\lambda(\alpha)$ *and* $\Lambda(\alpha)$ *respectively.*

We refer A. A. Бухштаб[14] and Wang Yuán[10] for the proof of Theorems C and E and Wang Yuán[10] for Theorem D.

IV. The Proof of Theorem 1

1. *The Estimation of* $\Gamma_\omega(\alpha, \beta)$.

Let $\beta > \alpha > 1$. Let

$$\Gamma_\omega(\alpha, \beta) = \sum_{\left(\frac{x^\eta}{q}\right)^{\frac{1}{\beta}} < p \leqslant \left(\frac{x^\eta}{q}\right)^{\frac{1}{\alpha}}} P_\omega\left(x, \ qp, \ \left(\frac{x^\eta}{q}\right)^{\frac{1}{\beta}}\right), \tag{4.1}$$

where p denotes prime. Then from Theorem B, we have

$$\Gamma_\omega(\alpha, \beta) \leqslant \left(\beta \int_{\beta(1-\alpha^{-1})}^{\beta-1} \frac{\Lambda(z)}{z(\beta - z)}\, dz\right) \frac{c_{qy}\, \mathrm{li}\, x}{\varphi(q) \log \frac{x^\eta}{q}} \left(1 + O\left(\frac{\log\log 3x}{\log x}\right)\right)$$

$$+ O\left(\frac{x}{\log^{A-2} x}\right). \tag{4.2}$$

2. *The Estimation of* Ω.

Let $P = \prod_{\substack{2 < p \leqslant x^{\frac{1}{4} - \frac{\varepsilon}{2}} \\ p \nmid x}} p$. Let $x = y$ be even integer. Further let $a = 1$, $q = 2$ and $d_i = x (i = 1, 2, \cdots)$. Then we have

$$\Omega \leqslant \sum_{(p_1, 2)} \sum_{\substack{n \leqslant \frac{x}{p_1 p_2} \\ (x - p_1 p_2 n, P) = 1}} a_n + O(x^{\frac{1}{4}}) \leqslant \sideset{}{'}\sum_{(p_1, 2)} \sideset{}{'}\sum_{n \leqslant \frac{x}{p_1 p_2}} a_n \left(\sum_{d \mid (x - p_1 p_2 n, P)} \lambda_d\right)^2$$

$$+ O(x^{\frac{1}{4}}) = \sum_{d_1 \mid P} \sum_{d_2 \mid P} \lambda_{d_1} \lambda_{d_2} \sum_{(p_1, 2)} \pi(x, p_1 p_2, [d_1, d_2], x) + O(x^{\frac{1}{4}}).$$

Obviously, $([d_1, d_2], p_1 p_2) = 1$ and $\lambda_d = O(\log x)$. Hence we have

$$\Omega \leqslant \sum_{(p_1, 2)} \sum_{d_1 \mid P} \sum_{d_2 \mid P} \lambda_{d_1} \lambda_{d_2} \frac{\mathrm{li}\, \frac{x}{p_1 p_2}}{\varphi([d_1, d_2])}$$

$$+ O\left(\log^2 x \cdot \sum_{\substack{d < x^\eta \\ (d, x) = 1}} |\mu(d)|\, 3^{\nu(d)} \left| \sum_{\substack{(p_1, 2) \\ (p_1 p_2, d) = 1}} \left(\pi(x, p_1 p_2, d, x) - \frac{\mathrm{li}\, \frac{x}{p_1 p_2}}{\varphi(d)}\right) \right| \right)$$

$$
+ O(x^{\frac{1}{4}}) \leqslant \sum_{(p_{1,2})} \sum_{d_1 \mid P} \sum_{d_2 \mid P} \lambda_{d_1} \lambda_{d_2} \frac{\operatorname{li} \dfrac{x}{p_1 p_2}}{\varphi([d_1, d_2])}
$$

$$
+ O\left(\log^2 x \cdot \sum_{\substack{d \leqslant x^7 \\ (d,x)=1}} |\mu(d)| \, 3^{\nu(d)} \left| \sum_{\substack{x^{\frac{13}{30}} < a \leqslant x^{\frac{2}{3}} \\ (a,d)=1}} f(a) \left(\pi(x, a, d, x) - \frac{\operatorname{li} \dfrac{x}{a}}{\varphi(d)} \right) \right| \right)
$$

$$
+ O(x^{\frac{1}{4}}),
$$

where

$$
f(a) = \begin{cases} 1, & \text{for } a = p_1 p_2 \text{ and } x^{\frac{1}{10}} < p_1 \leqslant x^{\frac{1}{3}} \leqslant p_2 \leqslant \left(\dfrac{x}{p_1} \right)^{\frac{1}{2}}, \\ 0, & \text{otherwise.} \end{cases}
$$

Therefore it follows from the Corollary of Theorem 2 (for $A = 5$) and Theorem D that

$$
\Omega \leqslant \left(8e^{\gamma} \sum_{(p_{1,2})} \frac{1}{p_1 p_2 \log \dfrac{x}{p_1 p_2}} + \delta \right) \frac{C_x x}{\log^2 x} \left(1 + O\left(\frac{1}{\log^{\frac{1}{2}} x} \right) \right)
$$

$$
\leqslant \left(8e^{\gamma} \int_3^{10} \frac{\log\left(2 - \dfrac{3}{y}\right)}{y - 1} \, dy + 2\delta \right) \frac{C_x x}{\log^2 x} \left(1 + O\left(\frac{1}{\log^{\frac{1}{2}} x} \right) \right)
$$

$$
< (7.01474 + 2\delta) \frac{C_x x}{\log^2 x} \left(1 + O \, \frac{1}{\log^{\frac{1}{2}} x} \right), \tag{4.3}
$$

for ε sufficiently small, where $\delta = o(1)$ as $\varepsilon \to 0$.

3) We have $\lambda_0(7) = 13.95578$ and $\Lambda_0(8) = 16.00624$ by (3.7) and (3.8) and $\Lambda_0(\alpha)(0 < \alpha \leqslant 6)$ is given by (3.13). Take $\beta - \alpha = 0.01$. Then we have $\lambda_0(3 + 0.01i)(0 \leqslant i \leqslant 400)$ by Theorem E. For examples, $\lambda_0(6) = 11.90332$, $\lambda_0(5) = 9.77058$, $\lambda_0(4) = 7.41296$ and $\lambda_0(3) = 4.44824$. Take also $\beta - \alpha = 0.01$. Then we have $\lambda_1(\alpha)$ and $\Lambda_1(\alpha)$ from $\lambda_0(\alpha)$ and $\Lambda_0(\alpha)$ by Theorem E, for example, $\lambda_1(5) = 9.87844$.

4) Let $x = y$ be even integer. Let $a = 1$, $q = 2$ and $d_i = x(i = 1, 2, \cdots)$. Further let

$$
M = \frac{1}{2} \sum_{x^{\frac{1}{10}} < p \leqslant x^{\frac{1}{3}}} P_\omega(x, 2p, x^{\frac{1}{10}}) + \frac{\Omega}{2} + O\left(x^{\frac{9}{10}} \right). \tag{4.4}
$$

Then from Theorem B, (4.2), (4.3) and 3), we have

$$
P_\omega(x, 2, x^{\frac{1}{10}}) - M \geqslant 2 \left(\lambda_1(5) - \frac{5}{2} \int_{\frac{1.5}{2.5}}^4 \frac{\Lambda_0(z)}{z(5 - z)} \, dz - 3.50737 - \delta \right) \frac{C_x x}{\log^2 x}
$$

$$
\times \left(1 + O \left(\frac{1}{\log^{\frac{1}{2}} x} \right) \right) = 2 \left(9.87844 - 10e^{\gamma} \int_1^2 \frac{dz}{(5 - 2z)(2z - 1 - z \log z)} \right.
$$

$$-2e^{\gamma}\log\frac{2}{1.5} - 3.50737 - \delta\Big) \times \frac{C_x x}{\log^2 x}\Big(1 + O\Big(\frac{1}{\log^{\frac{1}{2}} x}\Big)\Big)$$

$$> \frac{C_x x}{10\log^2 x}\Big(1 + O\Big(\frac{1}{\log^{\frac{1}{2}} x}\Big)\Big) > 1 \quad \text{(for sufficiently large } x). \tag{4.5}$$

Let p' denote prime. If $p'\,|\,x$ and $p'\,|\,(x-p)$, then $p'=p$. Since the number of prime divisors of x is $O(x^{\varepsilon})$, hence $P_{\omega}(x, 2, x^{\frac{1}{10}}) + O(x^{\frac{9}{10}})$ is equal to the number of primes p satisfying

$$2 < p < x, \text{ if } p'\,|\,(x-p), \text{ then } p' > x^{\frac{1}{10}}. \tag{4.6}$$

Since the number of primes p satisfying (4.6), such that $x-p$ has a prime divisor $p' > x^{\frac{1}{10}}$, is $P_{\omega}(x, 2p', x^{\frac{1}{10}})$ and the number of primes $p(\leqslant x)$, such that $x-p$ is divided by $p'^2(p' > x^{\frac{1}{10}})$, is at most

$$O\Big(\sum_{p>x^{\frac{1}{10}}}\Big(\frac{x}{p^2}+1\Big)\Big) = O(x^{\frac{9}{10}}),$$

hence the number of primes p satisfying (4.6), such that $x-p$ has at most 2 prime factors in $x^{\frac{1}{10}} < p' \leqslant x^{\frac{1}{3}}$ or 1 prime divisor in $x^{\frac{1}{10}} < p' \leqslant x^{\frac{1}{3}}$ and 2 prime divisors $> x^{\frac{1}{3}}$, is not exceeding M. Consequently, it follows from (4.5) that there exists a prime p such that $x-p$ has at most 2 prime factors for sufficiently large x. The theorem is proved.

References

[1]   Bombieri, E.: On the large sieve, *Mathematika*, **12** (1965), 201—225.

[2]   Chen, Jing-run: On the representation of a large even integer as the sum of a prime and the product of at most 2 primes, *Kexue Tongbao*, **17** (1966), 385—386.

[3]   Chen, Jing-run: On the representation of a larger even integer as the sum of a prime and the product of at most two primes, *Sci. Sin.*, **16** (1973), 157—176.

[4]   Estermann, T.: Eine neue Darstellung und neue Anwendung der Viggo-Brunschen Metode, *J. Rei. und Ang. Math.*, **168** (1932), 106—116.

[5]   Gallagher, P. X.: Bombieri's mean value theorem, *Mathematika*, **15** (1968), 1—6.

[6]   Pan, Cheng-dong: On the representation of large even integer as a sum of a prime and an almost prime, *Acta Math. Sin.*, **12** (1962), 95—106.

[7]   Prachar, K.: Primzahlverteilung, *Spr, Ver*, (1957).

[8]   Richert, H. E.: Selberg's sieve with weights, *Mathematika*, **16** (1969), 1—22.

[9]   Wang, Yuán: On the representation of large even integer as a sum of a prime and a product of at most 4 primes, *Acta Math. Sin.*, **6** (1956), 565—582.

[10]   Wang, Yuán: On the representation of large integer as a sum of a prime and an almost prime, *Sci. Sin.*, **11** (1962), 1033—1054.

[11]   Барбан, М. Б.: Новые применении большого решета Ю. В. Линника, *Тру. Ин. Мат. им. В. И. Романовского*, **22**, 1961.

[12]   Барбан, М. Б.: Плотность нулей L-рядов дирихре и задача о сложении простых и почти простых чисел, *Мат. сб*, **61**, (1963), 419—425.

[13]   Бухштаб, А. А.: Новые результаты в исследовании проблемы Гольдбаха-Эйлера и проблемы простых чисел близнецов, *ДАН СССР*, **162** (1965), 739—742.

[14]   Бухштаб, А. А.: Комбирнаторное усиление метода эратосфенова решета, *УМН СССР*, **22**, (1967), 199—226.

[15] Виноградов, А. И: О плотностной гипотезе для L-рядов дирихре, *ИАН СССР, сер. Мат.*, **23**, (1965), 903—934.

[16] Левин, Б. В.: Распределение "почти простых" чисел в целозначных полиномиальных, последовательностях, *Мат. сб.*, **61** (1963), 401—419.

[17] Пан, Чэн-дун: О представлении четных чисел в виде суммы простого и непревосходящего 4 простых произведения, *Sci. Sin.*, **12** (1963), 455—474.

[18] Реньи, А.: О представлении четных чисел в виде суммы простого и почти простого чисел, *ИАН СССР*, **2** (1948), 57—78.

REMARKS ON A THEOREM OF DAVENPORT*

WANG YUAN

Institute of Mathematics, Academia Sinica

Received 6 January 1975

We use $\mathbf{x}, \mathbf{a}, \mathbf{c}$ to denote the n-dimensional real vectors,

$$|\mathbf{x}| = |(x_1, \ldots, x_n)| = (x_1^2 + \cdots + x_n^2)^{1/2}$$

the modulus of $\mathbf{x}$, and Λ an n-dimensional lattice, i.e. a set formed by the vectors

$$u_1 \mathbf{a}_1 + \cdots + u_n \mathbf{a}_n,$$

where $\mathbf{a}_1, \ldots, \mathbf{a}_n$ are a given linearly independent vectors in n-dimensional Euclidean space and $u_1, \ldots, u_n$ are any integers. $\{\mathbf{a}_1, \ldots, \mathbf{a}_n\}$ is called a basis of Λ.

In this paper we shall use Brun's method to prove the following:

Theorem 1. *Let Λ be an n-dimensional lattice and*

$$\mathbf{c}_i(1 \leq i \leq n - 1)$$

be any given $n - 1$ vectors. Then there exists a basis $\{\mathbf{a}_1, \ldots, \mathbf{a}_n\}$ of Λ such that

$$|\mathbf{a}_i - N\mathbf{c}_i| = O(\log^3 N), \quad 1 \leq i \leq n - 1,$$

holds for any number $N(\geq 2)$, where the constant in "O" depends only on Λ and $\mathbf{c}_i$'s.

This gives an improvement of a result due to Davenport. In his original result, the error term should be replaced by $O(N^\varepsilon)$, where ε is any preassigned positive number and the constant in "O" depends on ε, Λ and $\mathbf{c}_i$'s. Besides its interest itself, Davenport's theorem is also useful in the study of diophantine approximations and geometry of numbers (see H. Davenport [1], J. W. S. Cassels [2] and C. G. Lekkerkerker [3]). We may also obtain some improvements on generalised Davenport's theorems by similar method.

Lemma 1. (Brun) *Let Q be a positive integer and $p_1 < p_2 < \cdots < p_r$ be a set of prime numbers not dividing Q. Let $P(Q; p_1, \ldots, p_r)$ be the sum of non-negative real*

---

*Acta Mathematica Sinica, **18**:4 (1975) 286–289.

numbers a_n's, where n satisfies the following conditions:

$$n \geq 1, \ n \equiv 0 \ (\mathrm{mod}\, Q), \ n \not\equiv 0 \ (\mathrm{mod}\, p_i), \quad 1 \leq i \leq r. \tag{1}$$

In particular, $P(Q)$ is the sum of a_n's with

$$n \geq 1, \quad n \equiv 0 \ (\mathrm{mod}\, Q). \tag{2}$$

If there are M and R depending only on a_n's such that

$$P(Q) = \frac{M}{Q} + O(R), \tag{3}$$

hereafter the constants in "O" are absolute constant, then there exists an absolute constant $\sigma \ (>0)$ such that

$$P(Q; p_1, \ldots, p_r) \geq \frac{\sigma M}{Q} \prod_{i=1}^{r} \left(1 - \frac{1}{p_i}\right) + O(p_r^{2.99} R). \tag{4}$$

We refer to G. Ricci [4] and Wang Yuan [5] for the proof, but the ind n in [5] should now be replaced by n.

Lemma 2. *Let q be an integer ≥ 2 and s, t be integers such that $(t, q) = 1$. Then there exists an absolute constant K such that there is a positive integer u satisfying $(tu + s, q) = 1$ in any interval with length not less than $K \log^3 q$.*

Proof. 1) Let G be a real number, $H = K \log^3 q$ and $q_1 = \prod_{\substack{p|q \\ p \leq \log^\delta q}} p$, where p denotes prime number, $\delta = \frac{3}{2.999}$ and K is a constant which will be defined later. Then

$$\Sigma = \sum_{\substack{G < u \leq G + H \\ (tu + s, q) = 1}} 1 \geq \Sigma_1 - \Sigma_2, \tag{5}$$

where

$$\Sigma_1 = \sum_{\substack{G < u \leq G + H \\ (tu + s, q_1) = 1}} 1 \quad \text{and} \quad \Sigma_2 = \sum_{\substack{p|q \\ p > \log^\delta q}} \sum_{\substack{G < u \leq G + H \\ tu + s \equiv 0 (\mathrm{mod}\, p)}} 1.$$

Since the number of prime divisors of q is at most $2 \log q$ and

$$\sum_{\substack{G < u \leq G + H \\ tu + s \equiv 0 (\mathrm{mod}\, Q)}} 1 = \left[\frac{H}{Q}\right] + \theta = \frac{H}{Q} + \theta \tag{6}$$

holds for $(t, q) = 1$, hereafter θ denotes a number satisfying $|\theta| \leq 1$ but not always the same in different occurrences, we have

$$\Sigma_2 = \sum_{\substack{p|q \\ p > \log^\delta q}} \left(\frac{H}{p} + \theta \right) \leq \frac{H}{\log^\delta q} \sum_{p|q} 1 + \sum_{p|q} 1$$

$$\leq 2H \log^{1-\delta} q + 2 \log q \leq 4K \log^{4-\delta} q. \tag{7}$$

Set

$$a_n = \begin{cases} 1, & \text{if } n = tu + s, \quad G < u \leq G + H; \\ 0, & \text{otherwise,} \end{cases}$$

in Lemma 1. Let $p_1 < p_2 < \cdots < p_r$ be all prime numbers satisfying $p|q$ and $p \leq \log^\delta q$. Then

$$\Sigma_1 = P(1; p_1, \ldots, p_r) \geq \sigma H \prod_{\substack{p|q \\ p \leq \log^\delta q}} \left(1 - \frac{1}{p} \right) + O(p_r^{2.99}),$$

by (6) and Lemma 1, and therefore

$$\Sigma_1 \geq \sigma H \frac{e^{-r}}{\delta \log \log q} \left(1 + O\left(\frac{1}{\log \log q} \right) \right) + O(\log^{2.99\delta} q) \geq \frac{2\tau K \log^3 q}{\log \log q}, \tag{8}$$

by Mertens theorem, where γ is Euler constant and τ is an absolute constant.

By (5), (7) and (8), we know that there exists an absolute constant q_0 such that

$$\Sigma > \frac{\tau K \log^3 q}{\log \log q} > 1, \quad q \geq q_0.$$

2) Suppose $q \leq q_0$. By Eratosthenes sieve, we obtain

$$\Sigma = \sum_{\substack{G < u \leq G + H \\ (tu + s, q) = 1}} 1 - \sum_{G < u \leq G + H} \sum_{d|(tu+s,q)} \mu(d)$$

$$= \sum_{d|q} \mu(d) \sum_{\substack{G < u \leq G + H \\ tu + s \equiv 0 (\text{mod } d)}} 1 = H \sum_{d|q} \frac{\mu(d)}{d} + \theta \sum_{d|q} 1$$

$$= H \prod_{p|q} \left(1 - \frac{1}{p} \right) + \theta q \geq \nu K \log^{\delta-1} q + \theta q,$$

where ν is an absolute constant. Since $q \leq q_0$, we may choose K sufficiently large such that $\Sigma > 1$. The lemma follows from 1) and 2).

Lemma 3. *Let* $\mathbf{b}_1,\ldots,\mathbf{b}_n$ *be a basis of* Λ *and*

$$\mathbf{a}_i = \sum_{j=1}^{n} v_{ij}\mathbf{b}_j, \quad 1 \le i \le m < n$$

be a set of vectors in Λ. *Then a necessary and sufficient condition that* $\mathbf{a}_1,\ldots,\mathbf{a}_m$ *can be extended to a basis* $\mathbf{a}_1,\ldots,\mathbf{a}_n$ *of* Λ *is that the determinants of all* $m \times m$ *submatrices of*

$$(v_{ij}), \quad 1 \le i \le m, \quad 1 \le j \le n$$

have no common factor.

See J. W. S. Cassels [2].

Proof of Theorem 1. Let $\mathbf{b}_1,\ldots,\mathbf{b}_n$ be a basis of Λ and

$$\mathbf{c}_i = \sum_{j=1}^{n} r_{ij}\mathbf{b}_j, \quad 1 \le i \le n, \tag{9}$$

where r_{ij}'s are real numbers. Now we proceed to show that we may choose a basis

$$\mathbf{a}_i = \sum_{j=1}^{n} v_{ij}\mathbf{b}_j, \quad 1 \le i \le j \tag{10}$$

of Λ such that

$$v_{ij} = Nr_{ij} + O(\log^3 N), \quad 1 \le i \le n-1, \quad 1 \le j \le n, \tag{11}$$

where N is any given number ≥ 2 and the constant in "O" depends only on n and r_{ij}'s.

We can choose v_{ij}'s with the property: For any integer $I < n$, the integers

$$R_I = \det(v_{ij}), \quad 1 \le i \le I, \ 1 \le j \le I$$

and

$$S_I = \det(v_{ij}), \quad 1 \le i \le I, \ 2 \le j \le I+1$$

are all nonzero and coprime.

In fact, we may prove it by induction. Suppose that $I = 1$. Let v_{11} be the integer which is nonzero and has the least distance with Nr_{11} and let v_{12} be the nonzero integer which is coprime with v_{11} and nearest to Nr_{12}. If $j > 2$, we choose v_{1j} to be the integer nearest to Nr_{1j}. Hence (11) holds for $i = 1, j \ne 2$. Suppose $i = 1$ and $j = 2$. If $v_{11} = \pm 1$, then v_{12} clearly satisfied the above conditions and (11). Otherwise, take $t = 1, s = 0, q = |v_{11}|$ and $G = Nr_{12}$, it follows from Lemma 2 that we may choose v_{12} such that the above conditions and (11) are satisfied. Hence $R_1 = v_{11}$ and $S_1 = v_{12}$ satisfy our requirements.

Now suppose that $I > 1$ and that v_{ij}'s are well defined for $i < I$. Choose v_{Ij} to be the integer nearest to Nr_{Ij} if $j \neq I, I+1$. Expanding R_I and S_I according to their last columns, we obtain

$$R_I = \pm v_{II} R_{I-1} + A \quad \text{and} \quad S_I = \pm v_{I,I+1} S_{I-1} + v_{II} B + C,$$

where A, B, C are definite integers. By the assumption of induction, R_{I-1} and S_{I-1} are nonzero, $(R_{I-1}, S_{I-1}) = 1$ and $S_{I-1} = O(N^{I-1})$, since S_{I-1} is a sum of $(I-1)!$ products of $I - 1$ v_{ij}'s. If $S_{I-1} \neq \pm 1$, then by Lemma 2 with $t = \pm R_{I-1}$, $S = A$, $q = |S_{I-1}|$ and $G = Nr_{II}$ we may choose v_{II} such that $R_I \neq 0$, $(R_I, S_{I-1}) = 1$ and

$$v_{II} - Nr_{II} = O(\log^3 |S_{I-1}|) = O(\log^3 N).$$

If $S_{I-1} = \pm 1$, we may clearly choose v_{II} satisfying the above requirements. If $R_{II} \neq \pm 1$ with this v_{II}, then by Lemma 2 with $t = \pm S_{I-1}$, $S = v_{II} B + C$, $q = |R_I|$ and $G = Nr_{I,I+1}$, there exists $v_{I,I+1}$ such that $S_I \neq 0$, $(S_I, R_I) = 1$ and

$$v_{I,I+1} - Nr_{I,I+1} = O(\log^3 |R_l|) = O(\log^3 N).$$

Otherwise, if $R_{II} \neq \pm 1$, we may choose evidently $v_{I,I+1}$ satisfying the above conditions, (11) follows by induction. Since $(R_{n-1}, S_{n-1}) = 1$, it follows from Lemma 3 that we may choose $\mathbf{a}_n$ such that $\mathbf{a}_1, \ldots, \mathbf{a}_n$ is a basis of Λ, and finally

$$|\mathbf{a}_i - N\mathbf{c}_i| = O(\log^3 N), \quad 1 \leq i \leq n - 1$$

by (10) and (11). The theorem is proved.

References

[1] H. Davenport, On a theorem of Furtwängler, *JLMS* **30** (1955) 186–195.
[2] J. W. S. Cassels, *An Introduction to the Geometry of Numbers* (Springer, 1959).
[3] C. G. Lekkerkerker, *Geometry of Numbers* (North-Holland, 1969).
[4] G. Ricci, Sur la Congettura di Goldbach e la costante di Schnirelmann, *Ann. del. R. Scu. Nor. Sup. di Pisa* **6** (1937) 70–115.
[5] Wang Yuan, On the least primitive root of a prime. *Acta Math. Sin.* **9**:4 (1959) 432–441.

Vol. XX No. 1   SCIENTIA SINICA   Jan. - Feb. 1977

ON ЛИННИК'S METHOD CONCERNING THE GOLDBACH NUMBER

WANG YUAN（王　元）

(*Institute of Mathematics, Academia Sinica*)

Received February 5, 1976.

ABSTRACT

In this paper, some conditional results concerning the Goldbach number are proved. For example, assuming that the density hypothesis of $\zeta(s)$ is true, the inequality $|N - p - p'| \leqslant c(\varepsilon)$ $(\ln N)^{\frac{148}{13}+\varepsilon}$ always has solutions; here ε is any pre-assigned positive number and p, p' are primes. It seems that there is a gap in Линник's original proof of the similar theorem.

I. INTRODUCTION

The even number that can be represented as a sum of two odd primes is called the Goldbach number. The binary Goldbach conjecture may be stated that every even integer > 4 is a Goldbach number.

Let $0 \leqslant \nu \leqslant 1/2$ and $T > 0$. Let $N(T, \nu)$ denote the number of zeros of Riemann's zeta function $\zeta(s)$ in the rectangle

$$\frac{1}{2} + \nu \leqslant \sigma \leqslant 1, \quad |t| \leqslant T. \tag{1.1}$$

Now, let us state the Riemann hypothesis (R) and the density hypothesis (D) as follows:

(R) $$N(T, \nu) = 0, \quad \left(0 < \nu \leqslant \frac{1}{2}, \quad T > 0\right),$$

and

(D) $$N(T, \nu) = O(T^{c(1-2\nu)}(\ln(T + 2))^r), \quad \left(0 < \nu \leqslant \frac{12}{37} + \varepsilon, \quad T > 0\right),$$

where r and $c(> 0)$ are constants, ε is any pre-assigned positive number and the constant implied by the symbol "O" depends on ε only.

Since

$$N_0(T) > cN(T, 0), \quad (T > T_0), \tag{1.2}$$

where c and T_0 are positive constants and $N_0(T)$ denotes the number of zeros of $\zeta(s)$

satisfying

$$\sigma = \frac{1}{2}, \quad |t| \leqslant T, \tag{1.3}$$

(cf. A. Selberg [4] and N. Levinson [2]), there is no need for us to consider the case $\nu = 0$ in (D). Since $\zeta(s)$ has no zero near the line $\sigma = 1$ (cf. И. М. Виноградов [5] and Н. М. Коробов [6]) and $N(T, \nu)$ is comparatively small for $\frac{12}{37} + \varepsilon \leqslant \nu \leqslant \frac{1}{2}$ (cf. M. N. Huxley [1]), we need not consider the case $\frac{12}{37} + \varepsilon \leqslant \nu \leqslant \frac{1}{2}$ either.

We use N to denote even integer, p, p' prime numbers, c positive constant, ε any pre-assigned positive number and $c(\varepsilon)$ positive constant depending on ε only. However, they are not always equal to the same values.

Theorem 1. *Suppose that*

$$N(T, \nu) \leqslant c(\varepsilon)T^{1-2\nu} \ln (T + 2), \quad \left(0 < \nu \leqslant \frac{12}{37} + \varepsilon\right). \tag{1.4}$$

Then for any given N, there exist p, p' such that

$$|N - p - p'| \leqslant c(\varepsilon)(\ln N)^{\frac{148}{13}+\varepsilon}. \tag{1.5}$$

Theorem 2. *Suppose that*

$$N(T, \nu) \leqslant c(\varepsilon)T^{1-2\nu}(\ln (T + 2))^{-\frac{72}{37}}, \quad \left(0 < \nu \leqslant \frac{12}{37} + \varepsilon\right). \tag{1.6}$$

Then for any given N, there exist p, p', such that

$$|N - p - p'| \leqslant c(\varepsilon)(\ln N)^{3+\varepsilon}. \tag{1.7}$$

In general, we may prove

Theorem 3. *Suppose that*

$$N(T, \nu) \leqslant \begin{cases} cT^{1-2\nu}(\ln (T + 2))^r, & \text{for } 0 < \nu \leqslant \alpha, \\ cT^{(1-2\nu)/(1+\varepsilon_1)} & \text{for } \alpha < \nu \leqslant \frac{1}{2}, \end{cases} \tag{1.8}$$

where α, r, ε_1 are constants $0 < \alpha < 1/2$, $r \geqslant -6\alpha$ and $\varepsilon_1 > 0$. Then for any given N, there exist p, p' such that

$$|N - p - p'| \leqslant c(\varepsilon)(\ln N)^{(3+r)(1-2\alpha)^{-1}+\varepsilon}. \tag{1.9}$$

We have also

Theorem 4. *Let q and N be two given positive integers such that $q \leqslant N/(\ln N)^{3+\varepsilon}$ and N is even if q is even. Then under the truth of the grand Riemann hypothesis, the equation*

$$N = p + p' + hq \tag{1.10}$$

always has solutions for $N > c(\varepsilon)$, where h satisfies $0 \leqslant h \leqslant (\ln N)^{3+\varepsilon}$.

The proofs of these theorems are based on Линник's method (cf. Ю. В. Линник [7]). Theorems 2 and 4 give some modifications of the corresponding theorems in [7], where Линник also stated a result that under the truth of (1.4) with $0 < \nu \leqslant 1/2$, the right-hand side of (1.5) might be replaced by $O((\ln N)^7)$. The present author, however, is unable to understand his arguments. It may have existed a gap in his proof. (See Remark in V.)

II.   SEVERAL LEMMAS

We use the notations

$$x = N^{-1} + 2\pi i \vartheta, \quad (0 \leqslant \vartheta \leqslant 1), \tag{2.1}$$

$$Q(N) = \sum_{p+p'=N} \ln p \ln p', \tag{2.2}$$

$$S(\vartheta, N) = \sum_{n=2}^{\infty} \Lambda(n) e^{-xn}, \tag{2.3}$$

$$S_1(\vartheta, N) = \sum_{\rho_k} x^{-\rho_k} \Gamma(\rho_k), \tag{2.4}$$

where $\sum_{\rho_k}$ denotes a sum over all zeros $\rho_k = \beta_k + it_k$ of $\zeta(s)$ in the critical domain $0 < \sigma < 1$.

$$S_2(\vartheta, N) = \sum_{\rho_k,\, t_k > 0} x^{-\rho_k} \Gamma(\rho_k), \tag{2.5}$$

$$S_3(\vartheta, N) = \sum_{\rho_k,\, t_k < 0} x^{-\rho_k} \Gamma(\rho_k). \tag{2.6}$$

Lemma 2.1. *For* $-1 \leqslant \sigma \leqslant 2$, *we have*

$$\Gamma(s) \leqslant c(|t| + 1)^{\sigma - \frac{1}{2}} e^{-\frac{\pi}{2}|t|}.$$

(cf. K. Prachar [3].)

Lemma 2.2.

$$\int_{-\infty}^{\infty} \frac{e^{2\pi i N_1 \vartheta}}{x^2} \, d\vartheta = N_1 e^{-N_1/N}.$$

Proof. Consider the integral

$$\int_C \frac{e^{2\pi i N_1 z}}{(N^{-1} + 2\pi i z)^2} \, dz,$$

where C denotes the closed contour which contains a segment $[-R, R]$ and a semi-circle $Re^{i\varphi}(0 \leqslant \varphi \leqslant \pi)$. Let $R \to \infty$. Then we have the lemma.

Lemma 2.3. $\sum_{n+n'=N} \Lambda(n)\Lambda(n') = Q(N) + O(\sqrt{N}(\ln N)^2).$

Lemma 2.4. *Suppose that* $N_1 = N + H$, *where* H *denotes an integer satisfying* $H = O(\sqrt{N}(\ln N)^2)$. *Then*

$$Q(N_1) = e \int_{-\frac{1}{2}}^{\frac{1}{2}} S(\vartheta, N)^2 e^{2\pi i N_1 \vartheta} d\vartheta + O(\sqrt{N}(\ln N)^2). \tag{2.7}$$

Lemma 2.5. *We have*

$$S(\vartheta, N) = x^{-1} - S_1(\vartheta, N) + O((\ln N)^2). \tag{2.8}$$

Proof. From Mellin's transform, we have

$$S(\vartheta, N) = -\frac{1}{2\pi i} \int_{2-i\infty}^{2+i\infty} \Gamma(s) x^{-s} \frac{\zeta'}{\zeta}(s) ds. \tag{2.9}$$

Obviously, we have

$$x^{-s} = \begin{cases} |x|^{-s} e^{\frac{\pi}{2}t - t\,\mathrm{arctg}\,\frac{1}{2\pi N \vartheta}} e^{-i\sigma\left(\frac{\pi}{2} - \mathrm{arctg}\,\frac{1}{2\pi N \vartheta}\right)}, & \text{for } \vartheta \geqslant 0, \\ |x|^{-s} e^{-\frac{\pi}{2}t - t\,\mathrm{arctg}\,\frac{1}{2\pi N|\vartheta|}} e^{i\sigma\left(\frac{\pi}{2} - \mathrm{arctg}\,\frac{1}{2\pi N|\vartheta|}\right)}, & \text{for } \vartheta < 0. \end{cases} \tag{2.10}$$

First, suppose that $\vartheta \geqslant 0$. Shifting the line of integration in (2.9) to the line $\sigma = -\frac{1}{2}$, we have

$$S(\vartheta, N) = x^{-1} - S_1(\vartheta, N) - \frac{1}{2\pi i} \int_{-\frac{1}{2}-i\infty}^{-\frac{1}{2}+i\infty} \Gamma(s) x^{-s} \frac{\zeta'}{\zeta}(s) ds. \tag{2.11}$$

Since

$$\frac{\zeta'}{\zeta}\left(-\frac{1}{2} + it\right) = O(\ln(|t| + 2)) \tag{2.12}$$

(cf. Prachar [3]), we have by (2.10), (2.11) and Lemma 2.1 that

$$\int_{-\frac{1}{2}-i\infty}^{-\frac{1}{2}+i\infty} \Gamma(s) x^{-s} \frac{\zeta'}{\zeta}(s) ds \ll \int_2^\infty t^{-1} e^{-t\,\mathrm{arctg}\,\frac{1}{2\pi N \vartheta}} \ln t\, dt$$

$$\ll \int_2^\infty t^{-1} e^{-t\,\mathrm{arctg}\,\frac{1}{2\pi}} \ln t\, dt \ll 1, \tag{2.13}$$

for the case $N\vartheta \leqslant 1$. Now, suppose that $N\vartheta \geqslant 1$. Then

$$\frac{1}{4\pi N \vartheta} \leqslant \mathrm{arctg}\,\frac{1}{2\pi N \vartheta} \leqslant \frac{1}{2\pi N \vartheta}, \tag{2.14}$$

and

$$\int_{-\frac{1}{2}-i\infty}^{-\frac{1}{2}+i\infty} \Gamma(s) x^{-s} \frac{\zeta'}{\zeta}(s) ds \ll \int_2^{4\pi N \vartheta} e^{-\frac{t}{4\pi N \vartheta}} t^{-1} \ln t\, dt$$

$$+ \int_{4\pi N \vartheta}^\infty e^{-\frac{t}{4\pi N \vartheta}} t^{-1} \ln t\, dt \ll (\ln N)^2. \tag{2.15}$$

Substituting (2.15) into (2.11), we have the lemma. The case $\vartheta < 0$ may be treated similarly. The lemma is proved.

Lemma 2.6. (Линник [7]). *Suppose that η satisfies $\frac{1}{4} \geqslant \eta \geqslant 4N^{-1}$. Then we have*

$$\int_{\eta}^{2\eta} |S_2(\vartheta, N)|^2 d\vartheta \ll \sum_{\rho_{k_1},\, t_{k_1}>0} \sum_{\rho_{k_2},\, t_{k_2}>0} (t_{k_1}+1)^{\beta_{k_1}-\frac{1}{2}}$$

$$\times (t_{k_2}+1)^{\beta_{k_2}-\frac{1}{2}}\, \eta^{1-\beta_{k_1}-\beta_{k_2}}\, \frac{1}{|t_{k_1}-t_{k_2}|+1}\, e^{-(t_{k_1}+t_{k_2})\operatorname{arctg}\frac{1}{4\pi N\eta}}, \qquad (2.16)$$

and

$$\int_{\eta}^{2\eta} |S_3(\vartheta, N)|^2 d\vartheta \ll \sum_{\rho_{k_1},\, t_{k_1}<0} \sum_{\rho_{k_2},\, t_{k_2}<0} (|t_{k_1}|+1)^{\beta_{k_1}-\frac{1}{2}}(|t_{k_2}|+1)^{\beta_{k_2}-\frac{1}{2}}$$

$$\times \frac{1}{|t_{k_1}-t_{k_2}|+1}\, e^{-(|t_{k_1}|+|t_{k_2}|)\left(\pi-\operatorname{arctg}\frac{1}{2\pi N\eta}\right)}. \qquad (2.17)$$

We need also some results concerning the critical zeros of $\zeta(s)$.

Lemma 2.7. *In any interval* $(t, t+1)$, *the number of critical zeros of* $\zeta(s)$ *does not exceed* $c \ln(|t|+2)$ *(cf. Prachar* [3].)

Lemma 2.8. (**Виноградов** [5]-**Коробов** [6].) $\zeta(s)$ *has no zero in the domain*

$$\sigma \geqslant 1 - \frac{c}{(\ln(|t|+2))^{2/3}}. \qquad (2.18)$$

Lemma 2.9. (**Huxley** [1].) *We have*

$$N(T, \nu) \ll \begin{cases} T^{\frac{12(1-2\nu)}{37\nu}+\varepsilon}, & for \ \dfrac{12}{37} \leqslant \nu \leqslant \dfrac{8}{21}, \\[3mm] T^{\frac{3(1-2\nu)}{2(1+2\nu)}+\varepsilon}, & for \ \dfrac{8}{21} \leqslant \nu \leqslant 1, \end{cases} \qquad (2.19)$$

where the constants implied by the symbol "$\ll$" *depend on* ε *only.*

From Lemmas 2.1 and 2.8, we have

Lemma 2.10. *Suppose that* $0 \leqslant \vartheta \leqslant 8N^{-1}$. *Then we have*

$$S_1(\vartheta, N) = O(Ne^{-c\,(\ln N)^{1/3}}). \qquad (2.20)$$

Lemma 2.11. *We have*

$$\int_0^1 |S(\vartheta, N)|^2 d\vartheta = \frac{1}{2} N \ln N + O(N \ln \ln 3N). \qquad (2.21)$$

III.   Линник's Method

Lemma A. *Suppose that* $N^{-\frac{1}{2}} \leqslant \aleph = \aleph(N) \leqslant \dfrac{1}{4}$. *If*

$$\int_0^{\aleph} |S_1(\vartheta, N)|^2 d\vartheta \ll N(\ln N)^{-\varepsilon/10}, \qquad (3.1)$$

hereafter the constants implied by the symbol "$\ll$" *depend only on* ε, *then for any given* N, *there exist* p, p' *such that*

$$|N - p - p'| \ll \aleph^{-1} (\ln N)^{\varepsilon/2}. \tag{3.2}$$

Proof. Let

$$J(N_1) = \int_{-\aleph}^{\aleph} S(\vartheta, N)^2 e^{2\pi i N_1 \vartheta} \, d\vartheta. \tag{3.3}$$

Then by Lemma 2.5 and the Schwarz inequality, we have

$$J(N_1) = \int_{-\aleph}^{\aleph} \frac{e^{2\pi i N_1 \vartheta}}{x^2} \, d\vartheta + R, \tag{3.4}$$

where

$$|R| \leqslant 2 \int_0^{\aleph} |S_1(\vartheta, N)|^2 d\vartheta + O(R_1), \tag{3.5}$$

in which

$$R_1 = \sqrt{\int_0^{\aleph} \frac{d\vartheta}{|x|^2} \int_0^{\aleph} |S_1(\vartheta, N)|^2 d\vartheta} + \sqrt{\aleph \int_0^{\aleph} \frac{d\vartheta}{|x|^2} (\ln N)^2}$$

$$+ \sqrt{\aleph \int_0^{\aleph} |S_1(\vartheta, N)|^2 d\vartheta \, (\ln N)^2} + (\ln N)^4. \tag{3.6}$$

From Lemma 2.2, we have

$$\int_{-\aleph}^{\aleph} \frac{e^{2\pi i N_1 \vartheta}}{x^2} \, d\vartheta = N_1 e^{-N_1/N} + O\left(\int_{\aleph}^{\infty} \frac{d\vartheta}{\vartheta^2}\right) = N_1 e^{-N_1/N} + O(\aleph^{-1}). \tag{3.7}$$

Since we have

$$\int_0^{\aleph} \frac{d\vartheta}{|x|^2} \ll \int_0^{N^{-1}} \frac{d\vartheta}{N^{-2}} + \int_{N^{-1}}^{\infty} \frac{d\vartheta}{\vartheta^2} \ll N, \tag{3.8}$$

it follows by $(3.1), (3.4), (3.5), (3.6), (3.7)$ and (3.8) that

$$J(N_1) = N_1 e^{-N_1/N} + O(N (\ln N)^{-\varepsilon/20}). \tag{3.9}$$

Denote

$$H = \aleph^{-1} (\ln N)^{\varepsilon/2}, \tag{3.10}$$

and

$$T(\vartheta) = \sum_{0 \leqslant y \leqslant H} e^{2\pi i y \vartheta}. \tag{3.11}$$

Then we have

$$|T(\vartheta)| \leqslant |\vartheta|^{-1} \leqslant \aleph^{-1} \leqslant H(\ln N)^{-\varepsilon/2}, \tag{3.12}$$

for $\vartheta \in [\aleph, 1/2]$ and $\vartheta \in [-1/2, -\aleph]$. Let $r = [4/\varepsilon] + 1$ and $x = x_1 + \cdots + x_r$, where $0 \leqslant x_j \leqslant H$, $(1 \leqslant j \leqslant r)$. Further let N_1 denote the integers of the form $N + x$. Then we have from (3.9), (3.12) and Lemmas 2.4 and 2.11 that

$$\sum_{N_1} Q(N_1) = e \sum_{N_1} \int_{-\frac{1}{2}}^{\frac{1}{2}} S(\vartheta, N)^2 e^{2\pi i N_1 \vartheta} d\vartheta + O(\sqrt{N} \, H^r (\ln N)^2)$$

$$= e \sum_{N_1} \int_{-\aleph}^{\aleph} S(\vartheta, N)^2 e^{2\pi i N_1 \vartheta} d\vartheta + O\left(\int_{\aleph}^{\frac{1}{2}} |S(\vartheta, N)|^2 |T(\vartheta)|^r d\vartheta\right)$$

$$+ O(\sqrt{N} H^r (\ln N)^2) = e \sum_{N_1} N_1 e^{-N_1/N} + O(N H^r (\ln N)^{-\varepsilon/20})$$

$$+ O\left(H'(\ln N)^{-\varepsilon r/2}\int_{-\frac{1}{2}}^{\frac{1}{2}}|S(\vartheta, N)|^2 d\vartheta\right)$$

$$= eN\sum_{N_1}{}' e^{-1}(1 + O(N^{-1}H)) + O(NH'(\ln N)^{-\varepsilon/20})$$

$$= NH'(1 + O(\ln N)^{-\varepsilon/20}) > 2, \tag{3.13}$$

for $N > c(\varepsilon)$. The lemma follows.

IV.   The Estimation of Integral

Denote

$$M = [2\ln N] \quad \text{and} \quad M_1 = [\ln N - c(\ln N)^{\frac{1}{3}}] + 1. \tag{4.1}$$

In this section, the constants implied by the symbol "$\ll$" are all absolute constants.

Lemma B. *Suppose that* $\dfrac{1}{4} \geqslant \eta \geqslant 4N^{-1}$ *and*

$$N(T, \nu) \ll T^{\mu(\nu)(1-2\nu)}(\ln(T+2))^r, \quad \left(0 < \nu \leqslant \frac{1}{2}\right), \tag{4.2}$$

where r *is a constant and* $\mu(\nu)$ *is a non-increasing function satisfying* $\dfrac{1}{2} \leqslant \mu(\nu) \leqslant 3$.
Then

$$\int_{\eta}^{2\eta}|S_1(\vartheta, N)|^2 d\vartheta \ll N\eta(\ln N)^3 + \sum_{m=2}^{M_1} N^{\frac{2m}{M} + \mu\left(\frac{m-1}{M}\right)\left(1-\frac{2m}{M}\right)}\eta^{\mu\left(\frac{m-1}{M}\right)\left(1-\frac{2m}{M}\right)}(\ln N)^{2+r}. \tag{4.3}$$

Proof. 1) Let

$$T = 2N\eta. \tag{4.4}$$

Then we have

$$N \geqslant T \geqslant 8, \quad \frac{1}{2\pi T} \geqslant \operatorname{arctg}\frac{1}{4\pi N\eta} \geqslant \frac{1}{4\pi T}. \tag{4.5}$$

Hence it follows by Lemma 2.6 that

$$\int_{\eta}^{2\eta}|S_2(\vartheta, N)|^2 d\vartheta \ll \Sigma_1 + \Sigma_2, \tag{4.6}$$

where

$$\Sigma_1 = \sum_{\rho_{k_1},\, t_{k_1}>0}\ \sum_{\substack{\rho_{k_2},\, \beta_{k_2}\leqslant\beta_{k_1} \\ t_{k_2}>0}}(t_{k_1}+1)^{\beta_{k_1}-\frac{1}{2}}(t_{k_2}+1)^{\beta_{k_2}-\frac{1}{2}}\eta^{1-\beta_{k_1}-\beta_{k_2}}$$

$$\times \frac{1}{|t_{k_1}-t_{k_2}|+1}e^{-(t_{k_1}+t_{k_2})/4\pi T}, \tag{4.7}$$

and Σ_2 is obtained if the condition $\beta_{k_2}\leqslant\beta_{k_1}$ in Σ_1 is replaced by $\beta_{k_2}\geqslant\beta_{k_1}$.

Clearly

$$\Sigma_1 \leqslant \Sigma_{11} + \Sigma_{12} + \Sigma_{13}, \tag{4.8}$$

where

$$\Sigma_{11} = \sum_{\substack{\rho_{k_1} \\ 0 < t_{k_1} \leqslant NT}} \sum_{\substack{\rho_{k_2},\, \beta_{k_2} \leqslant \beta_{k_1} \\ 0 < t_{k_2} \leqslant NT}} (t_{k_1} + 1)^{\beta_{k_1} - \frac{1}{2}} (t_{k_2} + 1)^{\beta_{k_2} - \frac{1}{2}} \eta^{1 - \beta_{k_1} - \beta_{k_2}}$$

$$\times \frac{1}{|t_{k_1} - t_{k_2}| + 1} \, e^{-(t_{k_1} + t_{k_2})/4\pi T}, \tag{4.9}$$

$$\Sigma_{12} = \sum_{\substack{\rho_{k_1} \\ t_{k_1} \geqslant NT}} \sum_{\substack{\rho_{k_2},\, \beta_{k_2} \leqslant \beta_{k_1} \\ 0 < t_{k_2} \leqslant t_{k_1}}} (t_{k_1} + 1)^{\beta_{k_1} - \frac{1}{2}} (t_{k_2} + 1)^{\beta_{k_2} - \frac{1}{2}} \eta^{1 - \beta_{k_1} - \beta_{k_2}}$$

$$\times \frac{1}{|t_{k_1} - t_{k_2}| + 1} \, e^{-(t_{k_1} + t_{k_2})/4\pi T}, \tag{4.10}$$

and Σ_{13} is obtained if the condition concerning t_{k_1} and t_{k_2} in Σ_{12} is replaced by $0 < t_{k_1} \leqslant t_{k_2}$ and $t_{k_2} \geqslant NT$.

2) Obviously, we have

$$\Sigma_{12} \ll \sum_{s=N}^{\infty} \sum_{\substack{\rho_{k_1} \\ sT \leqslant t_{k_1} < (s+1)T}} \sum_{l=1}^{(s+1)T} (sT) \eta^{-1} \sum_{\substack{\rho_{k_2} \\ l-1 \leqslant t_{k_1} - t_{k_2} \leqslant l}} \frac{1}{|t_{k_1} - t_{k_2}| + 1} \, e^{-s/4\pi}.$$

Hence it follows by Lemma 2.7 that

$$\Sigma_{12} \ll \eta^{-1} T \sum_{s=N}^{\infty} \sum_{\substack{\rho_{k_1} \\ sT \leqslant t_{k_1} \leqslant (s+1)T}} \sum_{l=1}^{(s+1)T} \frac{\ln sT}{l} \, s e^{-s/4\pi} \ll \eta^{-1} T^2 \sum_{s=N}^{\infty} s (\ln sT)^3 \, e^{-s/4\pi}$$

$$\ll N^2 (\ln N)^3 \, e^{-N/8\pi} \sum_{s=N}^{\infty} s (\ln s)^3 \, e^{-s/8\pi} \ll 1. \tag{4.11}$$

Since Σ_{13} satisfies the same relation as Σ_{12}, we have by (4.8) that

$$\Sigma_1 \ll \Sigma_{11} + 1. \tag{4.12}$$

3) Let I_m denote the interval $\left[\frac{1}{2} + \frac{m-1}{M}, \ \frac{1}{2} + \frac{m}{M} \right]$. Then from Lemma 2.8, we have

$$\Sigma_{11} = \sum_{m=1}^{M_1} \Sigma_{11}(m), \tag{4.13}$$

where

$$\Sigma_{11}(m) = \sum_{\substack{\rho_{k_1},\, \beta_{k_1} \in I_m \\ 0 < t_{k_1} \leqslant NT}} \sum_{\substack{\rho_{k_2},\, \beta_{k_2} \leqslant \beta_{k_1} \\ 0 < t_{k_2} \leqslant NT}} (t_{k_1} + 1)^{\beta_{k_1} - \frac{1}{2}} (t_{k_2} + 1)^{\beta_{k_2} - \frac{1}{2}} \eta^{1 - \beta_{k_1} - \beta_{k_2}}$$

$$\times \frac{1}{|t_{k_1} - t_{k_2}| + 1} \, e^{-(t_{k_1} + t_{k_2})/4\pi T}. \tag{4.14}$$

Obviously, we have

$$\Sigma_{11}(m) \ll \Sigma_{111}(m) + \Sigma_{112}(m), \tag{4.15}$$

where

$$\Sigma_{111}(m) = \sum_{s=0}^{N=1} \sum_{\substack{\rho_{k_1},\beta_{k_1}\in I_m \\ sT<t_{k_1}\leqslant(s+1)T}} \sum_{\substack{\rho_{k_2},\beta_{k_2}\leqslant\beta_{k_1} \\ t_{k_2}\leqslant(s+2)T}} (t_{k_1}+1)^{\beta_{k_1}-\frac{1}{2}} (t_{k_2}+1)^{\beta_{k_2}-\frac{1}{2}} \eta^{1-\beta_{k_1}-\beta_{k_2}}$$

$$\times \frac{1}{|t_{k_1}-t_{k_2}|+1}\, e^{-(t_{k_1}+t_{k_2})/4\pi T}, \tag{4.16}$$

and

$$\Sigma_{112}(m) = \sum_{s=0}^{N-1} \sum_{\substack{\rho_{k_1},\beta_{k_1}\in I_m \\ sT<t_{k_1}\leqslant(s+1)T}} \sum_{\substack{\rho_{k_2},\beta_{k_2}\leqslant\beta_{k_1} \\ (s+2)T<t_{k_2}}} (t_{k_1}+1)^{\beta_{k_1}-\frac{1}{2}} (t_{k_2}+1)^{\beta_{k_2}-\frac{1}{2}} \eta^{1-\beta_{k_1}-\beta_{k_2}}$$

$$\times \frac{1}{|t_{k_1}-t_{k_2}|+1}\, e^{-(t_{k_1}+t_{k_2})/4\pi T}. \tag{4.17}$$

4) Suppose that $m \geqslant 2$. Since $1 \ll (NT)^{1/M} \ll 1$, we have

$$\Sigma_{111}(m) \ll \sum_{s=0}^{N-1} \sum_{\substack{\rho_{k_1},\beta_{k_1}\in I_m \\ sT<t_{k_1}\leqslant(s+1)T}} ((s+1)T)^{2\beta_{k_1}-1}\eta^{1-2\beta_{k_1}} \sum_{l=1}^{(s+2)T} \frac{\ln(s+2)T}{l}\, e^{-s/4\pi}$$

$$\ll T^{2m/M}\eta^{-2m/M}(\ln N)^2 \sum_{s=0}^{N-1} (s+1)^{2m/M} e^{-s/4\pi} N\left((s+1)T, \frac{m-1}{M}\right)$$

$$\ll T^{\frac{2m}{M}+\mu\left(\frac{m-1}{M}\right)\left(1-\frac{2m}{M}\right)} \eta^{-2m/M} (\ln N)^{2+r} \sum_{s=0}^{\infty} (s+1)^{\frac{2m}{M}+\mu\left(\frac{m-1}{M}\right)\left(1-\frac{2m}{M}\right)} e^{-s/4\pi}$$

$$\ll N^{\frac{2m}{M}+\mu\left(\frac{m-1}{M}\right)\left(1-\frac{2m}{M}\right)} \eta^{\mu\left(\frac{m-1}{M}\right)\left(1-\frac{2m}{M}\right)} (\ln N)^{2+r}. \tag{4.18}$$

For the case $m = 1$, it follows by Lemma 2.7 that

$$\Sigma_{111}(1) = \sum_{s=0}^{N-1} (\ln(s+1)T)^2\, e^{-s/4\pi} N((s+1)T, 0)$$

$$\ll T(\ln N)^3 \sum_{s=0}^{N-1} (s+1) e^{-s/4\pi} \ll N\eta\,(\ln N)^3. \tag{4.19}$$

5) From Lemma 2.7, we have

$$\Sigma_{112}(m) \ll T^{-1} \sum_{s=0}^{N-1} \sum_{\substack{\rho_{k_1},\beta_{k_1}\in I_m \\ sT<t_{k_1}\leqslant(s+1)T}} ((s+1)T)^{\beta_{k_1}-\frac{1}{2}} \eta^{1-2\beta_{k_1}} e^{-s/4\pi}$$

$$\times \sum_{x=[(s+2)T]}^{\infty} \sum_{\substack{\rho_{k_2} \\ x<t_{k_2}\leqslant x+1}} (t_{k_2}+1)^{\beta_{k_1}-\frac{1}{2}} e^{-t_{k_2}/4\pi T}$$

$$\ll T^{-1} \sum_{s=0}^{N-1} \sum_{\substack{\rho_{k_1},\beta_{k_1}\in I_m \\ sT<t_{k_1}\leqslant(s+1)T}} ((s+1)T)^{\beta_{k_1}-\frac{1}{2}} \eta^{1-2\beta_{k_1}} e^{-s/4\pi}$$

$$\times \sum_{x=[(s+2)T]}^{\infty} (x+2)^{\beta_{k_1}-\frac{1}{2}} e^{-x/4\pi T} \ln (x+2)$$

$$\ll T^{-1} \sum_{s=0}^{N-1} \sum_{\substack{\rho_{k_1},\, \beta_{k_1} \in I_m \\ sT < t_{k_1} \leqslant (s+1)T}} ((s+1)T)^{\beta_{k_1}-\frac{1}{2}} \eta^{1-2\beta_{k_1}} e^{-s/4\pi}$$

$$\times \int_{[(s+2)T]}^{\infty} x^{\beta_{k_1}-\frac{1}{2}} e^{-x/4\pi T} \ln (x+2) dx$$

$$\ll \sum_{s=0}^{N-1} \sum_{\substack{\rho_{k_1},\, \beta_{k_1} \in I_m \\ sT < t_{k_1} \leqslant (s+1)T}} ((s+1)T)^{2\beta_{k_1}-1} \eta^{1-2\beta_{k_1}} e^{-s/4\pi} \ln N$$

$$\times \int_{s+1}^{\infty} y^{\beta_{k_1}-\frac{1}{2}} e^{-y/4\pi} \ln (y+2) dy$$

$$\ll \begin{cases} N^{\frac{2m}{M}+\mu\left(\frac{m-1}{M}\right)\left(1-\frac{2m}{M}\right)} \eta^{\mu\left(\frac{m-1}{M}\right)\left(1-\frac{2m}{M}\right)} (\ln N)^{1+r}, & \text{for } m \geqslant 2, \\ N\eta(\ln N)^2, & \text{for } m = 1. \end{cases} \qquad (4.20)$$

From (4.12), (4.13), (4.15), (4.18), (4.19) and (4.20), we have

$$\Sigma_1 \ll N\eta(\ln N)^3 + \sum_{m=2}^{M_1} N^{\frac{2m}{M}+\mu\left(\frac{m-1}{M}\right)\left(1-\frac{2m}{M}\right)} \eta^{\mu\left(\frac{m-1}{M}\right)\left(1-\frac{2m}{M}\right)} (\ln N)^{2+r}. \qquad (4.21)$$

Obviously, Σ_2 satisfies the same relation as Σ_1. Therefore we have by (4.6) that

$$\int_{\eta}^{2\eta} |S_2(\vartheta, N)|^2 d\vartheta \ll N\eta(\ln N)^3$$

$$+ \sum_{m=2}^{M_1} N^{\frac{2m}{M}+\mu\left(\frac{m-1}{M}\right)\left(1-\frac{2m}{M}\right)} \eta^{\mu\left(\frac{m-1}{M}\right)\left(1-\frac{2m}{M}\right)} (\ln N)^{2+r}. \qquad (4.22)$$

Similarly, we may prove that (4.22) holds also, if $S_2(\vartheta, N)$ is replaced by $S_3(\vartheta, N)$. Since

$$\int_{\eta}^{2\eta} |S_1(\vartheta, N)|^2 d\vartheta \leqslant 2 \int_{\eta}^{2\eta} |S_2(\vartheta, N)|^2 d\vartheta + 2 \int_{\eta}^{2\eta} |S_3(\vartheta, N)|^2 d\vartheta, \qquad (4.23)$$

hence we have the lemma.

V. THE PROOF OF THEOREM 1

Let

$$\aleph = \aleph(N) = (\ln N)^{-\frac{148}{13}-\frac{\varepsilon}{2}}. \qquad (5.1)$$

Let

$$\varepsilon_1 = 10^{-3}\varepsilon \quad \text{and} \quad M_2 = \left[2\left(\frac{12}{37} + \varepsilon_1\right) \ln N\right] + 1. \qquad (5.2)$$

Then it follows from Lemma 2.9 and the supposition of the theorem that there exists a positive constant $\varepsilon_2 = \varepsilon_2(\varepsilon)$ such that

$$N(T, \nu) \ll \begin{cases} T^{1-2\nu} \ln (T+2), & \text{for } \nu \in I_m \ (1 \leqslant m \leqslant M_2), \\ T^{\frac{1}{1+\varepsilon_2}(1-2\nu)}, & \text{for } \nu \in I_m \ (M_2 < m \leqslant M_1); \end{cases} \qquad (5.3)$$

hereafter the constants implied by the symbol "$\ll$" depend on ε only.

Let L be an integer such that

$$\frac{\aleph}{2^{L+1}} < \frac{4}{N} \leqslant \frac{\aleph}{2^{L}}. \tag{5.4}$$

Then

$$\sum_{m=1}^{M_2} N^{\frac{2m}{M}+\left(1-\frac{2m}{M}\right)}\left(\frac{\aleph}{2^l}\right)^{1-\frac{2m}{M}}(\ln N)^3 \ll \frac{N\aleph}{2^l}(\ln N)^3 \sum_{m=1}^{M_2}\left(\frac{\aleph}{2^l}\right)^{-\frac{2m}{M}}$$

$$\ll \frac{N\aleph}{2^l}\left(\frac{\aleph}{2^l}\right)^{-2M_2/M}(\ln N)^4 \ll N(\ln N)^{-\varepsilon/10}2^{-\frac{13}{38}l}, \tag{5.5}$$

for $1 \leqslant l \leqslant L$ and

$$\sum_{m=M_2+1}^{M_1} N^{\frac{2m}{M}+\frac{1}{1+\varepsilon_2}\left(1-\frac{2m}{M}\right)}\left(\frac{\aleph}{2^l}\right)^{\frac{1}{1+\varepsilon_2}\left(1-\frac{2m}{M}\right)}(\ln N)^2 \ll N^{\frac{2M_1}{M}+\frac{1}{1+\varepsilon_2}\left(1-\frac{2M_1}{M}\right)}(\ln N)^3$$

$$\ll N^{\frac{1}{1+\varepsilon_2}+\frac{\varepsilon_2}{1+\varepsilon_2}\cdot\frac{2M_1}{M}}(\ln N)^3 \ll N e^{-\frac{\varepsilon_2}{2}c(\ln N)^{1/3}} \ll N(\ln N)^{-2}. \tag{5.6}$$

Hence by Lemma B, we have

$$\int_{\aleph/2^l}^{\aleph/2^{l-1}} |S_1(\vartheta, N)|^2\, d\vartheta \ll N\aleph(\ln N)^3\, 2^{-l} + N(\ln N)^{-\varepsilon/10}\, 2^{-\frac{13}{38}l}$$

$$+ N(\ln N)^{-2} \ll N(\ln N)^{-\varepsilon/10}\, 2^{-\frac{13}{38}l} + N(\ln N)^{-2}. \tag{5.7}$$

Obviously, we have by Lemma 2.10 that

$$\int_0^{\aleph/2^L} |S_1(\vartheta, N)|^2\, d\vartheta \leqslant \int_0^{8/N} |S_1(\vartheta, N)|^2\, d\vartheta \ll N e^{-c(1,N)^{1/3}}. \tag{5.8}$$

Consequently, it follows by (5.7) and (5.8) that

$$\int_0^{\aleph} |S_1(\vartheta, N)|^2\, d\vartheta = \sum_{l=1}^{L}\int_{\aleph/2^l}^{\aleph/2^{l-1}} |S_1(\vartheta, N)|^2\, d\vartheta$$

$$+ \int_0^{\aleph/2^L} |S_1(\vartheta, N)|^2\, d\vartheta \ll N(\ln N)^{-\varepsilon/10}. \tag{5.9}$$

Hence, Theorem 1 follows from Lemma A.

We omit the proofs of Theorems 2 and 3. They are parallel to the proof of Theorem 1.

Remark. In his original proof of a theorem which is similar to Theorem 1, Линник used the following arguments:

"Hence, it is sufficient for us, if the inequality

$$N^{1+2\nu-\varphi(\nu)}\left(\frac{\aleph_N}{2^r}\right)^{1-\varphi(\nu)} \ll \frac{N}{(\ln N)^7}$$

or

$$\frac{\aleph_N}{2^r} \ll \frac{1}{(\ln N)^{\frac{7}{1-\varphi(\nu)}}} \cdot \frac{1}{N^{\frac{2\nu-\varphi(\nu)}{1-\varphi(\nu)}}}$$

holds for $0 \leqslant \nu \leqslant \dfrac{1}{2} - \dfrac{1}{(\ln N)^{10/11}}$" (cf. p. 515 of [7]). If we take $r = 1$, $\varphi(\nu) = 2\nu$

and $\aleph_N = (\ln N)^{-7}$, then the above inequality is not satisfied with $\nu = \dfrac{1}{2} - \dfrac{1}{(\ln N)^{10/11}}$.

Hence it seems to me that the conclusion $O((\ln N)^7)$ should be replaced by $O\left(e^{c(\ln N)^{10/11}\ln\ln 3N}\right)$.

VI.   The Proof of Theorem 4

1) Denote

$$E(\chi) = \begin{cases} 1, & \text{for } \chi = \chi_0 \\ 0, & \text{otherwise}, \end{cases} \tag{6.1}$$

$$\tau_\chi = \sum_{(l,q)=1} \chi(l) e^{2\pi i l/q}, \tag{6.2}$$

$$S(\vartheta, N, \chi) = \sum_{n=2}^{\infty} \Lambda(n)\chi(n)\, e^{-xn}, \tag{6.3}$$

$$S_1(\vartheta, N, \chi) = \sum_{\rho_\chi} x^{-\rho_\chi}\Gamma(\rho_\chi), \tag{6.4}$$

where $\displaystyle\sum_{\rho_\chi}$ denotes a sum over all critical zeros of $L(s, \chi)$. Then similar to Lemma

2.5, we have

$$S(\vartheta, N, \chi) = \frac{E(\chi)}{x} - S_1(\vartheta, N, \chi) - S_2(\vartheta, N, \chi), \tag{6.5}$$

where

$$S_2(\vartheta, N, \chi) = O((\ln N)^2). \tag{6.6}$$

Since

$$S\left(\frac{a}{q} + \vartheta, N\right) = \frac{1}{\varphi(q)} \sum_{\chi_q} \bar{\tau}_\chi \chi(a) S(\vartheta, N, \chi) + O((\ln N)^2), \tag{6.7}$$

for $(a, q) = 1$, where $\displaystyle\sum_{\chi_q}$ denotes a sum over all character mod q, and since $\tau_{\chi_0} =$

$\mu(q)$ and $|\tau_\chi| \leqslant \sqrt{q}$ for all χ mod q (cf. Prachar [3]), we have

$$S\left(\frac{a}{q} + \vartheta, N\right)^2 = \left(\frac{1}{\varphi(q)} \sum_{\chi_q} \bar{\tau}_\chi \chi(a)(S_1(\vartheta, N, \chi) + S_2(\vartheta, N, \chi))\right)^2 + O(N(\ln N)^2)$$

$$= \frac{\mu(q)^2}{\varphi(q)^2 x^2} + \frac{1}{\varphi(q)^2}\left(\sum_{\chi_q} \bar{\tau}_\chi \chi(a)(S_1(\vartheta, N, \chi) + S_2(\vartheta, N, \chi))\right)^2$$

$$- \frac{2\mu(q)}{\varphi(q)^2 x} \sum_{\chi_q} \bar{\tau}_\chi \chi(a)(S_1(\vartheta, N, \chi) + S_2(\vartheta, N, \chi)) + O(N(\ln N)^2). \tag{6.8}$$

Under the truth of the grand Riemann hypothesis, it is similar to Lemma B that

$$\int_{-\eta}^{\eta} |S_1(\vartheta, N, \chi)|^2 d\vartheta \ll N\eta(\ln N)^3, \tag{6.9}$$

for $\dfrac{1}{2} \geqslant \eta \geqslant 2N^{-1}$.

2) Suppose that N_1 is an integer satisfying $N/2 \leqslant N_1 \leqslant N$. Let

$$J_0(N_1, \eta) = \sum_{a=0}^{q-1} \int_{\frac{a}{q}-\eta}^{\frac{a}{q}+\eta} S(\vartheta, N)^2 e^{2\pi i N_1 \vartheta} d\vartheta, \tag{6.10}$$

where $\dfrac{1}{2q} \geqslant \eta \geqslant \dfrac{2}{N}$. Then it follows by (6.8) and (6.9) that

$$\begin{aligned}
J_0(N_1, \eta) &= \sum_{q_1|q} \sum_{(a,q_1)=1} e^{2\pi i a N_1/q} \int_{-\eta}^{\eta} S\left(\frac{a}{q_1} + \vartheta, N\right)^2 e^{2\pi i N_1 \vartheta} d\vartheta \\
&= \sum_{q_1|q} \frac{\mu(q_1)^2}{\varphi(q_1)^2} \sum_{(a,q)=1} e^{2\pi i a N_1/q} \int_{-\eta}^{\eta} \frac{e^{2\pi i N_1 \vartheta}}{x^2} d\vartheta \\
&\quad + O(R_1) + O(R_2) + O(N\eta(\ln N)^3),
\end{aligned} \tag{6.11}$$

where

$$R_1 = \sum_{q_1|q} \frac{1}{\varphi(q_1)^2} \sum_{(a,q_1)=1} \int_{-\eta}^{\eta} \left| \sum_{\chi_{q_1}} \bar{\tau}_\chi \chi(a)(S_1(\vartheta, N, \chi) + S_2(\vartheta, N, \chi)) \right|^2 d\vartheta, \tag{6.12}$$

and

$$R_2 = \sum_{q_1|q} \frac{1}{\varphi(q_1)^2} \sum_{(a,q_1)=1} \int_{-\eta}^{\eta} \left| x^{-1} \sum_{\chi_{q_1}} \bar{\tau}_\chi \chi(a)(S_1(\vartheta, N, \chi) + S_2(\vartheta, N, \chi)) \right| d\vartheta. \tag{6.13}$$

By (6.6) and (6.9), we have

$$\begin{aligned}
R_1 &= \sum_{q_1|q} \frac{1}{\varphi(q_1)^2} \int_{-\eta}^{\eta} \sum_{\chi_{q_1}} \sum_{\chi'_{q_1}} \bar{\tau}_\chi \tau_{\chi'} \sum_{(a,q_1)=1} \chi \bar{\chi}'(a)(S_1(\vartheta, N, \chi) + S_2(\vartheta, N, \chi)) \\
&\quad \times \overline{(S_1(\vartheta, N, \chi') + S_2(\vartheta, N, \chi'))} \, d\vartheta \ll \sum_{q_1|q} \frac{q_1}{\varphi(q_1)} \sum_{\chi_{q_1}} \int_{-\eta}^{\eta} \\
&\quad \times (|S_1(\vartheta, N, \chi)|^2 + O((\ln N)^4)) d\vartheta \ll \sum_{q_1|q} q_1 N\eta(\ln N)^3,
\end{aligned} \tag{6.14}$$

and

$$\begin{aligned}
R_2 &\leqslant \sum_{q_1|q} \frac{1}{\varphi(q_1)^2} \int_{-\eta}^{\eta} \left(\sum_{(a,q_1)=1} |x|^{-2} \right)^{1/2} \\
&\quad \times \left(\sum_{(b,q_1)=1} \left| \sum_{\chi_{q_1}} \bar{\tau}_\chi \chi(b)(S_1(\vartheta, N, \chi) + S_2(\vartheta, N, \chi)) \right|^2 \right)^{1/2} d\vartheta \\
&\leqslant \sum_{q_1|q} \frac{1}{\varphi(q_1)^2} \left(\int_{-\eta}^{\eta} \sum_{(a,q_1)=1} |x|^{-2} d\vartheta \right)^{1/2} \\
&\quad \times \left(\int_{-\eta}^{\eta} \sum_{(b,q_1)=1} \left| \sum_{\chi_{q_1}} \bar{\tau}_\chi \chi(b)(S_1(\vartheta, N, \chi) + S_2(\vartheta, N, \chi)) \right|^2 d\vartheta \right)^{1/2} \\
&\leqslant \sum_{q_1|q} \frac{1}{\varphi(q_1)^2} (\varphi(q_1)N)^{1/2} (q_1 \varphi(q_1)^2 N\eta(\ln N)^3)^{1/2} \\
&\leqslant \sum_{q_1|q} \frac{q_1^{1/2}}{\varphi(q_1)^{1/2}} N\eta^{1/2} (\ln N)^{3/2}.
\end{aligned} \tag{6.15}$$

Hence from $(6.11), (6.14)$ and (6.15), we have

$$
\begin{aligned}
J_0(N_1, \eta) &= \sum_{q_1 \mid q} \frac{\mu(q_1)^2}{\varphi(q_1)^2} \sum_{(a, q_1)=1} e^{2\pi i a N_1/q_1}(N_1 e^{-N_1/N} + O(\eta^{-1})) \\
&\quad + O\left(\sum_{q_1 \mid q} q_1 N\eta (\ln N)^3\right) + O\left(\sum_{q_1 \mid q} \frac{q_1^{1/2}}{\varphi(q_1)^{1/2}} N\eta^{1/2}(\ln N)^{3/2}\right) \\
&= N_1 e^{-N_1/N} \prod_{p \mid q} (1 + \Psi_{N_1}(p)) + O\left(\sum_{q_1 \mid q} \frac{1}{\eta\varphi(q_1)}\right) \\
&\quad + O\left(\sum_{q_1 \mid q} q_1 N\eta (\ln N)^3\right) + O\left(\sum_{q_1 \mid q} \frac{q_1^{1/2}}{\varphi(q_1)^{1/2}} N\eta^{1/2}(\ln N)^{3/2}\right), \quad (6.16)
\end{aligned}
$$

where

$$
\Psi_{N_1}(p) = \begin{cases} \dfrac{1}{p-1}, & \text{for } p \mid N_1, \\[2ex] -\dfrac{1}{(p-1)^2}, & \text{for } p \nmid N_1. \end{cases} \quad (6.17)
$$

3) Let

$$
\aleph = (\ln N)^{-3-\frac{\varepsilon}{2}}, \quad H = (\ln N)^{3+\varepsilon} \quad \text{and} \quad r = \left[\frac{4}{\varepsilon}\right] + 1. \quad (6.18)
$$

Let

$$
T(\vartheta) = \sum_{0 \leqslant h \leqslant \frac{H}{2r}} e^{-2\pi i q h \vartheta}. \quad (6.19)
$$

Let $\mathfrak{M}$ be the sum of intervals $\left[\dfrac{a}{q} - \dfrac{\aleph}{q}, \dfrac{a}{q} + \dfrac{\aleph}{q}\right]$ $(0 \leqslant a \leqslant q - 1)$ and m be the complementary set of $\mathfrak{M}$ in $\left[-\dfrac{\aleph}{q}, 1 - \dfrac{\aleph}{q}\right]$. Further let N_1 denote the integers of the form

$$
N_1 = N - q(h_1 + \cdots + h_r), \quad (6.20)
$$

where $0 \leqslant h \leqslant \dfrac{H}{2r}$. Then $\dfrac{N}{2} \leqslant N_1 \leqslant N$ and

$$
J = \sum_{N_1} \int_{-\frac{\aleph}{q}}^{1-\frac{\aleph}{q}} S(\vartheta, N)^2 e^{2\pi i N_1 \vartheta} d\vartheta = \sum_{N_1} J_0\left(N_1, \frac{\aleph}{q}\right) + \sum_{N_1} J_1\left(N_1, \frac{\aleph}{q}\right), \quad (6.21)
$$

where

$$
J_1\left(N_1, \frac{\aleph}{q}\right) = \int_m S(\vartheta, N)^2 e^{2\pi i N_1 \vartheta} d\vartheta. \quad (6.22)
$$

Since

$$
T(\vartheta) \ll \aleph^{-1}, \quad (6.23)
$$

for $\vartheta \in m$, therefore by (6.18) and Lemma 2.11, we have

$$
\sum_{N_1} J_1\left(N, \frac{\aleph}{q}\right) = \int_m S(\vartheta, N)^2 T(\vartheta)^r e^{2\pi i N\vartheta} d\vartheta \ll \aleph^{-r} \int_0^1 |S(\vartheta, N)|^2 d\vartheta
$$

$$\ll H^r (\ln N)^{-\varepsilon r/2} N \ln N \ll \frac{NH^r}{\ln N}. \tag{6.24}$$

Consequently, it follows by (6.16) (notice that $\Psi_{N_1}(p) = \Psi_N(p)$ for $p\,|\,q$), (6.21) and (6.24) that

$$J = \prod_{p\,|\,q} (1 + \Psi_N(p)) \sum_{N_1} N_1 e^{-N_1/N} + O\left(\frac{H^r q}{\aleph} \sum_{q_1\,|\,q} \frac{1}{\varphi(q_1)}\right)$$

$$+ O\left(NH^r \aleph (\ln N)^3 \sum_{q_1\,|\,q} \frac{1}{q_1}\right) + O\left(\frac{NH^r \aleph^{1/2}(\ln N)^{3/2}}{q^{1/2}} \sum_{q_1\,|\,q} \frac{q_1^{1/2}}{\varphi(q_1)^{1/2}}\right)$$

$$+ O\left(\frac{NH^r}{\ln N}\right) = \prod_{p\,|\,q} (1 + \Psi_N(p)) \sum_{N_1} N_1 e^{-N_1/N} + O(NH^r(\ln N)^{-\varepsilon/4}). \tag{6.25}$$

Since N is even if q is even, hence

$$\prod_{p\,|\,q} (1 + \Psi_N(p)) > c > 0, \tag{6.26}$$

where c is an absolute constant. Consequently, we have

$$J > c \sum_{N_1} N_1 e^{-N_1/N} + O(NH^r(\ln N)^{-\varepsilon/4}) > \frac{c}{3e^2(2r)^r} NH^r, \tag{6.27}$$

for $N > c(\varepsilon)$. The theorem follows.

<h3 style="text-align:center">REFERENCES</h3>

[1] Huxley, M. N.: Large values of Dirichlet polynomials, *Acta Arith.*, **24** (1973), 329—346; II, *Acta Arith.*, **27** (1975), 159—169; III, *Acta Arith.* (to appear.)

[2] Levinson, N.: More than one-third of zeros of Riemann's zeta function are on $\sigma = 1/2$, *Adv. in Math*; **4** (1974), 383—436.

[3] Prachar, K.: *Primzahlverteilung*, (1957) *Spr.-Ver.*

[4] Selberg, A.: On the zeros of Riemann's zeta function, *Skr. Nor. Vid. Akad. Oslo*; **10** (1942), 1—59.

[5] Виноградов, И. М.: Новая оценка функции $\zeta(1 + it)$, *ИАН СССР, сер. мат*; **22** (1958), 161—164

[6] Коробов, Н. М.: Оценки тригонометрических сумм и их приложения, *УМН СССР*, **4** (1958) 185—192.

[7] Линник, Ю. В.: Некоторые условные теоремы, касаюшиеся бинарных задач с простыми числами, *ДАН СССР*, **77** (1951), 15—18; Некоторые условные теоремы, касаюшиеся бинарных проблемы Гольдбах, *ИАН СССР, сер. мат*; **16** (1952), 503—520.

REMARKS CONCERNING A TRANSFERENCE THEOREM OF LINEAR FORMS*

WANG YUAN AND YU KUN-RUI

Institute of Mathematics, Academia Sinica

ZHU YAO-CHENG

Institute of Applied Mathematics, Academia Sinica

Received 1 June 1977
Revised 22 July 1977

Let $n \geq 2$ and $\theta_1, \ldots, \theta_n$ be a set of real numbers such that $1, \theta_1, \ldots, \theta_n$ are linearly independent over rational number field $\mathbb{Q}$. For a real number x, let

$$\|x\| = \min(x - [x], [x] + 1 - x) \quad \text{and} \quad \bar{x} = \max(1, |x|).$$

We use ε to denote a preassigned positive number and c_1, c_2 the positive constants depending only on $\varepsilon, n, \theta_1, \ldots, \theta_n$.

Proposition A. *There exists c_1 such that*

$$(\bar{x}_1 \cdots \bar{x}_n)^{1+\varepsilon} \|x_1 \theta_1 + \cdots + x_n \theta_n\| > 1$$

holds for any given set of integers $x_1, \ldots, x_n$ satisfying $x = \max_{1 \leq i \leq n} |x_i| > c_1$.

Proposition B. *There exists c_2 such that for any integer $y_{n+1} > c_2$, we have*

$$y_{n+1}^{1+\varepsilon} \|y_{n+1} \theta_1\| \cdots \|y_{n+1} \theta_n\| > 1.$$

In this paper, we shall use Mahler–Baker's method to prove the following transference theorem:

Theorem 1. *Propositions A and B are equivalent.*

A similar result may also be established if the ε in Propositions A and B is replaced respectively by certain functions $\varepsilon_1(x)$ and $\varepsilon_2(y_{n+1})$ which take positive values and tend to zero decreasingly.

---

*Acta Mathematica Sinica, **22**:2 (1979) 237–240.

Proof of Theorem 1. 1) Assume first the truth of Proposition A. We proceed to show that Proposition B holds. We may assume that $y_{n+1} \geq 2$. Let

$$\|y_{n+1}\theta_i\| = |y_{n+1}\theta_i - y_i|, \quad 1 \leq i \leq n. \tag{1}$$

Since $1, \theta_1, \ldots, \theta_n$ are linearly independent over $\mathbb{Q}$, we have

$$\|y_{n+1}\theta_i\| > 0, \quad 1 \leq i \leq n.$$

Let

$$\varphi_i = (y_{n+1}\|y_{n+1}\theta_1\| \cdots \|y_{n+1}\theta_n\|)^{\frac{1}{n}} \|y_{n+1}\theta_i\|^{-1}, \quad 1 \leq i \leq n. \tag{2}$$

Without loss of generality we may assume

$$\varphi_1 \geq \varphi_2 \geq \cdots \geq \varphi_n.$$

Since

$$\varphi_1\varphi_2 \cdots \varphi_n = y_{n+1} \geq 2, \tag{3}$$

it is impossible that $\varphi_i \leq 1$ for all i. If $\varphi_n > 1$, then we set $k = n$. Otherwise we define k to be the least integer satisfying $1 \leq k < n$ and $\varphi_{k+1} \cdots \varphi_n \leq 1$. The existence of such k follows by (3). Consider the system of linear inequalities of $x_1, \ldots, x_k, x_{n+1}$:

$$
\begin{aligned}
&|x_j| \leq \varphi_j, \quad 1 \leq j \leq k-1, \\
&|x_k| \leq \varphi_k \cdots \varphi_n, \\
&|y_1 x_1 + \cdots + y_k x_k + y_{n+1} x_{n+1}| < 1.
\end{aligned}
\tag{4}
$$

Since the absolute value of the determinant of linear forms on the left-hand sides of (4) is y_{n+1} and the product of right-hand sides of (4) is also y_{n+1} by (3), it follows by Minkowski's linear form theorem [3] that there are integers $x_1, \ldots, x_k, x_{n+1}$ not all zero and satisfying (4). The last inequality in (4) yields that

$$y_1 x_1 + \cdots + y_k x_k + y_{n+1} x_{n+1} = 0. \tag{5}$$

Hence $x_1, \ldots, x_k$ are not all zero. By the definition of k, we have $\varphi_1 \geq \cdots \geq \varphi_{k-1} > 1$, $\varphi_k \cdots \varphi_n > 1$, and

$$\bar{x}_1 \cdots \bar{x}_k \leq \varphi_1 \cdots \varphi_n = y_{n+1} \tag{6}$$

by (3) and (4). Therefore by the identity

$$y_1 x_1 + \cdots + y_k x_k + y_{n+1} x_{n+1} = (x_1\theta_1 + \cdots + x_k\theta_k + x_{n+1})y_{n+1}$$

$$- \sum_{i=1}^{k}(y_{n+1}\theta_i - y_i)x_i$$

and (1), (2), (4), (5), we obtain

$$|x_1\theta_1 + \cdots + x_k\theta_k + x_{n+1}|y_{n+1} \leq \sum_{i=1}^{k} |x_i| \|y_{n+1}\theta_i\|$$

$$\leq \sum_{i=1}^{k} \varphi_i \|y_{n+1}\theta_i\|$$

$$\leq k(y_{n+1}\|y_{n+1}\theta_1\| \cdots \|y_{n+1}\theta_n\|)^{\frac{1}{n}}. \qquad (7)$$

To prove Proposition B, it is sufficient to assume $\varepsilon < 1$. Suppose that $y_{n+1} \geq 2$ and

$$y_{n+1}^{1+\varepsilon}\|y_{n+1}\theta_1\| \cdots \|y_{n+1}\theta_n\| \leq 1. \qquad (8)$$

Then from Proposition A we may choose positive integer λ depending only on $\varepsilon, n, \theta_1, \ldots, \theta_n$ such that

$$(\overline{\lambda x_1} \cdots \overline{\lambda x_n})^{1+\frac{\varepsilon}{2n}} \|\lambda x_1\theta_1 + \cdots + \lambda x_n\theta_n\| > 1 \qquad (9)$$

holds for any not all zero integers $x_1, \ldots, x_n$, in particular, for the y_{n+1} with (8) and those $x_1, \ldots, x_k, x_{k+1} = \cdots = x_n = 0$, not all zero and satisfying (4). By the combination of (6)–(8) and $\varepsilon < 1$, we derive that

$$1 < (\overline{\lambda x_1} \cdots \overline{\lambda x_k})^{1+\frac{\varepsilon}{2n}} \|\lambda x_1\theta_1 + \cdots + \lambda x_k\theta_k\|$$

$$\leq \lambda^{k(1+\frac{\varepsilon}{2n})+1}(\bar{x}_1 \cdots \bar{x}_k)^{1+\frac{\varepsilon}{2n}} \|x_1\theta_1 + \cdots + x_k\theta_k\|$$

$$\leq \lambda^{n+2}\|x_1\theta_1 + \cdots + x_k\theta_k\|y_{n+1}y_{n+1}^{\frac{\varepsilon}{2n}}$$

$$\leq \lambda^{n+2}k(y_{n+1}\|y_{n+1}\theta_1\| \cdots \|y_{n+1}\theta_n\|)^{\frac{1}{n}} y_{n+1}^{\frac{\varepsilon}{2n}}$$

$$= \lambda^{n+2}k(y_{n+j}^{1+\varepsilon}\|y_{n+1}\theta_1\| \cdots \|y_{n+1}\theta_n\|)^{\frac{1}{n}} y_{n+1}^{-\frac{\varepsilon}{2n}}$$

$$\leq \lambda^{n+2}ny_{n+1}^{-\frac{\varepsilon}{2n}}.$$

This means that if (8) holds, then $y_{n+1} < n^{\frac{2n}{\varepsilon}}\lambda^{\frac{2n(n+2)}{\varepsilon}}$. The right-hand side of the above formula may be regarded as c_2 in Proposition B. Therefore Proposition B follows.

2) Assume now that Proposition B holds. We proceed to prove Proposition A. Let

$$x = \max_{1 \leq i \leq n} |x_i|,$$

where $x_1, \ldots, x_n$ are integers. Suppose that $x \geq 1$ and

$$(\bar{x}_1 \cdots \bar{x}_n \|x_1\theta_1 + \cdots + x_n\theta_n\|)^{\frac{1}{n}} < 1. \qquad (10)$$

Otherwise Proposition A is obviously true. Suppose

$$|x_1\theta_1 + \cdots + x_n\theta_n + x_{n+1}| = \|x_1\theta_1 + \cdots + x_n\theta_n\|. \qquad (11)$$

Set

$$\psi_i = (\bar{x}_1 \cdots \bar{x}_n \|x_1\theta_1 + \cdots + x_n\theta_n\|)^{\frac{1}{n}} \bar{x}_i^{-1}, \quad 1 \le i \le n. \tag{12}$$

Since $1, \theta_1, \ldots, \theta_n$ are linearly independent over $\mathbb{Q}$, we have

$$\psi_i > 0, \quad 1 \le i \le n.$$

Consider the system of linear inequalities of $y_1, \ldots, y_n, y_{n+1}$:

$$\begin{aligned} |y_i - y_{n+1}\theta_i| &\le \psi_i, \quad 1 \le i \le n, \\ |x_1 y_1 + \cdots + x_n y_n + x_{n+1} y_{n+1}| &< 1. \end{aligned} \tag{13}$$

Since the absolute value of the determinant of linear forms on the left-hand sides of (13) is $\|x_1\theta_1 + \cdots + x_n\theta_n\|$ by (11) which is equal to the product of the right-hand sides of (13) by (12), it follows by Minkowski's linear form theorem that there are integers $y_1, \ldots, y_n, y_{n+1}$ not all zero and satisfying (13). If $y_{n+1} = 0$, then $\psi_i < 1, 1 \le i \le n$ by (10), and therefore we have $y_i \ge 0, (1 \le i \le n)$ by the first n formulas in (13). This leads to a contradiction, and so $y_{n+1} \ne 0$. We may assume that $y_{n+1} > 0$. The last formula in (13) yields

$$x_1 y_1 + \cdots + x_n y_n + x_{n+1} y_{n+1} = 0. \tag{14}$$

Therefore from the identity

$$\begin{aligned} x_1 y_1 + \cdots + x_n y_n + x_{n+1} y_{n+1} = {}& (x_1\theta_1 + \cdots + x_n\theta_n + x_{n+1})y_{n+1} \\ & - \sum_{i=1}^{n} x_i(y_{n+1}\theta_i - y_i) \end{aligned}$$

and (10)–(14), we obtain

$$\begin{aligned} \|x_1\theta_1 + \cdots + x_n\theta_n\|y_{n+1} &\le \sum_{i=1}^{n} \bar{x}_i|y_{n+1}\theta_i - y_i| \le \sum_{i=1}^{n} \bar{x}_i\psi_i \\ &\le n(\bar{x}_1 \cdots \bar{x}_n \|x_1\theta_1 + \cdots + x_n\theta_n\|)^{\frac{1}{n}} < n. \end{aligned} \tag{15}$$

By (12) and (13), we derive

$$|y_{n+1}\theta_1 - y_1| \cdots |y_{n+1}\theta_n - y_n| \le \psi_1 \cdots \psi_n = \|x_1\theta_1 + \cdots + x_n\theta_n\|. \tag{16}$$

Now suppose that Proposition A is not true then there exists ε with $0 < \varepsilon < 1$ and a sequence of integer sets $(x_1, \ldots, x_n)$ with $x = \max_{1 \le i \le n} |x_i| \to \infty$ such that

$$(\bar{x}_1 \cdots \bar{x}_n)^{\frac{1+\varepsilon}{1-\varepsilon}} \|x_1\theta_1 + \cdots + x_n\theta_n\| \le 1. \tag{17}$$

It follows from Proposition B that there exists a positive integer τ such that

$$(\tau y_{n+1})^{1+\frac{\varepsilon}{n}} \|\tau y_{n+1}\theta_1\| \cdots \|\tau y_{n+1}\theta_n\| > 1 \tag{18}$$

holds for any positive integer y_{n+1}. Let $x_1, \ldots, x_n$ be a set of integers such that (17) holds and $x > n^{\frac{\varepsilon n}{\varepsilon}} \tau^{\frac{n(n+2)}{\varepsilon}}$ and let $y_1, \ldots, y_n, y_{n+1}$ $(y_{n+1} > 0)$ be a set of integers satisfying (13) which is induced by $x_1, \ldots, x_n$ as above. Then from (15)–(17) and $\varepsilon < 1$, we obtain

$$
\begin{aligned}
\tau y_{n+1} \| \tau y_{n+1} \theta_1 \| \cdots \| \tau y_{n+1} \theta_n \| &\le \tau^{n+1} y_{n+1} |y_{n+1}\theta_1 - y_1| \cdots |y_{n+1}\theta_n - y_n| \\
&\le \tau^{n+1} y_{n+1} \| x_1\theta_1 + \cdots + x_n\theta_n \| \\
&\le \tau^{n+1} n (\bar{x}_1 \cdots \bar{x}_n \| x_1\theta_1 + \cdots + x_n\theta_n \|)^{\frac{1}{n}} \\
&= n\tau^{n+1} ((\bar{x}_1 \cdots \bar{x}_n)^{\frac{1+\varepsilon}{1-\varepsilon}} \| x_1\theta_1 + \cdots + x_n\theta_n \|)^{\frac{1-\varepsilon}{n}} \\
&\quad \times (\bar{x}_1 \cdots \bar{x}_n)^{-\frac{\varepsilon}{n}} \| x_1\theta_1 + \cdots + x_n\theta_n \|^{\frac{\varepsilon}{n}} \\
&\le n\tau^{n+1} x^{-\frac{\varepsilon}{n}} n^{\frac{\varepsilon}{n}} y_{n+1}^{-\frac{\varepsilon}{n}} \\
&\le (n^{\frac{2n}{\varepsilon}} \tau^{\frac{n(n+2)}{\varepsilon}} x^{-1})^{\frac{\varepsilon}{n}} (\tau y_{n+1})^{-\frac{\varepsilon}{n}} < (\tau y_{n+1})^{-\frac{\varepsilon}{n}}.
\end{aligned}
$$

This gives a contradiction with (18), and thus Proposition A follows.

References

[1] K. Mahler, Ein Übertragungsprinzip für lineare Ungleichungen, *Čas. Pěst. Mat.* **68** (1939) 85–92.

[2] A. Baker, On some Diophantine inequalities involving the exponential function, *Can. J. Math.* **17** (1965) 616–626.

[3] Hua Loo-keng, *Introduction to Number Theory* (Science Press, 1957).

Vol. XXII No. 3   **SCIENTIA SINICA**   March 1979

A NOTE ON A TRANSFERENCE THEOREM
OF LINEAR FORMS

Wolfgang M. Schmidt   AND   Wang Yuan (王 元)

(*Department of Mathematics, University
of Colorado, USA*)   (*Institute of Mathematics, Academia Sinica*)

Received September 22, 1978.

ABSTRACT

In this paper, a transference theorem of linear forms in diophantine approximation is proved. The proof of the theorem depends on a theorem of Mahler.

1. Let m and n be integers $\geqslant 1$ and θ_{ij} $(1 \leqslant i \leqslant m, 1 \leqslant j \leqslant n)$ be a given set of mn real numbers. Let $\|x\|$ denote the distance from a real number x to the nearest integer and $\bar{x} = \max(1, |x|)$.

Property A. For any given $\varepsilon > 0$, the inequality

$$\left(\prod_{i=1}^{m} \|\theta_{i1}x_1 + \cdots + \theta_{in}x_n\| \right) (\bar{x}_1 \cdots \bar{x}_n)^{1+\varepsilon} < 1 \tag{1}$$

has only finitely many integral solutions $(x_1, \cdots, x_n)$.

Property B. For any given $\varepsilon > 0$, the inequality

$$\left(\prod_{j=1}^{n} \|\theta_{1j}y_1 + \cdots + \theta_{mj}y_m\| \right) (\bar{y}_1 \cdots \bar{y}_m)^{1+\varepsilon} < 1 \tag{2}$$

has only finitely many integral solutions $(y_1, \cdots, y_m)$.

In this note, we prove the

Theorem. *The properties A and B are equivalent.*

For the case $m = 1$ and $1, \theta_{11}, \cdots, \theta_{1n}$ are linearly independent over the rational field Q, we have the result of [1].

2. The proof of the Theorem depends on the following:

Lemma 1 (Mahler). *Let $f_k(\mathbf{x})$ $(1 \leqslant k \leqslant l)$ be l linearly independent homogeneous linear forms in the l variables $\mathbf{x} = (x_1, \cdots, x_l)$ and let $g_k(\mathbf{y})$ $(1 \leqslant k \leqslant l)$ be l linearly independent homogeneous linear forms in the l variables $\mathbf{y} = (y_1, \cdots, y_l)$ of determinant d. Suppose that all the products $x_i y_j$ $(1 \leqslant i, j \leqslant l)$ in*

$$\sum_{k=1}^{l} f_k(\mathbf{x}) g_k(\mathbf{y})$$

have integer coefficients. If the inequalities

$$|f_k(\mathbf{x})| \leqslant \lambda, \qquad 1 \leqslant k \leqslant l$$

are soluble with integral $\mathbf{x} \neq \mathbf{o}$, *then the inequalities*

$$|g_k(\mathbf{y})| \leqslant (l-1)|\lambda d|^{\frac{1}{l-1}}, \qquad 1 \leqslant k \leqslant l$$

are soluble with integral $\mathbf{y} \neq \mathbf{o}$.

See K. Mahler [2] and also J. W. S. Cassels [3].

3. **Proof of the Theorem.** If there is a $\theta_{ij} \in Q$, then Properties A and B are obviously both false. So we may suppose that all the $\theta'_{ij}s$ are irrational numbers. First, we shall prove that if there exists an i with $1 \leqslant i \leqslant m$ such that $1, \theta_{i1}, \cdots, \theta_{in}$ are linearly independent over Q, then Properties A and B are both false. Since there exist integers $a_1, \cdots, a_n$ not all zero and an integer a_0 such that

$$a_0 + a_1\theta_{i1} + \cdots + a_n\theta_{in} = 0, \tag{3}$$

it follows that (1) has infinitely many integral solutions $(ka_1, \cdots, ka_n)$ $(k = 1, 2, \cdots)$. Hence Property A is false.

Since $\theta_{ij} \bar{\in} Q$ $(1 \leqslant j \leqslant n)$, it is necessarily $n \geqslant 2$ by (3). Without loss of generality, we may assume that $a_n > 0$. Take $\varepsilon = \dfrac{1}{2(n-1)}$. Then it follows by Dirichlet's theorem that for any given integer $t > 1$, there exists an integer q such that

$$1 \leqslant q < t^{n-1} \text{ and } \|\theta_{ij}q\| \leqslant t^{-1}, \qquad 1 \leqslant j \leqslant n-1. \tag{4}$$

Take $y = a_n q$ and $c = n \max\limits_{1 \leqslant j \leqslant n} |a_j|$. Then

$$\|\theta_{ij}y\| < ct^{-1}, \qquad 1 \leqslant j \leqslant n-1,$$

and further by (3)

$$\|\theta_{in}y\| = \|\theta_{in}a_n q\| = \|(a_1\theta_{i1} + \cdots + a_{n-1}\theta_{i,n-1})q\|$$
$$\subseteq \|a_1\theta_{i1}q\| + \cdots + \|a_{n-1}\theta_{i,n-1}q\| < ct^{-1}.$$

Consequently, we have

$$\|\theta_{i1}y\| \cdots \|\theta_{in}y\| y^{1+\varepsilon} < (ct^{-1})^n (ct^{n-1})^{1+\varepsilon} < c^{n+2}t^{-\frac{1}{2}} < 1$$

if t is sufficiently large. Since $\|\theta_{ij}q\| \leqslant t^{-1}$ and $\theta_{ij} \bar{\in} Q$, there are infinitely many integers q satisfying (4) when t runs over all the integers > 1. Hence we have infinitely many integral solutions $(y_1, \cdots, y_m)$ of (2), where $y_i = a_n q$ and $y_j = 0$ $(j \neq i)$. So Property B is false.

Similarly, we can prove that if there exists a j with $1 \leqslant j \leqslant n$ such that $1, \theta_{1j}, \cdots, \theta_{mj}$ are linearly dependent over Q, then the Properties A and B are false also. So in the following, we may assume that for $1 \leqslant i \leqslant m$, $1, \theta_{i1}, \cdots, \theta_{in}$ are linearly independent over Q and for $1 \leqslant j \leqslant n$, $1, \theta_{1j}, \cdots, \theta_{mj}$ are linearly independent over Q.

By the symmetry of the two Properties, it is sufficient to prove the following assertion.

If Property A is false, then Property B is false also. $\tag{5}$

Suppose that there exists $\varepsilon > 0$ such that (1) has infinitely many integral solutions $(x_1, \cdots, x_n)$. Write

$$X = \bar{x}_1 \cdots \bar{x}_n. \tag{6}$$

For any integral solution of (1) with $X \geqslant 2$, we define $A_1, \cdots, A_m, A$ by

$$0 < \|\theta_{i1}x_1 + \cdots + \theta_{in}x_n\|$$
$$= X^{-A_i} \quad (1 \leqslant i \leqslant m) \quad \text{and} \quad A_1 + \cdots + A_m = A, \tag{7}$$

and $B_1, \cdots, B_n$ by

$$\bar{x}_j = X^{B_j}, \qquad (1 \leqslant j \leqslant n). \tag{8}$$

Then $A_i > 0 \ (1 \leqslant i \leqslant m)$, $B_j \geqslant 0 \ (1 \leqslant j \leqslant n)$ and $B_1 + \cdots + B_n = 1$. The exponents $A_1, \cdots, A_m$ and $B_1, \cdots, B_n$ depend on X. By (1), we have

$$A > 1 + \varepsilon. \tag{9}$$

Let

$$\overline{\lim} \ \frac{\min(A_1, \cdots, A_m)}{A} = 2\rho \quad (\text{as } X \to \infty). \tag{10}$$

Denote by $X_{n+i} \ (1 \leqslant i \leqslant m)$ the integers such that

$$|\theta_{i1}x_1 + \cdots + \theta_{in}x_n + x_{n+i}| = \|\theta_{i1}x_1 + \cdots + \theta_{in}x_n\|, \quad 1 \leqslant i \leqslant m. \tag{11}$$

1)　Suppose that $\rho > 0$. Then there are infinitely many integral solutions of (1) such that

$$\frac{\min(A_1, \cdots, A_m)}{A} > \rho > 0. \tag{12}$$

Write

$$\eta = \max(1, A - (n + m - 1)\min(A_1, \cdots, A_m)). \tag{13}$$

Then from (9) and (12), we have

$$\frac{\eta}{A} < \max\left(\frac{1}{1+\varepsilon}, \ 1 - (n + m - 1)\rho\right) < 1. \tag{14}$$

Let $\mathbf{x} = (x_1, \cdots, x_{n+m})$ and $\mathbf{y} = (y_1, \cdots, y_{n+m})$, and let

$$f_i(\mathbf{x}) = \begin{cases} X^{A_i}(\theta_{i1}x_1 + \cdots + \theta_{in}x_n + x_{n+i}), & 1 \leqslant i \leqslant m, \\ -X^{-\eta B_{i-m}}x_{i-m}, & m + 1 \leqslant i \leqslant m + n, \end{cases} \tag{15}$$

and

$$g_i(\mathbf{y}) = \begin{cases} X^{-A_i}y_i, & 1 \leqslant i \leqslant m, \\ X^{\eta B_{i-m}}(\theta_{1,i-m}y_1 + \cdots + \theta_{m,i-m}y_m + y_i), \\ \qquad m + 1 \leqslant i \leqslant m + n. \end{cases} \tag{16}$$

Since

$$\sum_{i=1}^{m+n} f_i(\mathbf{x})g_i(\mathbf{y}) = \sum_{i=1}^{m} x_{n+i}y_i - \sum_{j=1}^{n} x_j y_{m+j},$$

and since there exists an integral solution of (1) satisfying (12), it follows by Lemma 1 (with $\lambda = 1$ and $d = X^{\eta - A}$) that there exists an integral point $\mathbf{y} \neq \mathbf{o}$ such that

$$|y_i| \leqslant (n + m - 1) X^{\frac{\eta - A}{n+m-1} + A_i}, \qquad 1 \leqslant i \leqslant m, \tag{17}$$

and

$$|\theta_{1j}y_1 + \cdots + \theta_{mj}y_m + y_{m+j}|$$
$$\leqslant (n + m - 1) X^{\frac{\eta - A}{n+m-1} - \eta B_j}, \qquad 1 \leqslant j \leqslant n. \tag{18}$$

By (9) and (14), we have

$$|\theta_{1j}y_1 + \cdots + \theta_{mj}y_m + y_{m+j}| \leqslant (n + m - 1) X^{\frac{\eta - A}{n+m-1}}$$
$$\leqslant (n + m - 1) X^{\frac{A}{n+m-1}\left(\frac{\eta}{A}-1\right)}$$
$$\leqslant (n + m - 1) X^{-\frac{A}{n+m-1} \min\left(\frac{\varepsilon}{1+\varepsilon}, (n+m-1)\rho\right)}$$
$$< 1, \qquad 1 \leqslant j \leqslant n, \tag{19}$$

if X is sufficiently large. Hence $(y_1, \cdots, y_m) \neq \mathbf{o}$. Since the right-hand sides of (17) are all $\geqslant 1$ by (13), we may replace the y_i by $\bar{y}_i\, (1 \leqslant i \leqslant m)$ in (17). Take $\delta > 0$ such that

$$\frac{1 - (n - 1)\delta}{1 + m\delta} = \max\left(\frac{1}{1 + \varepsilon}, \; 1 - (n + m - 1)\rho\right) + \delta. \tag{20}$$

Then by (14), (17) and (18), we have

$$\left(\prod_{j=1}^{n} \|\theta_{1j}y_1 + \cdots + \theta_{mj}y_m\|\right) (\bar{y}_1 \cdots \bar{y}_m)^{1+\delta}$$
$$\leqslant (n + m - 1)^{n+m(1+\delta)} X^{\frac{1}{n+m-1}(\eta(1+m\delta)+A(-1+(n-1)\delta))}$$
$$- (n + m - 1)^{n+m(1+\delta)} X^{-\frac{(1+m\delta)A}{n+m-1}\left(\frac{1-(n-1)\delta}{1+m\delta}-\frac{\eta}{A}\right)}$$
$$< (n + m - 1)^{n+m(1+\delta)} X^{-\frac{(1+m\delta)\delta A}{n+m-1}}$$
$$< 1, \tag{21}$$

if X is sufficiently large.

Since $1, \theta_{1j}, \cdots, \theta_{mj}$ are linearly independent over Q for $1 \leqslant j \leqslant n$, it follows by (9) and (19) that there are infinitely many integral points $(y_1, \cdots, y_m)$ satisfying (21) when $(x_1, \cdots, x_n)$ runs over all the integral solutions of (1) satisfying (12). In other words, Property B is false. Thus the assertion (5) is proved.

2) We now proceed to prove the assertion (5) by induction on m. Since $\rho = 1/2$ for $m = 1$, the assertion holds for $m = 1$. Suppose that $m \geqslant 2$ and that the assertion holds for $m - 1$. If $\rho > 0$, then the assertion (5) is true by 1). In the following, we consider the case $\rho = 0$. Take

$$\eta = \frac{\varepsilon}{2(1 + \varepsilon)}. \tag{22}$$

Without loss of generality, we may assume that (1) has infinitely many integral solutions $(x_1, \cdots, x_n)$, such that

$$\frac{A_m}{A} < \eta, \tag{23}$$

which satisfy

$$\left(\prod_{i=1}^{m-1} \| \theta_{i1} x_1 + \cdots + \theta_{in} x_n \| \right) (\bar{x}_1 \cdots \bar{x}_n)^{1+\varepsilon/2}$$

$$< \left(\left(\prod_{i=1}^{m} \| \theta_{i1} x_1 + \cdots + \theta_{in} x_n \| \right) (\bar{x}_1 \cdots \bar{x}_n)^{1+\varepsilon} \right)^{1-\eta} < 1$$

by (7) and (22). Hence it follows by the hypothesis of the induction that there exists a $\tau > 0$ such that

$$\left(\prod_{j=1}^{n} \| \theta_{1j} y_1 + \cdots + \theta_{m-1, j} y_{m-1} \| \right) (\bar{y}_1 \cdots \bar{y}_{m-1})^{1+\tau} < 1$$

has infinitely many integral solutions $(y_1, \cdots, y_{m-1})$. Consequently, (2) has infinitely many integral solutions $(y_1, \cdots, y_{m-1}, 0)$ for $\varepsilon = \tau$. Thus we have the assertion (5), hence the Theorem is proved.

REFERENCES

[1]   Wang Yuan, Yu Kun-rui & Zhu Yao-cheng: A note on a transference theorem concerning the linear forms, *Acta Math. Sinica*, **22** (1979).

[2]   Mahler, K.: Ein Übertragungsprinzip für lineare Ungleichungen, *Čas. Pest Mat.*, **68** (1939), 85—92.

[3]   Cassels, J. W. S.: *An Introduction to Diophantine Approximation*, Camb. University Press, (1957).

Chin. Ann. of Math.
2 (1) 1981

A NOTE ON SOME METRICAL THEOREMS IN DIOPHANTINE APPROXIMATION

Wang Yuan   Yu Kunrui

(*Institute of Mathematics, Academia Sinica*)

§ 1. Introduction. Suppose that A is a set in n–dimensional Euclidean space R_n. If $(x_1, \cdots, x_n) \in A$ implies that $(x_1', \cdots, x_n') \in A$ for any $0 \leqslant x_i' \leqslant x_i (1 \leqslant i \leqslant n)$, then A is said to have property $\mathscr{P}$ which was introduced by Gallagher[1]. Denote by $|A|$ the measure of a measurable set A and by G_n the n–dimensional unit cube, i. e., the set of all points $(x_1, \cdots, x_n)$ with $0 \leqslant x_i < 1 (1 \leqslant i \leqslant n)$. We also use ε to denote any pre-assigned positive number.

Theorem 1. *Let $A_q (q = 1, 2, \cdots)$ be a sequence of measurable subsets of G_n. Suppose that each A_q has property $\mathscr{P}$ and $\psi(q) = |A_q|$ is a decreasing function of q. Let $N(h, \theta_1, \cdots, \theta_n)$ be the number of integers satisfying $1 \leqslant q \leqslant h$ and*

$$(\{q\theta_1\}, \cdots, \{q\theta_n\}) \in A_q. \tag{1}$$

Further let

$$\Psi(h) = \sum_{q=1}^{h} \psi(q) \quad and \quad \Omega(h) = \sum_{q=1}^{h} \psi(q) q^{-1}.$$

Then for almost all $(\theta_1, \cdots, \theta_n) \in R_n$, we have

$$N(h, \theta_1, \cdots, \theta_n) = \Psi(h) + O(\Psi(h)^{\frac{1}{2}} \Omega(h)^{\frac{1}{2}} (\log \Psi(h))^{2+\varepsilon}). \tag{2}$$

Let $\boldsymbol{q} = (q_1, \cdots, q_m)$ and $\boldsymbol{r} = (r_1, \cdots, r_m)$ denote the lattice points in R_m, where q_i's and r_i's are positive integers, $\boldsymbol{\theta} = (\theta_1, \cdots, \theta_m)$ a point in R_m, $\boldsymbol{q\theta} = q_1\theta_1 + \cdots + q_m\theta_m$ the scalar product of $\boldsymbol{q}$ and $\boldsymbol{\theta}$ and $d(\boldsymbol{q}) = \sum_{\substack{d|q_i \\ 1 \leqslant i \leqslant m}} 1$. We also use $\boldsymbol{q} \leqslant h$ to denote $q = \max(q_1, \cdots, q_m) \leqslant h$. Similarly, we may define $\boldsymbol{q} < h$, $\boldsymbol{q} \geqslant h$ and $\boldsymbol{q} > h$.

Theorem 2. *Let $\{A_q\}$ be a sequence of measurable subsets of G_n and each A_q have property $\mathscr{P}$. Put $\psi(\boldsymbol{q}) = |A_q|$ and denote by $N(h, \theta_1, \cdots, \theta_n)$ the number of lattice points $\boldsymbol{q}$ satisfying $\boldsymbol{q} \leqslant h$ and*

$$(\{\boldsymbol{q}\theta_1\}, \cdots, \{\boldsymbol{q}\theta_n\}) \in A_q \tag{3}$$

where $\boldsymbol{\theta}_i = (\theta_{i1}, \cdots, \theta_{im}) (1 \leqslant i \leqslant n)$. Set

$$\Psi(h) = \sum_{\boldsymbol{q} < h} \psi(\boldsymbol{q}) \quad and \quad \chi(h) = \sum_{\boldsymbol{q} < h} \psi(\boldsymbol{q}) d(\boldsymbol{q}).$$

Then

$$N(h, \boldsymbol{\theta}_1, \cdots, \boldsymbol{\theta}_n) = \Psi(h) + O(\chi(h)^{\frac{1}{2}} (\log \chi(h))^{\frac{3}{2}+\varepsilon}) \tag{4}$$

holds for almost all $(\boldsymbol{\theta}_1, \cdots, \boldsymbol{\theta}_n) = (\theta_{11}, \cdots, \theta_{1m}, \cdots, \theta_{n1}, \cdots, \theta_{nm}) \in R_{nm}$.

Manuscript received May 28, 1980.

Theorem 1 gives a modification of a theorem of Gallagher[1]. Take A_q to be the set of points $(x_1, \cdots, x_n)$ satisfying $0 \leqslant x_i < \psi_i(q)$ $(1 \leqslant i \leqslant n)$ and suppose that $\psi(q) = \prod\limits_{i=1}^{n} \psi_i(q)$ is a decreasing function of q in Theorem 1. Take A_q to be the set of points such that $0 \leqslant x_i < \psi_i(q)$ $(1 \leqslant i \leqslant n)$ in Theorem 2. Then we derive two theorems of Schmidt [2].

For any positive number τ, let $E(\tau)$ be the set of points $(x_1, \cdots, x_n)$ satisfying

$$0 \leqslant x_i < \frac{1}{2} \quad (1 \leqslant i \leqslant n) \quad \text{and} \quad \tau x_1 \cdots x_n < 1.$$

Then it is easily proved by mathematical induction that

$$|E(\tau)| = \begin{cases} 2^{-n}, \text{ for } \tau \leqslant 2^n \\ \tau^{-1} \sum\limits_{s=0}^{n-1} \frac{1}{s!} \Big(\log \frac{\tau}{2^n} \Big)^s, \quad \text{otherwise.} \end{cases}$$

Take $A_q = E(q(\log q)^n)$. Then $q(\log q)^n > 2^n$ and

$$\psi(q) = |A_q| = (q(\log q)^n)^{-1} \sum\limits_{s=0}^{n-1} \frac{1}{s!} \Big(\log \Big(\frac{q(\log q)^n}{2^n} \Big) \Big)^s$$

for $q \geqslant 8$. Hence

$$\Psi(h) = \sum\limits_{s=0}^{n-1} \frac{1}{s!} \sum\limits_{q=8}^{h} (q(\log q)^n)^{-1} \Big(\log \Big(\frac{q(\log q)^n}{2^n} \Big) \Big)^s + O(1)$$

$$= \frac{1}{(n-1)!} \sum\limits_{q=2}^{h} \frac{1}{q \log q} + O(1) = \frac{1}{(n-1)!} \log \log h + O(1).$$

Obviously

$$\Omega(h) = O(1).$$

Let $N(h, \theta_1, \cdots, \theta_n)$ denote the number of integers satisfying $1 \leqslant q \leqslant h$ and the inequalities

$$0 \leqslant \{q\theta_i\} < \frac{1}{2} \ (1 \leqslant i \leqslant n) \quad \text{and} \quad \Big(\prod\limits_{i=1}^{n} \{q\theta_i\} \Big) q(\log q)^n < 1.$$

Then it follows by Theorem 1 that

$$N(h, \theta_1, \cdots, \theta_n) = \frac{1}{(n-1)!} \log \log h + O((\log \log h)^{\frac{1}{2}} (\log \log \log h)^{2+\varepsilon}) \tag{5}$$

holds for almost all $(\theta_1, \cdots, \theta_n) \in R_n$. Let $\|x\|$ denote the distance from the real number x to the nearest integer. Then from (5) with some simple combinatorial considerations, we may derive

Theorem 3. *Let $N^*(h, \theta_1, \cdots, \theta_n)$ be the number of integers q satisfying*

$$\Big(\prod\limits_{i=1}^{n} \|q\theta_i\| \Big) q(\log q)^n < 1, \quad 1 \leqslant q \leqslant h. \tag{6}$$

Then for almost all $(\theta_1, \cdots, \theta_n) \in R_n$, we have

$$N^*(h, \theta_1, \cdots, \theta_n) = \frac{2^n}{(n-1)!} \log \log h + O((\log \log h)^{\frac{1}{2}} (\log \log \log h)^{2+\varepsilon}). \tag{7}$$

Similarly, we may derive from Theorem 1 that for almost all $(\theta_1, \cdots, \theta_n) \in R_n$, the inequality

$$\Big(\prod\limits_{i=1}^{n} \|q\theta_i\| \Big) q(\log q)^{n+\varepsilon} < 1 \tag{8}$$

has only finitely many positive integer solutions.

Let $f(1) = 1$ and $f(k) = k(\log k)^n$ or $k(\log k)^{n+s} (k > 1)$. Put $\bar{x} = \max(1, |x|)$. Take $A_q = E(f(q_1) \cdots f(q_m))$ in Theorem 2. Then we may obtain for almost all $(\theta_{11}, \cdots, \theta_{nm}) \in R_{nm}$ the asymptotic formula of the number of lattice points q satisfying $\max\limits_{1 \leqslant i \leqslant m} |q_i| \leqslant h$ and

$$\prod_{i=1}^{n} \|\theta_{i1}q_1 + \cdots + \theta_{im}q_m\| \prod_{j=1}^{m} f(\bar{q}_j) < 1. \tag{9}$$

From this formula we may derive

Theorem 4. *The inequality*

$$\prod_{i=1}^{n} \|\theta_{i1}q_1 + \cdots + \theta_{im}q_m\| \prod_{j=1}^{m} \overline{(q_j(\log \bar{q}_j)^{n+s})} < 1 \tag{10}$$

has only finitely many integral solutions for almost all $(\theta_{11}, \cdots, \theta_{nm}) \in R_{nm}$, *but*

$$\prod_{i=1}^{n} \|\theta_{i1}q_1 + \cdots + \theta_{im}q_m\| \prod_{j=1}^{m} \overline{(q_j(\log \bar{q}_j)^n)} < 1 \tag{11}$$

has infinitely many integral solutions for almost all $(\theta_{11}, \cdots, \theta_{nm}) \in R$.

Especially, it follows from Theorem 4 that property A in Schmidt and Wang's[4] transference theorem holds for almost all $(\theta_{11}, \cdots, \theta_{nm}) \in R_{nm}$.

The proofs of Theorems 1 and 2 are based on the method of Schmidt[1] and we may also treat the similar problems in non-linear diophantine approximation (Cf. Schmidt [3]).

Remark. We propose two conjectures.

$1°$ Suppose that $\alpha_{11}, \cdots, \alpha_{nm}$ are nm real algebraic numbers such that $1, \alpha_{11}, \cdots, \alpha_{nm}$ are linearly independent over rational field Q. Then

$$\prod_{i=1}^{n} \|\alpha_{i1}q_1 + \cdots + \alpha_{im}q_m\| \prod_{j=1}^{m} \bar{q}_j^{1+s} \gg 1.$$

$2°$ Suppose that $\beta_{ij} = e^{r_{ij}}$, where $r_{ij} (1 \leqslant i \leqslant n, 1 \leqslant j \leqslant m)$ are nm different rational numbers. Then

$$\prod_{i=1}^{n} \|\beta_{i1}q_1 + \cdots + \beta_{im}q_{im}\| \prod_{j=1}^{m} q_j^{1+s} \gg 1.$$

Put $n = 1$ or $m = 1$. Then $1°$ and $2°$ are Theorems of W. M. Schmidt[5] and A. Baker[6] respectively.

§ 2. **Proof of Theorem 1.** If $\sum \psi(q) < \infty$, the theorem can be proved by Borel-Cantelli's lemma (Cf. Gallagher [1]). Now we suppose that $\sum \psi(q)$ diverges. Evidently, we may confine ourselves to the case $(\theta_1, \cdots, \theta_n) \in G_n$. Let $\varphi(k, q)$ denote the number of integers l satisfying $0 \leqslant l < q$ and $(l, q) \leqslant k$. Put

$$\beta(q, \theta_1, \cdots, \theta_n) = \begin{cases} 1, & \text{if } (\theta_1, \cdots, \theta_n) \in A_q, \\ 0, & \text{otherwise,} \end{cases}$$

$$\gamma(q, \theta_1, \cdots, \theta_n) = \sum_{p_i, 1 \leqslant i \leqslant n} \beta(q, q\theta_1 - p_1, \cdots, q\theta_n - p_n),$$

$$\gamma(k, q, \theta_1, \cdots, \theta_n) = \sum_{\substack{p_i, (p_i, q) \leqslant k \\ 1 \leqslant i \leqslant n}} \beta(q, q\theta_1 - p_1, \cdots, q\theta_n - p_n),$$

$$I(q) = \int_0^1 \cdots \int_0^1 \gamma(q, \theta_1, \cdots, \theta_n) d\theta_1 \cdots d\theta_n,$$

$$I(k, q) = \int_0^1 \cdots \int_0^1 \gamma(k, q, \theta_1, \cdots, \theta_k) d\theta_1 \cdots d\theta_n,$$

$$I(k,\ q,\ r) = \int_0^1 \cdots \int_0^1 \gamma(k,\ q,\ \theta_1,\ \cdots,\ \theta_n)\, \gamma(k,\ r,\ \theta_1,\ \cdots,\ \theta_n)\, d\theta_1 \cdots d\theta_n,$$

$$\Psi(u,\ v) = \sum_{q=u+1}^{v} \psi(q),$$

$$N(k,\ u,\ v,\ \theta_1,\ \cdots,\ \theta_n) = \sum_{q=u+1}^{v} \gamma(k,\ q,\ \theta_1,\ \cdots,\ \theta_n)$$

and notice

$$N(v,\ \theta_1,\ \cdots,\ \theta_n) = \sum_{q=1}^{v} \gamma(q,\ \theta_1,\ \cdots,\ \theta_n).$$

Lemma 1.

$$I(q) = \psi(q),\quad I(k,\ q) = \psi(q)\varphi(k,\ q)^n q^{-n}, \tag{12}$$

$$I(k,\ q,\ r) \leqslant \psi(q)\psi(r) + (2^n - 1)\psi(q) A(k,\ q,\ r) q^{-1}, \tag{13}$$

where $A(k,\ q,\ r)$ is the number of integer pairs $(p,\ s)$ satisfying

$$qs - rp = 0,\quad 0 \leqslant p < q,\quad (p,\ q) \leqslant k,\quad (s,\ r) \leqslant k.$$

Proof

$$I(q) = \int_0^1 \cdots \int_0^1 \gamma(q,\ \theta_1,\ \cdots,\ \theta_n)\, d\theta_1 \cdots d\theta_n$$

$$= \int_0^q \cdots \int_0^q q^{-n}\gamma(q,\ q^{-1}\theta_1,\ \cdots,\ q^{-1}\theta_n)\, d\theta_1 \cdots d\theta_n$$

$$= q^{-n} \sum_{p_i,\ 1 \leqslant i \leqslant n} \int_0^q \cdots \int_0^q \beta(q,\ \theta_1 - p_1,\ \cdots,\ \theta_n - p_n)\, d\theta_1 \cdots d\theta_n$$

$$= q^{-n} q^n \int_0^1 \cdots \int_0^1 \beta(q,\ \theta_1,\ \cdots,\ \theta_n)\, d\theta_1 \cdots d\theta_n = \psi(q).$$

Similarly

$$I(k,\ q) = q^{-n} \sum_{\substack{p_i,\ (p_i,\ q) \leqslant k \\ 1 \leqslant i \leqslant n}} \int_0^q \cdots \int_0^q \beta(q,\ \theta_1 - p_1,\ \cdots,\ \theta_n - p_n)\, d\theta_1 \cdots d\theta_n$$

$$= \psi(q)\varphi(k,\ q)^n q^{-n}.$$

Now we proceed to prove (13). Divide the sum

$$I(k,\ q,\ r) = \sum_{\substack{p_i,\ (p_i,\ q) \leqslant k \\ s_i,\ (s_i,\ r) \leqslant k \\ 1 \leqslant i \leqslant n}} \int_0^1 \cdots \int_0^1 \beta(q,\ q\theta_1 - p_1,\ \cdots,\ q\theta_n - p_n)$$

$$\times \beta(r,\ r\theta_1 - s_1,\ \cdots,\ r\theta_n - s_n)\, d\theta_1 \cdots d\theta_n$$

into $n+1$ parts

$$I(k,\ q,\ r) = I_0 + \cdots + I_n, \tag{14}$$

where I_j is the sum of all the terms with exactly j indices $i_1,\ \cdots,\ i_j$ satisfying

$$qp_i - rs_i = 0.$$

We first estimate I_0.

$$I_0 \leqslant \sum_{\substack{p_i,\ s_i \\ qs_i - rp_i \neq 0 \\ 1 \leqslant i \leqslant n}} \int_0^1 \cdots \int_0^1 \beta(q,\ q\theta_1 - p_1,\ \cdots,\ q\theta_n - p_n)\beta(r,\ r\theta_1 - s_1,\ \cdots,\ r\theta_n - s_n)\, d\theta_1 \cdots d\theta_n$$

$$= \sum_{\substack{p_i,\ s_i \\ qs_i - rp_i \neq 0 \\ 1 \leqslant i \leqslant n}} \int_{-\frac{p_1}{q}}^{1-\frac{p_1}{q}} \cdots \int_{-\frac{p_n}{q}}^{1-\frac{p_n}{q}} \beta(q,\ q\theta_1',\ \cdots,\ q\theta_n')$$

$$\times \beta\left(r,\ r\theta_1' - \frac{qs_1 - rp_1}{q},\ \cdots,\ r\theta_n' - \frac{qs_n - rp_n}{q}\right) d\theta_1' \cdots d\theta_n'.$$

Write $(q,\ r) = d$, $q = dq'$ and $r = dr'$. Then we have $qs_i - rp_i = h_i d$. For given $h_i \neq 0$, p_i

is determined uniquely modulo q', so we have

$$I_0 \leqslant d^n \sum_{\substack{h_i \neq 0 \\ 1 \leqslant i \leqslant n}} \int_{-\infty}^{\infty} \cdots \int_{-\infty}^{\infty} \beta(q,\, q\theta_1,\, \cdots,\, q\theta_n)\, \beta\Big(r,\, r\theta_1 - \frac{h_1 d}{q},\, \cdots,\, r\theta_n - \frac{h_n d}{q}\Big)\, d\theta_1 \cdots d\theta_n.$$

$$(15)$$

Put

$$J(\lambda_1,\, \cdots,\, \lambda_n) = \int_{-\infty}^{\infty} \cdots \int_{-\infty}^{\infty} \beta(q,\, q\theta_1,\, \cdots,\, q\theta_n)\, \beta\Big(r,\, r\theta_1 - \frac{\lambda_1 d}{q},\, \cdots,\, r\theta_n - \frac{\lambda_n d}{q}\Big)$$
$$\times d\theta_1 \cdots d\theta_n.$$

We proceed to prove that J decreases as a function of λ_i when $\lambda_i \geqslant 0$ and increases when $\lambda_i \leqslant 0$ for $i = 1,\, \cdots,\, n$. Without loss of generality, we may suppose $i = 1$. In fact, for fixed $\theta_2,\, \cdots,\, \theta_n$ and $\lambda_2,\, \cdots,\, \lambda_n$, $\beta(q,\, q\theta_1,\, \cdots,\, q\theta_n)$ and $\beta\Big(r,\, r\theta_1 - \frac{\lambda_1 d}{q},\, \cdots,\, r\theta_n - \frac{\lambda_n d}{q}\Big)$ are the characteristic functions of two intervals on θ_1-axis which have fixed lengths and start from 0 and $\frac{\lambda_1 d}{qr}$ respectively. The measure of the overlap of the two intervals is

$$\int_{-\infty}^{\infty} \beta(q,\, q\theta_1,\, \cdots,\, q\theta_n)\, \beta\Big(r,\, r\theta_1 - \frac{\lambda_1 d}{q},\, \cdots,\, r\theta_n - \frac{\lambda_n d}{q}\Big)\, d\theta_1.$$

It is a decreasing function of λ_1 for $\lambda_1 \geqslant 0$ and an increasing function for $\lambda_1 \leqslant 0$. Hence J decreases for $\lambda_1 \geqslant 0$ and increases for $\lambda_1 \leqslant 0$. From this fact and (15) we have

$$I_0 \leqslant d^n \int_{-\infty}^{\infty} \cdots \int_{-\infty}^{\infty} \beta(q,\, q\theta_1,\, \cdots,\, q\theta_n)\, \beta\Big(r,\, r\theta_1 - \frac{\lambda_1 d}{q},\, \cdots,\, r\theta_n - \frac{\lambda_n d}{q}\Big)$$
$$\times d\theta_1 \cdots d\theta_n d\lambda_1 \cdots d\lambda_n = \psi(q)\psi(r).$$

$$(16)$$

As for $I_j\,(j \geqslant 1)$, since

$$\sum_{\substack{(p_i,\, q) \leqslant k \\ (s_i,\, r) \leqslant k \\ 1 \leqslant i \leqslant n}} \sum_{\substack{qs_i - rp_i \neq 0 \\ 1 \leqslant i \leqslant n-j}} \sum_{\substack{qs_i - rp_i = 0 \\ n-j < i \leqslant n}} \int_0^1 \cdots \int_0^1 \beta(q,\, q\theta_1 - p_1,\, \cdots,\, q\theta_n - p_n)$$

$$\times \beta(r,\, r\theta_1 - s_1,\, \cdots,\, r\theta_n - s_n)\, d\theta_1 \cdots d\theta_n$$

$$= \sum_{\substack{(p_i,\, q) \leqslant k \\ (s_i,\, r) \leqslant k \\ 1 \leqslant i \leqslant n}} \sum_{\substack{qs_i - rp_i \neq 0 \\ 1 \leqslant i \leqslant n-j}} \sum_{\substack{qs_i - rp_i = 0 \\ n-j < i \leqslant n}} \int_{-\frac{p_1}{q}}^{1 - \frac{p_1}{q}} \cdots \int_{-\frac{p_n}{q}}^{1 - \frac{p_n}{q}} \beta(q,\, q\theta_1',\, \cdots,\, q\theta_n')$$

$$\times \beta\Big(r,\, r\theta_1' - \frac{qs_1 - rp_1}{q},\, \cdots,\, r\theta_{n-j}' - \frac{qs_{n-j} - rp_{n-j}}{q},\, r\theta_{n-j+1}',\, \cdots,\, r\theta_n'\Big)\, d\theta_1' \cdots d\theta_n'$$

$$\leqslant A(k,\, q,\, r)^j \sum_{\substack{h_i \neq 0 \\ 1 \leqslant i \leqslant n-j}} d^{n-j} \int_{-\infty}^{\infty} \cdots \int_{-\infty}^{\infty} \beta(q,\, q\theta_1,\, \cdots,\, q\theta_n)$$

$$\times \beta\Big(r,\, r\theta_1 - \frac{dh_1}{q},\, \cdots,\, r\theta_{n-j} - \frac{dh_{n-j}}{q},\, r\theta_{n-j+1},\, \cdots,\, r\theta_n\Big)\, d\theta_1 \cdots d\theta_n$$

$$\leqslant A(k,\, q,\, r)^j d^{n-j} \int_{-\infty}^{\infty} \cdots \int_{-\infty}^{\infty} \beta(q,\, q\theta_1,\, \cdots,\, q\theta_n)$$

$$\times \beta\Big(r,\, r\theta_1 - \frac{d\lambda_1}{q},\, \cdots,\, r\theta_{n-j} - \frac{d\lambda_{n-j}}{q},\, r\theta_{n-j+1},\, \cdots,\, r\theta_n\Big)\, d\theta_1 \cdots d\theta_n d\lambda_1 \cdots d\lambda_{n-j}$$

$$\leqslant A(k,\, q,\, r)^j d^{n-j} \Big(\frac{q}{d}\Big)^{n-j} \int_{-\infty}^{\infty} \cdots \int_{-\infty}^{\infty} \beta(q,\, q\theta_1,\, \cdots,\, q\theta_n)\, d\theta_1 \cdots d\theta_n$$

$$= \psi(q)\, A(k,\, q,\, r)^j q^{-j},$$

hence

$$I_j \leqslant \binom{n}{j} \psi(q) A(k,\ q,\ r) q^{-1}. \tag{17}$$

(13) follows by combining (14), (16) and (17). The lemma is proved.

Lemma 2.

$$\int_0^1 \cdots \int_0^1 N(v,\ \theta_1,\ \cdots,\ \theta_n) d\theta_1 \cdots d\theta_n = \Psi(v), \tag{18}$$

$$\int_0^1 \cdots \int_0^1 N(k,\ u,\ v,\ \theta_1,\ \cdots,\ \theta_n) d\theta_1 \cdots d\theta_n = \sum_{q=u+1}^{v} \psi(q) \varphi(k,\ q)^n q^{-n}, \tag{19}$$

$$\int_0^1 \cdots \int_0^1 N(k,\ u,\ v,\ \theta_1,\ \cdots,\ \theta_n)^2 d\theta_1 \cdots d\theta_n \leqslant \Psi(u,\ v)^2 + 2(2^n-1) \sum_{q=u+1}^{v} \psi(q) d_k(q), \tag{20}$$

where $d_k(q) = \sum_{d|q, d \leqslant k} 1$.

Proof (18) and (19) follow from (12) immedeately. (20) follows from (13).

The rest part of the proof of Theorem 1 is similar to the proof of Theorem 1 in Schmidt [2], and we omit it here.

§ 3. Proof of Theorem 2. If $\sum \psi(q) < \infty$, the theorem can be proved by Borel-Cantelli's lemma (cf. Gallagher [1]). Now we suppose that $\sum \psi(q)$ diverges. We may also confine ourselves to the case $(\theta_1,\ \cdots,\ \theta_n) \in G_{nm}$. Let $\omega(0) = 0$ and $\omega(h)$, $h \geqslant 1$, be an increasing integral-valued function which tends to infinity. Set $S' = \{0\} \cup \{h > 0 \mid \omega(h-1) < \omega(h)\}$, $S'' = \{h \geqslant 0 \mid \omega(h) < \omega(h+1)\}$ and $S = \{\omega(h) \mid h \geqslant 0\}$. For integer $t > 0$, we define intervals of order t to be

$$(u2^t + v_1,\ (u+1)2^t + v_2],$$

where $u,\ v_1,\ v_2$ are non-negative integers such that $v_1 < 2^t$ and $v_1,\ v_2$ are the smallest non-negative integers satisfying $u2^t + v_1 \in S$, $(u+1)2^t + v_2 \in S$.

Lemma 3. *Every interval $(0,\ x]$ with $x \in S$ can be expressed as union of intervals $\cup I_i$ of the type described above, where no two of intervals I_i are of the same order.*

Proof (cf. [2]).

Put

$$\beta(q,\ \alpha_1,\ \cdots,\ \alpha_n) = \begin{cases} 1, & \text{if } (\alpha_1,\ \cdots,\ \alpha_n) \in A_q, \\ 0, & \text{otherwise.} \end{cases}$$

$$\gamma(q,\ \theta_1,\ \cdots,\ \theta_n) = \sum_{p_i,\ 1 \leqslant i \leqslant n} \beta(q,\ q\theta_1 - p_1,\ \cdots,\ q\theta_n - p_n),$$

$$I(q) = \int_{G_m} \cdots \int_{G_m} \gamma(q,\ \theta_1,\ \cdots,\ \theta_n) d\theta_1 \cdots d\theta_n,$$

$$I(q,\ r) = \int_{G_m} \cdots \int_{G_m} \gamma(q,\ \theta_1,\ \cdots,\ \theta_n) \gamma(r,\ \theta_1,\ \cdots,\ \theta_n) d\theta_1 \cdots d\theta_n,$$

$$\Psi(u,\ v) = \sum_{u < q \leqslant v} \psi(q),$$

$$N(u,\ v,\ \theta_1,\ \cdots,\ \theta_n) = \sum_{u < q \leqslant v} \gamma(q,\ \theta_1,\ \cdots,\ \theta_n),$$

and notice
$$N(v, \boldsymbol{\theta}_1, \cdots, \boldsymbol{\theta}_n) = \sum_{q \leqslant v} \gamma(\boldsymbol{q}, \boldsymbol{\theta}_1, \cdots, \boldsymbol{\theta}_n).$$

Lemma 4.
$$I(\boldsymbol{q}) = \psi(\boldsymbol{q}). \tag{21}$$
If $\boldsymbol{q}$ and $\boldsymbol{r}$ are linearly independent (this fact is abbreviated to $\boldsymbol{q}$, $\boldsymbol{r}$, l. i.), then
$$I(\boldsymbol{q}, \boldsymbol{r}) = \psi(\boldsymbol{q})\psi(\boldsymbol{r}). \tag{22}$$
If $\boldsymbol{q}$ and $\boldsymbol{r}$ are linearly dependent (this fact is abbreviated to $\boldsymbol{q}$, $\boldsymbol{r}$, l. d.), then
$$I(\boldsymbol{q}, \boldsymbol{r}) \leqslant \psi(\boldsymbol{q})\psi(\boldsymbol{r}) + (2^n - 1) A(q_1, r_1)\psi(\boldsymbol{q}) q_1^{-1}, \tag{23}$$
where $A(q_1, r_1)$ is the number of the integral solutions (p, s) of the equation
$$q_1 s - r_1 p = 0, \quad 0 \leqslant p < q_1.$$

Proof 1) Suppose that $\boldsymbol{q}$, $\boldsymbol{r}$, l. i. Without loss of generality we may assume that $\begin{vmatrix} q_1 & q_2 \\ r_1 & r_2 \end{vmatrix} \neq 0$. Let

$$T = \begin{pmatrix} q_1 & q_2 \cdots q_m \\ r_1 & r_2 \cdots r_m \\ & 0 \ I^{(m-2)} \end{pmatrix},$$

where $I^{(l)}$ is the $l \times l$ identity matrix. Obviously, det $T = q_1 r_2 - q_2 r_1$. Write $T\boldsymbol{\theta}_i = \boldsymbol{\xi}_i = (\xi_{i1}, \cdots, \xi_{im})$ $(1 \leqslant i \leqslant n)$ and $MG_m = \{(x_1, \cdots, x_m) \mid 0 \leqslant x_i < M, 1 \leqslant i \leqslant m\}$. Then

$$I(\boldsymbol{q}, \boldsymbol{r}) = M^{-nm} \int_{MG_m} \cdots \int_{MG_m} \gamma(\boldsymbol{q}, \boldsymbol{\theta}_1, \cdots, \boldsymbol{\theta}_n) \gamma(\boldsymbol{r}, \boldsymbol{\theta}_1, \cdots, \boldsymbol{\theta}_n) d\boldsymbol{\theta}_1 \cdots d\boldsymbol{\theta}_n$$

$$= M^{-nm} |q_1 r_2 - q_2 r_1|^{-n} \sum_{\substack{p_i, s_i \\ 1 \leqslant i \leqslant n}} \int_{T(MG_m)} \cdots \int_{T(MG_m)} \beta(\boldsymbol{q}, \xi_{11} - p_1, \cdots, \xi_{n1} - p_n)$$

$$\times \beta(\boldsymbol{r}, \xi_{12} - s_1, \cdots, \xi_{n2} - s_n) d\boldsymbol{\xi}_1 \cdots d\boldsymbol{\xi}_n. \tag{24}$$

Put $T(MG_m) \times \cdots \times T(MG_m) = D(M)$. Let P_1 and P_2 $(P_1 \subset D(M) \subset P_2)$ be two nm-dimensional parallelepipeds whose surfaces are parallel to the corresponding surfaces of $D(M)$ with distance $\sqrt{nm}$. Since

$$\int_{G_m} \cdots \int_{G_m} \beta(\boldsymbol{q}, \xi_{11}, \cdots, \xi_{n1}) \beta(\boldsymbol{r}, \xi_{12}, \cdots, \xi_{n2}) d\boldsymbol{\xi}_1 \cdots d\boldsymbol{\xi}_n = \psi(\boldsymbol{q})\psi(\boldsymbol{r}),$$

$$N(P_1)\psi(\boldsymbol{q})\psi(\boldsymbol{r}) \leqslant \sum_{\substack{p_i, s_i \\ 1 \leqslant i \leqslant n}} \int_{T(MG_m)} \cdots \int_{T(MG_m)} \beta(\boldsymbol{q}, \xi_{11} - p_1, \cdots, \xi_{n1} - p_n)$$

$$\times \beta(\boldsymbol{r}, \xi_{12} - s_1, \cdots, \xi_{n2} - s_n) d\boldsymbol{\xi}_1 \cdots d\boldsymbol{\xi}_n$$

$$\leqslant N(P_2)\psi(\boldsymbol{q})\psi(\boldsymbol{r}), \tag{25}$$

where $N(P_i)$ is the number of the lattice points in $P_i (i = 1, 2)$. Evidently, there exists a constant c which is independent of M such that

$$N(P_1) \geqslant |D(M - c)| = (M - c)^{nm} |q_1 r_2 - q_2 r_1|^n, \tag{26}$$
$$N(P_2) \leqslant |D(M + c)| = (M + c)^{nm} |q_1 r_2 - q_2 r_1|^n. \tag{27}$$

It follows from (24)—(27) that

$$\left(\frac{M - c}{M}\right)^{nm} \psi(\boldsymbol{q})\psi(\boldsymbol{r}) \leqslant I(\boldsymbol{q}, \boldsymbol{r}) \leqslant \left(\frac{M + c}{M}\right)^{nm} \psi(\boldsymbol{q})\psi(\boldsymbol{r}).$$

Let $M \to \infty$, then we have (22).

2) Take $\boldsymbol{r}=(1,\,0,\,\cdots,\,0)$ and $A_r=G_m$, then $\psi(\boldsymbol{r})=1$. Since all the components of $\boldsymbol{q}$ are positive integers, $\boldsymbol{q}$ and $\boldsymbol{r}$ are linearly independent. Hence (21) follows from (22). (Notice that the proof of (22) depends only on the linear independence of $\boldsymbol{q}$ and $\boldsymbol{r}$.)

3) Suppose $\boldsymbol{q}$, $\boldsymbol{r}$, $l.d.$. Put

$$W=\begin{pmatrix} 1 & \dfrac{q_2}{q_1} & \cdots & \dfrac{q_m}{q_1} \\[2mm] 0 & & I^{(m-1)} & \end{pmatrix},\quad \boldsymbol{\xi}_i=W\boldsymbol{\theta}_i\ (1\leqslant i\leqslant n).$$

Evidently, $\det W=1$. We have

$$I(\boldsymbol{q},\,\boldsymbol{r})=\sum_{\substack{p_i,\,s_i \\ 1\leqslant i\leqslant n}}\int_{W(G_m)}\cdots\int_{W(G_m)}\beta(\boldsymbol{q},\,q_1\xi_{11}-p_1,\,\cdots,\,q_1\xi_{n1}-p_n)$$

$$\times\beta(\boldsymbol{r},\,r_1\xi_{11}-s_1,\,\cdots,\,r_1\xi_{n1}-s_n)d\boldsymbol{\xi}_1\cdots d\boldsymbol{\xi}_n=I_0+\cdots+I_n,\qquad(28)$$

where I_j is the sum of all the terms with exactly j indices $i_1,\,\cdots,\,i_j$ having $q_1s_i-r_1p_i=0$.

For real $\alpha_1,\,\cdots,\,\alpha_n$, by the method similar to the proof of (13), we have

$$\sum_{\substack{p_i,\,s_i \\ q_1s_i-r_1p_i\neq0 \\ 1\leqslant i\leqslant n}}\int_{\alpha_1}^{1+\alpha_1}\cdots\int_{\alpha_n}^{1+\alpha_n}\beta(\boldsymbol{q},q_1\eta_1-p_1,\,\cdots,\,q_1\eta_n-p_n)\beta(\boldsymbol{r},\,r_1\eta_1-s_1,\,\cdots,\,r_1\eta_n-s_n)d\eta_1\cdots d\eta_n$$

$$\leqslant\sum_{\substack{p_i,\,s_i \\ q_1s_i-r_1p_i\neq0 \\ 1\leqslant i\leqslant n}}\int_{\alpha_1-\frac{p_1}{q_1}}^{1+\alpha_1-\frac{p_1}{q_1}}\cdots\int_{\alpha_n-\frac{p_n}{q_1}}^{1+\alpha_n-\frac{p_n}{q_1}}\beta(\boldsymbol{q},\,q_1\eta_1',\,\cdots,\,q_1\eta_n')$$

$$\times\beta\left(\boldsymbol{r},\,r_1\eta_1'-\frac{q_1s_1-r_1p_1}{q_1},\,\cdots,\,r_1\eta_n'-\frac{q_1s_n-r_1p_n}{q_1}\right)d\eta_1'\cdots d\eta_n'\leqslant\psi(\boldsymbol{q})\psi(\boldsymbol{r}).$$

Taking $\alpha_i=q_1^{-1}\sum_{j=2}^{m}q_j\xi_{ij}(1\leqslant i\leqslant n)$, we have

$$I_0\leqslant\int_0^1\cdots\int_0^1 d\xi_{12}\cdots d\xi_{1m}\cdots d\xi_{n2}\cdots d\xi_{nm}\sum_{\substack{p_i,\,s_i \\ q_1s_i-r_1p_i\neq0 \\ 1\leqslant i\leqslant n}}\int_{\alpha_1}^{1+\alpha_1}\cdots$$

$$\int_{\alpha_n}^{1+\alpha_n}\beta(\boldsymbol{q},\,q_1\xi_{11}-p_1,\,\cdots,\,q_1\xi_{n1}-p_n)\beta(\boldsymbol{r},\,r_1\xi_{11}-s_1,\,\cdots,\,r_1\xi_{n1}-s_n)d\xi_{11}\cdots d\xi_{n1}$$

$$\leqslant\psi(\boldsymbol{q})\psi(\boldsymbol{r}).\qquad(29)$$

As for $I_j(j\geqslant1)$, we first have

$$\sum_{\substack{p_i,\,s_i \\ q_1s_i-r_1p_i\neq0 \\ 1\leqslant i\leqslant n-j}}\sum_{\substack{p_i,\,s_i \\ q_1s_i-r_1p_i=0 \\ n-j<i\leqslant n}}\int_{\alpha_1}^{1+\alpha_1}\cdots\int_{\alpha_n}^{1+\alpha_n}\beta(\boldsymbol{q},\,q_1\eta_1-p_1,\,\cdots,\,q_1\eta_n-p_n)$$

$$\times\beta(\boldsymbol{r},\,r_1\eta_1-s_1,\,\cdots,\,r_1\eta_n-s_n)d\eta_1\cdots d\eta_n$$

$$=\sum_{\substack{p_i,\,s_i \\ q_1s_i-r_1p_i\neq0 \\ 1\leqslant i\leqslant n-j}}\sum_{\substack{p_i,\,s_i \\ q_1s_i-r_1p_i=0 \\ n-j<i\leqslant n}}\int_{\alpha_1-\frac{p_1}{q_1}}^{1+\alpha_1-\frac{p_1}{q_1}}\cdots\int_{\alpha_n-\frac{p_n}{q_1}}^{1+\alpha_n-\frac{p_n}{q_1}}\beta(\boldsymbol{q},\,q_1\eta_1',\,\cdots,\,q_1\eta_n')$$

$$\times\beta\left(\boldsymbol{r},\,r_1\eta_1'-\frac{q_1s_1-r_1p_1}{q_1},\,\cdots,\,r_1\eta_{n-j}'-\frac{q_1s_{n-j}-r_1p_{n-j}}{q_1},\,r_1\eta_{n-j+1}',\,r_1\eta_n'\right)d\eta_1'\cdots d\eta_n'$$

$$\leqslant\psi(\boldsymbol{q})A(q_1,\,r_1)^jq_1^{-j}\leqslant\psi(\boldsymbol{q})A(q_1,\,r_1)q_1^{-1}.$$

Taking $\alpha_i = q_1^{-1} \sum_{j=2}^{m} q_j \xi_{ij} \ (1 \leqslant i \leqslant n)$, we have

$$\sum_{\substack{p_i, \, s_i \\ q_1 s_i - r_1 p_i \neq 0 \\ 1 \leqslant i \leqslant n-j}} \sum_{\substack{p_i, \, s_i \\ q_1 s_i - r_1 p_i = 0 \\ n-j < i \leqslant n}} \int_{W(G_m)} \cdots \int_{W(G_m)} \beta(\boldsymbol{q}, \, q_1 \xi_{11} - p_1, \, \cdots, \, q_1 \xi_{n1} - p_n)$$

$$\times \beta(\boldsymbol{r}, \, r_1 \xi_{11} - s_1, \, \cdots, \, r_1 \xi_{n1} - s_n) d\boldsymbol{\xi}_1 \cdots d\boldsymbol{\xi}_n$$

$$= \int_0^1 \cdots \int_0^1 d\xi_{12} \cdots d\xi_{nm} \sum_{\substack{p_i, \, s_i \\ q_1 s_i - r_1 p_i \neq 0 \\ 1 \leqslant i \leqslant n-i}} \sum_{\substack{p_i, \, s_i \\ q_1 s_i - r_1 p_i = 0 \\ n-i < i \leqslant n}} \int_{\alpha_1}^{1+\alpha_1} \cdots$$

$$\int_{\alpha_n}^{1+\alpha_n} \beta(\boldsymbol{q}, \, q_1 \xi_{11} - p_1, \, \cdots, \, q_1 \xi_{n1} - p_n) \beta(\boldsymbol{r}, \, r_1 \xi_{11} - s_1, \, \cdots, \, r_1 \xi_{n1} - s_n) d\xi_{11} \cdots d\xi_{n1}$$

$$\leqslant \psi(\boldsymbol{q}) A(q_1, \, r_1) q_1^{-1}.$$

Hence

$$I_j \leqslant \binom{n}{j} \psi(\boldsymbol{q}) A(q_1, \, r_1) q_1^{-1}. \tag{30}$$

(23) follows from (28), (29) and (30). The lemma is proved.

Lemma 5.

$$\int_{G_m} \cdots \int_{G_m} N(u, \, v, \, \boldsymbol{\theta}_1, \, \cdots, \, \boldsymbol{\theta}_n) d\boldsymbol{\theta}_1 \cdots d\boldsymbol{\theta}_n = \Psi(u, \, v), \tag{31}$$

$$\int_{G_m} \cdots \int_{G_m} N(u, \, v, \, \boldsymbol{\theta}_1, \, \cdots, \, \boldsymbol{\theta}_n)^2 d\boldsymbol{\theta}_1 \cdots d\boldsymbol{\theta}_n \leqslant \Psi(u, \, v)^2 + 2(2^n - 1) \sum_{u < q \leqslant v} \psi(\boldsymbol{q}) d(\boldsymbol{q}). \tag{32}$$

Proof   (31) follows from (21) immediately. Let $\boldsymbol{r} \leqslant \boldsymbol{q}$ denote $r \leqslant q$. Then the left-hand side of (32) is equal to

$$\sum_{u < q, \, r \leqslant v} \int_{G_m} \cdots \int_{G_m} \gamma(\boldsymbol{q}, \, \boldsymbol{\theta}_1, \, \cdots, \, \boldsymbol{\theta}_n) \gamma(\boldsymbol{r}, \, \boldsymbol{\theta}_1, \, \cdots, \, \boldsymbol{\theta}_n) d\boldsymbol{\theta}_1 \cdots d\boldsymbol{\theta}_n$$

$$= \sum_{\substack{u < q, \, r \leqslant v \\ q, \, r, \, l.\, i.}} I(\boldsymbol{q}, \, \boldsymbol{r}) + \sum_{\substack{u < q, \, r \leqslant v \\ q, \, r, \, l.\, d.}} I(\boldsymbol{q}, \, \boldsymbol{r})$$

$$\leqslant \Psi(u, \, v)^2 + 2(2^n - 1) \sum_{\substack{u < r \leqslant q \leqslant v \\ q, \, r, \, l.\, d.}} \psi(\boldsymbol{q}) A(q_1, \, r_1) q_1^{-1} \tag{33}$$

by Lemma 4. If $\boldsymbol{r}$ is linearly dependent of $\boldsymbol{q}$ and satisfies $u < \boldsymbol{r} \leqslant \boldsymbol{q}$, we put

$$\frac{r_1}{q_1} = \cdots = \frac{r_m}{q_m} = \frac{a}{b}, \quad (a, \, b) = 1,$$

then $b \,|\, q_i \, (1 \leqslant i \leqslant m)$, $\dfrac{u}{q} < \dfrac{a}{b} \leqslant 1$. Conversely, if $(a, \, b) = 1$, $b \,|\, q_i \, (1 \leqslant i \leqslant m)$, $\dfrac{u}{q} < \dfrac{a}{b} \leqslant 1$,

then $\boldsymbol{r} = \left(\dfrac{a}{b} q_1, \, \cdots, \, \dfrac{a}{b} q_m \right)$ is a lattice point which is linearly dependent of $\boldsymbol{q}$ and satisfies $u < \boldsymbol{r} \leqslant \boldsymbol{q}$.

Hence

$$\sum_{\substack{u < r \leqslant q \leqslant v \\ q, \, r, \, l.\, d.}} \psi(\boldsymbol{q}) A(q_1, \, r_1) q_1^{-1} = \sum_{u < q \leqslant v} \psi(\boldsymbol{q}) \sum_{\substack{b \,|\, q_i \\ 1 \leqslant i \leqslant m}} \sum_{\substack{a \\ (a, \, b) = 1 \\ \frac{u}{q} < \frac{a}{b} \leqslant 1}} A\left(q_1, \, \frac{a}{b} q_1\right) q_1^{-1}$$

$$= \sum_{u < q \leqslant v} \psi(\boldsymbol{q}) \sum_{\substack{b \,|\, q_i \\ 1 \leqslant i \leqslant m}} \sum_{\substack{a \\ (a, \, b) = 1 \\ \frac{u}{q} < \frac{a}{b} \leqslant 1}} \frac{q_1}{b} q_1^{-1} \leqslant \sum_{u < q \leqslant v} \psi(\boldsymbol{q}) \sum_{\substack{b \,|\, q_i \\ 1 \leqslant i \leqslant m}} \frac{1}{b} b = \sum_{u < q \leqslant v} \psi(\boldsymbol{q}) d(\boldsymbol{q}).$$

Combining (33) and the above inequality, we obtain the lemma.

Take $\omega(h) = [\chi(h)]$. Let

$$L_s = \{(u,\ v)\ |\ u \in S',\ v \in S',\ (\omega(u),\ \omega(v)]\ \text{is an interval of any order}\ t,\ \omega(v) \leqslant 2^s\}$$

and let $h^* = h^*(s)$ be the greatest integer satisfying $\omega(h) \leqslant 2^s$.

Lemma 6.

$$\sum_{(u,\ v) \in L_s} \int_{G_m} \cdots \int_{G_m} (N(u,\ v,\ \theta_1,\ \cdots,\ \theta_n) - \Psi(u,\ v))^2 d\theta_1 \cdots d\theta_n = O(s2^s).$$

Proof By Lemma 5, every term in the above sum has upper bound

$$2(2^n - 1) \sum_{u < q \leqslant v} \psi(\boldsymbol{q}) d(\boldsymbol{q}).$$

We first sum over all the $(u,\ v) \in L_s$ for which $(\omega(u),\ \omega(v)]$ is an interval of fixed order t. Since intervals of order t cover the positive axis exactly once, we have the upper bound

$$2(2^n - 1)\chi(h) = O(2^s).$$

Summing over t, the lemma follows.

Lemma 7. *There exists a sequence of subsets* $\sigma_1,\ \sigma_2 \cdots$ *in* G_{nm} *with measures*

$$\mu_s = \int_{\sigma_s} d\theta_1 \cdots d\theta_n = O(s^{-1-\varepsilon})$$

such that $\qquad\qquad N(h,\ \boldsymbol{\theta}_1,\ \cdots,\ \boldsymbol{\theta}_n) = \Psi(h) + O(2^{\frac{s}{2}} s^{\frac{3}{2}+\varepsilon})$

holds for any $h \in S',\ \omega(h) \leqslant 2^s$ *and* $(\theta_1,\ \cdots,\ \theta_n) \in G_{nm} \backslash \sigma_s$.

Proof We define σ_s to be the set of $(\theta_1,\ \cdots,\ \theta_n) \in G_{nm}$ for which

$$\sum_{(u,\ v) \in L_s} (N(u,\ v,\ \theta_1,\ \cdots,\ \theta_n) - \Psi(u,\ v))^2 \leqslant s^{2+\varepsilon} 2^s \tag{34}$$

does not hold. By Lemma 6, we have

$$\mu_s = O(s^{-1-\varepsilon}).$$

If $h \in S',\ \omega(h) \leqslant 2^s$, then by Lemma 3, $(0,\ \omega(h)]$ can be expressed as union of at most s intervals $(\omega(u),\ \omega(v)]$, where $(u,\ v) \in L_s$. For $(\theta_1,\ \cdots,\ \theta_n) \in G_{nm} \backslash \sigma_s$, summing over these $(u,\ v)$, we obtain

$$(N(h,\ \boldsymbol{\theta}_1,\ \cdots,\ \boldsymbol{\theta}_n) - \Psi(h))^2 = (\sum_{(u,\ v)} (N(u,\ v,\ \theta_1,\ \cdots,\ \theta_n) - \Psi(u,\ v)))^2$$

$$\leqslant s \sum_{(u,\ v) \in L_s} (N(u,\ v,\ \theta_1,\ \cdots,\ \theta_n) - \Psi(u,\ v))^2 \leqslant s^{3+\varepsilon} 2^s$$

by (34) and Cauchy's inequality. The lemma follows.

Proof of Theorem 2. Since $\sum s^{-1-\varepsilon} < \infty$, there exists for almost all $(\theta_1,\ \cdots,\ \theta_n) \in G_{nm}$ a $s_0 = s_0(\theta_1,\ \cdots,\ \theta_n)$ such that $(\theta_1,\ \cdots,\ \theta_n) \overline{\in} \sigma_s$ for $s \geqslant s_0$. Suppose that $(\theta_1,\ \cdots,\ \theta_n)$ has such a s_0 and h is so large that $\omega(h) \geqslant 2^{s_0}$. Pick s satisfying $2^{s-1} \leqslant \omega(h) < 2^s$. Then $s > s_0$.

Suppose that $h \in S'$. Since $(\theta_1,\ \cdots,\ \theta_n) \overline{\in} \sigma_s$, Lemma 7 yields

$$N(h,\ \boldsymbol{\theta}_1,\ \cdots,\ \boldsymbol{\theta}_n) = \Psi(h) + O(2^{\frac{1}{2}s} s^{\frac{3}{2}+\varepsilon}) = \Psi(h) + O(\chi(h)^{\frac{1}{2}} (\log \chi(h))^{\frac{3}{2}+\varepsilon}),$$

that is, the theorem holds for $h \in S'$. We can prove similarly the theorem for $h \in S''$. For every h there exist $h' \in S'$ and $h'' \in S''$ such that

$$h' \leqslant h \leqslant h'',\qquad \omega(h') = \omega(h) = \omega(h'').$$

Then we have

$$|\Psi(h) - \Psi(h')| \leqslant |\chi(h) - \chi(h')| \leqslant 1$$

Similarly

$$|\Psi(h) - \Psi(h'')| \leqslant 1$$

Since

$$N(h', \theta_1, \cdots, \theta_n) \leqslant N(h, \theta_1, \cdots, \theta_n) \leqslant N(h'', \theta_1, \cdots, \theta_n),$$

the theorem follows.

References

[1]   Gallagher, P. X., Metric simultaneous Diophantine approximation, *JLMS*, **37** (1962), 387—390.

[2]   Schmidt, W. M., A metrical theorem in Diophantine approximation, *Can. J. of Math.*, **12** (1960), 619—631.

[3]   Schmidt, W. M., Metrical theorems on fractional parts of sequences, *TAMS*, **110** (1964), 493—518.

[4]   Schmidt W. M., and Wang Yuan, A note on a transference theorem of linear forms, *Sci. Sin.*, **22**: 3 (1979), 276—280.

[5]   Schmidt, W. M., Simultaneous Diophantine approximation to algebraic numbers by rationals, *Acta Math.*, **125** (1970), 189—201.

[6]   Baker, A., On some Diophantine inequalities involving the exponential function, *Can. J. Math.*, **17**: 4 (1965), 616—626.

关于丢番图逼近论中的若干测度定理

王　元　于坤瑞

（中国科学院数学研究所）

摘　　要

本文证明了两条关于丢番图逼近论中的测度定理．（详细叙述见本文 §1.）这些定理是 P. X. Gallagher 定理的改进，并包有 W. M. Schmidt 的测度定理．还可以导出，例如：

1°. 对于几乎所有的 $(\theta_1, \cdots, \theta_n) \in R_n$，适合于

$$\prod_{i=1}^{n} \|q\theta_i\| q (\log q)^n < 1, \quad 1 \leqslant q \leqslant h$$

的整数 q 的个数为

$$\frac{2^n}{(n-1)!} \log \log h + O\left((\log \log h)^{1/2+\varepsilon}\right),$$

此处 $\|X\|$ 表示实数 X 至最近整数的距离，ε 为任意正常数，而与 "O" 有关的常数依赖于 ε 与诸 θ_i．

2°. W. M. Schmidt 与王元的转换定理中的性质 A 对于几乎所有的 $(\theta_{11}, \cdots, \theta_{nm}) \in R_{nm}$ 都成立．

ACTA ARITHMETICA
XLVIII (1987)

Bounds for solutions of additive equations in an algebraic number field I

by

WANG YUAN * (Beijing, China)

1. Introduction. Let k be a rational integer $\geqslant 1$. Similar to Waring's problem, one can show by the Hardy–Littlewood's method that an equation

$$a_1 x_1^k + \ldots + a_s x_s^k = 0,$$

where $a_1, \ldots, a_s$ are given rational integers but not all of the same sign, has a nontrivial solution in nonnegative rational integers $x_1, \ldots, x_s$, provided only that $s \geqslant c_1(k)$. (See, e.g., H. Davenport [3]). Here we use $c(f, \ldots, g)$ to denote a positive constant depending on $f, \ldots, g$. As for a bound of these solutions, it was shown by J. Pitman [10] that if $s \geqslant c_2(k)$, then there exists a nontrivial solution in nonnegative integers such that

$$(1) \qquad \max_i x_i < c_3(k) \max(1, |a_1|, \ldots, |a_s|)^{c_4(k)}$$

where c_2 and c_4 are explicit. Under suitable conditions and if s is very large, the estimation can be considerably improved. (See, B. J. Birch [2] and W. M. Schmidt [11], [12].) In particular, Schmidt proved that if $s \geqslant c_5(k, \varepsilon)$, the equation

$$a_1 x_1^k + \ldots + a_s x_s^k = b_1 y_1^k + \ldots + b_s y_s^k$$

with positive rational integer coefficients has a nontrivial solution in nonnegative rational integers $x_1, \ldots, x_s, y_1, \ldots, y_s$ such that

$$(2) \qquad \max_{i,j}(x_i, y_j) \leqslant \max_{i,j}(a_i, b_j)^{1/k+\varepsilon}.$$

We use hereafter $\varepsilon, \varepsilon_1, \ldots$ to denote arbitrary preassigned positive numbers < 1. The number $1/k$ in (2) is best possible. Although the circle method is still used in the proof of (2), the treatment of the minor arcs is completely distinct from that in Waring's problem.

---

* Supported by the Institute for Advanced Study, Princeton, N. J. 08540.

                                      Wang Yuan

It was Siegel ([13], [14]) who succeeded in dealing with Waring's problem in an arbitrary algebraic number field by his generalized circle method, and he obtained the result corresponding to Hardy–Littlewood's estimation on $G(k)$. Siegel's result was improved by R. G. Ayoub [1], Y. Eda [4], O. Körner [8], R. M. Stemmler [15] and T. Tatuzawa [16], [17] in various aspects.

By the combination of the methods of Schmidt and Siegel, we can generalize Schmidt's theorem to an arbitrary algebraic number field.

Let K be an algebraic number field of degree n. Let $K^{(p)}$ ($1 \leqslant p \leqslant r_1$) be the real conjugates of K and let $K^{(q)}$ and $K^{(q+r_2)}$ ($r_1 + 1 \leqslant q \leqslant r_1 + r_2$) denote the complex conjugates of K, where $r_1 + 2r_2 = n$. Throughout this paper, the indices p and q are over the sets of integers cited above. For $\gamma \in K$, we denote by $\gamma^{(i)}$ ($1 \leqslant i \leqslant n$) the conjugates of γ and by $N(\gamma) = \prod_{i=1}^{n} \gamma^{(i)}$ the norm of γ. Let γ_j ($1 \leqslant j \leqslant n$) be numbers of K and x_j ($1 \leqslant j \leqslant n$) be real numbers. We set $\xi = \sum_{j=1}^{n} x_j \gamma_j$ and define $\xi^{(i)} = \sum_{j=1}^{n} x_j \gamma_j^{(i)}$ ($1 \leqslant i \leqslant n$). We use the notations

$$\|\xi\| = \max_{i} |\xi^{(i)}|, \quad S(\xi) = \sum_{i=1}^{n} \xi^{(i)} \quad \text{and} \quad E(\xi) = \exp(2\pi i S(\xi)),$$

where $\exp(x) = e^x$. A number γ of K is called *totally nonnegative* if $\gamma^{(p)} \geqslant 0$.

Let $\alpha_1, \ldots, \alpha_s, \beta_1, \ldots, \beta_s$ be $2s$ nonzero totally nonnegative integers of K. Consider the equation of the type

$$(3) \qquad \alpha_1 \lambda_1^k + \ldots + \alpha_s \lambda_s^k = \beta_1 \mu_1^k + \ldots + \beta_s \mu_s^k.$$

A set of numbers $\lambda_1, \ldots, \lambda_s, \mu_1, \ldots, \mu_s$ satisfying (3) is called a *nontrivial solution of* (3) if $\lambda_1, \ldots, \lambda_s, \mu_1, \ldots, \mu_s$ are totally nonnegative integers of K, not all zero. Set

$$(4) \qquad m = \max_{i,j} (N(\alpha_i), N(\beta_j)).$$

In this paper, we shall prove the following

THEOREM. *Suppose $s \geqslant c_6(k, n, \varepsilon)$. Then the equation* (3) *has a nontrivial solution such that*

$$(5) \qquad \max_{i,j} (N(\lambda_i), N(\mu_j)) \ll m^{1/k + \varepsilon}.$$

Here and below the constants implicit in $\ll$ or O may depend on k, K, ε, …, but not on m.

If $k = 1$, then $\lambda_i = \beta_i$, $\mu_i = \alpha_i$ ($1 \leqslant i \leqslant s$) is a nontrivial solution of (3) with (5). So we suppose $k \geqslant 2$ throughout this paper.

Suppose that $\alpha_1, \ldots, \alpha_s$ are given integers of K. In the second part of this investigation we will show that if $s \geqslant c_7(k, n, \varepsilon)$, the equations

$$\alpha_1 a_1 \lambda_1^k + \ldots + \alpha_s a_s \lambda_s^k = 0$$

has a solution in $a_1, \ldots, a_s, \lambda_1, \ldots, \lambda_s$, where each a_i is 1 or -1 and where λ_i $(1 \leqslant i \leqslant s)$ are totally nonnegative integers, not all zero, with

$$\max_i N(\lambda_i) \ll \max(1, |N(\alpha_1)|, \ldots, |N(\alpha_s)|)^\varepsilon.$$

2. Several lemmas.

LEMMA 1. *Let* $t_1, \ldots, t_{r_1+r_2}$ *be a set of real numbers satisfying*

$$(6) \qquad \sum_{p=1}^{r_1} t_p + 2 \sum_{q=r_1+1}^{r_1+r_2} t_q = 0.$$

Then there exists a totally nonnegative unit σ *of* K *such that*

$$c_8^{-1} e^{t_p} < \sigma^{(p)} < c_8 e^{t_p}, \qquad c_8^{-1} e^{t_q} < |\sigma^{(q)}| < c_8 e^{t_q},$$

where $c_8 = c_8(K)$.

See, e.g. Lemma 1.1 in Hua Loo Keng and Wang Yuan [6]. (Put $\sigma = \eta^2$.)

LEMMA 2. *There exists a rational integer* $c_9 = c_9(K)$ *such that for any integers* α, β *of* K, *where* $\beta \neq 0$, *there exist a rational integer* l *and an integer* ω *of* K *such that* $1 \leqslant l \leqslant c_9$ *and* $|N(l\alpha - \omega\beta)| < |N(\beta)|$.

See, e.g., K. Ireland and M. Rosen [7], p. 178.

LEMMA 3. *For any* t *integers* $\gamma_1, \ldots, \gamma_t$ *of* K, *not all zero, let* γ *be a nonzero element of the integral ideal* $\mathfrak{a} = (\gamma_1, \ldots, \gamma_t)$ *with the least norm in absolute value. Then*

$$c_9! \, \gamma_i/\gamma, \quad 1 \leqslant i \leqslant t$$

are integers.

Proof. Set $\alpha = \gamma_i$ and $\beta = \gamma$ in Lemma 2. Then there exist a rational integer l_i and an integer ω_i such that

$$|N(l_i \gamma_i - \omega_i \gamma)| < |N(\gamma)|, \quad 1 \leqslant l_i \leqslant c_9.$$

Since $l_i \gamma_i - \omega_i \gamma \in \mathfrak{a}$, it follows that $N(l_i \gamma_i - \omega_i \gamma) = 0$. Therefore $l_i \gamma_i - \omega_i \gamma = 0$, and $l_i \gamma_i/\gamma = \omega_i$ is an integer. Since $l_i | c_9!$,

$$\frac{c_9! \, \gamma_i}{\gamma} = \frac{c_9!}{l_i} \left(\frac{l_i \gamma_i}{\gamma} \right), \quad 1 \leqslant i \leqslant t,$$

are integers. The lemma is proved.

LEMMA 4. *For any* t *integral vectors* (γ_i, δ_i) $(1 \leqslant i \leqslant t)$ *of* K^2, *where* $\gamma_i \neq 0$

W a n g   Y u a n

$(1 \leqslant i \leqslant t)$, *if*

$$\frac{\delta_1}{\gamma_1} = \ldots = \frac{\delta_t}{\gamma_t},$$

then

$$(\gamma_i, \delta_i) = \frac{1}{c_9!} \chi_i (\gamma, \delta), \quad 1 \leqslant i \leqslant t,$$

where γ is defined in Lemma 3, and where δ and χ_i $(1 \leqslant i \leqslant t)$ are integers.

Proof. By Lemma 3, $\chi_i = c_9! \gamma_i/\gamma$ $(1 \leqslant i \leqslant t)$ are integers. Let

$$\frac{\delta_1}{\gamma_1} = \ldots = \frac{\delta_t}{\gamma_t} = \alpha.$$

Then $\delta_i = \alpha\gamma_i$ $(1 \leqslant i \leqslant t)$. Since $(\delta_1, \ldots, \delta_t) = \alpha(\gamma_1, \ldots, \gamma_t) = \alpha\mathfrak{a}$ is an integral ideal and $\gamma \in \mathfrak{a}$, $\delta = \alpha\gamma$ is an integer. Therefore

$$\frac{1}{c_9!} \chi_i \delta = \frac{1}{c_9!} \chi_i \alpha\gamma = \frac{1}{c_9!} \chi_i \frac{\delta_i}{\gamma_i} \gamma = \frac{1}{c_9!} \chi_i \gamma\delta_i \left(\frac{\chi_i \gamma}{c_9!}\right)^{-1} = \delta_i, \quad 1 \leqslant i \leqslant t.$$

The lemma is proved.

LEMMA 5. *For any nonzero integer σ, there exists a nonzero integer γ such that $\|\gamma\| \leqslant c_{10}(K)$ and $\gamma\sigma$ is totally nonnegative.*

Proof. If $r_1 = 0$, the lemma holds clearly. Now suppose that $r_1 > 0$. Let $\omega_1, \ldots, \omega_n$ be an integral basis of K. Let

$$c_{10} = 4 \max_i \sum_{j=1}^{n} |\omega_j^{(i)}| \quad \text{and} \quad N_p = \frac{\sigma^{(p)}}{2|\sigma^{(p)}|} c_{10}, \quad 1 \leqslant p \leqslant r_1.$$

Since the matrix $(\omega_j^{(p)})$ $(1 \leqslant p \leqslant r_1, 1 \leqslant j \leqslant n)$ has rank r_1, we may suppose $\det(\omega_j^{(p)}) \neq 0$ $(1 \leqslant p, j \leqslant r_1)$. The system of linear equations

$$\sum_{j=1}^{r_1} \omega_j^{(p)} x_j = N_p, \quad 1 \leqslant p \leqslant r_1$$

has a unique solution. Set $a_j = [x_j]$ $(1 \leqslant j \leqslant r_1)$, where $[x]$ denotes the integral part of x. Then we have an integer $\gamma = \sum_{j=1}^{r_1} a_j \omega_j$ satisfying $\gamma^{(p)} \sigma^{(p)} > 0$ and $\|\gamma\| \leqslant c_{10}$. The lemma is proved.

3. Reductions.

PROPOSITION 1. *Suppose that $x \geqslant 1/k$ and $s \geqslant c_{11}(k, n, x, \varepsilon)$. Then (3) has a nontrivial solution with*

$$\max_{i,j} \left(N(\lambda_i), N(\mu_j) \right) \ll m^{x+\varepsilon}.$$

The case $x = 1/k$ is the theorem.

One can prove by Siegel's method that if $s \geqslant c_{12}(k, n)$, then the equation of the type

$$\alpha_1 \lambda_1^k + \ldots + \alpha_t \lambda_t^k - \alpha_{t+1} \lambda_{t+1}^k - \ldots - \alpha_s \lambda_s^k = 0$$

has a nontrivial solution in totally nonnegative integers $\lambda_1, \ldots, \lambda_s$ such that

$$(7) \qquad\qquad \max_i N(\lambda_i) \ll \max N(\alpha_i)^{c_{13}(k, n)},$$

where $\alpha_1, \ldots, \alpha_s$ are given nonzero totally nonnegative integers and $1 \leqslant t \leqslant s - 1$.

It will suffice to prove Proposition 1 when m is large, say $m \geqslant c_{14}(k, K, x, \varepsilon)$. In fact, if $m < c_{14}$ and $s \geqslant c_{12}$, then it follows by (7) that (3) has a nontrivial solution such that

$$\max_{i,j} \left(N(\lambda_i), N(\mu_j) \right) \ll m^{c_{13}} \ll c_{14}^{c_{13}} \ll m^{x+\varepsilon}.$$

Let X be the set of x such that Proposition 1 holds. Then (7) shows that X is not empty. It is clear that X is a closed set. Hence the proof of Proposition 1 is reduced to proving that if $x > 1/k$ and $x \in X$, then there exists an $x' \in X$, where $x' < x$.

For $1 \leqslant j \leqslant s$, set

$$t_i = k^{-1} \left(\log N(\alpha_j)^{1/n} + \log |\alpha_j^{(i)}|^{-1} \right), \qquad 1 \leqslant i \leqslant n.$$

Then (6) holds, and therefore there exists a set of totally nonnegative units σ_j $(1 \leqslant j \leqslant s)$ such that

$$c_8^{-1} N(\alpha_j)^{1/nk} (\alpha_j^{(p)})^{-1/k} \leqslant \sigma_j^{(p)} \leqslant c_8 N(\alpha_j)^{1/nk} (\alpha_j^{(p)})^{-1/k},$$

$$c_8^{-1} N(\alpha_j)^{1/nk} |\alpha_j^{(q)}|^{-1/k} \leqslant |\sigma_j^{(q)}| \leqslant c_8 N(\alpha_j)^{1/nq} |\alpha_j^{(q)}|^{-1/k},$$

i.e.,

$$c_8^{-k} N(\alpha_j)^{1/n} \leqslant \alpha_j^{(p)} \sigma_j^{(p)k} \leqslant c_8^k N(\alpha_j)^{1/n},$$

$$c_8^{-k} N(\alpha_j)^{1/n} \leqslant |\alpha_j^{(q)} \sigma_j^{(q)k}| \leqslant c_8^k N(\alpha_j)^{1/n}, \qquad 1 \leqslant j \leqslant s.$$

Similarly, there exists a set of units τ_j $(1 \leqslant j \leqslant s)$ such that

$$c_8^{-k} N(\beta_j)^{1/n} \leqslant \beta_j^{(p)} \tau_j^{(p)k} \leqslant c_8^k N(\beta_j)^{1/n},$$

$$c_8^{-k} N(\beta_j)^{1/n} \leqslant |\beta_j^{(q)} \tau_j^{(q)k}| \leqslant c_8^k N(\beta_j)^{1/n}, \qquad 1 \leqslant j \leqslant s.$$

Let

$$\alpha_i' = \alpha_i \sigma_i^k, \qquad \beta_i' = \beta_i \tau_i^k, \qquad \lambda_i = \sigma_i \lambda_i', \qquad \mu_i = \tau_i \mu_i' \qquad (1 \leqslant i \leqslant s).$$

Then (3) becomes

$$(3)' \qquad\qquad \alpha_1' \lambda_1'^k + \ldots + \alpha_s' \lambda_s'^k = \beta_1' \mu_1'^k + \ldots + \beta_s' \mu_s'^k.$$

Wang Yuan

If Proposition 1 holds for x' and for the particular equation (3)', then we have a nontrivial solution of (3)' such that

$$\max_{i,j}\left(N(\lambda_i'), N(\mu_j')\right) \ll \max_{i,j}\left(N(\alpha_i'), N(\beta_j')\right)^{x'+\varepsilon}.$$

Since $N(\lambda_i') = N(\lambda_i)$, $N(\mu_i') = N(\mu_i)$, $N(\alpha_i') = N(\alpha_i)$, $N(\beta_i') = N(\beta_i)$ $(1 \leqslant i \leqslant s)$, we have a nontrivial solution of (3) with

$$\max_{i,j}\left(N(\lambda_i), N(\mu_j)\right) \ll m^{x'+\varepsilon}.$$

i.e., Proposition 1 holds for x' and for (3). Hence we may suppose without loss of generality that α_i and β_i satisfy

$$c_{15}^{-1} N(\alpha_i)^{1/n} < \alpha_i^{(p)} < c_{15} N(\alpha_i)^{1/n}, \quad c_{15}^{-1} N(\alpha_i)^{1/n} < |\alpha_i^{(q)}| < c_{15} N(\alpha_i)^{1/n},$$

(8)

$$c_{15}^{-1} N(\beta_i)^{1/n} < \beta_i^{(p)} < c_{15} N(\beta_i)^{1/n}, \quad c_{15}^{-1} N(\beta_i)^{1/n} < |\beta_i^{(q)}| < c_{15} N(\beta_i)^{1/n},$$

$$1 \leqslant i \leqslant s,$$

where $c_{15} = c_{15}(k, K)$.

In what follows, x will be a fixed number $> 1/k$ for which Proposition 1 holds. Take y sufficiently small such that

(9) $$1/k + 6c_{13}\, ny + 20ny < x \quad \text{and} \quad 22kny < 1,$$

and put

(10) $$x' = \max\left(x(1 - \tfrac{1}{2}y) + y/2kn,\ 1/k + 6c_{13}\,ny + 20ny\right),$$

so that $x' < x$. We proceed to prove that Proposition 1 holds for x'.

Let $\varepsilon_1 = \min(\varepsilon/8x', \varepsilon/4)$ and divide the interval $[0, 1]$ into a finite number of intervals $\{I\}$ of length $\leqslant \varepsilon_1$. If s is large, one of these intervals I will be such that many of the coefficients $\alpha_1, \ldots, \alpha_s$ are of the type

$$N(\alpha_i) = m^{a_i}, \quad a_i \in I.$$

We may therefore suppose without loss of generality that

$$\frac{N(\alpha_i)}{N(\alpha_j)} \leqslant m^{\varepsilon_1}, \quad 1 \leqslant i, j \leqslant s.$$

Similarly, we may suppose

$$\frac{N(\beta_i)}{N(\beta_j)} \leqslant m^{\varepsilon_1}, \quad 1 \leqslant i, j \leqslant s.$$

Let $a^n = m^{\varepsilon_1} \max_i N(\alpha_i)$ and $b^n = m^{\varepsilon_1} \max_i N(\beta_i)$. Let p_i and q_i be the largest rational integers such that

$$N(\alpha_i)\, p_i^{kn} \leqslant a^n \quad \text{and} \quad N(\beta_i)\, q_i^{kn} \leqslant b^n, \quad 1 \leqslant i \leqslant s.$$

Since $m \geqslant c_{14}$, $a^n/N(\alpha_i) \geqslant m^{\varepsilon 1}$ and $b^n/N(\beta_i) \geqslant m^{\varepsilon 1}$, we may suppose

$$p_i \geqslant 2^{-1/kn} \left(\frac{a^n}{N(\alpha_i)}\right)^{1/kn} \quad \text{and} \quad q_i \geqslant 2^{-1/kn} \left(\frac{b^n}{N(\beta_i)}\right)^{1/kn}, \quad 1 \leqslant i \leqslant s.$$

Hence

$$N(\alpha_i)\, p_i^{kn} \geqslant \tfrac{1}{2} a^n \quad \text{and} \quad N(\beta_i)\, q_i^{kn} \geqslant \tfrac{1}{2} b^n, \quad 1 \leqslant i \leqslant s.$$

Set $\alpha_i' = \alpha_i\, p_i^k$, $\beta_i' = \beta_i\, q_i^k$, $\lambda_i = p_i\, \lambda_i'$, $\mu_i = q_i\, \mu_i'$ $(1 \leqslant i \leqslant s)$. Then (3) becomes (3)′, and by (8), α_i' and β_i' satisfy

$$(2c_{15})^{-1} a < \alpha_i'^{(p)} < c_{15}\, a, \quad (2c_{15})^{-1} a < |\alpha_i'^{(q)}| < c_{15}\, a,$$

$$(2c_{15})^{-1} a < \beta_i'^{(p)} < c_{15}\, b, \quad (2c_{15})^{-1} b < |\beta_i'^{(q)}| < c_{15}\, b, \quad 1 \leqslant i \leqslant s.$$

Suppose that Proposition 1 holds for x' and for the particular equation (3)′. Then there exists a nontrivial solution of (3)′ with

$$\max_{i,j} \left(N(\lambda_i'), N(\mu_j')\right) \ll \max(a^n, b^n)^{x' + \varepsilon/4} \ll m^{(1+\varepsilon_1)(x' + \varepsilon/4)} \ll m^{x' + \varepsilon/2}.$$

Since

$$N(\alpha_i) = m^{\varepsilon 1}\, N(\alpha_i) \max_j N(\alpha_j)/m^{\varepsilon 1} \max_l N(\alpha_l)$$

$$= a^n m^{-\varepsilon 1}\, N(\alpha_i)/\max_l N(\alpha_l) \geqslant a^n m^{-2\varepsilon 1}, \quad 1 \leqslant i \leqslant s,$$

we have

$$p_i^n \leqslant p_i^{kn} \leqslant a^n/N(\alpha_i) \leqslant m^{2\varepsilon 1} \leqslant m^{\varepsilon/2}, \quad 1 \leqslant i \leqslant s,$$

and therefore

$$N(\lambda_i) \ll p_i^n\, N(\lambda_i') \ll m^{x' + \varepsilon}, \quad 1 \leqslant i \leqslant s.$$

Similarly

$$N(\mu_i) \ll m^{x' + \varepsilon}, \quad 1 \leqslant i \leqslant s,$$

i.e., Proposition 1 holds for x' and for (3). Thus in proving Proposition 1 for x', we may suppose that

$$(11) \quad c_{16}\, a < \alpha_i^{(p)} < c_{17}\, a, \quad c_{16}\, a < |\alpha_i^{(q)}| < c_{17}\, a, \quad c_{16}\, b < \beta_i^{(p)} < c_{17}\, b,$$

$$c_{16}\, b < |\beta_i^{(q)}| < c_{17}\, b, \quad 1 \leqslant i \leqslant s$$

for certain positive numbers a, b, where $c_{16} = c_{16}(k, K)$ and $c_{17} = c_{17}(k, K)$.

4. Continuation. In what follows, h will be the integer $c_{11}(k, n, x, \varepsilon)$ occurring in Proposition 1, and $s > h$. Set

$$(12) \qquad\qquad\qquad z = y/2kn^2.$$

124   Wang Yuan

We distinguish two cases.

A. There is a subset of h elements among $\alpha_1, \ldots, \alpha_s$, say $\alpha_1, \ldots, \alpha_h$ and there is a subset of h elements among $\beta_1, \ldots, \beta_s$, say $\beta_1, \ldots, \beta_h$, and there are totally nonnegative integers $\sigma_1, \ldots, \sigma_h, \tau_1, \ldots, \tau_h$ such that

$$(13) \qquad 0 < \|\sigma_i\| \leqslant m^z, \qquad 0 < \|\tau_i\| \leqslant m^z, \qquad 1 \leqslant i \leqslant h$$

and

$$|N(\sigma)| \geqslant m^y,$$

where σ is a nonzero element in the integral ideal $(\alpha_1 \sigma_1, \ldots, \alpha_h \sigma_h, \beta_1 \tau_1, \ldots, \beta_h \tau_h)$ with the least norm in absolute value.

By Lemma 5, we may choose a nonzero integer γ such that $\|\gamma\| \leqslant c_{10}$ and $\gamma\sigma$ is totally nonnegative. By Lemma 3,

$$\alpha'_i = \frac{c_9! \, \alpha_i \, \sigma_i^k \, \gamma^2}{\gamma\sigma} \quad \text{and} \quad \beta'_i = \frac{c_9! \, \beta_i \, \tau_i^k \, \gamma^2}{\gamma\sigma}, \qquad 1 \leqslant i \leqslant h$$

are all nonzero totally nonnegative integers. Therefore it follows from the case x of the Proposition 1 that the equation

$$\alpha'_1 \lambda'^k_1 + \ldots + \alpha'_h \lambda'^k_h = \beta'_1 \mu'^k_1 + \ldots + \beta'_h \mu'^k_h$$

has a nontrivial solution satisfying

$$\max_{i,j} \left(N(\lambda'_i), N(\mu'_j) \right) \ll \max_{i,j} \left(N(\alpha'_i), N(\beta'_j) \right)^{x+\varepsilon} \ll m^{(1+knz-y)(x+\varepsilon)}.$$

Let $\lambda_i = \sigma_i \lambda'_i$, $\mu_i = \tau_i \mu'_i$ $(1 \leqslant i \leqslant h)$ and $\lambda_i = \mu_i = 0$ $(h < i \leqslant s)$. Then by (10) and (12), the equation (3) has a nontrivial solution with

$$\max_{i,j} \left(N(\lambda_i), N(\mu_j) \right) \ll m^{(1+knz-y)(x+\varepsilon)+nz} \ll m^{(1-y/2)(x+\varepsilon)+nz} \ll m^{x'+\varepsilon}.$$

We are thus reduced to case

B. For any h elements, say $\alpha_1, \ldots, \alpha_h$, among $\alpha_1, \ldots, \alpha_s$, and for any h elements, say $\beta_1, \ldots, \beta_h$, among $\beta_1, \ldots, \beta_s$, and given any totally nonnegative integers $\sigma_1, \ldots, \sigma_h, \tau_1, \ldots, \tau_h$ satisfying (13), the integer σ defined as in the case A satisfies $|N(\sigma)| < m^y$.

Condition B depends on k, n, h, m, y, and it is denoted by $B(k, n, h, m, y)$.

PROPOSITION 2. *Let* $q = 1$ *or* -1. *Let*

$$(14) \qquad m = \max(a^n, b^n)$$

and let $\alpha_1, \ldots, \alpha_s, \beta_1, \ldots, \beta_s$ *be nonzero totally nonnegative integers satisfying* (11) *and* $B(k, n, h, m, y)$. *Then if* $s \geqslant c_{18}(k, n, h, y)$, *the equation*

$$\alpha_1 \lambda_1^k + \ldots + \alpha_s \lambda_s^k - \beta_1 \mu_1^k - \ldots - \beta_s \mu_s^k = q\chi$$

has a solution in totally nonnegative integers $\lambda_1, \ldots, \lambda_s, \mu_1, \ldots, \mu_s, \chi$, *not all zero, with*

$$\max_{i,j}\left(N(\lambda_i), N(\mu_j)\right) \ll m^{1/k+20ny}, \qquad \|\chi\| \leqslant m^{6y}.$$

Now we proceed to show that Proposition 2 implies that Proposition 1 is true for x'. Let x, x', y, z, h be as above. Suppose that c_{12} and c_{18} are integers. Let $s = uv$, where $u = c_{18}$ and $v = c_{12}$. Replace the indices $1 \leqslant l \leqslant s$ by double indices $1 \leqslant i \leqslant v$, $1 \leqslant j \leqslant u$. Then the equation (3) can be written as

$$(15) \qquad \sum_{i=1}^{v} \left(\alpha_{i1} \lambda_{i1}^k + \ldots + \alpha_{iu}^k \lambda_{iu}^k - \beta_{i1} \mu_{i1}^k - \ldots - \beta_{iu} \mu_{iu}^k\right) = 0.$$

For each i, $1 \leqslant i \leqslant v$, the coefficients $\alpha_{i1}, \ldots, \alpha_{iu}, \beta_{i1}, \ldots, \beta_{iu}$ satisfy the conditions in Proposition 2. Hence there are totally nonnegative integers $\lambda'_{i1}, \ldots, \lambda'_{iu}, \mu'_{i1}, \ldots, \mu'_{iu}, \chi_i$, not all zero, such that

$$\alpha_{i1} \lambda_{i1}'^k + \ldots + \alpha_{iu} \lambda_{iu}'^k - \beta_{i1} \mu_{i1}'^k - \ldots - \beta_{iu} \mu_{iu}'^k = q_i \chi_i$$

with

$$\max_{j,l}\left(N(\lambda'_{ij}), N(\mu'_{il})\right) \ll m^{1/k+20ny}, \qquad \|\chi_i\| \leqslant m^{6y}.$$

We may suppose that $\chi_i \neq 0$ $(1 \leqslant i \leqslant v)$. Otherwise we get a small solution straightaway. Take $q_1 = \ldots = q_{v-1} = 1$ and $q_v = -1$. Then by (7), the equation

$$\chi_1 \gamma_1^k + \ldots + \chi_{v-1} \gamma_{v-1}^k - \chi_v \gamma_v^k = 0$$

has a nontrivial solution satisfying

$$\max_{i} N(\gamma_i) \ll m^{6c_{13}ny}.$$

Let $\lambda_{ij} = \gamma_i \lambda'_{ij}$, $\mu_{ij} = \gamma_i \mu'_{ij}$ $(1 \leqslant i \leqslant v, 1 \leqslant j \leqslant u)$. Then we have a nontrivial solution of (15) having

$$\max_{i,j,t,l}\left(N(\lambda_{ij}), N(\mu_{il})\right) \ll m^{1/k+6c_{13}ny+20ny} \ll m^{x'}.$$

Thus Proposition 1 holds for x'.

5. Weyl's inequality. Let $\omega_1, \ldots, \omega_n$ be an integral basis of K, $\mathfrak{d}$ the different and D the absolute value of the discriminant of K. We can choose a basis $\varrho_1, \ldots, \varrho_n$ of $\mathfrak{d}^{-1}$ such that

$$S(\varrho_i \omega_j) = \begin{cases} 1, & \text{if } i = j, \\ 0, & \text{otherwise.} \end{cases}$$

126   W a n g   Y u a n

Set

$$\xi = x_1 \varrho_1 + \ldots + x_n \varrho_n \quad \text{and} \quad \eta = y_1 \omega_1 + \ldots + y_n \omega_n,$$

where x_i and y_i $(1 \leqslant i \leqslant n)$ are real numbers. We denote by $P(T)$ the set of $(y_1, \ldots, y_n)$ satisfying

$$0 \leqslant \eta^{(p)} \leqslant T, \quad |\eta^{(q)}| \leqslant T;$$

$\sum_{\lambda \in P(T)}$ a sum where λ runs over all integers such that $0 \leqslant \lambda^{(p)} \leqslant T, \; |\lambda^{(q)}| \leqslant T,$ and $\sum_{|\mu| \in P(T)}$ a sum of integers μ satisfying $\|\mu\| \leqslant T.$

LEMMA 6 (Siegel). *Let $h \geqslant 1$. Then for any ξ, there exist an integer α and a number β of $\mathfrak{d}^{-1}$ such that*

$$\|\alpha \xi - \beta\| < h^{-1}, \quad 0 < \|\alpha\| \leqslant h,$$

$$\max \left(h|\alpha^{(i)} \xi^{(i)} - \beta^{(i)}|, |\alpha^{(i)}| \right) \geqslant D^{-1/2}, \quad 1 \leqslant i \leqslant n$$

and

$$N\left((\alpha, \beta \mathfrak{d})\right) \leqslant D^{1/2}.$$

See Lemma 6 in Siegel [14]. Notice that the property of ξ belonging to supplementary domain is only used in the proof of his formula (41).

LEMMA 7 (Mitsui). *Let A, B, h be positive numbers satisfying $A \geqslant 1$, $h > 2^{5+r_2} D$ and $1 \leqslant B < 2^{-4-r_2} D^{-1/n} h$. Then for any ξ*

$$\sum_{|\mu| \in P(B)} \min \left(A, |1 - E(\xi \mu \omega_j)|^{-1} \; (1 \leqslant j \leqslant n) \right)$$

$$= O\left(AB^n \left(\frac{1}{\|\alpha\|} + \frac{1}{B} + \frac{h \log h}{AB} + \frac{\log h}{A} \right) \right),$$

here and also in Lemma 8, α denotes an integer satisfying the conditions in Lemma 6.

See Theorem 3.1 in Mitsui [9]. Notice that in the proof of his formula (3.42), we may use the estimation $|N(\alpha)| \geqslant c \|\alpha\|$ instead of $|N(\alpha)| \geqslant cT$ with $c = c(K)$.

LEMMA 8 (Weyl's inequality). *Let*

$$G = 2^{k-1} \quad \text{and} \quad L(\xi) = \sum_{\lambda \in P(T)} E(\lambda^k \xi), \quad \text{where} \quad T > k! \, 2^{5+k+r_2} D.$$

Let h be a number satisfying

$$k! \, 2^{4+k+r_2} D T^{k-1} < h \leqslant T^k.$$

Then

$$L(\xi) \ll T^{n+\varepsilon_2} \left(\frac{1}{\|\alpha\|} + \frac{1}{T} + \frac{h}{T^k} \right)^{1/G}.$$

Proof. By Hölder's inequality

$$|L(\xi)|^G = \left|\sum_\lambda \sum_{\lambda_1} E\big((\lambda+\lambda_1)^k \xi - \lambda^k \xi\big)\right|^{2^{k-2}}$$

$$\leqslant \left(\sum_{\lambda_1} \left|\sum_\lambda E(k\lambda_1 \lambda^{k-1}\xi + \ldots)\right|\right)^{2^{k-2}}$$

$$\ll T^{n(2^{k-2}-1)} \sum_{\lambda_1} \left|\sum_\lambda E(k\lambda_1 \lambda^{k-1}\xi + \ldots)\right|^{2^{k-2}}$$

$$\ll T^{n(2^{k-2}-1)} \sum_{\lambda_1} T^{n(2^{k-3}-1)} \sum_{\lambda_2} \left|\sum_\lambda E\big(k(k-1)\lambda_1 \lambda_2 \lambda^{k-2}\xi + \ldots\big)\right|^{2^{k-3}}.$$

$$\ll \ldots$$

$$\ll T^{n(G-k)} \sum_{\lambda_1} \sum_{\lambda_2} \cdots \sum_{\lambda_{k-1}} \left|\sum_\lambda E(\mu\lambda\xi)\right|,$$

where

(16) $$\mu = k!\,\lambda_1 \ldots \lambda_{k-1}, \qquad |\lambda_i| \in P(2T) \quad (1 \leqslant i \leqslant k-1),$$

and λ runs over all solutions of the conditions

$$\lambda + \lambda_{i_1} + \ldots + \lambda_{i_g} \in P(T) \quad (1 \leqslant i_1 < \ldots < i_g \leqslant k-1,\ G \leqslant g \leqslant k-1).$$

Let $A(\mu)$ denote the number of solutions of (16). Then by the well-known properties of the divisor function, we have

$$A(\mu) = \begin{cases} O(T^{n(k-2)}), & \text{if } \mu = 0, \\ O(T^{\varepsilon 2}), & \text{otherwise.} \end{cases}$$

Hence

$$|L(\xi)|^G \ll T^{n(G-2)} + T^{n(G-k)+\varepsilon 2} \sum_\mu \left|\sum_\lambda E(\mu\lambda\xi)\right|,$$

where the summation is extended over all μ, λ satisfying

$$\mu \in P(k!\,2^{k-1}\,T^{k-1}) \quad \text{and} \quad \lambda \in P(T).$$

Since

$$\sum_{\lambda \in P(T)} E(\mu\lambda\xi) = O\big(T^{n-1} \min(T, |1 - E(\mu\omega_j\xi)|^{-1}\ (1 \leqslant j \leqslant n))\big)$$

(cf. Siegel [14], p. 332), we have

$$|L(\xi)|^G \ll T^{n(G-2)} + T^{n(G-k)+\varepsilon 2} \sum_\mu T^{n-1} \min(T, |1 - E(\mu\omega_j\xi)|^{-1}\ (1 \leqslant j \leqslant n)).$$

Let $A = T$ and $B = k!\,2^{k-1}\,T^{k-1}$. Then by Lemma 7, we have

$$|L(\xi)|^G \ll T^{n(G-k)+\varepsilon 2 + n - 1 + 1 + n(k-1)} \left(\frac{1}{\|\alpha\|} + \frac{1}{T^{k-1}} + \frac{h \log h}{T^k} + \frac{\log h}{T}\right)$$

$$\ll T^{nG+2\varepsilon 2}\left(\frac{1}{\|\alpha\|} + \frac{1}{T} + \frac{h}{T^k}\right).$$

Wang Yuan

The lemma follows.

6. Schmidt's lemma. In this section, we shall generalize Schmidt's lemma to an arbitrary algebraic number field.

LEMMA 9 (Schmidt). *Suppose that* $T \geqslant c_{19}(k, K, \varepsilon_3)$, $C \geqslant T^{n-1/G+\varepsilon_3}$ *and* $|L(\xi)| \geqslant C$. *Then there exist a totally nonnegative integer* α *and an integer* β *such that*

$$\|\alpha\xi - \beta\| \ll \left(\frac{T^n}{C}\right)^G T^{-k+\varepsilon_3}$$

and

$$0 < \|\alpha\| \ll \left(\frac{T^n}{C}\right)^G T^{\varepsilon_3},$$

where $\alpha = \alpha'\gamma$ *and* $\beta = \beta'\gamma$ *in which* γ *is an integer satisfying* $\|\gamma\| \leqslant c_{20}(K)$ *and* α', β' *satisfy the conditions of Lemma 6 with* $h = T^{k-\varepsilon_3}(C/T^n)^G$.

Proof. We have

$$T^{k-\varepsilon_3}\left(\frac{C}{T^n}\right)^G \geqslant T^{k-\varepsilon_3}\left(\frac{T^{n-1/G+\varepsilon_3}}{T^n}\right)^G \geqslant T^{k-1+\varepsilon_3}$$

and

$$T^{k-\varepsilon_3}\left(\frac{C}{T^n}\right)^G \leqslant T^{k-\varepsilon_3}.$$

Let

$$h = T^{k-\varepsilon_3}\left(\frac{C}{T^n}\right)^G.$$

Then h satisfies the condition of Lemma 8 for $T \geqslant c_{19}$. By Lemma 6, there exist an integer α' and a number β' of $\mathfrak{d}^{-1}$ satisfying

$$\|\alpha'\xi - \beta'\| < h^{-1}, \qquad 0 < \|\alpha'\| \leqslant h$$

and the other conclusions in Lemma 6. Take $\varepsilon_2 = \varepsilon_3/2G$. Since

$$T^{n+\varepsilon_2}\left(\frac{h}{T^k}\right)^{1/G} = T^{n+\varepsilon_2-\varepsilon_3/G}\frac{C}{T^n} = CT^{-\varepsilon_2}$$

and

$$T^{n-1/G+\varepsilon_2} \leqslant CT^{\varepsilon_2-\varepsilon_3} < CT^{-\varepsilon_2},$$

we have by Lemma 8 that

$$C \leqslant |L(\xi)| \ll T^{n+\varepsilon_2}\|\alpha'\|^{-1/G},$$

i.e.,

$$0 < \|\alpha'\| \ll \left(\frac{T^n}{C}\right)^G T^{\varepsilon 2G} < \left(\frac{T^n}{C}\right)^G T^{\varepsilon 3}.$$

There exists a nonzero integer γ_1 such that $\|\gamma_1\| \leqslant c_{21}(K)$ and $\gamma_1 \beta'$ is an integer for any $\beta' \in \mathfrak{d}^{-1}$. (See, e.g., Hecke [5], p. 100.) By Lemma 5, there is a nonzero integer γ_2 with $\|\gamma_2\| < c_{10}$ such that $\gamma_1 \gamma_2 \alpha'$ is totally nonnegative. Let $\alpha = \gamma_1 \gamma_2 \alpha'$ and $\beta = \gamma_1 \gamma_2 \beta'$. The lemma follows.

7. The circle method. We denote by G_n the unit cube $\{(x_1, \ldots, x_n): 0 \leqslant x_i < 1 \ (1 \leqslant i \leqslant n)\}$. For any $\gamma \in K$, we can determine uniquely integral ideals $\mathfrak{a}, \mathfrak{b}$ such that

$$\gamma \mathfrak{d} = \mathfrak{b}/\mathfrak{a}, \quad (\mathfrak{a}, \mathfrak{b}) = 1.$$

We write $\gamma \to \mathfrak{a}$. Let $t > 1$ and $\Gamma(t)$ be the set consisting of $\gamma = x_1 \varrho_1 + \ldots$ $\ldots + x_n \varrho_n$ satisfying

$$(x_1, \ldots, x_n) \in G_n, \quad x_i \ (1 \leqslant i \leqslant n) \text{ rational numbers,}$$

$$\gamma \to \mathfrak{a} \quad \text{and} \quad N(\mathfrak{a}) \leqslant t^n.$$

Let

(17) $$h = abm^{20ky - y/n} \quad \text{and} \quad t = m^{y/n}.$$

For any $\gamma \in \Gamma(t)$, subject to $\gamma \to \mathfrak{a}$, we define the basic domain B_γ by

(18) $$\{(x_1, \ldots, x_n): (x_1, \ldots, x_n) \in G_n, \ \xi = x_1 \varrho_1 + \ldots + x_n \varrho_n$$

$$\text{such that } h\|\xi - \gamma_0\| < 1 \text{ for some } \gamma_0 \equiv \gamma \,(\mathrm{mod}\ \mathfrak{d}^{-1})\}.$$

We may prove that if $\gamma_1 \neq \gamma_2$, then $B_{\gamma_1} \cap B_{\gamma_2} = \emptyset$. In fact, suppose there is a $\xi \in B_{\gamma_1} \cap B_{\gamma_2}$, i.e., $h\|\xi - \gamma_{0i}\| < 1$, where $\gamma_{0i} \equiv \gamma_i \,(\mathrm{mod}\ \mathfrak{d}^{-1})$ $(i = 1, 2)$. For simplicity, we set $\gamma_{0i} = \gamma_i$ $(i = 1, 2)$. Write

$$\max(h|\xi^{(i)} - \gamma_j^{(i)}|, t^{-1}) = \sigma_j^{(i)}, \quad 1 \leqslant i \leqslant n, 1 \leqslant j \leqslant 2.$$

Then

$$\prod_{i=1}^{n} \sigma_j^{(i)} < 1, \quad \max(\sigma_j^{(i)})^{-1} \leqslant t, \quad j = 1, 2,$$

and thus

$$|\gamma_1^{(i)} - \gamma_2^{(i)}| \leqslant |\xi^{(i)} - \gamma_1^{(i)}| + |\xi^{(i)} - \gamma_2^{(i)}| \leqslant h^{-1}(\sigma_1^{(i)} + \sigma_2^{(i)})$$

$$= h^{-1} \sigma_1^{(i)} \sigma_2^{(i)} ((\sigma_1^{(i)})^{-1} + (\sigma_2^{(i)})^{-1}) \leqslant 2h^{-1} \sigma_1^{(i)} \sigma_2^{(i)} t, \quad 1 \leqslant i \leqslant n.$$

Suppose $\gamma_i \to \mathfrak{a}_i$ $(i = 1, 2)$. We have

$$N(\mathfrak{a}_1 \mathfrak{a}_2)|N(\gamma_1 - \gamma_2)| \leqslant (2h^{-1}t^3)^n < D^{-1},$$

130   Wang Yuan

since $m \geqslant c_{14}$. On the other hand, $\mathfrak{a}_1 \mathfrak{a}_2 (\gamma_1 - \gamma_2) \mathfrak{d}$ is an integral ideal, and thus

$$N(\mathfrak{a}_1 \mathfrak{a}_2) |N(\gamma_1 - \gamma_2)| \geqslant |N(\mathfrak{d}^{-1})| = D^{-1}.$$

This gives a contradiction, and therefore the assertion follows.

We define the supplementary domain E by

$$(19) \qquad E = G_n - \bigcup_{\gamma \in \Gamma(t)} B_\gamma.$$

We use the notations

$$\xi = x_1 \varrho_1 + \ldots + x_n \varrho_n, \qquad dx = dx_1 \ldots dx_n,$$

$$A = b^{1/k} m^{20y}, \qquad B = a^{1/k} m^{20y}, \qquad H = m^{6y},$$

$$(20) \qquad S_i(\xi) = \sum_{\lambda \in P(A)} E(\alpha_i \lambda^k \xi), \qquad T_i(\xi) = \sum_{\mu \in P(B)} E(-\beta_i \mu^k \xi), \qquad 1 \leqslant i \leqslant s,$$

$$S(\xi) = \prod_{i=1}^{s} S_i(\xi), \qquad T(\xi) = \prod_{i=1}^{s} T_i(\xi)$$

and

$$F(\xi) = \sum_{\chi \in P(H)} S(\xi) T(\xi) E(-q\chi\xi),$$

where $q = 1$ or -1. Let Z denote the number of solutions of the equation

$$\alpha_1 \lambda_1^k + \ldots + \alpha_s \lambda_s^k - \beta_1 \mu_1^k - \ldots - \beta_s \mu_s^k = q\chi$$

in totally nonnegative integers $\lambda_1, \ldots, \lambda_s, \mu_1, \ldots, \mu_s, \chi$ satisfying

$$\lambda_i \in P(A), \qquad \mu_i \in P(B) \quad (1 \leqslant i \leqslant s), \qquad \chi \in P(H).$$

Then

$$(21) \qquad Z = \sum_{\gamma \in \Gamma(t)} \int_B F(\xi)\, dx + \int_E F(\xi)\, dx.$$

We shall show that under the assumption made in Proposition 2, $Z > 1$.

8. Supplementary domain. Take ε_3 such that

$$(22) \qquad \varepsilon_3 < 1/2G, \qquad \varepsilon_3 (1 + 20y) < \tfrac{1}{2} z,$$

and s so large that

$$(23) \qquad s > \frac{10G}{z} + h.$$

LEMMA 10. *Suppose that* $(x_1, \ldots, x_n) \in G_n$ *and*

$$(24) \qquad |F(\xi)| \geqslant H^n (AB)^{ns} m^{-4}.$$

Then ξ lies in a basic domain.

Proof. We may suppose that

$$|S_1(\xi)| \geqslant \ldots \geqslant |S_s(\xi)|.$$

Then

$$F(\xi) \ll H^n A^{n(h-1)} B^{ns} |S_h(\xi)|^{s-h+1},$$

and thus by (24) and $m \geqslant c_{14}$, we have

$$|S_i(\xi)| \geqslant |S_h(\xi)| \geqslant A^n m^{-5/(s-h+1)} = C, \quad \text{say for } 1 \leqslant i \leqslant h.$$

By (20), (22) and (23), we have

$$m^{5/(s-h+1)} \leqslant A^{1/4y(s-h+1)} \leqslant A^{1/2G} < A^{1/G-\varepsilon_3},$$

and therefore

$$C \geqslant A^{n-1/G+\varepsilon_3}.$$

It follows by Lemma 9 that there are totally nonnegative integers σ_i $(1 \leqslant i \leqslant h)$ and integers φ_i $(1 \leqslant i \leqslant h)$ such that

$$0 < \|\sigma_i\| \ll m^{5G/(s-h+1)} A^{\varepsilon_3} < m^{z/2+z/2} = m^z$$

and

$$\|\xi\alpha_i\sigma_i - \varphi_i\| \ll m^{5G/(s-h+1)} A^{\varepsilon_3-k} < m^z A^{-k}, \quad 1 \leqslant i \leqslant h,$$

since $m \geqslant c_{14}$. After a recording of $\beta_1, \ldots, \beta_s$, we may also suppose that

$$|T_1(\xi)| \geqslant \ldots \geqslant |T_s(\xi)|.$$

Similarly, there are totally nonnegative integers τ_i $(1 \leqslant i \leqslant h)$ and integers ψ_i $(1 \leqslant i \leqslant h)$ having

$$0 < \|\tau_i\| < m^z \quad \text{and} \quad \|\xi\beta_i\tau_i - \psi_i\| < m^z B^{-k}, \quad 1 \leqslant i \leqslant h.$$

Hence by (11), (12), (20) and $m \geqslant c_{14}$, we have

$$\|\varphi_i\beta_j\tau_j - \psi_j\alpha_i\sigma_i\| = \|\varphi_i\beta_j\tau_j - \xi\alpha_i\sigma_i\beta_j\tau_j + \xi\alpha_i\sigma_i\beta_j\tau_j - \psi_j\alpha_i\sigma_i\|$$

$$\leqslant \|\beta_j\tau_j\|\|\xi\alpha_i\sigma_i - \varphi_i\| + \|\alpha_i\sigma_i\|\|\xi\beta_j\tau_j - \psi_j\|$$

$$\ll bm^{2z} A^{-k} + am^{2z} B^{-k} \ll m^{2z-20ky} < 1,$$

and thus

$$N(\varphi_i\beta_j\tau_j - \psi_j\alpha_i\sigma_i) = 0.$$

Since $\varphi_i\beta_j\tau_j - \psi_j\alpha_i\sigma_i$ is an integer, we have

$$\varphi_i\beta_j\tau_j - \psi_j\alpha_i\sigma_i = 0.$$

Thus by Lemma 4, the $2h$ integral vectors $c_9!(\alpha_i\sigma_i, \varphi_i)$ and $c_9!(\beta_i\tau_i, \psi_i)$

132   Wang Yuan

$(1 \leqslant i \leqslant h)$ are all integral multiples of an integral vector (σ, τ), where σ is a nonzero element of the integral ideal $(\alpha_1 \sigma_1, \ldots, \alpha_h \sigma_h, \beta_1 \tau_1, \ldots, \beta_h \tau_h)$ with the least norm in absolute value. Therefore the condition in case **B** yields that

$$0 < |N(\sigma)| < m^y.$$

Let

$$\sigma^{-1} \tau \mathfrak{d} = \mathfrak{b}/\mathfrak{a}, \quad (\mathfrak{a}, \mathfrak{b}) = 1.$$

Then $\mathfrak{a} | \sigma$, and thus

$$N(\mathfrak{a}) \leqslant |N(\sigma)| < m^y = t^n.$$

Since $\|\sigma_1\| < m^z$, we have

$$m^{(n-1)z} |\sigma_1^{(i)}| \geqslant N(\sigma_1) \geqslant 1, \quad 1 \leqslant i \leqslant n,$$

and by (11), (12), (17), (20) and $m \geqslant c_{14}$,

$$|\xi^{(i)} - (\sigma^{(i)})^{-1} \tau^{(i)}| = \frac{1}{|\alpha_1^{(i)} \sigma_1^{(i)}|} |\xi^{(i)} \alpha_1^{(i)} \sigma_1^{(i)} - \varphi_1^{(i)}|$$

$$\ll a^{-1} m^{nz} A^{-k} = a^{-1} b^{-1} m^{-20ky + nz}$$

$$= a^{-1} b^{-1} m^{-20ky + y/2nk} < h^{-1}, \quad 1 \leqslant i \leqslant n.$$

Therefore $\xi \in B_\gamma$, where $\gamma \equiv \sigma^{-1} \tau \pmod{\mathfrak{d}^{-1}}$. The lemma is proved.

9. Basic domain. We use the notations

$$\xi - \gamma = \zeta, \quad \eta = y_1 \omega_1 + \ldots + y_n \omega_n, \quad dy = dy_1 \ldots dy_n,$$

$$G_i(\gamma) = N(\mathfrak{a})^{-1} \sum_{\lambda(\mathrm{mod}\,\mathfrak{a})} E(\alpha_i \lambda^k \gamma), \quad H_i(\gamma) = N(\mathfrak{a})^{-1} \sum_{\mu(\mathrm{mod}\,\mathfrak{a})} E(-\beta_i \mu^k \gamma),$$

$$I_i(\zeta, A) = \int_{P(A)} E(\alpha_i \eta^k \zeta)\, dy, \quad J_i = (\zeta, B) = \int_{P(B)} E(-\beta_i \eta^k \zeta)\, dy, \quad 1 \leqslant i \leqslant s,$$

$$(25) \quad G(\gamma) = \prod_{i=1}^{s} G_i(\gamma), \quad H(\gamma) = \prod_{i=1}^{s} H_i(\gamma), \quad I(\zeta, A) = \prod_{i=1}^{s} I_i(\zeta, A) \quad \text{and}$$

$$J(\zeta, B) = \prod_{i=1}^{s} J_i(\zeta, B),$$

where $\gamma \to \mathfrak{a}$.

LEMMA 11. *Let $\mathfrak{a}$ be an integral ideal. Let $N(\mathfrak{a}, T)$ be the number of elements v of $\mathfrak{a}$ satisfying*

$$0 \leqslant v^{(p)} \leqslant T, \quad |v^{(q)}| \leqslant T.$$

Then

$$N(\mathfrak{a}, T) = \frac{(2\pi)^{r_2} T^n}{\sqrt{D} N(\mathfrak{a})} + O\left(\frac{T_0^{n-1}}{N(\mathfrak{a})^{1-1/n}}\right),$$

where $T_0 = \max(N(\mathfrak{a})^{1/n}, T)$.

See, e.g., Lemma 3.2 in Mitsui [9]. Notice that the conclusion is still true for the number of v satisfying $v + \mu \in \mathfrak{a}$, $0 \leqslant v^{(p)} + \mu^{(p)} \leqslant T$ and $|v^{(q)} + \mu^{(q)}| \leqslant T$, where μ is a given number in a residue class mod $\mathfrak{a}$.

Now we can prove the following lemma by the Siegel argument (see [14], pp. 328–330).

LEMMA 12. *Suppose that* $\xi \in B_y$. *Then*

(26) $$S_i(\xi) = G_i(\gamma) I_i(\zeta, A) + O(t^2 A^{n-1})$$

and

(27) $$T_i(\xi) = H_i(\gamma) J_i(\zeta, B) + O(t^2 B^{n-1}), \quad 1 \leqslant i \leqslant s.$$

Proof. Determine positive numbers $\theta^{(i)}$ $(1 \leqslant i \leqslant n)$, with $\theta^{(q)} = \theta^{(q+r_2)}$, such that

$$\theta^{(i)} \max(h|\zeta^{(i)}|, t^{-1} N(\mathfrak{a})^{1/n})$$

$$= D^{1/2n} \prod_{j=1}^{n} \max(h|\zeta^{(j)}|, t^{-1} N(\mathfrak{a})^{1/n})^{1/n} N(\mathfrak{a})^{1/n}, \quad 1 \leqslant i \leqslant n,$$

Then

$$\prod_{i=1}^{n} \theta^{(i)} = D^{1/2} N(\mathfrak{a}),$$

and it follows by Minkowski's linear form theorem that there exists $\alpha \in \mathfrak{a}$ such that $0 < |\alpha^{(i)}| \leqslant \theta^{(i)}$, $1 \leqslant i \leqslant n$. Hence $\alpha \mathfrak{a}^{-1} = \mathfrak{b}$ is an integral ideal and

$$N(\mathfrak{b}) = |N(\alpha)| N(\mathfrak{a})^{-1} \leqslant \left(\prod_{i=1}^{n} \theta^{(i)}\right) N(\mathfrak{a})^{-1} = \sqrt{D};$$

hence $\mathfrak{b}$ belongs to a finite set depending on K only. Let $\sigma_1, \ldots, \sigma_n$ be a basis of $\mathfrak{b}^{-1}$. Then $\mathfrak{a} = \alpha \mathfrak{b}^{-1}$ has a basis

$$\tau_i = \alpha \sigma_i, \quad 1 \leqslant i \leqslant n$$

satisfying

$$\|\tau_i\| = O(\|\alpha\|) = O(\max_i \theta^{(i)}) = O(t).$$

Let μ run over a complete residue system modulo $\mathfrak{a}$, and λ over all numbers

134   Wang Yuan

in $\mathfrak{a}$ such that $\lambda + \mu \in P(A)$. Then

$$(28) \qquad S_i(\xi) = \sum_{\mu(\mathrm{mod}\,\mathfrak{a})} E(\alpha_i \mu^k \gamma) \sum_{\substack{\mathfrak{a}|\lambda \\ \lambda + \mu \in P(A)}} E(\alpha_i (\lambda + \mu)^k \zeta).$$

Expressing λ in terms of τ_i $(1 \leqslant i \leqslant n)$, we obtain

$$\lambda = g_1 \tau_1 + \ldots + g_n \tau_n,$$

where $g_1, \ldots, g_n$ are rational integers. Let $G(\lambda)$ denote the cube

$$(s_1, \ldots, s_n): \sigma = s_1 \tau_1 + \ldots + s_n \tau_n, \qquad g_i \leqslant s_i < g_i + 1 \qquad (1 \leqslant i \leqslant n).$$

Then

$$\|\sigma - \lambda\| = O(t),$$

$$\|(\sigma + \mu)^k \zeta - (\lambda + \mu)^k \zeta\| \ll \|\sigma - \lambda\| \, \|\zeta\| \, (\|\sigma + \mu\|^{k-1} + \|\lambda + \mu\|^{k-1}) \ll th^{-1} A^{k-1},$$

and therefore by (11),

$$E\big(\alpha_i(\lambda + \mu)^k \zeta\big) = \int_{G(\lambda)} E\big(\alpha_i(\sigma + \mu)^k \zeta\big)\,ds + O(ath^{-1} A^{k-1}),$$

where $ds = ds_1 \ldots ds_n$. Since $N(\mathfrak{a}) \leqslant t^n \leqslant A$ by (17) and (20), it follows by Lemma 11 that the number of λ with $\mathfrak{a}|\lambda$ and $\lambda + \mu \in P(A)$ is $O(N(\mathfrak{a})^{-1} A^n)$. Therefore

$$\sum_{\substack{\mathfrak{a}|\lambda \\ \lambda + \mu \in P(A)}} E\big(\alpha_i(\lambda + \mu)^k \zeta\big)$$

$$= \sum_{\substack{\mathfrak{a}|\lambda \\ \lambda + \mu \in P(A)}} \int_{G(\lambda)} E\big(\alpha_i(\sigma + \mu)^k \zeta\big)\,ds + O\big(N(\mathfrak{a})^{-1} ath^{-1} A^{n+k-1}\big).$$

Let F denote the domain in the s-space defined by

$$0 \leqslant \sigma^{(p)} + \mu^{(p)} \leqslant A, \qquad |\sigma^{(q)} + \mu^{(q)}| \leqslant A.$$

Then the volume of the area belonging to exactly one of $\bigcup\limits_{\substack{\mathfrak{a}|\lambda \\ \lambda + \mu \in P(A)}} G(\lambda)$ and F is dominated by $O(N(\mathfrak{a})^{-1} t A^{n-1})$. (See, Siegel [14], p. 329.) Therefore by (17) and (20), we have

$$\sum_{\substack{\mathfrak{a}|\lambda \\ \lambda + \mu \in P(A)}} E\big(\alpha_i(\lambda + \mu)^k \zeta\big) = \int_F E\big(\alpha_i(\sigma + \mu)^k \zeta\big)\,ds + O\big(N(\mathfrak{a})^{-1} t^2 A^{n-1}\big).$$

Let $\sigma + \mu = \eta$. Since the Jacobian of $s_1, \ldots, s_n$ with respect to $y_1, \ldots, y_n$ is equal to

$$D^{1/2} |\det(\tau_j^{(i)})|^{-1} = N(\mathfrak{a})^{-1},$$

we have

$$\sum_{\substack{\mathfrak{a}\mid\lambda \\ \lambda+\mu\in P(A)}} E\big(\alpha_i(\lambda+\mu)^k\zeta\big) = N(\mathfrak{a})^{-1}I_i(\zeta, A) + O\big(N(\mathfrak{a})^{-1}t^2 A^{n-1}\big).$$

Substituting into (28), we have (26). The proof of (27) is similar.

10. Continuation. We use E_n to denote the whole n-dimensional Euclidean space.

LEMMA 13. *We have*

$$(29) \qquad I_i(\zeta, A) \ll \prod_{i=1}^{n} \min(A,\, a^{-1/k}\,|\zeta^{(i)}|^{-1/k})$$

and

$$(30) \qquad J_i(\zeta, B) \ll \prod_{i=1}^{n} \min(B,\, b^{-1/k}\,|\zeta^{(i)}|^{-1/k}).$$

See, Siegel [14], p. 335. The only difference between the proofs of (29) and the corresponding formula of Siegel is that we use $\alpha_i^{(p)}\tau^{(p)}$ and $|\alpha_i^{(q)}\tau^{(q)}|$ instead of his $\tau^{(p)}$ and $|\tau^{(q)}|$.

LEMMA 14.

$$\int_{B_\gamma} S(\xi)\,T(\xi)\,E(-q\chi\xi)\,dx$$

$$= G(\gamma)\,H(\gamma)\,E(-q\chi\gamma)\int_{E_n} I(\zeta, A)\,J(\zeta, B)\,dx + O\big((AB)^{ns}(ab)^{-n}m^{-20kny-17y}\big).$$

Proof. By Lemmas 12 and 13, we have

$$S(\xi)\,T(\xi) = G(\gamma)\,H(\gamma)\,I(\zeta, A)\,J(\zeta, B) + O\big((AB)^{ns}t^2\max(A^{-1}, B^{-1})\big).$$

Let

$$(31) \qquad \zeta^{(p)} = u_p, \qquad \zeta^{(q)} = u_q\,e^{i\varphi_q}.$$

The Jacobian of $x_1, \ldots, x_n$ with respect to u_p, u_q, φ_q is equal to the product of the Jacobian of $x_1, \ldots, x_n$ with respect to $\zeta^{(p)}, \zeta^{(q)}$ and the Jacobian of $\zeta^{(p)}, \zeta^{(q)}$ with respect to u_p, u_q, φ_q, i.e., it is equal to

$$2^{r_2}D^{1/2}\prod_{q} u_q.$$

It follows by (17) and (20) that

$$\int_{B_\gamma} dx \ll \prod_p \Big(\int_0^{h-1} du_p\Big)\prod_q\Big(\int_{-\pi}^{\pi}\int_0^{h-1} u_q\,du_q\,d\varphi_q\Big) \ll h^{-n} = (ab)^{-n}m^{-20kny+y}$$

136   Wang Yuan

and

$$\max(A^{-1}, B^{-1}) \ll m^{-20y}.$$

Therefore

$$(32) \quad \int_{B_\gamma} S(\xi)\,T(\xi)\,E(-q\chi\xi)\,dx$$

$$= G(\gamma)\,H(\gamma)\,E(-q\chi\gamma) \int_{B_\gamma} I(\zeta, A)\,J(\zeta, B)\,E(-q\chi\zeta)\,dx +$$

$$+ O\big((AB)^{ns}(ab)^{-n}m^{-20kny-17y}\big).$$

In the integral in the right-hand side of (32) we replace $E(-q\chi\zeta)$ by 1. Then by (20) and Lemma 13, the error is

$$(AB)^{ns} \int_{B_\gamma} \|\chi\zeta\|\,dx \ll (AB)^{ns} Hh^{-n-1} \ll (AB)^{ns}(ab)^{-n}m^{-20kny-17y}.$$

Hence

$$(33) \quad \int_{B_\gamma} S(\xi)\,T(\xi)\,E(-q\chi\xi)\,dx = G(\gamma)\,H(\gamma)\,E(-q\chi\gamma) \int_{B_\gamma} I(\zeta, A)\,J(\zeta, B)\,dx +$$

$$+ O\big((AB)^{ns}(ab)^{-n}m^{-20kny-17y}\big).$$

If $(x_1, \ldots, x_n)$ is a point in $E_n - B_\gamma$, then $h|\zeta^{(i)}| \geq 1$ is true for at least one index i. By Lemma 13 and (31), we have

$$\int_{E_n - B} I(\zeta, A)\,J(\zeta, B)\,dx$$

$$\ll \int_{E_n - B_\gamma} \Big(\prod_{i=1}^{n} \min(A, a^{-1/k}|\zeta^{(i)}|^{-1/k})\prod_{j=1}^{n} \min(B, b^{-1/k}|\zeta^{(j)}|^{-1/k})\Big)^s dx$$

$$\ll \Big(\int_{h^{-1}}^{\infty} (ab)^{-s/k} u^{-2s/k}\,du\Big)\Big(\int_{0}^{\infty} \min(A^2, a^{-s/k}v^{-s/k})\min(B^s, b^{-s/k}v^{-s/k})\,dv\Big)^{r_1-1} \times$$

$$\times \Big(\int_{-\pi}^{\pi}\int_{0}^{\infty} \min(A^{2s}, a^{-2s/k}w^{-2s/k})\min(B^{2s}, b^{-2s/k}w^{-2s/k})\,w\,dw\,d\varphi\Big)^{r_2} +$$

$$+ \Big(\int_{0}^{\infty} \min(A^s, a^{-s/k}u^{-s/k})\min(B^s, b^{-s/k}u^{-s/k})\,du\Big)^{r_1} \times$$

$$\times \Big(\int_{-\pi}^{\pi}\int_{h^{-1}}^{\infty} (ab)^{-2s/k}v^{-4s/k+1}\,dv\,d\varphi\Big)\Big(\int_{-\pi}^{\pi}\int_{0}^{\infty} \min(A^{2s}, a^{-2s/k}w^{-2s/k}) \times$$

$$\times \min(B^{2s}, b^{-2s/k}w^{-2s/k})\,w\,dw\,d\varphi\Big)^{r_2-1}.$$

Since

$$\int_0^\infty \min(A^s, a^{-s/k} u^{-s/k}) \min(B^s, b^{-s/k} u^{-s/k})\, du$$

$$\ll A^s \int_0^\Upsilon \min(B^s, b^{-s/k} u^{-s/k})\, du$$

$$\ll A^s \left(\int_0^{B^{-k} b^{-1}} B^s\, du + \int_{B^{-k} b^{-1}}^\infty b^{-s/k} u^{-s/k}\, du \right) \ll A^s B^{s-k} b^{-1}$$

and

$$\int_0^\infty \min(A^{2s}, a^{-2s/k} w^{-2s/k}) \min(B^{2s}, b^{-2s/k} w^{-2s/k})\, w\, dw \ll A^{2s} B^{2(s-k)} b^{-2},$$

we have by (9), (12), (17), (20) and (23),

$$\int_{E_n - B_\gamma} I(\zeta, A) J(\zeta, B)\, dx$$

$$\ll h^{2s/k-1} (ab)^{-s/k} A^{(r_1-1)s} b^{-(r_1-1)} B^{(r_1-1)(s-k)} A^{2r_2 s} b^{-2r_2} B^{2r_2(s-k)} +$$

$$+ A^{r_1 s} b^{-r_1} B^{r_1(s-k)} h^{4s/k-2} (ab)^{-2s/k} A^{2(r_2-1)s} b^{-2(r_2-1)} B^{2(r_2-1)(s-k)}$$

$$\ll h^{2s/k-1} (ab)^{-s/k} b^{-n+1} A^{(n-1)s} B^{(n-1)(s-k)} +$$

$$+ h^{4s/k-2} (ab)^{-2s/k} b^{-n+2} B^{(n-2)(s-k)}$$

$$\ll (AB)^{ns} (ab)^{-n} m^{-20kny} \left(m^{\frac{-2sy}{kn} + \frac{y}{n}} + m^{\frac{-4sy}{kn} + \frac{2y}{n}} \right)$$

$$\ll (AB)^{ns} (ab)^{-n} m^{-20kny - 17y}.$$

The lemma follows by substitution into (33).

11. The singular integral. Let $\eta' = y_1' \omega_1 + \ldots + y_n' \omega_n$, $\zeta' = x_1' \varrho_1 + \ldots + x_n' \varrho_n$, $dy' = dy_1' \ldots dy_n'$, $dx' = dx_1' \ldots dx_n'$, $\eta = A\eta'$ and $\zeta = a^{-1} b^{-1} m^{-20ky} \zeta'$. The Jacobians of $y_1, \ldots, y_n$ and $x_1, \ldots, x_n$ with respect to $y_1', \ldots, y_n'$ and $x_1', \ldots, x_n'$ are A^n and $(a^{-1} b^{-1} m^{-20ky})^n$ respectively. Set $\gamma_i = \alpha_i/a$ $(1 \leqslant i \leqslant s)$. Then

$$\alpha_i \eta^k \zeta = \gamma_i \eta'^k \zeta',$$

where by (11), γ_i $(1 \leqslant i \leqslant s)$ satisfy

$$(34) \qquad c_{16} < \gamma_i^{(p)} < c_{17}, \qquad c_{16} < |\gamma_i^{(q)}| < c_{17}, \qquad 1 \leqslant i \leqslant s.$$

Let us write η' and ζ' as η and ζ again and let

$$I_i(\zeta) = \int_P E(\gamma_i \eta^k \zeta)\, dy,$$

where $P = P(1)$. Then

$$I_i(\zeta, A) = A^n I_i(\zeta), \qquad 1 \leqslant i \leqslant s.$$

138   Wang Yuan

Similarly, we have

$$J_i(\zeta, B) = B^n J_i(\zeta), \quad 1 \leqslant i \leqslant s,$$

where

$$J_i(\zeta) = \int_P E(-\gamma_{s+i} \eta^k \zeta) \, dy, \quad 1 \leqslant i \leqslant s,$$

and $\gamma_{s+i} = \beta_i/b$ $(1 \leqslant i \leqslant s)$ which satisfy

(35) $$c_{16} < \gamma_{s+i}^{(p)} < c_{17}, \quad c_{16} < |\gamma_{s+i}^{(q)}| < c_{17}, \quad 1 \leqslant i \leqslant s.$$

Set

$$I(\zeta) = \prod_{i=1}^{s} I_i(\zeta), \quad J(\zeta) = \prod_{i=1}^{s} J_i(\zeta) \quad \text{and} \quad \Phi = \int_{E_n} I(\zeta) J(\zeta) \, dx.$$

Then we have

(36) $$\int_{E_n} I(\zeta, A) J(\zeta, B) \, dx = (AB)^{ns} (ab)^{-n} m^{-20kny} \Phi.$$

Now we shall treat the integral Φ by Tatuzawa's method. (See [17].)

If $F(x_1, \ldots, x_t)$ is nondecreasing for variables $x_{g_1}, \ldots, x_{g_r}$ and nonincreasing for other variables $x_{h_1}, \ldots, x_{h_s}$ $(r+s = t)$ over the rectangle

$$I = \{(x_1, \ldots, x_t): a_i \leqslant x_i \leqslant b_i \ (1 \leqslant i \leqslant t)\},$$

then F is said to be *monotonic over I*.

LEMMA 15 (Tatuzawa). *Let $F(x_1, \ldots, x_t)$ be a finite product of positive bounded monotonic functions over the rectangle*

$$\{(x_1, \ldots, x_t): 0 \leqslant x_i \leqslant c_i \ (1 \leqslant i \leqslant t)\}.$$

If we write

$$\chi_\lambda(x) = \frac{\sin 2\pi\lambda x}{\pi x},$$

then

$$\lim_{\substack{\lambda_i \to \infty \\ (1 \leqslant i \leqslant t)}} \int_0^{c_1} \cdots \int_0^{c_t} F(x_1, \ldots, x_t) \chi_{\lambda_1}(x_1) \ldots \chi_{\lambda_t}(x_t) \, dx_1 \ldots dx_t = (\tfrac{1}{2})^t F(+0, \ldots, +0).$$

See Tatuzawa [17], pp. 47–49.

LEMMA 16.

$$\Phi = D^{(1-2s)/2} k^{-2ns} N(\gamma_1 \ldots \gamma_s)^{-1/k} \prod_p F_p \prod_q H_q,$$

where

$$F_p = \int\limits_{U_p} \prod_{i=1}^{2s} w_i^{1/k-1}\, dw_1 \ldots dw_{2s-1}$$

in which U_p *denotes the domain*

$$0 \leqslant w_i \leqslant \gamma_i^{(p)}, \quad 1 \leqslant i \leqslant 2s, \quad w_{2s} = w_1 + \ldots + w_s - w_{s+1} - \ldots - w_{2s-1},$$

and where

$$H_q = \int\limits_{V_q} \prod_{i=1}^{2s} w_i^{1/k-1}\, dw_1 \ldots dw_{2s-1}\, d\varphi_1 \ldots d\varphi_{2s-1}$$

in which V_q *denotes the domain*

$$0 \leqslant w_i \leqslant |\gamma_i^{(q)}|^2, \quad 1 \leqslant i \leqslant 2s, \quad -\pi \leqslant \varphi_j \leqslant \pi, \quad 1 \leqslant j \leqslant 2s-1,$$

$$w_{2s} = |w_1^{1/2} e^{i\varphi_1} + \ldots + w_{2s-1}^{1/2} e^{i\varphi_{2s-1}}|^2.$$

Proof. By Lemma 13, we have

$$I_i(\zeta) \ll \prod_{j=1}^{n} \min(1, |\zeta^{(j)}|^{-1/k}), \quad 1 \leqslant i \leqslant s,$$

and $J_i(\zeta)$ $(1 \leqslant i \leqslant s)$ satisfy the same inequality. Then by the transformation (31), we have

$$\int\limits_{E_n} |I(\zeta) J(\zeta)|\, dx \ll \prod_p \left(\int_0^\infty \min(1, u_p^{-2s/k})\, du_p \right) \prod_q \left(\int_{-\pi}^{\pi} \int_0^\infty \min(1, u_q^{-4s/k}) u_q\, du_q\, d\varphi_q \right)$$

which converges for $s > k$. Therefore

$$\Phi = \lim_{\lambda_p, \lambda_q, \lambda_q' \to \infty} \Phi(\Omega), \quad \Phi(\Omega) = \int_\Omega I(\zeta) J(\zeta)\, dx,$$

where Ω denotes the closed region of x defined by

$$|v_p| \leqslant \lambda_p, \quad |v_q| \leqslant \lambda_q, \quad |v_q'| \leqslant \lambda_q'$$

in which

$$v_p = \zeta^{(p)}, \quad v_q = \frac{\zeta^{(q)} + \zeta^{(q+r_2)}}{\sqrt{2}}, \quad v_q' = \frac{\zeta^{(q)} - \zeta^{(q+r_2)}}{\sqrt{2}\, i}.$$

Consider $2ns$ real variables y_{ij} $(1 \leqslant i \leqslant 2s, 1 \leqslant j \leqslant n)$. Let

$$\eta_i = y_{i1}\omega_1 + \ldots + y_{in}\omega_n, \quad dY_i = dy_{i1} \ldots dy_{in}$$

and let P_i be the domain

$$0 \leqslant \eta_i^{(p)} \leqslant 1, \quad |\eta_i^{(q)}| \leqslant 1, \quad 1 \leqslant i \leqslant 2s.$$

140   Wang Yuan

Let

$$\gamma_1^{(i)}\eta_1^{(i)k} + \ldots + \gamma_s^{(i)}\eta_s^{(i)k} - \gamma_{s+1}^{(i)}\eta_{s+1}^{(i)k} - \ldots - \gamma_{2s}^{(i)}\eta_{2s}^{(i)k} = z_i, \qquad 1 \leqslant i \leqslant n,$$

and

$$u_p = z_p, \qquad u_q = \frac{z_q + z_{q+r_2}}{\sqrt{2}}, \qquad u'_q = -\frac{z_q - z_{q+r_2}}{\sqrt{2}\,i}.$$

Since

$$\zeta^{(q)} = \frac{v_q + iv'_q}{\sqrt{2}}, \qquad \zeta^{(q+r_2)} = \frac{v_q - iv'_q}{\sqrt{2}}, \qquad z_q = \frac{u_q - iu'_q}{\sqrt{2}}, \qquad z_{q+r_2} = \frac{u_q + iu'_q}{\sqrt{2}},$$

we have

$$\sum_{i=1}^{n} \zeta^{(i)} z_i = \sum_p u_p v_p + \sum_q u_q v_q + \sum_q u'_q v'_q.$$

The Jacobian of $x_1, \ldots, x_n$ with respect to v_p, v_q, v'_q is equal to

$$|\det(\varrho_i^{(j)})|^{-1}|-i|^{r_2} = D^{1/2}.$$

Set

$$dv = dv_1 \ldots dv_{r_1} dv_{r_1+1} \ldots dv_{r_1+r_2} dv'_{r_1+1} \ldots dv'_{r_1+r_2}.$$

We have

$$\begin{aligned}
\Phi(\Omega) &= \int_{P_1} \ldots \int_{P_{2s}} dY_1 \ldots dY_{2s} \int_{\Omega} \exp\Big(2\pi i \sum_{j=1}^{n} \zeta^{(j)} z_j\Big) dx \\
&= D^{1/2} \int_{P_1} \ldots \int_{P_{2s}} dY_1 \ldots dY_{2s} \int_{\Omega} \exp\Big(2\pi i\Big(\sum_p u_p v_p + \sum_q u_q v_q + \sum_q u'_q v'_q\Big)\Big) dv \\
&= D^{1/2} \int_{P_1} \ldots \int_{P_{2s}} \prod_p \chi_{\lambda_p}(u_p) \prod_q \big(\chi_{\lambda_q}(u_q)\,\chi_{\lambda'_q}(u'_q)\big) dY_1 \ldots dY_{2s}.
\end{aligned}$$

Let

$$z_i = t_1 \omega_1^{(i)} + \ldots + t_n \omega_n^{(i)}, \qquad 1 \leqslant i \leqslant n.$$

Then

$$-\gamma_{2s}^{(i)}\eta_{2s}^{(i)k} = z_i - \big(\gamma_1^{(i)}\eta_1^{(i)k} + \ldots + \gamma_s^{(i)}\eta_s^{(i)k} - \gamma_{s+1}^{(i)}\eta_{s+1}^{(i)k} - \ldots - \gamma_{2s-1}^{(i)}\eta_{2s-1}^{(i)k}\big).$$

The Jacobian of $y_{2s,1}, \ldots, y_{2s,n}$ with respect to $t_1, \ldots, t_n$ is equal to

$$|\det(k\gamma_{2s}^{(i)}\eta_{2s}^{(i)k-1}\omega_j^{(i)})|^{-1}|\det(\omega_r^{(l)})| = N(k^{-1}|\gamma_{2s}^{-1}\eta_{2s}^{1-k}|),$$

and the Jacobian of $t_1, \ldots, t_n$ with respect to u_p, u_q, u'_q is

$$|\det(\omega_r^{(l)})|^{-1}|i|^{r_2} = D^{-1/2}.$$

Therefore

$$\Phi(\Omega) = \int_Q \prod_p \chi_{\lambda_p}(u_p) \prod_q \left(\chi_{\lambda_q}(u_q)\, \chi_{\lambda'_q}(u'_q) \right) du \times$$

$$\times \int_{P_1} \dots \int_{P_{2s-1}} N(k^{-1} |\gamma_{2s}^{-1} \eta_{2s}^{1-k}|)\, dY_1 \dots dY_{2s-1},$$

where Q is a closed region containing the origin of u in its interior and

$$du = \prod_p du_p \prod_q (du_q\, du'_q).$$

Let

$$\eta_j^{(p)} = y_{j1} \omega_1^{(p)} + \dots + y_{jn} \omega_n^{(p)} = u_{jp}^{1/k},$$

$$\eta_j^{(q)} = y_{j1} \omega_1^{(q)} + \dots + y_{jn} \omega_n^{(q)} = u_{jq}^{1/2k} e^{i\psi_{jq}/k}, \qquad 1 \leqslant j \leqslant 2s.$$

The Jacobian of $y_{j1}, \dots, y_{jn}$ with respect to $u_{jp},\, u_{jq},\, u'_{jq}$ is

$$|\det(\omega_r^{(l)})|^{-1} N(k^{-1} |\eta_j^{1-k}|) = D^{-1/2}(k^{-1} |\eta_j^{1-k}|).$$

Then we have

$$\Phi(\Omega) = \int_Q \prod_p \chi_{\lambda_p}(u_p) \prod_q \left(\chi_{\lambda_q}(u_q)\, \chi_{\lambda'_q}(u'_q) \right) du \times$$

$$\times D^{(1-2s)/2} \int_R N(k^{-1} |\gamma_{2s}^{-1} \eta_{2s}^{1-k}|) \prod_{j=1}^{2s-1} N(k^{-1} |\eta_j^{1-k}|) \times$$

$$\times \prod_{j=1}^{2s-1} (du_{j1} \dots du_{j,r_1+r_2}\, d\psi_{j,r_1+1} \dots d\psi_{j,r_1+r_2}),$$

where R denotes the region

$$0 \leqslant u_{jl} \leqslant 1 \quad (1 \leqslant j \leqslant 2s-1,\ 1 \leqslant l \leqslant r_1+r_2),$$

$$-\pi \leqslant \psi_{rl} \leqslant \pi \quad (1 \leqslant r \leqslant 2s-1,\ r_1+1 \leqslant l \leqslant r_1+r_2)$$

$$-\gamma_{2s}^{(p)} \eta_{2s}^{(p)k} = z_p - (\gamma_1^{(p)} u_{1p} + \dots - \gamma_{2s-1}^{(p)} u_{2s-1,p}),$$

$$|\gamma_{2s}^{(q)} \eta_{2s}^{(q)k}| = |z_p - (\gamma_1^{(q)} u_{1q}^{1/2} e^{i\psi_{1q}} + \dots - \gamma_{2s-1}^{(q)} u_{2s-1,q}^{1/2} e^{i\psi_{2s-1,q}})|,$$

$$0 \leqslant \eta_{2s}^{(p)} \leqslant 1,\ |\eta_{2s}^{(q)}| \leqslant 1.$$

Therefore

$$\Phi = \lim_{\lambda_p, \lambda_q, \lambda'_q \to \infty} \Phi(\Omega) = D^{(1-2s)/2} \prod_p F'_p \prod_q H'_q,$$

where

$$F'_p = \lim_{\lambda_p \to \infty} k^{-2s} \gamma_{2s}^{(p)-1} \int_{Q_p} \chi_{\lambda_p}(u_p)\, du_p \int_{U'_p} \prod_{i=1}^{2s} w_i'^{1/k-1}\, dw'_1 \dots dw'_{2s-1}$$

142   Wang Yuan

in which w_i' is used instead of u_{ip}, Q_p denotes the range of u_p in Q and U_p' the domain

$$0 \leqslant w_i' \leqslant 1 \quad (1 \leqslant i \leqslant 2s), \qquad \gamma_1^{(p)} w_1' + \ldots - \gamma_{2s}^{(p)} w_{2s}' = z_p,$$

and where

$$H_q' = \lim_{\lambda_q, \lambda_q' \to \infty} k^{-4s} |\gamma_{2s}^{(q)}|^{-2} \int_{Q_q} \chi_{\lambda_q}(u_q) \chi_{\lambda_q'}(u_q') \, du_q \, du_q' \times$$

$$\times \int_{V_q'} \prod_{i=1}^{2s} w_i'^{1/k-1} \, dw_1' \ldots dw_{2s-1}' \, d\psi_1 \ldots d\psi_{2s-1}$$

in which w_i' stands for u_{iq}, ψ_i for ψ_{iq}, Q_q denotes the region of u_q and u_q' in Q and V_q' the domain

$$0 \leqslant w_i' \leqslant 1 \quad (1 \leqslant i \leqslant 2s), \qquad -\pi \leqslant \psi_j \leqslant \pi \quad (1 \leqslant j \leqslant 2s-1),$$

$$|\gamma_{2s}^{(q)}|^2 w_{2s}' = |z_q - (\gamma_1^{(q)} w_1'^{1/2} e^{i\psi_1} + \ldots - \gamma_{2s-1}^{(q)} w_{2s-1}'^{1/2} e^{i\psi_{2s-1}})|^2.$$

By Lemma 15 and the transformations $\gamma_i^{(p)} w_i' = w_i$ $(1 \leqslant i \leqslant 2s)$ for the integral F_p' and $|\gamma_i^{(q)}|^2 w_i' = w_i$ $(1 \leqslant i \leqslant 2s)$, $\varphi_j = \theta_j + \psi_j$ $(1 \leqslant j \leqslant s)$, $\varphi_l = \theta_l + \psi_l + \pi$ $(s+1 \leqslant l \leqslant 2s-1)$ for H_q', where $\theta_t = \arg \gamma_t^{(q)}$ $(1 \leqslant t \leqslant 2s-1)$, we have

$$F_p' = k^{-2s} \prod_{i=1}^{2s} \gamma_i^{(p)-1/k} F_p, \qquad H_q' = k^{-4s} \prod_{i=1}^{2s} |\gamma_i^{(q)}|^{-2/k} H_q,$$

and the lemma follows.

12. The proof of theorem. By Lemma 11, we have

$$(37) \qquad \sum_{\chi \in P(H)} 1 = \frac{(2\pi)^{r_2}}{\sqrt{D}} H^n + O(H^{n-1}).$$

Therefore by (9), (14), (21) and Lemma 10, we have

$$Z = \sum_{\gamma \in \Gamma(t)} \int_{B_\gamma} F(\xi) \, dx + O\left(H^n (AB)^{ns} (ab)^{-n} m^{-20kny-17y}\right).$$

For a given $\mathfrak{a}$, the number of γ in Γ, subject to $\gamma \to \mathfrak{a}$, is $O(N(\mathfrak{a}))$. By Theorems 35 and 76 in Hecke [5], it follows that the number of $\mathfrak{a}$ with $N(\mathfrak{a}) = d$ is $O\left(\sum_{d_1 \ldots d_n = d} 1\right) = O(d^{\varepsilon 4})$. Therefore

$$\sum_{\gamma \in \Gamma(t)} 1 \ll \sum_{N(\mathfrak{a}) \leqslant t^n} N(\mathfrak{a}) \ll \sum_{d \leqslant t^n} d^2 \ll t^{3n} = m^{3y},$$

and by (20), (36) and Lemmas 14 and 16, we have

$$Z = J_0 \, \mathfrak{S}(t, H) (AB)^{ns} (ab)^{-n} m^{-20kny} + O\left(H^n (AB)^{ns} (ab)^{-n} m^{-20kny-14y}\right),$$

where

$$J_0 = D^{(1-2s)/2} k^{-2ns} N(\gamma_1 \ldots \gamma_{2s})^{-1/k} \prod_p F_p \prod_q H_q$$

and

$$\mathfrak{S} = \mathfrak{S}(t, H) = \sum_{\chi \in P(H)} \sum_{\gamma \in \Gamma(t)} G(\gamma) H(\gamma) E(-q\chi\gamma).$$

Let $\sum^*$ denote a sum, where γ runs over a reduced residue system of $(\mathfrak{a}\mathfrak{d})^{-1}$, mod $\mathfrak{d}^{-1}$. Thus $\gamma \in (\mathfrak{a}\mathfrak{d})^{-1}$, $(\gamma, \mathfrak{d}^{-1}) = (\mathfrak{a}, \mathfrak{d}^{-1})$, and we take only one γ in each class modulo $\mathfrak{d}^{-1}$. Then

$$\mathfrak{S} = \sum_{N(\mathfrak{a})=1} \sum_{\gamma}^* G(\gamma) H(\gamma) \sum_{\chi \in P(H)} E(-q\chi\gamma) + \sum_{1 < N(\mathfrak{a}) \leqslant t^n} \sum_{\gamma}^* G(\gamma) H(\gamma) \sum_{\chi \in P(H)} E(-q\chi\gamma)$$

$$= \mathfrak{S}_1 + \mathfrak{S}_2, \quad \text{say}.$$

By (37) we have

$$\mathfrak{S}_1 = \sum_{\chi \in P(H)} 1 = \frac{(2\pi)^{r_2}}{\sqrt{D}} H^n + O(H^{n-1}).$$

If $N(\mathfrak{a}) > 1$, then

$$\sum_{\chi (\mathrm{mod}\, \mathfrak{a})} E(-q\chi\gamma) = 0.$$

(See, e.g., Hecke [5], p. 197.) For any given integer μ, it follows by (17), (20) and Lemma 11 that the number of $\nu \in \mathfrak{a}$ and $\nu + \mu \in P(H)$ is

$$\frac{(2\pi)^{r_2}}{\sqrt{D}\, N(\mathfrak{a})} H^n + O\left(\frac{H^{n-1}}{N(\mathfrak{a})^{1-1/n}}\right).$$

Hence if the domain $\chi \in P(H)$ is split up into a union of complete residue sets (mod $\mathfrak{a}$), plus a few others, remaining elements, say R elements, then

$$R \ll N(\mathfrak{a}) \frac{H^{n-1}}{N(\mathfrak{a})^{1-1/n}} = N(\mathfrak{a})^n H^{n-1},$$

and therefore by (17) and (20)

$$\mathfrak{S}_2 \ll \sum_{N(\mathfrak{a}) \leqslant t^n} \sum_{\gamma}^* R \ll H^{n-1} \sum_{N(\mathfrak{a}) \leqslant t^n} \sum_{\gamma}^* N(\mathfrak{a})^{1/n}$$

$$\ll H^{n-1} \sum_{N(\mathfrak{a}) \leqslant t^n} N(\mathfrak{a})^{1+1/n} \ll H^{n-1} \sum_{d \leqslant t^n} d^3 \ll H^{n-1} t^{4n} \ll H^n m^{-2y}.$$

Consequently, we have

$$\mathfrak{S} \geqslant c_{22}(K) H^n.$$

It follows by (23), (34) and (35) that $J_0 \geqslant c_{23}(k, K, h, y)$. Therefore

$$Z > c_{24}(k, K, h, y) H^n (AB)^{ns} (ab)^{-n} m^{-20kny} > 1$$

if $m \geqslant c_{14}(k, K, x', \varepsilon)$. The theorem is proved.

144   Wang Yuan

References

[1] R. G. Ayoub, *On the Waring–Siegel theorem*, Canad. J. Math. 5(1953), pp. 439–450.

[2] B. J. Birch, *Small zeros of diagonal forms of odd degree in many variables*, Proc. London Math. Soc. 21 (1970), pp. 12–18.

[3] H. Davenport, *Analytic methods for diophantine equations and diophantine inequalities*, Lecture Notes, Univ. of Michigan, 1962.

[4] Y. Eda, *On Waring's problem in algebraic number field*, Revista Colombiana de Mat., 1975, pp. 29–72.

[5] E. Hecke, *Lectures on Theory of Algebraic Numbers*, Springer-Verlag, 1980.

[6] Hua Loo Keng and Wang Yuan, *Applications of Number Theory to Numerical Analysis*, Springer-Verlag and Science Press (Beijing), 1981.

[7] K. Ireland and M. Rosen, *A Classical Introduction to Modern Number Theory*, Springer-Verlag, 1982.

[8] O. Körner, *Über das Waringsche Problem in algebraischen Zahlkörpern*, Math. Ann. 144 (1961), pp. 224–238.

[9] T. Mitsui, *On the Goldbach problem in an algebraic number field I*, J. Math. Soc. Japan 12 (1960), pp. 290–324.

[10] J. Pitman, *Bounds for solutions of diagonal equations*, Acta Arith. 19 (1971), pp. 223–247.

[11] W. M. Schmidt, *Small zeros of additive forms in many variables*, Trans. Amer. Math. Soc. 248 (1) (1979), pp. 121–133.

[12] — *Small zeros of additive forms in many variables II*, Acta Math. 143 (1979), pp. 219–232.

[13] C. L. Siegel, *Generalization of Waring's problem to algebraic number fields*, Amer. J. Math. 66 (1944), pp. 122–136.

[14] — *Sums of m-th powers of algebraic integers*, Ann. of Math. 46 (1945), pp. 313–339.

[15] R. M. Stemmler, *The easier Waring problem in algebraic number fields*, Acta Arith. 6 (1961), pp. 447–468.

[16] T. Tatuzawa, *On the Waring problem in an algebraic number field*, J. Math. Soc. Japan 10 (1958), pp. 322–341.

[17] — *On the Waring's problem in algebraic number fields*, Acta Arith. 24 (1973), pp. 37–60.

[18] Wang Yuan, *Bounds for solutions of additive equations in an algebraic number field II* (to appear).

INSTITUTE OF MATHEMATICS
ACADEMIA SINICA
Beijing, China

Received on 30. 5. 1984

and in revised form on 18. 11. 1985   (1428)

ACTA ARITHMETICA
XLVIII (1987)

Bounds for solutions of additive equations in an algebraic number field II

by

WANG YUAN* (Beijing, China)

1. Introduction. We use the conventions and notation introduced in [4] throughout this paper. Let $\alpha_1, \ldots, \alpha_s$ be a set of integers in K. Consider the additive form

$$(1) \qquad A(a, \lambda) = \alpha_1 a_1 \lambda_1^k + \ldots + \alpha_s a_s \lambda_s^k,$$

where $a = (a_1, \ldots, a_s)$ and $\lambda = (\lambda_1, \ldots, \lambda_s)$ are vectors. A set of numbers a, λ is called a *nontrivial solution of the equation*

$$(2) \qquad A(a, \lambda) = 0$$

if each a_i is 1 or -1, and the λ_i ($1 \leqslant i \leqslant s$) are totally nonnegative integers of K, not all zero. Write

$$\|\lambda\| = \max(\|\lambda_1\|, \ldots, \|\lambda_s\|) \quad \text{and} \quad |A| = \max(1, \|\alpha_1\|, \ldots, \|\alpha_s\|).$$

In this paper, we shall prove the following theorem by the combination of the methods of Schmidt [1] and Siegel [3].

THEOREM. *Suppose $s \geqslant c_1(k, n, \varepsilon)$. Then the equation (2) has a nontrivial solution with*

$$(3) \qquad \|\lambda\| \ll |A|^\varepsilon.$$

This gives a generalization of a theorem due to Schmidt [1]. He first established the estimation (3) for the case of rational field.

If a, λ is a nontrivial solution of (2) with (3) for the case of $k = 2$, then a, $\Lambda = (\Lambda_1, \ldots, \Lambda_s)$, with $\Lambda_i = \lambda_i^2$ ($1 \leqslant i \leqslant s$), is a nontrivial solution of the linear equation

$$\alpha_1 a_1 \Lambda_1 + \ldots + \alpha_s a_s \Lambda_s = 0,$$

having

$$\|\Lambda\| \ll |A|^2.$$

---

* Supported by the Institute for Advanced Study, Princeton, NJ 08540.

W a n g   Y u a n

Therefore we may suppose $k > 1$ throughout this paper.

If A is not identically zero, put

$$A' = \frac{c_2}{\sigma} A,$$

where σ is a nonzero element in the integral ideal $(\alpha_1, \ldots, \alpha_s)$ with the least norm in absolute value and $c_2 = c_2(K)$ is a rational integer such that $\sigma | c_2 \alpha_i$ $(1 \leqslant i \leqslant s)$. (See, e.g., Lemma 3 in [4].) We may suppose without loss of generality that

$$|N(\sigma)|^{1/n} \ll \|\sigma\| \ll |N(\sigma)|^{1/n},$$

because we can choose a unit η of K such that $\eta\sigma$ satisfies the above relation, and use $\eta\sigma$ instead of σ. (See, e.g., Lemma 1 in [4].) If $A \equiv 0$, then we put $A' = A$.

2. Reductions. One can prove by Siegel's method that if $s \geqslant c_3(k, n)$, then (2) has a nontrivial solution with

$$(4) \qquad \|\lambda\| \ll |A|^{c_4(k,n)}.$$

Let X be the set of x such that if $s \geqslant c_5(k, n, x)$, then (2) has a nontrivial solution with

$$\|\lambda\| \ll |A|^x.$$

(4) shows that X is not empty. Let x be the greatest lower bound of X. The conclusion of the theorem is $x = 0$. We will suppose that $x > 0$ and we will reach a contradiction.

We may choose y such that

$$(5) \qquad 0 < y < x \quad \text{and} \quad x + kx^2 - kxy - k^2 x^2 y < y.$$

(See Schmidt [1], p. 222.) Take z so small that

$$(6) \qquad y + 12xz < x, \quad z < y/10, \quad z < 1/10.$$

Then pick x' with

$$(7) \qquad \max\left(y + 12xz, \, x - xz/(2n)\right) < x' < x.$$

We proceed to prove that $x' \in X$. It will suffice to prove the assertion when $|A|$ is large, say $|A| \geqslant c_6(k, K, x')$. (See Wang Yuan [4].) And we may suppose clearly that $\alpha_i \neq 0$ $(1 \leqslant i \leqslant s)$. Finally, pick x'' with

$$(8) \qquad \max\left(y + 12xz, \, x - xz/(2n)\right) < x'' < x'$$

and choose ε_1 such that

$$(9) \qquad (1 + \varepsilon_1) x'' + 4\varepsilon_1/k < x'.$$

Since

$$1 \leqslant |N(\alpha_j)| = |\alpha_j^{(1)} \ldots \alpha_j^{(n)}| \leqslant |\alpha_j^{(i)}| |A|^{n-1}.$$

we have

$$\min_{i,j} |\alpha_j^{(i)}| \geqslant |A|^{-n+1}.$$

Divide the interval $[-n+1, 1]$ into a finite number of intervals $\{I\}$ of length $\leqslant \varepsilon_1$. One of these intervals I_1 will be such that there are not less than s_1 numbers among α_i's satisfying

$$|\alpha_j^{(1)}| = |A|^{e_1}, \quad e_1 \in I_1,$$

where $s_1 \geqslant s \Big/ \Big(\Big[\dfrac{n}{\varepsilon_1}\Big] + 1\Big)$. We may suppose without loss of generality that $\alpha_1, \ldots, \alpha_{s_1}$ satisfy the above relation. Similarly, there exists $I_2 \in \{I\}$ such that there are at least s_2 numbers in $\alpha_1, \ldots, \alpha_{s_1}$ satisfying

$$|\alpha_j^{(2)}| = |A|^{e_2}, \quad e_2 \in I_2,$$

where

$$s_2 \geqslant s_1 \Big/ \Big(\Big[\dfrac{n}{\varepsilon_1}\Big] + 1\Big) \geqslant s \Big/ \Big(\Big[\dfrac{n}{\varepsilon_1}\Big] + 1\Big)^2.$$

We may suppose that $\alpha_1, \ldots, \alpha_{s_2}$ have the above property. Continuing this process, we obtain t numbers among α_i's which we may suppose to be $\alpha_1, \ldots, \alpha_t$ such that

(10) $$|\alpha_j^{(i)}/\alpha_l^{(i)}| \leqslant |A|^{\varepsilon_1}, \quad 1 \leqslant j, l \leqslant t, 1 \leqslant i \leqslant n,$$

where $t \geqslant s \Big/ \Big(\Big[\dfrac{n}{\varepsilon_1}\Big] + 1\Big)^n$.

Suppose that if $\alpha_1, \ldots, \alpha_t$ satisfy (10) and $t \geqslant c_7(k, n, x')$, the equation

$$\alpha_1 a_1' \lambda_k'^h + \ldots + \alpha_t a_t' \lambda_t'^h = 0$$

has a nontrivial solution satisfying

$$\max_{i,j} |\lambda_j'^{(i)}| \ll |A|^{x'}.$$

Take $c_5(k, n, x') = \Big(\Big[\dfrac{n}{\varepsilon_1}\Big] + 1\Big)^n c_7(k, n, x')$ and set $a_i = a_i'$, $\lambda_i = \lambda_i'$ $(1 \leqslant i \leqslant t)$

$a_i = 1$, $\lambda_i = 0$ $(t < i \leqslant s)$. We have a nontrivial solution of (2) with

$$\|\lambda\| \ll |A|^{x'}.$$

Hence we may suppose that the coefficients of (2) satisfy

(11) $$|\alpha_j^{(i)}/\alpha_l^{(i)}| \leqslant |A|^{\varepsilon_1}, \quad 1 \leqslant j, l \leqslant s, 1 \leqslant i \leqslant n.$$

W a n g   Y u a n

Let σ_i $(1 \leqslant i \leqslant s)$ be a set of totally nonnegative units such that

$$(12) \qquad |N(\alpha_j)|^{1/n} \ll |\alpha_j^{(i)} \sigma_j^{(i)k}| \ll |N(\alpha_j)|^{1/n}, \qquad 1 \leqslant j \leqslant s, \, 1 \leqslant i \leqslant n.$$

(See, e.g., Lemma 1 in [4].) Let $a^n = |A|^{n\varepsilon_1} \max |N(\alpha_i)|$ and let P_i be the largest rational integer such that

$$|N(\alpha_i)| P_i^{kn} \leqslant a^n, \qquad 1 \leqslant i \leqslant s.$$

Since $|A| \geqslant c_6$ and $a^n/|N(\alpha_i)| \geqslant |A|^{n\varepsilon_1}$, we may suppose

$$P_i \geqslant 2^{-1/(kn)} \left(\frac{a^n}{|N(\alpha_i)|} \right)^{1/(kn)}, \qquad 1 \leqslant i \leqslant s.$$

Hence

$$|N(\alpha_i)| P_i^{kn} \geqslant \tfrac{1}{2} a^n.$$

Set

$$(13) \qquad \alpha_i' = \alpha_i \sigma_i^k P_i^k \quad \text{and} \quad \lambda_i = \sigma_1^{-1} \sigma_i P_i \lambda_i', \qquad 1 \leqslant i \leqslant s.$$

Then

$$A_1(a, \lambda') = \alpha_1' a_1 \lambda_1'^k + \ldots + \alpha_s' a_s \lambda_s'^k = \sigma_1^k A(a, \lambda),$$

where $\lambda' = (\lambda_1', \ldots, \lambda_s')$. By (11) and (12), we have

$$a \ll |\alpha_j^{(i)'}| \ll a, \qquad 1 \leqslant j \leqslant s, \, 1 \leqslant i \leqslant n$$

and

$$(14) \qquad |\sigma_1^{(i)-1} \sigma_j^{(i)}| = \left| \frac{\sigma_j^{(i)k} \alpha_j^{(i)} \alpha_j^{(i)-1}}{\sigma_1^{(i)k} \alpha_1^{(i)} \alpha_1^{(i)-1}} \right|^{1/k}$$

$$\ll |N(\alpha_j)|^{1/(kn)} |N(\alpha_1)|^{-1/(kn)} |A|^{\varepsilon_1/k} \ll |A|^{2\varepsilon_1/k},$$

$$1 \leqslant j \leqslant s, \, 1 \leqslant i \leqslant n.$$

Since

$$|N(\alpha_i)| = |A|^{n\varepsilon_1} |N(\alpha_i)| \max_j |N(\alpha_j)| / |A|^{n\varepsilon_1} \max_l |N(\alpha_l)|$$

$$= a^n |N(\alpha_i)| |A|^{-n\varepsilon_1} / \max |N(\alpha_l)| \geqslant a^n |A|^{-2n\varepsilon_1},$$

we have

$$(15) \qquad P_i^{kn} \leqslant \frac{a^n}{|N(\alpha_i)|} \leqslant |A|^{2n\varepsilon_1}.$$

Suppose that the equation $A_1(a, \lambda') = 0$ has a nontrivial solution such that

$$\|\lambda'\| \ll |A_1|^{x''} \ll a^{x''} \ll |A|^{(1+\varepsilon_1)x''}.$$

Then it follows by (9), (13), (14) and (15) that (2) has a nontrivial solution satisfying

$$\|\lambda\| \ll |A|^{(1+\varepsilon_1)x'' + 4\varepsilon_1/k} \ll |A|^{x'}.$$

Thus it will suffice to prove that if $s \geqslant c_8$ and if α_i $(1 \leqslant i \leqslant s)$ satisfy

$$(16) \qquad c_9 a < |\alpha_j^{(i)}| < c_{10} a, \qquad 1 \leqslant j \leqslant s, \, 1 \leqslant i \leqslant n,$$

where $c_9 = c_9(k, K)$, $c_{10} = c_{10}(k, K)$ and $a > c_6(k, K, x'')$, then (2) has a nontrivial solution with

$$(17) \qquad \qquad \qquad \|\lambda\| \ll a^{x''}.$$

Of course c_8 depends on k, n, x'', but since k, n, x, y, z, x', x'' are fixed, we will not indicate the dependency of c_8 (and of subsequent constants) on these parameters.

We shall first prove that this assertion can be derived by the following Proposition 1, and then give the proof of the proposition.

PROPOSITION 1. *Suppose that* α_i $(1 \leqslant i \leqslant s)$ *satisfy* (16). *If* $s \geqslant c_{11}$, *then either* (2) *has a nontrivial solution with* (17) *or there is a nonzero integer* χ *such that*

$$(18) \qquad A(a, \lambda) = \chi, \qquad \chi \in P(a^{6z}), \qquad \lambda_i \in P(a^y), \qquad 1 \leqslant i \leqslant s,$$

where each a_i *is* 1 *or* -1 *and* λ_j $(1 \leqslant j \leqslant s)$ *are totally nonnegative integers of* K, *not all zero.*

We may suppose that $c_5(k, n, 2x)$ and c_{11} are integers. Denote $v = c_5$, $u = c_{11}$ and $s = uv$. Replace the indices $1 \leqslant l \leqslant s$ by double indices $1 \leqslant i \leqslant v$, $1 \leqslant j \leqslant u$. Then (1) becomes

$$A(a, \lambda) = \sum_{i=1}^{v} A_i(a_i, \lambda_i),$$

where $a_i = (a_{i1}, \ldots, a_{iu})$, $\lambda_i = (\lambda_{i1}, \ldots, \lambda_{iu})$ and

$$A_i(a_i, \lambda_i) = \sum_{j=1}^{u} \alpha_{ij} a_{ij} \lambda_{ij}^k, \qquad 1 \leqslant i \leqslant v.$$

If there is an equation, say $A_i(a_i, \lambda_i) = 0$, which has a nontrivial solution having

$$\|\lambda_i\| \ll a^{x''},$$

then we have directly a nontrivial solution of (2) with (17). Otherwise it follows by Proposition 1 that there are nonzero integers $\chi_1, \ldots, \chi_v$ satisfying

$$A_i(u_i, \lambda_i) = \chi_i, \qquad \chi_i \in P(a^{6z}), \qquad \lambda_{ij} \in P(a^y), \qquad 1 \leqslant i \leqslant v, \, 1 \leqslant j \leqslant u,$$

where each a_{ij} is 1 or -1. Since the equation

$$B(b, \mu) = \chi_1 b_1 \mu_1^k + \ldots + \chi_v b_v \mu_v^k = 0$$

has a nontrivial solution with

$$\|\mu\| \ll \max \|\chi_i\|^{2x} \ll a^{12xz},$$

312   Wang Yuan

the equation (2) has a nontrivial solution $b_i\, a_{ij}$, $\mu_i\, \lambda_{ij}$ ($1 \leqslant i \leqslant v$, $1 \leqslant j \leqslant u$) satisfying

$$\max_{i,j} \|\mu_i\, \lambda_{ij}\| \leqslant \|\mu\| \max_i \|\lambda_i\| \ll a^{y + 12xz} \ll a^{x''}$$

by (8). Hence it remains only to prove Proposition 1.

3. The circle method. Let

$$(19) \qquad t = a^{z/n} \quad \text{and} \quad h = a^{1 + ky - z/n}.$$

Let $\Gamma(t)$ be the set consisting of $\gamma = x_1 \varrho_1 + \ldots + x_n \varrho_n$ satisfying $(x_1, \ldots, x_n) \in G_n$, x_i ($1 \leqslant i \leqslant n$) rational numbers, $\gamma \to \mathfrak{a}$ and $N(\mathfrak{a}) \leqslant t^n$. For any $\gamma \in \Gamma(t)$, subject to $\gamma \to \mathfrak{a}$, we define the basic domain B_γ by

$$\{(x_1, \ldots, x_n): (x_1, \ldots, x_n) \in G_n, \ \xi = x_1 \varrho_1 + \ldots + x_n \varrho_n$$

$$\text{such that } h\|\xi - \gamma_0\| < 1 \text{ for some } \gamma_0 \equiv \gamma (\mathrm{mod} \ \delta^{-1})\}.$$

We may prove that if $\gamma_1 \neq \gamma_2$, then $B_{\gamma_1} \cap B_{\gamma_2} = \emptyset$. In fact, suppose there is a $\xi \in B_{\gamma_1} \cap B_{\gamma_2}$, i.e., $h\|\xi - \gamma_{0i}\| < 1$, where $\gamma_{0i} \equiv \gamma_i (\mathrm{mod} \ \delta^{-1})$, $i = 1, 2$. For simplicity, we set $\gamma_{0i} = \gamma_i$ ($i = 1, 2$). Denote

$$\max(h|\xi^{(i)} - \gamma_j^{(i)}|, t^{-1}) = \sigma_j^{(i)}, \quad 1 \leqslant j \leqslant 2, 1 \leqslant i \leqslant n.$$

Then

$$\prod_{i=1}^{n} \sigma_j^{(i)} < 1, \quad \max_i \sigma_j^{(i)-1} \leqslant t, \quad j = 1, 2,$$

and thus

$$|\gamma_1^{(i)} - \gamma_2^{(i)}| \leqslant |\xi^{(i)} - \gamma_1^{(i)}| + |\xi^{(i)} - \gamma_2^{(i)}| \leqslant h^{-1}(\sigma_1^{(i)} + \sigma_2^{(i)})$$

$$= h^{-1} \sigma_1^{(i)} \sigma_2^{(i)}(\sigma_1^{(i)-1} + \sigma_2^{(i)-1}) \leqslant 2h^{-1} \sigma_1^{(i)} \sigma_2^{(i)} t.$$

Suppose $\gamma_i \to \mathfrak{a}_i$ ($i = 1, 2$). We have

$$N(\mathfrak{a}_1 \mathfrak{a}_2)|N(\gamma_1 - \gamma_2)| < (2h^{-1} t^3)^n < D^{-1},$$

since $a \geqslant c_6$. On the other hand, $\mathfrak{a}_1 \mathfrak{a}_2 (\gamma_1 - \gamma_2) \delta$ is an integral ideal, and thus

$$N(\mathfrak{a}_1 \mathfrak{a}_2)|N(\gamma_1 - \gamma_2)| \geqslant N(\delta^{-1}) = D^{-1}.$$

This gives a contradiction, and therefore the assertion follows.

We define the supplementary domain E by

$$(20) \qquad E = G_n - \bigcup_{\gamma \in \Gamma(t)} B_\gamma.$$

We use the notations

$$\xi = x_1 \varrho_1 + \ldots + x_n \varrho_n, \quad dx = dx_1 \ldots dx_n, \quad B = a^y, \quad H = a^{6z},$$

$$(21) \qquad S_i(\xi) = \sum_{\lambda \in P(B)} E(a_i \xi \alpha_i \lambda^k), \quad 1 \leqslant i \leqslant s, \quad S(\xi) = \prod_{i=1}^{s} S_i(\xi),$$

and

$$F(\xi) = \sum_{\chi \in P(H)} S(\xi) E(-\xi\chi),$$

where a_i $(1 \leqslant i \leqslant s)$ are defined by

(22)
$$a_{2p-1} = \frac{|\alpha_{2p-1}^{(p)}|}{\alpha_{2p-1}^{(p)}}, \qquad a_{2p} = -\frac{|\alpha_{2p}^{(p)}|}{\alpha_{2p}^{(p)}} \qquad (1 \leqslant p \leqslant r_1),$$

$$a_j = 1 \qquad (2r_1 + 1 \leqslant j \leqslant s).$$

Let Z denote the number of solutions of the equation

$$a_1 \alpha_1 \lambda_1^k + \ldots + a_s \alpha_s \lambda_s^k = \chi$$

in totally nonnegative integers $\lambda_1, \ldots, \lambda_s$, χ satisfying

$$\chi \in P(H), \qquad \lambda_i \in P(B), \qquad 1 \leqslant i \leqslant s.$$

Then

(23)
$$Z = \sum_{\gamma \in \Gamma(t)} \int_{B_\gamma} F(\xi)\, d\xi + \int_E F(\xi)\, d\xi.$$

We shall show that under the assumption made in Proposition 1, either (2) has a nontrivial solution with (17) or Z is > 1.

4. Supplementary domain. In this section $a_i = \pm 1$ $(1 \leqslant i \leqslant s)$ which are not restricted by (22).

LEMMA 1 (Schmidt). *Suppose that*

$$T \geqslant c_{12}(k, K, \varepsilon_2), \qquad C \geqslant T^{1-1/G+\varepsilon_2} \qquad and \qquad \left| \sum_{\lambda \in P(T)} E(\xi\lambda^k) \right| \geqslant C,$$

where $G = 2^{k-1}$. *Then there exist a totally nonnegative integer* α *and an integer* β *such that*

$$\|\alpha\zeta - \beta\| \ll \left(\frac{T^n}{C}\right)^G T^{\alpha_2 - h}$$

and

$$0 < \|\alpha\| \ll \left(\frac{T^n}{C}\right)^G T^{\varepsilon_2}.$$

See, e.g., Wang Yuan [3].

LEMMA 2. *Suppose that* $s \geqslant c_{13}$ *and* $\xi \in E$. *Then either*

(24)
$$|F(\xi)| < H^n B^{n(s-k)} a^{-2n}$$

or there is a nontrivial solution of (2) with (17).

Proof. Take ε_2 such that

(25)
$$0 < \varepsilon_2 < c_{14} < 1/(2G),$$

314   **Wang Yuan**

where c_{14} is a constant to be determined later. Set

$$(26) \qquad m = c_5(k, n, x + \varepsilon_2) \quad \text{and} \quad h = m^2.$$

Choose c_{13} $(> 8kn)$ sufficiently large such that if $s \geqslant c_{13}$, then

$$\frac{n(k + 2/y)}{s - h + 1} < \varepsilon_2,$$

and by (21), we have

$$(B^k a^2)^{n/(s-h+1)} = B^{n(k+2/y)/(s-h+1)} < B^{\varepsilon_2}.$$

If (24) fails to hold, then

$$H^n |S_1(\xi) \ldots S_s(\xi)| \geqslant F(\xi) \geqslant H^n B^{n(s-k)} a^{-2n}.$$

We may suppose without loss of generality that

$$|S_1(\xi)| \geqslant \ldots \geqslant |S_s(\xi)|.$$

Hence we have

$$|S_h(\xi)|^{s-h+1} B^{n(h-1)} \geqslant B^{n(s-k)} a^{-2n},$$

and thus by (25),

$$|S_i(\xi)| \geqslant B^{n(s-h-k+1)/(s-h+1)} a^{-2n/(s-h+1)}$$

$$\geqslant B^n (B^k a^2)^{-n/(s-h+1)} \geqslant B^{n-\varepsilon_2} > B^{n-1/G+\varepsilon_2}, \quad 1 \leqslant i \leqslant h.$$

Take $C = B^{n-\varepsilon_2}$ in Lemma 1. Then it follows that there are totally nonnegative integers σ_i $(1 \leqslant i \leqslant s)$ and integers β_j $(1 \leqslant j \leqslant s)$ such that

$$(27) \qquad \begin{aligned} \|\xi \alpha_i \sigma_i - \beta_i\| &\ll B^{-k+2G\varepsilon_2}, \\ 0 < \|\sigma_i\| &\ll B^{2G\varepsilon_2}, \quad 1 \leqslant i \leqslant h. \end{aligned}$$

Denote $\tau_i = \beta_i \sigma_i^{k-1}$ $(1 \leqslant i \leqslant h)$. Then

$$\|\xi \alpha_i \sigma_i^k - \tau_i\| \leqslant \|\sigma_i\|^{k-1} \|\xi \alpha_i \sigma_i - \beta_i\| \ll B^{-k+2kG\varepsilon_2}, \quad 1 \leqslant i \leqslant h.$$

Therefore by (16), we have

$$\|\alpha_i \sigma_i^k \tau_j - \alpha_j \sigma_j^k \tau_i\| \leqslant \|(\xi \alpha_j \sigma_j^k - \tau_j) \alpha_i \sigma_i^k\| + \|(\xi \alpha_i \sigma_i^k - \tau_i) \alpha_j \sigma_j^k\|$$

$$\ll a B^{-k+4kG\varepsilon_2}, \quad 1 \leqslant i, j \leqslant h,$$

i.e., the vectors $f_i = (\alpha_i \sigma_i^k, \tau_i)$ $(1 \leqslant i \leqslant h)$ satisfy

$$(28) \qquad \|\det(f_i, f_j)\| \ll a B^{-k+4kG\varepsilon_2}, \quad 1 \leqslant i, j \leqslant h.$$

Let β be a nonzero element in the integral idea $(\alpha_1 \sigma_1^k, \tau_1)$ with the least norm in absolute value. Then $\beta | c_2 \alpha_1 \sigma_1^k$ and $\beta | c_2 \tau_1$. (See § 1.) Set

$$c_2 \alpha_1 \sigma_1^k = \beta \sigma, \quad c_2 \tau_1 = \beta \tau \quad \text{and} \quad f = (\sigma, \tau).$$

Then

$$c_2 f_1 = \beta f.$$

We may choose β such that

(29) $$|N(\sigma)|^{1/n} \ll \|\sigma\| \ll |N(\sigma)|^{1/n}.$$

(See, e.g., Lemma 1 in [4].) We have also two integers σ' and τ' such that

$$\beta = \alpha_1 \sigma_1^k \tau' - \tau_1 \sigma',$$

therefore

$$c_2 \beta = \beta \sigma \tau' - \beta \tau \sigma',$$

i.e.,

(30) $$c_2 = \sigma \tau' - \tau \sigma'.$$

Set $g = (\sigma', \tau')$. Then by (30),

$$f_i = c_2^{-1} \varphi_i f + c_2^{-1} \psi_i g,$$

where

$$\varphi_i = \begin{vmatrix} \alpha_i \sigma_i^k & \sigma' \\ \tau_i & \tau' \end{vmatrix} \quad \text{and} \quad \psi_i = \begin{vmatrix} \sigma & \alpha_i \sigma_i^k \\ \tau & \tau_i \end{vmatrix}, \quad 1 \leqslant i \leqslant h$$

are integers. By (27), we have

$$1 \leqslant N(\sigma_j) \leqslant |\sigma_j^{(i)}| B^{2(n-1)G\varepsilon_2},$$

i.e.,

(31) $$|\sigma_j^{(i)}| \geqslant B^{-2(n-1)G\varepsilon_2}, \quad 1 \leqslant j \leqslant h, 1 \leqslant i \leqslant n.$$

Since $c_2 \alpha_1 \sigma_1^k = \beta \sigma$, we have by (16), (27), (29) and (31) that

(32) $$a\|\sigma\|^{-1} B^{2kG\varepsilon_2} \gg |\beta^{(i)}| \gg a\|\sigma\|^{-1} B^{-2k(n-1)G\varepsilon_2}, \quad 1 \leqslant i \leqslant n.$$

Therefore it follows by (28) that

(33) $$\|\psi_i\| = \|\det(f_i, f)\| \leqslant c_2 \max_i |\beta^{(i)}|^{-1} \|\det(f_i, f_1)\|$$

$$\ll \|\sigma\| B^{-k + 6knG\varepsilon_2} = M, \quad 1 \leqslant i \leqslant h.$$

1. Suppose $M \geqslant 1$. Replace the indices $1 \leqslant l \leqslant h$ by double indices $1 \leqslant i, j \leqslant m$. Define

$$A_i(a_i, \lambda_i) = \sum_{j=1}^{m} \psi_{ij} a_{ij} \lambda_{ij}^k, \quad 1 \leqslant i \leqslant m,$$

where $a_i = (a_{i1}, \ldots, a_{im})$ and $\lambda_i = (\lambda_{i1}, \ldots, \lambda_{im})$ $(1 \leqslant i \leqslant m)$. It follows by (26), (33) and the definition of the set X that the equation

$$A_i(a_i, \lambda_i) = 0$$

Wang Yuan

has a nontrivial solution having

$$(34) \qquad \|\lambda_i\| \ll M^{x+\varepsilon_2}, \qquad 1 \leqslant i \leqslant m.$$

Let

$$g_i = \sum_{j=1}^{m} a_{ij}\,\lambda_{ij}^k\,f_{ij} = c_2^{-1}\Big(\sum_{j=1}^{m} a_{ij}\,\lambda_{ij}^k\,\varphi_{ij}\Big)f, \qquad 1 \leqslant i \leqslant m.$$

The first coordinate of g_i is

$$\beta_i = c_2^{-1}\,\sigma \sum_{j=1}^{m} a_{ij}\,\lambda_{ij}^k\,\varphi_{ij}, \qquad 1 \leqslant i \leqslant m.$$

Therefore $c_2\,\beta_i/\sigma$ $(1 \leqslant i \leqslant m)$ are integers. If $\beta_1, \ldots, \beta_m$ are not all zero, then let χ be a nonzero element in the integral ideal $(\beta_1, \ldots, \beta_m)$ with least norm in absolute value and satisfying

$$|N(\chi)|^{1/n} \ll \|\chi\| \ll |N(\chi)|^{1/n}.$$

Then $\sigma|c_2\chi$ and by (29),

$$\|\sigma\| \ll \|\chi\|.$$

Consider the form

$$B(b,\,\mu) = \sum_{i=1}^{m} \beta_i\,b_i\,\mu_i^k,$$

where $b = (b_1, \ldots, b_m)$ and $\mu = (\mu_1, \ldots, \mu_m)$. Since

$$\beta_i = \sum_{j=1}^{m} a_{ij}\,\lambda_{ij}^k\,\alpha_{ij}\,\sigma_{ij}^k, \qquad 1 \leqslant i \leqslant m,$$

we have by (16), (27) and (34),

$$\|\beta_i\| \ll aB^{2kG\varepsilon_2}\,M^{(x+\varepsilon_2)k}, \qquad 1 \leqslant i \leqslant m,$$

and therefore the equation $B(b,\,\mu) = 0$ has a nontrivial solution satisfying

$$\|\mu\| \ll \max\left(1,\, \frac{aB^{2kG\varepsilon_2}\,M^{(x+\varepsilon_2)k}}{\|\sigma\|}\right)^{x+\varepsilon_2}.$$

Consequently (2) has a nontrivial solution

$$b_i\,a_{ij}, \qquad \mu_i\,\sigma_{ij}\,\lambda_{ij} \quad (1 \leqslant i,j \leqslant m),$$

$$a_l = 1, \qquad \lambda_l = 0 \quad (h < l \leqslant s)$$

satisfying

$$\|\lambda\| \ll B^{2G\varepsilon_2}\,M^{x+\varepsilon_2}\max\left(1,\, \frac{aB^{2kG\varepsilon_2}\,M^{(x+\varepsilon_2)k}}{\|\sigma\|}\right)^{x+\varepsilon_2}$$

$$\ll B^{2G\varepsilon_2} \max\left(M, \frac{aB^{2kG\varepsilon_2} M^{(x+\varepsilon_2)k+1}}{\|\sigma\|}\right)^{x+\varepsilon_2}$$

$$= B^{2G\varepsilon_2} I^{x+\varepsilon_2}, \quad \text{say}.$$

Since $c_2 \alpha_1 \sigma_1^k = \beta\sigma$, by (27), (29) and (33), we have

$$\|\sigma\| \ll aB^{2kG\varepsilon_2} (\max_i |\beta^{(i)}|)^{-1} \ll aB^{2kG\varepsilon_2},$$

$$M \ll aB^{-k+8knG\varepsilon_2},$$

and therefore

$$I \ll aB^{2kG\varepsilon_2} M^{(x+\varepsilon_2)k+1} \|\sigma\|^{-1}$$

$$= aB^{2kG\varepsilon_2} M^{(x+\varepsilon_2)k+1} M^{-1} B^{-k+6knG\varepsilon_2}$$

$$\ll aB^{-k+8knG\varepsilon_2} (aB^{-k+8knG\varepsilon_2})^{(x+\varepsilon_2)k}.$$

Since $a > c_6$, we have

$$\|\lambda\| \ll a^{x+kx^2-kxy-k^2x^2y+c_{15}(k,n)\varepsilon_2}.$$

Hence if c_{14} is sufficiently small, then by (5) and (8), we have the desired (17).

2. Suppose $M < 1$. We revert to the indices $1 \leqslant i \leqslant h$. By (33), we have $\psi_i = 0$ $(1 \leqslant i \leqslant h)$, i.e., $c_2 f_i$ $(1 \leqslant i \leqslant h)$ are integral multiples of the integral vector f. Let

$$B(b, \mu) = \sum_{i=1}^{h} \alpha_i \sigma_i^k b_i \mu_i^k,$$

where $b = (b_1, \ldots, b_h)$ and $\mu = (\mu_1, \ldots, \mu_h)$. Let χ be a nonzero element in the integral ideal $(\alpha_1 \sigma_1^k, \ldots, \alpha_h \sigma_h^k)$ with least norm in absolute value and satisfying $|N(\chi)|^{1/n} \gg \|\chi\| \gg |N(\chi)|^{1/n}$. Then $\sigma | c_2 \chi$, and thus $\|\sigma\| \ll \|\chi\|$. Hence the equation $B(b, \mu) = 0$ has a nontrivial solution satisfying

$$\|\mu\| \ll \max_i (\|\alpha_i \sigma_i^k\| \|\sigma\|^{-1})^{x+\varepsilon_2} \ll (aB^{2kG\varepsilon_2} \|\sigma\|^{-1})^{x+\varepsilon_2}.$$

It derives a nontrivial solution of (2):

$$a_i = b_i, \quad \lambda_i = \sigma_i \mu_i \quad (1 \leqslant i \leqslant h),$$

$$a_j = 1, \quad \lambda_j = 0 \quad (h < j \leqslant s).$$

If c_{14} is sufficiently small, then

$$\|\lambda\| \ll B^{2G\varepsilon_2} (aB^{2kG\varepsilon_2} \|\sigma\|^{-1})^{x+\varepsilon_2}$$

$$= a^{x+\varepsilon_2+2Gy\varepsilon_2+2kGy(x+\varepsilon_2)\varepsilon_2} \|\sigma\|^{-x-\varepsilon_2}$$

$$\leqslant a^{x+xz/(2n)} \|\sigma\|^{-x}.$$

Wang Yuan

If $\|\sigma\| \geqslant a^{z/n}$, then by (8), we have

$$\|\lambda\| \leqslant a^{x - xz/(2n)} \leqslant a^{x''},$$

i.e., (17) is true. Now we suppose that $\|\sigma\| < a^{z/n}$. Then by (27) and (31), we have

$$\|\xi - \sigma^{-1}\tau\| = \|\xi - \alpha_1^{-1}\sigma_1^{-1}\beta_1\| \leqslant \max_i \frac{1}{|\alpha_1^{(i)}\sigma_1^{(i)}|}\|\alpha_1\sigma_1\xi - \beta_1\|$$

$$\ll a^{-1}B^{2(n-1)G\varepsilon 2 - k + 2G\varepsilon 2} < a^{-1-ky+z/n} = h^{-1}$$

if c_{14} is sufficiently small. Let $\sigma^{-1}\tau\delta = \mathfrak{b}/\mathfrak{a}$, $(\mathfrak{a}, \mathfrak{b}) = 1$. Then $\mathfrak{a}|\sigma$, and thus

$$N(\mathfrak{a}) \leqslant |N(\sigma)| < a^z = t^n.$$

This means that $\xi \in B_\gamma$, where $\gamma \equiv \dot\sigma^{-1}\tau \,(\mathrm{mod}\ \delta^{-1})$. The lemma is proved.

5. Basic domain. We use the notations

$$(35) \qquad \xi - \gamma = \zeta, \qquad \eta = y_1\omega_1 + \ldots + y_n\omega_n, \qquad dy = dy_1 \ldots dy_n,$$

$$G_i(\gamma) = N(\mathfrak{a})^{-1} \sum_{\mu(\mathrm{mod}\,\mathfrak{a})} E(a_i\alpha_i\mu^k\gamma) \qquad (1 \leqslant i \leqslant s),$$

$$G(\gamma) = \prod_{i=1}^{s} G_i(\gamma),$$

$$I_i(\zeta, B) = \int_{P(B)} E(a_i\alpha_i\eta^k\zeta)\,dy \qquad (1 \leqslant i \leqslant s)$$

and

$$I(\zeta, B) = \prod_{i=1}^{s} I_i(\zeta, B),$$

where a_i $(1 \leqslant i \leqslant s)$ are defined by (22).

LEMMA 3. *Suppose that* $\xi \in B_\gamma$. *Then*

$$S_i(\xi) = G_i(\gamma)I_i(\zeta, B) + O(a^{2z/n}B^{n-1}).$$

See, e.g., Lemma 12 in Wang Yuan [4].

LEMMA 4.

$$I_i(\zeta, B) \ll \prod_{i=1}^{s} \min(B, a^{-1/k}|\zeta^{(i)}|^{-1/k}), \qquad 1 \leqslant i \leqslant s.$$

See Siegel [3], p. 335.

LEMMA 5.

$$\int_{B_\gamma} S(\xi)E(-\chi\xi)\,dx = G(\gamma)E(-\chi\gamma)\int_{E_n} I(\zeta, B)\,dx + O(B^{(s-k)n}a^{-n-7z}).$$

Proof. Let

$$\zeta^{(p)} = u_p \quad (1 \leqslant p \leqslant r_1), \qquad \zeta^{(q)} = u_q\, e^{i\varphi_q} \quad (r_1 + 1 \leqslant q \leqslant r_1 + r_2).$$

The Jacobian of $x_1, \ldots, x_n$ with respect to u_p, u_q, φ_q is

$$D^{1/2}\, 2^{r_2} \prod_q u_q.$$

Suppose $\xi \in B_\gamma$. Then by Lemmas 3 and 4, we have

$$S(\xi) = G(\gamma)\, I(\zeta, B) + O(B^{ns-1} a^{2z/n}).$$

Since by (19),

$$\int_{B_\gamma} dx \ll h^{-n} = a^{-n - kny + z},$$

and by (6),

$$(1 + 2/n)\,z - y < -.7z,$$

we have

$$(36) \quad \int_{B_\gamma} S(\xi)\, E(-\chi\xi)\, dx$$

$$= G(\gamma)\, E(-\chi\gamma) \int_{B_\gamma} I(\zeta, B)\, E(-\chi\zeta)\, dx + O(B^{m(s-k)} a^{-n - 7z}).$$

In the integral on the right-hand side of (36) we replace $E(-\chi\zeta)$ by 1. The error is

$$B^{ns} \int_{B_\gamma} \|\chi\zeta\|\, dx \ll B^{ns} H h^{-n-1} \ll B^{n(s-k)} a^{-n - 7z}$$

by (19) and (21). Hence

$$(37) \quad \int_{B_\gamma} S(\xi)\, E(-\chi\xi)\, dx = G(\gamma)\, E(-\chi\gamma) \int_{B_\gamma} I(\zeta, B)\, dx + O(B^{n(s-k)} a^{-n - 7z}).$$

If $(x_1, \ldots, x_n)$ is a point of $E_n - B_\gamma$, then the inequality $h|\zeta^{(i)}| \geqslant 1$ is true for at least one index i. Since $s \geqslant c_{13}$, it follows by Lemma 4 that

$$\int_{E_n - B_\gamma} I(\zeta, B)\, dx \ll \int_{E_n - B_\gamma} \Big(\prod_{i=1}^{n} \min(B, a^{-1/k} |\zeta^{(i)}|^{-1/k})\Big)^s dx$$

$$\ll \Big(\int_{h^{-1}}^{\infty} a^{-s/k} u^{-s/k}\, du\Big)\Big(\int_{0}^{\infty} \min(B^s, a^{-s/k} v^{-s/k})\, dv\Big)^{r_1 - 1}$$

$$\times \Big(\int_{-\pi}^{\pi} \int_{0}^{\infty} \min(B^{2s}, a^{-2s/k} w^{-2s/k})\, w\, dw\, d\varphi\Big)^{r_2}$$

$$+ \Big(\int_{0}^{\infty} \min(B^s, a^{-s/k} u^{-s/k})\, du\Big)^{r_1} \Big(\int_{-\pi}^{\pi} \int_{h^{-1}}^{\infty} a^{-2s/k} v^{-2s/k + 1}\, dv\, d\varphi\Big)$$

Wang Yuan

$$\times\Big(\int_{-\pi}^{\pi}\int_0^{\infty}\min(B^{2s},\,a^{-2s/k}w^{-2s/k})\,w\,dw\,d\varphi\Big)^{r_2-1}$$

$$\ll a^{-s/k}\,h^{s/k-1}\,a^{-(r_1-1)}\,B^{(s-k)(r_1-1)}\,a^{-2r_2}\,B^{2(s-k)r_2}$$

$$+a^{-r_1}\,B^{(s-k)r_1}\,a^{-2s/k}\,h^{2s/k-2}\,a^{-2(r_2-1)}\,B^{2(s-k)(r_2-1)}$$

$$\ll B^{(s-k)(n-1)}\,a^{-s/k-n+1+(1+ky-z/n)\left(\frac{s}{k}-1\right)}$$

$$+B^{(s-k)(n-2)}\,a^{-2s/k-n+2+(1+ky-z/n)\left(\frac{2s}{k}-2\right)}$$

$$\ll B^{n(s-k)}\,a^{-n}\big(a^{-\frac{zs}{kn}+\frac{z}{n}}+a^{-\frac{2zs}{kn}+\frac{2z}{n}}\big)$$

$$\ll B^{n(s-k)}\,a^{-n-7z}.$$

The lemma follows by substitution into (37).

6. Singular integral. Let

$$\eta' = y_1'\,\omega_1 + \ldots + y_n'\,\omega_n, \qquad \zeta' = x_1'\,\varrho_1 + \ldots + x_n'\,\varrho_n,$$

$$dy' = dy_1'\ldots dy_n', \qquad dx' = dx_1'\ldots dx_n',$$

$$\eta = B\eta' \qquad \text{and} \qquad \zeta = a^{-1-ky}\zeta'.$$

The Jacobians of $y_1, \ldots, y_n$ and $x_1, \ldots, x_n$ with respect to $y_1', \ldots, y_n'$ and $x_1', \ldots, x_n'$ are B^n and $(a^{-1-ky})^n$ respectively. Define $\gamma_i = \alpha_i/a$ $(1 \leqslant i \leqslant s)$. Then

$$\alpha_i\,\eta^k\,\zeta = \gamma_i\,\eta'^k\,\zeta', \qquad 1 \leqslant i \leqslant s,$$

and by (16), we have

$$(38) \qquad c_9 < |\gamma_j^{(i)}| < c_{10}, \qquad 1 \leqslant j \leqslant s,\ 1 \leqslant i \leqslant n.$$

Let us write η' and ζ' as η and ζ again and let

$$I_i(\zeta) = \int_P E(a_i\,\gamma_i\,\eta^k\,\zeta)\,dy \quad (1 \leqslant i \leqslant s) \qquad \text{and} \qquad I(\zeta) = \prod_{i=1}^s I_i(\zeta),$$

where $P = P(1)$. Then

$$I_i(\zeta, B) = B^n\,I_i(\zeta), \qquad 1 \leqslant i \leqslant s,$$

and thus

$$(39) \qquad \int_{E_n} I(\zeta, B)\,dx = B^{n(s-k)}\,a^{-n}\int_{E_n} I(\zeta)\,dx.$$

LEMMA 6.

$$\int_{E_n} I(\zeta)\,dx = D^{(1-s)/2}\,k^{-ns}\,|N(\gamma_1\ldots\gamma_s)|^{-1/k}\prod_p F_p\prod_q H_q,$$

where

$$F_p = \int_{U_p} \prod_{i=1}^{s} w_i^{1/k-1} \, dw_1 \dots dw_{s-1}$$

in which U_p denotes the domain

$$0 \leqslant w_i \leqslant |\gamma_i^{(p)}|, \quad 1 \leqslant i \leqslant s, \quad w_{2p} = w_{2p-1} \pm w_1 \pm \dots \pm w_s,$$

here the sign before w_i is the sign of $a_i \gamma_i^{(p)}$ (cf. (22)), and where

$$H_q = \int_{V_q} \prod_{i=1}^{s} w_i^{1/k-1} \, dw_1 \dots dw_{s-1} \, d\varphi_1 \dots d\varphi_{s-1}$$

in which V_q denotes the domain

$$0 \leqslant w_i \leqslant |\gamma_i^{(q)}|^2 \quad (1 \leqslant i \leqslant s), \qquad -\pi \leqslant \varphi_j \leqslant \pi \quad (1 \leqslant j \leqslant s-1),$$

$$w_s = |w_1^{1/2} e^{i\varphi_1} + \dots + w_{s-1}^{1/2} e^{i\varphi_{s-1}}|^2.$$

The proof is similar to that of Lemma 16 in Wang Yuan [4].

7. The proof of the theorem. We have

$$\sum_{\gamma \in \Gamma(t)} 1 \ll \sum_{N(\mathfrak{a}) \leqslant t^n} N(\mathfrak{a}) \ll \sum_{d \leqslant t^n} d^2 \ll t^{3n} = a^{3z}.$$

If (24) holds, then by (23), (39) and Lemmas 2, 5 and 6, we have

$$Z = \sum_{\gamma \in \Gamma(t)} \int_{B_\gamma} F(\xi) \, dx + O(H^n B^{n(s-k)} a^{-n-4z})$$

$$= J_0 \, \mathfrak{S}(t, H) B^{n(s-k)} a^{-n} + O(H^n B^{n(s-k)} a^{-n-4z}),$$

where

$$J_0 = D^{(1-s)/2} k^{-ns} |N(\gamma_1 \dots \gamma_s)|^{-1/k} \prod_p F_p \prod_q H_q$$

and

$$\mathfrak{S} = \mathfrak{S}(t, H) = \sum_{\chi \in P(H)} \sum_{\gamma \in \Gamma(t)} G(\gamma) E(-\chi\gamma).$$

Let $\sum^*$ denote a sum, where γ runs over a reduced residue system of $(\mathfrak{a}\delta)^{-1}$, mod δ^{-1}. Then

$$\mathfrak{S} = \sum_{N(\mathfrak{a})=1} \sum_{\gamma}^{*} G(\gamma) \sum_{\chi \in P(H)} E(-\chi\gamma) + \sum_{1 < N(\mathfrak{a}) \leqslant t^n} \sum_{\gamma}^{*} G(\gamma) \sum_{\chi \in P(H)} E(-\chi\gamma)$$

$$= \mathfrak{S}_1 + \mathfrak{S}_2, \quad \text{say}.$$

We have

$$\mathfrak{S}_1 = \sum_{\chi \in P(H)} 1 \gg H^n.$$

Wang Yuan

(See, e.g., [4].) If $N(\mathfrak{a}) > 1$, then

$$\sum_{\chi(\mathrm{mod}\,\mathfrak{a})} E(-\chi\gamma) = 0.$$

Therefore if the domain $\chi \in P(H)$ has to be split up into a union of complete residue set (mod $\mathfrak{a}$), plus a few other, remaining elements, say R elements, then

$$R \ll H^{n-1} N(\mathfrak{a})^{1/n},$$

and thus

$$\mathfrak{S}_2 \ll \sum_{N(\mathfrak{a}) \leqslant t^n} \sum_{\gamma}^{*} H^{n-1} N(\mathfrak{a})^{1/n} \ll H^{n-1} \sum_{N(\mathfrak{a}) \leqslant t^n} N(\mathfrak{a})^{1+1/n}$$

$$\ll H^{n-1} \sum_{d \leqslant t^n} d^3 \ll H^{n-1} t^{4n} \ll H^{n-2z}.$$

Hence

$$\mathfrak{S} \gg H^n.$$

It follows by (38) that $J_0 > c_{16}$, and therefore

$$Z > c_{17} H^n B^{n(s-k)} a^{-n} > 1$$

if $a > c_6(k, K, x'')$. The theorem is proved.

Remarks. 1. The inequality (3) can be replaced by

$$\max_i |N(\lambda_i)| \ll \max\left(1, |N(\alpha_1)|, \ldots, |N(\alpha_s)|\right)^{\varepsilon}.$$

(See [4].)

2. Consider the equation

$$(40) \qquad A(\lambda) = \sum_{i=1}^{s} \alpha_i \lambda_i^k = 0,$$

where $\alpha_1, .., \alpha_s$ are given integers in K. If k is an odd number, then

$$a_i \lambda_i^k = (a_i \lambda_i)^k \qquad (1 \leqslant i \leqslant s).$$

If $r_1 = 0$, i.e., K is totally complex, then the singular integral J_0 is always positive. Therefore in these two cases it follows by the theorem that if $s \geqslant c_{18}(k, n, \varepsilon)$, the equation (40) has a solution in integers $\lambda_1, \ldots, \lambda_s$, not all zero, satisfying

$$\|\lambda\| \ll |A|^{\varepsilon}.$$

3. We can further consider the problem of the estimation of bounds for solutions of certain diophantine inequalities in an algebraic number field (cf. [2]).

References

[1] W. M. Schmidt, *Small zeros of additive forms in many variables II*, Acta Math. 143 (1979), pp. 219–232.

[2] – *Diophantine inequalities for forms of odd degree*, Advances in Math. 38 (1980), pp. 128–151.

[3] C. L. Siegel, *Sums of m-th powers of algebraic integers*, Ann. of Math. 46 (1945), pp. 313–339.

[4] Wang Yuan, *Bounds for solutions of additive equations in an algebraic number field I*, Acta Arith. 48 (1987), pp.

INSTITUTE OF MATHEMATICS, ACADEMIA SINICA
Beijing, China

*Received on 30.5.1984
and in revised form on 7.1.1986*

(1429)

Reprinted from JOURNAL OF NUMBER THEORY
All Rights Reserved by Academic Press, New York and London
with permission from Elsevier

Vol. 29, No. 3, July 1988

Diophantine Inequalities for Forms in an Algebraic Number Field

WANG YUAN*

Institute of Mathematics, Academia Sinica, Beijing, China

Communicated by Hans Zassenhaus

Received March 26, 1986; revised March 30, 1987

1. INTRODUCTION

Given a vector with complex coefficients $\mathbf{x} = (x_1, ..., x_s)$ put

$$|\mathbf{x}| = \max |x_i|$$

and given a form (i.e., a homogeneous polynomial) F let

$$|F|$$

be the maximum absolute value of its coefficients. With every form F of degree k there is associated a form

$$\hat{F}(\mathbf{x}_1, ..., \mathbf{x}_k)$$

which is linear in each vector $\mathbf{x}_i$ $(1 \leqslant i \leqslant k)$ and symmetric in the k vectors $\mathbf{x}_1, ..., \mathbf{x}_k$ and such that

$$F(\mathbf{x}) = \hat{F}(\mathbf{x}, ..., \mathbf{x}).$$

Schmidt in 1980 [5] proved

THEOREM A (Schmidt).   *Given $h \geqslant 1$, $m \geqslant 1$, and odd numbers $k_1, ..., k_h$, and given a positive number E, however large, there is a constant*

$$c_1 = c_1(k_1, ..., k_h; m, E)$$

as follows. If $M \geqslant 1$ is real and if $F_1, ..., F_h$ are forms with real coefficients of

* Supported by National Science Foundation Grant MCS-8108814(A04).

respective degrees $k_1, ..., k_h$ in $\mathbf{x} = (x_1, ..., x_s)$ where $s \geqslant c_1$, then there are m linearly independent points $\mathbf{x}(1), ..., \mathbf{x}(m)$ in $\mathbb{Z}^s$ with

$$|\mathbf{x}(i)| \leqslant M, \qquad 1 \leqslant i \leqslant m$$

and

$$|\hat{F}(\mathbf{x}(i_1), ..., \mathbf{x}(i_{k_j}))| \ll M^{-E} |F_j|, \qquad 1 \leqslant j \leqslant h, \quad 1 \leqslant i_1, ..., i_{k_j} \leqslant m.$$

If the coefficients of all forms are rational integers, he derives from Theorem A the following.

THEOREM B (Schmidt). *Given $h \geqslant 1$, $m \geqslant 1$, and odd numbers $k_1, ..., k_h$, and given $\varepsilon > 0$, however small, there is a constant $c_2 = c_2(k_1, ..., k_h; m, \varepsilon)$ such that if $G_1, ..., G_h$ are forms of respective degrees $k_1, ..., k_h$ with integer coefficients in $\mathbf{x} = (x_1, ..., x_s)$, where $s \geqslant c_2$, then $G_1, ..., G_h$ vanish on an m-dimensional subspace which is spanned by integer points $\mathbf{x}(1), ..., \mathbf{x}(m)$ having*

$$|\mathbf{x}(i)| \ll G^\varepsilon, \qquad 1 \leqslant i \leqslant m,$$

where $G = \max(1, |G_1|, ..., |G_h|)$.

Remark 1. We will show that the conclusion of Theorem B is still true with suitable definition of G if $G_1, ..., G_h$ are forms of odd degrees with coefficients in integers in an algebraic number field K of degree n, provided only hat $s \geqslant c_3 (k_1, ..., k_h; n, m, \varepsilon)$.

In fact, if $\omega_1, ..., \omega_n$ is an integral basis of K, then G_i $(1 \leqslant i \leqslant h)$ can be written as

$$G_i = G_{i1}\omega_1 + \cdots + G_{in}\omega_n, \qquad 1 \leqslant i \leqslant h,$$

where $G_{ij}(1 \leqslant i \leqslant h, 1 \leqslant j \leqslant n)$ are forms with rational integer coefficients. Therefore we may consider the system of forms G_{ij} $(1 \leqslant i \leqslant h, 1 \leqslant j \leqslant n)$ instead of G_i $(1 \leqslant i \leqslant h)$, and the assertion follows.

The first step of the proof of Theorem A is to prove a result on additive equations, namely

PROPOSITION A (Schmidt [4]). *Suppose $\varepsilon > 0$ and suppose*

$$D(\mathbf{x}) = d_1 x_1^k + \cdots + d_s x_s^k$$

is an additive form of odd degree k with $d_i \in Z$ $(1 \leqslant i \leqslant s)$ and with $s \geqslant c_4(k, \varepsilon)$. Then there is a nonzero point $\mathbf{x}$ in $\mathbb{Z}^s$ with

$$D(\mathbf{x}) = 0 \qquad and \qquad |\mathbf{x}| \leqslant \max(1, |D|^\varepsilon).$$

326   WANG YUAN

The next step is to prove a special case of Theorem A, namely when there is only one additive form. And the final step is to prove the theorem by an inductive argument.

To consider the problem of diophatine inequalities for forms of arbitrary degrees we shall need a proposition for additive equations of arbitrary degree. If the solutions of the equations belong to a totally complex algebraic number field (i.e., a field with no isomorphic-embedding into the reals), such a result can be proved by the combination of the methods of Schmidt [4] and Siegel [6]. (See Proposition 1.)

Let K be an algebraic number field of degree n over rationals. $K^{(1)}, ..., K^{(r_1)}$ are real conjugates and $K^{(r_1+1)}, ..., K^{(r_1+2r_2)}$ are the complex conjugates of K. Here $r_1 + 2r_2 = n$. The complex conjugates are so arranged that, for $r_1 + 1 \leqslant q \leqslant r_1 + r_2$, the fields $K^{(q)}$ and $K^{(q+r_2)}$ are obtained from one another by interchanging $i(= \sqrt{-1})$ and $-i$. If $r_1 = 0$, K is said to be a totally complex algebraic number field. The ring of integers of K is denoted by J.

Elements of K are denoted by Greek letters. For $\gamma \in K$, we denote by $\gamma^{(i)}(1 \leqslant i \leqslant n)$ the conjugates of γ and $N(\gamma) = \prod_{i=1}^{n} \gamma^{(i)}$ the norm of γ. Let γ_j $(1 \leqslant j \leqslant n)$ be numbers of K and x_j $(1 \leqslant j \leqslant n)$ be real numbers. We set $\xi = \sum_{j=1}^{n} x_j \gamma_j$ and define $\xi^{(i)} = \sum_{j=1}^{n} x_j \gamma_j^{(i)}$ $(1 \leqslant i \leqslant n)$. So the definition of $\xi^{(i)}$'s is extended to the numbers not necessarily elements of K. We use the notations

$$\|\xi\| = \max_{1 \leqslant i \leqslant n} |\xi^{(i)}|$$

and

$$\|\lambda\| = \max_{1 \leqslant j \leqslant s} \|\lambda_j\| = \max_{i,j} |\lambda_j^{(i)}|$$

for a vector $\lambda = (\lambda_1, ..., \lambda_s)$.

THEOREM 1.   *Let K be a totally complex algebraic number field of degree $2r$. Given positive integers h, m, and $k_1, ..., k_h$, and given a positive number E, however large, there is a constant*

$$c_5 = c_5(k_1, ..., k_h; r, m, E)$$

as follows. If $M \geqslant 1$ is real and $F_1, ..., F_h$ are forms with complex coefficients of respective degrees $k_1, ..., k_h$ in $\lambda = (\lambda_1, ..., \lambda_s)$ where $s \geqslant c_5$, then there are m linearly independent points $\lambda(1), ..., \lambda(m)$ in J^s with

$$\|\lambda(i)\| \leqslant M, \qquad 1 \leqslant i \leqslant m,$$

and

$$|\hat{F}(\lambda(i_1), ..., \lambda(i_{k_j}))| \ll M^{-E} |F_j|, \qquad 1 \leqslant j \leqslant h, \quad 1 \leqslant i_1, ..., i_{k_j} \leqslant m,$$

here and below the constant implicit in $\ll$ or 0 may depend on $k_1, ..., k_h$, K, m, E, ε but not on M, $F_1, ..., F_h$, and G.

In particular, it follows that

$$|F_j(\lambda(i))| \ll M^{-E}|F_j|, \qquad 1 \leqslant j \leqslant h, \quad 1 \leqslant i \leqslant m.$$

Suppose now that $G_1, ..., G_h$ are forms of respective degrees $k_1, ..., k_h$ with coefficients in J. Let

$$\|G_i\|$$

denote the maximum absolute value of its coefficients and their conjugates. Further let

$$G = \max(1, \|G_1\|, ..., \|G_h\|).$$

Suppose that $s \geqslant c_5(k_1, ..., k_h; r, m, 4rk_1 \cdots k_h \varepsilon^{-1}) = c_6(k_1, ..., k_h; r, m, \varepsilon)$, say, where $0 < \varepsilon < 1$. Apply Theorem 1 with $M = M_0 G^\varepsilon$, where $M_0 = M_0(k_1, ..., k_h; K, m, \varepsilon)$ is to be chosen in a moment. We obtain m linearly independent points $\lambda(1), ..., \lambda(m)$ in J^s with

$$\|\lambda(i)\| \leqslant M_0 G^\varepsilon, \qquad 1 \leqslant i \leqslant m$$

and

$$|\hat{G}_j(\lambda(i_1), ..., \lambda(i_{k_j}))| \ll GM^{-4rk_1 \cdots k_h \varepsilon^{-1}} \ll M_0^{-4rk_1 \cdots k_h} G^{-4rk_1 \cdots k_h + 1}. \tag{1}$$

On the other hand, $k_j! \hat{G}_j(\lambda(i_1), ..., \lambda(i_{k_j}))$ is an integer in K and

$$\|k_j! \hat{G}_j(\lambda(i_1), ..., \lambda(i_{k_j}))\| \ll GM^{k_j} \ll M_0^{k_j} G^{1 + k_j}.$$

If $\hat{G}_j(\lambda(i_1), ..., \lambda(i_{k_j})) \neq 0$, then

$$|N(k_j! \, \hat{G}_j(\lambda(i_1), ..., \lambda(i_{k_j})))| \geqslant 1,$$

and therefore

$$|\hat{G}_j(\lambda(i_1), ..., \lambda(i_{k_j}))| \gg (M_0^{k_j} G^{1 + k_j})^{-2r + 1} \gg M_0^{-2rk_j} G^{-4rk_j + 1}$$

which leads to a contradiction with (1) if M_0 is sufficiently large. Therefore

$$\hat{G}_j(\lambda(i_1), ..., \lambda(i_{k_j})) = 0,$$

and we have the following

THEOREM 2. *Given positive integers $k_1, ..., k_h$ and m, and given ε, however small, there is a constant $c_6 = c_6(k_1, ..., k_h; r, m, \varepsilon)$ such that if $G_1, ..., G_k$ are*

328   WANG YUAN

forms of respective degrees $k_1, ..., k_h$ in $\lambda = (\lambda_1, ..., \lambda_s)$ with coefficients in J, where $s \geqslant c_6$, then $G_1, ..., G_h$ vanish on an m-dimensional subspace which is spanned by m points $\lambda(1), ..., \lambda(m)$ in J^s having

$$\|\lambda(i)\| \ll G^\varepsilon, \qquad 1 \leqslant i \leqslant m.$$

Remark 2. Theorem 2 gives an improvement for a result due to Peck [3]. He first established by the combination of the methods of Brauer [1] and Siegel [6] the existence of $\lambda(i)$ $(1 \leqslant i \leqslant m)$ with $c_6'(k_1, ..., k_h; K, m)$ instead of c_6, but the estimation of $\|\lambda(i)\|$'s was not considered. As we have shown in Remark 1, a similar result can be obtained if the coefficients of $G_1, ..., G_h$ are integers in any given algebraic number field K_1.

The proof of Theorem 1 relies on the Schmidt's method [5]. In this paper, we shall prove a special case of Theorem 1, namely when there is only one additive form, and give only a sketch on the inductive argument from a single additive form to a system of forms, since it is easily treated by Schmidt's method.

2. ADDITIVE FORMS

We suppose that K is a given totally complex algebraic number field of degree $2r$ throughout this paper.

PROPOSITION 1. *Suppose $\varepsilon > 0$, $k \geqslant 1$, and $s \geqslant c_7(k, r, \varepsilon)$. Then given an additive form*

$$A(\lambda) = \alpha_1 \lambda_1^k + \cdots + \alpha_s \lambda_s^k$$

with coefficients in the integer of K, there is a nonzero point $\lambda \in J^s$ with

$$A(\lambda) = 0 \qquad and \qquad \|\lambda\| \ll \max(1, \|A\|^\varepsilon).$$

This proposition was proved in [8]. (See Remark 2 in [8].) Now we shall prove the following proposition which is a very special case of Theorem 1.

PROPOSITION 2. *Given $k \geqslant 1$ and given E, however large, there is a constant $c_8 = c_8(k, r, E)$ with the following property. Let $A(\lambda)$ be an additive form of degree k with complex coefficients in at least c_8 variables. Then for real $M \geqslant 1$ there is a nonzero point $\lambda \in J^s$ with*

$$\|\lambda\| \leqslant M \qquad and \qquad |A(\lambda)| \ll M^{-E} |A|.$$

We begin with simple reductions. Suppose we can prove the conclusion for $s \geq c_8(k, r, E)$ and $M \geq c_9(k, K, E)$. Then in the case of $1 \leq M < c_9$, we put

$$\lambda_1 = 1 \qquad \text{and} \qquad \lambda_2 = \cdots = \lambda_s = 0,$$

and therefore

$$\|\lambda\| = 1 \qquad \text{and} \qquad |A(\lambda)| \leq |A| = c_9^E c_9^{-E} |A| \ll c_9^{-E} |A|.$$

Hence it will suffice to prove Proposition 2 for large values of M, say for $M \geq c_9(k, K, E)$.

Choose $\delta > 0$ so small that

$$2\delta + \frac{E + 2\delta}{E + \frac{1}{8}} < 1.$$

Proposition 2 is obvious if there is an α_i with $|\alpha_i| < M^{-E} |A|$. In fact, we may take $\lambda_i = 1$, $\lambda_j = 0$ $(j \neq i)$ in this case. So we may suppose that

$$M^{-E} |A| \leq |\alpha_i| \leq |A|, \qquad 1 \leq i \leq s.$$

Cover the interval $[-E, 0]$ by a finite number of intervals $\{I\}$ of length δ. One of these intervals I will be such that at least $[s/([E/\delta] + 1)]$ of the α_i are of the type $|\alpha_i| = M^{e_i} |A|$ with $e_i \in I$. We may suppose without loss of generality that this holds for all i. Put

$$L = M^\delta |A| = M^\delta \max |\alpha_i|$$

and choose natural numbers $q_1, ..., q_s$, each as large as possible, with

$$|\alpha_i| q_i^k \leq L, \qquad 1 \leq i \leq s.$$

Since $L/|\alpha_i| > M^\delta$, we have

$$\tfrac{1}{2} L \leq |\alpha_i| q_i^k \leq L \tag{2}$$

if M is sufficiently large. Further, $q_i^k \leq L/|\alpha_i| = M^\delta |A|/|\alpha_i| \leq M^{2\delta}$. Suppose Proposition 2 is true for

$$B(\mu) = \alpha_1 q_1^k \mu_1^k + \cdots + \alpha_s q_s^k \mu_s^k$$

with

$$M_0 = M^{(E + 2\delta)/(E + 1/8)}, \qquad E_0 = E + \tfrac{1}{8}$$

330   WANG YUAN

in place of M, E, i.e., there is a nonzero point $\mu \in J^s$ with

$$\|\mu\| \leqslant M_0 \qquad \text{and} \qquad |B(\mu)| \ll M_0^{-E_0} |B|.$$

Let $\lambda_i = q_i \mu_i$ $(1 \leqslant i \leqslant s)$. Since $|B| \leqslant L = M^\delta |A|$, we have

$$|A(\lambda)| \ll M^{-((E+2\delta)/(E+1/8))(E+1/8)} M^\delta |A| \ll M^{-E} |A|$$

and

$$\|\lambda\| \ll (\max q_i) \|\mu\| \ll M^{2\delta + ((E+2\delta)/(E+1/8))} < M$$

if M is large, ie., Proposition 2 is true for $A(\lambda)$.

What is special about B is that by (2) each of its coefficients is at least $\frac{1}{2}|B|$ in absolute value. By what we said it is clear that if Proposition 2 is true with $E + \frac{1}{8}$ in place of E for form A with

$$\tfrac{1}{2} |A| \leqslant |\alpha_i| \leqslant |A|, \qquad 1 \leqslant i \leqslant s,$$

then it is true with E for general forms. By homogeneity we may replace the above relation by

$$|A| = 1 \qquad \text{and} \qquad \tfrac{1}{2} \leqslant |\alpha_i| \leqslant 1, \qquad 1 \leqslant i \leqslant s. \tag{3}$$

It now will suffice to prove the following statements.

(i)   The conclusion of Proposition 2 is true for $0 \leqslant E \leqslant \frac{1}{2}$ for forms A with (3), provided only that $s \geqslant c_{10}(k, r, E)$.

(ii)   The conclusion of Proposition 2 is true for E for forms A with (3), provided only that $s \geqslant c_{10}(k, r, E)$ and that Proposition 2 is true for $E - \frac{1}{4}$ for general additive forms.

3. ANALYTIC METHOD

Let $\omega_1, \ldots, \omega_{2r}$ be a basis of J with $\omega_i^{(q+r)} = \bar{\omega}_i^{(q)}$ $(1 \leqslant i \leqslant 2r, 1 \leqslant q \leqslant r)$, δ the different and D the absolute value of the discriminant of K. We choose a basis $\rho_1, \ldots, \rho_{2r}$ of δ^{-1} such that

$$T(\rho_i \omega_j) = \begin{cases} 1, & \text{if} \quad i = j \\ 0, & \text{if} \quad i \neq j, \end{cases}$$

where $T(\gamma) = \sum_{i=1}^{2r} \gamma^{(i)}$. Given a set of complex numbers $c_1, \ldots, c_{2r}$ with $c_{q+r} = \bar{c}_q$ $(1 \leqslant q \leqslant r)$, there is a unique set of real numbers $y_1, \ldots, y_{2r}$ such that

$$c_i = y_1 \omega_1^{(i)} + \cdots + y_{2r} \omega_{2r}^{(i)}, \qquad 1 \leqslant i \leqslant 2r.$$

Suppose that $\alpha_1, ..., \alpha_s$ is a set of complex numbers satisfying (3). We define for each j,

$$\alpha_j^{(q)} = \alpha_j, \qquad \alpha_j^{(q+r)} = \bar{\alpha}_j, \qquad 1 \leqslant q \leqslant r.$$

Then α_j has a unique representation

$$\alpha_j^{(i)} = z_1 \omega_1^{(i)} + \cdots + z_{2r} \omega_{2r}^{(i)} \ (1 \leqslant i \leqslant 2r), \qquad z_l \in \mathbb{R} \ (1 \leqslant l \leqslant 2r).$$

Set $\xi = x_1 \rho_1 + \cdots + x_{2r} \rho_{2r}$, $\eta = y_1 \omega_1 + \cdots + y_{2r} \omega_{2r}$, $dx = dx_1 \cdots dx_{2r}$, and $dy = dy_1 \cdots dy_{2r}$. We denote by $P(Q)$ the set of $(y_1, ..., y_{2r})$ satisfying

$$\|\eta\| \leqslant Q$$

and $\sum_{\lambda \in P(Q)}$ a sum where λ runs over all integers in K with $\|\lambda\| \leqslant Q$. Set

$$S_i(\xi) = \sum_{\lambda \in P(M)} E(\alpha_i \xi \lambda^k), \qquad I_i(\xi) = \int_{P(M)} E(\alpha_i \xi \eta^k)\, dy, \qquad 1 \leqslant i \leqslant s,$$

$$S(\xi) = \prod_{i=1}^{s} S_i(\xi) \qquad \text{and} \qquad I(\xi) = \prod_{i=1}^{s} I_i(\xi),$$

where $E(\gamma) = e(T(\gamma))$ with $e(x) = e^{2\pi i x}$. We use the notations

$$K(x) = \left(\frac{\sin \pi x M^{-E}}{\pi x} \right)^2 \qquad \text{and} \qquad K(x) = \prod_{j=1}^{2r} K(x_j),$$

where x is a real variable and $\mathbf{x} = (x_1, ..., x_{2r})$ is a real vector variable.

We wish to estimate the number Z of solutions of

$$|A(\lambda) \ll M^{-E} \tag{4}$$

in points $\lambda \in J^s$ subject to

$$\|\lambda\| \leqslant M. \tag{5}$$

Our plan is to show that either $Z > 1$ or (i) and (ii) holds. Therefore (i) and (ii) hold in every case. Put

$$\theta = (rk + (r+1)E + 3)^{-1}(4k + 5E)^{-1}, \tag{6}$$

$$m = \begin{cases} 1 & \text{in the case (i),} \\ c_8(k, r, E - \tfrac{1}{4}) & \text{in the case (ii),} \end{cases} \tag{7}$$

$$n = c_7(k, r, \theta), \tag{8}$$

$$h = mn, \tag{9}$$

WANG YUAN

and choose $\eta > 0$ so small that

$$41 r G h \eta < 1, \tag{10}$$

where $G = 2^{k-1}$. Now let s be so large that

$$(2r(r+1)(k+E) + 2r + 2)(s - h + 1)^{-1} \leqslant \eta. \tag{11}$$

In what follows, the constants in $\ll$ or 0 may depend on s (in addition to k, K, E). But observe that if Proposition 2 is true for a particular value of s, then it also holds for larger values of s. We assume M to be large, i.e., $M \geqslant c_9$.

LEMMA 1. *We have for real Q*

$$M^E \int_{-\infty}^{\infty} e(\beta Q) \, K(\beta) \, d\beta = \begin{cases} 1 - M^E |Q|, & \text{if} \quad |Q| < M^{-E}, \\ 0, & \text{if} \quad |Q| \geqslant M^{-E}. \end{cases} \tag{12}$$

Proof. Make substitutions $\beta = M^E \alpha$, and $M^E Q$ for Q in Lemma 50 in [2].

Let E_n denote the n-dimensional Euclidean space. Expand $\sum_{j=1}^{s} \alpha_j \lambda_j^k$ as

$$\sum_{j=1}^{s} \alpha_j^{(i)} \lambda_j^{(i)k} = A_1 \omega_1^{(i)} + \cdots + A_{2r} \omega_{2r}^{(i)}, \qquad 1 \leqslant i \leqslant 2r. \tag{13}$$

Then by Lemma 1, we have

$$M^{2rE} \int_{E_{2r}} S(\xi) \, K(\mathbf{x}) \, dx$$

$$= M^{2rE} \sum_{\lambda_1 \in P(M)} \cdots \sum_{\lambda_s \in P(M)} \prod_{j=1}^{2r} \int_{-\infty}^{\infty} e(x_j A_j) \, K(x_j) \, dx_j$$

$$= \sum_{\substack{\lambda_1 \in P(M) \\ |A_i| < M^{-E}}} \cdots \sum_{\lambda_s \in P(M)} (1 - |A_1| \, M^E) \cdots (1 - |A_{2r}| \, M^E). \tag{14}$$

Let

$$\eta_j = \gamma_{j1} \omega_1 + \cdots + y_{j,2r} \omega_{2r}, \qquad dY_j = dy_{j1} \cdots dy_{j,2r}, \qquad 1 \leqslant j \leqslant s,$$

and

$$\sum_{j=1}^{s} \alpha_j^{(i)} \eta_j^{(i)k} = B_1 \omega_1^{(i)} + \cdots + B_{2r} \omega_{2r}^{(i)}, \qquad 1 \leqslant i \leqslant 2r.$$

Then by Lemma 1, we have

$$M^{2rE} \int_{E_{2r}} I(\xi)\, K(\mathbf{x})\, dx$$

$$= \int_{\substack{P(M) \\ |B_i| < M^{-E}}} \cdots \int_{P(M)} (1 - |B_1|\, M^E) \cdots (1 - |B_{2r}|\, M^E)\, dY_1 \cdots dY_s. \quad (15)$$

If $|A_i| < M^{-E}$, $1 \leqslant i \leqslant 2r$, then it follows by (13) that $A(\lambda) \ll M^{-E}$, and thus the right-hand side of (14) gives a lower bound for Z. The general idea now will be to show that the right-hand side of (15) is large and to show that the left-hand sides of (14) and (15) differ little.

LEMMA 2. *The right-hand side of (15) is*

$$\gg M^{2r(s-k-E)}.$$

Proof. Let

$$\eta_j^{(q)} = u_{jq}^{1/k}\, e^{i\varphi_{jq}/k}, \qquad 1 \leqslant j \leqslant s, \quad 1 \leqslant q \leqslant r.$$

The Jacobian of y_{jl} $(1 \leqslant l \leqslant 2r)$ with respect to u_{jq}, φ_{jq} $(1 \leqslant q \leqslant r)$ is

$$2^r k^{-2} D^{-1/2} \prod_{q=1}^{r} u_{jq}^{2/k-1},$$

and so the right-hand side of (15) is equal to

$$2^{rs} k^{-2rs} D^{-s/2} \Phi, \qquad (16)$$

where

$$\Phi = \int_{(U)} \prod_{i=1}^{2r} (1 - M^E\, |B_i|) \prod_{j=1}^{s} \prod_{q=1}^{r} u_{jq}^{2/k-1}\, du\, d\varphi \qquad (17)$$

in which $du = \prod_{j=1}^{s} \prod_{q=1}^{r} du_{jq}$, $d\varphi = \prod_{j=1}^{s} \prod_{q=1}^{r} d\varphi_{jq}$, and (U) denotes the domain $0 \leqslant u_{jq} \leqslant M^k$, $-\pi \leqslant \varphi_{jq} \leqslant \pi$ $(1 \leqslant j \leqslant s, \; 1 \leqslant q \leqslant r)$, $|B_i| < M^{-E}$ $(1 \leqslant i \leqslant 2r)$.

Let

$$u_{jq} = |\alpha_j|^{-1}\, M^{-E} v_{jq}, \qquad 1 \leqslant j \leqslant s, \quad 1 \leqslant q \leqslant r.$$

Then

$$\Phi = |\alpha_1 \cdots \alpha_s|^{-(2r)/k}\, M^{-(2rEs)/k} \Psi, \qquad (18)$$

WANG YUAN

where

$$\Psi = \int_{(V)} \prod_{i=1}^{2r} (1 - M^E |B_i|) \prod_{j=1}^{s} \prod_{q=1}^{r} v_{jq}^{2/k-1} \, dv \, d\varphi \qquad (19)$$

in which $dv = \prod_{j=1}^{s} \prod_{q=1}^{r} dv_{jq}$, (V) denotes the domain $0 \leqslant v_{jq} < |\alpha_j| M^{E+k}$, $-\pi \leqslant \varphi_{jq} \leqslant \pi (1 \leqslant j \leqslant s, \ 1 \leqslant q \leqslant r)$, $|B_i| < M^{-E} \ (1 \leqslant i \leqslant 2r)$.

Let

$$\theta_{jq} = \varphi_{jq} + \arg \alpha_j^{(q)} \ (1 \leqslant j \leqslant s, \ 1 \leqslant q \leqslant 2r) \quad \text{and} \quad M^E B_i = D_i (1 \leqslant i \leqslant 2r).$$

Then

$$\sum_{j=1}^{s} v_{jq} e^{i\theta_{jq}} = M^E \sum_{j=1}^{s} \alpha_j^{(q)} u_{jq} e^{i\varphi_{jq}} = M^E \sum_{j=1}^{s} \alpha_j^{(q)} \eta_j^{(q)k}$$

$$= D_1 \omega_1^{(q)} + \cdots + D_{2r} \omega_{2r}^{(q)}, \qquad 1 \leqslant q \leqslant 2r,$$

and

$$\Psi = \int_{(W)} \prod_{i=1}^{2r} (1 - |D_i|) \prod_{j=1}^{s} \prod_{q=1}^{r} v_{jq}^{2/k-1} \, dv \, d\theta,$$

where $d\theta = \prod_{j=1}^{s} \prod_{q=1}^{r} d\theta_{jq}$, and (W) denotes the domain $0 \leqslant v_{jq} \leqslant |\alpha_j| M^{E+k}$, $-\pi \leqslant \theta_{jq} \leqslant \pi \ (1 \leqslant j \leqslant s, \ 1 \leqslant q \leqslant r)$, $|D_i| < 1 \ (1 \leqslant i \leqslant 2r)$.

Since $\frac{1}{2} \leqslant |\alpha_j| \leqslant 1 \ (1 \leqslant j \leqslant s)$, the domain

$$\frac{1}{8} M^{E+k} \leqslant v_{1q} \leqslant \frac{1}{4} M^{E+k} \quad (1 \leqslant q \leqslant r),$$

$$\frac{M^{E+k}}{32s} \leqslant v_{jq} \leqslant \frac{M^{E+k}}{16s} \quad (2 \leqslant j \leqslant s-1, 1 \leqslant q \leqslant r), \qquad (w)$$

$$-\pi \leqslant \theta_{jq} \leqslant \pi \quad (1 \leqslant j \leqslant s-1, 1 \leqslant q \leqslant r),$$

$$|D_l| < \frac{1}{2} \quad (1 \leqslant l \leqslant 2r)$$

is contained in (W). The volume of (w) is

$$\gg M^{(E+k)r(s-2)}.$$

In fact, for any given v_{jq}, $\theta_{jq} (1 \leqslant j \leqslant s-1)$ in (w), we let $v_{1q} e^{i\theta_{1q}} + \cdots + v_{s-1,q} e^{i\theta_{s-1,q}} = V e^{i\theta}$. Then $M^{E+k}/16 \leqslant V \leqslant (5M^{E+k})/16$, and v_{sq} and θ_{sq} satisfy $1 \ll v_{sq} - V \ll 1$ and $\theta_{sq} - \theta \ll M^{-E-k}$. The assertion follows. The integrand of the integral Ψ in (w) is

$$\gg M^{(E+k)rs(2/k-1)}.$$

Hence

$$\Psi \gg M^{(E+k)r(s-2)+(E+k)rs(2/k-1)} \gg M^{2r(s-k-E)+2rEs/k},$$

and the lemma follows by (16)–(19).

LEMMA 3. *If $\|\xi\| \leqslant M^{9/10-k}$, then*

$$\sum_{\lambda \in P(M)} E(\xi\lambda^k) = \int_{P(M)} E(\xi\eta^k)\, dy + O(M^{2r-1/10}).$$

Proof. Expressing an integer λ in terms of the ω_i $(1 \leqslant i \leqslant 2r)$, we obtain

$$\lambda = g_1\omega_1 + \cdots + g_{2r}\omega_{2r},$$

where the g_i $(1 \leqslant i \leqslant 2r)$ are rational integers. Let $G(\lambda)$ denote the cube

$$(y_1, \ldots, y_{2r})\colon \eta = y_1\omega_1 + \cdots + y_{2r}\omega_{2r}, \qquad g_i \leqslant y_i < g_i + 1, \qquad 1 \leqslant i \leqslant 2r.$$

Then

$$\|\eta - \lambda\| \ll 1,$$

$$\|\xi\eta^k - \xi\lambda^k\| \ll \|\xi\|\, \|\eta - \lambda\|\, (\|\eta\|^{k-1} + \|\lambda\|^{k-1}) \ll M^{9/10-k+k-1} \ll M^{-1/10}$$

and so

$$E(\xi\lambda^k) = \int_{G(\lambda)} E(\xi\eta^k)\, dy + O(M^{-1/10}).$$

Since the volume of the area belonging to exactly one of $\bigcup_{\lambda \in P(M)} G(\lambda)$ and $P(M)$ is dominated by $O(M^{2r-1})$, we have

$$\sum_{\lambda \in P(M)} E(\xi\lambda^k) \sum_{\lambda \in P(M)} \int_{G(\lambda)} E(\xi\eta^k)\, dy + O(M^{2r-1/10})$$

$$= \int_{P(M)} E(\xi\eta^k)\, dy + O(M^{2r-1/10}),$$

and the lemma is proved.

LEMMA 4. *We have*

$$\int_{P(M)} E(\xi\eta^k)\, dy \ll \prod_{j=1}^{2r} \min(M, |\xi|^{(i)}|^{-1/k}).$$

See, e.g., Siegel [6], p. 335.

336   WANG YUAN

Lemma 5. *We have*

$$\int_{\|\xi\| \leq M^{9/10-k}} S(\xi)\, K(\mathbf{x})\, dx \gg M^{2r(s-k-2E)}. \tag{20}$$

Proof. By Lemma 4, we have for any q,

$$\int_{|\xi^{(q)}| > M^{9/10-k}} I(\xi)\, K(\mathbf{x})\, dx$$

$$\ll M^{-4rE} \int_{|\xi^{(q)}| > M^{9/10-k}} \prod_{j=1}^{2r} \min(M^s, |\xi^{(j)}|^{-s/k})\, dx.$$

Let $\xi^{(q)} = u_q e^{i\varphi_q}$ $(1 \leq q \leq r)$. The Jacobian of x_i $(1 \leq i \leq 2r)$ with respect to u_q, φ_q $(1 \leq q \leq r)$ is $2^r D^{1/2} \prod_{q=1}^{r} u_q$. Therefore the above expression is

$$\ll M^{-4rE} \left(\int_{M^{9/10-k}}^{\infty} \int_{-\pi}^{\pi} u^{-(2s)/k+1}\, du\, d\varphi \right)$$

$$\times \left(\int_{0}^{\infty} \int_{-\pi}^{\pi} \min(M^{2s}, w^{-(2s)/k})\, w\, dw\, d\theta \right)^{r-1}$$

$$\ll M^{-4rE + (9/10-k)(-(2s)/k+2) + 2(r-1)(s-k)}$$

$$\ll M^{2r(s-k-2E) - (9s)/(5k) + 9/5} \ll M^{2r(s-k-2E)-1}$$

if s is sufficiently large, i.e., $s > 2k$. Therefore it follows by Lemma 2 that

$$\int_{\|\xi\| \leq M^{9/10-k}} I(\xi)\, K(\mathbf{x})\, dx \gg M^{2r(s-k-2E)}. \tag{21}$$

It remains to compare the integral (21) with the one in (20). By Lemma 3 we have for $\|\xi\| \leq M^{9/10-k}$,

$$|S(\xi) - I(\xi)| = \left| \prod_{i=1}^{s} (I_i(\xi) + O(M^{2r-1/10})) - I(\xi) \right|$$

$$\ll M^{2r-1/10} \prod_{i=1}^{2r} \min(M^{s-1}, |\xi^{(i)}|^{-(s-1)/k}) + M^{s(2r-1/10)},$$

and there fore the left-hand sides of (20), (21) have a difference

$$\ll M^{-4rE + 2r - 1/10} \left(\int_{0}^{M^{9/10-k}} \int_{-\pi}^{\pi} \min(M^{2(s-1)}, u^{-(2(s-1))/k})\, u\, du\, d\varphi \right)^{r}$$

$$+ M^{-4rE} \int_{\|\xi\| \leq M^{9/10-k}} M^{s(2r-1/10)}\, dx$$

$$\ll M^{-4rE+2r-1/10}\left(\int_0^{M^{-k}} M^{2s-2}u\,du + \int_{M^{-k}}^{\infty} u^{-(2(s-1))/k+1}\,du\right)^r$$

$$+\ M^{2r(s-k-2E)-s/10+(9r)/5}$$

$$\ll M^{2r(s-k-2)-1/10}$$

if s is large, i.e., $s \geqslant 20r$. The lemma follows.

LEMMA 6. *We have for any q*

$$\int_{|\xi^{(q)}| > M^{rk+(r+1)E+1}} S(\xi)\,K(\mathbf{x})\,dx \ll M^{2r(s-k-2E)-1}.$$

Proof. It follows from $|\xi^{(q)}| > M^{rk+(r+1)E+1}$ that at least one of the x_l in the expression $\xi = x_1\rho_1 + \cdots + x_{2r}\rho_{2r}$ satisfies

$$|x_l| \gg M^{rk+(r+1)E+1}.$$

Since

$$|K(x)| \leqslant \min(M^{-E}, \pi^{-1}|x|^{-1})^2,$$

we have

$$\int_{|\xi^{(q)}| > M^{rk+(r+1)E+1}} S(\xi)\,K(\mathbf{x})\,dx$$

$$\ll M^{2rs}\left(\int_{M^{rk+(r+1)E+1}}^{\infty} \frac{dx}{x^2}\right)^2\left(\int_0^{\infty} \min(M^{-2E}, x^{-2})\,dx\right)^{2r-2}$$

$$\ll M^{2rs-2rk-2(r+1)E-2-2(r-1)E}$$

$$\ll M^{2r(s-k-2E)-1}.$$

The lemma is proved.

For $1 \leqslant q \leqslant r$, let (U_q) denote the domain

$$\|\xi\| \leqslant M^{rk+(r+1)E+1}, \qquad |\xi^{(q)}| > M^{9/10-k}.$$

Suppose for a moment that the r integrals satisfy

$$\int_{(U_q)} |S(\xi)|\,dx \leqslant M^{2r(s-k)-1}, \qquad 1 \leqslant q \leqslant r. \tag{22}$$

338   WANG YUAN

Then it follows from Lemmas 5 and 6 that

$$\int_{E_{2r}} S(\xi)\, K(\mathbf{x})\, dx = \int_{\|\xi\| \leqslant M^{9/10-k}} S(\xi)\, K(\mathbf{x})\, dx + O(M^{2r(s-k-2E)-1})$$

$$\gg M^{2r(s-k-2E)}$$

and in view of (14) it follows that

$$Z \gg M^{2r(s-k-E)} > 1.$$

4. The Proof of Proposition 2

We may thus suppose (22) is false, i.e., there is a (U_q) such that

$$\int_{(U_q)} |S(\xi)|\, dx > M^{2r(s-k)-1}, \qquad 1 \leqslant q \leqslant r. \tag{23}$$

Let $\sigma(\mathbf{i})$ be a transformation such that

$$\gamma^{(j)} \to \gamma^{(i_j)}, \qquad 1 \leqslant j \leqslant r,$$

where $\mathbf{i} = (i_1, \ldots, i_r)$ is a permutation of $(1, \ldots, r)$. Since $S(\xi)$ is invariant under the group $\{\sigma(\mathbf{i})\}$, (23) is true for any q. In particular, it holds for $q = 1$, i.e.,

$$\int_{(U_1)} |S(\xi)|\, dx > M^{2r(s-k)-1}. \tag{24}$$

There is a $\xi(= \xi^{(1)})$ satisfying

$$\|\xi\| \leqslant M^{rk+(r+1)E+1}, \qquad |\xi| > M^{9/10-k} \tag{25}$$

and

$$|S(\xi)| \geqslant M^{2r(s-(r+1)(E+k)-1)-2}. \tag{26}$$

Otherwise, we have

$$\int_{(U_1)} |S(\xi)|\, dx \ll M^{2r(s-(r+1)(E+k)-1)-2} \int_{\|\xi\| \leqslant M^{rk+(r+1)E+1}} dx$$

$$\ll M^{2r(s-k)-2}$$

which leads to a contradiction with (24) if M is large.

We may suppose without loss of generality that

$$|S_1(\xi)| \geqslant \cdots \geqslant |S_s(\xi)|.$$

The left-hand side of (26) is

$$\ll |S_h(\xi)|^{s-h+1} M^{2r(h-1)},$$

and thus we have by (11)

$$|S_i(\xi)| \geqslant M^{2r-(2r(r+1)(E+k)+2r+2)/(s-h+1)} \geqslant M^{2r-\eta}, \qquad 1 \leqslant i \leqslant h. \quad (27)$$

LEMMA 7. *Suppose that* $\eta > 0$, $M \geqslant c_{11}(k, K, \eta)$, *and* $C \geqslant M^{2r-1/G+\eta}$, *where* $G = 2^{k-1}$. *If* $|\sum_{\lambda \in P(M)} E(\zeta\lambda^k)| \geqslant C$, *then there are two integers* α, β *of* K *such that*

$$\|\alpha\zeta - \beta\| \ll \left(\frac{M^{2r}}{C}\right)^G M^{\eta-k}$$

and

$$0 < \|\alpha\| \ll \left(\frac{M^{2r}}{C}\right)^G M^{\eta}.$$

See, e.g., [7].

In view of (10) and (27) we can apply Lemma 7 with $C = M^{2r-\eta}$ and $\zeta = \xi\alpha_i$ $(1 \leqslant i \leqslant h)$. We obtain integers $\sigma_1, ..., \sigma_h$ and $\beta_1, ..., \beta_h$ of K with

$$0 < \|\sigma_i\| \ll M^{2G\eta} \qquad \text{and} \qquad \|\alpha_i\sigma_i\xi - \beta_i\| \ll M^{-k+2G\eta}, \qquad 1 \leqslant i \leqslant h.$$

Setting $\sigma = \sigma_1 \cdots \sigma_h$ and $(\sigma\beta_i)/\sigma_i = \tau_i$ $(1 \leqslant i \leqslant h)$ we have

$$0 < \|\sigma\| \ll M^{2Gh\eta} \qquad \text{and} \qquad \|\alpha_i\sigma\xi - \tau_i\| \ll M^{-k+2Gh\eta}, \qquad 1 \leqslant i \leqslant h. \quad (28)$$

Set

$$\alpha_i = \frac{\tau_i}{\sigma\xi} + \gamma_i, \qquad 1 \leqslant i \leqslant h.$$

Then

$$|\sigma| = \frac{|N(\sigma)|}{|\sigma^{(2)} \cdots \sigma^{(2r)}|} \gg M^{-2(2r-1)Gh\eta},$$

and by (10), (25), (28) we have

$$|\gamma_i| \ll |\sigma\xi|^{-1} M^{-k+2Gh\eta} \ll M^{k-9/10+4rGh\eta-k} < M^{-4/5} \quad (29)$$

340   WANG YUAN

if M is large. So with $\lambda = (\lambda_1, ..., \lambda_s) = (\mu_1, ..., \mu_h, 0, ..., 0) = (\mu, 0)$ we have

$$A(\lambda) = |\sigma\xi|^{-1} B(\mu) + C(\mu),$$

where

$$B(\mu) = \tau_1 \mu_1^k + \cdots + \tau_h \mu_h^k$$

is a form with coefficients in J and

$$C(\mu) = \gamma_1 \mu_1^k + \cdots + \gamma_h \mu_h^k.$$

From (25), (28) we obtain

$$\|\tau_i\| \ll M^{rk + (r+1)E + 1 + 2Gh\eta}, \qquad 1 \leqslant i \leqslant h,$$

and therefore

$$\|B\| < M^{rk + (r+1)E + 2} \tag{30}$$

by (10), if M is large.

(i)   First, suppose that $0 \leqslant E \leqslant \frac{1}{2}$. In view of Proposition 1, and since

$$h = mn = n = c_7(k, r, \theta)$$

by (7), (8), (9), there is a nonzero vector $\mu \in J^h$ with

$$B(\mu) = 0 \qquad \text{and} \qquad \|\mu\| \ll \max(1, \|B\|^\theta) \ll M^{(rk + (r+1)E + 2)\theta} \ll M^{1/(4k)}$$

by (6) and (30). With $\lambda = (\mu, 0)$ we have for large M, $\|\lambda\| \leqslant M$ and

$$|A(\lambda)| = |C(\mu)| \leqslant h\,|C|\,\|\mu\|^k \ll M^{-4/5} M^{1/4} \leqslant M^{-E}$$

by (29).

(ii)   Next, suppose that Proposition 2 holds for $E - \frac{1}{4}$ in place of E. We have $h = mn$ by (9). Write

$$\mu = (\mu_1, ..., \mu_m), \qquad B(\mu) = B_1(\mu_1) + \cdots + B_m(\mu_m),$$

$$C(\mu) = C_1(\mu_1) + \cdots + C_m(\mu_m),$$

where each μ_i has n components. Since $n = c_7(k, r, \theta)$ by (8), there are nonzero vectors $\mu_1, ..., \mu_m$ in J^n with

$$B_i(\mu_i) = 0 \qquad \text{and} \qquad \|\mu_i\| \ll \max(1, \|B\|^\theta) \ll M^{(rk + (r+1)E + 2)\theta}$$

$$\ll \min(M^{1/(4k)}, M^{1/(5E)}), \qquad 1 \leqslant i \leqslant m. \tag{31}$$

Setting $\lambda = v_1 \mu_1 + \cdots + v_m \mu_m$ we have

$$A(v_1 \mu_1 + \cdots + v_m \mu_m) = C_1(\mu_1) v_1^\kappa + \cdots + C_m(\mu_m) v_m^k = D(v),$$

say. Since $m = c_8(k, r, E - \frac{1}{4})$ and

$$|D| \leqslant \max |C_i(\mu_i)| \ll M^{-4/5} M^{1/4} \ll M^{-1/2},$$

by (7), (29), (31), we have a point $v \in J^m$ such that

$$\|v\| \leqslant M^{1 - 1/(4E)}$$

and

$$|D(v)| \ll M^{-(1 - 1/(4E))(E - 1/4)} |D| \ll M^{-E + 1/2 - 1/(16E) - 1/2} < M^{-E}$$

if M is large. Consequently, in view of $\lambda = v_1 \mu_1 + \cdots + v_m \mu_m$ we have

$$|A(\lambda)| = |D(v)| < M^{-E}$$

and by (31)

$$\|\lambda\| \leqslant \|v\| \, \|\mu\| \ll M^{1 - 1/(4E) + 1/(5E)} < M$$

if M is large, i.e., Proposition 2 holds for E for forms with (3), and thus the proposition is proved.

5. The Proof of Theorem 1

Theorem 1 will be proved by induction on the values of

$$k = \max(k_1, \ldots, k_h).$$

The case $k = 1$ is proved by the box principle. We may suppose that $|F_1| = \cdots = |F_h| = 1$. Given $M \geqslant 2$, consider the points $\lambda \in I^s$ with $\lambda_i \in P(M/2)$ $(1 \leqslant i \leqslant s)$. Then there are

$$c_{12}(K) M^{2rs}$$

such points λ. (See, e.g., [7, Lemma 11].) Given such a λ, the point

$$\mathbf{p}(\lambda) = (F_1(\lambda), \ldots, F_h(\lambda))$$

lies in the domain $D: (p_1, \ldots, p_h)$, where $p_j = x_j + y_j$, $x_j, y_j \in \mathbb{R}$, $|x_j| \leqslant (sM)/2$ and $|y_j| \leqslant (sM)/2$ $(1 \leqslant j \leqslant h)$. Let v be the rational integer in

$$(c_{12} M^{2rs})^{1/(2h)} - 1 \leqslant v < (c_{12} M^{2rs})^{1/(2h)}$$

and divide each region of p_j in D into v^2 subsquares of side $(sM)/v$. Since $v^{2h} < c_{12} M^{2rs}$, there are two points $\mathbf{p}(\lambda)$ and $\mathbf{p}(\lambda')$ such that for $j = 1, ..., h$, $F_j(\lambda)$ and $F_j(\lambda')$ will lie in the same subsquare. Then $\Lambda = \lambda - \lambda'$ has

$$0 < \|\Lambda\| \leqslant \|\lambda\| + \|\lambda'\| \leqslant M$$

and for $j = 1, ..., h$,

$$|F_j(\Lambda)| = |F_j(\lambda) - F_j(\lambda')| \leqslant \sqrt{\left(\frac{sM}{v}\right)^2 + \left(\frac{sM}{v}\right)^2} < \frac{2sM}{v} \ll M^{-E}$$

if

$$s \geqslant c_{13}(h, r, E) = c_5(1, ..., 1; r, 1, E) = u,$$

say. So Theorem 1 is true for $k = m = 1$. Let

$$c_5(1, ..., 1; r, m, E) = mu.$$

Replace the indices $1 \leqslant l \leqslant mu$ by double indices $1 \leqslant i \leqslant m$, $1 \leqslant j \leqslant u$. If $s \geqslant mu$, we let

$$\lambda(i) = (\underbrace{0, ..., 0}_{u(i-1)}, \lambda_j, ..., \lambda_{iu}, \underbrace{0, ..., 0}_{s - ui}), \qquad 1 \leqslant i \leqslant m.$$

They are linearly independent and satisfy

$$|F_j(\lambda(i)| \ll M^{-E}, \qquad \|\lambda(i)\| \leqslant M, \qquad 1 \leqslant j \leqslant h, \quad 1 \leqslant i \leqslant m.$$

So Theorem 1 is true for $k = 1$.

If $F = F(\lambda) = F(\lambda_1, ..., \lambda_s)$ and $G = G(\mu) = G(\mu_1, ..., \mu_t)$ are forms, write

$$F \to G$$

if there are t linearly independent points $\lambda_1, ..., \lambda_t$ in J^s with

$$G(\mu_1, ..., \mu_t) = F(\mu, \lambda_1 + \cdots + \mu_t \lambda_t). \tag{32}$$

Put

$$\psi(F, G) = \min \max(\|\lambda_1\|, ..., \|\lambda_t\|),$$

where the minimum is over $\lambda_1, ..., \lambda_t$ with (32). Put $\psi(F, G) = \infty$ if $F \nrightarrow G$. We use $s(F)$ to denote the number of variables of F.

Suppose that $k > 1$ and that Theorem 1 has already been proved for forms of degrees less than k. First we treat the case of a single form F of degree k.

LEMMA 8. *Suppose $k \geqslant 1$, $l \geqslant 1$, $E > 0$. If F is a form of degree k with $s(F) \geqslant c_{14}(k, r, l, E)$, and if $M \geqslant 1$, then there is a form G with $F \to G$ and $\psi(F, G) \leqslant M$, and where G is a form in $l + 1$ variables, of the type*

$$G = \sigma\mu^k + F_1(v_1, ..., v_l) + H(\mu, v_1, ..., v_l) \tag{33}$$

with

$$|H| \ll M^{-E} |F|.$$

Proof. Let $\mathbf{e}(1), ..., \mathbf{e}(l+1)$ be the first $l + 1$ unit vectors in E_s. Consider the forms

$$K_{p_1, ..., p_{k-u}}(\lambda) = \hat{F}(\underbrace{\lambda, ..., \lambda}_{u}, \mathbf{e}(p_1), ..., \mathbf{e}(p_{k-u})), \tag{34}$$

where u takes all the values from 1 to $k - 1$, and $p_1, ..., p_{k-u}$ all the values from 1 to $l + 1$. The total number of the forms (34) is less than $k(l+1)^k$, and each is of degree $\leqslant k - 1$ in λ. Each form (34) has

$$|K_{p_1, ..., p_{k-u}}(\lambda)| \leqslant |F|.$$

So by the part of Theorem 1 which we already know, we see that there exists $c_{14}(k, r, l, E)$ such that if $s \geqslant c_{14}$ there is a point $\lambda_0 \neq \mathbf{0}$ in J^s with

$$\|\lambda_0\| \leqslant M$$

and

$$|K_{p_1, ..., p_{k-u}}(\lambda_0)| \ll M^{-E} |F| \tag{35}$$

for all the forms (34). We may suppose without loss of generality that $\lambda_0, \mathbf{e}(1), ..., \mathbf{e}(l)$ are linearly independent. Writing

$$G(\mu, v_1, ..., v_l) = F(\mu\lambda_0 + v_1\mathbf{e}(1) + \cdots + v_l\mathbf{e}(l)),$$

we have $F \to G$ and

$$\psi(F, G) \leqslant \max(\|\lambda_0\|, \|\mathbf{e}(1)\|, ..., \|\mathbf{e}(l)\|) \leqslant M.$$

G is of the type (33) with

$$\sigma = F(\lambda_0), \qquad F_1(v_1, ..., v_l) = F(v_1\mathbf{e}(1) + \cdots + v_l\mathbf{e}(l))$$

and

$$H(\mu, v_1, ..., v_l)$$

$$= \sum_{u=1}^{k-1} \binom{k}{u} \hat{F}(\underbrace{\mu\lambda_0, ..., \mu\lambda_0}_{u}, \underbrace{v_1\mathbf{e}(1) + \cdots + v_l\mathbf{e}(l), ..., v_1\mathbf{e}(1) + \cdots + v_l\mathbf{e}(l)}_{k-u}).$$

344   WANG YUAN

It follows from (35) that

$$|H| \ll M^{-E} |F|,$$

and the lemma is proved.

We omit the remaining part of the proof, since it is parallel to the corresponding part of the proof of Theorem A.

ACKNOWLEDGMENT

I am grateful to Professor W. M. Schmidt for his suggestion to consider the present problem and for all of his help.

REFERENCES

1. R. BRAUER, A note on systems of homogeneous algebraic equations, *Bull. Amer. Math. Soc.* **51** (1945), 749–755.
2. H. DAVENPORT, "Analytic Methods for Diophantine Equations and Diophantine Inequalities," Lecture Notes, University of Michigan, 1962.
3. L. G. PECK, Diophantine equations in algebraic number fields, *Amer. J. Math.* **71** (1949), 387–402.
4. W. M. SCHMIDT, Small zeros of additive forms in many variables II, *Acta Math.* **143** (1979), 219–232.
5. W. M. SCHMIDT, Diophantine inequalities for forms of odd degree, *Adv. in Math.* **38** (1980), 128–151.
6. C. L. SIEGEL, Sums of m-th powers of algebraic integers, *Ann. of Math.* **46** (1945), 313–339.
7. WANG YUAN, Bounds for solutions of additive equations in an algebraic number field I, *Acta Arith.* **48** (1987), 21–48.
8. WANG YUAN, Bounds for solutions of additive equations in an algebraic number field II, *Acta Arith.* **48** (1987), to appear.

Vol. 32 No. 5   SCIENCE IN CHINA (Series A)   May 1989

ON HOMOGENEOUS ADDITIVE CONGRUENCES

WANG Yuan (王　元)

(Institute of Mathematics, Academia Sinica, Beijing)

Received August 26, 1988.

ABSTRACT

In this paper, the additive equations of the type $\alpha_1\lambda_1^k + \cdots + \alpha_s\lambda_s^k = 0$ are studied, $\alpha_i's$ being integers of an algebraic number field K of degree n. The main result is as follows: If $s \geq (2k)^{n+1}$ (or $s \geq ckn\log k$ for $2 + k$), the equation is solved nontrivially in any $\mathfrak{P}$-adic field, where $\mathfrak{P}$ is a prime ideal of K.

Key words: additive equation, prime ideal, congruence, singular series.

I. INTRODUCTION

Let K be an algebraic number field of degree n. For any $\lambda \in K$, we use $\lambda^{(l)}(1 \leq l \leq r)$ to denote the real conjugates and $\lambda^{(m)}(r + 1 \leq m \leq n)$ the complex conjugates of λ. Let k be a rational integer ≥ 1 and suppose $\alpha_1, \cdots, \alpha_s$ is a set of integers in K. Consider the additive equation[1]

$$\alpha_1\lambda_1^k + \cdots + \alpha_s\lambda_s^k = 0. \tag{1}$$

One can apply Siegel's generalized circle method[5,6] on Waring's problem in algebraic number fields to prove that if $s \geq c_1(k, n)$ (hereafter we use $c(f, \cdots, g)$ to denote a positive constant depending on $f, \cdots, g$), then there is an asymptotic formula for the number of solutions $\lambda_1, \cdots, \lambda_s$ of (1), where $\lambda_1, \cdots, \lambda_s$ are integers of K oat istying

$$0 \leq \lambda_i^{(l)} \leq P, \ |\lambda_i^{(m)}| \leq P, \ 1 \leq i \leq s. \tag{2}$$

The main term in the asymptotic formula is essentially a power of P multiplied by the so-called singular series connected to $\alpha_1, \cdots, \alpha_s$.

Ler $\mathfrak{P}$ denote a prime ideal of K and r be a rational integer ≥ 1. Consider the congruence

$$\alpha_1\lambda_1^k + \cdots + \alpha_1\lambda_s^k \equiv 0(\mathrm{mod}\mathfrak{P}^r). \tag{3}$$

A set of numbers $\lambda_1, \cdots, \lambda_s$ of K satisfying (3) is called a nontrivial solution of (3) if $\lambda_1, \cdots, \lambda_s$ are integers, not all divisible by $\mathfrak{P}$.

---

1) We suppose that the coefficients of additive equations and forms are nonzero integers in K throughout this paper.

In order to derive from the asymptotic formula for the number of solutions to Eq. (1) in (2) that (1) either has infinitely many solutions or has a small solution in (2) with P depending on $k, K, \alpha_1, \cdots, \alpha_s$ (cf. [8, 9]), we need to know that the value of singular series is positive. This is equivalent to the congruence condition: For any prime ideal power $\mathfrak{P}^r$, the congruence (3) has a nontrivial solution.

Let $\Gamma^*(k, K, \mathfrak{P})$ be the least number such that if $s \geqslant \Gamma^*(k, K, \mathfrak{P}^r)$ and $\alpha_1, \cdots,$ α_s are any given integers in K, the congruence (3) has a nontrivial solution. Write

$$\Gamma^*(k, K) = \max_{\mathfrak{P}, r} \Gamma^*(k, K, \mathfrak{P}^r).$$

Peck[4] first established that $\Gamma^*(k, K) \leqslant 4k^{2n+3} + 1$. However for the case of rational field Q, Davenport and Lewis[2] proved that $\Gamma^*(k, Q) \leqslant k^2 + 1$, with equality whenever $k + 1$ is a prime. Chowla and Shimula[1] showed that $\Gamma^*(k, Q) \leqslant ck \log k$ if k is odd, which is best possible apart from some possible improvements on the constant c. Their methods can be generalized to treat the problem for the estimation of $\Gamma^*(k, K)$, and we shall prove the following

Theorem 1. *We have*

$$\Gamma^*(k, K) \leqslant \begin{cases} ckn \log k, & \text{if } k \text{ is odd,} \\ (2k)^{n+1}, & \text{if } k \text{ is an integer } \geqslant 1. \end{cases}$$

II.   Preliminary Lemmas

For $\gamma \in K$, we denote $S(\gamma) = \sum_{i=1}^{n} \gamma^{(i)}$ and $E(r) = e(S(\gamma))$, where $e(x) = e^{2\pi i x}$. Let δ be the different of K. Suppose that $\alpha \in K$ and $\alpha\delta = \mathcal{G}/\mathfrak{P}$, where $\mathcal{G}$ is an integral ideal and $\mathfrak{P} \nmid \mathcal{G}$. Consider the exponential sum

$$S(\alpha\lambda^k, \mathfrak{P}) = \sum_{\lambda \,(\mathrm{mod}\,\mathfrak{P})} E(\alpha\lambda^k),$$

where $\lambda(\mathrm{mod}\,\mathfrak{P})$ denotes λ running over a complete residue system.

Lemma 1. *Let* $d = (k, N(\mathfrak{P}) - 1)$. *Then*

$$|S(\alpha\lambda^k, \mathfrak{P})| \leqslant (d - 1)\sqrt{N(\mathfrak{P})}.$$

Proof. Choose η to be a primitive root $\mathrm{mod}\,\mathfrak{P}$. Then $\mu \equiv \eta^{\mathrm{ind}\mu}(\mathrm{mod}\,\mathfrak{P})$ for any μ with $\mathfrak{P} \nmid \mu$. The necessary and sufficient condition for the solubility of the congruence $\lambda^k \equiv \mu(\mathrm{mod}\,\mathfrak{P})$ is $k\,\mathrm{ind}\lambda \equiv \mathrm{ind}\mu(\mathrm{mod}N(\mathfrak{P}) - 1)$ or $d \mid \mathrm{ind}\mu$. If the congruence is soluble, it has d incongruent solutions $\mathrm{mod}\,\mathfrak{P}$.

Suppose $d = 1$. Then

$$S(\alpha\lambda^k, \mathfrak{P}) = \sum_{\mu\,(\mathrm{mod}\,\mathfrak{P})} E(\alpha\mu) = 0,$$

(see Hecke [3], p.197.) Now suppose that $d > 1$. Then

$$S(\alpha\lambda^k, \mathfrak{P}) = 1 + \sum_{\mu\,(\mathrm{mod}\,\mathfrak{P})}^{*} E(\alpha\mu) \sum_{m=0}^{d-1} e\left(\frac{m\,\mathrm{ind}\mu}{d}\right),$$

where $\sum^*$ denotes a sum in which μ runs over a reduced residue system $\mathrm{mod}\mathfrak{B}$. Since $\sum\limits_{\mu(\mathrm{mod}\mathfrak{B})}^* E(\alpha\mu) = -1$, we have

$$S(\alpha\lambda^k, \mathfrak{B}) = \sum_{m=1}^{d-1} \sum_{\mu(\mathrm{mod}\mathfrak{B})}^* e\left(\frac{m\,\mathrm{ind}\mu}{d}\right) E(\alpha\mu),$$

and by the Schwarz inequality we admit

$$|S(\alpha\lambda^k, \mathfrak{B})|^2 \leqslant (d-1)\sum_{m=1}^{d-1} \sum_{\mu(\mathrm{mod}\mathfrak{B})}^* \sum_{\mu_1(\mathrm{mod}\mathfrak{B})}^* e\left(\frac{m(\mathrm{ind}\mu - \mathrm{ind}\mu_1)}{d}\right) E(\alpha(\mu - \mu_1)).$$

For any given μ and μ_1, let τ be the number such that

$$\mu \equiv \mu_1\tau(\mathrm{mod}\mathfrak{B}).$$

Then

$$|S(\alpha\lambda^k, \mathfrak{B})|^2 \leqslant (d-1)\sum_{m=1}^{d-1} \sum_{\mu(\mathrm{mod}\mathfrak{B})}^* \sum_{\mu_1(\mathrm{mod}\mathfrak{B})}^* e\left(\frac{m\,\mathrm{ind}\tau}{d}\right) E(\alpha(\tau-1)\mu_1)$$

$$= (d-1)\sum_{m=1}^{d-1}\left(N(\mathfrak{B}) - 1 + \sum_{(\tau\,\mathrm{mod}\mathfrak{p})}^{**} e\left(\frac{m\,\mathrm{ind}\tau}{d}\right)\right.$$

$$\left. \cdot \sum_{\mu_1(\mathrm{mod}\mathfrak{p})}^* E(\alpha(\tau-1)\mu_1)\right)$$

$$= (d-1)\sum_{m=1}^{d-1}\left(N(\mathfrak{B}) - 1 - \sum_{\tau(\mathrm{mod}\mathfrak{p})}^{**} e\left(\frac{m\,\mathrm{ind}\tau}{d}\right)\right),$$

where $\sum^{**}$ denotes the sum $\sum^*$ with $\tau = 1$ deleted. Since

$$\sum_{\tau(\mathrm{mod}\mathfrak{p})}^{**} e\left(\frac{m\,\mathrm{ind}\tau}{d}\right) = -1 + \sum_{\tau(\mathrm{mod}\mathfrak{p})}^* e\left(\frac{m\,\mathrm{ind}\tau}{d}\right)$$

$$= -1 + \frac{N(\mathfrak{B}) - 1}{d}\sum_{q=1}^{d} e\left(\frac{mq}{d}\right) = -1,$$

we have

$$|S(\alpha\lambda^k, \mathfrak{B})| \leqslant (d-1)\sqrt{N(\mathfrak{B})}.$$

The lemma is proved.

Let π be an integer of K such that $\mathfrak{B}\|\pi$. If S is a set of complete residue system $\mathrm{mod}\mathfrak{B}$, then any integer α of K has a unique $\mathfrak{B}$-adic representation

$$\alpha = \sum_{i=0}^{\infty} \mu_i\pi^i, \quad \mu_i \in S.$$

In fact, μ_0 is uniquely determined by $\mu_0 \equiv \alpha(\mathrm{mod}\mathfrak{B})$, μ_1 is then determined by $\pi\mu_1 \equiv \alpha - \mu_0(\mathrm{mod}\mathfrak{B}^2)$ and so on.

Consider two additive forms:

$$A = \sum_{i=1}^{s} \alpha_i \lambda_i^k \quad \text{and} \quad B = \sum_{i=1}^{s} \beta_i \mu_i^{sk}$$

with integer coefficients. We shall say that A is $\mathfrak{P}$-adically equivalent to B if A can be changed into A' by a transformation $\lambda_i = \tau_i \mu_i (1 \leqslant i \leqslant s)$ such that $A' \equiv \gamma B$ (mod $\mathfrak{P}^r$) for any $r \geqslant 1$, where $\tau_i (1 \leqslant i \leqslant s)$ are nonzero integers of K. It is easily seen that the relationship is reflexive and transitive. Now we proceed to prove that the relationship is symmetrical. In fact, suppose that

$$\sum_{i=1}^{s} \alpha_i (\tau_i \mu_i)^k \equiv r \sum_{i=1}^{s} \beta_i \mu_i^k \pmod{\mathfrak{P}^r} \tag{4}$$

holds for any $r \geqslant 1$. Then set $\mu_i = \gamma \tau_1 \cdots \tau_{i-1} \tau_{i+1} \cdots \tau_s \lambda_i (1 \leqslant i \leqslant s)$. We have

$$\sum_{i=1}^{s} \beta_i \gamma^k (\tau_1 \cdots \tau_{i-1} \tau_{i+1} \cdots \tau_s)^k \lambda_i^k \equiv \gamma^{k-1} \sum_{i=1}^{s} \alpha_i \tau_i^k (\tau_1 \cdots \tau_{i-1} \tau_{i+1} \cdots \tau_s)^k \lambda_i^k$$

$$\equiv \gamma^{k-1} (\tau_1 \cdots \tau_s)^k \sum_{i=1}^{s} \alpha_i \lambda_i^k \pmod{\mathfrak{P}^r}.$$

If the congruence (3) has a nontrivial solution for every r, then it is said that the equation $A = 0$ is solved nontrivially in $\mathfrak{P}$-adic field or that the form A represents zero $\mathfrak{P}$-adically. We may prove that if A and B are $\mathfrak{P}$-adically equivalent, then they can or cannot represent zero simultaneously in the $\mathfrak{P}$-adic field. In fact, from (4) the solubility of $A \equiv 0 (\text{mod} \mathfrak{P}^r)$ is found to follow from the solubility of $B \equiv 0$ (mod $\mathfrak{P}^{r'}$) for some sufficiently large r'.

Lemma 2. *Any additive form A is equivalent $\mathfrak{P}$-adically to a form of the type*

$$F = F^{(0)} + \pi F^{(1)} + \cdots + \pi^{k-1} F^{(k-1)}, \tag{5}$$

where $F^{(j)}$ is an additive form in v_j variables (the variables in distinct forms $F^{(j)}$ being distinct) with all coefficients $\not\equiv 0$ (mod $\mathfrak{P}$), and where $v_0 \geqslant s/k$.

Proof. Represent every α_i by a series of π. By a substitution $\lambda_i = \pi^{r_i} \mu_i$ with suitable $r_i (1 \leqslant i \leqslant s)$, we can ensure that A is equivalent $\mathfrak{P}$-adically to a form of the type (5), where $F^{(j)} (0 \leqslant j \leqslant k-1)$ satisfy the conditions stated in the lemma except the condition that $v_0 \geqslant s/k$. Put $\lambda_i = \pi \mu_i$ for the variables in $F^{(0)}$. Then F is equivalent $\mathfrak{P}$-adically to the form

$$F^{(1)} + \pi F^{(2)} + \cdots + \pi^{k-2} F^{(k-1)} + \pi^{k-1} F^{(0)}.$$

Hence we can choose a cyclic permutation of $F^{(0)}, \cdots, F^{(k-1)}$ to ensure that the number of variables in $F^{(0)}$ is $\geqslant s/k$. The lemma is proved.

III.　The Proof of Theorem 1 (Odd Degree k)

We use the notations: $\Pi = \prod_{i=1}^{s} \alpha_i$, ε means any pre-assigned positive number and k means an odd integer $\geqslant c(\varepsilon)$.

Lemma 3. *If $\mathfrak{P} \nmid \Pi$, then*

$$\Gamma^*(k, K, \mathfrak{P}) \leqslant \left(\frac{2}{\log 2} + \varepsilon\right) \log k.$$

Proof. 1) Suppose $N(\mathfrak{P}) \leqslant k^{2+\varepsilon/2}$. Consider the $2^s - 1$ sums

$$\alpha_i(1 \leqslant i \leqslant s), \ \alpha_i + \alpha_j(1 \leqslant i < j \leqslant s), \ \cdots, \alpha_1 + \cdots + \alpha_s.$$

If $2^s - 1 > N(\mathfrak{P})$, then at least two of the sums are congruent mod$\mathfrak{P}$. Since -1 is a kth power residue mod$\mathfrak{P}$, the congruence $A \equiv 0(\text{mod}\mathfrak{P})$ has a nontrivial solution if $2^s > k^{2+\varepsilon/2}(1 + k^{-2-\varepsilon/2})$ or

$$s \log 2 > \left(2 + \frac{\varepsilon}{2}\right) \log k + \log(1 + k^{-2-\varepsilon/2}).$$

The lemma follows.

2) Suppose $N(\mathfrak{P}) > k^{2+\varepsilon/2}$. We shall prove the following stronger inequality

$$\Gamma^*(k, K, \mathfrak{P}) \leqslant \left[\frac{8}{\varepsilon}\right] + 4.$$

Let $M(\mathfrak{P})$ denote the number of solutions of the congruence

$$\alpha_1\lambda_1^k + \cdots + \alpha_{s-1}\lambda_{s-1}^k \equiv -\alpha_s(\text{mod}\mathfrak{P}), \ \lambda_i(\text{mod}\mathfrak{P}), \ 1 \leqslant i \leqslant s-1.$$

Then

$$M(\mathfrak{P}) = \frac{1}{N(\mathfrak{P})} \sum_\mu \sum_{\lambda_1(\text{mod}p)} \cdots \sum_{\lambda_{s-1}(\text{mod}p)} E(\mu(\alpha_1\lambda_1^k + \cdots + \alpha_{s-1}\lambda_{s-1}^k + \alpha_s)),$$

where μ runs over a complete residue system of $(\mathfrak{P}\delta)^{-1}$, mod δ^{-1}. Therefore

$$M(\mathfrak{P}) = N(\mathfrak{P})^{s-2} + N(\mathfrak{P})^{-1} \sum_\mu{}^* E(\mu\alpha_s) \sum_{j=1}^{s-1} S(\mu\alpha_j\lambda_i^k, \mathfrak{P}).$$

By Lemma 1, we have

$$\left| N(\mathfrak{P})^{-1} \sum_\mu{}^* E(\mu\alpha_s) \sum_{j=1}^{s-1} S(\mu\alpha_j\lambda_i^k, \mathfrak{P}) \right| \leqslant (d-1)^{s-1} N(\mathfrak{P})^{\frac{s-1}{2}}.$$

Consequently, if

$$(d-1)^{s-1} N(\mathfrak{P})^{\frac{s-1}{2}} < N(\mathfrak{P})^{s-2}, \tag{6}$$

then

$$M(\mathfrak{P}) > 0.$$

If s satisfies

$$N(\mathfrak{P}) > k^{\frac{2s-2}{s-3}} = k^{2+\frac{4}{s-3}}, \tag{7}$$

then (6) holds. (7) follows from $N(\mathfrak{P}) > k^{2+\varepsilon/2}$ if $s - 3 > 8/\varepsilon$. Hence if $s \geqslant [8/\varepsilon] + 4$, we can get a nontrivial solution

$$\lambda_1, \cdots, \lambda_{s-1}, 1$$

of the congruence $A \equiv 0(\text{mod}\mathfrak{P})$. The lemma is proved.

Lemma 4. *Suppose that* $\mathfrak{P}\nmid\Pi, \mathfrak{P}^e\|p$ *and* $p^b\|k$, *where* p *is a rational prime,*

$e \geqslant 1$ *and* $b \geqslant 1$. *Then if* $w \geqslant (b+2)e$, *we have*

$$\Gamma^*(k, K, \mathfrak{P}^{w-e}) \leqslant \left(\frac{2}{\log 2} + \varepsilon\right) n \log k.$$

Proof. Suppose $w = (b+2)e$. Let s be an integer such that

$$2^s - 1 > N(k)^2 \geqslant N(p)^{2b} \geqslant N(\mathfrak{P})^{2be} \geqslant N(\mathfrak{P})^{(b+1)e}. \tag{8}$$

Then at least two of the sums

$$\alpha_i(1 \leqslant i \leqslant s), \quad \alpha_i + \alpha_j(1 \leqslant i < j \leqslant s), \quad \cdots, \quad \alpha_1 + \cdots + \alpha_s$$

are congruent $\mathrm{mod}\,\mathfrak{P}^{(b+1)e}$, i.e. the congruence

$$A \equiv 0(\mathrm{mod}\,\mathfrak{P}^{(b+1)e})$$

has a nontrivial solution. From (8) it follows that the condition of s is

$$s \log 2 > \log(k^{2n} + 1)$$

or

$$s \geqslant \left(\frac{2}{\log 2} + \varepsilon\right) n \log k.$$

The lemma holds for the case $w = (b+2)e$. Suppose that $w \geqslant (b+2)e$ and that the congruence

$$A \equiv 0(\mathrm{mod}\ (\mathfrak{P}^{w-e})$$

has a nontrivial solution $\lambda_1, \cdots, \lambda_s$. Without loss of generality, we may assume that $\mathfrak{P} \nmid \lambda_1$. We now proceed to prove that

$$A \equiv 0(\mathrm{mod}\ \mathfrak{P}^w)$$

also has a nontrivial solution. Set

$$\mu_i = \lambda_i + \nu_i \pi^{w-be-e}, \quad 1 \leqslant i \leqslant s.$$

We start to show that

$$\sum_{i=1}^{s} \alpha_i \mu_i^k \equiv \sum_{i=1}^{s} \alpha_i \lambda_i^k + k\pi^{w-be-e} \sum_{i=1}^{s} \alpha_i \nu_i \lambda_i^{k-1}(\mathrm{mod}\,\mathfrak{P}^w). \tag{9}$$

In fact, consider the $(h+1)$-th term $\dbinom{k}{h} \lambda_i^{k-h}(\nu_i \pi^{w-be-e})^h$ in the binomial expansion of

$$\mu_i^k = (\lambda_i + \nu_i \pi^{w-be-e})^k,$$

where $h \geqslant 2$. Suppose $p^a \| h$. Since $\dbinom{k}{h} = \dfrac{k}{h}\dbinom{k-1}{h-1}$, the $(h+1)$-th term is divisible by $\mathfrak{P}^g$, where

$$g = h(w - be - e) + be - ae.$$

To prove (9) it suffices to show that $g \geqslant w$ or

$$(h-1)(w - be) \geqslant he + ae.$$

Since $w \geqslant (b+2)e$, it suffices to show that

     SCIENCE IN CHINA (Series A)   Vol. 32

$$2(h-1)e \geqslant (h+a)e,$$

i. e.,

$$h \geqslant a + 2. \tag{10}$$

Since $2 \nmid k$ and $p \mid k$, we have $3^a \leqslant p^a \leqslant h$, and therefore (10) follows by $3^a = (1 + 2)^a \geqslant 1 + 2a \geqslant a + 2$ for the case $a \geqslant 1$. If $a = 0$, (10) follows immediately from $h \geqslant 2$. Write

$$k = \pi^{be}\varphi \quad \text{and} \quad \sum_{i=1}^{s} \alpha_i \lambda_i^k = \pi^{w-e}\psi$$

in their $\mathfrak{P}$-adic representations, where $\mathfrak{P} \nmid \varphi$. Then from (9), we have

$$\sum_{i=1}^{s} \alpha_i \mu_i^k \equiv \pi^{w-e}\left(\psi + \varphi \sum_{i=1}^{s} \alpha_i \nu_i \lambda_i^{k-1}\right)(\mathrm{mod}\,\mathfrak{P}^w).$$

Since $(\varphi \alpha_1 \lambda_1^{k-1}, \mathfrak{P}) = 1$, the congruence

$$\psi + \varphi \alpha_1 \lambda_1^{k-1} \nu_1 \equiv 0(\mathrm{mod}\,\mathfrak{P}^e)$$

has a unique solution $\nu_1(\mathrm{mod}\,\mathfrak{P}^e)$. Put $\nu_i = 0(2 \leqslant i \leqslant s)$. Then we have a solution $\mu_1, \cdots, \mu_s$ of

$$\sum_{i=1}^{s} \alpha_i \mu_i^k \equiv 0(\mathrm{mod}\,\mathfrak{P}^w),$$

where $\mathfrak{P} \nmid \mu_1$. The lemma follows by induction.

Proof of Theorem 1 (odd degree k). By Lemma 2, it suffices to consider only the forms F of the type (5).

1) Suppose $\mathfrak{P} \mid k$. Then it follows by Lemma 4 that the equation $F^{(0)} = 0$ can be solved nontrivially in $\mathfrak{P}$-adic field if

$$v_0 \geqslant \left(\frac{2}{\log 2} + \varepsilon\right) n \log k,$$

and therefore $F = 0$ can be solved if

$$s \geqslant \left(\frac{2}{\log 2} + \varepsilon\right) nk \log k.$$

The theorem follows.

2) Suppose $\mathfrak{P} \nmid k$ and $\mathfrak{P} \nmid \Pi$. If $s \geqslant \left(\frac{2}{\log 2} + \varepsilon\right)\log k$ and $w \geqslant 2$, then we may assume that

$$F \equiv 0(\mathrm{mod}\,\mathfrak{P}^{w-1})$$

has a nontrivial solution with $\mathfrak{P} \nmid \lambda_1$. Now we proceed to show that $F \equiv 0(\mathrm{mod}\,\mathfrak{P}^w)$ also has a nontrivial solution. Set

$$\mu_i = \lambda_i + \nu_i \pi^{w-1}, \quad 1 \leqslant i \leqslant s.$$

Then

$$\sum_{i=1}^{s} \alpha_i \mu_i^k \equiv \sum_{i=1}^{s} \alpha_i \lambda_i^k + k\pi^{w-1} \sum_{i=1}^{s} \alpha_i \nu_i \lambda_i^{k-1} (\mathrm{mod}\,\mathfrak{P}^w).$$

Since $\sum_{i=1}^{s} \alpha_i \lambda_i^k$ is divided by $(k\pi^{w-1}\alpha_1\lambda_1^{k-1}, \mathfrak{P}^w) = \mathfrak{P}^{w-1}$, the congruence

$$\sum_{i=1}^{s} \alpha_i \lambda_i^k + k\pi^{w-1}\alpha_1\lambda_1^{k-1}\nu_1 \equiv 0(\mathrm{mod}\,\mathfrak{P}^w)$$

has a unique solution $\nu_1(\mathrm{mod}\,\mathfrak{P})$. Put $\nu_i = 0(2 \leqslant i \leqslant s)$. We have a solution of $F \equiv 0 \ (\mathrm{mod}\,\mathfrak{P}^w)$ with $\mu_1 \not\equiv 0(\mathrm{mod}\,\mathfrak{P})$. The theorem follows by induction.

3) Suppose $\mathfrak{P}\nmid k$ and $\mathfrak{P}\,|\,\Pi$. It follows by 2) that $F^{(0)} = 0$ is solved nontrivially in $\mathfrak{P}$-adic field if $v_0 \geqslant \left(\dfrac{2}{\log 2} + \varepsilon\right) \log k$, and therefore $F = 0$ is solved if

$$s \geqslant \left(\frac{2}{\log 2} + \varepsilon\right) k \log k.$$

The theorem is proved.

<h3 style="text-align:center">IV. The Proof of Theorem 1 (Arbitrary Degree k)</h3>

Let p denote a rational prime and $\omega(\alpha)$ the exponent with which $\mathfrak{P}$ enters into the canonical factorization of α, i.e. $\mathfrak{P}^{\omega(\alpha)}\|\alpha$ and

$$l_0 = \left[\frac{e}{p-1}\right] + be + 1, \tag{11}$$

where $p^b\|k$ and $\mathfrak{P}^e\|p$.

Lemma 5. *The series*

$$\sum_{i=1}^{\infty} \frac{\alpha^i}{i!} \quad and \quad \sum_{i=1}^{\infty} (-1)^i \frac{\alpha^i}{i}$$

are convergent provided $\omega(\alpha) > \dfrac{e}{p-1}$, *and they are denoted by* $\exp \alpha$ *and* $\log(1 + \alpha)$ *respectively.*

See Ref. [7].

Lemma 6. *Suppose $\mathfrak{P}\nmid\beta$. If the congruence*

$$\beta\xi^k \equiv \alpha(\mathrm{mod}\,\mathfrak{P}^{l_0})$$

has a nontrivial solution, then so has the congruence

$$\beta\eta^k \equiv \alpha(\mathrm{mod}\,\mathfrak{P}^l)$$

for any $l > l_0$.

Proof. Since $\mathfrak{P}\nmid\beta$, there exists $\bar\beta$ such that

$$\beta\bar\beta \equiv 1(\mathrm{mod}\,\mathfrak{P}^l).$$

Set $\gamma = \alpha\bar{\beta}$ and $\gamma - \xi^k = \mathfrak{P}^{l_0}\mathscr{Q}$, where $\mathscr{Q}$ is an integral ideal. Since $(\xi, \mathfrak{P}) = 1$, it follows from $\xi^k\left(\dfrac{\gamma}{\xi^k} - 1\right) = \mathfrak{P}^{l_0}\mathscr{Q}$ that $\mathfrak{P}^{l_0}\Big|\left(\dfrac{\gamma}{\xi^k} - 1\right)$. Therefore by Lemma 5, there exists

$$\log\left(1 + \left(\frac{\gamma}{\xi^k} - 1\right)\right) = \log\frac{\gamma}{\xi^k}.$$

Since

$$\omega\left(\frac{1}{k}\log\frac{\gamma}{\xi^k}\right) = \omega\left(\log\frac{\gamma}{\xi^k}\right) - \omega(k) \geqslant l_0 - be > \frac{e}{p-1},$$

there exists also

$$\exp\left(\frac{1}{k}\log\frac{\gamma}{\xi^k}\right) = A, \quad \text{say,}$$

by Lemma 5. Then

$$\gamma = (A\xi)^k$$

and therefore $A\xi$ is an integer. Choose an integer η such that

$$\omega(A\xi - \eta) \geqslant l.$$

Then

$$\omega(\eta^k - \gamma) = \omega(\eta^k - (A\xi)^k) \geqslant l,$$

i. e.

$$\eta^k \equiv \gamma(\mathrm{mod}\mathfrak{P}^l).$$

The lemma follows.

Lemma 7. *Let* $d = (k, N(\mathfrak{P}) - 1)$. *Suppose that*

$$\alpha_1 \cdots \alpha_{d+1} \not\equiv 0(\mathrm{mod}\mathfrak{P}).$$

Then

$$\alpha_1\lambda_1^k + \cdots + \alpha_{d+1}\lambda_{d+1}^k \equiv 0(\mathrm{mod}\mathfrak{P})$$

has a solution with $\lambda_1 \not\equiv 0 \ (\mathrm{mod}\mathfrak{P})$.

Proof. It is well known that the congruences

$$y^k \equiv m \ (\mathrm{mod}\mathfrak{P}) \quad \text{and} \quad y^d \equiv m \ (\mathrm{mod}\mathfrak{P})$$

are soluble or not soluble simultaneously. Hence if the conclusion of the lemma does not hold, then for any $\lambda_2, \cdots, \lambda_{d+1}$, we have

$$\alpha_1 + \alpha_2\lambda_2^d + \cdots + \alpha_{d+1}\lambda_{d+1}^k \not\equiv 0 \ (\mathrm{mod}\mathfrak{P}),$$

i.e.,

$$(\alpha_1 + \alpha_2\lambda_2^d + \cdots + \alpha_{d+1}\lambda_{d+1}^d)^{N(\mathfrak{P})-1} \equiv 1 \ (\mathrm{mod}\mathfrak{P}),$$

which should be an identity. But the expansion of the left-hand side contains a term

$$\alpha(\lambda_2^d \cdots \lambda_{d+1}^d)^{\frac{N(\mathfrak{P})-1}{d}} = \alpha(\lambda_2 \cdots \lambda_{d+1})^{N(\mathfrak{P})-1}$$

where $\mathfrak{P} \nmid \alpha$, and every other term contains at least one of the variables $\lambda_2, \cdots, \lambda_{d+1}$ with a power less than $N(\mathfrak{P}) - 1$, which leads to a contradiction, and the lemma follows.

To prove the theorem it is sufficient to consider only the forms F of the type (5). First we define an operation of contraction as follows. Consider a sum of $d + 1$ terms

$$\alpha_1 \lambda_1^k + \cdots + \alpha_{d+1} \lambda_{d+1}^k \tag{12}$$

in $F^{(0)}$. Since $\alpha_1 \cdots \alpha_{d+1} \not\equiv 0 \pmod{\mathfrak{P}}$, the congruence

$$\alpha_1 \mu_1^k + \cdots + \alpha_{d+1} \mu_{d+1}^k \equiv 0 \pmod{\mathfrak{P}}$$

has a solution with $\mu_1 \not\equiv 0 \pmod{\mathfrak{P}}$ by Lemma 7. We can assume that the left-hand side of the congruence is equal to a nonzero number α, or otherwise it follows directly that $F = 0$ is solved nontrivially in the $\mathfrak{P}$-adic field if $s \geq d + 1$. Set $\lambda_i = \mu_i \lambda$ ($1 \leq i \leq d + 1$). Then there holds

$$\alpha_1 \lambda_1^k + \cdots + \alpha_{d+1} \lambda_{d+1}^k = (\alpha_1 \mu_1^k + \cdots + \alpha_{d+1} \mu_{d+1}^k) \lambda^k = \alpha \lambda^k, \tag{13}$$

where $\mathfrak{P}^i \| \alpha$, $i \geq 1$. Thus α has a $\mathfrak{P}$-adic representation

$$\alpha = \pi^i \nu_i + \cdots, \quad \mathfrak{P} \nmid \nu_i. \tag{14}$$

Therefore the operation of contraction consists in replacing the sum (12) by a single term (13). Notice that $\lambda \not\equiv 0 \pmod{\mathfrak{P}}$ implies $\lambda_1 \not\equiv 0 \pmod{\mathfrak{P}}$, where λ_1 is called the distinguished variable in (12).

Operations of contraction are applied to groups of $d + 1$ terms in $F^{(0)}$ and here any one of the variables can be chosen as the distinguished variable. The number of remaining variables in $F^{(0)}$ is then $\leq d$, and we put them to be equal to zero. Hence there results a form of the type

$$\pi G^{(1)} + \pi^2 G^{(2)} + \cdots,$$

where $G^{(j)}$ contains the original v_j terms of $F^{(j)}$ together possibly with some additional terms, each arising in the form (13) from one of the contractions. The variables in these additional terms are called the derived variables. If $i \geq k$, we define $v_i = 0$.

The next step is to apply contractions to groups of $d + 1$ terms in $G^{(1)}$, subject to the condition that each group contains at least one term with a derived variable; such a variable is chosen the distinguished variable in the contraction. The process is continued, and at each stage we take care that at least one of the variables in a group is derived, either directly or indirectly, from a variable in $F^{(0)}$.

Suppose that, after a series of permissible contraction, we get a form H such that $H \equiv 0 \pmod{\mathfrak{P}^{l_0}}$ is soluble with at least one of the derived variables in H, not divisible by $\mathfrak{P}$. This implies a solution of $F \equiv 0 \pmod{\mathfrak{P}^{l_0}}$, and on tracing back the derived variable to its ancester in $F^{(0)}$, we see that the solution is nontrivial.

Finally, if we get a form

$$H = \pi^u H^{(u)} + \pi^{u+1} H^{(u+1)} + \cdots \tag{15}$$

in which any one of the forms $H^{(l_0)}$, $H^{(l_0+1)}$, $\cdots$ contains a derived variable, then we take the derived variable to be 1 and all the other variables equal to zero. Thus a nontrivial solution of $F \equiv 0 (\mathrm{mod}\mathfrak{P}^{l_0})$ is derived.

Lemma 8. *We can obtain a form H of the type* (15) *from a form of the type* (5) *by repeated contractions, where $H^{(u)}$, $H^{(u+1)}$, $\cdots$ are additive forms in distinct variables with all coefficients $\not\equiv 0$ $(\mathrm{mod}\mathfrak{P})$. For $j \leqslant k-1$, the form $H^{(j)}$ includes the terms of $F^{(j)}$, and possibly some additional terms contain derived variables; for $j \geqslant k$ the form $H^{(j)}$ can contain derived variables only. Further, if S_u denote the total number of derived variables in H, then*

$$S_u > \frac{v_0}{(d+1)^u} - 1.$$

Proof. Suppose first that $u = 1$. We divide the v_0 terms of $F^{(0)}$ into m sets of $d+1$ terms, where

$$m = \left[\frac{v_0}{d+1}\right] \geqslant \frac{v_0}{d+1} - \frac{d}{d+1} > \frac{v_0}{d+1} - 1.$$

The lemma holds. Suppose that $u \geqslant 1$ and the lemma is true for u. We proceed to show that the lemma holds for $u+1$. Let w denote the number of derived variables in $H^{(u)}$. Then the total number of derived variables in $H^{(u+1)}$, $\cdots$ is $S_u - w$. We divide the $v_u + w$ terms in $H^{(a)}$ into as many sets of $d+1$ terms as possible, subject to he condition that each set contains at least one of the w derived variables. The number of sets formed is allowed to be

$$\begin{cases} w, & \text{if } v_u \geqslant dw, \\ \left[\dfrac{v_u + w}{d+1}\right], & \text{if } v_u < dw. \end{cases}$$

The remaining variables in $H^{(u)}$ are equal to zero. Each set of $d+1$ terms can be contracted into a single term

$$\pi^g \alpha \lambda^k, \quad g \geqslant u+1, \quad \alpha \not\equiv 0 \ (\mathrm{mod}\mathfrak{P}).$$

Adding such a term to the corresponding form $\pi^g H^{(g)}$, we obtain a form of the type

$$\pi^{u+1}P^{(u+1)} + \pi^{u+2}P^{(u+2)} + \cdots,$$

where each $P^{(j)}$ contains the variables in $H^{(j)}$ and derived new variables of the form $\pi^j \alpha \lambda^j$, $\mathfrak{P} \nmid \alpha$. The total number of derived variables in $P^{(u+1)}, \cdots$ is expressed as

$$(S_u - w) + \begin{cases} w, & \text{if } v_u \geqslant dw, \\ \left[\dfrac{v_u + w}{d+1}\right], & \text{if } v_u < dw. \end{cases}$$

In the first case, we find

$$S_{u+1} = S_u > \frac{v_0}{(d+1)^u} - 1 > \frac{v_0}{(d+1)^{u+1}} - 1.$$

In the second case, we find

$$S_{u+1} = S_u - w + \left[\frac{v_u + w}{d+1}\right]$$

$$\geq S_u - w + \frac{v_u + w}{d+1} - \frac{d}{d+1}.$$

Since $w \leq S_u$, it follows that

$$S_{u+1} \geq \frac{S_u}{d+1} + \frac{dw}{d+1} - w + \frac{v_\mu + w}{d+1} - \frac{d}{d+1}$$

$$= \frac{S_u}{d+1} + \frac{v_u}{d+1} - \frac{d}{d+1} \geq \frac{S_u}{d+1} - \frac{d}{d+1}$$

$$> \frac{v_0}{(d+1)^{u+1}} - \frac{1}{d+1} - \frac{d}{d+1} = \frac{v_0}{(d+1)^{u+1}} - 1,$$

and the lemma follows by induction.

Lemma 9. *If* $s \geq (2k)^{n+1}$, *the congruence*

$$F \equiv 0 \pmod{\mathfrak{P}^{l_0}}$$

always has a nontrivial solution.

Proof. Let $k = p^b k_0$, where $p \nmid k_0$. If $N(\mathfrak{P}) = p^f$, then $fe \leq n$,

$$d = (k, N(\mathfrak{P}) - 1) = (k, p^f - 1) = (k_0, p^f - 1),$$

$$d + 1 \leq \min(2k_0, p^f)$$

and

$$(d+1)^{l_0} \leq (d+1)^{\frac{e}{p-1} + be + 1}$$

$$\leq (p^{\frac{1}{p-1}})^{fe} p^{fbe} \cdot (2k_0)$$

$$\leq 2^{n+1}(k_0 p^b)^{fe} \leq 2^{n+1} k^n.$$

Therefore, Lemmas 2 and 8 lead to

$$S_{l_0} > \frac{v_0}{(d+1)^{l_0}} - 1 > \frac{s}{k(d+1)^{l_0}} - 1 \geq \frac{s}{(2k)^{n+1}} - 1 \geq 0.$$

The lemma is proved.

Proof of Theorem 1 (arbitrary degree k). Suppose that $\lambda_1, \cdots, \lambda_s$ is a solution of

$$\alpha_1 \lambda_1^k + \cdots + \alpha_s \lambda_s^k \equiv 0 \pmod{\mathfrak{P}^{l_0}}$$

with $\mathfrak{P} \nmid \lambda_1$. Let $l > l_0$ and

$$\mu_i = \lambda_i + \pi^{l_0} v, \quad 2 \leq i \leq s,$$

where v runs over a complete residue system mod $\mathfrak{P}^{l-l_0}$. By Lemma 6, the congruence

$$\alpha_1 \lambda^k \equiv -\alpha_2 \mu_2^k - \cdots - \alpha_s \mu_s^k \pmod{\mathfrak{P}^l}$$

has a nontrivial solution λ for any given $\mu_2, \cdots, \mu_s$, i.e., the congruence

$$\alpha_1 \lambda_1^k + \cdots + \alpha_s \lambda_s^k \equiv 0 \pmod{\mathfrak{P}^l}$$

has no less then $N(\mathfrak{P})^{(l-l_0)(s-1)}$ nontrivial solutions. The theorem is proved.

REFERENCES

[1] Chowla, S. & Shimula, G., On the representation of zero by a linear combination of k-th powers, *Norske Vid. Sels. Forh.*, **36**(1963), 169—176.

[2] Davenport, H. & Lewis, D. J., Homogeneous additive equations, *Proc. Roy. Soc.*, A **274** (1963) 443—460.

[3] Hecke, E., *Lectures on Theory of Algebraic Numbers*, Springer-Verlag, 1980.

[4] Peck, L. G., Diophantine equations in algebraic number fields, *Amer. J. Math.*, **71** (1949), 387—402.

[5] Siegel, C. L., Generalization of Waring's problem to algebraic number fields, *Amer. J. Math.*, **66** (1944), 122—136.

[6] ————, Sums of m-th powers of algebraic integers, *Ann. of Math.*, **46** (1945), 313—339.

[7] Tatuzawa, T., On Waring's problem in algebraic number fields, *Acta Arith.*, **24**(1973), 37—60.

[8] Wang Yuan, Bounds for solutions of additive equations in an algebraic number field I, *Acta Arith.*, **48**(1987), 117—144.

[9] ————, Bonuds for soluitons of additive equations in an algebraic number field II, *ibid.*, **48**(1987), 307—323.

Reprinted from JOURNAL OF NUMBER THEORY
All Rights Reserved by Academic Press, New York and London
with permission from Elsevier

Vol. 45, No. 3, November 1993

Small Solutions of Congruences

WANG YUAN

Institute of Mathematics, Academia Sinica,
Beijing 100 080, China

Communicated by Alan C. Woods

Received June 10, 1990; revised February 28, 1992

Let $Q_j(\lambda) = Q_j(\lambda_1, ..., \lambda_s)$ $(1 \leqslant j \leqslant h)$ be a system of quadratic forms with coefficients in integers of an algebraic number field K of degree n, and let $\mathfrak{a}$ be an integral ideal of K. The purpose of the paper is to prove that the system of congruences $Q_j(\lambda) \equiv 0$ $(\mathrm{mod}\ \mathfrak{a})$ $(1 \leqslant j \leqslant h)$ has a nonzero solution λ satisfying $\max_{i,j} |\lambda_j^{(i)}| \ll N(\mathfrak{a})^{1/2+\varepsilon}$ provided that $s \geqslant c(h, n, \varepsilon)$. This improves a result of T. Cochrane (1987, *Illinois J. Math.* **31**, 618–625) and also gives a generalisation of a result of R. C. Baker (1980, *Mathematika* **27**, 30–45). The small solutions of additive congruences are also considered. © 1993 Academic Press, Inc.

1. INTRODUCTION

Let K be an algebraic number field of degree n over $\mathbb{Q}$ with a ring of integers J. Let $Q_j(\lambda) = \sum_{1 \leqslant q, r \leqslant s} \alpha_{qr}(j)\, \lambda_q \lambda_r$ $(\alpha_{qr}(j) = \alpha_{rq}(j),\ 1 \leqslant j \leqslant h)$ be a system of h quadratic forms in $\lambda = (\lambda_1, ..., \lambda_s)$ with coefficients in J. Using a generalised geometrical method of Schinzel, Schlickewei, and Schmidt [5] on the small solutions of quadratic congruences, Cochrane [3] proved that if $\mathfrak{a}$ is a nonzero integral ideal of K and $h < s/2$, then the system of congruences

$$Q_j(\lambda) \equiv 0 \qquad (\mathrm{mod}\ \mathfrak{a}), \qquad 1 \leqslant j \leqslant h \tag{1}$$

has a solution λ in J with

$$0 < \max_i |N(\lambda_i)| \leqslant c_1(K)\, N(\mathfrak{a})^\delta, \tag{2}$$

where

$$\delta = \begin{cases} \frac{1}{2} + 3/2(s-1), & \text{if } h = 1, 2, \\ h/(h+1) + h/(h+1)(s-r), & \text{if } h > 2. \end{cases}$$

$c(f, ..., g)$ denotes hereafter a positive constant depending on $f, ..., g, N(\lambda_i)$

262   WANG YUAN

and $N(\mathfrak{a})$ are norms of λ_i and $\mathfrak{a}$, and r is the remainder on dividing $s-h$ by $h+1$. However, in the rational case, Baker [1] has given a result in which $\delta \to \frac{1}{2}$ rather than $h/(h+1)$, as $s \to \infty$. Baker [2] proved also that for any given $\varepsilon > 0$ and integers $k, q \geqslant 2$, if $s \geqslant c_2(k, h, \varepsilon)$, then the system of h congruences

$$\sum_{j=1}^{s} a_{ij} x_j^k \equiv 0 \qquad (\bmod\, q),\ 1 \leqslant i \leqslant h$$

with coefficients in $\mathbb{Z}$ has a solution in nonnegative integers $x_1, ..., x_s$ satisfying

$$0 < \max_i x_i \leqslant q^{1/k + \varepsilon}. \tag{3}$$

We may use the form $x_1^k + \cdots + x_s^k$ $(k \geqslant 2)$ to show that the limiting exponents $\frac{1}{2}$ and $1/k$ in (2) and (3) are best possible, even when $K = \mathbb{Q}$ and $h = 1$.

The proof of Baker's results is based on a discrete version of Schmidt's method on small solutions of additive equations in many variables [6]. In Baker's argument, the relation

$$\sum_{u=1}^{q} \exp\left(2\pi i\, \frac{au}{q}\right) = \begin{cases} 1, & \text{if } q \,|\, a, \\ 0, & \text{if } q \nmid a, \end{cases}$$

is used instead of the integral over $[0, 1)$ in the circle method, and the case $h = 1$ of Baker's estimation (3) is derived immediately from the following Schmidt theorem: Let $a_1, ..., a_s, b_1, ..., b_s$ be natural numbers. Then there is a constant $c_3(k, \varepsilon)$ such that if $s \geqslant c_3(k, \varepsilon)$, the equation

$$a_1 x_1^k + \cdots + a_s x_s^k = b_1 y_1^k + \cdots + b_s y_s^k$$

has a solution in nonnegative integers $x_1, ..., x_s, y_1, ..., y_s$ with

$$0 < \max_{i,j} (x_i,\, y_j) \leqslant \max_{i,j} (a_i,\, b_j)^{1/k + \varepsilon}.$$

(See Schmidt [6].)

Since Schmidt's theorem has been generalised to algebraic numer fields by the combination of his method and Siegel's version of the circle method [7, 8], we could generalise the above Baker theorems to algebraic number fields by a similar discrete method.

Let $K^{(l)}$ $(1 \leqslant l \leqslant r_1)$ be the real conjugate of K, and $K^{(m)}$ and $K^{(m+r_2)}$ $(r_1 + 1 \leqslant m \leqslant r_1 + r_2)$ the complex conjugates of K, where $r_1 + 2r_2 = n$ is the degree of K. Throughout this paper, the indices l and m are over the sets of integers cited above. For $\gamma \in K$ we denote by $\gamma^{(i)}$ $(1 \leqslant i \leqslant n)$ the conjugates of γ. A number γ of K is called totally nonnegative if $\gamma^{(l)} \geqslant 0$, and we denote by P the set of all totally nonnegative integers of J. We use $\mathfrak{a}, \mathfrak{q}, \ldots$ to denote the nonzero integral ideals, $\wp, \wp_1, \ldots$ the prime ideals, δ the different (ground ideal) of K, and ε any pre-assigned positive number. The constants implicit in the symbols $\ll$ and 0 may depend on $k, h, K, \varepsilon, \ldots$ but not on $\mathfrak{a}, \mathfrak{q}, \ldots$.

THEOREM 1. *Let*

$$F_i(\lambda) = \sum_{j=1}^{s} \alpha_{ij} \lambda_j^k, \qquad 1 \leqslant i \leqslant h \tag{4}$$

be forms with coefficients in J. Then if $s \geqslant c_4(k, h, n, \varepsilon)$, there is a $\lambda \in P^s$ such that

$$F_i(\lambda) \equiv 0 \qquad (\text{mod } \mathfrak{a}), \qquad 1 \leqslant i \leqslant h \tag{5}$$

and

$$0 < \max_j N(\lambda_j) \ll N(\mathfrak{a})^{1/k + \varepsilon}.$$

THEOREM 2. *Suppose $s \geqslant c_s(h, n, \varepsilon)$. Then the system of quadratic congruences (1) has a solution λ in J^s with*

$$0 < \max_j |N(\lambda_j)| \ll N(\mathfrak{a})^{1/2 + \varepsilon}.$$

We prove Theorem 1 first and then derive Theorem 2 from Theorem 1 by the method of algebraic diagonalization with the aid of the box principle. Since the conclusions of Theorems 1 and 2 can be improved to $N(\mathfrak{a})^\varepsilon$ if K is a totally complex algebraic number field, we assume in the following that r_1 is positive. See Wang [8, Chap. 11].

2. LEMMAS

Let $\omega_1, \ldots, \omega_n$ be a basis of J and D the absolute value of the discriminant of K; i.e., $D = |\det(\omega_j^{(i)})|^2$ $(1 \leqslant i, j \leqslant n)$. Let γ_j $(1 \leqslant j \leqslant n)$ be numbers of K and x_j $(1 \leqslant j \leqslant n)$ be real numbers. We set $\xi = \sum_{j=1}^{n} x_j \gamma_j$ and define $\xi^{(i)} = \sum_{j=1}^{n} x_j \gamma_j^{(i)}$ $(1 \leqslant i \leqslant n)$. We use the notations

$$\|\xi\| = \max_i |\xi^{(i)}|, \qquad S(\xi) = \sum_{i=1}^{n} \xi^{(i)}, \qquad E(\xi) = \exp(2\pi i S(\xi)),$$

and $P(T)$ the set of integers in P satisfying $\|\lambda\| \leqslant T$.

264   WANG YUAN

LEMMA 1. *Suppose $\alpha \in J$. Then*

$$\sum_{\xi} E(\xi\alpha) = \begin{cases} N(\mathfrak{a}), & \text{if} \quad \mathfrak{a} \mid \alpha, \\ 0, & \text{otherwise}, \end{cases}$$

where ξ runs over a complete residue system of $(\mathfrak{a}\delta)^{-1}$, mod δ^{-1}.

See, e.g., Wang Yuan [8, Chap. 2].

LEMMA 2. *Let T_i $(1 \leqslant i \leqslant n)$ be positive numbers such that $T_{m+r_2} = T_m$. Let $\mu \in J$ and $N(\mathfrak{a}, \mathbf{T})$ denote the number v of P satisfying*

$$v \equiv \mu \pmod{\mathfrak{a}}, \qquad 0 \leqslant v^{(l)} \leqslant T_l, \qquad |v^{(m)}| \leqslant T_m.$$

Then

$$N(\mathfrak{a}, \mathbf{T}) = \frac{(2\pi)^{r_2}}{\sqrt{D}\, N(\mathfrak{a})}\, T_1 \cdots T_n + O\left(\frac{T_0^{n-1}}{N(\mathfrak{a})^{1-1/n}}\right),$$

where $T_0 = \max(N(\mathfrak{a})^{1/n}, (T_1 \cdots T_n)^{1/n})$.

See, e.g., Wang Yuan [8, Chap. 3].

Remark. Take $T_i = c_6(K)\, N(\mathfrak{a})^{1/n}$ $(1 \leqslant i \leqslant n)$, where $c_6(K)$ is sufficiently large. Then it follows from Lemma 2 that $N(\mathfrak{a}, \mathbf{T}) \gg 1$; i.e., there is a $v \in P$ satisfying $v \equiv \mu \pmod{\mathfrak{a}}$ and $0 < \|v\| \ll N(\mathfrak{a})^{1/n}$. Therefore we may choose a complete residue system $R(\mathfrak{a})$, modulo $\mathfrak{a}$, such that $\gamma \in P$ and $0 < \|\gamma\| \ll N(\mathfrak{a})^{1/n}$ for each $\gamma \in R(\mathfrak{a})$.

LEMMA 3. *Let $f(\lambda) = \alpha_k \lambda^k + \cdots + \alpha$, λ be a polynomial with coefficients in J and $\alpha_k \neq 0$ and let $S(f, T) = \sum_{\lambda \in P(T)} E(f(\lambda)\, \xi)$, where λ runs over all integers in $P(T)$. Suppose that $T \geqslant c_7(k, K, \varepsilon)$, $C \geqslant T^{n-1/2^{\varepsilon-1}+\varepsilon}$, and $|S(f, T)| \geqslant C$. Then there exist $\alpha \in P$ and $\beta \in J$ such that*

$$\|\alpha\alpha_k \xi - \beta\| \ll \left(\frac{T^n}{C}\right)^{2^{k-1}} T^{-k+\varepsilon}$$

and

$$0 < \|\alpha\| \ll \left(\frac{T^n}{C}\right)^{2^{k-1}} T^{\varepsilon}.$$

See, e.g., Wang Yuan [8, Chap. 3].

LEMMA 4. *Let $F(\lambda) = \sum_{i=1}^{s} \alpha_i \lambda_i^k$ be an additive form with nonzero integer* *coefficients in* J. *Suppose that* $s \geqslant c_8(k, n)$ *and that* $\alpha_i^{(l)}$ $(1 \leqslant i \leqslant s)$ *have different signs for each* l. *Then the equation*

$$F(\lambda) = 0$$

has a solution $\lambda \in P^s$ *with*

$$0 < \max_i \|\lambda_i\| \ll \max_i \|\alpha_i\|^{c_9(k,n)}$$

See, e.g., Wang Yuan [8, Chap. 8].

LEMMA 5. *Let* $c = c_8(k, n)$ *and* r *be a natural number. Then there is a matrix* $G_r = (g_{ij})$ $(1 \leqslant i \leqslant r, \ 1 \leqslant j \leqslant c^r)$ *whose entries are* ± 1 *having the following property. Given integers* β_{ij} *with* $\|\beta_{ij}\| \leqslant B$, *there is a* λ *in* P^{c^r} *satisfying*

$$\sum_{j=1}^{c^r} g_{ij}\beta_{ij}\lambda_j^k = 0, \qquad 1 \leqslant i \leqslant r \tag{6}$$

and

$$0 < \max_i \|\lambda_i\| \ll B^{c_{10}(k,r,n)}. \tag{7}$$

Proof. We take $c \geqslant 2r_1$, and for each j, $1 \leqslant j \leqslant r_1$, we choose $g_{1,2j-1}$ and $g_{1,2j}$ such that $g_{1,2j-1}$, $\beta_{1,2j-1}^{(j)}$ and $g_{1,2j}\beta_{1,2j}^{(j)}$ have different signs. Note that if there is a $\beta_{1,j} = 0$, then the assertion of the lemma is obvious for $r = 1$. So the case $r = 1$ of the lemma follows from Lemma 4 by taking $c_{10}(k, 1, n) = c_9(k, n)$. Now suppose that $r \geqslant 2$ and that the lemma is true for $r - 1$. Let G_r be formed by G_{r-1} as follows:

$$G_r = \begin{pmatrix} G_{r-1}^{(1)} & & & G_{r-1}^{(1)} \\ & & \cdots & \\ g_{r1}, \ \ldots, \ g_{rc^{r-1}} & & & g_{r,(c-1)c^{r-1}+1}, \ \ldots, \ g_{rc^r} \end{pmatrix}.$$

The vertical lines here divide G_r into c blocks. The variables λ_j $(1 \leqslant j \leqslant c^r)$ are also divided into c blocks with indices $B_1 = \{1, \ldots, c^{r-1}\}, \ldots, B_c = \{(c-1)c^{r-1}+1, \ldots, c^r\}$. By induction we can choose g_{ij} $(1 \leqslant i \leqslant r-1, j \in B_v)$ such that there are μ_j $(j \in B_v)$ in P satisfying

$$\sum_{j \in B_v} g_{ij}\beta_{ij}\mu_j^k = 0, \qquad i \leqslant i \leqslant r-1 \tag{8}$$

and

$$0 < \max_j \|\mu_j\| \ll B^{c_{10}(k,r-1,n)}. \tag{9}$$

266   WANG YUAN

For $1 \leqslant v \leqslant c$, let α_v be the integers given by

$$\alpha_v = \sum_{j \in B_v} \beta_{rj} \mu_j^k.$$

Then we may choose $g_{rj} = g_r(B_v)$ $(j \in B_v)$ such that $g_r(B_v) \alpha_v^{(l)}$ $(1 \leqslant v \leqslant c)$ have different signs for each l. Consider the equation

$$g_r(B_1) \alpha_1 v_1^k + \cdots + g_r(B_r) \alpha_c v_c^k = 0. \tag{10}$$

It follows from Lemma 4 that (10) has a solution $\mathbf{v}$ in P^c satisfying

$$0 < \max_i \|v_i\| \ll \max_i \|\alpha_i\|^{c_q(k,n)} \ll B^{(1 + kc_{10}(k,r-1,n))c_q(k,n)}. \tag{11}$$

Define $\lambda_i = v_v \mu_i$ $(i \in B_v)$. It follows from (8) that (6) holds for $i = 1, ..., r - 1$. From (10), we deduce that (6) also holds for $j = r$. Equation (7) follows from (9) and (11) and by taking

$$c_{10}(k, n, r) = (1 + (k + 1) c_{10}(k, r - 1, n)) c_9(k, n).$$

The lemma is proved.

Remark. Lemma 5 was established by Baker [2] for $K = \mathbb{Q}$ and $\beta_{ij} \geqslant 0$.

LEMMA 6. *For any given 2s nonzero integers* $\alpha_1, ..., \alpha_s, \beta_1, ..., \beta_s$ *in P, the equation*

$$\alpha_1 \lambda_1^k + \cdots + \alpha_s \lambda_s^k = \beta_1 \mu_1^k + \cdots + \beta_s \mu_s^k$$

has a solution in integers $\lambda_1, ..., \lambda_s, \mu_1, ..., \mu_s$ *of P, not all zero, such that*

$$\max_{i,j} (N(\lambda_i), N(\mu_j)) \ll \max_{i,j} (N(\alpha_i), N(\beta_j))^{1/k + \varepsilon}$$

provided that $s \geqslant c_{11}(k, n, \varepsilon)$.

See Wang Yuan [7, 8].

LEMMA 7 (The case $h = 1$ of Theorem 1). *Suppose that* $s \geqslant c_4(k, 1, n, \varepsilon)$ *and* $\alpha_i \in J$ $(1 \leqslant i \leqslant s)$. *Then the congruence*

$$\sum_{i=1}^s \alpha_i \lambda_i^k \equiv 0 \qquad (\mathrm{mod}\ \mathfrak{a}) \tag{12}$$

has a solution $\boldsymbol{\lambda} \in P^s$ *such that*

$$0 < \max_i N(\lambda_i) \ll N(\mathfrak{a})^{1/k + \varepsilon}. \tag{13}$$

Proof. By the remark of Lemma 2, we may suppose that $\alpha_i \in P$ and $N(\alpha_i) \ll N(\mathfrak{a})$ $(1 \leqslant i \leqslant s)$. Take $\alpha \in \mathfrak{a}$ such that $\alpha \in P$ and $N(\mathfrak{a}) \leqslant N(\alpha) \ll N(\mathfrak{a})$ and consider the equation

$$\alpha_1 \lambda_1^k + \cdots + \alpha_s \lambda_s^k = \alpha(\mu_1^k + \cdots + \mu_s^k). \tag{14}$$

Take $c_4(k, 1, n, \varepsilon) = c_{11}(k, n, \varepsilon)$. Then it follows from Lemma 6 that Eq. (14) has a solution in integers $\lambda_1, ..., \lambda_s, \mu_1, ..., \mu_s$ in P, not all zero, such that

$$\max_{i,j} (N(\lambda_i), N(\mu_j)) \ll \max(\max_i N(\alpha_i), N(\alpha))^{1/k + \varepsilon} \ll N(\mathfrak{a})^{1/k + \varepsilon}.$$

It is clear that $\lambda \neq \mathbf{0}$, since $r_1 > 0$. Hence from (14) we have a solution λ of (12) satisfying (13). The lemma is proved.

3. Reduction

PROPOSITION 1. *Let* $1/k \leqslant x \leqslant 1$ *and* $s \geqslant c_{12}(k, h, n, x, \varepsilon)$. *Let* $F_j(\lambda)$ $(1 \leqslant j \leqslant h)$ *be forms as in* (4). *Then for each ideal* $\mathfrak{a}$, *there are integers* $\lambda_1, ..., \lambda_s$ *in* P *satisfying* (5) *and*

$$0 < \max_i N(\lambda_i) \ll N(\mathfrak{a})^{x + \varepsilon}.$$

The case $x = 1/k$ is Theorem 1.

It is obvious that Proposition 1 is true for $x = 1$, because we can choose by Lemma 2 an integer α satisfying $\alpha \in P$, $\mathfrak{a} \mid \alpha$, and $0 < N(\alpha) \ll N(\mathfrak{a})$ and take $\lambda_1 = \cdots = \lambda_s = \alpha$.

It will suffice to prove Proposition 1 when $N(\mathfrak{a})$ is large, say $N(\mathfrak{a}) \geqslant c_{13}(k, h, K, x, \varepsilon)$. In fact, if $N(\mathfrak{a}) < c_{13}$ and $s \geqslant c_{12}$, then (5) has a solution $\lambda_1 = \cdots = \lambda_s = \alpha$ with α stated above such that

$$N(\alpha) \ll N(\mathfrak{a}) \ll c_{13} \ll N(\mathfrak{a})^{x + \varepsilon}.$$

Let X be the set of x such that Proposition 1 holds. X contains 1, and so X is not empty. It is clear that X is a closed set. Hence the proof of Proposition 1 is reduced to proving that if $1 \geqslant x > 1/k$ and $x \in X$, then there exists an $x' \in X$ with $x' < x$.

We assume that Theorem 1 holds for $h - 1$ forms $(h \geqslant 2)$ by the induction, since Theorem 1 is true for $h = 1$ by Lemma 7, and that Proposition 1 is true for a number x of X. We proceed to show that Proposition 1 is true for a number x' with $x' < x$.

Suppose that j is a rational integer $\geqslant 2$ and that there is no prime ideal $\wp$ such that $\wp^j \mid \mathfrak{a}$. Then $\mathfrak{a}$ is called a *j-free* ideal. Now we first show that

268

WANG YUAN

if Proposition 1 is true for square-free moduli in at least $c_{14}(k, h, n, x', \varepsilon)$ variables, then it is also true for j-free moduli, provided that $s \geqslant c_{14}^{j-1}$. The rational case of this assertion was established by Baker [2]. Now we use the induction and assume that the assertion is true for $(j-1)$-free moduli, where $j \geqslant 3$, and we proceed to prove the assertion for j-free moduli.

Write a j-free ideal $\mathfrak{a}$ as $\mathfrak{a} = \prod_{i=1}^{q} \wp_i^{a_i}$, where $1 \leqslant a_i \leqslant j-1$ $(1 \leqslant i \leqslant q)$ and where $\wp_i$, $\wp_r$ $(i \neq r)$ are distinct prime ideals. Let $b_i = a_i$ if $1 \leqslant a_i < j-1$ and $b_i = a_i - 1$ otherwise. Let $\mathfrak{a}_1 = \prod_{i=1}^{q} \wp_i^{b_i}$ and $\mathfrak{a}_2 = \prod_{i=1}^{q} \wp_i^{a_i - b_i}$. Then $\mathfrak{a} = \mathfrak{a}_1 \mathfrak{a}_2$, where $\mathfrak{a}_1$ is $(j-1)$-free and $\mathfrak{a}_2$ is square-free. We may choose an integer $\pi^{(i)}$ such that $(\pi^{(i)}, \mathfrak{a}) = \wp_i$ $(i = 1, ..., q)$. See, e.g., Hecke [4, Theorem 74]. Set $\pi_1 = \prod_{i=1}^{q} \pi^{(i)b_i}$ and $\pi_2 = \prod_{i=1}^{q} \pi^{(i)a_i - b_i}$. Then $\mathfrak{a}_1 | \pi_1$, $\mathfrak{a}_2 | \pi_2$ and $\pi_2 | \pi_1$. Any α in K which is a $\wp_i$-adic integer for all i has a unique representation

$$\alpha = \sum_{i=0}^{\infty} \mu_i \pi_1^i, \qquad \mu_i \in R(\mathfrak{a}_1),$$

where $R(\mathfrak{a}_1)$ is a given set of complete residue systems mod $\mathfrak{a}_1$. In fact, there is a unique μ_0 such that $\alpha \equiv \mu_0 \pmod{\mathfrak{a}_1}$ by the Chinese Remainder Theorem, and μ_1 is then determined by $\pi_1 \mu_1 \equiv \alpha - \mu_0 \pmod{\mathfrak{a}_1^2}$, and so on.

Set $u = C_{14}^{j-2}$, $v = c_{14}$, and $s = uv$. Rewrite the forms (4) by

$$F_j(\lambda_{11}, ..., \lambda_{vu}) = \sum_{i=1}^{v} (\alpha_{i1}(j) \lambda_{j1}^k + \cdots + \alpha_{iu}(j) \lambda_{iu}^k), \qquad 1 \leqslant j \leqslant h.$$

By the assumption of induction, we may choose integers $\lambda'_{i1}, ..., \lambda'_{iu}$ in P such that

$$\alpha_{i1}(j) \lambda_{i1}^{\prime k} + \cdots + \alpha_{iu}(j) \lambda_{iu}^{\prime k} \equiv 0 \qquad (\mathrm{mod}\ \mathfrak{a}), \ 1 \leqslant j \leqslant h$$

and

$$0 < \max_t N(\lambda'_{it}) \ll N(\mathfrak{a}_1)^{x' + \varepsilon}.$$

Write

$$\alpha_i(j) = \alpha_{i1}(j) \lambda_{i1}^{\prime k} + \cdots + \alpha_{iu}(j) \lambda_{iu}^{\prime k}, \qquad 1 \leqslant i \leqslant v, \ 1 \leqslant j \leqslant h.$$

Expand $\alpha_i(j)$ in a power series of π_1,

$$\alpha_i(j) = \pi_1 \beta_i(j),$$

where

$$\beta_i(j) = \sum_{l=0}^{\infty} \mu_{lj}^{(i)} \pi_1^l, \qquad \mu_{lj}^{(i)} \in R(\mathfrak{a}_1).$$

Since $\pi_2 | \pi_1$, $\beta_i(j)$ is also a power series of π_2, that is,

$$\beta_i(j) = \sum_{l=0}^{\infty} v_{lj}^{(i)} \pi_2^l, \qquad v_{lj}^{(i)} \in R(\mathfrak{a}_2).$$

Since $\mathfrak{a}_2$ is square-free, we derive by the inductive hypothesis that the congruences

$$\sum_{i=1}^{v} \beta_i(j) \mu_i^k \equiv \sum_{i=1}^{v} v_{0j}^{(i)} \mu_i^k \equiv 0 \qquad (\text{mod } \mathfrak{a}_2),\ 1 \leqslant j \leqslant h$$

have a solution in P^v satisfying

$$0 < \max_i N(\mu_i) \ll N(\mathfrak{a}_2)^{x'+\varepsilon}.$$

Set $\lambda_{it} = \mu_i \lambda_{it}'$ $(1 \leqslant i \leqslant v,\ 1 \leqslant t \leqslant u)$. Then we have

$$F_j(\lambda_{11}, \ldots, \lambda_{vu}) \equiv 0 \qquad (\text{mod } \mathfrak{a}),\ 1 \leqslant j \leqslant h$$

and

$$0 < \max_{i,t} N(\lambda_{it}) \ll N(\mathfrak{a})^{x'+\varepsilon}.$$

This completes the proof of the induction step.

In particular, Proposition 1 holds for x' if $s \geqslant c_{14}^{k-1}$ and $\mathfrak{a}$ is k-free. Now we may write any ideal $\mathfrak{a}$ as $\mathfrak{a} = \mathfrak{B}^k \mathfrak{C}$, where $\mathfrak{C}$ is k-free. So if we set $s = c_{14}^{k-1}$, we have integers $\lambda_1', \ldots, \lambda_s'$ in P such that

$$F_j(\lambda') = F_j(\lambda_1', \ldots, \lambda_s') \equiv 0 \qquad (\text{mod } \mathfrak{C}),\ 1 \leqslant j \leqslant h$$

and

$$0 < \max_i N(\lambda_i') \ll N(\mathfrak{C})^{x'+\varepsilon}.$$

We take an integer β by Lemma 2 such that $\beta \in P$, $\mathfrak{B}/\beta$, and $N(\mathfrak{B}) \leqslant N(\beta) \ll N(\mathfrak{B})$. Then $\lambda = \beta\lambda' = (\beta\lambda_1', \ldots, \beta\lambda_s')$ satisfies

$$F_j(\lambda) = \beta^k F_j(\lambda') \equiv 0 \qquad (\text{mod } \mathfrak{a}),\ 1 \leqslant j \leqslant h$$

and

$$0 < \max_i N(\lambda_i) = N(\beta) \max_i N(\lambda_i') \ll (N(\mathfrak{B})^k N(\mathfrak{C}))^{x'+\varepsilon} \ll N(\mathfrak{a})^{x'+\varepsilon}.$$

Therefore we may assume that $\mathfrak{a}$ is square-free in Proposition 1.

270   WANG YUAN

4. Division into Two Cases

Set

$$y = \left(x - \frac{1}{k} \right) c_{10}(k, h, n)^{-1} (2(h+6)\, n)^{-1}, \tag{15}$$

$$z = \left(x - \frac{1}{k} \right) y(2h+6)^{-1} \tag{16}$$

$$x' = x - z. \tag{17}$$

We have

$$\frac{1}{k} + c_{10}(k, h, n)\, 2y(h+5)\, n$$

$$= \frac{1}{k} + \left(x - \frac{1}{k} \right) 2(h+5)\, n(2(h+6)\, n)^{-1}$$

$$= \frac{1}{k} + \left(x - \frac{1}{k} \right) (h+5)(h+6)^{-1}$$

$$= x - \left(x - \frac{1}{k} \right) (h+6)^{-1} < x - z = x'. \tag{18}$$

Write $\mathbf{a}_i = (\alpha_{1i}, ..., \alpha_{hi})$ $(1 \leqslant i \leqslant s)$ for the coefficients of λ_i^k in (4). For any two vectors $\mathbf{\alpha} = (\alpha_1, ..., \alpha_h)$ and $\mathbf{\beta} = (\beta_1, ..., \beta_h)$, we denote by $\mathbf{\alpha\beta} = \sum_{i=1}^{h} \alpha_i \beta_i$ the inner product of $\mathbf{\alpha}$ and $\mathbf{\beta}$. A vector $\mathbf{\gamma} = (\gamma_1, ..., \gamma_h) \in J^h$ is said to be prime to an ideal q, if the greatest common divisor $(\gamma_1, ..., \gamma_h, q) = 1$, where γ_i is regarded as the principal ideal generated by γ_i.

Condition $D(k, h, t, \mathfrak{a}, y)$. For any t coefficient vectors, say $\mathbf{a}_1, ..., \mathbf{a}_t$, any divisor q of $\mathfrak{a}$, and any $\mathbf{\gamma}$ in J^h prime to q such that

$$\mathbf{\gamma a}_i \equiv 0 \qquad (\mathrm{mod}\ q),\ 1 \leqslant i \leqslant t,$$

we have

$$N(q) \leqslant N(\mathfrak{a})^y.$$

Proposition 2. *Let t be an integer $\geqslant 1$. Let the forms $F_j(\lambda)$ $(1 \leqslant j \leqslant h)$ have the property $D(k, h, t, \mathfrak{a}, y)$. Then if $s \geqslant c_{15}(k, h, n, x, t)$, the system of congruences*

$$F_j(\lambda) \equiv \mu_j \qquad (\mathrm{mod}\ \mathfrak{a}),\ 1 \leqslant j \leqslant h,$$

has a solution $\lambda_1, ..., \lambda_s, \mu_1, ..., \mu_h$ *in* P^{s+h} *satisfying*

$$0 < \max_i \|\lambda_i\| \ll N(\mathfrak{a})^{(1/k + y)(1/n)} \qquad and \qquad \max_i \|\mu_i\| \ll N(\mathfrak{a})^{2y(h+4)}.$$

Let $d = c_4(k, h-1, n, z)$, $w = c_{12}(k, h, n, x, z)$, and $t = dw$. We proceed to show that Proposition 1 is true if F_j $(1 \leqslant j \leqslant h)$ satisfy the condition $D(k, h, t, \mathfrak{a}, y)$. Write $c = c_g$, $u = c_{15}$, and $s = c^h u$. After a change of notation, (4) becomes

$$\sum_{r=1}^{c^h} (\alpha_{r1}(j) \lambda_{r1}^k + \cdots + \alpha_{ru}(j) \lambda_{ru}^k) \equiv 0 \qquad (\mathrm{mod}\ \mathfrak{a}), \ 1 \leqslant j \leqslant h. \qquad (19)$$

For each r, $1 \leqslant r \leqslant c^h$, consider the system of h forms

$$g_{jr}(\alpha_{r1}(j) \lambda_{r1}^k + \cdots + \alpha_{ru}(j) \lambda_{ru}^k) \qquad (1 \leqslant j \leqslant h)$$

in u variables. These forms satisfy the condition $D(k, h, t, \mathfrak{a}, y)$. Note that the introduction of g_{jr} $(= \pm 1)$ does not affect the condition $(\gamma_1, ..., \gamma_h, \mathfrak{q}) = 1$. We choose g_{jr} $(1 \leqslant j \leqslant h, \ 1 \leqslant r \leqslant c^h)$ for the case such that β_{jr} are all totally nonnegative in Lemma 5. It follows from Proposition 2 that the system of congruences

$$g_{jr}(\alpha_{r1}(j) \lambda_{r1}'^k + \cdots + \alpha_{ru}(j) \lambda_{ru}'^k) \equiv v_{rj} \qquad (\mathrm{mod}\ \mathfrak{a}), \ 1 \leqslant j \leqslant h, \qquad (20)$$

has a solution $\lambda_{r1}', ..., \lambda_{ru}', v_{r1}, ..., v_{rh}$ in P^{u+h} satisfying

$$0 < \max_i \|\lambda_{ri}'\| \ll N(\mathfrak{a})^{(1/k + y)(1/n)}, \qquad \max_j \|v_{rj}\| \ll N(\mathfrak{a})^{2y(h+4)}. \qquad (21)$$

By Lemma 5, the system of equations

$$\sum_{r=1}^{c^h} g_{jr} v_{rj} \mu_r^k = 0, \qquad 1 \leqslant j \leqslant h \qquad (22)$$

has a solution μ_r $(1 \leqslant r \leqslant c^h)$ satisfying

$$0 < \max_r \|\mu_r\| \ll N(\mathfrak{a})^E, \qquad (23)$$

where $E = c_{10}(k, h, n)\, 2y(h+4)$.

By (18), (20)–(23), we see that $\lambda_{ri} = \mu_r \lambda_{ri}'$ $(1 \leqslant r \leqslant c^k, \ 1 \leqslant i \leqslant u)$ is a solution of (19) satisfying $\lambda_{ri} \in P$ and

$$0 < \max_{r,i} N(\lambda_{ri}) \ll N(\mathfrak{a})^{1/k + y + En} \ll N(\mathfrak{a})^{x'}, \qquad 1 \leqslant r \leqslant c^h, \ 1 \leqslant i \leqslant u.$$

We are left with the case where the condition $D(k, h, t, \mathfrak{a}, y)$ fails; i.e.,

272 WANG YUAN

there is a divisor $\mathfrak{q}$ of $\mathfrak{a}$, t coefficient vectors $\mathbf{a}, ..., \mathbf{a}_t$ (say), and an integral vector $\gamma = (\gamma_1, ..., \gamma_h)$ such that

$$N(\mathfrak{q}) > N(\mathfrak{a})^v, \tag{24}$$

$$(\gamma_1, ..., \gamma_h, \mathfrak{q}) = 1 \tag{25}$$

and

$$\gamma \mathbf{a}_i \equiv 0 \qquad (\bmod \ \mathfrak{q}), \ 1 \leqslant i \leqslant t. \tag{26}$$

We set all the variables, except the first $t = dw$, equal to zero, and so

$$F_j(\lambda) = \sum_{u=1}^{d} (\alpha_{u1}(j)\,\lambda_{u1}^k + \cdots + \alpha_{uw}(j)\,\lambda_{uw}^k), \qquad 1 \leqslant j \leqslant h. \tag{27}$$

A consequence of (26) is that the congruence

$$\gamma_1 F_1(\lambda) + \cdots + \gamma_h F_h(\lambda) \equiv 0 \qquad (\bmod \ \mathfrak{q}) \tag{28}$$

holds for every integer vector λ. Now we show that there is a divisor $\mathfrak{q}'$ of $\mathfrak{q}$ such that

$$N(\mathfrak{q}) \geqslant N(\mathfrak{q})^{1/h}, \tag{29}$$

and some j, $1 \leqslant j \leqslant h$, for which

$$(\gamma_j, \mathfrak{q}') = 1. \tag{30}$$

In fact, since $\mathfrak{q}$ is square-free, we write $\mathfrak{q} = \wp_1 \cdots \wp_a$, where $\wp_i$ are distinct prime ideals. Because of (25), no $\wp_j$ divides every element of $\gamma_1, ..., \gamma_h$, hence

$$\prod_{j=1}^{h} (\gamma_j, \mathfrak{q}) \mid \wp_1^{h-1} \cdots \wp_a^{h-1} - \mathfrak{q}^{h-1}.$$

Thus $N((\gamma_j, \mathfrak{q})) \leqslant N(\mathfrak{q})^{1-1/h}$ for some j in $1 \leqslant j \leqslant h$, and (29) and (30) follow by taking $\mathfrak{q}' = \mathfrak{q}/(\gamma_j, \mathfrak{q})$. We may suppose without loss of generality that j in (30) is $j = 1$. From (28), we deduce that if

$$F_2(\lambda) \equiv \cdots \equiv F_h(\lambda) \equiv 0 \qquad (\bmod \ \mathfrak{q}'), \tag{31}$$

then

$$F_1(\lambda) \equiv 0 \qquad (\bmod \ \mathfrak{q}'). \tag{32}$$

For each $u \leqslant d$, it follows by the definition of $w = c_{12}(k, h, n, x, z)$ that the system of congruences

$$\alpha_{u1}(j)\,\lambda_{u1}'^k + \cdots + \alpha_{uw}(j)\,\lambda_{uw}'^k \equiv 0 \qquad \left(\bmod \frac{\mathfrak{a}}{\mathfrak{q}'}\right),\ 1 \leqslant j \leqslant h, \qquad (33)$$

has a solution $\lambda_{u1}', \ldots, \lambda_{uw}'$ in P^w such that

$$0 < \max_i N(\lambda_{ui}') \ll N\left(\frac{\mathfrak{a}}{\mathfrak{q}'}\right)^{x+z}. \qquad (34)$$

We denote by $\alpha_u(j)$ the expression on the left-hand side of (33). Since $d = c_4(k, h-1, n, z)$ and Theorem 1 is assumed to be true for $h - 1$ forms by induction, i.e., the system of congruences

$$\alpha_1(j)\,\mu_1^k + \cdots + \alpha_d(j)\,\mu_d^k \equiv 0 \qquad (\bmod\ \mathfrak{q}'),\ 2 \leqslant j \leqslant h, \qquad (35)$$

has a solution $\mu_1, \ldots, \mu_d$ in P^d satisfying

$$0 < \max_i N(\mu_i) \ll N(\mathfrak{q}')^{1/k + z}. \qquad (36)$$

Write $\lambda_{ui} = \mu_u \lambda_{ui}'$ $(1 \leqslant u \leqslant d,\ 1 \leqslant i \leqslant w)$. From (27) and (35), it follows that (31) holds. So by (32), we have

$$F_1(\lambda) \equiv F_2(\lambda) \equiv \cdots \equiv F_h(\lambda) \equiv 0 \qquad (\bmod\ \mathfrak{q}') \qquad (37)$$

and by (33), we have

$$F_1(\lambda) \equiv F_2(\lambda) \equiv \cdots \equiv F_h(\lambda) \equiv 0 \qquad \left(\bmod \frac{\mathfrak{a}}{\mathfrak{q}'}\right). \qquad (38)$$

By the combination of (37) and (38), we deduce that

$$F_1(\lambda) \equiv F_2(\lambda) \equiv \cdots \equiv F_h(\lambda) \equiv 0 \qquad (\bmod\ \mathfrak{a}),$$

since $\mathfrak{a}$ is square-free. And by (34) and (36), we have

$$0 < \max_{i,j} N(\lambda_{ij}) \ll N(\mathfrak{q}')^{1/k + z} \left(\frac{N(\mathfrak{a})}{N(\mathfrak{q}')}\right)^{x+z}$$

$$\ll N(\mathfrak{a})^{x+z} N(\mathfrak{q}')^{-(x - 1/k)}. \qquad (39)$$

Now $N(\mathfrak{q}') \geqslant N(\mathfrak{q})^{1/h}$ and $N(\mathfrak{q}) > N(\mathfrak{a})^y$ from (24) and (29). Therefore the right-hand side of (39) is

$$\ll N(\mathfrak{a})^{x + z - (x - 1/k)y/h} \ll N(\mathfrak{a})^{x + z - 2z} \ll N(\mathfrak{a})^{x'}$$

by (16), and Proposition 1 is proved for x'.

WANG YUAN

5. PROOF OF PROPOSITION 2

Let

$$T = N(\mathfrak{a})^{(1/k + y)(1/n)}, \qquad Z = N(\mathfrak{a})^{2y(h+4)}, \qquad s \geqslant t^2(1 + z^{-1}2^k(h+2)). \qquad (40)$$

Let W denote the number of solutions of the congruences

$$F_j(\lambda) \equiv \mu_j \qquad (\text{mod } \mathfrak{a}), \ 1 \leqslant j \leqslant h,$$

satisfying $\lambda_i \in P(T)$, $1 \leqslant i \leqslant s$, and $\mu_j \in P(Z)$, $1 \leqslant j \leqslant h$, where $F_j(\lambda)$ $(1 \leqslant j \leqslant h)$ are forms stated in (4). By Lemma 1, it follows that

$$N(\mathfrak{a})^{-1} \sum_\beta E((F_j(\lambda) - \mu_j)\,\beta) = \begin{cases} 1, & \text{if } F_j(\lambda) \equiv \mu_j \,(\text{mod } \mathfrak{a}), \\ 0, & \text{otherwise}, \end{cases}$$

where β runs over a complete residue system of $(\mathfrak{a}\delta)^{-1}$, mod δ^{-1}. Hence

$$W = N(\mathfrak{a})^{-h} \sum_{\substack{\lambda_i \in P(T) \\ 1 \leqslant i \leqslant s}} \sum_{\substack{\mu_j \in P(Z) \\ 1 \leqslant j \leqslant h}} \sum_{\substack{\beta_q \\ 1 \leqslant q \leqslant h}} E\left(\sum_{r=1}^{h} (F_r(\lambda) - \mu_r)\,\beta_r \right),$$

where each β_q runs over a complete residue system of $(\mathfrak{a}\delta)^{-1}$, mod δ^{-1}, $1 \leqslant q \leqslant h$. We may write W as

$$W = N(\mathfrak{a})^{-h} \sum_{\substack{\lambda_i \in P(T) \\ 1 \leqslant i \leqslant s}} \sum_{\substack{\mu_j \in P(Z) \\ 1 \leqslant j \leqslant h}} \sum_{q \mid \mathfrak{a}} \sideset{}{^*}\sum_{\substack{\gamma_q \\ 1 \leqslant q \leqslant h}} E$$
$$\times \left(\sum_{r=1}^{h} (F_r(\lambda) - \mu_r)\,\gamma_r \right),$$

where $\sum^*$ denotes a sum such that each γ_q runs over a complete residue system of $(\mathfrak{q}\delta)^{-1}$, mod δ^{-1}, $1 \leqslant q \leqslant h$, and satisfies $((\mathfrak{q}\delta)\,\gamma_1, ..., (\mathfrak{q}\delta)\,\gamma_h, \mathfrak{q}) = 1$. We write

$$N(\mathfrak{a})^h\,W = \sum_{\mathfrak{q} \mid \mathfrak{a}} S(\mathfrak{q}), \qquad (41)$$

where

$$S(\mathfrak{q}) = \sum_{\substack{\lambda_i \in P(T) \\ 1 \leqslant i \leqslant s}} \sum_{\substack{\mu_j \in P(Z) \\ 1 \leqslant j \leqslant h}} \sideset{}{^*}\sum_{\substack{\gamma_q \\ 1 \leqslant g \leqslant h}} E$$
$$\times \left(\sum_{r=1}^{h} (F_r(\lambda) - \mu_r)\,\gamma_r \right).$$

By Lemma 2, it follows that

$$\sum_{\lambda \in P(T)} 1 = \frac{(2\pi)^{r_2}}{\sqrt{D}} T^n + O(T^{n-1}),$$

and so

$$S(1) \gg T^{sn} Z^{hn}. \tag{42}$$

Now suppose that $N(\mathfrak{q}) > 1$. By Lemma 1 we have

$$\sum_{\mu \,(\mathrm{mod}\, \mathfrak{q})} E(-\mu\gamma) = 0,$$

where μ runs over a complete residue system, mod $\mathfrak{q}$, and where $\gamma \in (\mathfrak{q}\delta)^{-1}$, $\gamma \notin \delta^{-1}$. By Lemma 2 we see that the number of integers μ satisfying $\mu \in P(Z)$ and $\mu \equiv v \;(\mathrm{mod}\, \mathfrak{q})$ is equal to

$$\frac{(2\pi)^{r_2}}{\sqrt{D}\, N(\mathfrak{q})} Z^n + O\left(\frac{Z^{n-1}}{N(\mathfrak{q})^{1-1/n}}\right).$$

Hence if the domain $\mu \in P(Z)$ is split up into a union of a complete residue system, mod $\mathfrak{q}$, plus a few other, remaining elements, say R elements, then

$$R \ll N(\mathfrak{q}) \frac{Z^{n-1}}{N(\mathfrak{q})^{1-1/n}} \ll N(\mathfrak{q})^{1/n} Z^{n-1}.$$

Therefore

$$S(\mathfrak{q}) \ll T^{sn} Z^{hn-1} N(\mathfrak{q})^{h+1/n}.$$

Since the number of ideals $\mathfrak{q}$ with $N(\mathfrak{q}) = d$ is $O(d^{\varepsilon})$ (see, e.g., Hecke [4, Theorems 35, 76]), we have

$$\sum_{\substack{\mathfrak{q}\,|\,\mathfrak{a} \\ 1 < N(\mathfrak{q}) \leqslant N(\mathfrak{a})^{2y}}} |S(\mathfrak{q})| \ll T^{sn} Z^{hn-1} \sum_{1 < N(\mathfrak{q}) \leqslant N(\mathfrak{a})^{2y}} N(\mathfrak{q})^{h+1/n}$$

$$\ll T^{sn} Z^{hn-1} \sum_{d \leqslant N(\mathfrak{a})^{2y}} d^{h+2}$$

$$\ll T^{sn} Z^{hn-1} N(\mathfrak{a})^{2y(h+3)}$$

$$\ll T^{sn} Z^{hn} N(\mathfrak{a})^{-y_n}. \tag{43}$$

We claim that

$$\sum_{\substack{\mathfrak{q}\,|\,\mathfrak{a} \\ N(\mathfrak{q}) > N(\mathfrak{a})^{2y}}} |S(\mathfrak{q})| \ll T^{sn} Z^{hn} N(\mathfrak{a})^{-1}. \tag{44}$$

276   WANG YUAN

Otherwise there is some q satisfying $q \mid \mathfrak{a}$ and $N(q) > N(\mathfrak{a})^{2y}$ such that

$$|S(q)| \gg T^{sn} Z^{hn} N(\mathfrak{a})^{-2}.$$

Then we have

$$\left| \sum_{\substack{\lambda_i \in P(T) \\ 1 \leqslant i \leqslant s}} E\left(\sum_{j=1}^{s} \gamma_j F_j(\lambda) \right) \right| \gg T^{sn} N(\mathfrak{a})^{-2} N(q)^{-h}$$

for some γ with $((q\delta)\gamma_1, ..., (q\delta)\gamma_h, q) = 1$, that is,

$$\prod_{i=1}^{s} |S_i| \gg T^{sn} N(\mathfrak{a})^{-2} N(q)^{-h}, \tag{45}$$

where

$$S_i = \sum_{\lambda \in P(T)} E\left(\sum_{j=1}^{h} \gamma_j \alpha_{ji} \lambda^k \right)$$

$$= \sum_{\lambda \in P(T)} E(\gamma \mathfrak{a}_i \lambda^k).$$

We may suppose without loss of generality that

$$|S_1| \geqslant \cdots \geqslant |S_s|.$$

Then the left-hand side of (45) is

$$\ll T^{(t-1)n} |S_t|^{(s+1-t)}.$$

It follows that for $i = 1, ..., t$,

$$|S_i| \geqslant |S_t| \gg T^n (N(\mathfrak{a})^2 N(q)^h)^{-1/(s+1-t)}.$$

Take t sufficiently large. We have by (40),

$$(N(\mathfrak{a})^2 N(q)^h)^{1/(s+1-t)} \leqslant N(\mathfrak{a})^{(h+2)/(s+1-t)} \leqslant N(\mathfrak{a})^{z/t^2 2^k}$$

Set $C = T^n N(\mathfrak{a})^{-z/t^2 2^k}$. We have $C \geqslant T^{n - 1/2^{k-1} + z/t^2 2^k}$. Hence by Lemma 3, we have $\sigma_i \in P$ and $\tau_i \in J$ $(1 \leqslant i \leqslant t)$ such that

$$\|(\gamma \mathfrak{a}_i)\sigma_i - \tau_i\| \ll N(\mathfrak{a})^{z/2t^2} T^{-k + z/2t^2} \ll N(\mathfrak{a})^{-(1 + ky)/n + z/t^2}$$

and

$$0 < \|\sigma_i\| \ll N(\mathfrak{a})^{z/2t^2} N(\mathfrak{a})^{z/2t^2} \ll N(\mathfrak{a})^{z/t^2}, \qquad 1 \leqslant i \leqslant t.$$

Therefore

$$N((\gamma\mathbf{a}_i)\,\sigma_i - \tau_i) \ll N(\mathfrak{a})^{-1-ky+nz/t^2} < (DN(\mathfrak{a}))^{-1}$$

$$\leqslant (DN(\mathfrak{q}))^{-1}, \qquad 1 \leqslant i \leqslant t,$$

because $N(\mathfrak{a}) \geqslant c_{13}(k, h, K, x', \varepsilon)$. This means that

$$(\gamma\mathbf{a}_i)\,\sigma_i = \tau_i, \qquad 1 \leqslant i \leqslant t.$$

Since $((\mathfrak{q}\delta)\,\gamma_1, ..., (\mathfrak{q}\delta)\,\gamma_h, \mathfrak{q}) = 1$, we may choose $\beta \in \mathfrak{q}\delta$ such that

$$(\beta\gamma_1, ..., \beta\gamma_h, \mathfrak{q}) = 1.$$

Since

$$\mathfrak{q} \mid (\beta\gamma\mathbf{a}_i)\,\sigma_i, \qquad 1 \leqslant i \leqslant t,$$

we have

$$\mathfrak{q}' \mid \beta\gamma\mathbf{a}_i, \qquad 1 \leqslant i \leqslant t,$$

where

$$\mathfrak{q}' = \frac{\mathfrak{q}}{(\mathfrak{q}, \sigma_1 \cdots \sigma_t)}.$$

Therefore we have

$$N(\mathfrak{q}') \geqslant \frac{N(\mathfrak{q})}{N(\sigma_1 \cdots \sigma_t)} \gg N(\mathfrak{q})\,N(\mathfrak{a})^{-nz/t} \gg N(\mathfrak{a})^{2y-nz/t} \gg N(\mathfrak{a})^y.$$

This gives a contradiction, and we have proved (44). By (41)–(44), we have

$$W \gg N(\mathfrak{a})^{-h}\,T^{sn}Z^{hn} \gg 1.$$

Thus Proposition 2 and also Theorem 1 are proved.

6. Proof of Theorem 2

With a quadratic form

$$Q(\lambda) = \sum_{1 \leqslant i,j \leqslant s} \alpha_{ij}\lambda_i\lambda_j \qquad (\alpha_{ij} = \alpha_{ji})$$

we associate the bilinear form

$$\hat{Q}(\lambda, \mu) = \sum_{1 \leqslant i,j \leqslant s} \alpha_{ij}\lambda_i\mu_j.$$

278   WANG YUAN

Then

$$\hat{Q}(\lambda, \mu) = \hat{Q}(\mu, \lambda) \qquad \text{and} \qquad \hat{Q}(\lambda, \lambda) = Q(\lambda).$$

By the remark of Lemma 2, we may take $R(\mathfrak{a})$ as a complete residue system, $\bmod\ \mathfrak{a}$, such that for any $\alpha \in R(\mathfrak{a})$, we have $0 < \|\alpha\| \leqslant c_{16}(K) N(\mathfrak{a})^{1/n}$. We assume that the coefficients of $Q_i(\lambda)$ $(1 \leqslant i \leqslant h)$ belong to $R(\mathfrak{a})$.

Set $u = c_4(2, h, n, \varepsilon/2)$, $v = c_{17}(h, n, \varepsilon)$, and $s = uv$. Let

$$\mu_i = (\underbrace{0, ..., 0}_{(i-1)v}, \mu_{i1}, ..., \mu_{iv}, \underbrace{0, ..., 0}_{(u-i)v}), \qquad 1 \leqslant i \leqslant u.$$

Now we define $\mu_1, ..., \mu_u$ by induction such that

(1) μ_{ij} $(1 \leqslant i \leqslant u,\ 1 \leqslant j \leqslant v)$ are integers in J,
(2) $\|\mu_{ij}\| \leqslant N(\mathfrak{a})^{\varepsilon/2n}$ $(1 \leqslant i \leqslant u,\ 1 \leqslant j \leqslant v)$,
(3) $\hat{Q}_j(\mu_r, \mu_q) = 0$ $(1 \leqslant j \leqslant h,\ 1 < r < q \leqslant u)$.

We define μ_1 by $\mu_{11} = \cdots = \mu_{1v} = 1$, which satisfies conditions (1), (2), and (3). Suppose now that $\mu_1, ..., \mu_{q-1}$ have been defined. Consider the $h(q-1)(r_1 + r_2)$ linear forms

$$Q_j^{(l)}(\mu_r, \mu_q), \qquad Q_j^{(m)}(\mu_r, \mu_q), \qquad 1 \leqslant j \leqslant h,\ 1 \leqslant r < q.$$

Consider the points $\mu \in P^v$ with components $\mu_i \in P(N(\mathfrak{a})^{\varepsilon/2n}/2)$, $1 \leqslant i \leqslant v$. Then by Lemma 2 it follows that there are $c_{18}(K) N^{v\varepsilon/2}$ such points. Given such a μ, the point

$$\chi(\mu) = (\hat{Q}_1^{(1)}(\mu_1, \mu), ..., \hat{Q}_h^{(r_1)}(\mu_{q-1}, \mu), \hat{Q}_1^{(r_1+1)}(\mu_1, \mu), ..., \hat{Q}_h^{(r_1+r_2)}(\mu_{q-1}, \mu))$$

lies in the domain D_i $(f_1, ..., f_{h(q-1)r_1}, g_1, ..., g_{h(q-1)r_2})$, where

$$-c_{16}N(\mathfrak{a})^{1/n+\varepsilon/2n} v \leqslant f_j \leqslant c_{16}N(\mathfrak{a})^{1/n+\varepsilon/2n} v, \qquad 1 \leqslant j \leqslant h(q-1)r_1,$$
$$g_j = x_j + iy_j, \qquad\qquad x_j, y_j \in \mathbb{R},$$
$$|x_j| \leqslant c_{16}N(\mathfrak{a})^{1/n+\varepsilon/2n} v,$$
$$|y_j| \leqslant c_{16}N(\mathfrak{a})^{1/n+\varepsilon/2n} v, \qquad 1 \leqslant j \leqslant h(q-1)r_2.$$

Let t be the rational integer in

$$(c_{18}N(\mathfrak{a})^{v\varepsilon/2})^{1/h(q-1)n} - 1 \leqslant t < (c_{18}N(\mathfrak{a})^{v\varepsilon/2})^{1/h(q-1)n}$$

and divide each interval of f_j in D into t subintervals of length

$2c_{16}N(\mathfrak{a})^{1/n+\varepsilon/2n}vt^{-1}$ and each square g_j in D into t^2 subsquares of side $2c_{16}N(\mathfrak{a})^{1/n+\varepsilon/2n}vt^{-1}$. Since

$$t^{h(q-1)n}<c_{18}N(\mathfrak{a})^{v\varepsilon/2},$$

it follows by the Box principle that there are two distinct points $\chi(\mu')$ and $\chi(\mu'')$ such that $\hat{Q}_j^{(l)}(\mu_r,\mu')$ and $\hat{Q}_j^{(l)}(\mu_r,\mu'')$ will lie in the same subinterval for each j, r, l and $\hat{Q}_j^{(m)}(\mu_r,\mu')$ and $\hat{Q}_j^{(m)}(\mu_r,\mu'')$ in the same subsquare for each j, r, m. Set

$$\mu_q=\mu'-\mu''.$$

Then

$$0<\|\mu_q\|\leqslant\|\mu'\|+\|\mu''\|\leqslant N(\mathfrak{a})^{\varepsilon/2n}$$

and

$$|\hat{Q}_j^{(l)}(\mu_r,\mu_q)|\leqslant 2c_{16}N(\mathfrak{a})^{1/n+\varepsilon/2n}vt^{-1}<1,$$
$$|\hat{Q}_j^{(m)}(\mu_r,\mu_q)|\leqslant\sqrt{2(2c_{16}N(\mathfrak{a})^{1/n+\varepsilon/2n}vt^{-1})^2}$$
$$<3c_{16}N(\mathfrak{a})^{1/n+\varepsilon/2n}vt^{-1}<1,\qquad 1\leqslant j\leqslant h,\ 1\leqslant r<q,$$

if we take $v=c_{17}$ sufficiently large. Hence

$$|N(\hat{Q}_j(\mu_r,\mu_q))|<1,\qquad 1\leqslant j\leqslant h,\ 1\leqslant r<q.$$

Since $\hat{Q}_j(\mu_r,\mu_q)\in J$, we have

$$\hat{Q}_j(\mu_r,\mu_q)=0,\qquad 1\leqslant j\leqslant h,\ 1\leqslant r<q.$$

So μ_q is well defined, and thus we obtain $\mu_1,...,\mu_u$ by induction. Set

$$\lambda=v_1\mu_1+\cdots+v_u\mu_u.$$

Then

$$Q_j(\lambda)=\sum_{1\leqslant q,r\leqslant u}\hat{Q}_j(\mu_r,\mu_q)v_rv_q$$
$$=\sum_{q=1}^{u}\hat{Q}_j(\mu_q,\mu_q)v_q^2,\qquad 1\leqslant j\leqslant h.$$

By Theorem 1 it follows that the diagonal congruences

$$\sum_{q=1}^{u}\hat{Q}_j(\mu_q,\mu_q)v_q^2\equiv 0\qquad(\text{mod }\mathfrak{a}),\qquad 1\leqslant j\leqslant h,$$

WANG YUAN

has a solution $\mathbf{v}$ in P^u satisfying

$$0 < \max_i N(v_i) \ll N(\mathfrak{a})^{1/2 + \varepsilon/2},$$

and therefore the system of quadratic congruences

$$Q_j(\lambda) \equiv 0 \qquad (\bmod \ \mathfrak{a}), \ 1 \leqslant j \leqslant h,$$

has a solution $\lambda = v_1 \mu_1 + \cdots + v_u \mu_u$ satisfying

$$0 < \max_i |N(\lambda_i)| = \max_{i,j} |N(v_i \mu_{ij})| \ll N(\mathfrak{a})^{1/2 + \varepsilon/2 + \varepsilon/2} \ll N(\mathfrak{a})^{1/2 + \varepsilon}.$$

Theorem 2 is proved

ACKNOWLEDGMENTS

I am grateful to the referee for his useful suggestions and help, and to the NSF of the People's Republic of China and a Glorious Sun Fellowship through the Committee for Educational Exchange with China for financial support.

REFERENCES

1. R. C. BAKER, Small solutions of quadratic and quartic congruences, *Mathematika* **27** (1980), 30–45.
2. R. C. BAKER, "Diophantine Inequalities," Oxford Sci. Pub., Oxford, 1986.
3. T. COCHRANE, Small solutions of congruences over algebraic number fields, *Illinois J. Math.* **31** (1987), 618–625.
4. E. HECKE, "Lectures on Theory of Algebraic Numbers," Springer-Verlag, New York/Berlin, 1980.
5. A. SCHINZEL, H.-P. SCHLICKEWEI, AND W. M. SCHMIDT, Small solutions of quadratic congruences and small fractional parts of quadratic forms, *Acta Arith.* **37** (1980), 241–248.
6. W. M. SCHMIDT, Small zeros of additive forms in many variables, *Trans. Amer. Math. Soc.* **248** (1979), 121–133.
7. WANG YUAN, Bounds for solutions of additive equations in an algebraic number field, I, *Acta Arith.* **48** (1987), 21–48.
8. WANG YUAN, "Diophantine Equations and Inequalities in Algebraic Number Fields," Springer-Verlag, New York/Berlin, 1992.

Acta Mathematica Sinica, New Series
1993, Vol.9, No.4, pp.382–389

On Small Zeros of Quadratic Forms
over Finite Fields (II)

Wang Yuan（王　元）

Institute of Mathematics, Academia Sinica, Beijing, China

Received October 24, 1991

Abstract. Let $Q(\underline{x}) = Q(x_1, \cdots, x_n)$ be a quadratic form with integer coefficients and let p denote a prime. Cochrane[1] proved that if $n \geq 4$ then $Q(\underline{x}) \equiv 0 \pmod{p}$ has a solution $\underline{x} \neq \underline{0}$ satisfying $|\underline{x}| \ll \sqrt{p}$, where $|\underline{x}| = \max |x_i|$. The aim of the present paper is to generalize the above result to finite fields.

§1. Introduction

Let $\mathbb{F}(= \mathbb{F}_q)$ be a finite field of order $q = p^m$, where p is an odd prime. Let $\mathbb{Z}_p = \mathbb{Z}/p\mathbb{Z} = \left\{ 0, \pm 1, \cdots, \pm \dfrac{p-1}{2} \right\}$ be a complete residue system modulo p. Let $\omega_1, \cdots, \omega_m$ be a basis of $\mathbb{F}$. The elements of $\mathbb{F}$ can be represented by

$$x = a_1\omega_1 + \cdots + a_m\omega_m, \quad a_i \in \mathbb{Z}_p, \quad 1 \leq i \leq m.$$

Set

$$|x| = \max |a_i|. \tag{1}$$

If $\underline{x} = (x_1, \cdots, x_n)$ is a vector with components in $\mathbb{F}$, then we define

$$|\underline{x}| = \max |x_i|. \tag{2}$$

Note that $|x|$ and $|\underline{x}|$ depend on the basis of $\mathbb{F}$. A quadratic form over $\mathbb{F}$ is a polynomial in $\mathbb{F}[\underline{x}]$ of the type,

$$Q(\underline{x}) = \underline{x}Q\underline{x}' = \sum_{1 \leq i,j \leq n} q_{ij} x_i x_j,$$

where $q_{ij} = q_{ji}, \underline{x} = (x_1, \cdots, x_n)$ and $Q = (q_{ij})(1 \leq i, j \leq n)$ is the matrix associated with $Q(\underline{x})$.

For the case $\mathbb{F} = \mathbb{Z}_p$, Heath-Brown[2] first established by his celebrated method based on Fourier analysis that if n is an even integer ≥ 4, then the congruence $Q(\underline{x}) \equiv 0 (\mathrm{mod}\, p)$ has a solution satisfying

$$0 < |\underline{x}| \ll \sqrt{p}\log p. \tag{3}$$

The right hand side of (3) is of best possible apart from a possible improvement on the second order $\log p$. Cochrane[1] proved that

$$|\underline{x}| \ll \sqrt{p} \tag{4}$$

under the same condition by the use of finite Fourier analysis. Wang Yuan[3] has generalized Heath-Brown's result to all finite fields with the norms defined by (1) and (2). In this note we shall show that (4) is also true for all finite fields.

Theorem 1.   *If n is an even integer ≥ 4, then $Q(\underline{x})$ with coefficients in $\mathbb{F}$ has a zero $\underline{x}$ in $\mathbb{F}^n$ satisfying*

$$0 < |\underline{x}| \ll \sqrt{p},$$

where the constant implicit in $\ll$ depends only on n and m.

§2. Preliminaries

We use the notations: $\mathbb{F}^*$ denotes the multiplicative group of $\mathbb{F}$, $\left(\frac{a}{q}\right)$ the Legendre symbol, $\Delta = \left(\frac{(-1)^{n/2}\det Q}{q}\right)$, $e_p(x) = e^{2\pi i x/p}$, V the set of zeros of $Q(\underline{x})$ in $\mathbb{F}^n$, $\Sigma_{\underline{x}}$ a sum where $\underline{x}$ runs over $\mathbb{F}^n$, and $Q^*(\underline{x})$ the quadratic form associated with matrix Q^{-1}, Let $\mathcal{I}$ be the trace from $\mathbb{F}$ to $\mathbb{Z}_p$ and

$$\tau\left(\frac{a}{q}\right) = \sum_{y \in F} e_p(\mathcal{I}(ay))\left(\frac{y}{q}\right).$$

A region of points S in $\mathbb{R}^{mn}$ is said to be star-shaped about the origin O if for any point P in S the line segment joining P and O is also contained in S.

Lemma 1 (Cochrane [1]).   *Let S be a star-shaped region about O with $|\underline{x}| < p/2$ for all $\underline{x} \in S$. For $0 < r < 1$, let $rS = \{r\underline{x} | \underline{x} \in S\}$. Let V be the set of zeros modulo p of any given form in $\underline{x}$. Then*

$$|rS \cap V| \leq 1 + \frac{r}{1-r}|S \cap V|,$$

where $|T|$ denotes the number of points in T.

Lemma 2.   *If $\underline{y} \neq 0$ and $\Delta \neq 0$, then*

$$\sum_V e_p(\mathcal{I}(\underline{x}\,\underline{y}')) = \begin{cases} q^{\frac{n}{2}-1}(q-1)\Delta, & \text{if } Q^*(\underline{y}) = 0, \\ -q^{\frac{n}{2}-1}\Delta, & \text{otherwise} \end{cases} \tag{5}$$

and

$$\sum_V 1 = |V| = q^{n-1} + q^{\frac{n}{2}-1}(q-1)\Delta. \tag{6}$$

Proof. Suppose that $\underline{y} \neq \underline{0}$. Then

$$\sum_V e_p(\mathcal{I}(\underline{\underline{x}}\ \underline{y}')) = q^{-1}\Lambda,$$

where

$$\Lambda = \sum_{\underline{\underline{x}}} e_p(\mathcal{I}(\underline{\underline{x}}\ \underline{y}')) \sum_{a \in \mathbf{F}^*} e_p(aQ(\underline{x})).$$

Let $Q = TDT'$, where $D = [d_1, \cdots, d_n]$ is a diagonal matrix. Let $\underline{x}T = \underline{z}$ and $\underline{y}T^{-1\prime} = \underline{u}$. Since

$$\tau\left(\frac{1}{q}\right) = \left(\frac{-1}{q}\right)^{1/2}\sqrt{q},$$

we have

$$
\begin{aligned}
\Lambda &= \sum_{a \in \mathbf{F}^*} \sum_{\underline{\underline{z}}} e_p(\mathcal{I}(a\underline{z}D\underline{z}' + \underline{\underline{z}}\underline{u}')) \\
&= \sum_{a \in \mathbf{F}^*} \prod_{i=1}^{n} \left(\sum_{z_i \in \mathbf{F}} e_p(\mathcal{I}(ad_i z_i^2 + u_i z_i)) \right) \\
&= \sum_{a \in \mathbf{F}^*} \prod_{i=1}^{n} e_p\left(\mathcal{I}\left(-\frac{u_i^2}{4ad_i}\right)\right) \sum_{z_i \in \mathbf{F}} e_p\left(\mathcal{I}\left(ad_i\left(z_i + \frac{u_i}{2ad_i}\right)^2\right)\right) \\
&= \sum_{a \in \mathbf{F}^*} \prod_{i=1}^{n} e_p\left(\mathcal{I}\left(-\frac{u_i^2}{4ad_i}\right)\right) \sum_{v_i \in \mathbf{F}} e_p(\mathcal{I}(ad_i v_i))\left(1 + \left(\frac{v_i}{q}\right)\right) \\
&= \sum_{a \in \mathbf{F}^*} \prod_{i=1}^{n} e_p\left(\mathcal{I}\left(\frac{-u_i^2}{4ad_i}\right)\right)\left(\frac{ad_i}{q}\right)\tau\left(\frac{1}{q}\right) \\
&= q^{n/2}\Delta \sum_{a \in \mathbf{F}^*} e_p(\mathcal{I}(-(4a)^{-1}Q^*(\underline{y}))),
\end{aligned}
$$

and thus (5) follows. The proof of (6) is similar.

Let $\alpha(\underline{x})$ be a complex function of $\mathbf{F}^n$ with Fourier expansion

$$\alpha(\underline{x}) = \sum_{\underline{\underline{y}}} c(\underline{y}) e_p(\mathcal{I}(\underline{\underline{x}}\ \underline{y}')).$$

Then

$$\sum_{\underline{\underline{x}}} \alpha(\underline{\underline{x}}) e_p(-\mathcal{I}(\underline{\underline{x}}\ \underline{y}')) = \sum_{\underline{\underline{x}}} \sum_{\underline{\underline{z}}} c(\underline{z}) e_p(\mathcal{I}(\underline{\underline{x}}(\underline{z}' - \underline{y}'))) = q^n c(\underline{y}). \tag{7}$$

Lemma 3 (Basic identity). *If $\Delta \neq 0$, then*

$$\sum_V \alpha(\underline{\underline{x}}) = q^{-1}\sum_{\underline{\underline{x}}} \alpha(\underline{\underline{x}}) - \Delta q^{\frac{n}{2}-1}\alpha(\underline{0}) + \Delta q^{\frac{n}{2}} \sum_{\substack{\underline{\underline{y}} \\ Q^*(\underline{y})=0}} c(\underline{y}).$$

Proof. By Lemma 2 and (7) we have

$$\sum_V \alpha(\underline{\underline{x}}) = \sum_V \sum_{\underline{\underline{y}}} c(\underline{y}) e_p(\mathcal{I}(\underline{\underline{x}}\ \underline{y}'))$$

$$
\begin{aligned}
&= c(\underline{0})\sum_{V}1 + \sum_{\underline{y}\neq\underline{0}}c(\underline{y})\sum_{V}e_p(\mathcal{I}(\underline{\underline{x}}\ \underline{y}'))\\
&= c(\underline{0})(q^{n-1}+q^{\frac{n}{2}-1}(q-1)\Delta)+q^{\frac{n}{2}-1}(q-1)\Delta\sum_{\substack{\underline{y}\neq\underline{0}\\ Q^{*}(\underline{y})=0}}c(\underline{y})\\
&\qquad -q^{\frac{n}{2}-1}\Delta\sum_{\substack{\underline{y}\neq\underline{0}\\ Q^{*}(\underline{y})\neq0}}c(\underline{y})\\
&= q^{-1}\sum_{\underline{\underline{x}}}\alpha(\underline{\underline{x}})-\Delta q^{\frac{n}{2}-1}\sum_{\underline{\underline{x}}}c(\underline{\underline{x}})+\Delta q^{\frac{n}{2}}\sum_{\substack{\underline{y}\\ Q^{*}(\underline{y})=0}}c(\underline{y}).
\end{aligned}
$$

The lemma is proved

Let B be a box of points in $\mathbb{F}^n$ of the type

$$
B = \left\{\underline{\underline{x}}\,\big|\,\underline{\underline{x}}\in\mathbb{F}^n,\ x_i=\sum_{j=1}^{m}x_{ij}\omega_j,\ a_{ij}\leq x_{ij}<a_{ij}+m_{ij},\ 1\leq i\leq n,\ 1\leq j\leq m\right\},\quad (8)
$$

where $a_{ij},m_{ij}\in\mathbb{Z}$ and $0<a_{ij}+m_{ij}<p$, $1\leq i\leq n$, $1\leq j\leq m$. Let $\chi_B(\underline{\underline{x}})$ be the characteristic function of B and

$$
\chi_B(\underline{\underline{x}}) = \sum_{\underline{y}}c_B(\underline{y})e_p(\mathcal{I}(\underline{\underline{x}}\ \underline{y}')).
$$

Set $y_i=\sum_{j=1}^{m}y_{ij}\omega_j'$, where $\{\omega_1',\cdots,\omega_m'\}$ is the dual basis of $\{\omega_1,\cdots,\omega_m\}$ i.e;

$$
\mathcal{I}(\omega_i\omega_j') = \begin{cases} 1, & \text{if } i=j,\\ 0, & \text{otherwise}.\end{cases}
$$

Then

$$
\begin{aligned}
\mathcal{I}(\underline{\underline{x}}\ \underline{y}') &= \mathcal{I}\left(\sum_{i=1}^{n}x_iy_i\right) = \sum_{i=1}^{n}\sum_{j=1}^{m}\sum_{k=1}^{m}x_{ij}y_{ik}\mathcal{I}(\omega_j\omega_k')\\
&= \sum_{i=1}^{n}\sum_{j=1}^{m}x_{ij}y_{ij}
\end{aligned}
$$

and by (7), we have

$$
\begin{aligned}
c_B(\underline{y}) &= q^{-n}\sum_{\underline{\underline{x}}}\chi_B(\underline{\underline{x}})e_p(-\mathcal{I}(\underline{\underline{x}}\ \underline{y}'))\\
&= q^{-n}\prod_{i=1}^{n}\prod_{j=1}^{m}e_p\left(-\left(a_{ij}+\frac{m_{ij}}{2}-\frac{1}{2}\right)y_{ij}\right)\frac{\sin\frac{\pi m_{ij}y_{ij}}{p}}{\sin\frac{\pi y_{ij}}{p}}.
\end{aligned}\qquad (9)
$$

Let $\alpha(\underline{\underline{x}})$ be the convolution of $\chi_B(\underline{\underline{x}})$ with itself, i.e;

$$
\alpha(\underline{\underline{x}}) = \chi_B * \chi_B = \sum_{\underline{u}}\chi_B(\underline{u})\chi_B(\underline{\underline{x}}-\underline{u}).
$$

Then the Fourier coefficient $c(\underline{y})$ of $\alpha(\underline{x})$ is

$$
\begin{aligned}
c(\underline{y}) \ &= q^{-n} \sum_{\underline{\underline{x}}} \sum_{\underline{\underline{u}}} \chi_B(\underline{\underline{u}}) \chi_B(\underline{\underline{x}} - \underline{\underline{u}}) e_p(-\mathcal{I}(\underline{\underline{x}}\ \underline{y}')) \\
&= q^{-n} \sum_{\underline{\underline{x}}} \sum_{\underline{\underline{u}}} \sum_{\underline{\underline{z}}} \sum_{\underline{\underline{v}}} c_B(\underline{\underline{z}}) e_p(\mathcal{I}(\underline{\underline{u}}\ \underline{z}')) c_B(\underline{\underline{v}}) e_p(\mathcal{I}(\underline{\underline{v}}(\underline{\underline{x}} - \underline{\underline{u}})')) e_p(-\mathcal{I}(\underline{\underline{x}}\ \underline{y}')) \\
&= \sum_{\underline{\underline{x}}} \sum_{\underline{\underline{v}}} c_B(\underline{\underline{v}})^2 e_p(\underline{\underline{x}}(\underline{v} - \underline{y})') = q^n c_B(\underline{y})^2 .
\end{aligned}
\tag{10}
$$

Lemma 4. *Suppose that $\Delta = -1$. Let B be a box of the type*

$$
B = \{ \underline{\underline{x}} \mid \underline{\underline{x}} \in \mathbb{F}^n, \ |x_{ij}| \le B_{ij}, 1 \le i \le n, 1 \le j \le m \}
\tag{11}
$$

for some nonnegative integers B_{ij}. Let t be a positive integer satisfying $(tm + 2) \log 2 \le \left(\frac{n}{2} - 1 \right) \log q$. If $B_{ij} < 2^{-(n+t+3)m} p$ for all i, j or $|B| \ge 2^{-(n+t+2)nm^2} q^{\frac{n}{2}}$, then

$$
|B \cap V| \le 2^{(n+t+3)nm^2+1} q^{-1} |B| + 2^{-tm} q^{\frac{n}{2}-1} .
$$

Proof. Take $a_{ij} = -B_{ij}$ and $m_{ij} = 2B_{ij} + 1$. It follows from (9) and (10) that $c_B(\underline{y})$ is real and $c(\underline{y})$ is positive. Since

$$
\sum_{\underline{\underline{x}}} \alpha(\underline{\underline{x}}) = \sum_{\underline{\underline{u}}} \chi_B(\underline{\underline{u}}) \sum_{\underline{\underline{x}}} \chi_B(\underline{\underline{x}} - \underline{\underline{u}}) = |B|^2
$$

and

$$
\alpha(\underline{0}) = \sum_{\underline{\underline{u}}} \chi_B(\underline{\underline{u}}) \chi_B(-\underline{\underline{u}}) = |B|,
$$

we derive from Lemma 3 that

$$
\sum_{\underline{\underline{x}} \in V} \alpha(\underline{\underline{x}}) \le q^{-1} |B|^2 + q^{\frac{n}{2}-1} |B|.
$$

On the other hand we have

$$
\sum_{\underline{\underline{x}} \in V} \alpha(\underline{\underline{x}}) \ge \sum_{\underline{\underline{x}} \in B \cap V} 2^{-nm} |B| = 2^{-nm} |B| |B \cap V|
$$

and

$$
|B \cap V| \le 2^{nm} (q^{-1} |B| + q^{\frac{n}{2}-1}).
\tag{12}
$$

Set $r = 2^{-(n+t+2)m}$. Suppose first that $|B| \ge r^{nm} q^{\frac{n}{2}}$. Then

$$
q^{\frac{n}{2}-1} \le 2^{(n+t+2)nm^2} q^{-1} |B|,
$$

and so

$$
\begin{aligned}
|B \cap V| \ &\le 2^{nm} q^{-1} |B| + 2^{nm+(n+t+2)nm^2} q^{-1} |B| \\
&\le 2^{(n+t+3)nm^2+1} q^{-1} |B|.
\end{aligned}
$$

The lemma follows. Suppose now that $|B| < r^{nm}q^{\frac{n}{2}}$ and $B_{ij} < \frac{1}{2}rp$ for all i,j. Set

$$r^{-1}B = \{\underline{x}|\underline{x} \in I\!\!F^n, \ |x_{ij}| \le r^{-1}B_{ij}, 1 \le i \le n, 1 \le j \le m\}.$$

Then for any $\underline{x} \in r^{-1}B$ we have $|\underline{x}| < p/2$ and thus by (12),

$$|r^{-1}B \cap V| \le 2^{nm}(r^{-nm}q^{-1}|B| + q^{\frac{n}{2}-1}) \le 2^{nm+1}q^{\frac{n}{2}-1}$$

We obtain by applying Lemma 1 to the region $r^{-1}B$ that

$$|B \cap V| \le 1 + \frac{r}{1-r}2^{nm+1}q^{\frac{n}{2}-1} \le 1 + \frac{2^{-tm-1}}{1-r}q^{\frac{n}{2}-1} \le 2^{-tm}q^{\frac{n}{2}-1}$$

since $(tm + 2)\log 2 \le \left(\frac{n}{2} - 1\right)\log q$. The lemma is proved.

§3. Proof of Theorem 1

If $\Delta = 0$ or $\Delta = 1$, then the theorem is proved in Wang[3]. So we assume that $\Delta = -1$. Theorem 1 is a consequence of the following

Lemma 5. *Suppose that* $p > 12^{2nm} \cdot 2^{10nm+12m}$ *and that* B *is a box of the type* (8) *satisfying*

$$m_{ij} \ge 2^{(5n+7)m}12^{nm+1} \quad \text{and} \quad |B| > 2^{(3n+4)nm^2+2}12^{nm+1}q^{n/2}. \tag{13}$$

Then B *contains a nonzero solution of* $Q(\underline{x}) = 0$.

Proposition 1. *Let* $S = B$ *be a box of the type* (8) *and* T *be any translation of* S. *Suppose that* $p > 12^{2nm}2^{10nm+12m}$,

$$m_{ij} \ge 2^{(5n+6)m}12^{nm+1} \tag{14}$$

and

$$2^{(3n+3)nm^2+2}12^{nm+1}q^{\frac{n}{2}} \le |S| < 2^{(3n+3)nm^2+3}12^{nm+1}q^{\frac{n}{2}}. \tag{15}$$

Then the set $S + T$ *contains a nontrivial zero of* $Q(\underline{x})$.

We first derive Lemma 5 from Proposition 1 and then prove Proposition 1.

Proof of Lemma 5. Let

$$S = \left\{\underline{x}|\underline{x} \in I\!\!F^n, 0 \le x_{ij} < \left[\frac{m_{ij} + 1}{2}\right], 1 \le i \le n, 1 \le j \le m\right\}$$

and $T = S + \underline{a}$ with $\underline{a} = (a_{11}, \cdots, a_{nm})$. Then $|S| \ge 2^{-nm}|B| > 2^{(3n+3)nm^2+2}12^{nm+1}q^{n/2}$ and each side of S has length exceeding $2^{(5n+6)m}12^{nm+1}$. We may assume that S satisfies (14) and (15), since we may shorten the sides of S if $|S|$ is large. By our assumption $p > 12^{2nm}2^{10nm+12m}$ we obtain

$$\left(2^{(5n+6)m}12^{nm+1}\right)^{nm} < 2^{(3n+3)nm+2}12^{nm+1}p^{\frac{nm}{2}}.$$

Hence there exists a subset of S containing a cube with edge length $2^{(5n+6)m}12^{nm+1}$ and satisfying (14) and (15). Since $0 \le x_{ij} < \left[\frac{m_{ij}+1}{2}\right]$, we have $a_{ij} \le 2x_{ij} + a_{ij} \le a_{ij} +$

$2\left(\dfrac{m_{ij}+1}{2}\right) - 2 \le a_{ij} + m_{ij} - 1, 1 \le i \le n, 1 \le j \le m$. Therefore $S+T$ is a subset of B and $S+T$ contains a nontrivial zero of $Q(\underline{x})$ by Proposition 1. The lemma is proved.

Proof of Proposition 1.   Let χ_S and χ_T be the characteristic functions of S and T with Fourier coefficients $c_S(\underline{y})$ and $c_T(\underline{y})$ respectively. Let $\alpha(\underline{x}) = \chi_S * \chi_T$ be the convolution of χ_S and χ_T. Then similar to (9) and (10) we derive that the Fourier coefficient $c(\underline{y})$ of $\alpha(\underline{x})$ satisfies

$$|c(\underline{y})| \le q^{-n} \prod_{i=1}^{n} \prod_{j=1}^{m} \min\left(m_{ij}^2, \frac{p^2}{4y_{ij}^2}\right).$$

Let $1 \le j \le n$ and $1 \le l \le m$ and let π and τ be injections of $\{1, \cdots, j\}$ into $\{1, \cdots, n\}$ and $\{1, \cdots, l\}$ into $\{1, \cdots, m\}$ respectively. Let $k_1, \cdots, k_j, \zeta_1, \cdots, \zeta_l$ be nonnegative integers and

$$B' = \left\{\underline{y}\,\middle|\,\underline{y} \in \mathbb{F}^n, |y_{\pi(i)\tau(s)}| \le \frac{2^{k_i+\zeta_s}p}{m_{\pi(i)\tau(s)}}, 1 \le i \le j, \quad 1 \le s \le l, \right. \tag{16}$$
$$\left. |y_{uv}| \le \frac{p}{2m_{uv}} \text{ otherwise}\right\}.$$

If $\dfrac{2^{k_i+\zeta_s}p}{m_{\pi(i)\tau(s)}} < 2^{-(3n+3)m}p, 1 \le i \le j, 1 \le s \le l$ and $\dfrac{p}{2m_{uv}} < 2^{-(3n+3)m}p$, then Lemma 4 applies to B' with $t = 2n$. Otherwise $2^{k_i+\zeta_s} \ge 2^{-(3n+3)m}m_{\pi(i)\tau(s)}$, for some i, s, then

$$\begin{aligned}
|B'| &\ge \prod_{i=1}^{j}\prod_{s=1}^{l} \frac{2^{k_i+\zeta_s}p}{m_{\pi(i)\tau(s)}} \prod_{\substack{u,v \\ \text{otherwise}}} \frac{p}{2m_{uv}} \ge \frac{p^{nm}}{2^{nm}|B|} \prod_{i=1}^{j}\prod_{s=1}^{l} 2^{k_i+\zeta_s} \\
&\ge \frac{p^{nm}}{2^{nm} \cdot 2^{(3n^2+3n+3)m^2} 12^{nm+1} p^{\frac{nm}{2}}} 2^{-(3n+3)m} 2^{(5n+6)m} 12^{nm+1} \\
&= 2^{-(3n+2)nm^2} q^{\frac{n}{2}},
\end{aligned}$$

and so Lemma 4 again applies to B' with $t = 2n$. Let V^* be the set of zeros of $Q^*(\underline{x})$ in $\mathbb{F}^n$. Then by Lemma 4

$$|B' \cap V^*| \le c_1 q^{-1}|B'| + c_2 q^{\frac{n}{2}-1},$$

where $c_1 = 2^{(3n+3)nm^2+1}$ and $c_2 = 2^{-2nm}$.

  Set $\displaystyle\sum_{\substack{\underline{y} \\ Q^*(\underline{y})=0}} = \sum_{V^*} = \sum{}^*$. Let π and τ run through the sets of all injections of $\{1, \cdots, j\}$ into $\{1, \cdots, n\}$ and $\{1, \cdots, l\}$ into $\{1, \cdots, m\}$ respectively such that $\pi(h) < \pi(i)$ and $\tau(h) < \tau(i)$ for $h < i$. Then

$$\begin{aligned}
\sum{}^* |c(\underline{y})| &= \sum_{j=0}^{n}\sum_{l=0}^{m}\sum_{\pi}\sum_{\tau} \sum_{\substack{|y_{\pi(i)\tau(s)}| \ge \frac{p}{2m_{\pi(i)\tau(s)}} \\ |y_{uv}| \le \frac{p}{2m_{uv}} \text{ otherwise}}}^{*} |c(\underline{y})| \\
&\le q^{-n} \sum_{j=0}^{n}\sum_{l=0}^{m}\sum_{\pi}\sum_{\tau}\sum_{k_1=0}^{\infty}\cdots\sum_{\zeta_l=0}^{\infty} \sum_{\frac{2^{k_i+\zeta_s}p}{2m_{\pi(i)\tau(s)}} \le |y_{\pi(i)\tau(s)}| \le \frac{2^{k_i+\zeta_s+1}p}{2m_{\pi(i)\tau(s)}}}^{*}
\end{aligned}$$

$$\prod_{i=1}^{j}\prod_{s=1}^{l}\frac{m_{\pi(i)\tau(s)}^2}{2^{2(k_i+\zeta_s)}}\prod_{\substack{u,v\\ \text{otherwise}}} m_{uv}^2$$

$$\leq q^{-n}|B|^2\sum_{j=0}^{n}\sum_{l=0}^{m}\sum_{\pi}\sum_{\tau}\sum_{k_1=0}^{\infty}\cdots\sum_{\zeta_s=0}^{\infty}\prod_{i=1}^{j}\prod_{s=1}^{l}\frac{1}{2^{2(k_i+\zeta_s)}}\Big(c_1 q^{-1}\prod_{i=1}^{j}$$

$$\cdot\prod_{s=1}^{l}\frac{2^{k_i+\zeta_s+2}p}{m_{\pi(i)\tau(s)}}\prod_{\substack{u,v\\ \text{otherwise}}}\frac{2p}{m_{uv}}+c_2 q^{\frac{n}{2}-1}\Big)$$

$$= q^{-n}|B|^2\sum_{j=0}^{n}\sum_{l=0}^{m}\sum_{\pi}\sum_{\tau}\sum_{k_1=0}^{\infty}\cdots\sum_{\zeta_s=0}^{\infty}\Big(c_1 q^{n-1}|B|^{-1}2^{nm+jl}\prod_{i=1}^{j}\prod_{s=1}^{l}\frac{1}{2^{k_i+\zeta_s}}$$

$$+c_2 q^{\frac{n}{2}-1}\prod_{i=1}^{j}\prod_{s=1}^{l}\frac{1}{4^{k_i+\zeta_s}}\Big)$$

$$= c_1 2^{2nm}q^{-1}|B|\sum_{j=0}^{n}\binom{n}{j}2^j\sum_{l=0}^{m}\binom{m}{l}2^l+c_2 q^{-\frac{n}{2}-1}|B|^2\sum_{j=0}^{n}\binom{n}{j}$$

$$\cdot\Big(\frac{4}{3}\Big)^j\sum_{l=0}^{m}\binom{m}{l}\Big(\frac{4}{3}\Big)^l$$

$$= c_1 2^{2nm}q^{-1}3^{n+m}|B|+c_2\Big(\frac{7}{3}\Big)^{n+m}q^{-\frac{n}{2}-1}|B|^2$$

$$< c_1 12^{nm+1}q^{-1}|B|+c_2\Big(\frac{7}{3}\Big)^{nm+1}q^{-\frac{n}{2}-1}|B|^2.$$

Thus by Lemma 3 we obtain

$$\sum_{V}\alpha(\underline{x}) \geq q^{-1}|B|^2-2^{(3n+3)nm^2+1}12^{nm+1}q^{\frac{n}{2}-1}|B|-4\Big(\frac{7}{12}\Big)^{nm+1}q^{-1}|B|^2$$

$$\geq 0.6 q^{-1}|B|^2-2^{(3n+3)nm^2+1}12^{nm+1}q^{\frac{n}{2}-1}|B|$$

$$= q^{-1}|B|\Big(0.6|B|-2^{(3n+3)nm^2+1}12^{nm+1}q^{\frac{n}{2}}\Big)>1.$$

Therefore $(S+T)\cap V$ contains a nonzero solution of $Q(\underline{x})=0$. The proposition is proved.

Acknowlegment.   This work was supported by the NSF of P.R. China and by a Glorious Sun Fellowship through the committee for Educational Exchange with China for financial Support.

References

[1] Todd Cochrane, *Small zeros of quadratic forms modulo p, III*, J. of Number Theory, **37** (1991), 92–99.

[2] Heath-Brown, D.R., *Small solutions of quadratic congruences*, Glasgov Math. J., **27** (1985), 87–93.

[3] Wang Yuan, *On small zeros of quadratic forms over finite fields*, J. of Number Theory, **31** (1989), 272–284.

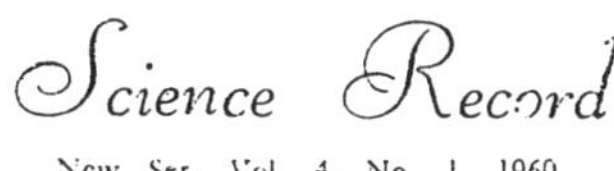
New Ser. Vol. 4, No. 1, 1960

Mathematics

Remarks Concerning Numerical Integration*

HUA LOO-KENG** (華罗庚) AND WANG YUAN (王　元)

Institute of Mathematics, Academia Sinica

Let $f(x_1, \cdots, x_s)$ be a function of $x_1, \cdots, x_s$ with period 1; it has a Fourier expansion

$$f(x_1, \cdots, x_s) = \sum_{m_1=-\infty}^{\infty} \cdots \sum_{m_s=-\infty}^{\infty} C(m_1, \cdots, m_s)e^{2\pi i(m_1 x_1 + \cdots + m_s x_s)}, \tag{1}$$

where

$$C(m_1, \cdots, m_s) = \int_0^1 \cdots \int_0^1 f(x_1, \cdots, x_s)e^{-2\pi i(m_1 x_1 + \cdots + m_s x_s)}dx_1 \cdots dx_s. \tag{2}$$

For a given set of natural numbers q and a_i $(1 \leqslant a_i < q), i = 1, 2, \cdots, s$, we have

$$\sum_{t=1}^{q} f\left(\frac{a_1 t}{q}, \cdots, \frac{a_s t}{q}\right) = q \sum_{\substack{m_1=-\infty \\ a_1 m_1 + \cdots + a_s m_s \equiv 0 (\mathrm{mod}\ q)}}^{\infty} \cdots \sum_{m_s=-\infty}^{\infty} C(m_1, \cdots, m_s).$$

Therefore, we have

$$\left| \int_0^1 \cdots \int_0^1 f(x_1, \cdots, x_s)\, dx_1 \cdots dx_s - \frac{1}{q}\sum_{t=1}^{q} f\left(\frac{a_1 t}{q}, \cdots, \frac{a_s t}{q}\right)\right|$$

$$= \left| \sideset{}{'}\sum_{a_1 m_1 + \cdots + a_s m_s \equiv 0 (\mathrm{mod}\ q)} C(m_1, \cdots, m_s)\right| = R, \tag{3}$$

where Σ' denotes the omitting of the term with $m_1 = m_2 = \cdots = m_s = 0$.

We assume that

$$|C(m_1, \cdots, m_s)| \leqslant \frac{C}{[(|m_1| + 1)\cdots(|m_s| + 1)]^a}, \tag{4}$$

---

* Received Oct. 27, 1959.

** Member of Academia Sinica.

where $\alpha > 1$ and $C > 0$ are constants. It can be easily shown that

$$R \leqslant \underset{\substack{a_1 m_1 + \cdots + a_s m_s \equiv 0 (\mathrm{mod}\, q)}}{\sum}{}' \frac{1}{[(|m_1| + 1)\cdots(|m_s| + 1)]^\alpha}$$

$$\leqslant \underset{\substack{a_1 m_1 + \cdots + a_s m_s \equiv 0 (\mathrm{mod}\, q) \\ |m_i| \leqslant \frac{1}{2} q}}{\sum}{}' \frac{1}{[(|m_1| + 1)\cdots(|m_s| + 1)]^\alpha} + O\left(\frac{1}{q^\alpha}\right).$$

The last sum is equivalent to $q^{-s\alpha}\Omega$ with

$$\Omega = \underset{\substack{a_1 m_1 + \cdots + a_s m_s \equiv 0 (\mathrm{mod}\, q) \\ |m_i| \leqslant \frac{1}{2} q}}{\sum}{}' \frac{1}{\left[\left(\left\langle \frac{m_1}{q} \right\rangle + \frac{1}{q}\right)\cdots\left(\left\langle \frac{m_s}{q} \right\rangle + \frac{1}{q}\right)\right]^\alpha}, \tag{5}$$

where $\langle \xi \rangle$ denotes the distance from ξ to its nearest integer.

It was proved by Коробов[1] that for q being a prime p, there exist integers $a_1, \cdots, a_s$ such that

$$R = O\left(\frac{\log^{\alpha s} p}{p^\alpha}\right). \tag{6}$$

The constant implied by O in his paper can all be made explicit. He also pointed out the counter part that there exist functions satisfying (1) and (4) with

$$R \geqslant \frac{C}{q^\alpha} \tag{7}$$

for any choice of $a_1, \cdots, a_s$ and q.

Therefore the explicit choice of $a_1, \cdots, a_s$, q seems to be the central problem in the numerical integration. We prove

Theorem 1. For $a_1 = 1$, $a_2 = p_n = \frac{1}{2}[(1 + \sqrt{2})^n + (1 - \sqrt{2})^n]$ and $q = q_n = \frac{1}{2\sqrt{2}}[(1 + \sqrt{2})^n - (1 - \sqrt{2})^n]$, we have

$$R = O\left(\frac{\log q_n}{q_n^\alpha}\right). \tag{8}$$

Proof. It is sufficient to prove that

$$\Omega = 2 \underset{1 \leqslant x \leqslant \frac{1}{2} q_n}{\sum} \frac{1}{\left(\left\langle \frac{x}{q_n} \right\rangle \left\langle \frac{p_n}{q_n} x \right\rangle\right)_\varepsilon^\alpha} = O(q_n^\alpha \log q_n).$$

10

We divide Ω into n subsums:

$$J_m = \sum_{q_{m-1} \leqslant x < q_m} \frac{1}{\left(\left\langle \frac{x}{q_n}\right\rangle\left\langle \frac{p_n}{q_n}x\right\rangle\right)^a}, \quad m = 2, \cdots, n;$$

however, for the last one, J_n, we replace q_n by $\frac{q_n}{2}$.

Let y be the integer satisfying

$$\left|y - \frac{p_n}{q_n}x\right| \leqslant \frac{1}{2}. \tag{9}$$

The system of equations

$$\left.\begin{aligned} x &= q_m u + p_m v \\ y &= p_m u + 2q_m v \end{aligned}\right\} \tag{10}$$

has integral solution in u and v, since its determinant $-p_m^2 + 2q_m^2 = (-1)^{m+1}$. In case $q_{m-1} \leqslant x < q_m$, we have

$$uv < 0. \tag{11}$$

From $p_n + \sqrt{2}\,q_n = (1 + \sqrt{2})^n$, we deduce

$$\begin{aligned} q_n y - p_n x &= (q_n p_m - p_n q_m)u - (p_n p_m - 2q_n q_m)v \\ &= (-1)^m (- p_{n-m}u + q_{n-m}v). \end{aligned}$$

Then, we have, by (9) and (11),

$$|q_{n-m}v| < |-p_{n-m}u + q_{n-m}v| \leqslant \frac{1}{2}q_n. \tag{12}$$

To each x satisfying $q_{m-1} \leqslant x < q_m$, we have a unique v satisfying (9) and (10). Therefore

$$J_m \leqslant q_n^{2a} \sum_{q_{m-1} \leqslant x < q_m} \frac{1}{(x|-p_{n-m}u + q_{n-m}v|)^a}$$

$$\leqslant \frac{q_n^{2a}}{q_{m-1}^a} \sum_v \frac{1}{q_{n-m}^a|v|^a}$$

$$= O\left(\frac{q_n^{2a}}{q_{m-1}^a q_{n-m}^a}\right) = O(q_n^a).$$

Consequently, we have

$$\Omega = O(nq_n^a) = O(q_n^a \log q_n);$$

the theorem is now proved.

Remark. The constant $1 + \sqrt{2}$ is used merely for the sake of simplicity. In fact, the unit of any real quadratic field can be used instead. For example, the field $R(\sqrt{5})$ suggests the use of Fibonacci numbers as well. It suggests also the possibility to treat the high dimensional case by means of the units of a totally real field of degree s.

For the counter part, we have the following improvement of (7).

Theorem 2. There exist functions satisfying (1) and (4); for any q and $a_1, \cdots, a_s$ we have

$$R \geqslant \frac{C' \log q}{q^a}, \tag{6'}$$

where $C' > 0$ is a constant.

Therefore, Theorem 1 gives a best possible choice for $s = 2$.

Besides, under the condition that

$$\left| \frac{\partial^r f(x_1, \cdots, x_s)}{\partial x_1^{i_1} \cdots \partial x_s^{i_s}} \right| \leqslant K \quad i_1 + \cdots + i_s = r \tag{4'}$$

for a fixed $r > s$, we have

Theorem 3. Let q be a positive integer and $p_1 < \cdots < p_s$ be s pairwise prime integers lying between $q^{1/s}$ and $2^s q^{1/s}$. Then we have the following formula for approximating the s-ple integral by a single integral

$$\left| \int_0^1 \cdots \int_0^1 f(x_1, \cdots, x_s)\, dx_1 \cdots dx_s - \int_0^1 f(a_1 x, \cdots, a_s x)\, dx \right| = O\left(q^{-\frac{r}{s}}\right),$$

where $a_i = p_1 \cdots p_s / p_i$.

Consequently, we have

Theorem 4. Under the same assumption of Theorem 3, we have

$$\left| \int_0^1 \cdots \int_0^1 f(x_1, \cdots, x_s)\, dx_1 \cdots dx_s - \frac{1}{q} \sum_{t=1}^{q} f\left(\frac{a_1 k}{q}, \cdots, \frac{a_s k}{q}\right) \right| = O\left(\frac{1}{q^{r/s}}\right).$$

REFERENCE

[1] Коробов, Н. М. 1959 *ДАН СССР*, **124**(6), 1207—1210.

SCIENTIA   SINICA

Vol. X, No. 6, 1961

MATHEMATICS

A NOTE ON INTERPOLATION OF A CERTAIN CLASS OF FUNCTIONS*

WANG YUAN (王　元)

(Institute of Mathematics, Academia Sinica)

Let E_s^a be the class of functions

$$f(x_1, \cdots, x_s) = \sum \cdots \sum_{-\infty}^{\infty} C(m_1, \cdots, m_s) e^{2\pi i(m_1 x_1 + \cdots + m_s x_s)}, \qquad (1)$$

in which the Fourier coefficients satisfy

$$|C(m_1, \cdots, m_s)| \leqslant \frac{1}{(\bar{m}_1 \cdots \bar{m}_s)^a}, \qquad (2)$$

where $\bar{m} = \max(1, |m|)$ and $\alpha > 1$ is an absolute constant.

Let N be an odd prime number and N_1 be an integer satisfying $3 < N_1 < \dfrac{N}{ln^s N}$.   Let

$$\widetilde{C}(m_1, \cdots, m_s) = \frac{1}{N} \sum_{k=1}^{N} f\left(\frac{ka_1}{N}, \cdots, \frac{ka_s}{N}\right) e^{-2\pi i \frac{a_1 m_1 + \cdots + a_s m_s}{N} k}, \qquad (3)$$

$$P(x_1, \cdots, x_s) = \sum_{\bar{m}_1 \cdots \bar{m}_s < N_1} \widetilde{C}(m_1, \cdots, m_s) e^{2\pi i(m_1 x_1 + \cdots + m_s x_s)}, \qquad (4)$$

and

$$\Delta = \min_a \sup_{f \in E_s^a} \int_0^1 \cdots \int_0^1 |f(x_1, \cdots, x_s) - P(x_1, \cdots, x_s)|^2 dx_1, \cdots, dx_s, \qquad (5)$$

where $a_1, \cdots, a_s$ are integers.

Рябенький[1] first proved that

$$\Delta \leqslant A_0(N_1^{2a+1} N^{-2a} ln^{2as+s-1} N + N_1^{-(2a-1)} ln^{s-1} N_1)^{1)} \qquad (6)$$

(Cf. also [2]).

---

* Received June 5, 1961.

1) In this paper, A_ν and B_ν denote positive constants, depending on a and s only.

The aim of the present note is to give the following improvement:

Theorem 1.

$$A_1(N_1^{2a}N^{-2a}+N_1^{-(2a-1)}ln^{s-1}N_1) \leqslant \Delta \leqslant A_2(N_1^{2a}N^{-2a}ln^{2a(s-1)}N+N_1^{-(2a-1)}ln^{s-1}N_1). \quad (7)$$

Evidently, the theorem given above does not permit further essential improvements. There follows immediately

Theorem 2. *Let* $N_1 = [N^{\frac{2a}{4a-1}} ln^{-\frac{(2a-1)(s-1)}{4a-1}} N]$. *Then there exist integers* $a_i = a_i(N)$ $(1 \leqslant i \leqslant s)$ *such that the inequality*

$$\int_0^1 \cdots \int_0^1 |f(x_1, \cdots, x_s) - P(x_1, \cdots, x_s)|^2 dx_1 \cdots dx_s \leqslant$$

$$\leqslant A_3 N^{-\frac{2a(2a-1)}{4a-1}} ln^{\frac{4a^2}{4a-1}(s-1)} N \quad (8)$$

holds for any $f(x_1, \cdots, x_s) \in E_s^a$.

Proof of Theorem 1. 1) The right-hand side of the inequality (7) may be deduced from the following two Lemmas.

Lemma 1. (Бахвалов[3]) *There exist* $a_i = a_i(N)$, $(1 \leqslant i \leqslant s)$, *such that*

$$\sum_{\substack{a_1 l_1 + \cdots + a_s l_s \equiv 0 (mod\ N)}}^{\infty} \cdots \sum_{-\infty}' \frac{1}{(\bar{l}_1, \cdots, \bar{l}_s)^a} \leqslant A_4 \frac{ln^{a(s-1)}N}{N^a}, \quad (9)$$

where $\sum'$ *denotes the omitting of the term with* $l_1 = \cdots = l_s = 0$.

Lemma 2. *Let* $l_1, \cdots, l_s$ *be non-negative integers and* $1 < N_1 \leqslant \frac{\bar{l}_1, \cdots, \bar{l}_s}{3^s}$. *Then*

$$\sum_{\substack{\bar{m}_1 \cdots \bar{m}_s < N_1 \\ m_j \geqslant 0}} \frac{1}{[(\overline{l_1 - m_1}) \cdots (\overline{l_s - m_s})]^a} \leqslant B_s \frac{N_1^u}{(\bar{l}_1, \cdots, \bar{l}_s)^a}. \quad (10)$$

Proof. For $s = 1$, we have

$$\sum_{0 \leqslant m_1 < N_1} \frac{1}{(\overline{l_1 - m_1})^a} < \left(\frac{3}{2}\right)^a \frac{N_1}{\bar{l}_1^a} \leqslant B_1 \frac{N_1^a}{\bar{l}_1^a}.$$

Assume that the lemma holds for $s \leqslant k$. Then

$$\sum_{\substack{\bar{m}_1 \cdots \bar{m}_{k+1} < N_1 \\ m_j \geqslant 0}} \frac{1}{[(\overline{l_1 - m_1}) \cdots (\overline{l_{k+1} - m_{k+1}})]^a} \leqslant \Sigma_1 + \Sigma_2 + \cdots + \Sigma_{k+1},$$

634

where

$$\Sigma_i = \sum_{\substack{\overline{m}_1 \cdots \overline{m}_{k+1} < N_1 \\ m_i < \frac{\overline{l}_i}{2} \\ m_j \geqslant 0}} \frac{1}{[\,(\overline{l_1 - m_1}) \cdots (\overline{l_{k+1} - m_{k+1}})\,]^\alpha}.$$

(i) Suppose $N_1 < \dfrac{\overline{l}_2 \cdots \overline{l}_{k+1}}{3^k}$. Then

$$\Sigma_1 \leqslant \sum_{0 \leqslant m_1 < \frac{\overline{l}_1}{2}} \frac{1}{(\overline{l_1 - m_1})^\alpha} \sum_{\substack{\overline{m}_2 \cdots \overline{m}_{k+1} < \frac{N_1}{\overline{m}_1} \\ m_j \geqslant 0}} \frac{1}{[\,(\overline{l_2 - m_2}) \cdots (\overline{l_{k+1} - m_{k+1}})\,]^\alpha} \leqslant$$

$$\leqslant B_k \frac{N_1^\alpha}{(\overline{l}_2 \cdots \overline{l}_{k+1})^\alpha} \sum_{0 \leqslant m_1 < \frac{\overline{l}_1}{2}} \frac{1}{\overline{m}_1^\alpha (\overline{l_1 - m_1})^\alpha} < 2^{\alpha+1} B_k \zeta(\alpha) \frac{N_1^\alpha}{(\overline{l}_1 \cdots \overline{l}_{k+1})^\alpha}.$$

(ii) Suppose $N_1 \geqslant \dfrac{\overline{l}_2 \cdots \overline{l}_{k+1}}{3^k}$. Then

$$\Sigma_1 \leqslant \sum_{0 \leqslant m_1 < \frac{3^k N_1}{\overline{l}_2 \cdots \overline{l}_{k+1}}} \frac{1}{(\overline{l_1 - m_1})^\alpha} \sum_{\substack{\overline{m}_2 \cdots \overline{m}_{k+1} < \frac{N_1}{\overline{m}_1} \\ m_j \geqslant 0}} \frac{1}{[\,(\overline{l_2 - m_2}) \cdots (\overline{l_{k+1} - m_{k+1}})\,]^\alpha} +$$

$$+ \sum_{\frac{3^k N_1}{\overline{l}_2 \cdots \overline{l}_{k+1}} \leqslant m_1 < \frac{\overline{l}_1}{2}} \frac{1}{(\overline{l_1 - m_1})^\alpha} \sum_{\substack{\overline{m}_2 \cdots \overline{m}_{k+1} < \frac{N_1}{\overline{m}_1} \\ m_j \geqslant 0}} \frac{1}{[\,(\overline{l_2 - m_2}) \cdots (\overline{l_{k+1} - m_{k+1}})\,]^\alpha} \leqslant$$

$$\leqslant (3\zeta(\alpha))^k \sum_{0 \leqslant m_1 < \frac{3^k N_1}{\overline{l}_2 \cdots \overline{l}_{k+1}}} \frac{1}{(\overline{l_1 - m_1})^\alpha} +$$

$$+ \sum_{\frac{3^k N_1}{\overline{l}_2 \cdots \overline{l}_{k+1}} \leqslant m_1 < \frac{\overline{l}_1}{2}} \frac{1}{(\overline{l_1 - m_1})^\alpha} \cdot \frac{B_k N_1^\alpha}{\overline{m}_1^\alpha (\overline{l}_2 \cdots \overline{l}_{k+1})^\alpha}$$

$$\leqslant 3^{\alpha+k} \zeta^k(\alpha) \frac{3^k N_1}{\overline{l}_2 \cdots \overline{l}_{k+1} \overline{l}_1^\alpha} + \frac{2^\alpha B_k N_1^\alpha \zeta(\alpha)}{(\overline{l}_1 \cdots \overline{l}_{k+1})^\alpha} \leqslant$$

$$\leqslant (3^{\alpha+k+\alpha k} \zeta^k(\alpha) + 2^\alpha B_k \zeta(\alpha)) \frac{N_1^\alpha}{(\overline{l}_1 \cdots \overline{l}_{k+1})^\alpha}.$$

Hence

$$\Sigma_1 \leqslant \left(\frac{B_{k+1}}{k+1} \right) \frac{N_1^\alpha}{(\overline{l}_1 \cdots \overline{l}_{k+1})^\alpha}.$$

Similarly, we have

$$\Sigma_i \leqslant \left(\frac{B_{k+1}}{k+1}\right)\frac{N_1^\alpha}{(\bar{l}_1 \cdots \bar{l}_{k+1})^\alpha}, \quad (2 \leqslant i \leqslant k+1).$$

Therefore

$$\sum_{\substack{\overline{m}_1 \cdots \overline{m}_{k+1} < N_1 \\ m_j \geqslant 0}} \frac{1}{[(\overline{l_1 - m_1}) \cdots (\overline{l_{k+1} - m_{k+1}})]^\alpha} \leqslant B_{k+1} \frac{N_1^\alpha}{(\bar{l}_1 \cdots \bar{l}_{k+1})^\alpha}.$$

Hence the lemma follows by mathematical induction.

2) Take

$$f(x_1, \cdots, x_s) = \sum_{-\infty}^{\infty} \cdots \sum \frac{1}{(\overline{m}_1 \cdots \overline{m}_s)^\alpha} e^{2\pi i(m_1 x_1 + \cdots + m_s x_s)}.$$

Then we have, for any $a_1, \cdots, a_s$,

$$\int_0^1 \cdots \int_0^1 |f(x_1, \cdots, x_s) - P(x_1, \cdots, x_s)|^2 dx_1 \cdots dx_s =$$

$$= \sum_{\overline{m}_1 \cdots \overline{m}_s < N_1} \left(\sum_{\substack{-\infty \\ a_1 l_1 + \cdots + a_s l_s \equiv 0 (\mathrm{mod}\ N)}}^{\infty} \cdots \sum{}' \frac{1}{[(\overline{l_1 - m_1}) \cdots (\overline{l_{k+1} - m_{k+1}})]^\alpha} \right)^2 +$$

$$+ \sum_{\overline{m}_1 \cdots \overline{m}_s \geqslant N_1} \frac{1}{(\overline{m}_1 \cdots \overline{m}_s)^{2\alpha}} \geqslant$$

$$\geqslant \sum_{m_1 m_2 < N_1} \sum_{a_1 l_1 + a_2 l_2 \equiv 0 (\mathrm{mod}\ N)}{}' \frac{1}{[(\overline{l_1 - m_1})(\overline{l_2 - m_2})]^{2\alpha}} + A_5 N_1^{-(2\alpha-1)} ln^{s-1} N_1.$$

Suppose first that $(a_i, N) = 1$ $(i = 1, 2)$. Let the solution of the congruence $a_1 x \equiv -a_2 (\mathrm{mod}\ N)$ be $x = a(|a| < N)$, and $\frac{p_t}{q_t}$ be the t-th convergence of $\frac{a}{N}$, $\frac{p_n}{q_n} = \frac{a}{N}$ and $1 = q_0 < q_1 < \cdots < q_r < N_1 \leqslant q_{r+1} < \cdots < q_n = N$. Since for any integer b,

$$|p_n q_t - q_n b| \leqslant \frac{q_n}{q_{t+1}}, \quad (0 \leqslant t \leqslant n-1),$$

therefore

$$\int_0^1 \cdots \int_0^1 |f(x_1, \cdots, x_s) - P(x_1, \cdots, x_s)|^2 dx_1 \cdots dx_s \geqslant$$

$$\geqslant \sum_{m_1 m_2 < N_1} \sum_{l_1 \equiv a l_2 (\mathrm{mod}\ N)} \frac{1}{[(\overline{l_1 - m_1})(\overline{l_2 - m_2})]^{2\alpha}} + A_5 N_1^{-(2\alpha-1)} ln^{s-1} N_1 \geqslant$$

636

$$\geqslant \sum_{m_1 \, m_2 < N_1} \sum_{t=0}^{n-1} \frac{1}{\left[(q_t - m_1)\left(\left[\dfrac{q_n}{q_{t+1}}\right] - m_2\right)\right]^{2a}} + A_5 N_1^{-(2a-1)} ln^{s-1} N_1 \geqslant$$

$$\geqslant \left(\frac{q_{r+1}}{q_n}\right)^{2a} + A_5 N_1^{-(2a-1)} ln^{s-1} N_1 \geqslant$$

$$\geqslant A_2(N_1^{2a} N^{-2a} + N_1^{-(2a-1)} ln^{s-1} N_1).$$

Evidently, this is true also for the case $(a_1, N) > 1$ or $(a_2, N) > 1$. Hence we have the left-hand side of the inequality (7).

Similarly, we have

Theorem 3. *Let* $N_1 = [N^{\frac{a}{2a-1}} ln^{\frac{(s-1)(1-a)}{2a-1}} N]$. *Then there exist* $a_i = a_i(N)$ $(1 \leqslant i \leqslant s)$ *such that*

$$\left| f(x_1, \cdots, x_s) - \right.$$

$$\left. - \sum_{\overline{m}_1 \cdots \overline{m}_s < N_1} \left[\frac{1}{N} \sum_{k=1}^{N} f\left(\frac{a_1 k}{N}, \cdots, \frac{a_s k}{N}\right) e^{-2\pi i \frac{(m_1 a_1 + \cdots + m_s a_s)}{N} k}\right] e^{2\pi i (m_1 x_1 + \cdots + m_s x_s)} \right| \leqslant$$

$$\leqslant A_6 N^{-\frac{a(a-1)}{2a-1}} ln^{\frac{a^2}{2a-1}(s-1)} N \tag{11}$$

holds for all $f(x_1, \cdots, x_s) \in E_s^a$.

Remark. This method may be used to improve the theorem of Шахов[4] (i.e., the error term $O(N^{-\frac{a-1}{2}+\varepsilon})$ may be repaced by $O(N^{-\frac{a}{2}+\varepsilon})$).

REFERENCES

[1] Рябенький, В. С. 1960 О Таблицах и интерполяции функций из некоторого класса, *ДАН СССР*, **131**, № 5, 1025—1027.

[2] Смоляк, С. А. 1960 Интерполяционные и квадратурные формулы на классах W_s^a и E_s^a, *ДАН СССР*, **131**, № 5, 1028—1031.

[3] Бахвалов, Н. С. 1959 О приближенном вычислении кратных интегралов, *Вестник ГМУ*, **4**, 3—18.

[4] Шахов, Ю. Н. 1961 О приближенном решении уравнений Вольтерра II рода методом итераций, *ДАН СССР*, **136**, № 6, 1302—1305.

Vol. XIII, No. 6   SCIENTIA SINICA (Notes)   1007—8

MATHEMATICS

On Diophantine Approximations and Numerical Integrations (I)

Let $f(x_1, \cdots, x_s)$ be a function with absolute convergent Fourier series

$$f(x_1, \cdots, x_s) = \sum_{-\infty}^{\infty} \cdots \sum C(m_1, \cdots, m_s) \times$$
$$\times e^{2\pi i (m_1 x_1 + \cdots + m_s x_s)},$$

where

$$|C(m_1, \cdots, m_s)| \leqslant \frac{1}{(\overline{m}_1 \cdots \overline{m}_s)^\alpha},$$

$\overline{m} = \max(1, |m|)$, and $\alpha > 1$ is a constant. The class of the functions with this property is denoted by E_s^α. For $\alpha = 2$ and for a set of integers q; $a_1, \cdots, a_s$, we have

$$\sup_{f \in E_s^2} \left| \int_0^1 \cdots \int_0^1 f(x_1, \cdots, x_s)\, dx_1 \cdots dx_s - \right.$$
$$\left. - \frac{1}{q} \sum_{k=1}^q f\left(\frac{a_1 k}{q}, \cdots, \frac{a_s k}{q} \right) \right| \leqslant$$
$$\leqslant \left(\frac{\pi^2}{6} \right)^s H(q; a_1, \cdots, a_s),$$

where
$$H(q; a_1, \cdots, a_s) =$$
$$= \begin{cases} \dfrac{3^s}{q}\left[1 + 2 \displaystyle\sum_{k=1}^{\frac{q-1}{2}} \prod_{\nu=1}^{s} \left(1 - 2\left\{\dfrac{a_\nu k}{q}\right\}\right)^2 \right] - 1, & \text{if } 2 \nmid q; \\[2mm] \dfrac{3^s}{q}\left[1 + \displaystyle\prod_{\nu=1}^{s} \left(1 - 2\left\{\dfrac{a_\nu}{2}\right\}\right)^2 + \right. \\[2mm] \left. + 2 \displaystyle\sum_{k=1}^{\frac{q}{2}-1} \prod_{\nu=1}^{s} \left(1 - 2\left\{\dfrac{a_\nu k}{q}\right\}\right)^2 \right] - 1, & \text{if } 2 \mid q. \end{cases}$$

The main problem in the study of numerical integrations is to find a set of integers $q; a_1, \cdots, a_s$ such that the error $H(q; a_1, \cdots, a_s)$ is comparatively small. Коробов[1] proved that for $q = p$ being a prime, there exists an integer a such that

$$H(p; 1, a, \cdots, a^{s-1}) = H(p, a) \ll \frac{\log^{2s} p}{p^2}.$$

For a given p, in order to find the integer a such that $H(p, a)$ takes its minimum, we have to perform $O(p^2)$ elementary operations (add., sub., mult., and div.). It was improved later to $O(q^{4/3})$.

The aim of this note and the succeeding one is to give two direct methods. From the numerical view of consideration, the results are as sharp as those obtained by the method of Коробов. (They have the same significant figures.)[2]

Let $p_1, \cdots, p_t$ be t distinct primes. We consider the algebraic field $R(\sqrt{p_1}, \cdots, \sqrt{p_t})$. Let

$$\varepsilon_1, \cdots, \varepsilon_{2^t-1}$$

be a set of independent units, which are the solution of Pell's equation

$$x^2 - p_{i_1} \cdots p_{i_k} y^2 = \pm 4 \quad (x > 0,\ y > 0)$$

with least $\dfrac{x}{2} + \dfrac{\sqrt{p_{i_1} \cdots p_{i_k}}\, y}{2}$, where $k \geqslant 1$ and $1 \leqslant i_1 < \cdots < i_k \leqslant t$ is any choice of $1, 2, \cdots, t$. We may take

$$\varepsilon_i = \begin{cases} \dfrac{x_i}{2} + \dfrac{\sqrt{d_i}\, y_i}{2}, \\[2mm] \qquad x_i \equiv y_i \equiv 1 \ (\mathrm{mod}\ 2) \quad (1 \leqslant i \leqslant m); \\[2mm] x_i + \sqrt{d_i}\, y_i \quad (m + 1 \leqslant i \leqslant 2^t - 1). \end{cases}$$

We take positive integers $n_1, \cdots, n_t$ such that for any $i \not\equiv j$, we have

$$c_1 \varepsilon_j^{n_i} < \varepsilon_i^{n_i} < c_2 \varepsilon_j^{n_i},$$

where c_1 and c_2 are two positive constants. Consider the expression

$$\varepsilon_1^{n_1} \cdots \varepsilon_{2^t-1}^{n_{2^t-1}} = q_0^{(n)} + q_1^{(n)} \varepsilon_1 + \cdots + q_m^{(n)} \varepsilon_m +$$
$$+ q_{m+1}^{(n)} \sqrt{d_{m+1}} + \cdots + q_{2^t-1}^{(n)} \sqrt{d_{2^t-1}} \tag{1}$$

and $2^t - 1$ transformations:

$$(\sigma_{i_1 \cdots i_k}) \quad \sqrt{p_\nu} \to \begin{cases} \sqrt{p_\nu}, & \text{if } \nu \not\equiv i_j \ (1 \leqslant j \leqslant k); \\ -\sqrt{p_\nu}, & \text{if } \nu = i_j \ (1 \leqslant j \leqslant k). \end{cases} \tag{2}$$

Applying (2) to Eq. (1), we have $2^t - 1$ equations. Combining them with (1), we have a system of 2^t equations, of which we have the solutions:

$$2q_0^{(n)} + q_1^{(n)} + \cdots +$$
$$+ q_m^{(n)} = \varepsilon_1^{n_1} \cdots \varepsilon_{2^t-1}^{n_{2^t-1}} \Big/ 2^{t-1} + O(|\varepsilon_1|^{-n_1}),$$
$$q_i^{(n)} = \varepsilon_1^{n_1} \cdots \varepsilon_{2^t-1}^{n_{2^t-1}} \Big/ 2^{t-1} \sqrt{d_i} + O(|\varepsilon_1|^{-n_1})$$
$$(1 \leqslant i \leqslant m),$$
$$q_i^{(n)} = \varepsilon_1^{n_1} \cdots \varepsilon_{2^t-1}^{n_{2^t-1}} \Big/ 2^t \sqrt{d_i} + O(|\varepsilon_1|^{-n_1})$$
$$(m + 1 \leqslant i \leqslant 2^t - 1). \tag{3}$$

For $2 \mid (q_1^{(n)}, \cdots, q_m^{(n)})$, we define $q = q_0^{(n)} + \frac{1}{2}(q_1^{(n)} + \cdots + q_m^{(n)})$, $a_1 = 1$, $a_{i+1} \equiv d_i q_i^{(n)}/2 \ (\mathrm{mod}\ q)$ $(1 \leqslant i \leqslant m)$, $a_{i+1} \equiv d_i q_i^{(n)} \ (\mathrm{mod}\ q)$ $(m+1 \leqslant i \leqslant 2^t-1)$.

For $2 \nmid (q_1^{(n)}, \cdots, q_m^{(n)})$, we define $q = 2q_0^{(n)} + (q_1^{(n)} + \cdots + q_m^{(n)})$, $a_1 = 1$, $a_{i+1} \equiv d_i q_i^{(n)} \ (\mathrm{mod}\ q)$ $(1 \leqslant i \leqslant m)$, $a_{i+1} \equiv 2 d_i q_i^{(n)} \ (\mathrm{mod}\ q)$ $(m+1 \leqslant i \leqslant 2^t - 1)$.

From (3) we have the simultaneous diophantine approximations:

$$\left|\frac{a_i}{q} - \sqrt{d_{i-1}}\right| = O\left(\frac{1}{q^{1+\frac{1}{2^t+1}}}\right) \quad (2 \leqslant i \leqslant 2^t).$$

We take these $q; a_1, \cdots, a_{2^t}$ to evaluate the value of $H(q; a_1, \cdots, a_{2^t})$. The effectiveness of the present method is illustrated by the following examples:

Example 1. The field $R(\sqrt{2})$ has $\varepsilon = 1 + \sqrt{2}$ as the fundamental unit. Let $\varepsilon^n = q_0^{(n)} + \sqrt{2} q_1^{(n)}$. Then

$$H(q_0^{(n)}; 1, 2q_1^{(n)}) \ll \frac{\log q_0^{(n)}}{q_0^{(n)2}} \quad (\text{cf. } [3]).$$

Example 2. We take $R(\sqrt{2}, \sqrt{5})$ and

$$\varepsilon_1 = \frac{1 + \sqrt{5}}{2}, \quad \varepsilon_2 = 1 + \sqrt{2}, \quad \varepsilon_3 = 3 + \sqrt{10}.$$

From

$$\varepsilon_1^6 \varepsilon_2^4 \varepsilon_3^2 = (9 + 4\sqrt{5})(17 + 12\sqrt{2})(19 + 6\sqrt{10})$$

$$= 5787 + 4092\sqrt{2} + 2588\sqrt{5} + 1830\sqrt{10},$$

we have

$$q = 5787, \quad a_1 = 1, \quad a_2 = 2397,$$
$$a_3 = 1366, \quad a_4 = 939.$$

By numerical calculations, we have

$$H(5787; 1, 2397, 1366, 939) \leqslant 0.001498.$$

Remark. Here we suggest the following method to calculate numerically the integration of $2^t - 1$-dimensional space

$$\int_0^1 \cdots \int_0^1 f(x_1, \cdots, x_s)\, dx_1, \cdots, dx_s \approx$$

$$\approx \frac{1}{(2q+1)^l} \sum_{j=-ql}^{ql} \mu_{q,l,j}\, f(\varepsilon_1 j, \cdots, \varepsilon_{2^t-1} j),$$

where $f \in E_{2^t-1}^{\alpha}$, l is the least integer $\geqslant \alpha$, $\mu_{q,l,j}$ is defined by

$$\left(\sum_{j=-q}^{q} z^j\right)^l = \sum_{j=-ql}^{ql} \mu_{q,l,j} z^j,$$

and $\varepsilon_1, \cdots, \varepsilon_{2^t-1}$ is a set of independent units of $R(\sqrt{p_1}, \cdots, \sqrt{p_t})$ (cf. [4]).

Hua Loo Keng (华罗庚)

Wang Yuan (王　元)

*University of Science and
  Technology of China;
Institute of Mathematics,
  Academia Sinica;*
  April 14, 1964

References

[1] Коробов, Н. М. 1960 *ДАН СССР*, **132** (5), 1009—1012.

[2] Салтыков, А. И. 1963 *Жур. Выч. Мат. и Мат. Физ.*, **3** (1), 181—186.

[3] Hua, Loo Keng & Wang, Yuan 1960 *Sci. Rec.*, New Ser., **4** (1), 8—11.

[4] Бахвалов, Н. С. 1959 *Вес. МГУ*, **4**, 3—18.

MATHEMATICS

On Diophantine Approximations and Numerical Integrations (II)

We use the same notations as those in the previous note. Let p be a prime $\geqslant 5$ and $r = \dfrac{p-3}{2}$. For the algebraic field $R\left(\cos\dfrac{2\pi}{p}\right)$, we take the following $r+1$ units:

$$2\cos\frac{2\pi}{p}, \quad 2\cos\frac{4\pi}{p}, \quad \cdots, \quad 2\cos\frac{2(r+1)\pi}{p}. \tag{1}$$

They are arranged according to their absolute values as follows:

$$|\varepsilon_1^{(1)}| > |\varepsilon_2^{(1)}| > \cdots > |\varepsilon_{r+1}^{(1)}|.$$

For $1 \leqslant \nu \leqslant r+1$, we introduce the transformation

$$(\sigma_\nu) \qquad 2\cos\frac{2\pi\mu}{p} \to 2\cos\frac{2\pi\mu\nu}{p} \quad (1 \leqslant \mu \leqslant r+1).$$

The transformation (σ_ν) carries the set (1) onto itself. We use the notation

$$(\sigma_\nu) \qquad \varepsilon_\mu^{(1)} \to \varepsilon_\mu^{(\nu)} \quad (1 \leqslant \mu \leqslant r+1).$$

Let

$$\Delta = \begin{pmatrix} 1, & \varepsilon_1^{(1)}, & \cdots, & \varepsilon_r^{(1)} \\ 1, & \varepsilon_1^{(2)}, & \cdots, & \varepsilon_r^{(2)} \\ \cdots\cdots\cdots\cdots\cdots\cdots\cdots \\ 1, & \varepsilon_1^{(r+1)}, & \cdots, & \varepsilon_r^{(r+1)} \end{pmatrix},$$

$$\widetilde{\Delta} = \frac{1}{p}\begin{pmatrix} 2-\varepsilon_{r+1}^{(1)}, & 2-\varepsilon_{r+1}^{(2)}, & \cdots, & 2-\varepsilon_{r+1}^{(r+1)} \\ \varepsilon_1^{(1)}-\varepsilon_{r+1}^{(1)}, & \varepsilon_1^{(2)}-\varepsilon_{r+1}^{(2)}, & \cdots, & \varepsilon_1^{(r+1)}-\varepsilon_{r+1}^{(r+1)} \\ \cdots\cdots\cdots\cdots\cdots\cdots\cdots\cdots\cdots\cdots \\ \varepsilon_r^{(1)}-\varepsilon_{r+1}^{(1)}, & \varepsilon_r^{(2)}-\varepsilon_{r+1}^{(2)}, & \cdots, & \varepsilon_r^{(r+1)}-\varepsilon_{r+1}^{(r+1)} \end{pmatrix}.$$

We have the following properties:

(i) $\displaystyle\sum_{\mu=1}^{r+1} \varepsilon_\mu^{(\nu)} = -1,$

(ii) $\displaystyle\sum_{\mu=1}^{r+1}{}' \varepsilon_\mu^{(\nu)} = (-1)\, H\!\left[\frac{\nu}{2}\right],$

(iii) $|\det \Delta| = p^{\frac{r}{2}},$

(iv) $\Delta\,\widetilde{\Delta} = I.$

Let

$$A = \begin{pmatrix} \log|\varepsilon_1^{(2)}|, & \cdots, & \log|\varepsilon_r^{(2)}| \\ \cdots\cdots\cdots\cdots\cdots\cdots\cdots \\ \log|\varepsilon_1^{(r+1)}|, & \cdots, & \log|\varepsilon_r^{(r+1)}| \end{pmatrix}.$$

We have $\det A \neq 0$[1]. We may take

---

1) For the case $\det A = 0$, we may use the following set of independent units:

$$\eta_e = \sin\frac{\pi}{p}\, g^{l+1} \Big/ \sin\frac{\pi}{p}\, g^l \quad (l = 0, 1, \cdots, r-1),$$

where g is a primitive root mod p.

$$\det\begin{pmatrix} \log\left|\dfrac{\varepsilon_1^{(3)}}{\varepsilon_1^{(2)}}\right|, & \cdots, & \log\left|\dfrac{\varepsilon_{r-1}^{(3)}}{\varepsilon_{r-1}^{(2)}}\right| \\ \cdots\cdots\cdots\cdots\cdots\cdots\cdots\cdots\cdots \\ \log\left|\dfrac{\varepsilon_1^{(r+1)}}{\varepsilon_1^{(2)}}\right|, & \cdots, & \log\left|\dfrac{\varepsilon_{r-1}^{(r+1)}}{\varepsilon_{r-1}^{(2)}}\right| \end{pmatrix} \neq 0.$$

Hence the system of equations

$$|\varepsilon_1^{(i)^{a_1}}, \cdots, \varepsilon_r^{(i)^{a_r}}| = |\varepsilon_1^{(2)^{a_1}}, \cdots, \varepsilon_r^{(2)^{a_r}}|$$
$$(3 \leqslant i \leqslant r+1)$$

has a unique nonzero solution

$$\left(\frac{\alpha_1}{\alpha_r}, \cdots, \frac{\alpha_{r-1}}{\alpha_r}\right).$$

From $\det A \neq 0$, we may assume that

$$|\varepsilon_1^{(2)^{a_1}}, \cdots, \varepsilon_r^{(2)^{a_r}}| < 1.$$

Take positive integer n_r to be sufficiently large and the system of integers $n_1, \cdots, n_{r-1}$ such that

$$\left|\frac{n_i}{n_r} - \frac{\alpha_i}{\alpha_r}\right| < \frac{1}{n} \quad (1 \leqslant i \leqslant r-1),$$

and

$$|\varepsilon_1^{(i)^{n_1}}, \cdots, \varepsilon_r^{(i)^{n_r}}| < 1 \quad (2 \leqslant i \leqslant r+1).$$

Therefore we have

$$|\varepsilon_1^{(i)^{n_1}}, \cdots, \varepsilon_r^{(i)^{n_r}}| = O\big(|\varepsilon_1^{(1)^{n_1}}, \cdots, \varepsilon_r^{(1)^{n_r}}|^{-\frac{1}{r}}\big)$$
$$(2 \leqslant i \leqslant r+1).$$

The system of equations

$$\varepsilon_1^{(i)^{n_1}}, \cdots, \varepsilon_r^{(i)^{n_r}} = q_0^{(n)}(2 - \varepsilon_{r+1}^{(i)})/p +$$
$$+ \sum_{i=1}^{r} q_j^{(n)}(\varepsilon_j^{(i)} - \varepsilon_{r+1}^{(i)})/p \quad (1 \leqslant i \leqslant r+1)$$

gives

$$q_0^{(n)} = \varepsilon_1^{(1)^{n_1}}, \cdots, \varepsilon_r^{(1)^{n_r}} +$$
$$+ O\big(|\varepsilon_1^{(1)^{n_1}}, \cdots, \varepsilon_r^{(1)^{n_r}}|^{-\frac{1}{r}}\big),$$

$$q_i^{(n)} = \varepsilon_1^{(1)^{n_1}}, \cdots, \varepsilon_r^{(1)^{n_r}}\varepsilon_i^{(1)} +$$
$$+ O\big(|\varepsilon_1^{(1)^{n_1}}, \cdots, \varepsilon_r^{(1)^{n_r}}|^{-\frac{1}{r}}\big) \quad (1 \leqslant i \leqslant r). \tag{2}$$

Let

$$q = |q_0^{(n)}|, \quad a_1 = 1,$$
$$a_{i+1} \equiv |q_i^{(n)}| \pmod{q} \quad (1 \leqslant i \leqslant r).$$

From (2) we have the following simultaneous diophantine approximations:

$$\left|\frac{a_i}{q} - \varepsilon_{i-1}^{(1)}\right| \ll \frac{1}{q^{1+\frac{1}{r}}} \quad (2 \leqslant i \leqslant r+1).$$

The practical numerical method for finding $q_i^{(n)}(0 \leqslant i \leqslant r)$ can be described as follows: From the expression

$$\varepsilon_1^{(1)^{n_1}}, \cdots, \varepsilon_r^{(1)^{n_r}} = h_1^{(n)}\varepsilon_1^{(1)} + \cdots + h_{r+1}^{(n)}\varepsilon_{r+1}^{(1)},$$

we have

$$q_0^{(n)} = - \sum_{i=1}^{r+1} h_i^{(n)},$$

$$q_i^{(n)} = p h_i^{(n)} - 2 \sum_{j=1}^{r+1} h_j^{(n)} \quad (1 \leqslant i \leqslant r).$$

Example 1. Take $R\left(\cos\dfrac{2\pi}{5}\right)$ and

$$\varepsilon_1 = 2\cos\frac{4\pi}{5}, \qquad \varepsilon_2 = 2\cos\frac{2\pi}{5}.$$

From $2 - \varepsilon_2 = \sqrt{5}\,\varepsilon_2$, $\varepsilon_1 - \varepsilon_2 = \sqrt{5}$, and the expression $\varepsilon_1^n = q_0^{(n)}\varepsilon_2 - q_1^{(n)}$, we have

$$H(|q_0^{(n)}|; 1, |q_1^{(n)}|) \ll \frac{\log|q_0^{(n)}|}{|q_0^{(n)}|^2} \quad \text{(cf. [1]).}$$

Example 2. Take $R\left(\cos\dfrac{2\pi}{7}\right)$ and

$$\varepsilon_1 = 2\cos\frac{6\pi}{7}, \qquad \varepsilon_2 = 2\cos\frac{2\pi}{7},$$

$$\varepsilon_3 = 2\cos\frac{4\pi}{7}.$$

From the equation $|\varepsilon_2^\alpha \varepsilon_3^\beta| = |\varepsilon_3^\alpha \varepsilon_1^\beta|$, we have $\dfrac{\alpha}{\beta} \doteq 1.356 \doteq \dfrac{4}{3}$. From the expression

$$\varepsilon_1^4 \varepsilon_2^6 = -227\,\varepsilon_1 - 45\,\varepsilon_2 - 146\,\varepsilon_3,$$

we have

$$q = 418;\ a_1 = 1,\ a_2 = 335,\ a_3 = 103.$$

By numerical calculation, we have

$$H(418; 1, 335, 103) \leqslant 0.0108146.$$

Example 3. Considering the field $R\left(\cos\dfrac{2\pi}{11}\right)$ and $R\left(\cos\dfrac{2\pi}{13}\right)$ respectively, we have

$$H(9389; 1, 8628, 6408, 2908, 7800) \leqslant$$
$$\leqslant 0.0081175,$$

$$H(41204; 1, 38810, 31766, 20480, 5610,$$
$$29223) \leqslant 0.009425.$$

Further numerical examples will be given elsewhere later.

Remarks. The following numerical method of evaluating the r-ple integral is suggested

$$\int_0^1 \cdots \int_0^1 f(x_1, \cdots, x_s)\, dx_1 \cdots dx_s \approx$$

$$\approx \frac{1}{(2q+1)^l} \sum_{j=-ql}^{ql} \mu_{q,l,j}\, f(\varepsilon_1^{(1)} j, \cdots, \varepsilon_r^{(1)} j).$$

Hua Loo Keng (华罗庚)

Wang Yuan (王　元)

*University of Science and
Technology of China;
Institute of Mathematics,
Academia Sinica;*
April 14, 1964

Reference

[1] Hua, Loo Keng & Wang, Yuan 1960 *Sci. Rec.*, New Ser., 4(1), 8—11.

SCIENTIA SINICA

Vol. XIV, No. 7, 1965

<u>*MATHEMATICS*</u>

ON NUMERICAL INTEGRATION OF PERIODIC FUNCTIONS
OF SEVERAL VARIABLES*

HUA LOO KENG (华罗庚) AND WANG YUAN (王　元)

(University of Science and Technology of China; Institute of Mathematics, Academia Sinica)

ABSTRACT

By means of the algebraic theory of numbers the authors suggest a method for evaluating multiple integrals. A numerical example of eleven dimensions shows the advantage of the present method.

I. INTRODUCTION

It is the Monte Carlo method to approximate a multiple integral by a single sum over a random variable. More precisely, let $f(x_1, \cdots, x_s)$ be a continuous periodic function of s-variables, each with period 1. If we identify the opposite faces of the unit cube $0 \leqslant x_\nu \leqslant 1$ $(\nu = 1, 2, \cdots, s)$ as an s-ring R, a periodic continuous function may be considered as a continuous function defined on the s-ring. Let

$$(x_1^{(i)}, \cdots, x_s^{(i)}), \quad i = 1, 2, 3, \cdots \tag{1.1}$$

be a variable random uniformly distributed on the ring R; then the average

$$\frac{1}{n} \sum_{i=1}^{n} f(x_1^{(i)}, \cdots, x_s^{(i)}) \tag{1.2}$$

of the values of $f(x_1, \cdots, x_s)$ for the sequence of numbers (1.1) is approximately equal to the value of the integral

$$\int_0^1 \cdots \int_0^1 f(x_1, \cdots, x_s) dx_1 \cdots dx_s \tag{1.3}$$

with a probability error

$$O\left(\frac{1}{\sqrt{n}}\right).$$

The method of theory of numbers is to construct explicitly a sequence (1.1) so that the difference of (1.3) and (1.2) has an absolute error. Some theoretic results in this respect have been collected in a monograph of the authors[2], and we shall not discuss them here. However, it is worth mentioning that both Коробов[5] and Halton[6] have made important contributions to the subject considered.

---

* Received Feb. 8, 1965.

Here, we aim at the specific computational methods and numerical results. In the beginning Коробов applied the method of complete exponential sum in the analytic theory of numbers (see, e.g., Hua Loo Keng[1]) and found a method of using a simple sum to approximate a multiple integral with an absolute error which is the same as the probability error $O\left(\dfrac{1}{\sqrt{n}}\right)$ obtained by the Monte Carlo method, where n is the number of points of division. Later, he proved a theorem which even improves the accuracy; but it is a theorem similar to that of existence, and for computational purpose, it is necessary to compare all "divisions" of a certain type so as to find the one with minimal error.

In the light of the theory of algebraic numbers and Diophantine approximations, we suggest a set of points of division. Our division is essentially as good as the best division of Коробов (the example given below shows that they have the same significant figures), yet the amount of computation involved is considerably reduced. To be more definite, if his amount is denoted by P, our amount is only $O(\sqrt{P})$. Naturally, using a computer of the same speed, we may find a method treating the integrals of higher dimensions and obtain results of better accuracy

To show the efficiency, we give the following numerical example:

For a periodic function $f(x_1, \cdots, x_{11})$ satisfying certain conditions mentioned in the text, we have

$$\left| \int_0^1 \cdots \int_0^1 f(x_1, \cdots, x_{11}) dx_1 \cdots dx_{11} - \right.$$

$$\left. - \frac{1}{q} \sum_{t=1}^{q} f\left(\frac{a_1 t}{q}, \cdots, \frac{a_{11} t}{q}\right) \right| \leqslant c \left(\frac{\pi^2}{6}\right)^{11} \times 0.233543,$$

where

$$q = 698047, \quad a_1 = 1, \qquad a_2 = 685041,$$
$$a_3 = 646274, \quad a_4 = 582461, \quad a_5 = 494796,$$
$$a_6 = 384914, \quad a_7 = 254860, \quad a_8 = 107051,$$
$$a_9 = 642292, \quad a_{10} = 467527, \quad a_{11} = 284044.$$

Further examples of 5- and 6-ple integrals shall be given in the text.

It is worth while to mention that the division points so obtained have applications to the theory of probability.

II.   Multiple Integrals and Simple Sums

First of all, let us construct some auxiliary functions:

Lemma 2.1. *Let* $\{x\}$ *be the distance from* x *to its nearest integer. Then we have*

$$3(1 - 2\{x\})^2 = \sum_{m=-\infty}^{\infty} \frac{e^{2\pi i m x}}{\dfrac{\pi^2}{6} \bar{m}^2},$$

where $\bar{n} = \max(1, |n|)$.

Proof. It is sufficient to prove that

$$c_m = \int_0^1 3(1 - 2\{x\})^2 e^{-2\pi i m x} dx = \begin{cases} 1 & \text{for } m = 0, \\[2mm] \dfrac{6}{\pi^2 m^2} & \text{for } m \neq 0. \end{cases}$$

Evidently we have

$$c_m = 2 \int_0^{\frac{1}{2}} 3(1 - 2x)^2 \cos 2\pi m x \, dx.$$

For $m = 0$,

$$c_0 = 2 \int_0^{\frac{1}{2}} 3(1 - 2x)^2 dx = -(1 - 2x)^3 \Big|_0^{\frac{1}{2}} = 1.$$

For $m \neq 0$, integrating by parts twice, we have

$$c_m = 6(1 - 2x)^2 \frac{\sin 2\pi m x}{2\pi m} \Big|_0^{\frac{1}{2}} + 24 \int_0^{\frac{1}{2}} \frac{(1 - 2x)\sin 2\pi m x}{2\pi m} dx =$$

$$= -\frac{12}{\pi m}(1 - 2x)\frac{\cos 2\pi m x}{2\pi m} \Big|_0^{\frac{1}{2}} - 48 \int_0^{\frac{1}{2}} \frac{\cos 2\pi m x}{(2\pi m)^2} dx =$$

$$= \frac{6}{\pi^2 m^2}.$$

Consequently we have
Lemma 2.2.

$$g(x_1, \cdots, x_s) = 3^s \prod_{\nu=1}^{s}(1 - 2\{x_\nu\}) = \sum_{-\infty}^{\infty} \cdots \sum \frac{e^{2\pi i(m_1 x_1 + \cdots + m_s x_s)}}{\dfrac{\pi^2}{6}m_1^2 \cdots \dfrac{\pi^2}{6}m_s^2}.$$

Let $q > 0$, $a_1, \cdots, a_s$ be integers; we have

$$\frac{1}{q} \sum_{t=1}^{q} g\left(\frac{a_1 t}{q}, \cdots, \frac{a_s t}{q}\right) = \sum_{\substack{-\infty \\ m_1 a_1 + \cdots + m_s a_s \equiv 0 (\bmod q)}}^{\infty} \cdots \sum \frac{1}{\dfrac{\pi^2}{6}m_1^2 \cdots \dfrac{\pi^2}{6}m_s^2}. \tag{2.1}$$

The main idea is to find $q, a_1, \cdots, a_s$ so that

$$H_1(q; a_1, \cdots, a_s) = \frac{1}{q}\sum_{t=1}^{q} g\left(\frac{a_1 t}{q}, \cdots, \frac{a_s t}{q}\right) \tag{2.2}$$

as small as possible.

Naturally, the best way is, for a fixed q among all possible $0 \leqslant a_\nu < q$, to find a set of $(a_1, \cdots, a_s)$ for minimizing $H_1(q, a_1, \cdots, a_s)$. However, it is too complicated to do so; hence Коробов suggested the following method.

To find the smallest value among the $p - 1$ sums

$$H_1(p; 1, z, \cdots, z^{s-1}),$$

where $p = q$ is a prime, and z runs over $1, 2, \cdots, p - 1$. This method requires $O(q^2)$ elementary operations; for a larger q, it can be improved to $O(q^{4/3})$.

With the aid of algebraic theory of numbers, we give a set of $q, a_1, \cdots, a_s,$ without very complicated computations (i.e., the ordinary calculating machine is good enough for our purpose). To estimate the value of $H_1(q, a_1, \cdots, a_s)$, we require only $O(q)$ elementary operations.

Let $f(x_1, \cdots, x_s)$ be a periodic function with period 1 for each variable and having the Fourier expansion

$$f(x_1, \cdots, x_s) = \sum_{-\infty}^{\infty} \cdots \sum c(m_1, \cdots, m_s) e^{2\pi i(m_1 x_1 + \cdots + m_s x_s)},$$

where

$$|c(m_1, \cdots, m_s)| < \frac{c}{(\overline{m_1} \cdots \overline{m_s})^2},$$

and c is a positive number. For example, the function satisfying

$$\left| \frac{\partial^{2s}}{\partial x_1^2 \cdots \partial x_s^2} f(x_1, \cdots, x_s) \right| < (2\pi)^{2s} c$$

belongs to the class considered.

Consider

$$\frac{1}{q} \sum_{t=1}^{q} f\left(\frac{a_1 t}{q}, \cdots, \frac{a_s t}{q}\right) = \sum_{-\infty}^{\infty} \cdots \sum c(m_1, \cdots, m_s) \frac{1}{q} \sum_{t=1}^{q} e^{2\pi i(a_1 m_1 + \cdots + a_s m_s)\frac{t}{q}} =$$

$$= \sum_{\substack{-\infty \\ a_1 m_1 + \cdots + a_s m_s \equiv 0 (\mathrm{mod}\ q)}}^{\infty} \cdots \sum c(m_1, \cdots, m_s).$$

Consequently

$$\int_0^1 \cdots \int_0^1 f(x_1, \cdots, x_s) dx_1 \cdots dx_s - \frac{1}{q} \sum_{t=1}^{q} f\left(\frac{a_1 t}{q}, \cdots, \frac{a_s t}{q}\right) =$$

$$= - \sum_{\substack{a_1 m_1 + \cdots + a_s m_s \equiv 0(\mathrm{mod}\ q)}} \cdots \sum{}' c(m_1, \cdots, m_s), \tag{2.3}$$

where $\sum \cdots \sum'$ indicates the omitting of the term $m_1 = \cdots = m_s = 0$ since

$$c(0, \cdots, 0) = \int_0^1 \cdots \int_0^1 f(x_1, \cdots, x_s) dx_1 \cdots dx_s.$$

Consequently, the difference between the multiple integral

$$\int_0^1 \cdots \int_0^1 f(x_1, \cdots, x_s) dx_1 \cdots dx_s$$

and the simple sum

$$\frac{1}{q} \sum_{t=1}^{q} f\left(\frac{a_1 t}{q}, \cdots, \frac{a_s t}{q}\right)$$

is dominated by

$$\left| \sum \cdots \sum_{a_1 m_1 + \cdots + a_s m_s \equiv 0 (\bmod\ q)}' c(m_1, \cdots, m_s) \right| \leqslant \sum \cdots \sum' \frac{1}{m_1^2 \cdots m_s^2} =$$

$$= \left(\frac{\pi^2}{6}\right)^s [H_1(q; a_1, \cdots, a_s) - 1]$$

by (2.1) and (2.2).

III. Some Properties of the Totally Real Algebraic Numbers

Let $\mathscr{F}$ denote a totally real algebraic number field of the nth degree. For a number η of $\mathscr{F}$, let $\eta^{(1)}(=\eta)$, $\eta^{(2)}$, $\cdots$, $\eta^{(n)}$ be its conjugates. Assume that

$$\varpi_1, \cdots, \varpi_n$$

be an integral basis of $\mathscr{F}$. Form a matrix

$$\varOmega = \begin{pmatrix} \varpi_1^{(1)}, \cdots, \varpi_n^{(1)} \\ \cdots\cdots\cdots \\ \varpi_1^{(n)}, \cdots, \varpi_n^{(n)} \end{pmatrix}. \tag{3.1}$$

The matrix

$$S = \varOmega' \varOmega = \left(\sum_{k=1}^{n} \varpi_i^{(k)} \varpi_j^{(k)} \right) \tag{3.2}$$

is called the fundamental matrix of the field $\mathscr{F}$. Clearly, it is a symmetric matrix with rational integral elements. (The invariants of the fundamental matrix under the modular group are characteristic properties of the algebraic number field $\mathscr{F}$.) From (3.2), we deduce immediately

$$\varOmega^{-1} = S^{-1} \varOmega'. \tag{3.3}$$

The determinant of S is the discriminant of the field.

Let η be a unit of $\mathscr{F}$ with absolute value >1 and its conjugates (beside itself) satisfy

$$|\eta^{(j)}| \leqslant c |\eta|^{-\frac{1}{n-1}}, \quad j = 2, \cdots, n, \tag{3.4}$$

where c is a constant. We express η as follows:

$$\eta = h_1 \varpi_1 + \cdots + h_n \varpi_n, \tag{3.5}$$

where $h_1, \cdots, h_n$ are rational integers. From (3.5) and its conjugates we have

$$(\eta^{(1)}, \cdots, \eta^{(n)}) = (h_1, \cdots, h_n)\varOmega'. \tag{3.6}$$

Hence

$$(\eta^{(1)}, \cdots, \eta^{(n)})\varOmega = (h_1, \cdots, h_n)S = (k_1, \cdots, k_n)$$

$$\text{(by definition)}.$$

Or

$$\sum_{i=1}^{n} \eta^{(i)}\tilde{\omega}_j^{(i)} = k_j.$$

Therefore, we have

$$\left| \eta_i \tilde{\omega}_j - k_j \right| \leqslant \sum_{i=2}^{n} |\eta^{(i)}| \, |\tilde{\omega}_j^{(i)}| \leqslant (n-1)c\tilde{\omega} |\eta|^{-\frac{1}{n-1}}, \tag{3.7}$$

where

$$\tilde{\omega} = \max_{i,j} |\tilde{\omega}_j^{(i)}|.$$

Let

$$1 = a_1\tilde{\omega}_1 + \cdots + a_n\tilde{\omega}_n \tag{3.8}$$

and

$$k = a_1 k_1 + \cdots + a_n k_n.$$

Then, from (3.7), we deduce that

$$|\eta - k| \leqslant \sum_{j=1}^{n} |a_j| \, |\eta\tilde{\omega}_j - k_j| \leqslant A|\eta|^{-\frac{1}{n-1}}, \tag{3.9}$$

where the constant A depends only on $\mathscr{F}$ and C. (A will appear later but may be different in each occurrence.)

Hence, we obtain the simultaneous Diophantine approximation

$$\left| \tilde{\omega}_j - \frac{k_j}{k} \right| \leqslant \frac{A}{k^{n/n-1}}, \quad 1 \leqslant j \leqslant n. \tag{3.10}$$

This is not new. Our purpose is to suggest a computational method for obtaining $k_j's$.

For $n = 2$, we can use continued fractions to treat the present problem. In case of $n > 2$, the situation is entirely different. Various classical methods can only prove the existence of an infinitely many sets of $k, k_1, \cdots, k_n$ satisfying (3.10), but this does not suggest any effective way for finding $k, k_1, \cdots, k_n$. It is shown in this section that the problem for finding $k, k_1, \cdots, k_n$ is equivalent to the problem for finding an η in the totally real algebraic field so that (3.4) is satisfied.

If we know a set of independent units, then by the following lemma, we can find a sequence of units $\{\eta_j\}$ satisfying (3.4) and

$$|\eta_j| \to \infty, \quad \text{as } j \to \infty.$$

Consequently, we have infinitely many sets

$$k, k_1, \cdots, k_n$$

satisfying (3.10). (It is practically known.)

Lemma 3.1. *Let*

$$L_i(x_1, \cdots, x_m) = a_{i1}x_1 + \cdots + a_{im}x_m, \quad 1 \leqslant i \leqslant m.$$

If $\det(a_{ij}) \neq 0$, *there is a point* $(x_1, \cdots, x_m)$ *such that*

$$L_i = L_j < K, \quad 1 \leqslant i, j \leqslant m, \tag{3.11}$$

where K is any given real number. Further let

$$N = \max(|a_{11}| + \cdots + |a_{1m}|, \cdots, |a_{m1}| + \cdots + |a_{mm}|). \tag{3.12}$$

Then there is a set of integers $(y_1^{(t)}, \cdots, y_m^{(t)})$ *such that*

$$L_i(y_1^{(t)}, \cdots, y_m^{(t)}) < -(2t - 1)N. \tag{3.13}$$

and

$$|L_i(y_1, \cdots, y_m) - L_j(y_1, \cdots, y_m)| \leqslant 2N. \tag{3.14}$$

Proof. We solve the system of equations

$$L_1 = L_2 = \cdots = L_m = K - 1$$

for $x_1, \cdots, x_m$. Obviously, the solution will satisfy (3.11).

Now, we take $K = -2Nt$ and $y_i^{(t)} = [x_i]$; then

$$L_i(y_1^{(t)}, \cdots, y_m^{(t)}) \leqslant L_i(x_1, \cdots, x_m) + \sum_{k=1}^{m} |a_{ik}| < -2tN + N = -(2t - 1)N$$

and

$$|L_i(y_1^{(t)}, \cdots, y_m^{(t)}) - L_j(y_1^{(t)}, \cdots, y_m^{(t)})| = \left| \sum_{k=1}^{m} (a_{ik} - a_{jk}) y_k^{(t)} \right| \leqslant$$

$$\leqslant \left| \sum_{k=1}^{m} (a_{ik} - a_{jk}) x_k \right| + \sum_{k=1}^{m} |a_{ik} - a_{jk}| =$$

$$= \sum_{k=1}^{m} |a_{ik} - a_{jk}| \leqslant 2N.$$

The lemma follows.

By means of the lemma, we are going to find the unit η.

Theorem 3.1. *In the totally real algebraic number field $\mathscr{F}$, we have a sequence of units $\eta_t(= \eta_t^{(1)})$ $t = 1, 2, 3, \cdots$, whose conjugates $\eta_t^{(2)}, \cdots, \eta_t^{(n)}$ satisfy*

$$|\eta_t^{(i)}| < e^{-(2t-1)a}, \quad 2 \leqslant i \leqslant n \tag{3.15}$$

and

$$e^{-2a}|\eta_t^{(j)}| \leqslant |\eta_t^{(i)}| \leqslant e^{2a}|\eta_t^{(j)}|, \quad 2 \leqslant i, j \leqslant n, \tag{3.16}$$

where a is a constant depending only on the field $\mathscr{F}$.

Proof. **Let**

$$\varepsilon_1, \cdots, \varepsilon_r \quad (r = n - 1)$$

be a set of independent units of $\mathscr{F}$.

Let

$$\xi^{(i)} = \varepsilon_1^{(i)l_1} \cdots \varepsilon_r^{(i)l_r}, \quad 2 \leqslant i \leqslant n.$$

Taking the logarithm of the absolute value of $\xi^{(i)}$, we have

$$\log|\xi^{(i)}| = \sum_{i=1}^{r} l_i \log|\varepsilon_j^{(i)}|, \quad 2 \leqslant i \leqslant n.$$

Let

$$a = \max_{2 \leqslant i \leqslant n} \left(\sum_{j=1}^{r} \log|\varepsilon_j^{(i)}| \right).$$

Since

$$\det\left(\log|\varepsilon_j^{(i)}|\right) \neq 0,$$

the solution of

$$\log|\xi^{(2)}| = \cdots = \log|\xi^{(n)}| = -2at - 1$$

is denoted by $l_1, \cdots, l_r$. Let

$$[l_i] = a_i^{(t)}, \quad 1 \leqslant t \leqslant r.$$

By Lemma 3, the units

$$\eta_t = \varepsilon_1^{a_1{}^{(t)}} \cdots \varepsilon_r^{a_r{}^{(t)}}$$

satisfy the requirement of our theorem.

IV. The Cyclotomic Field

Let p be a prime $\geqslant 5$. $n = \dfrac{1}{2}(p-1)$, and $r = \dfrac{1}{2}(p-3) = n - 1$. The

real cyclotomic field $R\left(2\cos\dfrac{2\pi}{p}\right)$ is an algebraic field of degree n. The field has a set

of integral bases

$$2\cos\frac{2\pi}{p}, \quad 2\cos\frac{4\pi}{p}, \cdots, \quad 2\cos\frac{2n\pi}{p}. \tag{4.1}$$

It is well known that

$$\sum_{l=1}^{n} 2\cos\frac{2\pi l}{p} = -1 \tag{4.2}$$

and

$$\prod_{l=1}^{n} 2\cos\frac{2\pi l}{p} = (-1)^{\left[\frac{n+1}{2}\right]} = (-1)^{\frac{p^2-1}{8}}. \tag{4.3}$$

We use g to denote a primitive root, mod p. Since

$$2\cos\frac{2\pi}{p}g^{l\pm n} = 2\cos\left(-\frac{2\pi}{p}g^l\right) = 2\cos\frac{2\pi}{p}g^l, \tag{4.4}$$

the integral base (4.1) can be expressed as

$$\tilde{\omega}_l = 2\cos\frac{2\pi}{p}\, g^{l+m} \quad 1 \leqslant l \leqslant n, \tag{4.5}$$

where

$$|\tilde{\omega}_l| > |\tilde{\omega}_n| \quad (1 \leqslant l \leqslant n-1).$$

From (4.4), we define $\tilde{\omega}_{l\pm n} = \tilde{\omega}_l$.

The transformation

$$\sigma:\quad \tilde{\omega}_l \rightarrow \tilde{\omega}_{l+1}$$

is an automorphism of the cyclotomic field, and n automorphisms

$$\sigma, \sigma^2, \cdots, \sigma^{n-1}, \sigma^n \,(= \sigma^0 = 1)$$

form the group of automorphism of the field. A number η of the cyclotomic field has n conjugates under the automorphisms.

Since

$$\sum_{t=1}^{n} 2\cos\frac{2\pi lt}{p} \cdot 2\cos\frac{2\pi kt}{p} = \sum_{t=1}^{n}\left(2\cos\frac{2\pi(l+k)t}{p} + 2\cos\frac{2\pi(l-k)t}{p}\right) =$$

$$= -2 + \begin{cases} p, & \text{for } l = k, \\ 0, & \text{for } l \neq k, \end{cases}$$

we have

$$S = \begin{pmatrix} \tilde{\omega}_1, & \tilde{\omega}_2, & \cdots, & \tilde{\omega}_n \\ \tilde{\omega}_2, & \tilde{\omega}_3, & \cdots, & \tilde{\omega}_1 \\ \multicolumn{4}{c}{\cdots\cdots\cdots\cdots} \\ \tilde{\omega}_n, & \tilde{\omega}_1, & \cdots, & \tilde{\omega}_{n-1} \end{pmatrix}^2 = pI - 2M, \tag{4.6}$$

where $M = (m_{ij})$, $m_{ij} = 1$.

Taking a set of independent units $\varepsilon_1, \cdots, \varepsilon_r$ of $R\left(2\cos\frac{2\pi}{p}\right)$, then from the method of § III, we take

$$\eta = \varepsilon_1^{l_1} \cdots \varepsilon_r^{l_r} \tag{4.7}$$

so that the magnitudes of the absolute values of its $n-1$ conjugates are about the same, and all less than 1. If we let

$$\eta = \sum_{j=1}^{n} h_j \tilde{\omega}_j, \tag{4.8}$$

then by § III, we have

$$\left|\frac{k_i}{k} - \tilde{\omega}_i\right| = 0\left(\frac{1}{|k|^{n/n-1}}\right), \quad 1 \leqslant i \leqslant n, \tag{4.9}$$

where

$$
\begin{cases}
k = -\sum_{j=1}^{n} h_j, \\
k_i = ph_i - 2\sum_{i=1}^{n} h_j, \quad 1 \leqslant i \leqslant n.
\end{cases}
\tag{4.10}
$$

It is known (Fricke[7]) that

$$
\rho_l = \frac{\sin \dfrac{\pi}{p} g^{l+1}}{\sin \dfrac{\pi}{p} g^l}, \quad 1 \leqslant l \leqslant r
\tag{4.11}
$$

is a set of independent units of the cyclotomic field $R\left(2\cos\dfrac{2\pi}{p}\right)$.   It is more convenient to use the set of units

$$
2\cos\frac{2\pi l}{p}, \quad 1 \leqslant l \leqslant r.
\tag{4.12}
$$

However, they do not always form an independent set.   In the next section, we shall find a necessary and sufficient condition for which (4.12) are independent units.

V.   Units of the Totally Real Cyclotomic Field

Theorem 5.1. *Let*

$$
\zeta_l = 2\cos\frac{2\pi}{p} g^l, \quad l = 1, \cdots, r.
$$

A necessary and sufficient condition that $\zeta_1, \cdots, \zeta_r$ *form a set of independent units is that*

(i)   *2 is a primitive root*, mod p, *or*

(ii)   *2 belongs to the exponent* $\dfrac{1}{2}(p-1) = n$, mod p *and* $p = 7 \pmod 8$.

Proof. If $2 \equiv g^l \pmod p$, then

$$
\zeta_\mu = 2\cos\frac{2\pi}{p} g^\mu = 2\cos\frac{\pi}{p} g^{\mu+l} = \frac{\sin \dfrac{\pi}{p} g^{\mu+2l}}{\sin \dfrac{\pi}{p} g^{\mu+l}} = \rho_{\mu+l}\cdots\rho_{\mu+2l-1}.
$$

1)   If 2 is a primitive root, mod p, we take $g = 2$, so that $l = 1$, and hence

$$
\zeta_\mu = \rho_{\mu+1}.
$$

Therefore $\zeta_1, \cdots, \zeta_r$ are independent.

2)   If 2 is not a primitive root and belongs to the exponent S, then let $p - 1 = ls$

and take the primitive root g, mod p such that

$$g^l \equiv 2 \pmod{p}.$$

(i)   *If $l|n$, then let $n = lt$, hence*

$$\zeta_1 \zeta_{l+1} \cdots \zeta_{(t-1)l+1} = (\rho_{l+1} \cdots \rho_{2l})(\rho_{2l+1} \cdots \rho_{3l}) \cdots (\rho_1 \cdots \rho_l) = \prod_{j=1}^{n} \rho_j = -1.$$

Since $l > 1$, therefore $\zeta_1, \cdots, \zeta_r$ are not independent.

(ii)   Let $l \,|\, 2n$, but $l \nmid n$, i.e., $l = 2m$ and $n = mt$.
For $m \neq 1$, we have

$$\zeta_1 \zeta_{m+1} \cdots \zeta_{(t-1)m+1} = (\rho_{2m+1} \cdots \rho_{4m})(\rho_{3m+1} \cdots \rho_{5m}) \cdots (\rho_{m+1} \cdots \rho_{3m}) = \prod_{j=1}^{n} \rho_j^2 = 1.$$

Therefore $\zeta_1, \cdots, \zeta_r$ are not independent.

For $m = 1$, if we have integers $l_1, \cdots, l_n$, not all zero, such that

$$\pm 1 = \zeta^{l_1} \cdots \zeta_n^{l_n} = (\rho_3 \rho_4)^{l_1}(\rho_4 \cdot \rho_5)^{l_2} \cdots (\rho_n \rho_1)^{l_{n-2}}(\rho_1 \rho_2)^{l_{n-1}}(\rho_2 \rho_3)^{l_n} =$$
$$= \rho_1^{l_{n-2}+l_{n-1}} \rho_2^{l_{n-1}+l_n} \rho_3^{l_n+l_1} \cdots \rho_n^{l_{n-3}+l_{n-2}},$$

owing to the independence of $\rho_1, \cdots, \rho_n$, it follows that

$$l_{n-2} + l_{n-1} = l_{n-1} + l_n = l_n + l_1 = \cdots = l_{n-3} + l_{n-2}.$$

Consequently we have

$$l_1 = l_2 = \cdots = l_n.$$

In other words, $\zeta_1, \cdots, \zeta_r$ are independent, and 2 belongs to the exponent n, where $2 \nmid n$. From

$$2^n \equiv \left(\frac{2}{p}\right) = (-1)^{\frac{p^2-1}{8}} \pmod{p},$$

we have $p \equiv 7 \pmod 8$.   This completes the proof.

VI.   Selection of the Set of Points of Divisions and Numerical Results

Let $\varpi_1, \cdots, \varpi_n$ be an integral basis of the totally real field $R\left(2\cos\dfrac{2\pi}{p}\right)$ satisfying

$$|\varpi_i| > |\varpi_n| \quad (1 \leqslant i \leqslant n-1).$$

We define

$$k, k_1, \cdots, k_n$$

as in (4.10).

Define

$$q = |k|, \quad a_1 = 1, \quad a_{j+1} = |k_j| \quad (1 \leqslant j \leqslant r). \tag{6.1}$$

We use the following approximation formula for the multiple integral:

$$\int_0^1 \cdots \int_0^1 f(x_1, \cdots, x_n) dx_1 \cdots dx_n \approx \frac{1}{q} \sum_{t=1}^{q} f\left(\frac{a_1 t}{q}, \cdots, \frac{a_n t}{q}\right). \tag{6.2}$$

Our method is summarized as follows:

Take a set of independent units $\varepsilon_1, \cdots, \varepsilon_r$ of $R\left(2\cos\frac{2\pi}{p}\right)$ and take $l_1, \cdots, l_r$ so that the absolute values of the $n - 1$ conjugates of

$$\eta = \varepsilon_1^{l_1} \cdots \varepsilon_r^{l_r}$$

are about the same, and less than 1. Express η in terms of the integral basis

$$\eta = h_1 \bar{\omega}_1 + \cdots + h_n \bar{\omega}_n. \tag{6.3}$$

Then we take

$$q = \left|\sum_{j=1}^{n} h_j\right|, \quad a_1 = 1, \quad a_{j+1} = \left|ph_j - 2\sum_{j=1}^{n} h_j\right|, \quad 1 \leqslant j \leqslant r, \tag{6.4}$$

as the set of points of division.

Now, let us give some numerical examples:

Example 1. Take $R\left(2\cos\frac{2\pi}{5}\right)$ and

$$\varepsilon_1 = 2\cos\frac{4\pi}{5}, \quad \varepsilon_2 = 2\cos\frac{2\pi}{5}.$$

From

$$\varepsilon_1^n = q_0^{(n)}\varepsilon_2 - q_1^{(n)},$$

we have

$$H_1(\,|q_0^{(n)}|; 1, |q_1^{(n)}|) = 1 + O\left(\frac{\log|q_0^{(n)}|}{|q_0^{(n)}|^2}\right).$$

This is the result of [2].

Example 2. Take $R\left(2\cos\frac{2\pi}{7}\right)$ and

$$\varepsilon_1 = 2\cos\frac{6\pi}{7}, \quad \varepsilon_2 = 2\cos\frac{2\pi}{7}, \quad \varepsilon_3 = 2\cos\frac{4\pi}{7}.$$

From the equation

$$|\varepsilon_2^\alpha \varepsilon_3^\beta| = |\varepsilon_3^\alpha \varepsilon_1^\beta|,$$

we have approximately

$$\frac{\alpha}{\beta} = 1.356\cdots \doteq \frac{4}{3}.$$

From the expansion

$$\varepsilon_1^8 \varepsilon_2^6 = -227\,\varepsilon_1 - 45\,\varepsilon_2 - 146\,\varepsilon_3,$$

we have

$$q = 418, \quad a_1 = 1, \quad a_2 = 335, \quad a_3 = 103.$$

After numerical computation, we have

$$H_1(418; 1, 335, 103) \leqslant 1.0108146.$$

Example 3. Take $R\left(2\cos\dfrac{2\pi}{11}\right)$ and

$$\varepsilon_1 = 2\cos\frac{10\pi}{11}, \quad \varepsilon_2 = 2\cos\frac{2\pi}{11}, \quad \varepsilon_3 = 2\cos\frac{8\pi}{11}, \quad \varepsilon_4 = 2\cos\frac{4\pi}{11}, \quad \varepsilon_5 = 2\cos\frac{6\pi}{11}.$$

Solving the system of equations

$$|\varepsilon_2^\alpha \varepsilon_4^\beta \varepsilon_5^\gamma \varepsilon_3^\delta| = |\varepsilon_3^\alpha \varepsilon_5^\beta \varepsilon_2^\gamma \varepsilon_1^\delta| = |\varepsilon_4^\alpha \varepsilon_3^\beta \varepsilon_1^\gamma \varepsilon_5^\delta| = |\varepsilon_3^\alpha \varepsilon_1^\beta \varepsilon_4^\gamma \varepsilon_2^\delta|,$$

we have

$$\frac{\alpha}{\delta} \doteq 1.412 \doteq \frac{7}{5}, \quad \frac{\beta}{\delta} \doteq 1.584 \doteq \frac{8}{5}, \quad \frac{\gamma}{\delta} \doteq 0.944 \doteq \frac{5}{5}.$$

From the expansion

$$\varepsilon_1^7 \varepsilon_2^8 \varepsilon_3^5 \varepsilon_4^5 = -3345\,\varepsilon_1 - 271\,\varepsilon_2 - 2825\,\varepsilon_3 - 998\,\varepsilon_4 - 1950\,\varepsilon_5,$$

we have

$$q = 9389, \quad a_1 = 1, \quad a_2 = 8628, \quad a_3 = 6408, \quad a_4 = 2908, \quad a_5 = 7800.$$

After numerical computation, we have

$$H_1(q; a_1, \cdots, a_5) \leqslant 1.0081175.$$

Example 4. Take $R\left(2\cos\dfrac{2\pi}{23}\right)$ and

$$\varepsilon_1 = 2\cos\frac{12\pi}{23}, \quad \varepsilon_2 = 2\cos\frac{2\pi}{23}, \quad \varepsilon_3 = 2\cos\frac{10\pi}{23},$$

$$\varepsilon_4 = 2\cos\frac{4\pi}{23}, \quad \varepsilon_5 = 2\cos\frac{8\pi}{23}, \quad \varepsilon_6 = 2\cos\frac{6\pi}{23}.$$

From the expansion

$$\varepsilon_1^7 \varepsilon_2^8 \varepsilon_3^5 \varepsilon_4^6 \varepsilon_5^4 = -12494\,\varepsilon_1 - 726\,\varepsilon_2 - 11084\,\varepsilon_3 - 2738\,\varepsilon_4 - 8587\,\varepsilon_5 - 5575\,\varepsilon_6,$$

we have

$$q = 41204, \quad a_1 = 1, \quad a_2 = 33810, \quad a_3 = 31766,$$
$$a_4 = 20480, \quad a_5 = 5610, \quad a_6 = 29223.$$

After computation, we have

$$H_1(q; a_1, \cdots, a_6) \leqslant 1.0094250.$$

Example 5. Take $R\left(2\cos\dfrac{2\pi}{23}\right)$ and

$$\varepsilon_1 = 2\cos\frac{22\pi}{23}, \quad \varepsilon_2 = 2\cos\frac{2\pi}{23}, \quad \varepsilon_3 = 2\cos\frac{20\pi}{23}, \quad \varepsilon_4 = 2\cos\frac{4\pi}{23},$$

$$\varepsilon_5 = 2\cos\frac{18\pi}{23}, \quad \varepsilon_6 = 2\cos\frac{6\pi}{23}, \quad \varepsilon_7 = 2\cos\frac{16\pi}{23}, \quad \varepsilon_8 = 2\cos\frac{8\pi}{23},$$

$$\varepsilon_9 = 2\cos\frac{14\pi}{23}, \quad \varepsilon_{10} = 2\cos\frac{10\pi}{23}, \quad \varepsilon_{11} = 2\cos\frac{12\pi}{23}.$$

From the expansion

$$\varepsilon_1^5\varepsilon_2^6\varepsilon_3^4\varepsilon_4^7\varepsilon_5^4\varepsilon_6^3\varepsilon_7^4\varepsilon_8^4\varepsilon_9^3\varepsilon_{10}^2 = -120834\,\varepsilon_1 - 2251\,\varepsilon_2 - 116374\,\varepsilon_3 - 8837\,\varepsilon_4 - 107785\,\varepsilon_5 -$$
$$- 19269\,\varepsilon_6 - 95704\,\varepsilon_7 - 32774\,\varepsilon_8 - 81027\,\varepsilon_9 - 48350\,\varepsilon_{10} - 64842\,\varepsilon_{11},$$

we have

$$q = 698047, \quad a_1 = 1, \quad a_2 = 685041, \quad a_3 = 646274,$$
$$a_4 = 582461, \quad a_5 = 494796, \quad a_6 = 384914, \quad a_7 = 254860,$$
$$a_8 = 107051, \quad a_9 = 642292, \quad a_{10} = 467527, \quad a_{11} = 284044.$$

After computation, we have

$$H_1(q; 1, a_1, \cdots, a_{11}) \leqslant 1.2333543.$$

Remark. We can also use as well other totally real fields for finding the division points for evaluating multiple integrals, for example, the Dirichlet field, i.e., the field obtained by successive quadratic extensions to the rational field (Hua and Wang)[3,4]. However, we conjecture that the cyclotomic field gives the best result among the fields of the same degree. Another advantage of the cyclotomic field is the convenience for calculation.

Example. We use the Dirichlet field $R(\sqrt{2}, \sqrt{5})$, and take

$$\varepsilon_1 = \frac{1+\sqrt{5}}{2}, \quad \varepsilon_2 = 1 + \sqrt{2}, \quad \varepsilon_3 = 3 + \sqrt{10}.$$

Expanding

$$\varepsilon_1^6\varepsilon_2^4\varepsilon_3^2 = (9 + 4\sqrt{5})(17 + 12\sqrt{2})(19 + 6\sqrt{10}),$$
$$= 5787 + 4092\sqrt{2} + 2588\sqrt{5} + 1830\sqrt{10},$$

we have

$$H_1(5787; 1, 2397, 1366, 939) \leqslant 1.001498.$$

REFERENCES

[1] Hua, Loo Keng 1957 *Additive Theory of Numbers*, Peking, China: Science Press.

[2] Hua, Loo Keng & Wang, Yuan 1963 *Theory of Numerical Integrations and Their Applications*, Peking, China: Science Press.

[3] Hua, Loo Keng & Wang, Yuan 1964 On diophantine approximations and numerical integrations (I), *Scientia Sinica*, 13, 1007—1009.

[4] Hua, Loo Keng & Wang, Yuan 1964 On diophantine approximations and numerical integrations (II), *Scientia Sinica*, 13, 1009—1010.

[5] Коробов, Н. М. 1963 *Теоретикочисловые методы в приближенном анализе*, ГИФМЛ.

[6] Halton, J. H. 1960 On the efficiency of certain quasi-random sequences of points in evaluating multi-dimensional integrals, *Num. Math.*, 27(2), 73—79.

[7] Fricke, R. 1928 *Lehrbuch der Algebra*, Bd. III, s. 225.

ON INTERPOLATION OF A CERTAIN CLASS OF FUNCTIONS*

WANG YUAN

Institute of Mathematics, Academia Sinica

I. Let E_s^α (C) be a class of functions:

$$f(x_1, \ldots, x_s) = \sum_{-\infty}^{\infty} \cdots \sum C(m_1, \ldots, m_s) e^{2\pi i (m_1 x_1 + \cdots + m_s x_s)},$$

where

$$|C(m_1, \ldots, m_s)| \leq \frac{C}{(\bar{m}_1 \cdots \bar{m}_s)^\alpha},$$

in which $\bar{m} = \max(1, |m|), \alpha > 1$ and $C > 0$ are absolute constants. If $f \in E_s^\alpha(C)$, then we define v by

$$v(x_1, \ldots, x_s) = \sum_{-\infty}^{\infty} \cdots \sum B(m_1, \ldots, m_s) C(m_1, \ldots, m_s) e^{2\pi i (m_1 x_1 + \cdots + m_s x_s)},$$

where $C(m_1, \ldots, m_s)$ is the Fourier coefficient of f and $B(m_1, \ldots, m_s)$ satisfies

$$|B(m_1, \ldots, m_s)| \leq \frac{1}{(\bar{m}_1 \cdots \bar{m}_s)^\omega}, \quad 0 \leq \omega \leq 1.$$

Let N be a prime number and

$$N_1 = \left[N^{\frac{\alpha}{2\alpha-1}} (\ln N)^{\frac{-(s-1)(\alpha-1)}{2\alpha-1}} \right] + 1.$$

Let

$$\tilde{C}(m_1, \ldots, m_s) = \frac{1}{N} \sum_{k=1}^{N} f\left(\frac{a_1 k}{N}, \ldots, \frac{a_s k}{N} \right) e^{-2\pi i \frac{a_1 m_1 + \cdots + a_s m_s}{N} k},$$

$$Q(x_1, \ldots, x_s) = \sum_{\bar{m}_1 \cdots \bar{m}_s \leq N_1} B(m_1, \ldots, m_s) \tilde{C}(m_1, \ldots, m_s) e^{2\pi i (m_1 x_1 + \cdots + m_s x_s)}$$

and

$$\Delta = \min_a \sup_{f \in E_s^a(C)} |v(x_1, \ldots, x_s) - Q(x_1, \ldots, x_s)|,$$

where $a = (a_1, \ldots, a_s)$ is an integral vector.

---

*Kexue Tongbao, **9** (1996) 389–391.

                                      Y. Wang

The aim of this paper is to prove the following two theorems:

Theorem 1. *The estimation*

$$\Delta \leq c_1(\alpha,s)CN^{-\frac{\alpha(\alpha+\omega-1)}{2\alpha-1}}(\ln N)^{\frac{(s-1)(\alpha^2+\alpha\omega-\omega)}{2\alpha-1}},$$

holds, where $c_1(\alpha,s)$ is a constant depending on α and s.

Suppose that $f \in E_s^\alpha(C)$ and f is an odd function with respect to each variable, and that $u(x_1,\ldots,x_s)$ satisfies Poisson equation in unit cube G_s and takes value 0 on its boundary. Then from Theorem 1, it follows that

Theorem 2. *Let $s \geq 2$. Then for any given prime number N, there exists an integral vector $a = (a_1,\ldots,a_s)$ such that*

$$\left| u(x_1,\ldots,x_s) - \sum_{k=1}^N f\left(\frac{a_1 k}{N},\ldots,\frac{a_s k}{N}\right)\psi_k(x_1,\ldots,x_s)\right|$$

$$\leq c_2(\alpha,s)N^{\frac{-\alpha(\alpha s+2-s)}{(2\alpha-1)s}}(\ln N)^{\frac{(s-1)(\alpha^2 s+2\alpha-2)}{(2\alpha-1)s}},$$

where

$$\psi_k(x_1,\ldots,x_s) = -\frac{1}{4\pi N}\sideset{}{'}\sum_{\bar{m}_1\cdots\bar{m}_s \leq N_1}\frac{e^{2\pi i\left(m_1\left(x_1-\frac{a_1 k}{N}\right)+\cdots+m_s\left(x_s-\frac{a_s k}{N}\right)\right)}}{m_1^2+\cdots+m_s^2},$$

in which $\sum'$ denotes that the term $m_1 = \cdots = m_s = 0$ is omitted.

These two theorems give improvements for the previous results due to Korobov [1, 2] and the author [3], for example, the error term in Theorem 2 is $O(N^{-4/3+\varepsilon})$ for the case $\alpha = s = 2$, while the original results of Korobov and the author are $O(N^{-1+\varepsilon})$ and $O(N^{-6/5+\varepsilon})$ respectively.

II. To prove Theorem 1 we need the following two lemmas

Lemma 1. [4] *Let $l_i(1 \leq i \leq s)$ and n_1 be integers satisfying $\bar{l}_1\cdots\bar{l}_s \geq 3^s$ and $1 \leq n_1 \leq \bar{l}_1\cdots\bar{l}_s/3^s$ and let α be a real number > 1. Then*

$$\sum_{\bar{m}_1\cdots\bar{m}_s \leq n_1}\frac{1}{((\bar{l}_1+m_1)\cdots(\bar{l}_s+m_s))^\alpha} \leq c_3(\alpha,s)\frac{n_1^\alpha}{(\bar{l}_1\cdots\bar{l}_s)^\alpha}.$$

Lemma 2. [5] *Let N be a prime number and $\alpha > 1$. Then there is an integral vector $a = (a_1,\ldots,a_s)$ such that*

$$\sideset{}{'}\sum_{a_1 m_1+\cdots+a_s m_s \equiv 0 \,(\mathrm{mod}\,N)}\frac{1}{(\bar{m}_1\cdots\bar{m}_s)^\alpha} \leq c_4(\alpha,s)\frac{(\ln N)^{\alpha(s-1)}}{N^\alpha}.$$

Proof of Theorem 1.

1)

$$|v - Q| \leq \sum_{\bar{m}_1 \cdots \bar{m}_s \leq N_1} |B(m_1, \ldots, m_s)| \, |C(m_1, \ldots, m_s) - \tilde{C}(m_1, \ldots, m_s)|$$

$$+ \sum_{\bar{m}_1 \cdots \bar{m}_s > N_1} |B(m_1, \ldots, m_s)| \, |C(m_1, \ldots, m_s)| = \Sigma_1 + \Sigma_2.$$

2) Since

$$C(m_1, \ldots, m_s) - \tilde{C}(m_1, \ldots, m_s) = - \sideset{}{'}\sum_{a_1 l_1 + \cdots + a_s l_s \equiv 0 \,(\mathrm{mod}\, N)} C(l_1 + m_1, \ldots, l_s + m_s),$$

we have

$$|\Sigma_1| \leq \sum_{\bar{m}_1 \cdots \bar{m}_s \leq N_1} \frac{1}{(\bar{m}_1 \cdots \bar{m}_s)^\omega} \sideset{}{'}\sum_{a_1 l_1 + \cdots + a_s l_s \equiv 0 \,(\mathrm{mod}\, N)} \frac{C}{((\bar{l}_1 + m_1) \cdots (\bar{l}_s + m_s))^\alpha}.$$

It follows from Lemma 2 that the congruence

$$a_1 l_1 + \cdots + a_s l_s \equiv 0 \,(\mathrm{mod}\, N)$$

has a nonzero solution satisfying

$$\bar{l}_1 \cdots \bar{l}_s \geq c_6(\alpha, s) \frac{N}{(\ln N)^{s-1}} > \frac{N_1}{3^s}$$

whenever $N > c_5(\alpha, s)$. Note that the theorem holds evidently if $N \leq c_5(\alpha, s)$.
Therefore by Lemmas 1 and 2, we obtain

$$|\Sigma_1| \leq C \sum_{k=0}^{[\log_2 N_1]} \sum_{2^{-k-1} N_1 < \bar{m}_1 \cdots \bar{m}_s \leq 2^{-k} N_1} \frac{1}{(\bar{m}_1 \cdots \bar{m}_s)^\omega}$$

$$\times \sideset{}{'}\sum_{a_1 l_1 + \cdots + a_s l_s \equiv 0 \,(\mathrm{mod}\, N)} \frac{1}{((\bar{l}_1 + m_1) \cdots (\bar{l}_s + m_s))^\alpha}$$

$$\leq C \sum_{k=0}^{[\log_2 N_1]} (2^{-k-1} N_1)^{-\omega}$$

$$\times \sideset{}{'}\sum_{a_1 l_1 + \cdots + a_s l_s \equiv 0 \,(\mathrm{mod}\, N)} \sum_{\bar{m}_1 \cdots \bar{m}_s \leq 2^{-k} N_1} \frac{1}{((\bar{l}_1 + m_1) \cdots (\bar{l}_s + m_s))^\alpha}$$

$$\leq c_3(\alpha, s) C \sum_{k=0}^{[\log_2 N_1]} (2^{-k-1} N_1)^{-\omega} \sideset{}{'}\sum_{a_1 l_1 + \cdots + a_s l_s \equiv 0 \,(\mathrm{mod}\, N)} \frac{(2^{-k} N_1)^\alpha}{(\bar{l}_1 \cdots \bar{l}_s)^\alpha}$$

$$\leq c_7(\alpha, s) C N_1^{-\omega + \alpha} N^{-\alpha} (\ln N)^{\alpha(s-1)}$$

$$\times \sum_{k=0}^{\infty} (2^{\alpha-1})^{-k} \leq c_8(\alpha, s) C N^{\frac{-\alpha(\alpha+\omega-1)}{2\alpha-1}} (\ln N)^{\frac{(s-1)(\alpha^2+\alpha\omega-\omega)}{2\alpha-1}}.$$

　　　　　　　　　　Y. Wang

3)

$$|\Sigma_2| \le \sum_{\bar{m}_1\cdots\bar{m}_s > N_1} \frac{C}{(\bar{m}_1\cdots\bar{m}_s)^{\alpha+\omega}}$$

$$\le c_9(\alpha, s) C N_1^{-\alpha-\omega+1}(\ln N_1)^{s-1}$$

$$\le c_{10}(\alpha, s) C N^{\frac{-\alpha(\alpha+\omega-1)}{2\alpha-1}}(\ln N)^{\frac{(s-1)(\alpha^2+\alpha\omega-\omega)}{2\alpha-1}}.$$

The theorem follows from 1), 2) and 3).

References

[1] N. M. Korobov, *Problems in Comp. Math. and Comp. Tech.*, MASGES (1963).
[2] N. M. Korobov, Number theoretic methods in asymptotic analysis, GEFML (1963).
[3] Wang Yuan, *Sci. Sinica* **14**[4] (1965) 629–631.
[4] Wang Yuan, *Sci. Sinica* **10**[6] (1961) 632–636.
[5] N. S. Bachvalov, *Vestnik Moscow Univ.* **4** (1959) 3–18.

Vol. XVI No. 4   **SCIENTIA SINICA**   November 1973

ON UNIFORM DISTRIBUTION AND NUMERICAL ANALYSIS (I)

(NUMBER-THEORETIC METHOD)

Hua Loo-keng (华罗庚) and Wang Yuan (王 元)

Received June 13, 1973.

Abstract

In this paper, the uniformly distributed sequences of sets defined by means of a real cyclotomic field have been dealt with. We have obtained the estimations of their discrepancies and applied them to the problem of numerical integration.

I. Introduction

Let G_s be a unit cube

$$0 \leqslant x_1 \leqslant 1, \cdots, 0 \leqslant x_s \leqslant 1$$

of s-dimensional space, and $n_1 < n_2 < \cdots$ be a sequence of positive integers. Let

$$P_{n_l}(j) = (x_1^{(n_l)}(j), \cdots, x_s^{(n_l)}(j)), \quad (1 \leqslant j \leqslant n_l)$$

be a set of points in G_s. For any $(\gamma_1, \cdots, \gamma_s) \in G_s$, let $N_{n_l}(\gamma_1, \cdots, \gamma_s)$ denote the number of points of $P_{n_l}(j)(1 \leqslant j \leqslant n_l)$ satisfying the inequalities

$$0 \leqslant x_1^{(n_l)}(j) < \gamma_1, \cdots, 0 \leqslant x_s^{(n_l)}(j) < \gamma_s.$$

If

$$\lim_{l \to \infty} \frac{N_{n_l}(\gamma_1, \cdots, \gamma_s)}{n_l} = \gamma_1 \cdots \gamma_s,$$

then the sequence of sets $(P_{n_l}(j))(n_1 < n_2 < \cdots)$ is called uniformly distributed on G_s. Futhermore, if we have sharper condition

$$\left| \frac{N_{n_l}(\gamma_1, \cdots, \gamma_s)}{n_l} - \gamma_1 \cdots \gamma_s \right| < \varphi(n_l),$$

where $\varphi(n_l) = o(1)$, then the sequence of sets $(P_{n_l}(j))(n_1 < n_2 < \cdots)$ is uniformly distributed with discrepency $\varphi(n)$. In case $n_l = l$, $x_1^{(l)}(j) = x_1(j), \cdots$, and $x_s^{(l)}(j) = x_s(j)(l = 1, 2, \cdots)$, the sequence $P(j) = (x_1(j), \cdots, x_s(j))$ $(j = 1, 2, \cdots)$ will be called uniformly distributed on G_s.

Let p be a prime $\geqslant 5$. Set $r = \dfrac{1}{2}(p-1)$ and $q = r - 1 = \dfrac{1}{2}(p-3)$. The real cyclotomic field $\Re_r = R\left(2 \cos \dfrac{2\pi}{p}\right)$ is an algebraic field of degree r. Let

$$\omega_1 = 2 \cos \frac{2\pi}{p}, \quad \omega_2 = 2 \cos \frac{4\pi}{p}, \cdots, \quad \omega_q = 2 \cos \frac{2\pi q}{p}$$

and let

$$\left|\frac{h_i^{(l)}}{n_l} - \omega_i\right| \leqslant c(\Re_r) n_l^{-1-\frac{1}{q}}, \quad (1 \leqslant i \leqslant q)$$

be the simultaneous approximations of ω's by rationals, where we use $c(f, \cdots, g)$ to denote the positive constants depending on $f, \cdots, g$ only, but not always with the same value.

Theorem 1. *Let*

$$P(j) = (\{\omega_1 j\}, \cdots, \{\omega_q j\}) \quad (j = 1, 2, \cdots), \tag{1.1}$$

where $\{x\}$ denotes the fractional part of x. Then the sequence $P(j)(j = 1, 2, \cdots)$ has discrepancy

$$\varphi(n) = c(\Re_r, \ \varepsilon)n^{-1+\varepsilon},$$

ε being any pre-assigned positive number. (The convention will be adopted throughout the present paper).

Theorem 2. *Set*

$$P_{n_l}(j) = \left(\left\{\frac{j}{n_l}\right\}, \ \left\{\frac{h_1^{(l)} j}{n_l}\right\}, \cdots, \left\{\frac{h_q^{(l)} j}{n_l}\right\}\right) \quad (1 \leqslant j \leqslant n_l). \tag{1.2}$$

Then the sequence of sets $(P_{n_l}(j))$ has discrepancy

$$\varphi(n) = c(\Re_r, \ \varepsilon)n^{-\frac{1}{2}-\frac{1}{2q}+\varepsilon}.$$

Let $E_s^a(C)$ be the class of functions

$$f(x_1, \cdots, x_s) = \sum_{-\infty}^{\infty} \cdots \sum C(m_1, \cdots, m_s)e^{2\pi i(m_1 x_1 + \cdots + m_s x_s)},$$

in which the Fourier coefficients satisfy

$$|C(m_1, \cdots, m_s)| \leqslant \frac{C}{(\overline{m}_1 \cdots \overline{m}_s)^a},$$

where α and C are positive constants and $\overline{m} = \max(1, |m|)$.

Suppose $\alpha > 1$ throughout the present paper.

Applying the sequences of sets (1.1) and (1.2) to the problem of numerical integration, we have

Theorem 3. *Let l be the least integer $\geqslant \alpha$ and let $\mu_{n, l, j}$ be a set of integers defined by*

$$\left(\sum_{j=-n}^{n} z^j\right)^l = \sum_{j=-n_l}^{n_l} \mu_{n, l, j} z^j.$$

Then

$$\underset{f \in E_q^a(C)}{\text{Sup}} \left| \int_0^1 \cdots \int_0^1 f(x_1, \cdots, x_q)dx_1 \cdots dx_q - \frac{1}{(2n+1)^l} \sum_{j=-n_l}^{n_l} \mu_{n, l, j} f(\omega_1 j, \cdots, \omega_q j) \right|$$

$$\leqslant C \cdot c(\Re_r, \alpha, \varepsilon)n^{-\alpha+\varepsilon}. \tag{1.3}$$

Theorem 4. *We have*

$$\mathop{\text{Sup}}_{f \in E_r^a(C)} \left| \int_0^1 \cdots \int_0^1 f(x_1, \cdots, x_r)\, dx_1 \cdots dx_r - \frac{1}{n_l} \sum_{j=1}^{n_l} f\left(\frac{j}{n_l}, \frac{h_1^{(l)} j}{n_l}, \cdots, \frac{h_q^{(l)} j}{n_l}\right) \right|$$

$$\leqslant C \cdot c(\Re_r, \alpha, \varepsilon) n_l^{-\frac{a}{2} - \frac{a}{2q} + \varepsilon}. \tag{1.4}$$

It is well known that under certain conditions, the integral of non-periodic function may be calculated by an integral of periodic function[14]. We may also use the following simple formula instead of formula (1.3) for the case $\alpha = 2$.

$$\mathop{\text{Sup}}_{f \in E_q^2(C)} \left| \int_0^1 \cdots \int_0^1 f(x_1, \cdots, x_q)\, dx_1 \cdots dx_q - \frac{1}{n} \sum_{j=-n}^{n} \left(1 - \frac{|j|}{n}\right) f(\omega_1 j, \cdots, \omega_q j) \right|$$

$$\leqslant C \cdot c(\Re_r, \varepsilon) n^{-2+\varepsilon}. \tag{1.5}$$

It is followed immediately by the Ω-result of Roth[9] on the discrepancy of uniformly distributed sequence of sets that the estimation given in Theorem 1 is the best possible one of its kinds apart from some possible improvements about the order n^ε. This type of distribution may be recognized as the best distributed sequence of sets. The other known best distributed sequences of sets are those proposed by Коробов[14] and Halton[2] in 1959 and 1960 respectively. We can prove also that the error term given in Theorem 3 does not allow essential improvements[15]. Perhaps, it is worth mentioning that only $c(\Re_r) \log n_l$ elementary operations are required for obtaining the sequence of integers $(h_1^{(l)}, \cdots, h_q^{(l)}; n_l)$ in (1.2).

The classical method in numerical integration is established by means of the sequence of sets

$$\left(\frac{j_1}{m}, \cdots, \frac{j_s}{m}\right) \quad (0 \leqslant j_1, \cdots, j_s \leqslant m - 1). \tag{1.6}$$

That is, the integral over G_s is calculated approximately by the sum

$$\frac{1}{m^s} \sum_{j_1, \cdots, j_s} f\left(\frac{j_1}{m}, \cdots, \frac{j_s}{m}\right). \tag{1.7}$$

The number of points of (1.6) is $n = m^s$. We can prove easily that the discrepancy of (1.6) is $\geqslant n^{-1/s}$ and the error term in classical quadrature formula of the functions belonging to $E_s^a(C)$ over G_s is $\geqslant 2Cn^{-a/s}$. $\Big($ Take $f(x_1, \cdots, x_s) = C(e^{2\pi i m_1 x_1} + e^{-2\pi i m_1 x_1})/m^a$. Then we have $\int_0^1 \cdots \int_0^1 f(x_1, \cdots, x_s)\, dx_1 \cdots dx_s = 0$ and $\frac{1}{n} \sum_{j_1, \cdots, j_s} f\left(\frac{j_1}{m}, \cdots, \frac{j_s}{m}\right) = \frac{2C}{n^{a/s}}. \Big)$

In 1959, Бахвалов[12] and the authors[6] proved independently the formula (1.4) for the case $q = 1$ with the error term $O(F_l^{-a} \log 3F_l)$ by means of the Fibonacci sequence (F_l), where

$$F_l = \frac{1}{\sqrt{5}}\left(\left(\frac{1+\sqrt{5}}{2}\right)^{l+1} - \left(\frac{1-\sqrt{5}}{2}\right)^{l+1}\right) \quad (l \geqslant 1).$$

Since $\Re_2 = R\left(2\cos\dfrac{2\pi}{5}\right) = R(\sqrt{5}\,)$, the sequences of sets (1.1) and (1.2) may be recognized as the generalizations of the sequences of sets

$$\left\{\frac{\sqrt{5}-1}{2}\,j\right\}\ (j = 1,\ 2,\ \cdots)\ \text{and}\ \left(\left\{\frac{j}{F_l}\right\},\ \left\{\frac{F_{l-1}\,j}{F_l}\right\}\right)\ (1 \leqslant j \leqslant F_l) \qquad (1.8)$$

obtained by Golden section and Fibonacci sequence respectively. Hence we suggested also the possibility to treat the high dimensional case of numerical integration by means of a real cyclotomic field and gave some numerical examples of the formula (1.4) for the case $2 \leqslant q \leqslant 10^{[6-8]}$. Indeed, other totally real fields can be used as well instead of the cyclotomic field, for example, the Dirichlet field $R(\sqrt{P_1}, \cdots, \sqrt{P_h})$, where $P's$ are h distinct primes, but it seems that the cyclotomic field often gives best result among the fields of the same degree. Another advantage of cyclotomic field is convenience for calculation.

The proofs of the above theorems depend on the following important result of Wolfgang M. Schmidt[10] concerning the simultaneous approximations of algebraic numbers by rationals:

Lemma 1.1. *Let* $\alpha_1, \cdots, \alpha_s$ *be a set of real algebraic numbers such that* $1, \alpha_1, \cdots, \alpha_s$ *are linearly independent over rational field* R. *Then we obtain*

$$\langle \alpha_1 m_1 + \cdots + \alpha_s m_s \rangle \geqslant c(\alpha_1, \cdots, \alpha_s, \varepsilon)(\overline{m}_1 \cdots \overline{m}_s)^{-1-\varepsilon},$$

$m_1, \cdots, m_s$ *denoting any set of integers which are not all equal to zeros and* $\langle x \rangle = \min(\{x\}, 1 - \{x\})$.

We have also used the number-theoretic methods in numerical analysis introduced by Бахвалов[12], Коробов[14], Haselgrove[3] and Hlawka[5].

If a similar result of A. Baker[1] is used instead of Lemma 1.1, we may also obtain the corresponding results on numerical analysis. For example, we may prove that the sequence

$$P(j) = (\{ej\},\ \{e^2 j\},\ \cdots,\ \{e^s j\}),\ (j = 1, 2, \cdots) \qquad (1.9)$$

has discrepancy $\varphi(n) = c(s, \varepsilon)n^{-1+\varepsilon}$.

By Wang 520-calculator, we obtain the following example of the formula (1.5)

$$\operatorname*{Sup}_{f \in E_4^2(C)}\left| \int_0^1 \cdots \int_0^1 f(x_1, \cdots, x_4)dx_1 \cdots dx_4 - \frac{1}{3000}\sum_{j=-3000}^{3000}\left(1 - \frac{|j|}{3000}\right)\right.$$

$$\left. \times f\left(2j\cos\frac{2\pi}{11},\ 2j\cos\frac{4\pi}{11},\ 2j\cos\frac{6\pi}{11},\ 2j\cos\frac{8\pi}{11}\right)\right| \leqslant 0.065C.$$

II. The Totally Real Algebraic Field

Let $\mathscr{F}_s$ denote a totally real algebraic field of the degree s. For a number η of $\mathscr{F}_s$, let $\eta^{(1)}(=\eta), \eta^{(2)}, \cdots, \eta^{(s)}$ be its conjugates. Assume that

$$\omega_1,\ \cdots,\ \omega_s$$

is an integral basis of $\mathscr{F}_s$. Form a matrix

$$\Omega = (\omega_j^{(i)})\ (1 \leqslant i,\ j \leqslant s).$$

The matrix

$$S = \Omega'\Omega = \left(\sum_{k=1}^{s} \omega_i^{(k)}\omega_j^{(k)} \right), \quad (1 \leqslant i, j \leqslant s)$$

is called the fundamental matrix of $\mathscr{F}_s$. Clearly, it is a symmetric matrix with rational integer elements. The invariants of the fundamental matrix under the modular groups are characteristic properties of the algebraic number field. The determinant $\det S$ of S is called the discriminant of the field.

Let η be a unit of $\mathscr{F}_s$ satisfying

$$|\eta| > 1, \quad |\eta^{(j)}| \leqslant c(\mathscr{F}_s)|\eta|^{-\frac{1}{s-1}} \quad (2 \leqslant j \leqslant s). \tag{2.1}$$

We express η as follows:

$$\eta = \sum_{i=1}^{s} k_i \omega_i, \tag{2.2}$$

where $k's$ are rational integers. From (2.2) and its conjugates, it appears

$$(\eta^{(1)}, \cdots, \eta^{(s)}) = (k_1, \cdots, k_s)\Omega'. \tag{2.3}$$

Hence

$$(\eta^{(1)}, \cdots, \eta^{(s)})\Omega = (k_1, \cdots, k_s)S = (h_1, \cdots, h_s) \quad \text{(by definition)}$$

or

$$h_j = \sum_{i=1}^{s} \eta^{(i)} \omega_j^{(i)}, \quad (1 \leqslant j \leqslant s).$$

Therefore, we have

$$|\eta\omega_j - h_j| \leqslant \sum_{i=2}^{s} |\eta^{(i)}|\,|\omega_j^{(i)}| \leqslant c(\mathscr{F}_s)|\eta|^{-\frac{1}{s-1}}$$

by (2.1). Suppose

$$1 = \sum_{i=1}^{s} a_i \omega_i$$

and

$$n = \sum_{i=1}^{s} a_i h_i .$$

Then we deduce that

$$|\eta - n| = \left| \sum_{i=1}^{s} a_i\omega_i\eta - \sum_{i=1}^{s} a_i h_i \right| \leqslant \sum_{i=1}^{s} |a_i|\,|\omega_i\eta - h_i| \leqslant c(\mathscr{F}_s)|\eta|^{-\frac{1}{s-1}}.$$

Hence

$$\left| \omega_j - \frac{h_j}{n} \right| \leqslant c(\mathscr{F}_s)|n|^{-1-\frac{1}{s-1}}, \quad (1 \leqslant j \leqslant s). \tag{2.4}$$

In the case of $s > 2$, various classical methods can only prove the existence of an infinitely many sets of integers $(h_1, \cdots, h_s; n)$ satisfying (2.4), but this does not suggest effective way to finding $(h_1, \cdots, h_s; n)$. It is shown in this section that the

488   SCIENTIA SINICA   Vol. XVI

problem for finding $(h_1, \cdots, h_s; n)$ is equivalent to the problem for finding a unit η in the totally real algebraic field $\mathscr{F}_s$ so that (2.1) is satisfied.

If a complete set of independent units of $\mathscr{F}_s$ is known, then by the following theorem, we can find a sequence of units (η_l) satisfying (2.1) and

$$|\eta_l| \to \infty \quad (\text{as } l \to \infty).$$

As a result, we have infinitely many sets of integers

$$(h_1, \cdots, h_s; n)$$

satisfying (2.4).

Theorem 2.1. *In the totally real algebraic field $\mathscr{F}_s$, we have a sequence of units $\eta_l \ (= \eta_l^{(1)}) \ (l = 1, 2, \cdots)$, whose conjugates satisfy*

$$|\eta_l^{(i)}| < e^{-(2l-1)c} \quad (2 \leqslant i \leqslant s)$$

and

$$e^{-2c}|\eta_l^{(j)}| \leqslant |\eta_l^{(i)}| \leqslant e^{2c}|\eta_l^{(j)}| \quad (2 \leqslant i, \ j \leqslant s),$$

where $c = c(\mathscr{F}_s) > 0$.

Proof. Let

$$\varepsilon_1, \cdots, \varepsilon_{s-1}$$

be a complete set of independent units of $\mathscr{F}_s$. Set

$$\xi^{(i)} = \varepsilon_1^{(i)l_1} \cdots \varepsilon_{s-1}^{(i)l_{s-1}} \quad (2 \leqslant i \leqslant s)$$

and

$$c = \max_{2 \leqslant i \leqslant s} \left(\sum_{j=1}^{s-1} |\log|\varepsilon_j^{(i)}|| \right).$$

Since

$$\det (\log|\varepsilon_j^{(i)}|) \neq 0, \quad (2 \leqslant i \leqslant s, \ 1 \leqslant j \leqslant s-1),$$

we denote the solution of the system of linear equations

$$\log|\xi^{(2)}| = \cdots = \log|\xi^{(s)}| = -2cl - 1$$

by

$$l_1 = l_1^{(l)}, \cdots, l_{s-1} = l_{s-1}^{(l)}.$$

Set

$$[l_i^{(l)}] = a_i^{(l)}, \quad (1 \leqslant i \leqslant s-1),$$

where $[x]$ denotes the integer part of x. Define

$$\eta_l = \varepsilon_1^{a_1^{(l)}} \cdots \varepsilon_{s-1}^{a_{s-1}^{(l)}}.$$

Then it follows

$$\log|\eta_l^{(i)}| = \sum_{j=1}^{s-1} a_j^{(l)} \log|\varepsilon_j^{(i)}| \leqslant \sum_{j=1}^{s-1} l_j^{(l)} \log|\varepsilon_j^{(i)}| + \sum_{j=1}^{s-1} |\log|\varepsilon_j^{(i)}||$$

$$= \log|\xi^{(i)}| + \sum_{j=1}^{s-1} |\log|\varepsilon_j^{(i)}|| < -2cl + c = -(2l-1)c \quad (2 \leqslant i \leqslant s)$$

and

$$\left|\log|\eta_I^{(i)}| - \log|\eta_I^{(j)}|\right| \leqslant \left|\log|\xi^{(i)}| - \log|\xi^{(j)}|\right|$$

$$+ \sum_{k=1}^{s-1} \left(\left|\log|\varepsilon_k^{(i)}|\right| + \left|\log|\varepsilon_k^{(j)}|\right|\right) \leqslant 2c \quad (2 \leqslant i,\ j \leqslant s).$$

The theorem is proved.

Remark. It follows immediately from Theorem 2.1 that there requires only $c(\mathscr{F}_s)\cdot$ $\log|n|$ elementary operations for finding the set of integers $(h_1, \cdots, h_s; n)$ satisfying (2.4).

III. Some Examples

1) Let p be a prime $\geqslant 5$, $r = \dfrac{1}{2}(p-1)$ and $q = r - 1 = \dfrac{1}{2}(p-3)$. The real cyclotomic field $\mathfrak{R}_r = R\left(2\cos\dfrac{2\pi}{p}\right)$ is an algebraic field of degree r. The field has a set of integral basis

$$\omega_1 = 2\cos\frac{2\pi}{p},\quad \omega_2 = 2\cos\frac{4\pi}{p},\ \cdots,\quad \omega_r = 2\cos\frac{2\pi r}{p}. \tag{3.1}$$

Let σ be a cyclic permutation of $(\omega_1, \cdots, \omega_r)$ $(\omega_1 \to \omega_2, \cdots, \omega_r \to \omega_1)$. Then from a number $\eta\ (=\eta^{(1)})$ of $\mathfrak{R}_r$, we have its q conjugates $\eta^{(2)}, \cdots, \eta^{(r)}$ under the q transformations $\sigma, \sigma^2, \cdots, \sigma^q$. Hence

$$S = \Omega'\Omega = pI - 2M,$$

where I is the identical matrix and $M = (m_{ij})$, in which $m_{ij} = 1$.

Taking a set of complete independent units $\varepsilon_1, \cdots, \varepsilon_q$ of $\mathfrak{R}_r$, then by Theorem 2.1, we construct a unit

$$\eta = \varepsilon_1^{a_1} \cdots \varepsilon_q^{a_q},$$

satisfying

$$|\eta| > 1, \quad e^{-2c}|\eta^{(j)}| \leqslant |\eta^{(i)}| \leqslant e^{2c}|\eta^{(j)}|, \quad (2 \leqslant i,\ j \leqslant r),$$

where $c = c(\mathfrak{R}_r) > 0$.

If

$$\eta = \sum_{i=1}^{r} k_i\,\omega_i,$$

then from

$$(h_1, \cdots, h_r) = (k_1, \cdots, k_r)S = (k_1, \cdots, k_r)(pI - M),$$

we have

$$h_i = pk_i - 2\sum_{j=1}^{r} k_j, \quad (1 \leqslant i \leqslant r).$$

Since

$$\sum_{i=1}^{r} \omega_i = -1,$$

hence it is found that

$$n = - \sum_{j=1}^{r} h_j = - \sum_{j=1}^{r} k_j.$$

Therefore, we have the simultaneous Diophantine approximations

$$\left| \frac{h_i}{n} - \omega_i \right| \leqslant c(\Re_r) n^{-1-\frac{1}{q}}, \quad (1 \leqslant i \leqslant r). \tag{3.2}$$

It is well-known that

$$\rho_l = \frac{\sin \dfrac{\pi}{p} g^{l+1}}{\sin \dfrac{\pi}{p} g^l}, \quad (1 \leqslant l \leqslant q)$$

is a set of complete independent units of the cyclotomic field $\Re_r$, where g denotes a primitive root mod p. It is more convenient to use the set of units

$$\omega_l = 2 \cos \frac{2\pi l}{p}, \quad (1 \leqslant l \leqslant q).$$

However, they do not always form an independent set. In order that they form a set of complete independent units, it is necessary and sufficient that (i) 2 is a primitive root mod p, or (ii) 2 belongs to the exponent r mod p and $p \equiv 7 \pmod 8$[8].

2) Let $p_1, \cdots, p_h$ be h distinct primes, $m = 2^h$ and $l = m - 1$. The real Dirichlet field $\mathscr{D}_m = R(\sqrt{p_1}, \cdots, \sqrt{p_h})$ is an algebraic field of degree m. Let

$$\varepsilon_1, \cdots, \varepsilon_l$$

be a set of complete independent units of $\mathscr{D}_m$. Take ε's as the solutions of Pell's equations

$$x^2 - p_{i_1} \cdots p_{i_k} y^2 = \pm 4$$

with the least of $\dfrac{x}{2} + \dfrac{\sqrt{p_{i_1} \cdots p_{i_k}}}{2} y$, where $k \geqslant 1$ and $1 \leqslant i_1 < \cdots < i_k \leqslant h$ is any choice of $1, 2, \cdots, h$. Suppose

$$\varepsilon_i = \begin{cases} \dfrac{x_i}{2} + \dfrac{\sqrt{d_i} y_i}{2}, & x_i \equiv y_i \equiv 1 \pmod 2 \quad (1 \leqslant i \leqslant \tau), \\ x_i + \sqrt{d_i}\, y_i, & (\tau + 1 \leqslant i \leqslant l), \end{cases}$$

where x_i and y_i are rational integers.

The integral basis of $\mathscr{D}_m$ is

$$\omega_1 = 1, \quad \omega_2 = \varepsilon_1, \quad \cdots, \quad \omega_{\tau+1} = \varepsilon_\tau, \quad \omega_{\tau+2} = \sqrt{d_{\tau+1}}, \quad \cdots, \quad \omega_m = \sqrt{d_l}.$$

Consider l transformations

$$(\sigma_{i_1, \cdots, i_k}) \quad \sqrt{p_\nu} \to \begin{cases} -\sqrt{p_\nu} & \text{for } \nu = i_j \quad (1 \leqslant j \leqslant k) \\ \sqrt{p_\nu}, & \text{otherwise}, \end{cases}$$

where $k \geqslant 1$ and $1 \leqslant i_1 < \cdots < i_k \leqslant h$ is any choice of $1, 2, \cdots, h$. Applying these transformations to any element $\eta \; (= \eta^{(1)})$ of $\mathscr{D}_m$, we have its l conjugates $\eta^{(2)}, \cdots, \eta^{(m)}$. Form a matrix

$$\Omega = (\omega_i^{(j)}) \quad (1 \leqslant i, \; j \leqslant m),$$

then we obtain

$$S = \Omega'\Omega = \begin{pmatrix} A & 0 \\ 0 & B \end{pmatrix},$$

where $A = (a_{ij}) \; (1 \leqslant i, \; j \leqslant \tau + 1)$ and $B = (b_{\mu\nu}) \; (1 \leqslant \mu, \; \nu \leqslant l - \tau)$ in which $a_{11} = 2^h$, $a_{ii} = 2^{h-2}(x_{i-1}^2 + d_{i-1}y_{i-1}^2) \; (2 \leqslant i \leqslant \tau + 1)$, $a_{1j} = a_{j1} = 2^{h-1}x_{j-1} \; (2 \leqslant j \leqslant \tau + 1)$, $a_{ij} = 2^{h-2}x_{i-1}x_{j-1}(2 \leqslant i, \; j \leqslant \tau + 1, \; i \neq j)$, $b_{\mu\mu} = 2^h d_{\tau+\mu} \; (1 \leqslant \mu \leqslant l - \tau)$ and $b_{\mu\nu} = 0 \; (\mu \neq \nu)$.

Construct a unit

$$\eta = \varepsilon_1^{a_1} \cdots \varepsilon_l^{a_l}$$

of $\mathscr{D}_m$ according to Theorem 2.1 such that

$$|\eta| > 1, \quad e^{-2c}|\eta^{(j)}| \leqslant |\eta^{(i)}| \leqslant e^{2c}|\eta^{(j)}| \quad (2 \leqslant i, j \leqslant m),$$

where $c = c(\mathscr{D}_m) > 0$.

If

$$\eta = \sum_{i=1}^{m} k_i \omega_i,$$

then from

$$(k_1, \; \cdots, \; k_m) S = (h_1, \; \cdots, \; h_m)$$

we have

$$n = h_1 = 2^{h-1}\left(2k_1 + \sum_{i=1}^{\tau} x_i k_{i+1}\right),$$

$$h_j = 2^{h-2}\left(2x_{j-1}k_1 + d_{j-1}y_{j-1}^2 k_j + \sum_{i=2}^{\tau+1} x_{i-1}x_{j-1}k_i\right) \quad (2 \leqslant j \leqslant \tau + 1),$$

$$h_j = 2^h k_j d_{j-1} \quad (\tau + 2 \leqslant j \leqslant m).$$

Hence the simultaneous Diophantine approximations

$$\left| \omega_i - \frac{h_i}{n} \right| \leqslant c(\mathscr{D}_m) n^{-1-\frac{1}{l}}, \quad (1 \leqslant i \leqslant m).$$

IV.   Uniform Distribution

In this paper, we use $\boldsymbol{\gamma} = (\gamma_1, \cdots, \gamma_s)$ to denote the vector with real components and $\mathbf{m} = (m_1, \cdots, m_s)$ the vector with integral components. We also use the notations $\|\boldsymbol{\gamma}\| = \bar{\gamma}_1 \cdots \bar{\gamma}_s$, $|\boldsymbol{\gamma}| = |\gamma_1 \cdots \gamma_s|$ and $(\boldsymbol{\alpha}, \boldsymbol{\beta}) = \sum_{i=1}^{s} \alpha_i \beta_i$ (the scalar product of $\boldsymbol{\alpha}$ and $\boldsymbol{\beta}$). In this section, we shall prove the formula of Erdös-Turan-Koksma.

 SCIENTIA SINICA Vol. XVI

Theorem 4.1. *Let η be a number satisfying $\dfrac{1}{6} > \eta > 0$ and let h be an integer $> \dfrac{1}{\eta}$. Then for any $\boldsymbol{\gamma} \in G_s$, we have*

$$\left| \frac{1}{n_l} N_{n_l}(\boldsymbol{\gamma}) - |\boldsymbol{\gamma}| \right| < 2^{s+2}\, \varphi(n_l),$$

where

$$\varphi(n_l) = \sideset{}{'}\sum_{|m_i| \leqslant h} \frac{1}{\|\pi \mathbf{m}\|} \left| \frac{1}{n} \sum_{j=1}^{n_l} e^{2\pi i (P_{n_l}(j),\, \mathbf{m})} \right| + \frac{\log^s 9h}{\eta h} + 2^{s+1}\, \eta,$$

in which Σ' denotes a sum with an exception $\mathbf{m} = \mathbf{o} = (0, \cdots, 0)$.

Lemma 4.1.[13] *Let r be a positive integer, α, β be real numbers and Δ satisfy*

$$0 < \Delta < \frac{1}{2}, \quad \Delta \leqslant \beta - \alpha \leqslant 1 - \Delta.$$

Then there exists a function $\psi(x)$ with periodic 1 such that

(i) $\psi(x) = 1$, *for* $\alpha + \dfrac{1}{2}\Delta \leqslant x \leqslant \beta - \dfrac{1}{2}\Delta$,

(ii) $0 \leqslant \psi(x) \leqslant 1$, *for* $\alpha - \dfrac{1}{2}\Delta \leqslant x \leqslant \alpha + \dfrac{1}{2}\Delta$ *and* $\beta - \dfrac{1}{2}\Delta \leqslant x \leqslant \beta + \dfrac{1}{2}\Delta$,

(iii) $\psi(x) = 0$, *for* $\beta + \dfrac{1}{2}\Delta \leqslant x \leqslant 1 + \alpha - \dfrac{1}{2}\Delta$,

(iv) $\psi(x)$ *possesses a Fourier expansion*

$$\psi(x) = \beta - \alpha + \Sigma' C(m) e^{2\pi i m x},$$

of which the coefficients satisfy

$$|C(m)| \leqslant \min\left(\beta - \alpha,\ \frac{1}{\pi|m|},\ \left(\frac{1}{\pi|m|}\right)^{r+1} \left(\frac{r}{\Delta}\right)^r \right).$$

Lemma 4.2. *Set $0 < \delta \leqslant \psi(n_l)$. If a uniformly distributed sequence of sets $(P_{n_l}(j))$ $(n_1 < n_2 < \cdots)$ satisfies*

$$\left| \frac{1}{n_l} N_{n_l}(\boldsymbol{\gamma}) - |\boldsymbol{\gamma}| \right| < \varphi(n_l)$$

for $\delta \leqslant \gamma_i \leqslant 1 - \delta$ $(1 \leqslant i \leqslant s)$, then

$$\left| \frac{1}{n_l} N_{n_l}(\boldsymbol{\gamma}) - |\boldsymbol{\gamma}| \right| < 2^{s+2}\varphi(n_l)$$

holds for all $\boldsymbol{\gamma} \in G_s$.

Proof. For simplicity, hereafter we omit the index l of n_l and $h^{(l)}$.

1) Set $\delta \leqslant \alpha_i < \beta_i \leqslant 1 - \delta$ $(1 \leqslant i \leqslant s)$ and let $N_n(\boldsymbol{\alpha}, \boldsymbol{\beta})$ be the number of points $P_n(j)$ $(1 \leqslant j \leqslant n)$ satisfying

$$\alpha_i \leqslant x_i^{(n)}(j) < \beta_i \quad (1 \leqslant i \leqslant s).$$

We can prove easily that

$$\left| \frac{1}{n} N_n(\boldsymbol{\alpha}, \boldsymbol{\beta}) - |\boldsymbol{\beta} - \boldsymbol{\alpha}| \right| < 2^s \varphi(n).$$

2) Let $\mathscr{D}$ be the domain

$$\delta \leqslant x_i < 1 - \delta \quad (1 \leqslant i \leqslant s)$$

and $\overline{\mathscr{D}}$ denote the complementary set of $\mathscr{D}$ in G_s. The measure of $\overline{\mathscr{D}}$ is $\leqslant 2^s \delta \leqslant 2^s \varphi(n)$. By 1), the number of points $P_n(j)$ belonging to $\mathscr{D}$ equals

$$(1 - 2\delta)^s n + \vartheta 2^s n \varphi(n),$$

where ϑ and ϑ's that will appear later are not the same, but with absolute value $\leqslant 1$. Therefore, the number of points of $P_n(j)$ belonging to $\overline{\mathscr{D}}$ does not exceed

$$(1 - (1 - 2\delta)^s) n + 2^s \varphi(n) n \leqslant 2^{s+1} \varphi(n) n.$$

3) Combining 1) with 2), we deduce that for an arbitrary $\boldsymbol{\gamma} \in G_s$ we have

$$\left| \frac{1}{n} N_n(\boldsymbol{\gamma}) - |\boldsymbol{\gamma}| \right| < (2^s + 2^s + 2^{s+1}) \varphi(n) = 2^{s+2} \varphi(n).$$

The lemma is proved.

The Proof of Theorem 4.1. By Lemma 4.2, it is sufficient to prove that

$$\left| \frac{1}{n} N_n(\boldsymbol{\gamma}) - |\boldsymbol{\gamma}| \right| < \varphi(n)$$

under the conditions

$$3\eta < \gamma_i \leqslant 1 - 3\eta \quad (1 \leqslant i \leqslant s).$$

Introducing function

$$G_x(y) = \begin{cases} 1, & \text{for } 0 \leqslant y < x, \\ 0, & \text{for } x \leqslant y < 1, \end{cases}$$

we have evidently

$$\frac{1}{n} N_n(\mathbf{x}) = \frac{1}{n} \sum_{j=1}^{n} G_{x_1}(x_1^{(n)}(j)) \cdots G_{x_s}(x_s^{(n)}(j)). \tag{4.1}$$

For $3\eta \leqslant x \leqslant 1 - 3\eta$, we construct according to Lemma 4.1 two auxiliary functions $G_x^{(1)}(y)$ and $G_x^{(2)}(y)$. Function $G_x^{(1)}(y)$ satisfies

(i) $G_x^{(1)}(y) = 1$, for $2\eta \leqslant y \leqslant x - \eta$,

(ii) $0 \leqslant G_x^{(1)}(y) \leqslant 1$, for $\eta \leqslant y \leqslant 2\eta$ and $x - \eta \leqslant y \leqslant x$,

(iii) $G_x^{(1)}(y) = 0$, for $x \leqslant y \leqslant 1 + \eta$,

(iv) $G_x^{(1)}(y)$ has a Fourier expansion

$$G_x^{(1)}(y) = x - 2\eta + \Sigma' C_1(m) e^{2\pi i m y},$$

where

$$|C_1(m)| \leqslant \min\left(x - 2\eta, \frac{1}{\pi |m|}, \frac{1}{\eta \pi^2 m^2} \right).$$

Function $G_x^{(2)}(y)$ satisfies

(i)′ $G_x^{(2)}(y) = 1$, for $-\eta \leqslant y \leqslant x$,

(ii)′ $0 \leqslant G_x^{(2)}(y) \leqslant 1$, for $-2\eta \leqslant y \leqslant -\eta$ and $x \leqslant y \leqslant x + \eta$,

(iii)′ $G_x^{(2)}(y) = 0$, for $x + \eta \leqslant y \leqslant 1 - 2\eta$,

(iv)′ $G_x^{(2)}(y)$ has a Fourier expansion

$$G_x^{(2)}(y) = x + 2\eta + \Sigma' C_2(m) e^{2\pi i m y},$$

where

$$|C_2(m)| \leqslant \min\left(x + 2\eta, \ \frac{1}{\pi|m|}, \ \frac{1}{\eta\pi^2 m^2}\right).$$

From (iv) it follows

$$G_x^{(1)}(y) = x - 2\eta + \sum_{|m|\leqslant h}' C_1(m) e^{2\pi i m y} + \vartheta \sum_{|m|>h} \frac{1}{\eta\pi^2 m^2}$$

$$= x - 2\eta + \sum_{|m|\leqslant h}' C_1(m) e^{2\pi i m y} + \frac{2\vartheta}{\pi^2 \eta h},$$

where $\displaystyle\sum_{m>h} \frac{1}{m^2} \leqslant \int_h^\infty \frac{dt}{t^2} = h^{-1}$. Put $x = x_i$ and $y = y_i$ in the above formula and write $C_0^{(1)} = x_i - 2\eta$ for all i. Then we have s equations. Multiplying separately the left hand sides and the right hand sides of these equations, we have

$$G_{x_1}^{(1)}(y_1)\cdots G_{x_s}^{(1)}(y_s) = \sum_{|m_i|\leqslant h} C_1(m_1)\cdots C_1(m_s) e^{2\pi i(m_1 y_1 + \cdots + m_s y_s)}$$

$$+ \frac{2\vartheta}{\pi^2 \eta h}\left(1 + \sum_{|m|\leqslant h} |C_1(m)|\right)^s = (x_1 - 2\eta)\cdots(x_s - 2\eta)$$

$$+ \sum_{|m_i|\leqslant h}' C_1(m_1)\cdots C_1(m_s) e^{2\pi i(m_1 y_1 + \cdots + m_s y_s)} + \frac{\vartheta \log^s 9h}{\eta h}$$

$$\cdot \left(1 + \sum_{|m|\leqslant h} |C_1(m)| \leqslant 2 + \frac{2}{\pi} + \frac{2}{\pi}\int_1^h \frac{dt}{t} < 2 + \log h < \log 9h\right).$$

Consequently, it is found that

$$G_{x_1}^{(1)}(y_1)\cdots G_{x_s}^{(1)}(y_s) = x_1 \cdots x_s + \sum_{|m_i|\leqslant h}' C_1(m_1)\cdots C_1(m_s) e^{2\pi i(m_1 y_1 + \cdots + m_s y_s)}$$

$$+ \vartheta\left(\frac{\log^s 9h}{\eta h} + 2^{s+1}\eta\right). \tag{4.2}$$

Similarly, we find

$$G_x^{(2)}(y_1)\cdots G_{x_s}^{(2)}(y_s) = x_1 \cdots x_s + \sum_{|m_i|\leqslant h}' C_2(m_1)\cdots C_2(m_s) e^{2\pi i(m_1 y_1 + \cdots + m_s y_s)}$$

$$+ \vartheta\left(\frac{\log^s 9h}{\eta h} + 2^{s+1}\eta\right). \tag{4.3}$$

From the definition of $G_x^{(1)}(y)$ and $G_x^{(2)}(y)$, we have

$$G_{x_1}^{(1)}(y_1)\cdots G_{x_s}^{(1)}(y_s) \leqslant G_{x_1}(y_1)\cdots G_{x_s}(y_s) \leqslant G_{x_1}^{(2)}(y_1)\cdots G_{x_s}^{(2)}(y_s). \tag{4.4}$$

Hence the theorem follows from (4.1), (4.2), (4.3) and (4.4).

V. The Proof of Theorem 1

Since $1, \omega_1, \cdots, \omega_q$ are linearly independent over rational field R, Theorem 1 follows from Lemma 1.1 and the following

Theorem 5.1. *If for any vector* $\mathbf{m} \neq \mathbf{o}$, *the inequality*

$$\langle (\mathbf{m}, \boldsymbol{\gamma}) \rangle > b \|\mathbf{m}\|^{-a} \tag{5.1}$$

holds, where a *and* b *are constants satisfying* $s + 1 \geq a > 1$ *and* $1 \geq b > 0$, *then the sequence*

$$P(j) = (\{\gamma_1 j\}, \cdots, \{\gamma_s j\}) \quad (j = 1, 2 \cdots)$$

has discrepancy

$$\varphi(n) = c(a, b, s) n^{-1+2s(a-1)} (\log 3n)^{1+s\delta_1, a},$$

where $\delta_{a,\beta}$ *denotes Kronecker symbol.*

Lemma 5.1. *Let* δ *be any real number. Then*

$$\left| \sum_{j=1}^{n} e^{2\pi i \delta j} \right| \leq \min \left(n, \frac{1}{2\langle \delta \rangle} \right).$$

Proof. If δ is not an integer, then we have

$$\left| \sum_{j=1}^{n} e^{2\pi i \delta j} \right| = \left| \frac{e^{2\pi i \delta (n+1)} - e^{2\pi i \delta}}{e^{2\pi i \delta} - 1} \right| \leq \frac{1}{|\sin \pi \delta|} \leq \frac{1}{2\langle \delta \rangle}.$$

Hence we have the lemma.

Lemma 5.2. *Let* $g(\mathbf{m})$ *be a non-negative function of* $\mathbf{m}$. *Then*

$$\sideset{}{'}\sum_{|m_i| \leq h} \frac{g(\mathbf{m})}{\|\mathbf{m}\|} \leq \sum_{l=0}^{s} \sum_{\mathbf{i}} \frac{1}{h^l} \sum_{m_{i_{l+1}}=1}^{h} \cdots \sum_{m_{i_s}=1}^{h} \frac{1}{m_{i_{l+1}}^2 \cdots m_{i_s}^2}$$

$$\times \sum_{|k_{i_1}| \leq h+1} \cdots \sum_{|k_{i_l}| \leq h+1} \sum_{|k_{i_{l+1}}| \leq m_{i_{l+1}}} \cdots \sideset{}{'}\sum_{|k_{i_s}| \leq m_{i_s}} g(\mathbf{k}),$$

where $\displaystyle\sum_{\mathbf{i}}$ *denotes a sum in which* $\mathbf{i} = (i_1, \cdots, i_s)$ *runs over all permutations of* $(1, 2, \cdots, s)$.

Proof. Since

$$\sum_{|m| \leq h} \frac{g(m)}{\overline{m}} = g(0) + \sum_{m=1}^{h} \frac{1}{m} (g(m) + g(-m))$$

$$= g(0) + \sum_{m=1}^{h} \left(\frac{1}{m} - \frac{1}{m+1} \right) \sum_{1 \leq k \leq m} (g(k) + g(-k))$$

$$+ \frac{1}{h+1} \sideset{}{'}\sum_{|k| \leq h+1} g(k) \leq \sum_{m=1}^{h} \frac{1}{m^2} \sum_{|k| \leq m} g(k) + \frac{1}{h} \sum_{|k| \leq h+1} g(k),$$

we have

$$
\sideset{}{'}\sum_{|m_i|\leqslant h} \frac{g(\mathbf{m})}{\|\mathbf{m}\|} \leqslant \sideset{}{'}\sum_{|m_i|\leqslant h} \frac{1}{\overline{m}_1\cdots\overline{m}_{s-1}} \left(\sum_{m_s=1}^{h} \frac{1}{m_s^2} \sum_{|k_s|\leqslant m_s} g(m_1,\cdots,m_{s-1},k_s) \right.
$$

$$
\left. + \frac{1}{h} \sum_{|k_s|\leqslant h+1} g(m_1,\cdots,m_{s-1},k_s) \right) + \sum_{|m_i|\leqslant h} \frac{1}{\overline{m}_1\cdots\overline{m}_{s-1}}
$$

$$
\times \left(\sum_{m_s=1}^{h} \frac{1}{m_s^2} \sideset{}{'}\sum_{|k_s|\leqslant m_s} g(m_1,\cdots,m_{s-1},k_s) + \frac{1}{h} \sideset{}{'}\sum_{|k_s|\leqslant h+1} g(m_1,\cdots,m_{s-1},k_s) \right)
$$

$$
\leqslant \cdots \leqslant \sum_{l=0}^{s} \sum_{i} \frac{1}{h^l} \sum_{m_{i_{l+1}}=1}^{h} \cdots \sum_{m_{i_s}=1}^{h} \frac{1}{m_{i_{l+1}}^2 \cdots m_{i_s}^2}
$$

$$
\times \sum_{|k_{i_1}|\leqslant h+1} \cdots \sum_{|k_{i_l}|\leqslant h+1} \sum_{|k_{i_{l+1}}|\leqslant m_{i_{l+1}}} \cdots \sideset{}{'}\sum_{|k_{i_s}|\leqslant m_{i_s}} g(\mathbf{k}).
$$

The lemma is proved.

Lemma 5.3. *Set* $\mathbf{m} \neq \mathbf{o}$ *and* $Q = [2^{sa}\|\mathbf{m}\|^a b^{-1}] + 1$. *If* (5.1) *holds for any* $\mathbf{m} \neq \mathbf{o}$, *then in any interval* $(P, P + Q^{-1}]$, *there contains at most a point* $(\mathbf{k}, \boldsymbol{\gamma})$ $= \sum_{i=1}^{s} k_i \gamma_i$, *where* $\mathbf{k}$ *is a vector with integral components which satisfy* $|k_i| \leqslant |m_i|$ $(1 \leqslant i \leqslant s)$.

Proof. If there contain two points $(\mathbf{k}', \boldsymbol{\gamma})$ and $(\mathbf{k}'', \boldsymbol{\gamma})$ in the interval $(P, P + Q^{-1}]$, where $\mathbf{k}' \neq \mathbf{k}''$, $|k_i'| \leqslant |m_i|$ and $|k_i''| \leqslant |m_i|$ $(1 \leqslant i \leqslant s)$, then we obtain

$$
\langle (\mathbf{k}' - \mathbf{k}'', \boldsymbol{\gamma}) \rangle \leqslant Q^{-1}.
$$

On the other hand, from (5.1), we have

$$
\langle (\mathbf{k}' - \mathbf{k}'', \boldsymbol{\gamma}) \rangle > b\|\mathbf{k}' - \mathbf{k}''\|^{-a} \geqslant 2^{-sa} b\|\mathbf{m}\|^{-a} > Q^{-1},
$$

which leads to a contradiction. Hence we have the lemma.

Lemma 5.4. *Set* $\mathbf{m} \neq \mathbf{o}$ *and* $Q = [2^{sa}\|\mathbf{m}\|^a b^{-1}] + 1$. *If* (5.1) *holds for any* $\mathbf{m} \neq \mathbf{o}$, *then*

$$
\sideset{}{'}\sum_{|k_i|\leqslant|m_i|} \frac{1}{\langle (\mathbf{k}, \boldsymbol{\gamma}) \rangle} \leqslant 4Q \log 3Q.
$$

Proof. Divide the interval $(0, 1]$ into Q subintervals

$$
I_j = \left(\frac{j}{Q}, \frac{j+1}{Q} \right] \quad (j = 0, 1, \cdots, Q-1).
$$

By (5.1), we deduce that none of the points $(\mathbf{k}, \boldsymbol{\gamma})$ lies in the interval I_0, where $\mathbf{k} \neq \mathbf{o}$ and $|k_i| \leqslant |m_i|$ $(1 \leqslant i \leqslant s)$. It follows by Lemma 5.3 that there contains at most a point $(\mathbf{k}, \boldsymbol{\gamma})$ in any interval I_j, where $j \geqslant 1$. Hence

$$
\sideset{}{'}\sum_{|k_i|\leqslant|m_i|} \frac{1}{\langle (\mathbf{k}, \boldsymbol{\gamma}) \rangle} \leqslant 4 \sum_{j=1}^{Q-1} \frac{Q}{j} \leqslant 4Q \log 3Q.
$$

The lemma follows.

Lemma 5.5. *Let h be an integer ≥ 2. If (5.1) holds for any $\mathbf{m} \neq \mathbf{o}$, then*

$$\sideset{}{'}\sum_{|m_i|\leqslant h} \frac{1}{\|\mathbf{m}\|\langle(\mathbf{m}, \boldsymbol{\gamma})\rangle} \leqslant c(a, b, s)h^{s(a-1)}(\log h)^{1+s\delta_1, a}.$$

Proof. From Lemmas 5.2 and 5.4, we have

$$\sideset{}{'}\sum_{|m_i|\leqslant h} \frac{1}{\|\mathbf{m}\|\langle(\mathbf{m}, \boldsymbol{\gamma})\rangle} \leqslant \sum_{l=0}^{s} \sum_{\mathbf{i}} \frac{1}{h^l} \sum_{m_{i_{l+1}}=1}^{h} \cdots \sum_{m_{i_s}=1}^{h} \frac{1}{m_{i_{l+1}}^2 \cdots m_{i_s}^2}$$

$$\times \sum_{|k_{i_1}|\leqslant h+1} \cdots \sum_{|k_{i_l}|\leqslant h+1} \sum_{|k_{i_{l+1}}|\leqslant m_{i_{l+1}}} \cdots \sideset{}{'}\sum_{|k_{i_s}|\leqslant m_{i_s}} \frac{1}{\langle(\mathbf{k}, \boldsymbol{\gamma})\rangle}$$

$$\leqslant \sum_{l=0}^{s} \sum_{\mathbf{i}} \frac{1}{h^l} \sum_{m_{i_{l+1}}=1}^{h} \cdots \sum_{m_{i_s}=1}^{h} \frac{1}{m_{i_{l+1}}^2 \cdots m_{i_s}^2} c(a, b, s)h^{la}(m_{i_{l+1}} \cdots m_{i_s})^a \log h$$

$$\leqslant c(a, b, s)h^{s(a-1)}(\log h)^{1+s\delta_1, a}.$$

The lemma is proved.

The Proof of Theorem 5.1. By Lemmas 5.1 and 5.5, we have

$$\sideset{}{'}\sum_{|m_i|\leqslant h} \frac{1}{\|\mathbf{m}\|} \left| \frac{1}{n} \sum_{j=1}^{n} e^{2\pi i(\mathbf{m}, \boldsymbol{\gamma})j} \right| \leqslant \frac{1}{n} \sideset{}{'}\sum_{|m_i|\leqslant h} \frac{1}{2\|\mathbf{m}\|\langle(\mathbf{m}, \boldsymbol{\gamma})\rangle}$$

$$\leqslant c(a, b, s)n^{-1}h^{s(a-1)}(\log h)^{1+s\delta_1, a}.$$

Take $\eta = \dfrac{1}{7n}$ and $h = 8n^2$. Then we have the theorem by Theorem 4.1.

VI. The Proof of Theorem 2

We use P_M^s to denote the s-dimensional parallelopiped with edges parallel to coordinate axes and volume $\leqslant M$, and $\mathbf{u} = (u_0, \cdots, u_s)$ and $\mathbf{h} = (1, h_1, \cdots, h_s)$ the $(s + 1)$-dimensional vectors with integral components.

Evidently, we may deduce Theorem 2 from Lemma 1.1 and the following two theorems.

Theorem 6.1. *Let n be an integer > 1 and M be a number $\geqslant 1$. Further let $\mathbf{a} = (a_1, \cdots, a_s)$ be a vector with integral components. If the congruence*

$$(\mathbf{a}, \mathbf{m}) = \sum_{i=1}^{s} a_i m_i \equiv 0 \pmod{n} \tag{6.1}$$

has no solution in the domain

$$\|\mathbf{m}\| \leqslant M, \quad \mathbf{m} \neq \mathbf{o}, \tag{6.2}$$

then the set

$$\left(\left\{ \frac{a_1 j}{n} \right\}, \cdots, \left\{ \frac{a_s j}{n} \right\} \right), \quad (1 \leqslant j \leqslant n),$$

has discrepancy

$$\varphi(n) = c(s, \varepsilon)M^{-1+\varepsilon}.$$

Theorem 6.2. *Let*

$$\left| \frac{h_i}{n} - \gamma_i \right| \leqslant dn^{-1-\frac{1}{s}}, \quad (1 \leqslant i \leqslant s), \tag{6.3}$$

be the simultaneous approximations of γ's by rationals, where d is a positive constant. If (5.1) holds for any $\mathbf{m} \neq \mathbf{o}$, then there exists constant $c(a, b, d, s)$ (<1) such that the congruence

$$(\mathbf{h}, \mathbf{u}) = u_0 + \sum_{i=1}^{s} h_i u_i \equiv 0 \pmod{n} \tag{6.4}$$

has no solution in the domain

$$\|\mathbf{u}\| \leqslant c(a, b, d, s)n^{\left(1+\frac{1}{s}\right)/(a+1)}, \quad \mathbf{u} \neq \mathbf{o}. \tag{6.5}$$

Lemma 6.1. *Let l be an integer $\geqslant 1$. Then the s-dimensional domain*

$$\|\mathbf{m}\| < lM \tag{6.6}$$

can be covered by at most $c(\varepsilon)^s l^{1+\varepsilon} M^\varepsilon$ parallelopipeds of the type P_M^s.

Proof. Take

$$c(\varepsilon) = 2^{2+\varepsilon} \sum_{j=0}^{\infty} \left(\bar{j}^{-(1+\varepsilon)} + 2^{-\varepsilon j} \right).$$

1) For $s = 1$, since the domain (6.6) is the interval $(-lM, lM)$, it can be covered by at most

$$\frac{2lM}{M} = 2l$$

intervals of the type $[c, c + M]$, where c is a real number. Hence the lemma is true for $s = 1$.

2) Suppose k is a positive integer and assume that the lemma holds for $s = 1, \cdots, k$. Now we proceed to prove the validity of the lemma for $s = k + 1$.

Divide the domain

$$\bar{m}_1 \cdots \bar{m}_{k+1} < lM \tag{6.7}$$

into $2[\log_2 M] + 3$ sub-domains

    (i) $m_{k+1} = j$, $\bar{j} \leqslant l$,

    (ii) $2^i l < |m_{k+1}| \leqslant 2^{i+1} l$, $(i = 0, 1, \cdots, [\log_2 M])$

by the hyperplanes

$$m_{k+1} = 0, \quad \pm 2^i l, \quad (i = 0, 1, \cdots, [\log_2 M]).$$

3) Suppose that $m_{k+1} = j$. Then

$$\bar{m}_1 \cdots \bar{m}_k < \frac{lM}{\bar{j}} < \left(\left[\frac{l}{\bar{j}} \right] + 1 \right) M.$$

Hence by our inductive hypothesis, we see that the above k-dimensional domain can be covered by at most

$$Q = c(\varepsilon)^k \left(\left[\frac{l}{\bar{j}} \right] + 1 \right)^{1+\varepsilon} M^\varepsilon$$

parallelopipeds of the type P_M^k. Using these P_M^k as bases and 1 as height, we construct P_M^{k+1}. Hence the sub-domain $m_{k+1} = j$ can be covered by at most Q parallelopipeds of the type P_M^{k+1}. Consequently, the domain defind by (i) can be covered by at most

$$2 \sum_{j=0}^{l} c(\varepsilon)^k \left(\left[\frac{l}{j} \right] + 1 \right)^{1+\varepsilon} M^\varepsilon \tag{6.8}$$

parallelopipeds of the type P_M^{k+1}.

4) Consider the sub-domain of (6.7)

$$2^i l < m_{k+1} \leqslant 2^{i+1} l. \tag{6.9}$$

Then for $m_{k+1} = 2^i l + 1$, we have

$$\overline{m}_1 \cdots \overline{m}_k < \frac{M}{2^i}.$$

Hence according to our inductive hypothesis, the above domain can be covered by at most

$$c(\varepsilon)^k \left(\frac{M}{2^i} \right)^\varepsilon \tag{6.10}$$

parallelopipeds of the type $P_{M/2^i}^k$. Using the $P_{M/2^i}^k$ as base and 2^i as height, we construct P_M^{k+1}. Since $2^i l + 2^i l + 1 > 2^{i+1} l$, the domain (6.9) can be covered by at most $c(\varepsilon)^k l \left(\frac{M}{2^i} \right)^\varepsilon$ parallelopipeds of the type P_M^{k+1}. Consequently, the domain defined by (ii) can be covered by at most

$$2c(\varepsilon)^k \sum_{i=0}^{[\log_2 M]} l \left(\frac{M}{2^i} \right)^\varepsilon \tag{6.11}$$

parallelopipeds of the type P_M^{k+1}.

5) It follows by (6.8) and (6.11) that the domain (6.7) can be covered by at most

$$c(\varepsilon)^k l^{1+\varepsilon} M^\varepsilon \sum_{j=0}^{\infty} \left(\frac{2^{2+\varepsilon}}{j^{1+\varepsilon}} + \frac{2}{2^{\varepsilon j}} \right) \leqslant c(\varepsilon)^{k+1} l^{1+\varepsilon} M^\varepsilon$$

parallelopipeds of the type P_M^{k+1}. Hence the lemma follows by mathematical induction.

Remark. The term $c(\varepsilon)^s l^{1+\varepsilon} M^\varepsilon$ in the above lemma may be replaced by $c(s) l \log^{s-1} \cdot 3lM$[12].

Lemma 6.2. *Let T_M^l be the number of solutions of the congruence* (6.1) *in the domain* (6.6). *If the congruence* (6.1) *has no solution in the domain* (6.2), *then we have*

$$T_M^l \leqslant c(\varepsilon)^s l^{1+\varepsilon} M^\varepsilon.$$

Proof. By Lemma 6.1, it is sufficient to prove that the congruence (6.1) has at most 1 solution in any parallelopiped of the type P_M^s. Suppose that the congruence (6.1) has two solutions $\mathbf{m}'$ and $\mathbf{m}''$ in certain P_M^s, where $\mathbf{m}' \neq \mathbf{m}''$. Let $\mathbf{m} = \mathbf{m}' - \mathbf{m}''$. Then $\|\mathbf{m}\| \leqslant M$ and

$$(\mathbf{a}, \mathbf{m}) = (\mathbf{a}, \mathbf{m}') - (\mathbf{a}, \mathbf{m}'') \equiv 0 \pmod{n}.$$

This leads to a contradiction. Hence we have the lemma.

The Proof of Theorem 6.1. Take $3\delta s = \varepsilon$. By Lemma 6.2, we have

$$\sum_{|m_i|\leqslant h}{}' \frac{1}{\|\mathbf{m}\|} \left| \frac{1}{n} \sum_{j=1}^{n} e^{2\pi i(\mathbf{a},\,\mathbf{m})j/n} \right| = \sum_{\substack{|m_i|\leqslant h \\ (\mathbf{a},\,\mathbf{m})\equiv 0 \ (\mathrm{mod}\ n)}}{}' \frac{1}{\|\mathbf{m}\|} \leqslant \sum_{l=1}^{h^s} \frac{(T_M^{l+1} - T_M^l)}{lM}$$

$$= \frac{1}{M} \sum_{l=1}^{h^s} T_M^{l+1}\left(\frac{1}{l} - \frac{1}{l+1} \right) + \frac{T_M^{h^s+1}}{(h^s+1)M} \leqslant c(\delta)^s M^{-1+\delta} h^{s\delta}.$$

$(T_M^1 = 0)$. Take $\eta = \dfrac{1}{7M}$ and $h = 7([M]+1)^2$. Then the theorem follows by Theorem 4.1.

The Proof of Theorem (6.2). Let $\mathbf{u} \neq \mathbf{o}$ be a solution of the congruence (6.4). If $u_i = 0\ (1 \leqslant i \leqslant s)$, we obtain $u_0 \neq 0$. From (6.4), we have $u_0 \equiv 0\ (\mathrm{mod}\ n)$. Hence the $\|\mathbf{u}\| \geqslant n$. Consequently, $\mathbf{u}$ is not belonging to the domain (6.5). Therefore we may suppose $(u_1, \cdots, u_s) \neq (0, \cdots, 0)$. If

$$\bar{u}_1 \cdots \bar{u}_s \geqslant \left(\frac{b}{2ds} \right)^{\frac{1}{a+1}} n^{\left(1+\frac{1}{s}\right)/(a+1)},$$

then it appears

$$\|\mathbf{u}\| \geqslant \left(\frac{b}{2ds} \right)^{\frac{1}{a+1}} n^{\left(1+\frac{1}{s}\right)/(a+1)},$$

hence we have the theorem. Now suppose

$$\bar{u}_1 \cdots \bar{u}_s < \left(\frac{b}{2ds} \right)^{\frac{1}{a+1}} n^{\left(1+\frac{1}{s}\right)/(a+1)}.$$

Since

$$\langle \alpha - \beta \rangle \geqslant \langle \alpha \rangle - \langle \beta \rangle,$$

from (5.1) and (6.3), we have

$$\frac{|u_0|}{n} \geqslant \left\langle \frac{u_0}{n} \right\rangle = \left\langle \frac{1}{n} \sum_{i=1}^{n} h_i u_i \right\rangle \geqslant \left\langle \sum_{i=1}^{s} \gamma_i u_i \right\rangle - \left\langle \sum_{i=1}^{s} \left(\frac{h_i}{n} - \gamma_i \right) u_i \right\rangle$$

$$\geqslant \frac{b}{(\bar{u}_1 \cdots \bar{u}_s)^a} - \frac{ds}{n^{1+\frac{1}{s}}} (\bar{u}_1 \cdots \bar{u}_s).$$

Therefore

$$|u_0| \bar{u}_1 \cdots \bar{u}_s > \frac{1}{2} (2ds)^{\frac{a-1}{a+1}} b^{\frac{2}{a+1}} n^{\left(1+\frac{1}{s}\right)/(a+1)}.$$

The theorem follows.

VII. THE PROOF OF THEOREM 3

We use the notations

$$I(f) = \int_0^1 \cdots \int_0^1 f(x_1, \cdots, x_s)\ dx_1 \cdots dx_s$$

and

$$P_n(f) = I(f) - \frac{1}{(2n+1)^l} \sum_{j=-nl}^{nl} \mu_{n,l,j}\, f(j\boldsymbol{\gamma}).$$

Clearly, Theorem 3 is the consequence of Lemma 1.1 and the following
Theorem 7.1. *If* (5.1) *holds for any* $\mathbf{m} \neq \mathbf{o}$, *then*

$$\operatorname*{Sup}_{f \in E_s^\alpha(C)} |P_n(f)| \leqslant C \cdot c(a, b, \alpha, s)\, n^{(-a+sa(a-1))/(a-1)} (\log 3n)^{a+sa\delta_1, a}.$$

Proof. Since
$\left| \dfrac{\sin h\pi\delta}{h \sin \pi\delta} \right| \leqslant 1$ (where h is a positive integer and δ is a real number but not an integer),

$$\frac{1}{(2n+1)^l} \sum_{j=-nl}^{nl} \mu_{n,l,j}\, f(j\boldsymbol{\gamma}) = \frac{1}{(2n+1)^l} \sum C(\mathbf{m}) \sum_{j=-nl}^{nl} \mu_{n,l,j}\, e^{2\pi i(\mathbf{m},\,\boldsymbol{\gamma})j}$$

$$= C(\mathbf{o}) + \frac{1}{(2n+1)^l} \sum{}' C(\mathbf{m}) \left(\sum_{j=-n}^{n} e^{2\pi i(\mathbf{m},\,\boldsymbol{\gamma})j} \right)^l$$

$$= C(\mathbf{o}) + \sum{}' C(\mathbf{m}) \left(\frac{\sin(2n+1)\pi(\mathbf{m},\,\boldsymbol{\gamma})}{(2n+1)\sin\pi(\mathbf{m},\,\boldsymbol{\gamma})} \right)^l$$

and

$$C(\mathbf{o}) = I(f),$$

then we have

$$\operatorname*{Sup}_{f \in E_s^\alpha(C)} |P_n(f)| \leqslant C \sum{}' \frac{1}{\|\mathbf{m}\|^a} \left| \frac{\sin(2n+1)\pi(\mathbf{m},\,\boldsymbol{\gamma})}{(2n+1)\sin\pi(\mathbf{m},\,\boldsymbol{\gamma})} \right|^a = C(\Sigma_1 + \Sigma_2), \qquad (7.1)$$

where Σ_1 denotes a sum in which the $\mathbf{m}$'s satisfy the conditions $|m_i| \leqslant n^{\frac{a}{a-1}}$ $(1 \leqslant i \leqslant s)$ and $\mathbf{m} \neq \mathbf{o}$, while Σ_2 the remaining part.

For $\alpha > 0$, $a_i > 0$ and $\sum_i a_i < \infty$, we have

$$\sum_i a_i^\alpha = \sum_i \left(\frac{a_i}{\sum_j a_j} \right)^\alpha \left(\sum_k a_k \right)^\alpha \leqslant \sum_i \frac{a_i}{\sum_j a_j} \left(\sum_k a_k \right)^\alpha = \left(\sum_k a_k \right)^\alpha.$$

Hence by Lemmas 5.1 and 5.5 it is found that

$$\Sigma_1 \leqslant \frac{1}{2^a(2n+1)^a} \sum_{\substack{|m_i| \leqslant n^{\frac{a}{a-1}}}}{}' \frac{1}{\|\mathbf{m}\|^a \langle(\mathbf{m},\,\boldsymbol{\gamma})\rangle^a}$$

$$\leqslant \frac{1}{2^a(2n+1)^a} \left(\sum_{\substack{|m_i| \leqslant n^{\frac{a}{a-1}}}}{}' \frac{1}{\|\mathbf{m}\| \langle(\mathbf{m},\,\boldsymbol{\gamma})\rangle} \right)^a$$

$$\leqslant c(a, b, \alpha, s)\, n^{-a+sa(a-1)/(a-1)} (\log 3n)^{a+sa\delta_1,\,a}. \qquad (7.2)$$

It is evidently

$$\sum_2 \leqslant \sum_{i=1}^{s} \sum_{\substack{a \\ |m_i|>n^{\frac{a}{a-1}}}} \frac{1}{|m_i|^a} \sum \frac{1}{\|\mathbf{m}\|^a} \leqslant c(\alpha, s)\, n^{-a}. \tag{7.3}$$

Substituting (7.2) and (7.3) into (7.1), we have the theorem.

VIII. The Proof of Theorem 4

Use the notation

$$Q_n(f) = I(f) - \frac{1}{n} \sum_{j=1}^{n} f\left(\frac{j\mathbf{a}}{n}\right).$$

Theorem 4 follows immediately from Lemma 1.1, Theorem 6.2 and the following

Theorem 8.1. *If the congruence (6.1) has no solution in the domain (6.2), then we have*

$$\operatorname{Sup}_{f \in E_s^a(C)} |Q_n(f)| \leqslant C \cdot c(\alpha, \varepsilon)^s M^{-a+\varepsilon}.$$

Proof. Clearly, we may suppose that $\varepsilon < \alpha - 1$. Since

$$\frac{1}{n} \sum_{j=1}^{n} f\left(\frac{j\mathbf{a}}{n}\right) = \frac{1}{n} \sum_{j=1}^{n} \sum C(\mathbf{m}) e^{2\pi i(\mathbf{a},\, \mathbf{m})j/n}$$

$$= C(\mathbf{o}) + \sum{}' C(\mathbf{m}) \frac{1}{n} \sum_{j=1}^{n} e^{2\pi i(\mathbf{a},\, \mathbf{m})j/n} = C(\mathbf{o}) + \sum_{(\mathbf{a},\mathbf{m})\equiv 0 \ (\mathrm{mod}\ n)}{}' C(\mathbf{m}),$$

from Lemma 6.2, it is known

$$\operatorname{Sup}_{f \in E_s^a(C)} |Q_n(f)| \leqslant C \sum_{(\mathbf{a},\mathbf{m})\equiv 0 \ (\mathrm{mod}\ n)}{}' \frac{1}{\|\mathbf{m}\|^a} \leqslant C \sum_{l=1}^{\infty} \frac{(T_M^{l+1} - T_M^l)}{(lM)^a}$$

$$= C \sum_{l=1}^{\infty} T_M^{l+1} \left(\frac{1}{l^a} - \frac{1}{(l+1)^a}\right) \leqslant C \cdot c(\varepsilon)^s M^{-a+\varepsilon}$$

$$\times \sum_{l=1}^{\infty} \frac{\alpha}{l^{a-\varepsilon}} \leqslant C \cdot c(\alpha, \varepsilon)^s M^{-a+\varepsilon}$$

$$\left(\frac{1}{l^a} - \frac{1}{(l+1)^a} = \alpha \int_{l}^{l+1} x^{-a-1}\, dx \leqslant \frac{\alpha}{l^{a+1}}\right). \quad \text{The theorem is proved.}$$

IX. Examples

Denote

$$P_n^*(f) = I(f) - \frac{1}{n} \sum_{j=-(n-1)}^{n-1} \left(1 - \frac{|j|}{n}\right) f(j\boldsymbol{\gamma}).$$

First of all, we prove the following

Theorem 9.1. *Let* $\gamma_1, \cdots, \gamma_s$ *be a set of real number such that* $1, \gamma_1, \cdots, \gamma_s$ *are linearly independent over the rational field* R. *Then we have*

$$\sup_{f \in E_s^2(C)} |P_n^*(f)| \leqslant C \left(\frac{\pi^2}{6}\right)^s (W(n; \gamma_1, \cdots, \gamma_s) - 1),$$

where

$$W(n; \gamma_1, \cdots, \gamma_s) = \frac{3^s}{n} + 2 \frac{3^s}{n} \sum_{j=1}^{n-1} \left(1 - \frac{j}{n}\right) \prod_{\nu=1}^{s} (1 - 2\{\gamma_\nu j\})^2.$$

Lemma 9.1. *We have*

$$\sum_{m=-\infty}^{\infty} \frac{e^{2\pi i m x}}{\frac{\pi^2}{6} m^2} = 3(1 - 2\{x\})^2.$$

Proof. Since

$$3 \int_0^1 (1 - 2x)^2 e^{2\pi i m x}\, dx = \begin{cases} 0, & \text{for } m = 0, \\ \dfrac{6}{\pi^2 m^2}, & \text{for } m \neq 0, \end{cases}$$

therefore we have the lemma.

The Proof of Theorem 9.1. Since

$$\sum_{j=-(n-1)}^{n-1} (n - |j|) = \sum_{k=0}^{n-1} \sum_{j=-k}^{k} 1 = n^2$$

and

$$\sum_{j=-(n-1)}^{n-1} (n - |j|) e^{2\pi i j \delta} = \sum_{k=0}^{n-1} \sum_{j=-k}^{k} e^{2\pi i j \delta}$$

$$= \frac{1}{\sin \pi \delta} \sum_{k=0}^{n-1} \sin (2k + 1)\pi \delta = \left(\frac{\sin n\pi\delta}{\sin \pi\delta}\right)^2,$$

where δ is a real number but not an integer, we find

$$\frac{1}{n} \sum_{j=-(n-1)}^{n-1} \left(1 - \frac{|j|}{n}\right) f(j\boldsymbol{\gamma}) = \frac{1}{n^2} \sum_{j=-(n-1)}^{n-1} (n - |j|) \sum C(\mathbf{m}) e^{2\pi i (\mathbf{m}, \gamma) j}$$

$$= \frac{1}{n^2} \sum C(\mathbf{m}) \sum_{j=-(n-1)}^{n-1} (n - |j|) e^{2\pi i (\mathbf{m}, \gamma) j}$$

$$= C(\mathbf{o}) + \frac{1}{n^2} {\sum}' C(\mathbf{m}) \left(\frac{\sin n\pi(\mathbf{m}, \boldsymbol{\gamma})}{\sin \pi(\mathbf{m}, \boldsymbol{\gamma})}\right)^2.$$

Hence by Lemma 9.1, we have

$$\sup_{f \in E_s^2(C)} |P_n^*(f)| \leqslant \frac{C}{n^2} {\sum}' \frac{1}{\|\mathbf{m}\|^2} \left(\frac{\sin n\pi(\mathbf{m}, \boldsymbol{\gamma})}{\sin \pi(\mathbf{m}, \boldsymbol{\gamma})}\right)^2 = \frac{C}{n^2} {\sum}' \frac{1}{\|\mathbf{m}\|^2}$$

$$\times \sum_{k=0}^{n-1} \sum_{j=-k}^{k} e^{2\pi i (\mathbf{m}, \gamma) j} \leqslant \frac{C}{n^2} \left(\frac{2\pi}{6}\right)^s \sum_{k=0}^{n-1} \sum_{j=-k}^{k} {\sum}' \frac{e^{2\pi i (\mathbf{m}, \gamma) j}}{\left\|\frac{\pi^2}{6} \mathbf{m}\right\|}$$

$$= \frac{C}{n^2}\left(\frac{\pi^2}{6}\right)^s \sum_{k=0}^{n-1}\sum_{j=-k}^{k}\left(3^s\prod_{\nu=1}^{s}(1-2\{\gamma_\nu j\})^2 - 1\right)$$

$$= C\left(\frac{\pi^2}{6}\right)^s\left(\sum_{j=-(n-1)}^{n-1}(n-|j|)\frac{3^s}{n^2}\prod_{\nu=1}^{s}(1-2\{\gamma_\nu j\})^2 - 1\right)$$

$$= C\left(\frac{\pi^2}{6}\right)^s(W(n;\gamma_1,\cdots,\gamma_s)-1).$$

The theorem is proved.

By Wang 520 calculator, we obtain the following two tables:

Table 1: $s=3$

n	$W\left(n;\ \dfrac{\sqrt{5}-1}{2},\ \sqrt{2},\ \sqrt{10}\right)$	$W(n;e,e^2,e^3)$
100	1.08877	1.10689
500	1.01351	1.00914
1000	1.00572	1.00294

Table 2: $s=4$

n	$W\left(n;\ 2\cos\dfrac{2\pi}{11},\ 2\cos\dfrac{4\pi}{11},\ 2\cos\dfrac{6\pi}{11},\ 2\cos\dfrac{8\pi}{11}\right)$	$W(n;e,e^2,e^3,e^4)$
1000	1.03263	1.13899
1500	1.02139	1.11848
3000	1.00887	

References

[1] Baker, A. 1965 On some Diophantine inequalities involving the exponential function, *Can. J. Math.*, **17**(4), 616—626.

[2] Halton, J. H. 1960 On the efficiency of certain quasirandom sequences of points in evaluating multi-dimensional integrals, *Num. Math.*, **27**(2), 84—90.

[3] Haselgrove, C. B. 1961 A method for numerical integration, *Math. Comp.*, **15**(76), 323—337.

[4] Hlawka, E. 1962 Zur angenäherten Berechnung mehrfacher Integrale, *Mon. Math.*, **66**(2), 140—151.

[5] ———— 1964 Uniform distribution modulo 1 and numerical analysis, *Comp. Math.*, **16**(1—2), 92—105.

[6] Hua, Loo-keng & Wang Yuan 1960 Remarks concerning numerical integration, *Sci. Rec.*, **4**(1), 8—11.

[7] ———— 1964 On Diophantine approximations and numerical integrations (I) (II), *Sci. Sin.*, **13**(6), 1007—1010.

[8] ———— 1965 On numerical integration of periodic functions of several variables, *Sci. Sin.*, **14**(7), 964—978.

[9] Roth, K. F. 1954 On irregularities distribution, *Math.*, **1**(2), 73—79.

[10] Schmidt, Wolfgang M. 1970 Simultaneous approximation to algebraic numbers by rationals, *Acta Math.*, **125**, 189—201.

[11] Wyel, H. 1913 Über die Gleichverteilung von Zahlen mod Eins, *Math. Ann.*, **77**, 313—352.

[12] Бахбалов Н. С. 1959 О приближенном вычислении кратных интегралов, *Вес. Мос. Ун-та.*, **4**, 3—18.

[13] Виноградов И. М. 1971 Метод тригонометрических Сумм в теории чисел, *Физмат. Лит.*, Изд. «Наука».

[14] Коробов Н. М. 1963 Теоретико-числовые методы в приближенном анализе, *Физмат. Лит. Мос.*.

[15] Шарыгин И. Ф. 1963 Оценки снизу погрешиости квадратурных формул на классах функций, *Жур. Мат. и Мат. Физ.* **3**(2), 370—376.

[16] *Applications of Number Theory to Numerical Analysis*, Zaremba, S. K. (Ed.), Acad. Press, 1972.

[17] 津田孝夫 1973 多变数问题的数值解析，株式会社（日文）.

Note added on the 19th of May, 1973. A theorem similar to Theorem 5.1 was proved independently by H. Niederreiter and some results given in [8] were improved by S. Haber (See [16]). In the course of publication, we add two more new books, [16] and [17].

Vol. XVII No. 3   SCIENTIA SINICA   June 1974

ON UNIFORM DISTRIBUTION AND NUMERICAL ANALYSIS (II) (NUMBER-THEORETIC METHOD)

Hua Loo-keng (华罗庚) and Wang Yuan (王　元)

Received June 13, 1973.

Abstract

In this paper, we shall give some applications of the sequences of sets defined in a previous paper to the problems of numerical integration, interpolation, and the approximate solution of Fredholm integral equation of the second type. We have studied in addition the well-known sequences of sets introduced by Коробов.

I. Statement of Results

Let $f(\boldsymbol{x}) = f(x_1, \cdots, x_s)$ be a periodic function of s-variables, each with periodic 1. Let $\boldsymbol{a} = (\alpha_1, \cdots, \alpha_s)$ be a vector and let $\rho_k = \beta_k = 0$ for $\alpha_k = 0$, where ρ_k represents a non-negative integer and $0 < \beta_k \leqslant 1$ for $\alpha_k = \rho_k + \beta_k > 0$ $(1 \leqslant k \leqslant s)$. Define

$$\delta_h^k f = (2i)^{-1}(f(x_1, \cdots, x_k + h, \cdots, x_s) - f(x_1, \cdots, x_k - h, \cdots, x_s)).$$

Suppose that the partial derivatives

$$\frac{\partial^{\tau_1 + \cdots + \tau_s} f}{\partial x_1^{\tau_1} \cdots \partial x_s^{\tau_s}} = f(\boldsymbol{x})^{(\tau_1, \cdots, \tau_s)}, \quad (0 \leqslant \tau_i \leqslant \rho_i, \; 1 \leqslant i \leqslant s),$$

exist, which all are the periodic functions of s-variables, each with periodic 1. Denote

$$\|f^a\| = \sup_{\substack{0 < h_k \leqslant \infty \\ \boldsymbol{x} \in G_s}} \left| \prod_{a_k > 0} ((h_k^{-\beta_k} \delta_{h_k}^k) f)^{(\rho_1, \cdots, \rho_s)} \right|.$$

The class of functions f satisfying the following conditions

$$\|f^{(\theta_1 \alpha_1, \cdots, \theta_s \alpha_s)}\| \leqslant A$$

is denoted by $H_s^a(A)$, where $\theta_1, \cdots, \theta_s = 0$ or 1 and $A(> 0)$ is an absolute constant. For the special case $\alpha_1 = \cdots = \alpha_s = \alpha$, we denote the class of functions by $H_s^\alpha(A)$.

The inventions and notations introduced in [1] are also used in this paper. Applying the sequences of sets (1.1) and (1.2) of [1] to the problem of numerical integration, we have

Theorem 1. *Let α be a number satisfying $1 \geqslant \alpha > 0$. Then*

$$\sup_{f \in H_q^\alpha(A)} \left| \int_0^1 \cdots \int_0^1 f(x_1, \cdots, x_q) dx_1 \cdots dx_q - \frac{1}{n} \sum_{j=1}^n f(\omega_1 j, \cdots, \omega_q j) \right|$$
$$\leqslant Ac(\mathfrak{R}_r, \alpha, \varepsilon) n^{-\alpha+\varepsilon}. \tag{1.1}$$

Theorem 2. *Let α be a number satisfying $1 \geqslant \alpha > 0$. Then*

$$\sup_{f \in H_r^\alpha(A)} \left| \int_0^1 \cdots \int_0^1 f(x_1, \cdots, x_r) dx_1 \cdots dx_r - \frac{1}{n} \sum_{j=1}^n f\left(\frac{j}{n}, \frac{h_1 j}{n}, \cdots, \frac{h_q j}{n}\right) \right|$$
$$\leqslant Ac(\mathfrak{R}_r, \alpha, \varepsilon) n^{-\frac{\alpha}{2}-\frac{\alpha}{2q}+\varepsilon}. \tag{1.2}$$

Next, we use the sequences of sets (1.1) and (1.2) of [1] to treat the problem of interpolation and the problem of approximation solution of Fredholm integral equation of the second type.

Moreover, we have studied the uniformly distributed sequences of sets

$$\left(\left\{\frac{j}{p}\right\}, \left\{\frac{j^2}{p}\right\}, \cdots, \left\{\frac{j^s}{p}\right\}\right), \quad (1 \leqslant j \leqslant p), \tag{1.3}$$

and

$$\left(\left\{\frac{j}{p}\right\}, \left\{\frac{aj}{p}\right\}, \cdots, \left\{\frac{a^{s-1}j}{p}\right\}\right), \quad (1 \leqslant j \leqslant p), \tag{1.4}$$

where p is a prime number and $a = a(p)$ is an integer depending on p. The two sequences of sets were proposed first by Коробов in 1959 and 1960 respectively[5,6].

Theorem 3. *Let p be a prime. Then*

$$\sup_{f \in H_s^\alpha(A)} \left| \int_0^1 \cdots \int_0^1 f(x_1, \cdots, x_s) dx_1 \cdots dx_s - \frac{1}{p} \sum_{j=1}^p f\left(\frac{j}{p}, \frac{j^2}{p}, \cdots, \frac{j^s}{p}\right) \right|$$
$$\leqslant \begin{cases} Ac(\alpha, s)p^{-1/2}, & \text{for } 1 \geqslant \alpha > \dfrac{1}{2}, \\[2mm] Ac(\alpha, s)p^{-\alpha} (\log p)^{s-1+\delta_{1/2,\,\alpha}}, & \text{for } \dfrac{1}{2} \geqslant \alpha > 0. \end{cases} \tag{1.5}$$

Besides, we propose the sequence of set

$$\left(\left\{\frac{a}{p}\right\}, \left\{\frac{aj}{p}\right\}, \cdots, \left\{\frac{aj^{s-1}}{p}\right\}\right) \quad (1 \leqslant a, \; j \leqslant p) \tag{1.6}$$

and apply it to the problem of numerical integration.

Consider the Volterra integral equation of the second type

$$\varphi(x) = \int_0^x K(x, y)\varphi(y)dy + f(x), \tag{1.7}$$

where $f(x) \in H_1^\alpha(A)$ and $K(x, y) \in H_2^\alpha(A)$. We use the notations

$$\nu(\alpha) = \begin{cases} \dfrac{1}{2}, & \text{for } \alpha > 1, \\[3mm] \dfrac{2\alpha}{1 + 4\alpha - \alpha^2}, & \text{for } 1 \geqslant \alpha > 0, \end{cases} \tag{1.8}$$

$$g = [p^{\nu(\alpha)}], \tag{1.9}$$

$$Q = \left[\nu(\alpha)\alpha \, \frac{\log_2 p}{\log_2 \log_2 3p}\right] \tag{1.10}$$

and

$$B_{j,\,s,\,p}(a) = \sum_{\overline{m}_1 \cdots \overline{m}_s < g} e^{-2\pi i(m_1 + m_2 a + \cdots + m_s a^{s-1})j/p}$$

$$\times \int_0^x \int_0^{x_1} \cdots \int_0^{x_{s-1}} e^{2\pi i(m_1 x_1 + \cdots + m_s x_s)} \, dx_1 \cdots dx_s. \tag{1.11}$$

Theorem 4. *Let p be a prime and α be a positive number. Then there exists an integer a such that the solution of Eq. (1.7) can be written as*

$$\varphi(x) = f(x) + \frac{1}{p} \sum_{j=1}^{p} \sum_{s=1}^{Q} B_{j,\,s,\,p}(a) K\left(x, \frac{j}{p}\right)$$

$$\times K\left(\frac{j}{p}, \frac{aj}{p}\right) \cdots K\left(\frac{a^{s-2}j}{p}, \frac{a^{s-1}j}{p}\right) f\left(\frac{a^{s-1}j}{p}\right) + O(p^{-\nu(\alpha)\alpha + \varepsilon}), \tag{1.12}$$

where the constant implied by the symbol "O" depends only on α, A and ε.

In order to obtain the integer $a = a(p)$, we require $c(s)p^2$ elementary operations. For the proof of Theorem 3, we require the well-known result of A. Weil[2] concerning the estimation of exponential sums.

Lemma 1.1. *Let p be a prime and let $m_1, \cdots, m_s$ be a set of integers, at least one of which is not dividing by p. Then it is seen that*

$$\left|\sum_{j=1}^{p} e^{2\pi i(m_1 j + \cdots + m_s j^s)/p}\right| \leqslant (s-1)\sqrt{p}\,.$$

The estimations given in Theorems 1 and 3 for the case $1/2 \geqslant \alpha > 0$ are the best possible kinds, apart from some possible improvements about the order $(\log p)^{s-1+\delta_{1,\,\alpha}}$ (cf [1]).

Previously, the formula (1.5) was proved under more restrictive condition $\alpha > 1/2$ (see Коробов[5] and Шахов[9]). The Theorem 4 gives a slight modification to the result obtained by Шахов[8,9]. In his original result, the error term is $O(p^{-\frac{\alpha}{2}+\frac{1}{2}+\varepsilon})$, α being an integer $\geqslant 2$.

II. THE CLASSES OF FUNCTIONS

Let c be a constant such that $0 < c < 1$. Let

$$\mu(x) = \begin{cases} \cos^2\left(\frac{\pi}{2} \log_2 \left|\frac{x}{c}\right|\right), & \text{for } \frac{c}{2} \leqslant |x| \leqslant 2c, \\ 0, & \text{otherwise} \end{cases}$$

and let

$$\mu_0(x) = 1 - \sum_{t=1}^{\infty} \mu_t(x), \tag{2.1}$$

where $\mu_t(x) = \mu(2^{1-t}x)$ $(t \geqslant 1)$.

Let $f(\boldsymbol{x})$ be a periodic function of s-variables, each with periodic 1. Suppose that $f(\boldsymbol{x})$ has the Fourier expansion

$$f(\boldsymbol{x}) \sim \sum C(\boldsymbol{m})e^{2\pi i(\boldsymbol{m},\,\boldsymbol{x})}.$$

For a vector $\boldsymbol{t} = (t_1, \cdots, t_s)$ with non-negative integral components, we define

$$\varphi_{\boldsymbol{t}}(\boldsymbol{x}) = \sum C_{\boldsymbol{t}}(\boldsymbol{m})e^{2\pi i(\boldsymbol{m},\,\boldsymbol{x})},$$

where

$$C_{\boldsymbol{t}}(\boldsymbol{m}) = C(\boldsymbol{m})\mu_{t_1}(m_1) \cdots \mu_{t_s}(m_s). \tag{2.2}$$

Let $\boldsymbol{a} = (\alpha_1, \cdots, \alpha_s)$ be a vector with non-negative real components. The class of functions $f(\boldsymbol{x})$ satisfying

$$\|\varphi_{\boldsymbol{t}}\| = \sup_{\boldsymbol{x}\in G_s} |\varphi_{\boldsymbol{t}}(\boldsymbol{x})| \leqslant B2^{-(\boldsymbol{a},\,\boldsymbol{t})}$$

is denoted by $Q_s^a(B)$, $B(>0)$ being an absolute constant; it is so especially, when denoted by $Q_s^a(B)$ for the case $\alpha_1 = \cdots = \alpha_s = \alpha$.

Lemma 2.1. *Let* $\alpha > 0$. *Then*

$$H_s^a(A) \subset Q_s^a(A \cdot c(\alpha)^s) \subset E_s^a(A \cdot c(\alpha)^s).$$

Lemma 2.2. *Let* $\alpha > 0$ *and* $f(\boldsymbol{x}) \in Q_s^a(B)$. *Then it is inferred*

$$f(\boldsymbol{x}) = \sum{}'' \varphi_{\boldsymbol{t}}(\boldsymbol{x}),$$

where $\sum''$ *denotes a sum in which* $\boldsymbol{t}$ *runs over all vectors with non-negative components.* See Бахвалов[3, 4].

III.   THE PROOF OF THEOREM 1

Denote

$$R_n(f) = I(f) - \frac{1}{n}\sum_{j=1}^{n} f(j\boldsymbol{\gamma}).$$

Evidently, Theorem 1 is the consequence of Lemma 1.1 of [1], Lemma 2.1 and the following

Theorem 3.1. *Suppose that* $1 \geqslant \alpha > 0$. *If Eq.* (5.1) *of* [1] *holds for any* $\boldsymbol{m} \neq \boldsymbol{o}$, *then*

$$\sup_{f\in Q_s^a(B)} |R_n(f)| \leqslant Bc(a, b, \alpha, s)n^{-a+s(a-1)}(\log 3n)^{s+s\delta_1,\,a}.$$

Proof. By Lemma 2.2, we have

$$R_n(f) = \sum{}'' R_n(\varphi_{\boldsymbol{t}}) \tag{3.1}$$

for $f \in Q_s^a(B)$. It follows by the definition of $\varphi_{\boldsymbol{t}}$ that

$$|R_n(\varphi_{\boldsymbol{t}})| \leqslant 2\|\varphi_{\boldsymbol{t}}\| \leqslant 2B \cdot 2^{-at_0}, \tag{3.2}$$

where

$$t_0 = t_1 + \cdots + t_s. \tag{3.3}$$

From Lemma 5.1 of [1], we have

$$|R_n(\varphi_t)| \leqslant \Sigma' |C_t(\boldsymbol{m})| \frac{1}{n} \left| \sum_{j=1}^{n} e^{2\pi i(\boldsymbol{m},\, \boldsymbol{\gamma})j} \right|$$

$$\leqslant n^{-1} \Sigma' |C_t(\boldsymbol{m})| \frac{1}{\langle (\boldsymbol{m}, \boldsymbol{\gamma}) \rangle}. \tag{3.4}$$

Hence from Eqs. (3.1), (3.2), (3.3) and (3.4), we have

$$\sup_{f \in Q_s^\alpha(B)} |R_n(f)| \leqslant \sup_{f \in Q_s^\alpha(B)} \Sigma'' |R_n(\varphi_t)| \leqslant \Sigma_1 + \Sigma_2, \tag{3.5}$$

where

$$\Sigma_1 = \sup_{f \in Q_s^\alpha(B)} \sum_{t_0 \leqslant \log_2 n}'' \frac{1}{n} \Sigma' \frac{|C_t(\boldsymbol{m})|}{\langle (\boldsymbol{m}, \boldsymbol{\gamma}) \rangle}$$

and

$$\Sigma_2 = 2B \sum_{t_0 > \log_2 n}'' 2^{-\alpha t_0}.$$

Since it follows from Eq. (2.2) that $C_t(\boldsymbol{m}) = 0$ for $\|\boldsymbol{m}\| \geqslant 2^{t_0}$, therefore by Lemma 5.5 of [1], Lemma 2.1 and Eq. (2.1), we have

$$\Sigma_1 \leqslant \sup_{f \in Q_s^\alpha(B)} n^{-1} \sum_{\|\boldsymbol{m}\| \leqslant n}' \frac{1}{\langle (\boldsymbol{m}, \boldsymbol{\gamma}) \rangle} \Sigma'' |C_t(\boldsymbol{m})|$$

$$\leqslant \sup_{f \in Q_s^\alpha(B)} n^{-1} \sum_{\|\boldsymbol{m}\| \leqslant n}'' \frac{|C(\boldsymbol{m})|}{\langle (\boldsymbol{m}, \boldsymbol{\gamma}) \rangle}$$

$$\leqslant Bc(\alpha, s)n^{-1} \sum_{\|\boldsymbol{m}\| \leqslant n}' \frac{1}{\|\boldsymbol{m}\|^\alpha \langle (\boldsymbol{m}, \boldsymbol{\gamma}) \rangle}$$

$$\leqslant Bc(\alpha, s)n^{-1} \sum_{\|\boldsymbol{m}\| \leqslant n}' \frac{n^{1-\alpha}}{\|\boldsymbol{m}\| \langle (\boldsymbol{m}, \boldsymbol{\gamma}) \rangle}$$

$$\leqslant Bc(a, b, \alpha, s)n^{-\alpha+s(a-1)}(\log 3n)^{1+s\delta_1,\, \alpha}. \tag{3.6}$$

Since the number of non-negative solutions $(t_1, \cdots, t_s)$ of the Diophantine equation (3.3) is

$$\binom{t_0 + s - 1}{s - 1} = \frac{(t_0 + s - 1)!}{t_0!(s - 1)!} \leqslant c(s) t_0^{s-1},$$

therefore

$$\Sigma_2 \leqslant Bc(s) \sum_{t=[\log_2 n]+1}^{\infty} 2^{-\alpha t} t^{s-1} \leqslant Bc(\alpha, s) n^{-\alpha}(\log 3n)^{s-1}. \tag{3.7}$$

Substituting (3.6) and (3.7) into (3.5), we have the theorem.

IV. THE PROOF OF THEOREM 2

We use the notation

$$S_n(f) = I(f) - \frac{1}{n} \sum_{j=1}^{n} f\left(\frac{j\boldsymbol{a}}{n} \right),$$

where $\boldsymbol{a} = (a_1, \cdots, a_s)$ is a vector with integral components.

Evidently, Theorem 2 follows immediately from Lemma 1.1 of [1], Theorem 6.2 of [1] and the following

Theorem 4.1. *Let α be a number satisfying $1 \geqslant \alpha > 0$. Let n be an integer $\geqslant 2$ and M be a number $\geqslant 1$. If the congruence*

$$(\boldsymbol{a}, \boldsymbol{m}) = \sum_{i=1}^{s} a_i m_i \equiv 0 \ (\mathrm{mod} \ n) \tag{4.1}$$

has no solution in the domain

$$\|\boldsymbol{m}\| \leqslant M, \quad \boldsymbol{m} \neq \boldsymbol{o}, \tag{4.2}$$

then we have

$$\sup_{f \in Q_s^\alpha(B)} |S_n(f)| \leqslant Bc(\alpha, \varepsilon)^s M^{-\alpha+\varepsilon}.$$

Proof. Clearly, we may suppose that $\varepsilon < \alpha$. In a similar manner to (3.5), we have

$$\sup_{f \in Q_s^\alpha(B)} |S_n(f)| \leqslant \Sigma_1 + \Sigma_2, \tag{4.3}$$

where

$$\Sigma_1 = \sup_{f \in Q_s^\alpha(B)} \sum_{t_0 > \log_2 M}'' |S_n(\varphi_t)|,$$

and

$$\Sigma_2 = \sup_{f \in Q_s^\alpha(B)} \sum_{t_0 \leqslant \log_2 M}'' |S_n(\varphi_t)|.$$

By (3.2), we have

$$\Sigma_1 \leqslant 2B \sum_{t_0 > \log_2 M}'' 2^{-\alpha t_0} \leqslant 2B \sum_{t_0 > \log_2 M}'' 2^{-(\alpha-\varepsilon)t_0 - \varepsilon t_0}$$

$$\leqslant Bc(\alpha, \varepsilon)^s M^{-\alpha+\varepsilon}. \tag{4.4}$$

Since $C_t(\boldsymbol{m}) = 0$ for $\|\boldsymbol{m}\| \geqslant 2^{t_0}$, and

$$|S_n(\varphi_t)| = \left| \Sigma' C_t(\boldsymbol{m}) \frac{1}{n} \sum_{j=1}^{n} e^{2\pi i(\boldsymbol{a},\boldsymbol{m})j/n} \right| \leqslant \sum_{(\boldsymbol{a},\boldsymbol{m}) \equiv 0 \ (\mathrm{mod} \ n)}' |C_t(\boldsymbol{m})|,$$

therefore it appears

$$\Sigma_2 \leqslant \sup_{f \in Q_s^\alpha(B)} \sum_{(\boldsymbol{a},\boldsymbol{m}) \equiv 0 \ (\mathrm{mod} \ n)}' \sum_{t_0 \leqslant \log_2 M}'' |C_t(\boldsymbol{m})| = 0. \tag{4.5}$$

Hence the theorem follows from (4.3), (4.4) and (4.5).

V.   The Proof of Theorem 3

Denote

$$T_p(f) = I(f) - \frac{1}{p} \sum_{j=1}^{p} f\left(\frac{j}{p}, \frac{j^2}{p}, \cdots, \frac{j^s}{p}\right).$$

By Lemmas 2.1 and 2.2, we have

$$\sup_{f \in H_s^\alpha(A)} |T_p(f)| \leqslant \Sigma_1 + \Sigma_2, \tag{5.1}$$

where

$$\Sigma_1 = \sup_{f \in H_s^\alpha(A)} \sum_{t_0 > \log_2 p}'' |T_p(\varphi_t)|,$$

and

$$\Sigma_2 = \sup_{f \in H_s^\alpha(A)} \sum_{t_0 < \log_2 p}'' |T_p(\varphi_t)|.$$

It is similar to (3.7), that

$$\Sigma_1 \leqslant 2 \sup_{f \in H_s^\alpha(A)} \sum_{t_0 > \log_2 p}'' \|\varphi_t\| \leqslant Ac(\alpha, s) \sum_{t=[\log_2 p]}^\infty 2^{-\alpha t} t^{s-1}$$

$$\leqslant Ac(\alpha, s) p^{-\alpha} (\log p)^{s-1}. \tag{5.2}$$

Let $\boldsymbol{m} \neq \boldsymbol{o}$. If one of the relations $|m_i| < 2^{t_i}$ $(1 \leqslant i \leqslant s)$ is not satisfying, then it is known that $C_t(\boldsymbol{m}) = 0$. If $|m_i| < 2^{t_i}$ for all i, where $1 \leqslant i \leqslant s$, and $t_0 < \log_2 p$, then p can not divide all the $m_i's$. Hence by Lemma 1.1, we have

$$\Sigma_2 \leqslant \sup_{f \in H_s^\alpha(A)} \sum_{t_0 < \log_2 p}'' \sum' |C_t(\boldsymbol{m})| \left| \frac{1}{p} \sum_{j=1}^p e^{2\pi i(m_1 j + \cdots + m_s j^s)/p} \right|$$

$$\leqslant (s-1) p^{-\frac{1}{2}} \sup_{f \in H_s^\alpha(A)} \sum_{t_0 < \log_2 p}'' \sum_{|m_i| < 2^{t_i}}' |C_t(\boldsymbol{m})|.$$

Consequently, by Schwarz inequality and

$$\|\varphi_t\|_{L_2} \leqslant \|\varphi_t\| \leqslant Ac(\alpha, s) 2^{-\alpha t_0},$$

where

$$\|\varphi_t\|_{L_2} = \left(\int_{G_s} |\varphi_t|^2 d\boldsymbol{x} \right)^{\frac{1}{2}},$$

we have

$$\Sigma_2 \leqslant (s-1) p^{-1/2} \sup_{f \in H_s^\alpha(A)} \sum_{t_0 < \log_2 p}'' \left(\sum_{|m_i| < 2^{t_i}} 1 \right)^{\frac{1}{2}} \left(\sum_{|m_i| < 2^{t_i}} |C_t(\boldsymbol{m})|^2 \right)^{\frac{1}{2}}$$

$$\leqslant (s-1) p^{-1/2} \sup_{f \in H_s^\alpha(A)} \sum_{t_0 < \log_2 p}'' 2^{\frac{s+t_0}{2}} \|\varphi_t\|_{L_2}$$

$$\leqslant Ac(\alpha, s) p^{-1/2} \sum_{t_0 < \log_2 p}'' 2^{-\left(\alpha - \frac{1}{2}\right) t_0}$$

$$= Ac(\alpha, s) p^{-1/2} \sum_{t=0}^{[\log_2 p]} 2^{-\left(\alpha - \frac{1}{2}\right) t} (t+1)^{s-1}$$

$$\leqslant \begin{cases} Ac(\alpha, s) p^{-1/2}, & \text{for } 1 \geqslant \alpha > \dfrac{1}{2}, \\[2ex] Ac(\alpha, s) p^{-\alpha} (\log p)^{s-1+\delta_{\frac{1}{2}, \alpha}}, & \text{for } \dfrac{1}{2} \geqslant \alpha > 0. \end{cases} \tag{5.3}$$

By substituting (5.2) and (5.3) into (5.1), the theorem is immediate.

Remark. For the case $\alpha > 1$, we propose to use the uniformly distributed sequence of sets (1.6) and the following result holds true: Let p be a prime and $n = p^2$. Then we obtain

$$\sup_{f \in E_s^\alpha(C)} \left| I(f) - \frac{1}{n} \sum_{a=1}^{p} \sum_{j=1}^{p} f\left(\frac{a}{p}, \frac{aj}{p}, \cdots, \frac{aj^{s-1}}{p}\right) \right| \leqslant C \cdot c(\alpha, s) n^{-\frac{1}{2}}.$$

Proof. Evidently, we have for $f \in E_s^\alpha(C)$ that

$$\frac{1}{n} \sum_{a=1}^{p} \sum_{j=1}^{p} f\left(\frac{a}{p}, \frac{aj}{p}, \cdots, \frac{aj^{s-1}}{p}\right)$$

$$= \frac{1}{p^2} \sum C(\boldsymbol{m}) \sum_{a=1}^{p} \sum_{j=1}^{p} e^{2\pi i(m_1 + m_2 j + \cdots + m_s j^{s-1})a/p}$$

$$= I(f) + p^{-1} \sum{}' C(\boldsymbol{m}) \sum_{\substack{m_1 + \cdots + m_s j^{s-1} \equiv 0 \ (\mathrm{mod}\ p) \\ 1 \leqslant j \leqslant p}} 1.$$

If at least one of m's is not dividing by p, then the number of solutions of the congruence

$$m_1 + \cdots + m_s j^{s-1} \equiv 0 (\mathrm{mod}\ p) \quad (1 \leqslant j \leqslant p)$$

is at most $s - 1$. Therefore

$$\sup_{f \in E_s^\alpha(C)} \left| I(f) - \frac{1}{n} \sum_{a=1}^{p} \sum_{j=1}^{p} f\left(\frac{a}{p}, \frac{aj}{p}, \cdots, \frac{aj^{s-1}}{p}\right) \right|$$

$$\leqslant C p^{-1} \sum{}' \frac{1}{\|\boldsymbol{m}\|^a} \sum_{\substack{m_1 + \cdots + m_s j^{s-1} \equiv 0 \ (\mathrm{mod}\ p) \\ 1 \leqslant j \leqslant p}} 1$$

$$\leqslant C \cdot (s - 1) p^{-1} \sum \frac{1}{\|\boldsymbol{m}\|^a} + C \sum{}' \frac{1}{\|p\boldsymbol{m}\|^a}$$

$$\leqslant C \cdot 2s p^{-1} \sum \frac{1}{\|\boldsymbol{m}\|^a} \leqslant C \cdot c(\alpha, s) n^{-\frac{1}{2}}.$$

The assertion is proved.

VI.   INTERPOLATIONS

We use the notations

$$\Delta = \sup_{f \in \varOmega_s^\alpha(B)} \left\| f(\boldsymbol{x}) - \frac{1}{n} \sum_{j=1}^{n} f\left(\frac{j\boldsymbol{a}}{n}\right) \sum_{\|\boldsymbol{m}\| \leqslant n_1} e^{2\pi i\left(\boldsymbol{m}, \boldsymbol{x} - \frac{j\boldsymbol{a}}{n}\right)} \right\|_{L_2}.$$

Theorem 6.1. *If the congruence (4.1) has no solution in the domain (4.2), then it yields*

$$\Delta \leqslant B \cdot s!^{1/2} c(\alpha, \varepsilon)^s M^{-\nu(a)a + \varepsilon},$$

where

$$n_1 = [M^{\nu(a)}],$$

in which $\nu(\alpha)$ is defined by (1.8).

Lemma 6.1. *Let $l = (l_1, \cdots, l_s)$ be a vector with integral components such that $\|l\| \geqslant 3^s$. Further let N satisfy $1 \leqslant N \leqslant \dfrac{\|l\|}{3^s}$. Then*

$$\sum_{\|m\| \leqslant N} \frac{1}{\|l + m\|^\alpha} < \begin{cases} s! \, c(\alpha, \, \varepsilon)^s \dfrac{N^{1+\varepsilon}}{\|l\|^\alpha}, & \text{for } 1 \geqslant \alpha > 0, \\[3mm] s! \, c(\alpha)^s \dfrac{N^\alpha}{\|l\|^\alpha}, & \text{for } \alpha > 1. \end{cases}$$

Proof. Suppose first that $1 \geqslant \alpha > 0$. Since $N \leqslant \dfrac{l_1}{3}$ for $s = 1$, we have

$$\sum_{\overline{m}_1 \leqslant N} \frac{1}{(\overline{l_1 + m_1})^\alpha} \leqslant \left(\frac{3}{2}\right)^\alpha \frac{3N}{l_1^\alpha}.$$

Thus the lemma is true for $s = 1$. We suppose that k is a positive integer and that the lemma holds for $s = 1, \cdots, k$. Now we proceed to prove the validity of the lemma for $s = k + 1$.

Clearly, it follows from $\overline{m}_1 \cdots \overline{m}_{k+1} \leqslant N \leqslant \dfrac{l_1 \cdots l_{k+1}}{3^{k+1}}$ that there exists at least an m_i such that $\overline{m}_i < \dfrac{l_i}{2}$, where $1 \leqslant i \leqslant k + 1$. Hence

$$\sum_{\overline{m}_1 \cdots \overline{m}_{k+1} \leqslant N} \frac{1}{((\overline{l_1 + m_1}) \cdots (\overline{l_{k+1} + m_{k+1}}))^\alpha} \leqslant \Sigma_1 + \cdots + \Sigma_{k+1},$$

where

$$\Sigma_i = \sum_{\substack{\overline{m}_1 \cdots \overline{m}_{k+1} \leqslant N \\ \overline{m}_i < \frac{l_i}{2}}} \frac{1}{((\overline{l_1 + m_1}) \cdots (\overline{l_{k+1} + m_{k+1}}))^\alpha}, \quad (1 \leqslant i \leqslant k + 1).$$

1) Suppose that $N \leqslant \dfrac{l_2 \cdots l_{k+1}}{3^k}$. Then by our inductive hypothesis, we have

$$\begin{aligned}
\Sigma_1 &\leqslant \sum_{\overline{m}_1 \leqslant N} \frac{2^\alpha}{l_1^\alpha} \sum_{\overline{m}_2 \cdots \overline{m}_{k+1} \leqslant \frac{N}{\overline{m}_1}} \frac{1}{((\overline{l_2 + m_2}) \cdots (\overline{l_{k+1} + m_{k+1}}))^\alpha} \\[2mm]
&\leqslant \frac{k! \, c(\alpha, \, \varepsilon)^k N^{1+\varepsilon}}{(l_1 \cdots l_{k+1})^\alpha} \sum_{\overline{m}_1 \leqslant N} \frac{1}{(\overline{m}_1)^{1+\varepsilon}} \\[2mm]
&\leqslant \frac{k! \, c(\alpha, \, \varepsilon)^{k+1} N^{1+\varepsilon}}{(l_1 \cdots l_{k+1})^\alpha}.
\end{aligned}$$

2) Suppose that $N > \dfrac{l_2 \cdots l_{k+1}}{3^k}$. Then

$$\Sigma_2 \leqslant \sigma_1 + \sigma_2,$$

where

$$\sigma_1 = \frac{2^\alpha}{l_1^\alpha} \sum_{\overline{m}_1 \leqslant \frac{3^k N}{l_2 \cdots l_{k+1}}} \sum_{\overline{m}_2 \cdots \overline{m}_{k+1} \leqslant \frac{N}{\overline{m}_1}} \frac{1}{((\overline{l_2 + m_2}) \cdots (\overline{l_{k+1} + m_{k+1}}))^\alpha},$$

and

$$\sigma_2 = \frac{2^{\alpha}}{\overline{l}_1^{\alpha}} \sum_{\frac{3^k N}{\overline{l}_2 \cdots \overline{l}_{k+1}} < \overline{m}_1 \leqslant \frac{\overline{l}_1}{2}} \sum_{\overline{m}_2 \cdots \overline{m}_{k+1} \leqslant \frac{N}{\overline{m}_1}} \frac{1}{((\overline{l}_2 + m_2) \cdots (\overline{l}_{k+1} + m_{k+1}))^{\alpha}}.$$

It is evidently known that

$$\sigma_1 \leqslant \frac{2^{\alpha}}{\overline{l}_1^{\alpha}} \sum_{\overline{m}_1 \leqslant \frac{3^k N}{\overline{l}_2 \cdots \overline{l}_{k+1}}} \sum_{\overline{m}_2 \cdots \overline{m}_{k+1} \leqslant \frac{N}{\overline{m}_1}} \frac{(\overline{m}_2 \cdots \overline{m}_{k+1})^{1-\alpha+\varepsilon}}{(\overline{m}_2 \cdots \overline{m}_{k+1})^{1+\varepsilon}}$$

$$\leqslant \frac{c(\alpha, \varepsilon)^k N^{1-\alpha+\varepsilon}}{\overline{l}_1^{\alpha}} \sum_{\overline{m}_1 \leqslant \frac{3^k N}{\overline{l}_2 \cdots \overline{l}_{k+1}}} \frac{(\overline{m}_1)^{\alpha}}{(\overline{m}_1)^{1+\varepsilon}}$$

$$\leqslant c(\alpha, \varepsilon)^k \frac{N^{1+\varepsilon}}{(\overline{l}_2 \cdots \overline{l}_{k+1})^{\alpha}} \sum_{m_1=-\infty}^{\infty} \frac{1}{(\overline{m}_1)^{1+\varepsilon}}$$

$$\leqslant c(\alpha, \varepsilon)^{k+1} \frac{N^{1+\varepsilon}}{(\overline{l}_1 \cdots \overline{l}_{k+1})^{\alpha}}.$$

By our inductive hypothesis, we have

$$\sigma_2 \leqslant \frac{k! \, c(\alpha, \varepsilon)^k N^{1+\varepsilon}}{(\overline{l}_1 \cdots \overline{l}_{k+1})^{\alpha}} \sum_{m_1=-\infty}^{\infty} \frac{1}{(\overline{m}_1)^{1+\varepsilon}}$$

$$\leqslant \frac{k! \, c(\alpha, \varepsilon)^{k+1} N^{1+\varepsilon}}{(\overline{l}_1 \cdots \overline{l}_{k+1})^{\alpha}}.$$

3)　From 1) and 2) we obtain

$$\Sigma_1 \leqslant \frac{k! \, c(\alpha, \varepsilon)^{k+1} N^{1+\varepsilon}}{(\overline{l}_1 \cdots \overline{l}_{k+1})^{\alpha}}.$$

Since the Σ's satisfy the same relation, hence the lemma follows by mathematical induction immediately.

We may treat the case $\alpha > 1$ similarly. The lemma is thus proved.

Remark. For the case $1 \geqslant \alpha > 0$, the conclusion of the above Lemma may be replaced by

$$\sum_{\|\boldsymbol{m}\| \leqslant N} \frac{1}{\|\boldsymbol{l} + \boldsymbol{m}\|^{\alpha}} < s! \, c(\alpha)^s \frac{N(\log 3N)^{s-1}}{\|\boldsymbol{l}\|^{\alpha}}.$$

Lemma 6.2. *Suppose that $Q \geqslant 1$ and $1 \geqslant \alpha > 0$. If the congruence (4.1) has no solution in the domain (4.2), then*

$$\sideset{}{'}\sum_{\substack{(\boldsymbol{a}, \boldsymbol{m}) \equiv 0 \,(\mathrm{mod}\ n) \\ \|\boldsymbol{m}\| \leqslant Q}} \frac{1}{\|\boldsymbol{m}\|^{\alpha}} \leqslant c(\varepsilon)^s Q^{1-\alpha+\varepsilon} M^{-1}.$$

Proof. By Theorem 8.1 of [1], we have

$$\sideset{}{'}\sum_{(\boldsymbol{a}, \boldsymbol{m}) \equiv 0 \,(\mathrm{mod}\ n)} \frac{1}{\|\boldsymbol{m}\|^{1+\varepsilon}} \leqslant c(\varepsilon)^s M^{-1}.$$

Consequently we note

$$\sideset{}{'}\sum_{\substack{(\boldsymbol{a},\,\boldsymbol{m})\equiv 0\ (\mathrm{mod}\ n)\\ \|\boldsymbol{m}\|\leqslant Q}} \frac{1}{\|\boldsymbol{m}\|^{a}} \leqslant Q^{1-a+\varepsilon} \sideset{}{'}\sum_{(\boldsymbol{a},\,\boldsymbol{m})\equiv 0\ (\mathrm{mod}\ n)} \frac{1}{\|\boldsymbol{m}\|^{1+\varepsilon}}$$

$$\leqslant c(\varepsilon)^{s}\, Q^{1-a+\varepsilon}\, M^{-1}.$$

The lemma is proved.

The Proof of Theorem 6.1.   Clearly, we may suppose that $\varepsilon < a\nu(\alpha)$. Take

$$T = \begin{cases} [\log_2 M] + 1, & \text{for } \alpha > 1, \\[2mm] [\log_2 M n_1^{-1+\frac{a}{2}}] + 1, & \text{for } 1 \geqslant \alpha > 0. \end{cases}$$

By Minkowski inequality, we have

$$\Delta \leqslant \Delta_1 + \Delta_2 + \Delta_3, \tag{6.1}$$

where

$$\Delta_1 = \sup_{f \in \mathcal{Q}_s^a(B)} \left\| f(\boldsymbol{x}) - \sideset{}{''}\sum_{t_0 \leqslant T} \varphi_t(\boldsymbol{x}) \right\|_{L_2},$$

$$\Delta_2 = \sup_{f \in \mathcal{Q}_s^a(B)} \left\| \sideset{}{''}\sum_{t_0 \leqslant T} \left(\varphi_t(\boldsymbol{x}) - \frac{1}{n} \sum_{j=1}^{n} \varphi_t\left(\frac{j\boldsymbol{a}}{n}\right) \sum_{\|\boldsymbol{m}\|\leqslant n_1} e^{2\pi i\left(\boldsymbol{m},\,\boldsymbol{x}-\frac{j\boldsymbol{a}}{n}\right)} \right) \right\|_{L_2},$$

and

$$\Delta_3 = \sup_{f \in \mathcal{Q}_s^a(B)} \left\| \frac{1}{n} \sum_{j=1}^{n} f\left(\frac{j\boldsymbol{a}}{n}\right) \sum_{\|\boldsymbol{m}\|\leqslant n_1} e^{2\pi i\left(\boldsymbol{m},\,\boldsymbol{x}-\frac{j\boldsymbol{a}}{n}\right)} \right.$$

$$\left. - \sideset{}{''}\sum_{t_0 \leqslant T} \frac{1}{n} \sum_{j=1}^{n} \varphi_t\left(\frac{j\boldsymbol{a}}{n}\right) \sum_{\|\boldsymbol{m}\|\leqslant n_1} e^{2\pi i\left(\boldsymbol{m},\,\boldsymbol{x}-\frac{j\boldsymbol{a}}{n}\right)} \right\|_{L_2}.$$

1)   By Minkowski inequality and Lemma 2.2, we have

$$\Delta_1 \leqslant \sup_{f \in \mathcal{Q}_s^a(B)} \sideset{}{''}\sum_{t_0 > T} \|\varphi_t\|_{L_2} \leqslant B \sideset{}{''}\sum_{t_0 > T} 2^{-(a-\varepsilon)t_0 - \varepsilon t_0}$$

$$\leqslant Bc(\alpha, \varepsilon)^{s}\, 2^{-(a-\varepsilon)T} \leqslant Bc(\alpha, \varepsilon)^{s}\, M^{-\nu(a)a+\varepsilon}. \tag{6.2}$$

2) $\Delta_2 \leqslant \sigma_1 + \sigma_2$, where

$$\sigma_1 = \sup_{f \in \mathcal{Q}_s^a(B)} \left\| \sideset{}{''}\sum_{t_0 \leqslant T} \sum_{\|\boldsymbol{m}\|\leqslant n_1} \left(C_t(\boldsymbol{m}) - \frac{1}{n} \sum_{j=1}^{n} \varphi_t\left(\frac{j\boldsymbol{a}}{n}\right) e^{-2\pi i\left(\boldsymbol{m},\,\frac{j\boldsymbol{a}}{n}\right)} \right) e^{2\pi i(\boldsymbol{m},\,\boldsymbol{x})} \right\|_{L_2},$$

and

$$\sigma_2 = \sup_{f \in \mathcal{Q}_s^a(B)} \left\| \sideset{}{''}\sum_{t_0 \leqslant T} \sum_{\|\boldsymbol{m}\|> n_1} C_t(\boldsymbol{m})\, e^{2\pi i(\boldsymbol{m},\,\boldsymbol{x})} \right\|_{L_2}.$$

Since

$$C_t(\boldsymbol{m}) - \frac{1}{n} \sum_{j=1}^{n} \varphi_t\left(\frac{j\boldsymbol{a}}{n}\right) e^{-2\pi i\left(\boldsymbol{m},\,\frac{j\boldsymbol{a}}{n}\right)} = - \sum_{(\boldsymbol{a},\,\boldsymbol{l})\equiv 0\ (\mathrm{mod}\ n)} C_t(\boldsymbol{l}+\boldsymbol{m})$$

and

$$\|\boldsymbol{l}\| \leqslant 2^{s}\,\|\boldsymbol{m}\| \cdot \|\boldsymbol{l}+\boldsymbol{m}\|,$$

therefore we have

$$\sigma_1^2 \leqslant \sup_{f \in Q_s^a(B)} \left\| \sideset{}{''}\sum_{t_0 \leqslant T} \sum_{\|m\| \leqslant n_1} \sideset{}{'}\sum_{(a,\,l)\equiv 0 \ (\mathrm{mod}\ n)} C_t(l+m)\, e^{2\pi i (m,\,x)} \right\|_{L_2}^2$$

$$= \sup_{f \in Q_s^a(B)} \sum_{\|m\| \leqslant n_1} \left(\sideset{}{''}\sum_{t_0 \leqslant T} \sideset{}{'}\sum_{(a,\,l)\equiv 0 \ (\mathrm{mod}\ n)} |C_t(l+m)| \right)^2$$

$$\leqslant B^2\, c(\alpha)^s \sum_{\|m\| \leqslant n_1} \left(\sideset{}{'}\sum_{\substack{(a,\,l)\equiv 0 \ (\mathrm{mod}\ n) \\ \|l\| \leqslant 2^{s+T} n_1}} \frac{1}{\|l+m\|^\alpha} \right)^2$$

$$\leqslant B^2\, c(\alpha)^s n_1^a \sum_{\|m\| \leqslant n_1} \sideset{}{'}\sum_{\substack{(a,\,l)\equiv 0 \ (\mathrm{mod}\ n) \\ \|l\| \leqslant 2^{s+T} n_1}} \frac{1}{\|l+m\|^\alpha} \sideset{}{'}\sum_{\substack{(a,\,l')\equiv 0 \ (\mathrm{mod}\ n) \\ \|l'\| \leqslant 2^{s+T} n_1}} \frac{1}{\|l'\|^\alpha} .$$

Evidently, we may suppose that $n_1 \leqslant \dfrac{M}{3^s}$. Hence it follows by Theorem 8.1 of [1], Lemma 6.1 and Lemma 6.2 that

$$\sigma_1^2 \leqslant \begin{cases} B^2 s!\, c(\alpha, \varepsilon)^s n_1^{2a} M^{-2a+2\varepsilon}, & \text{for } \alpha > 1, \\ B^2 s!\, c(\alpha, \varepsilon)^s n_1^{3-a+\varepsilon} M^{-2} 2^{2T(1-a)+T\varepsilon}, & \text{for } 1 \geqslant \alpha > 0. \end{cases}$$

By Schwarz inequality, we have

$$\sigma_2^2 = \sup_{f \in Q_s^a(B)} \sum_{\|m\| > n_1} \left(\sideset{}{''}\sum_{t_0 \leqslant T} |C_t(m)| \right)^2$$

$$\leqslant \sup_{f \in Q_s^a(B)} \sum_{\|m\| > n_1} \sideset{}{''}\sum_{t_0 \leqslant T} 2^{-\frac{\varepsilon t_0}{2}} \sideset{}{''}\sum_{t_0 \leqslant T} 2^{\frac{\varepsilon t_0}{2}} |C_t(m)|^2$$

$$\leqslant c(\varepsilon)^s \sup_{f \in Q_s^a(B)} \sideset{}{''}\sum_{t_0 \leqslant T} \sum_{\|m\| > n_1} 2^{\frac{\varepsilon t_0}{2}} |C_t(m)|^2$$

$$\leqslant c(\varepsilon)^s \sup_{f \in Q_s^a(B)} \sideset{}{''}\sum_{t_0 > \log_2 n_1} 2^{\frac{\varepsilon t_0}{2}} \|\varphi_t\|_{L_2}^2$$

$$\leqslant B^2 c(\varepsilon)^s \sideset{}{''}\sum_{t_0 > \log_2 n_1} 2^{-2a t_0 + \frac{\varepsilon t_0}{2}} \leqslant B^2 c(\alpha, \varepsilon)^s n_1^{-2a+\varepsilon} .$$

Therefore

$$\Delta_2 \leqslant B \cdot s!^{1/2} c(\alpha, \varepsilon)^s M^{-\nu(a)a+\varepsilon} . \tag{6.3}$$

3) $$\quad \Delta_3 = \sup_{f \in Q_s^a(B)} \left\| \sideset{}{''}\sum_{t_0 > T} \frac{1}{n} \sum_{j=1}^n \varphi_t\!\left(\frac{ja}{n}\right) \sum_{\|m\| \leqslant n_1} e^{2\pi i \left(m,\, x - \frac{ja}{n}\right)} \right\|_{L_2}$$

$$\leqslant \sideset{}{''}\sum_{t_0 > T} \sup_{f \in Q_s^a(B)} \|\varphi_t\| \left(\sum_{\|m\| \leqslant n_1} 1 \right)^{\frac{1}{2}} \leqslant B \sideset{}{''}\sum_{t_0 > T} 2^{-a t_0} \left(\sum_{\|m\| \leqslant n_1} \frac{n_1^{1+\varepsilon}}{\|m\|^{1+\varepsilon}} \right)^{\frac{1}{2}}$$

$$\leqslant B\, c(\alpha, \varepsilon)^s n_1^{\frac{1}{2}+\frac{\varepsilon}{2}} 2^{-aT+\frac{\varepsilon T}{2}} \leqslant B\, c(\alpha, \varepsilon)^s M^{-\nu(a)a+\varepsilon} . \tag{6.4}$$

The theorem follows from (6.1), (6.2), (6.3) and (6.4).

Remark. The conclusion of the above theorem may be replaced by

$$\Delta \leqslant B\, c(\alpha, s)\, M^{-\nu(a)a} (\log 3M)^{\nu_1(a)},$$

where

$$\nu_1(\alpha) = \begin{cases} s - 1, & \text{for } \alpha > 1, \\[2mm] \dfrac{(1 - 5\alpha + \alpha^2)(s - 1) + 2\alpha^2 \delta_{1,a}}{1 + 4\alpha - \alpha^2}, & \text{for } 1 \geq \alpha > 0. \end{cases}$$

Now we apply the sequences of sets (1.2) of [1] to the problem of interpolation. We use the notations $\boldsymbol{h} = (1, h_1, \cdots, h_q)$ where $(h_1, \cdots, h_q; n)$ denotes the set of integers defined by (1.2) of [1]. Then we have

Theorem 6.2. *We have*

$$\Delta = \sup_{f \in H_r^\alpha(A)} \left\| f(\boldsymbol{x}) - \frac{1}{n} \sum_{j=1}^n f\left(\frac{j\boldsymbol{h}}{n}\right) \sum_{\|\boldsymbol{m}\| \leq n_1} e^{-2\pi i\left(\boldsymbol{m}, \frac{j\boldsymbol{h}}{n}\right)} e^{2\pi i(\boldsymbol{m}, \boldsymbol{x})} \right\|_{L_2}$$

$$\leq A\, c(\mathfrak{R}_r, \varepsilon)\, n^{\frac{-\nu(a)\alpha(q+1)}{2q} + \varepsilon},$$

where

$$n_1 = \left[n^{\frac{\nu(a)(q+1)}{2q}} \right].$$

VII. CONTINUATION

Theorem 7.1. *Let p be a prime and $n_1 = [p^{\nu(a)}]$. Then there exists a vector $\boldsymbol{a} = (1, a, \cdots, a^{s-1})$, where a is an integer, such that*

$$\sup_{f \in Q_s^\alpha(B)} \left\| f(\boldsymbol{x}) - \frac{1}{p} \sum_{j=1}^p f\left(\frac{j\boldsymbol{a}}{n}\right) \sum_{\|\boldsymbol{m}\| \leq n_1} e^{-2\pi i\left(\boldsymbol{m}, \frac{j\boldsymbol{a}}{n}\right)} e^{2\pi i(\boldsymbol{m}, \boldsymbol{x})} \right\|_{L_2}$$

$$\leq B \cdot s!^{\frac{1}{2}} c(\alpha, \varepsilon)^s p^{-\nu(a)a + \varepsilon}.$$

Proof. By Theorem 6.1, it is sufficient to prove that there exists an integer a such that the congruence

$$(\boldsymbol{a}, \boldsymbol{m}) = m_1 + m_2 a + \cdots + m_s a^{s-1} \equiv 0 \pmod{p} \tag{7.1}$$

has no solution in the domain

$$\|\boldsymbol{m}\| \leq c(s, \varepsilon) p^{1-\varepsilon}, \quad \boldsymbol{m} \neq \boldsymbol{o}, \tag{7.2}$$

where

$$c(s, \varepsilon) = \left(2 \sum_{m=-\infty}^{\infty} \frac{1}{(\bar{m})^{1+\varepsilon}} \right)^{-s}.$$

Evidently, we may suppose that $c(s, \varepsilon) p^{1-\varepsilon} \geq 1$. Take a vector belonging to the domain (7.2). Then the number of solutions of the congruence (7.1) is at most $s - 1$ in the interval $1 \leq a \leq p$. Consequently, the total number of solutions of the congruence (7.1) in domain (7.2) and $1 \leq a \leq p$ is at most

$$\sideset{}{'}\sum_{\|\boldsymbol{m}\| \leq c(s, \varepsilon) p^{1-\varepsilon}} \sum_{\substack{(\boldsymbol{a}, \boldsymbol{m}) \equiv 0 \pmod{p} \\ 1 \leq a \leq p}} 1 \leq (s - 1) \sideset{}{'}\sum_{\|\boldsymbol{m}\| \leq c(s, \varepsilon) p^{1-\varepsilon}} \frac{\|\boldsymbol{m}\|^{1+\varepsilon}}{\|\boldsymbol{m}\|^{1+\varepsilon}}$$

$$\leq (s - 1) c(s, \varepsilon) p \left(\sum_{m=-\infty}^{\infty} \frac{1}{(\bar{m})^{1+\varepsilon}} \right)^s < \frac{p}{2}.$$

Hence, there exists an integer a satisfying $1 \leqslant a \leqslant p$ such that the congruence (7.1) has no solution in (7.2). The theorem follows.

Theorem 7.2. *Let n be an integer $\geqslant 3$. Then for any integer satisfying $n > n_1 \geqslant 1$ and any vector with integral components $\boldsymbol{a} = (a_1, \cdots, a_s)$, we have*

$$\Delta = \sup_{f \in H_s^\alpha(A)} \left\| f(\boldsymbol{x}) - \frac{1}{n} \sum_{j=1}^n f\left(\frac{j\boldsymbol{a}}{n}\right) \sum_{\|\boldsymbol{m}\| \leqslant n_1} e^{-2\pi i \left(\boldsymbol{m}, \frac{j\boldsymbol{a}}{n}\right)} e^{2\pi i (\boldsymbol{m}, \boldsymbol{x})} \right\|_{L_2} \geqslant \frac{A}{(4\pi)^{\alpha+1}} n^{-\alpha/2}.$$

Proof. Evidently, we may suppose that $a_1 = -1$ and $(a_2, n) = 1$. Let $\frac{p_t}{q_t}$ denote the t-th convergence of $\frac{p_h}{q_h} = \frac{a_2}{n}$. Suppose that n_1 satisfies

$$1 = q_0 < \cdots < q_k \leqslant n_1 < q_{k+1} \leqslant \cdots \leqslant q_h = n.$$

Let

$$K = (p_h q_k - q_h p_k).$$

It then follows

$$|K| \leqslant \frac{q_h}{q_{k+1}} \leqslant \frac{n}{n_1}.$$

Take a function of $H_s^\alpha(A)$:

$$f(\boldsymbol{x}) = \frac{A}{4(2\pi)^{\alpha+1}} \left(\frac{e^{2\pi i(Kx_1 + x_2)}}{|K|^\alpha} + \frac{e^{-2\pi i(Kx_1 + x_2)}}{|K|^\alpha} + \frac{e^{2\pi i(n_1+1)x_1}}{(n_1 + 1)^\alpha} + \frac{e^{-2\pi i(n_1+1)x_1}}{(n_1 + 1)^\alpha} \right).$$

Then we have

$$\Delta^2 \geqslant \sum_{\overline{m}_1 \overline{m}_2 \leqslant n_1} \left| \sum_{l_1 \equiv a_2 l_2 \ (\mathrm{mod} \ n)}' C(l_1 + m_1, l_2 + m_2) \right|^2 + 2 \left(\frac{A}{4(2\pi)^{\alpha+1}} \right)^2 (n_1 + 1)^{-2\alpha}$$

$$\geqslant \sum_{\overline{m}_1 \overline{m}_2 \leqslant n_1} (C(K + m_1, q_k + m_2) + C(-K + m_1, -q_k + m_2))^2$$

$$+ 2 \left(\frac{A}{4(2\pi)^{\alpha+1}} \right)^2 (n_1 + 1)^{-2\alpha} \geqslant C(K, 1)^2 + C(-K, -1)^2$$

$$+ 2 \left(\frac{A}{4(2\pi)^{\alpha+1}} \right)^2 (n_1 + 1)^{-2\alpha} \geqslant 2 \left(\frac{A}{4(2\pi)^{\alpha+1}} \right)^2 (n^{-2\alpha} n_1^{2\alpha} + (n_1 + 1)^{-2\alpha})$$

$$\geqslant 4 \left(\frac{A}{4(2\pi)^{\alpha+1}} \right)^2 n^{-\alpha} \left(\frac{n_1}{n_1 + 1} \right)^\alpha \geqslant \left(\frac{A}{(4\pi)^{\alpha+1}} \right)^2 n^{-\alpha}.$$

The theorem follows.

Remark. It follows from Theorem 7.2 that Theorem 7.1 does not permit further substantial improvements, apart from some possible improvements about the order p^ε for the case $\alpha > 1$.

VIII.   THE APPROXIMATE SOLUTION OF FREDHOLM INTEGRAL
EQUATION OF THE SECOND TYPE

For simplicity, we use capital Latin letter to denote a vector of s-dimensional space. Now we study the approximate solution of Fredholm integral equation of the second type

$$\varphi(P) = \lambda \int_{G_s} K(P, Q)\varphi(Q)dQ + f(P), \tag{8.1}$$

where $f(P) \in H_s^a(A)$ and $K(P, Q) \in H_{2s}^a(A)$.

Let

$$D(\lambda) = 1 + \sum_{\nu=1}^{\infty} \frac{(-1)^\nu}{\nu!} \lambda^\nu \int_{G_{\nu s}} K\begin{pmatrix} P_1, & \cdots, & P_\nu \\ P_1, & \cdots, & P_\nu \end{pmatrix} dP_1 \cdots dP_\nu$$

denote the Fredholm kernel, where

$$K\begin{pmatrix} P_1, & \cdots, & P_\nu \\ Q_1, & \cdots, & Q_\nu \end{pmatrix} = \det\left(K(P_i, Q_j)\right), \quad (1 \leqslant i, j \leqslant \nu).$$

Further let

$$\Delta(\lambda) = \det\left(\delta_{ij} - \frac{\lambda}{n} K(M_i, M_j)\right), \quad (1 \leqslant i, j \leqslant n).$$

We assume, moreover, that $D(\lambda) \neq 0$.

Theorem 8.1. *If the quadrature formula*

$$\sup_{F \in H_s^a(A)} \left| \int_{G_s} F(P)dP - \frac{1}{n} \sum_{j=1}^{n} F(M_j) \right| \leqslant Ac(\alpha, s)\varphi(n) \tag{8.2}$$

holds with $\varphi(n) = o(1)$, *and if* $\tilde{\varphi}(M_k)$ *denotes the solution of the system of linear equations,*

$$\tilde{\varphi}(M_j) = \frac{\lambda}{n} \sum_{k=1}^{n} K(M_j, M_k)\tilde{\varphi}(M_k) + f(M_j) \quad (1 \leqslant j \leqslant n). \tag{8.3}$$

Then the solution of the equation (8.1) *can be written as*

$$\varphi(P) = f(P) + \sum_{j=1}^{n} K(P, M_j)\tilde{\varphi}(M_j) + O(\varphi(n)),$$

where the constant implied by the symbol "O" depends only on λ, K *and* f.

Lemma 8.1. *Let* a_{ij} *be real numbers and let*

$$\tilde{A}_n(k) = \det(a'_{ij}), \quad (1 \leqslant i, j \leqslant n),$$

where $a'_{ii} = 1 + a_{ii}(1 \leqslant i \leqslant k)$ *and otherwise* $a'_{ij} = a_{ij}$. *Further let* $A_n(k)$ *denote the least upper bound of the absolute value of* $\tilde{A}_n(k)$ *under the condition*

$$|a_{ij}| \leqslant \frac{\gamma}{n},$$

where γ *is a constant. Then there exist two constants* γ_1 *and* γ_2 *such that for any positive integer* n, *we have*

$$A_n(n-1) \leqslant \frac{\gamma_1}{n} \text{ and } A_n(n) \leqslant \gamma_2.$$

Lemma 8.2. *If the integral formula* (8.2) *holds, then there exists constant* $n_0 = n_0(\lambda, A, \alpha, s)$, *such that for* $n > n_0$, *we have*

$$|\Delta(\lambda)| \geqslant \frac{1}{2}\,|D(\lambda)|.$$

See Коробов[7].

Lemma 8.3. *The solution of Eq.* (8.1) *belongs to* $H_s^\alpha(A')$, A' *being a constant which depends on* λ, K *and* f *only.*

Proof. From

$$\varphi(P) - f(P) = \lambda \int_{G_s} K(P, Q)\varphi(Q)dQ,$$

we have

$$|(\varphi(P)-f(P))^{(a\vartheta_1, \cdots, a\vartheta_s)}| \leqslant \sup_{Q \in G_s} |K(P, Q)^{(a\vartheta_1, \cdots, a\vartheta_s)}|\,|\lambda| \int_{G_s} |\varphi(Q)|dQ \leqslant A'',$$

where $\vartheta_1, \cdots, \vartheta_s = 0$ or 1. Hence $\varphi(P) - f(P) \in H_s^\alpha(A'')$. Consequently

$$\varphi(P) = f(P) + (\varphi(P) - f(P)) \in H_s^\alpha(A').$$

The lemma is proved.

The Proof of Theorem 8.1. By Lemma 8.3, we have $K(P, Q)\varphi(Q) \in H_s^\alpha(A')$. Therefore we have

$$\varphi(P) = \frac{\lambda}{n} \sum_{k=1}^{n} K(P, M_k)\varphi(M_k) + f(P) + O(\varphi(n)).$$

Consequently

$$\varphi(M_j) = \frac{\lambda}{n} \sum_{k=1}^{n} K(M_j, M_k)\varphi(M_k) + f(M_j) + O(\varphi(n)) \quad (1 \leqslant j \leqslant n).$$

From (8.3), we have the system of linear equations

$$z_j = \sum_{k=1}^{n} a_{jk}z_k + b_j \quad (1 \leqslant j \leqslant n),$$

where

$$z_j = \varphi(M_j) - \tilde{\varphi}(M_j), \quad a_{jk} = \frac{\lambda}{n} K(M_j, M_k), \quad b_j = O(\varphi(n)).$$

Let $\Delta_k(\lambda)$ be the determinant obtained from $\Delta(\lambda)$ by replacing its k column

$$\left(-\frac{\lambda}{n} K(M_1, M_k), \cdots, 1 - \frac{\lambda}{n} K(M_k, M_k), \cdots, -\frac{\lambda}{n} K(M_n, M_k)\right)'$$

with

$$(b_1, \cdots, b_n)'.$$

Then we have

$$z_j = \frac{\Delta_j(\lambda)}{\Delta(\lambda)}, \quad (1 \leqslant j \leqslant n).$$

As n sufficiently large, we have

$$|\Delta(\lambda)| > \frac{1}{2}\,|D(\lambda)| > 0$$

by Lemma 8:2. Further since

$$\left|\frac{\lambda}{n}\,K(M_j,\,M_k)\right| \leqslant \frac{|\lambda|\,A}{n},$$

therefore by Lemma 8.1, we have

$$|\Delta_j| \leqslant |b_j B_j| + \sum_{\substack{1 \leqslant k \leqslant n \\ k \neq j}} |b_k B_k| \leqslant \gamma_2 |b_j| + \frac{\gamma_1}{n} \sum_{k=1}^{n} |b_k| = O(\varphi(n)),$$

where B_k is the cofactor of b_k in Δ_j.   Hence

$$z_j = O(\varphi(n)), \quad (1 \leqslant j \leqslant n).$$

The theorem follows.

Let

$$M_j = \left(\left\{\frac{j}{n}\right\},\ \left\{\frac{h_1 j}{n}\right\},\ \cdots,\ \left\{\frac{h_q j}{n}\right\}\right) \quad (1 \leqslant j \leqslant n)$$

be the sequence of sets defined by (1.2) of [1].   Then from Theorem 4 of [1] and Theorem 8.1, we have

Theorem 8.2. *Suppose that $s = r$.   Then the solution of the equation (8.1) can be written as*

$$\varphi(P) = f(P) + \frac{\lambda}{n} \sum_{j=1}^{n} K(P,\,M_j)\tilde{\varphi}(M_j) + O(n^{-\frac{\alpha}{2}-\frac{\alpha}{2q}+\varepsilon}),$$

where $\tilde{\varphi}(M_j)$ denotes the solutions of the system of linear equations (8.3) and the constant implied by the symbol "O" depends on $\lambda, K, f, \Re_r$ and ε only.

IX.   The Proof of Theorem 4

The solution of the equation

$$\varphi(x) = \int_0^x K(x,\,y)\varphi(y)dy + f(x) \tag{9.1}$$

is given by the Neumann series

$$\varphi(x) = f(x) + \sum_{\nu=1}^{\infty} \varphi_\nu(x), \tag{9.2}$$

where

$$\varphi_\nu(x) = \int_0^x \int_0^{x_1} \cdots \int_0^{x_{\nu-1}} G_\nu(x;\,x_1,\,\cdots,\,x_\nu)dx_1 \cdots dx_\nu,$$

$$G_\nu(x;\,x_1,\,\cdots,\,x_\nu) = K(x,\,x_1)K(x_1,\,x_2)\cdots K(x_{\nu-1},\,x_\nu)f(x_\nu).$$

Since $f \in H_1^\alpha(A)$ and $K(x,\,y) \in H_2^\alpha(A)$, we have

$$G_\nu(x;\,x_1,\,\cdots,\,x_\nu) \in H_{\nu+1}^\alpha(2^{(\alpha+1)(\nu+1)}A^{\nu+1}).$$

Hence

$$|\varphi_\nu(x)| \leqslant 2^{(a+1)(\nu+1)} A^{\nu+1} \int_0^x \cdots \int_0^{x_{\nu-1}} dx_1 \cdots dx_\nu \leqslant \frac{2^{(a+1)(\nu+1)} A^{\nu+1}}{\nu!}.$$

Consequently there exists

$$\left|\sum_{\nu=Q+1}^\infty \varphi_\nu(x)\right| \leqslant \sum_{\nu=Q+1}^\infty \frac{2^{(a+1)(\nu+1)} A^{\nu+1}}{\nu!} \leqslant \frac{c(A,\varepsilon)^Q}{Q!} \leqslant c(A,\alpha,\varepsilon)p^{-\nu(a)a+\varepsilon}. \qquad (9.3)$$

Let

$$\widetilde{G}(x; x_1, \cdots, x_\nu) = \frac{1}{p} \sum_{j=1}^p K\left(x, \frac{j}{p}\right) K\left(\frac{j}{p}, \frac{aj}{p}\right) \cdots K\left(\frac{a^{\nu-2}j}{p}, \frac{a^{\nu-1}j}{p}\right)$$

$$\times f\left(\frac{a^{\nu-1}j}{p}\right) \sum_{\overline{m}_1\cdots\overline{m}_\nu\leqslant g} e^{-2\pi i(m_1+m_2 a+\cdots+m_\nu a^{\nu-1})j/p}\, e^{2\pi i(m_1 x_1+\cdots+m_\nu x_\nu)}.$$

Then it follows by Theorem 7.1 that there exists an integer a such that

$$\left|\varphi_\nu(x) - \int_0^x \cdots \int_0^{x_{\nu-1}} \widetilde{G}_\nu dx_1 \cdots dx_\nu\right| \leqslant \int_0^x \cdots \int_0^{x_{\nu-1}} |G_\nu - \widetilde{G}_\nu| dx_1 \cdots dx_\nu$$

$$\leqslant \left(\int_0^x \cdots \int_0^{x_{\nu-1}} dx_1 \cdots dx_\nu\right)^{\frac{1}{2}} \|G_\nu - \widetilde{G}_\nu\|_{L_2}$$

$$\leqslant A^{\nu+1} c(\alpha,\varepsilon)^\nu p^{-\nu(a)a+\varepsilon} \quad (1 \leqslant \nu \leqslant Q). \qquad (9.4)$$

The theorem follows immediately from (9.2), (9.3) and (9.4).

Remarks. 1. The present method can be used to treat more general integral equation of the type

$$\varphi(x_1, \cdots, x_{s+l}) = \int_0^1 \cdots \int_0^1 \int_0^{x_{s+1}} \cdots \int_0^{x_{s+l}} K(x_1, \cdots, x_{s+l}, y_1, \cdots, y_{s+l})$$

$$\times \varphi(y_1, \cdots, y_{s+l}) dy_1 \cdots dy_{s+l} + f(x_1, \cdots, x_{s+l}),$$

where $s \geqslant 0$, $l \geqslant 1$, $f \in H_{s+l}^a(A)$ and $K \in H_{2s+2l}^a(A)$.

2. The present method can also be used to treat the Fredholm integral equation of the second type when λ is sufficiently small.

3. The sequences of sets defined by (1.2) of [1] cannot be used here, since we don't know the explicit relation between r and $c(\Re_r, \varepsilon)$, where $c(\Re_r, \varepsilon)$ is the constant appeared in Theorem 2 of [1].

REFERENCES

[1] Hua, L. K. & Wang, Y.: On uniform distribution and numerical analysis, (I), *Sci. Sin.*, **16** (1973), 483—505.

[2] Weil, A.: On some exponential sums, *PNAS*, USA, **34** (1948), 204—207.

[3] Бахвалов, Н. С.: Теоремы вложения для классов функций с несколькими ограниченными производными, *Вес. Мос. ун-та*, **3** (1963), 7—16.

[4] Бахвалов, Н. С.: Об оптимальных оценках сходимости к вадратурных процессов и методов интегрирования типа Монте-Карло на классах функций, *Доп к жур. выч. мат. и мат. физ; изд, «Наука». Мос.* (1964), 5—63.

[5] Коробов, Н. М.: Приближенное вычисление кратных интегралов с помощью методов теории чисел, *ДАН СССР*, **115** (1957), 1062—1065.

[6] Коробов, Н. М.: О приближенном вычислении кратных интегралов, *ДАН СССР*, **124** (1959), 1207—1210.

[7] Коробов, Н. М.: Теоретико-числовые методы в приближенном анализе, *Физмат. Лит. Мос.* (1963).

[8] Шахов, Ю. Н.: О приближенном решении много-мерных линейных уравнений Вольтера II рода методом итераций, доп. к жур. выч. мат. и мат. физ; *Изд. «Наука». Мос.* (1964), 75—100.

[9] Шахов, Ю. Н.: О вычислении интегралов растущей кратности, *Жур. выч. мат. и мат. физ;* **5** (1965), 911—916.

Vol. XVIII No. 2   **SCIENTIA SINICA**   Mar. - Apr. 1975

ON UNIFORM DISTRIBUTION AND NUMERICAL ANALYSIS (III)

(NUMBER-THEORETIC METHOD)

HUA LOO-KENG (华罗庚) AND WANG YUAN (王 元)

Received Jan. 3, 1974.

ABSTRACT

We present in this paper, a study of the uniformly distributed sequences of sets defined by elementary symmetric functions of roots of $f(x) = 0$, where $f(x)$ is the minimal polynomial of a Pisot-Vijayaraghavan number (cf. Jacobi-Perron algorithm). We obtain the estimations of their discrepancies and then apply them to the problem of numerical integration. For practical purposes, two suggestions concerning the polynomials $f(x)$ are given.

I. INTRODUCTION

The Pisot-Vijayaraghavan numbers (PV number) are those algebraic integers $\omega > 1$, whose conjugates other than ω itself lie in the open unit circle (Cf. [2] Chap. VIII). Let ω be a PV number of degree $s(\geqslant 2)$ with conjugates $\omega(= \omega^{(1)} = \omega_s^{(1)})$, $\omega^{(2)}, \cdots, \omega^{(s)}$ so that

$$\omega > 1, \quad |\omega^{(2)}| \leqslant |\omega^{(3)}| \leqslant \cdots \leqslant |\omega^{(s)}| < 1. \tag{1.1}$$

Further let ω satisfy the irreducible equation

$$f(\omega) = 0, \quad f(x) = x^s - a_{s-1}x^{s-1} - \cdots - a_1 x - a_0, \tag{1.2}$$

where $a_0, a_1, \cdots, a_{s-1}$ are integers.

The conventions and notations introduced in [3, 4] are also used here.

Theorem 1. *Let* $\mathbf{b} = (b_0, b_1, \cdots, b_{s-1})$ *be an integral vector* $\neq \mathbf{o}$. *Let* $(Q_n) \cdot$ $(= (Q_n^{(s)}))$ *be a sequence of integers defined by the following recurrent formula*

$$Q_i = b_i \quad (0 \leqslant i \leqslant s - 1),$$
$$Q_n = a_{s-1}Q_{n-1} + a_{s-2}Q_{n-2} + \cdots + a_1 Q_{n-s+1} + a_0 Q_{n-s}(n \geqslant s). \tag{1.3}$$

Then for any given large number N, *there exists an* $n_0 = n_0(\omega, \mathbf{b}, N)$. *When* $n > n_0$, *it holds true that*

$$|Q_n| > N \tag{1.4}$$

and

$$\left| \frac{Q_{n+1}}{Q_n} - \omega \right| < c(\omega, \mathbf{b})|Q_n|^{-1-\rho}, \tag{1.5}$$

where

$$\rho = -\frac{\log|\omega^{(s)}|}{\log \omega}. \tag{1.6}$$

It follows from Theorem 1 immediately that

$$\left|\frac{Q_{n+k}}{Q_n} - \omega^k\right| < c(\omega, \mathbf{b})|Q_n|^{-1-\rho} \qquad (1 \leqslant k \leqslant s-1, n > n_0). \qquad (1.7)$$

Since $1, \omega, \cdots, \omega^{s-1}$ is linearly independent over rational field R, hence by the Theorems 6.1, 6.2 (with a slight modification) and 8.1 of [3] and Theorem 4.1 of [4] we have

Theorem 2. *The sequence of sets*

$$\left(\left\{\frac{k}{Q_n}\right\}, \ \left\{\frac{Q_{n+1}k}{Q_n}\right\}, \ \cdots, \ \left\{\frac{Q_{n+s-1}k}{Q_n}\right\}\right) \ \ (1 \leqslant k \leqslant |Q_n|, \ n > n_0) \qquad (1.8)$$

is uniformly distributed and has discrepancy

$$\varphi(|Q_n|) = c(\omega, \mathbf{b}, \varepsilon)|Q_n|^{-\frac{1}{2}-\frac{\rho}{2}+\varepsilon}, \qquad (1.9)$$

where ε is any pre-assigned positive number.

Theorem 3. *Suppose that $\alpha > 1$. Then*

$$\underset{f \in E_s^\alpha(C)}{\mathrm{Sup}} \left|\int_0^1 \cdots \int_0^1 f(x_1, \cdots, x_s)dx_1 \cdots dx_s\right.$$

$$\left. - \frac{1}{|Q_n|}\sum_{k=1}^{|Q_n|} f\left(\frac{k}{Q_n}, \frac{Q_{n+1}k}{Q_n}, \ \cdots, \ \frac{Q_{n+s-1}k}{Q_n}\right)\right|$$

$$\leqslant C \cdot c(\omega, \mathbf{b}, \alpha, \varepsilon)|Q_n|^{-\frac{\alpha}{2}-\frac{\rho\alpha}{2}+\varepsilon}, \quad (n > n_0). \qquad (1.10)$$

Theorem 4. *Suppose that $1 \geqslant \alpha > 0$. Then*

$$\underset{f \in H_s^\alpha(A)}{\mathrm{Sup}} \left|\int_0^1 \cdots \int_0^1 f(x_1, \cdots, x_s)dx_1 \cdots dx_s\right.$$

$$\left. - \frac{1}{|Q_n|}\sum_{k=1}^{|Q_n|} f\left(\frac{k}{Q_n}, \frac{Q_{n+1}k}{Q_n}, \ \cdots, \ \frac{Q_{n+s-1}k}{Q_n}\right)\right|$$

$$\leqslant A \cdot c(\omega, \mathbf{b}, \alpha, \varepsilon)|Q_n|^{-\frac{\alpha}{2}-\frac{\rho\alpha}{2}+\varepsilon}, \quad (n > n_0). \qquad (1.11)$$

For practical purposes, we give two suggestions.

1) Take $a_0 = a_1 = \cdots = a_{s-1} = 1$. We denote the largest real root of the equation

$$F(x) = x^s - x^{s-1} - \cdots - x - 1 = 0, \qquad (1.12)$$

by $\eta(= \eta^{(1)} = \eta_s^{(1)})$ and its other roots by $\eta^{(2)}, \cdots, \eta^{(s)}$. Let $(F_n)(= (F_n^{(s)}))$ be the sequence of integers defined by the recurrent formula

$$F_0 = F_1 = \cdots = F_{s-2} = 0, \quad F_{s-1} = 1,$$

$$F_{n+s} = F_{n+s-1} + F_{n+s-2} + \cdots + F_{n+1} + F_n, \quad (n \geqslant 0). \qquad (1.13)$$

As usual, (F_n) is called the generalized Fibonacci sequence of dimension s. From the theory of Jacobi-Perron algorithm, we have

$$\lim_{n \to \infty} \frac{F_{n+1}}{F_n} = \eta, \tag{1.14}$$

(Cf. [1] Chap. 7). In this paper, we shall prove that η is a PV number and

$$\left| \frac{F_{n+1}}{F_n} - \eta \right| < c(\eta) F_n^{-1 - \frac{1}{2^s \log 2} - \frac{1}{2^{2s+1}}}, \quad (n \geqslant s). \tag{1.15}$$

Consequently, if we take $(Q_n) = (F_n)$ in Theorems 2, 3, 4, then the right hand sides of (1.9), (1.10) and (1.11) become $c(\eta) F_n^{-\frac{1}{2} - \frac{1}{2^{s+1} \log 2}}$, $C \cdot c(\eta, \alpha) F_n^{-\frac{\alpha}{2} - \frac{\alpha}{2^{s+1} \log 2}}$ and $A \cdot c(\eta, \alpha) F_n^{-\frac{\alpha}{2} - \frac{\alpha}{2^{s+1} \log 2}}$ respectively.

2) Take $a_0 = 1, a_1 = \cdots = a_{s-2} = 0$ and $a_{s-1} = L$, where L is an integer $\geqslant 2$. We denote the largest real root of the equation

$$G(x) = x^s - L x^{s-1} - 1 = 0 \tag{1.16}$$

by $\tau (= \tau^{(1)} = \tau_s^{(1)})$ and its other roots by $\tau^{(2)}, \cdots, \tau^{(s)}$. Let $(G_n)(= (G_n^{(s)}))$ be the sequence of integers defined by the recurrent formula

$$G_0 = G_1 = \cdots = G_{s-2} = 0, \quad G_{s-1} = 1,$$
$$G_{n+s} = L G_{n+s-1} + G_n, \quad (n \geqslant 0). \tag{1.17}$$

In this paper, we shall prove that τ is a PV number and

$$\left| \frac{G_{n+1}}{G_n} - \tau \right| < c(\tau) G_n^{-1 - \frac{1}{s-1} + \frac{2}{(s-1)L \log L} - \frac{1}{(s-1)L^{s+3}}}, \quad (n \geqslant s). \tag{1.18}$$

Consequently, if we take $(Q_n) = (G_n)$ in Theorems 2, 3, 4, then the right hand sides of (1.9), (1.10), (1.11) become $c(\tau) G_n^{-\frac{1}{2} - \frac{1}{2(s-1)} + \frac{1}{(s-1)L \log L}}$, $C \cdot c(\tau, \alpha) G_n^{-\frac{\alpha}{2} - \frac{\alpha}{2(s-1)} + \frac{\alpha}{(s-1)L \log L}}$ and $A \cdot c(\tau, \alpha) G_n^{-\frac{\alpha}{2} - \frac{\alpha}{2(s-1)} + \frac{\alpha}{(s-1)L \log L}}$ respectively.

The results given here are slightly rougher than the corresponding results obtained by the real cyclotomic field $\mathfrak{R}_s$ in the previous papers [3, 4]. However, compared with $(h_1, \cdots, h_s; n)$ in [3, 4], the magnitude of elementary operations required for finding (Q_n) is decreased.

Finally, we shall give some numerical results obtained by Wang 520 calculator as an example

$$\operatorname*{Sup}_{f \in E_4^2(C)} \left| \int_0^1 \cdots \int_0^1 f(x_1, \cdots, x_4) dx_1 \cdots dx_4 \right.$$
$$\left. - \frac{1}{F_{19}} \sum_{k=1}^{F_{19}} f\left(\frac{k}{F_{19}}, \frac{F_{20} k}{F_{19}}, \frac{F_{21} k}{F_{19}}, \frac{F_{22} k}{F_{19}} \right) \right| < 0.0054 C, \tag{1.19}$$

where $F_{19} = 10671$, $F_{20} = 20569$, $F_{21} = 39648$ and $F_{22} = 76424$.

II. The Proof of Theorem 1

1) *Elementary Symmetric Functions.* Let

$$S_l = \omega^{(1)l} + \omega^{(2)l} + \cdots + \omega^{(s)l}, \quad l = 1, 2, \cdots. \tag{2.1}$$

It is known that S_l can be evaluated by the following recurrent formula of Newton

$$\begin{cases} S_1 = a_{s-1}, \\ S_2 = a_{s-1}S_1 + 2a_{s-2}, \\ \quad\cdots\cdots\cdots\cdots\cdots \\ S_s = a_{s-1}S_{s-1} + a_{s-2}S_{s-2} + \cdots + a_1 S_1 + s a_0 \end{cases} \tag{2.2}$$

and

$$S_n = a_{s-1}S_{n-1} + a_{s-2}S_{n-2} + \cdots + a_1 S_{n-s+1} + a_0 S_{n-s} \tag{2.3}$$

for $n > s$.

It is evident that

$$|S_n - \omega^n| \leqslant \sum_{i=2}^{s} |\omega^{(i)}|^n < s - 1.$$

Hence we have

$$|S_n| < \omega^n + s - 1 < s\omega^n.$$

Consequently, we have the Diophantine approximation

$$\left| \frac{S_{n+1}}{S_n} - \omega \right| = |S_{n+1} - \omega S_n| \, |S_n|^{-1} = \left| \sum_{i=2}^{s} \omega^{(i)n}(\omega^{(i)} - \omega) \right| |S_n|^{-1}$$

$$< (\omega + 1)(s - 1)|\omega^{(s)}|^n |S_n|^{-1} = (\omega + 1)(s - 1)\omega^{-\rho n}|S_n|^{-1}$$

$$< (\omega + 1)(s - 1)s^\rho |S_n|^{-1-\rho}, \quad (n > n_0). \tag{2.4}$$

2) *Suppose That the Initial Vector* $\mathbf{b} = (0, \cdots, 0, 1)$. Let $l_1, \cdots, l_s$ be a set of non-negative integers and

$$\Delta(l_1, \cdots, l_s) = \begin{vmatrix} \omega^{(1)l_1}, & \omega^{(2)l_1}, & \cdots, & \omega^{(s)l_1} \\ \cdots\cdots\cdots\cdots\cdots\cdots \\ \omega^{(1)l_s}, & \omega^{(2)l_s}, & \cdots, & \omega^{(s)l_s} \end{vmatrix}. \tag{2.5}$$

Let

$$P_l(s - 1) = \frac{\Delta(l, s - 2, \cdots, 1, 0)}{\Delta(s - 1, s - 2, \cdots, 1, 0)}. \tag{2.6}$$

Then $P_l(s - 1)$ is an algebraic integer since $\Delta(l, s - 2, \cdots, 1, 0)$ may be divided by $\Delta(s - 1, s - 2, \cdots, 1, 0)$. Consequently $P_l(s - 1)$ is a rational integer since $P_l(s - 1)$ is a symmetric function of $\omega^{(1)}, \cdots, \omega^{(s)}$.

It is obvious that

$$P_0(s - 1) = P_1(s - 1) = \cdots = P_{s-2}(s - 1) = 0, \quad P_{s-1}(s - 1) = 1. \tag{2.7}$$

Since

$$\omega^{(i)n} = \omega^{(i)n-s}\omega^{(i)s} = \omega^{(i)n-s}(a_{s-1}\omega^{(i)s-1} + a_{s-2}\omega^{(i)s-2} + \cdots + a_1\omega^{(i)} + a_0)$$

$$= a_{s-1}\omega^{(i)n-1} + a_{s-2}\omega^{(i)n-2} + \cdots + a_1\omega^{(i)n-s+1} + a_0\omega^{(i)n-s}, \quad (0 \leqslant i \leqslant s-1),$$

for $n \geqslant s$, we have

$$P_n(s-1) = a_{s-1}P_{n-1}(s-1) + a_{s-2}P_{n-2}(s-1) + \cdots$$
$$+ a_1P_{n-s+1}(s-1) + a_0P_{n-s}(s-1), \quad (n \geqslant s). \tag{2.8}$$

It follows from (2.6) that

$$P_n(s-1) = \sum_{i=1}^{s} \frac{\omega^{(i)n}}{\displaystyle\prod_{j \neq i} (\omega^{(i)} - \omega^{(j)})}, \quad (n \geqslant 0). \tag{2.9}$$

Consequently, we have the Diophantine approximation

$$\left| \frac{P_{n+1}(s-1)}{P_n(s-1)} - \omega \right| < c(\omega)|P_n(s-1)|^{-1-\rho}, \quad (n > n_0). \tag{2.10}$$

3) *The Proof of Theorem 1.* Let

$$P_l(i) = \frac{\Delta(s-1, \cdots, i+1, l, i-1, \cdots, 0)}{\Delta(s-1, \cdots, 0)}, \quad (0 \leqslant i \leqslant s-1). \tag{2.11}$$

Then we have evidently that

$$P_l(i) = \delta_{i,l}, \quad (0 \leqslant i, l \leqslant s-1), \tag{2.12}$$

where $\delta_{i,l}$ denotes Kronecker symbol and

$$P_n(i) = a_{s-1}P_{n-1}(i) + a_{s-2}P_{n-2}(i) + \cdots$$
$$+ a_1P_{n-s+1}(i) + a_0P_{n-s}(i), \quad (0 \leqslant i \leqslant s-1), \tag{2.13}$$

for $n \geqslant s$. Let

$$Q_l = b_0P_l(0) + b_1P_l(1) + \cdots + b_{s-1}P_l(s-1), \quad (l = 0, 1, \cdots). \tag{2.14}$$

Then $Q_l (l = 0, 1, \cdots)$ satisfy

$$Q_i = b_i \quad (0 \leqslant i \leqslant s-1), \tag{2.15}$$

and

$$Q_n = a_{s-1}Q_{n-1} + a_{s-2}Q_{n-2} + \cdots + a_1Q_{n-s+1} + a_0Q_{n-s}, \tag{2.16}$$

for $n \geqslant s$. Let $W_i^{(j)}$ denote the cofactor of $\omega^{(j)i}$ in $\Delta(s-1, \cdots, 0)$. Then

$$P_l(i) = \frac{1}{\Delta(s-1, \cdots, 0)} \sum_{j=1}^{s} W_i^{(j)}\omega^{(j)l}, \quad (0 \leqslant i \leqslant s-1). \tag{2.17}$$

Hence we have from (2.14) that

$$Q_l = \frac{1}{\Delta(s-1, \cdots, 0)} \sum_{j=1}^{s} \omega^{(j)l} \sum_{i=0}^{s-1} b_iW_i^{(j)}, \quad (l = 0, 1, \cdots). \tag{2.18}$$

Since

$$W_i = W_i^{(1)} = (-1)^{s-i+1} \begin{vmatrix} \omega^{(2)s-1}, & \cdots, & \omega^{(s)s-1} \\ \cdots \cdots \cdots \cdots \\ \omega^{(2)i+1}, & \cdots, & \omega^{(s)i+1} \\ \omega^{(2)i-1}, & \cdots, & \omega^{(s)i-1} \\ \cdots \cdots \cdots \cdots \\ 1, & \cdots, & 1 \end{vmatrix}$$

$$= (-1)^{s-i+1}\sigma_{s-1-i}(\omega^{(2)}, \cdots, \omega^{(s)})W_{s-1}, \quad (0 \leqslant i \leqslant s-1), \tag{2.19}$$

where $\sigma_l(\omega^{(2)}, \cdots, \omega^{(s)}) = \sum\limits_{2 \leqslant i_1 < \cdots < i_l \leqslant s} \omega^{(i_1)} \cdots \omega^{(i_l)}$ denotes the elementary symmetric function of $\omega^{(2)}, \cdots, \omega^{(s)}$, therefore we can prove easily that

$$\sigma_l(\omega^{(2)}, \cdots, \omega^{(s)}) = g_l(\omega), \quad (0 \leqslant l \leqslant s-1), \tag{2.20}$$

where $g_l(\omega)$ is a polynomial of degree l and has rational coefficients. Its leading coefficient is equal to ± 1. Hence $W_0, W_1, \cdots, W_{s-1}$ is linearly independent over rational field R, since $1, \omega, \cdots, \omega^{s-1}$ is a set of basis of the algebraic number field $R(\omega)$. Especially, we have

$$b_0 W_0 + \cdots + b_{s-1} W_{s-1} \neq 0. \tag{2.21}$$

It follows from (2.18) that

$$Q_l = \frac{(b_0 W_0 + \cdots + b_{s-1} W_{s-1})\omega^l}{\Delta(s-1, \cdots, 0)} + O(|\omega^{(s)}|^l), \tag{2.22}$$

where the constant implied by the symbol "O" depends on ω and $\mathbf{b}$ only. Hence the theorem follows from (2.21) and (2.22).

Remark 1. For practical purposes, we suggest to take the initial values $Q_0 = Q_1 = \cdots = Q_{s-2} = 0$ and $Q_{s-1} = 1$ as usual (Cf. [1])

Remark 2. For the case $s = 2$, $a_0 = a_1 = 1$ and $\mathbf{b} = (0, 1)$, we have from (2.18) the well-known formula

$$F_l^{(2)} = \frac{1}{\sqrt{5}}\left(\left(\frac{1+\sqrt{5}}{2}\right)^l - \left(\frac{1-\sqrt{5}}{2}\right)^l\right), \quad (l = 0, 1, 2\cdots). \tag{2.23}$$

Hence the formula (2.18) may be recognized as the generalization of (2.23).

III.   The Estimation of η

Lemma 3.1. *We have*

$$2 - \frac{1}{2^{s-1}} < \eta < 2 - \frac{1}{2^s} \tag{3.1}$$

and

$$|\eta^{(i)}| \leqslant \eta - 1, \quad (2 \leqslant i \leqslant s). \tag{3.2}$$

To prove Lemma 3.1, we shall need

Lemma 3.2. *If the coefficients of the polynomial*

$$g(x) = a_s x^s + a_{s-1} x^{s-1} + \cdots + a_1 x + a_0$$

satisfy $a_s \geqslant a_{s-1} \geqslant \cdots \geqslant a_1 \geqslant a_0 > 0$, *then no root of the equation* $g(x) = 0$ *has modulus greater than* 1 (Cf [7]).

Proof. Since

$$|(1 - x)g(x)| \geqslant a_s|x|^{s+1} - ((a_s - a_{s-1})|x|^s + (a_{s-1}$$
$$- a_{s-2})|x|^{s-1} + \cdots + (a_1 - a_0)|x| + a_0) > a_s|x|^s(|x| - 1) > 0$$

for $|x| > 1$. The lemma follows.

The proof of Lemma 3.1. 1) Denote

$$Q(x) = (x - 1)F(x) = x^{s+1} - 2x^s + 1.$$

Then we have

$$Q\left(2 - \frac{1}{2^s}\right) = \left(2 - \frac{1}{2^s}\right)^{s+1} - 2\left(2 - \frac{1}{2^s}\right)^s + 1$$

$$= 1 - \left(2 - \frac{1}{2^s}\right)^s \frac{1}{2^s} = 1 - \left(1 - \frac{1}{2^{s+1}}\right)^s > 0$$

and

$$Q\left(2 - \frac{1}{2^{s-1}}\right) = \left(2 - \frac{1}{2^{s-1}}\right)^{s+1} - 2\left(2 - \frac{1}{2^{s-1}}\right)^s + 1$$

$$= 1 - \left(2 - \frac{1}{2^{s-1}}\right)^s \frac{1}{2^{s-1}} = 1 - 2\left(1 - \frac{1}{2^s}\right)^s$$

$$= 1 - \left(2^{\frac{1}{s}} - \frac{1}{2^{s-\frac{1}{s}}}\right)^s.$$

Let

$$g(s) = 2^s - 1 - 2^{s-\frac{1}{s}}.$$

Then

$$g'(s) = 2^s \log 2 - 2^{s-\frac{1}{s}}\left(1 + \frac{1}{s^2}\right)\log 2$$

$$= 2^s\left(1 - 2^{-\frac{1}{s}}\left(1 + \frac{1}{s^2}\right)\right)\log 2.$$

Since

$$2^s \geqslant \left(1 + \frac{1}{s^2}\right)^{s^2},$$

that is $g'(s) > 0$ for $s \geqslant 2$, therefore $g(s)$ is increasing for $s \geqslant 2$. Consequently

$$2^s - 1 - 2^{s-\frac{1}{s}} > 0,$$

$$2^{\frac{1}{s}} - \frac{1}{2^{s-\frac{1}{s}}} > 1.$$

Hence $Q\left(2 - \dfrac{1}{2^{s-1}}\right) < 0$.   (3.1) is thus proved.

2) Let

$$F(x) = (x - \eta)R(x)$$

and

$$R(x) = x^{s-1} + \beta_{s-2}x^{s-2} + \cdots + \beta_1 x + \beta_0.$$

Then

$$\begin{cases} -\eta\beta_0 = -1, \\ \beta_0 - \eta\beta_1 = -1, \\ \cdots\cdots\cdots \\ \beta_{s-2} - \eta = -1. \end{cases} \tag{3.3}$$

Hence we have

$$\beta_0 = \frac{1}{\eta}, \quad \beta_1 = \frac{\eta + 1}{\eta^2}, \quad \cdots, \quad \beta_{s-3} = \frac{\eta^{s-3} + \eta^{s-4} + \cdots + \eta + 1}{\eta^{s-2}},$$

$$\beta_{s-2} = \frac{\eta^{s-2} + \eta^{s-3} + \cdots + \eta + 1}{\eta^{s-1}} = \eta - 1.$$

Let

$$\gamma_j = \begin{cases} \beta_j/\beta_{j+1}, & \text{for } 0 \leqslant j \leqslant s - 3; \\ \beta_j, & \text{for } j = s - 2. \end{cases}$$

Since $\dfrac{x}{x + 1}$ is an increasing function of x for $x \geqslant 0$ and $\gamma_j = \dfrac{\eta^{j+1} + \eta^j + \cdots + \eta}{\eta^{j+1} + \eta^j + \cdots + \eta + 1}$
$(0 \leqslant j \leqslant s - 2)$, we have

$$\gamma_{s-2} > \gamma_{s-3} > \cdots > \gamma_0. \tag{3.4}$$

Let $x = \beta_{s-2}y$. Then

$$\tilde{R}(y) = R(\beta_{s-2}y) = \beta_{s-2}^{s-1}y^{s-1} + \beta_{s-2}^{s-1}y^{s-2} + \beta_{s-3}\beta_{s-2}^{s-3}y^{s-3} + \cdots + \beta_1\beta_{s-2}y + \beta_0.$$

From (3.4) we have

$$\beta_{s-2}^{s-1} > \beta_{s-3}\beta_{s-2}^{s-3} > \cdots > \beta_1\beta_{s-2} > \beta_0. \tag{3.5}$$

Hence it follows by Lemma 3.2 that the moduli of roots of $\tilde{R}(y) = 0$ are all $\leqslant 1$.
That is, the moduli of roots of $R(x) = 0$ are all $\leqslant \beta_{s-2} = \eta - 1$. The lemma is
proved.

Lemma 3.3. *We have* $|\eta^{(2)}| = \eta^{-1}$ *for* $s = 2$ *and* $|\eta^{(2)}| = |\eta^{(3)}| = \eta^{-\frac{1}{2}}$ *for*
$s = 3$.

Proof. Obviously, for $s = 2$ we have $|\eta^{(2)}| = \eta^{-1}$. For $s = 3$, since

$$x^3 - x^2 - x - 1 = (x - \eta)\left(x^2 + (\eta - 1)x + \frac{1}{\eta}\right)$$

and

$$(\eta - 1)^2 - \frac{4}{\eta} = \frac{\eta^3 - 2\eta^2 + \eta - 4}{\eta} = \frac{-\eta^2 + 2\eta - 3}{\eta} < 0,$$

hence $\eta^{(2)}$, $\eta^{(3)}$ is a pair of conjugate complex numbers. Therefore $|\eta^{(2)}| = |\eta^{(3)}| = \eta^{-\frac{1}{2}}$. The lemma is proved.

IV. The Estimation of τ

Lemma 4.1. *We have*

$$L < \tau < L + L^{-(s-1)} \tag{4.1}$$

and

$$(L + L^{-\frac{1}{s-1}})^{-\frac{1}{s-1}} < |\tau^{(i)}| < (L - (L-1)^{-\frac{1}{s-1}})^{-\frac{1}{s-1}}, \quad (2 \leqslant i \leqslant s). \tag{4.2}$$

Proof. 1) (4.1) follows immediately from

$$G(L) = -1 < 0$$

and

$$G(L + L^{-(s-1)}) = L^{-(s-1)}(L + L^{-(s-1)})^{s-1} - 1 > 0.$$

2) Let χ be the real root of the equation

$$g(x) = x^s + Lx^{s-1} - 1 = 0$$

in the interval $(0, 1)$. Since $g'(x) \neq 0$ in $(0, 1)$, therefore χ is the only root of $g(x) = 0$ in $(0, 1)$. It follows from

$$g(0) = -1 < 0$$

and

$$g(1) = L > 0$$

that $g(\chi - \delta) < 0$ for any δ satisfying $0 < \delta < \chi$. Hence on the circle $|x| = \chi - \delta$, we have

$$1 > |x^s + Lx^{s-1}|.$$

It follows by Rouché's theorem that 1 and $G(x)$ have the same number of zero in the domain $|x| < \chi - \delta$. Therefore $G(x)$ has no zero in the domain $|x| < \chi - \delta$. Let $\delta \to 0$. Then the moduli of roots of $G(x) = 0$ are all $\geqslant \chi$. Since

$$g((L + L^{-\frac{1}{s-1}})^{-\frac{1}{s-1}}) = ((L + L^{-\frac{1}{s-1}})^{-\frac{1}{s-1}} + L)(L + L^{-\frac{1}{s-1}})^{-1} - 1$$

$$< (L^{-\frac{1}{s-1}} + L)(L^{-\frac{1}{s-1}} + L)^{-1} - 1 = 0,$$

we have

$$|\tau^{(i)}| \geqslant \chi > (L + L^{-\frac{1}{s-1}})^{-\frac{1}{s-1}}, \quad (2 \leqslant i \leqslant s).$$

3) Suppose that $L > 2$. Let Ω be the only real root of the equation

$$h(x) = x^s - Lx^{s-1} + 1 = 0$$

in the interval $(0, 1)$. Since

$$h(0) = 1 > 0$$

and

$$h(1) = -L + 2 < 0,$$

we have $h(\Omega + \delta) < 0$ for any δ satisfying $0 < \delta < 1 - \Omega$. Hence on the circle $|x| = \Omega + \delta$, we have

$$|Lx^{s-1}| > |x^s + 1|.$$

It follows by Rouché's theorem that x^{s-1} and $G(x)$ have the same number of zeros in the domain $|x| < \Omega + \delta$. Therefore $G(x)$ has $s - 1$ zeros in the domain $|x| < \Omega + \delta$. Let $\delta \to 0$. Then $G(x) = 0$ has $s - 1$ roots in the region $|x| \leqslant \Omega$. Since

$$h((L - (L - 1)^{-\frac{1}{s-1}})^{-\frac{1}{s-1}})$$

$$= ((L - (L - 1)^{-\frac{1}{s-1}})^{-\frac{1}{s-1}} - L)(L - (L - 1)^{-\frac{1}{s-1}})^{-1} + 1$$

$$= (L - (L - 1)^{-\frac{1}{s-1}})^{-1}((L - (L - 1)^{-\frac{1}{s-1}})^{-\frac{1}{s-1}} - (L - 1)^{-\frac{1}{s-1}}) < 0,$$

we have

$$|\tau^{(i)}| \leqslant \Omega < (L - (L - 1)^{-\frac{1}{s-1}})^{-\frac{1}{s-1}}, \quad (2 \leqslant i \leqslant s).$$

4) Suppose that $L = 2$. We proceed to prove that the equation

$$G(x) = x^s - 2x^{s-1} - 1 = 0$$

has $s - 1$ roots with moduli < 1. It is equivalent to prove that

$$H(y) = y^s + 2y - 1 = 0$$

has $s - 1$ roots with moduli > 1. Let $\delta > 0$. Then it follows from Rouché's theorem that $H_\delta(y) = y^s + (2 + \delta)y - 1$ and y have the same number of zero in the domain $|y| < 1$. Therefore $H_\delta(y)$ has $s - 1$ roots with moduli $\geqslant 1$. Let $\delta \to 0$. Then $H(y) = 0$ also has $s - 1$ roots with moduli $\geqslant 1$. Since $H(y) = 0$ has no root satisfying $|y| = 1$, $H(y) = 0$ has $s - 1$ roots with moduli > 1. The lemma is proved.

Lemma 4.2.   *We have* $|\tau^{(2)}| = \tau^{-1}$ *for* $s = 2$ *and* $|\tau^{(2)}| = |\tau^{(3)}| = \tau^{-\frac{1}{2}}$ *for* $s = 3$.

Proof.   Clearly, for $s = 2$ we have $|\tau^{(2)}| = \tau^{-1}$. For $s = 3$, since

$$x^3 - Lx^2 - 1 = (x - \tau)\left(x^2 + (\tau - L)x + \frac{1}{\tau}\right)$$

and

$$(\tau - L)^2 - \frac{4}{\tau} = \frac{\tau^3 - 2L\tau^2 + L^2\tau - 4}{\tau} = \frac{-L\tau^2 + L^2\tau - 3}{\tau} < 0,$$

hence $\tau^{(2)}$, $\tau^{(3)}$ is a pair of conjugate complex numbers. Therefore $|\tau^{(2)}| = |\tau^{(3)}| = \tau^{-\frac{1}{2}}$. The lemma is proved.

V.   IRREDUCIBILITY OF POLYNOMIALS

In this section, we prove two theorems concerning the irreducibility of polynomials.

Let

$$w(x) = x^s + a_{s-1}x^{s-1} + \cdots + a_1 x + a_0, \tag{5.1}$$

where $a_0, a_1, \cdots, a_{s-1}$ are integers.

Theorem 5.1. (Xie Ting-fan and Pei Ding-yi [8]) *If*

$$|a_1| > |a_0^{s-1}| + |a_{s-1}a_0^{s-2}| + \cdots + |a_2 a_0| + 1, \quad a_0 \neq 0, \tag{5.2}$$

then $w(x)$ is irreducible over rational field R.

Proof. From (5.2), we have

$$|a_1 a_0| > |a_0^s| + |a_{s-1}a_0^{s-1}| + \cdots + |a_2 a_0^2| + |a_0|.$$

It follows by Rouché's theorem that $w(x)$ and x have the same number of zero in the domain $|x| < |a_0|$. Therefore $w(x)$ has only one zero ϑ in the domain $|x| < |a_0|$.

It follows easily from (5.2) that the equation $w(x) = 0$ has no root with modulus $|a_0|$. If $w(x) = u(x)v(x)$, where $u(x)$ and $v(x)$ are polynomials with integral coefficients and with degrees ≥ 1 and if $u(\vartheta) = 0$, then the moduli of roots of $v(x) = 0$ are all $> |a_0|$. Hence

$$|a_0| = |w(0)| = |u(0)v(0)| \geq |v(0)| > |a_0|,$$

which leads to a contradiction. Thus we have the theorem.

Theorem 5.2. *If $|a_0| = 1$ and $w(x) = 0$ has only one root ϑ with modulus ≥ 1, then $w(x)$ is irreducible over rational field R.*

Proof. If $w(x) = u(x)v(x)$, where $u(x)$ and $v(x)$ are polynomials with integral coefficients and with degrees ≥ 1 and if $u(\vartheta) = 0$, then the moduli of roots of $v(x) = 0$ are all < 1. Hence $|v(0)| < 1$, which leads to a contradiction. The theorem follows.

It follows from Lemmas 3.1 and 4.1 immediately that $F(x)$ and $G(x)$ are irreducible over rational field R. Hence we have

Corollary 5.1. *η and τ are PV numbers.*

Remark 1. Theorem 5.1 improves a result of Perron [5] and a result of Bernstein (Cf. [1] Theorem 12) too.

Remark 2. The polynomials

$$x^s - x^{s-1} - 1 \quad (s = 2, 3, \cdots)$$

are not all irreducible over rational field R. For example,

$$x^5 - x^4 - 1 = (x^2 - x + 1)(x^3 - x - 1)$$

(Cf. Theorem 11 of [1]).

VI. Rational Approximations of η and τ

Lemma 6.1. *If $n \geqslant s$, then*

$$\left| \frac{F_{n+1}}{F_n} - \eta \right| < c(\eta) F_n^{-1 - \frac{1}{2^s \log 2} - \frac{1}{2^{2s+1}}}. \tag{6.1}$$

Proof. From Lemma 3.1, we have

$$\rho = - \frac{\log |\eta^{(s)}|}{\log \eta} \geqslant - \frac{\log (\eta - 1)}{\log \eta} \geqslant \frac{- \log \left(1 - \dfrac{1}{2^s} \right)}{\log 2}$$

$$\geqslant \frac{1}{2^s \log 2} + \frac{1}{2^{2s+1}}.$$

Hence the lemma follows from Theorem 1 and Corollary 5.1.

Lemma 6.2. *If $n \geqslant s$, then*

$$\left| \frac{G_{n+1}}{G_n} - \tau \right| < c(\tau) G_n^{-1 - \frac{1}{s-1} + \frac{2}{(s-1)L \log L} - \frac{1}{(s-1)L^{s+3}}}. \tag{6.2}$$

Proof. Since

$$\log (L - (L - 1)^{-\frac{1}{s-1}}) = \log L + \log (1 - L^{-1}(L - 1)^{-\frac{1}{s-1}})$$

$$\geqslant \log L - L^{-1}(L - 1)^{-\frac{1}{s-1}} - L^{-2}(L - 1)^{-\frac{2}{s-1}}$$

$$\geqslant \log L - L^{-1} - L^{-2}$$

and

$$\log (L + L^{-(s-1)}) = \log L + \log (1 + L^{-s}) \leqslant \log L + L^{-s},$$

we have

$$\rho = - \frac{\log |\tau^{(s)}|}{\log \tau} \geqslant \frac{\log L - L^{-1} - L^{-2}}{(s - 1)(\log L + L^{-s})}$$

$$\geqslant \frac{1}{s - 1} \left(1 - \frac{1}{L \log L} - \frac{1}{L^2 \log L} \right) \left(1 - \frac{1}{L^s \log L} \right)$$

$$\geqslant \frac{1}{s - 1} \left(1 - \frac{1}{L \log L} - \frac{1}{L^2 \log L} - \frac{1}{L^s \log L} + \frac{1}{L^{s+1} \log^2 L} \right)$$

$$\geqslant \frac{1}{s - 1} \left(1 - \frac{2}{L \log L} + \frac{1}{L^{s+3}} \right).$$

Hence the lemma follows from Theorem 1 and Corollary 5.1.

Remark. If we use Lemmas 3.3 and 4.2 to replace Lemmas 3.1 and 4.1 in the proofs of Lemmas 6.1 and 6.2 respectively, then the corresponding right hand sides of (6.1) and (6.2) may be improved to $c(\eta) F_n^{(2)-2}$ and $c(\tau) G_n^{(2)-2}$ for $s = 2$ and $c(\eta) F_n^{(3)-3/2}$ and $c(\tau) G_n^{(3)-3/2}$ for $s = 3$.

VII. Examples

Let n be an integer ≥ 1 and let $a_1, \cdots, a_s$ be integers. Denote

$$Q_n(f) = \int_0^1 \cdots \int_0^1 f(x_1, \cdots, x_s)dx_1 \cdots dx_s - \frac{1}{n}\sum_{k=1}^{n} f\left(\frac{a_1 k}{n}, \cdots, \frac{a_s k}{n}\right). \quad (7.1)$$

Theorem 7.1. *We have*

$$\operatorname*{Sup}_{f \in E_s^2(C)} |Q_n(f)| \leq C(H(n; a_1, \cdots, a_s) - 1), \quad (7.2)$$

where

$$H(n; a_1, \cdots, a_s) = \begin{cases} \frac{1}{n}\left(\left(1 + \frac{\pi^2}{3}\right)^s + 2\sum_{k=1}^{\frac{n-1}{2}} \prod_{\nu=1}^{s}\left(1 - \frac{\pi^2}{6} \right. \right. \\ \left. \left. + \frac{\pi^2}{2}\left(1 - 2\left\{\frac{a_\nu k}{n}\right\}\right)^2\right)\right), \quad \text{when } 2 \nmid n; \\ \frac{1}{n}\left(\left(1 + \frac{\pi^2}{3}\right)^s + \left(1 - \frac{\pi^2}{6}\right)^\mu\left(1 + \frac{\pi^2}{3}\right)^{s-\mu} \right. \\ \left. + 2\sum_{k=1}^{\frac{n}{2}-1} \prod_{\nu=1}^{s}\left(1 - \frac{\pi^2}{6} + \frac{\pi^2}{2}\left(1 - 2\left\{\frac{a_\nu k}{n}\right\}\right)^2\right)\right), \\ \quad \text{when } 2 \mid n, \end{cases} \quad (7.3)$$

in which μ denotes the number of odd integers in $a_\nu (1 \leq \nu \leq s)$. (*Cf.* [9])

Lemma 7.1. *We have*

$$\sum_{m=-\infty}^{\infty} \frac{e^{2\pi i m x}}{\overline{m}^2} = 1 - \frac{\pi^2}{6} + \frac{\pi^2}{2}(1 - 2\{x\})^2.$$

Proof. Since

$$\int_0^1 \left(1 - \frac{\pi^2}{6} + \frac{\pi^2}{2}(1 - 2x)^2\right) e^{-2\pi i m x}\,dx = \overline{m}^{-2},$$

the lemma follows.

The Proof of Theorem 7.1. From Lemma 7.1, we have

$$\operatorname*{Sup}_{f \in E_s^2(C)} |Q_n(f)| \leq \frac{C}{n}\sum_{k=1}^{n}\sideset{}{'}\sum \frac{e^{2\pi i(a_1 m_1 + \cdots + a_s m_s)k/n}}{(\overline{m}_1 \cdots \overline{m}_s)^2}$$

$$= C\left(\frac{1}{n}\sum_{k=1}^{n}\prod_{\nu=1}^{s}\left(\sum_{m_\nu=-\infty}^{\infty} \frac{e^{2\pi i a_\nu m_\nu k/n}}{\overline{m}_\nu^2}\right) - 1\right)$$

$$= C\left(\frac{1}{n}\sum_{k=1}^{n}\prod_{\nu=1}^{s}\left(1 - \frac{\pi^2}{6} + \frac{\pi^2}{2}\left(1 - 2\left\{\frac{a_\nu k}{n}\right\}\right)^2\right) - 1\right)$$

$$= C(H(n; a_1, \cdots, a_s) - 1).$$

The theorem is proved.

Take $n = F_m^{(s)}$, $a_1 = 1$ and $a_\nu = F_{m+\nu-1}^{(s)}$ $(2 \leqslant \nu \leqslant s)$. We have the following two tables by Wang 520 calculator.

Table 1

$s = 3,\ 4$

n	$H(n; a_1, a_2, a_3)$	n	$H(n; a_1, a_2, a_3, a_4)$
149	1.17442	401	1.36254
927	1.01102	2872	1.02416
1705	1.00480	10671	1.00540

Table 2

$s = 5,\ 6$

n	$H(n; a_1, \cdots, a_5)$	n	$H(n; a_1, \cdots, a_6)$
13624	1.07428	29970	1.14458

Remark. In addition to the two generalizations of Fibonacci sequence in our papers, there is another generalization given by G. N. Raney [6]. Let $Q = Q_n = (a_{ij})$, where $a_{ij} = 1$ for $i + j \leqslant n + 1$ and $a_{ij} = 0$ for otherwise condition. Let $\phi_{n,0} = (1, 0, \cdots, 0)'$ and $\phi_{n, d+1} = Q\phi_{n, d}(d = 0, 1, 2, \cdots)$. The sequence of column vectors $\phi_{n, d}$ is called the generalized Fibonacci sequence.

Let $D_n(\lambda) = \det (Q - \lambda I)$ be the characteristic equation of Q. The $D_n(\lambda)$ is irreducible over rational field R for $n = \dfrac{p - 1}{2}$, where p is a prime $\geqslant 5$, but not always irreducible for other n. When $D_n(\lambda)$ is irreducible, we may obtain by the given method of this paper, n sequences of integers $\phi_{n, d}(r)(r = 1, 2, \cdots, n; d = 0, 1, 2, \cdots)$ satisfying the recurrent formula

$$\sum_{j=0}^{n} \begin{pmatrix} n - j + \left[\dfrac{1}{2} j\right] \\ \left[\dfrac{1}{2} j\right] \end{pmatrix} (-1)^{\frac{(n-j)(n-j-1)}{2}+j} \phi_{n, k+j}(r) = 0,$$

where the initial values are the row vectors of the matrix $(\phi_{n,0}, \cdots, \phi_{n,n-1})$ respectively. Then we have

$$(\phi_{n, d}(1), \cdots, \phi_{n, d}(n))' = \phi_{n, d}.$$

Since $D_n(\lambda)$ is not the minimum polynomial of a Pisot-Vijayaraghavan number in general, it seems that we can not expect the rapid convergence of $\phi_{n, d+1}(r)/\phi_{n, d}(r)$ to λ_1, where λ_1 is the characteristic root of the matrix Q whose absolute value is the greatest.

Corrigendum. In § 3, 1) of [3], we may write the ω's as

$$\omega_l = 2\cos\frac{2\pi g^l}{p} \quad (1 \leqslant l \leqslant r),$$

where g is a primitive root mod p. The transformations $\sigma^j (1 \leqslant j \leqslant q)$ should be defined as

$$(\sigma^j) \qquad \omega_l \to \omega_{l+j} \quad (1 \leqslant l \leqslant r).$$

<h2 style="text-align:center">References</h2>

[1] Bernstein, L.: *The Jacobi-Perron Algorithm, Its Theory and Application, Lec. Not. in Math*; Spr. Ver. Pre; (1971), 207.

[2] Cassels, J. W. S.: *An Introduction to Diophantine Approximation*, Camb. Univ. Pre., (1957).

[3] Hua, Loo-keng & Wang Yuan: On uniform distribution and numerical analysis (I) (Number-theoretic method), *Sci. Sin.*, **4**, (1973), 483—505.

[4] Hua Loo-keng & Wang Yuan: On uniform distribution and numerical analysis (II) (Number-theoretic method), *Sci. Sin.*, **17** (1974), 331—348.

[5] Perron, O.: Grundlagen fur eine Theorie des Jacobischen Kettenbruchalgorithmus, *Math. Ann*, **64**, (1906), 1—76.

[6] Raney, G. M.: Generalization of the Fibonacci sequence to n dimensions, *Can. J. of Math.*, **18** (1966), 332—349.

[7] Uspensky, J. V.: *Theory of equations*, MH book Com. Inc; (1948).

[8] Xie Ting-fan & Pei Ding-yi: On irreducibility of polynomials, *Kexue Tongbao* (to appear).

[9] See *"Applications of Number Theory to Numerical Analysis"*, edited by S. K. Zaremba, Acad. Pre., (1972).

A NOTE ON UNIFORM DISTRIBUTION AND EXPERIMENTAL DESIGN

Wang Yuan (王 元) and Fang Kaitai (方开泰)

(Institute of Mathematics, Academia Sinica)

Received December 29, 1979.

I. Introduction

If there are s factors of which each factor has q levels in an experiment, where $q > 1$, then the orthogonal array is often used. The number of experiments is $0(q^2)$ which is too large when q is comparatively large, for example, $q \geqslant 9$. The number of experiments may be decreased by BIB (balanced incomplete blocks) method for the case $s = 2$ only (cf. [1]). Hence, it is necessary to find a method of which the number of experiments is much lower than $0(q^2)$.

If there are s factors of which each has q levels, then the number of all possible experiments is q^s. The orthogonal array is to choose $0(q^2)$ experiments by the theory of orthogonal Latin square. In this note, a design of q experiments is proposed by the theory of uniform distribution which will be called the uniform design. Some applications to industry show the advantages of the uniform design as compared with the orthogonal array.

II. Methods

1. Suppose that the integers $a_i \, |(1 \leqslant i \leqslant s)$ satisfy $a_1 = 1$, $1 < a_i < q(2 \leqslant i \leqslant s)$, $a_i \not\equiv a_j (i \not\equiv q)$ and g.c.d. $(a_i, q) = 1(1 \leqslant i \leqslant s)$, then we use the set of points

$$P_q(k) = (ka_1, ka_2, \cdots, ka_s) \qquad (\text{mod } q), \, k = 1, 2, \cdots, q \qquad (2.1)$$

to arrange the experiments by the usual way. Obviously, s needs to be $\leqslant \varphi(q)$, where $\varphi(q)$ denotes the Euler function, i.e. the number of reduced residue classes modulo q. In other words, if the level of the experiments is q, then the number of the factors needs to be $\leqslant \varphi(q)$. We use the notation $\boldsymbol{a} = (a_1, \cdots a_s)$.

Lemma 2.1. *Suppose that* $q = p_1^{l_1} p_2^{l_2} \cdots p_m^{l_m}$, *where* p_1, p_2, $\cdots p_m$ *are different primes, then*

$$\varphi(q) = q \prod_{i=1}^{m} \left(1 - \frac{1}{p_i}\right).$$

We always use p to denote a prime. From Lemma 2.1, we have $\varphi(q) \leqslant q - 1$, where the equality holds for $q = p$ only. The choice of $\boldsymbol{a}$'s is based on the theory of uniform distribution which will be stated in the next section. Since $a_1 = 1$ and there are $\begin{pmatrix} \varphi(q) - 1 \\ s - 1 \end{pmatrix}$ vectors $\boldsymbol{a}$, hence the number of possible choices of $\boldsymbol{a}$'s is too large when $\varphi(q)$ is comparatively large.

2. Suppose that $q = p$. We suggest to use the set of points

$$Q_p(k) = (k, kb, \cdots, kb^{s-1}) \qquad (\mathrm{mod}\ p), \ k = 1, 2, \cdots, p \qquad (2.2)$$

to arrange the experiments, where b is an integer satisfying $1 < b < p$ and $b^i \not\equiv b^j(\mathrm{mod}\ p)(i \neq j)$. We use the notation $\boldsymbol{b} = (1, b, \cdots, b^{s-1})$. It is noted that b is often chosen among the primitive roots modulo p.

3. Suppose that $q = p - 1$. Then the set of points may be obtained by omitting the last coordinate of the points of (2.2).

III. Uniform Distribution

Let G_s be the s-dimensional unit cube. Let $1 < n_1 < n_2 < \cdots$ be a sequence of integers and $R_{n_l}(k) = (x_1^{(n_l)}(k), \cdots, x_s^{(n_l)}(k))\,(1 \leqslant k \leqslant n_l)$ be a set of points G_s. For any given $\boldsymbol{r} = (r_1, r_2, \cdots, r_s) \in G_s$, let $N_{n_l}(\boldsymbol{r})$ denote the number of points $R_{n_l}(k)$ $(1 \leqslant k \leqslant n_l)$ satisfying

$$0 \leqslant x_i^{(n_l)}(k) < r_i, \qquad 1 \leqslant i \leqslant s.$$

If

$$\sup_{r \in G_s} \left| \frac{N_{n_l}(\boldsymbol{r})}{n_l} - |\boldsymbol{r}| \right| = O(1),$$

as $n_l \to \infty$, where $|\boldsymbol{r}| = \prod_{i=1}^{s} r_i$, then the sequence of sets $\{R_{n_l}(k)\}(1 < n_1 < n_2 < \cdots)$ is called uniformly distributed on G_s.

For simplicity, we omit the index l. Let

$$\sup_{r \in G_s} \left| \frac{N_n(\boldsymbol{r})}{n} \right| - |\boldsymbol{r}| \right| = D(n).$$

Then $D(n)$ is called the discrepancy of $R_n(k)(1 \leqslant k < n)$. We will choose $\boldsymbol{a}$ and $\boldsymbol{b}$ such that the sets of points

$$\left(\left\{ \frac{a_1 k}{q} \right\}, \cdots, \left\{ \frac{a_s k}{q} \right\} \right) \qquad (1 \leqslant k \leqslant g) \qquad (3.1)$$

and

$$\left(\left\{ \frac{k}{p} \right\}, \left\{ \frac{kb}{p} \right\}, \cdots, \left\{ \frac{kb^{s-1}}{p} \right\} \right) \qquad (1 \leqslant k \leqslant p) \qquad (3.2)$$

have lower discrepancies. We use $D(q, \boldsymbol{a})$ and $D(p, \boldsymbol{b})$ to denote the discrepancies of (3.1) and (3.2) respectively. The set (3.1) with $q = p$ was first introduced by H. M. Коробов[2] and E. Hlawka[3] independently in their studies on the theory of numerical integration over G_s and the set (3.2) was proposed by H. M. Коробов[4].

We use the notations $\bar{x} = \max(1, |x|)$, $\|\boldsymbol{x}\| = \bar{x}_1, \cdots, \bar{x}_s$ and

$$(\boldsymbol{x}, \boldsymbol{y}) = \sum_{i=1}^{s} x_i y_i$$

the scalar product of two vectors $\boldsymbol{x}$ and $\boldsymbol{y}$.

Lemma 3.1 (*Erdös-Turen-Koksma*). *Suppose that* η *is a positive number and* h *is an integer* $\geqslant 2$, *then*

$$D(n) = \sum_{|m_i| < h}{}' \frac{1}{\|\pi\boldsymbol{m}\|} \left| \frac{1}{n} \sum_{k=1}^{n} e^{2\pi i(\boldsymbol{m}), R_n(k)} \right| + O(\eta) + O(\eta^{-1}h^{-1}(\ln h)^{s-1}),$$

where $\sum'$ *denotes a sum with an exception* $m_1 = \cdots = m_s = 0$ (cf. [5]).

Lemma 3.2. *Suppose that* g. c. d. $(a_i, q) = 1 (1 \leqslant i \leqslant s)$, *then for any integer* $r \geqslant 1$,

$$\sum_{\substack{(\boldsymbol{a},\boldsymbol{m}) \equiv 0(\mathrm{mod} q) \\ -qr < m_i \leqslant qr}}{}' \frac{1}{\|\pi\boldsymbol{m}\|} - \sum_{\substack{(\boldsymbol{a},\boldsymbol{m}^{(0)}) \equiv 0(\mathrm{mod} q) \\ -q/2 < m_i^{(0)} \leqslant q/2}}{}' \frac{1}{\|\pi\boldsymbol{m}^{(0)}\|} = O(q^{-1}(\ln r q)^{s-1})$$

(cf. [5]).

It derives from Lemmas 3.1 and 3.2 that

Lemma 3.3.

$$D(q, \boldsymbol{a}) = \sum_{\substack{(\boldsymbol{a},\boldsymbol{m}) \equiv 0(\mathrm{mod} q) \\ |m_i| \leqslant q/2}}{}' \frac{1}{\|\pi\boldsymbol{m}\|} + O(q^{-1}(\ln q)^s).$$

Lemma 3.4. *Suppose that* $x \in (0, 1)$, *then*

$$\sum_{|m| < q} \frac{e^{2\pi i m x}}{\pi m} = 1 - \frac{2}{\pi} \ln(2 \sin \pi x) + \frac{\vartheta}{\pi q \langle x \rangle} ,$$

where $|\vartheta| \leqslant 1$ *and* $\langle x \rangle = \min(\{x\}, 1 - \{x\})$ *in which* $\{x\}$ *denotes the fractional part of* x (cf. [5]).

Theorem 3.1.

$$D(q, \boldsymbol{a}) = \frac{1}{q} \sum_{k=1}^{q-1} \prod_{\nu=1}^{s} \left(1 - \frac{2}{\pi} \ln \left(2 \sin \pi \left\{ \frac{a_\nu k}{q} \right\} \right) \right) - 1 + O(q^{-1}(\ln q)^s). \qquad (3.3)$$

The theorem follows from Lemmas 3.1, 3.3 and 3.4.

Since the second term and third term of the right-hand side of (3.3) are -1 and $O(q^{-1}(\ln q)^s)$ respectively, hence the comparison of the quantity $D(q, \boldsymbol{a})$ may be reduced to the comparison of the quantity

$$D_1(q, a) = \frac{1}{q} \sum_{k=1}^{q-1} \prod_{\nu=1}^{s} \left(1 - \frac{2}{\pi} \ln \left(2 \sin \pi \left\{ \frac{a_\nu k}{q} \right\} \right) \right).$$

Sometimes $D_1(q, \boldsymbol{a})'s$ are difficult to make a difference, hence we introduce

$$D_2(q, \boldsymbol{a}) = \frac{1}{q} \sum_{k=1}^{q} \prod_{\nu=1}^{s} \left(1 - \frac{2}{\pi} \ln \left(2 \sin \pi \frac{a_{\nu k}}{q+1} \right) \right),$$

where $a_{\nu k} \equiv a_\nu k(\mathrm{mod}\ q)$, $1 \leqslant a_{\nu k} < q$ *for* $1 \leqslant k < q$, *and* $a_{\nu q} = q$.

Theorem 3.2.

$$D_2(q, \boldsymbol{a}) = D_1(q, \boldsymbol{a}) + O(q^{-1}(\ln q)^s).$$

Proof. Clearly

$$D_2(q, \boldsymbol{a}) = J + O(q^{-1}(\ln q)^s),$$

where

$$J = \frac{1}{q} \sum_{k=1}^{q-1} \prod_{v=1}^{s} \left(1 - \frac{2}{\pi} \ln\left(2 \sin \pi \frac{a_{vk}}{q+1} \right) \right).$$

And

$$|D_2(q, \boldsymbol{a}) - J| = O\left(q^{-1} \sum_{k=1}^{q-1} \sum_{r=1}^{s} \frac{a_{rk}}{q^2 \langle a_{rk}/q \rangle} (\ln q)^{s-1} \right)$$

$$= O\left(q^{-1} \sum_{l=1}^{q} (\ln q)^{s-}l^{l-1} \right) = O(q^{-1}(\ln q)^s).$$

The theorem follows.

The integral vector $\boldsymbol{a}$ such that $D_2(q, \boldsymbol{a})$ attains its minimum with respect to $\boldsymbol{a}$ may be used to define the set of points (2.1) which is called a uniform design. Similarly, the set (2.2) with minimal $D_2(p, \boldsymbol{b})$ is also called a uniform design.

IV. Tables

Since the congruences $(\boldsymbol{a}, \boldsymbol{m}) \equiv 0 \pmod{q}$ and $(\boldsymbol{b}, \boldsymbol{m}) \equiv 0 \pmod{p}$ may be written by $a_\mu^{-1}(\boldsymbol{a}, \boldsymbol{m}) \equiv 0 \pmod{q}$ $(2 \leqslant \mu \leqslant s)$ and $b^{-s+1}(\boldsymbol{b}, \boldsymbol{m}) \equiv 0 \pmod{p}$ respectively, hence the calculations of $D_2(a_\mu^{-1}\boldsymbol{a}, q)$ $(2 \leqslant \mu \leqslant s)$ and $D_2(b^{-(s-1)}\boldsymbol{b}, p)$ may be neglected.

1. For $q = p$, we use the set of points (2.2). If $q = p - 1$, then we have the set of points by omitting the least coordinate of corresponding set of points of the level p. By the calculation on computer DJS-2, we have the tables of the uniform design for $q = 4, 5, 6, 7, 10, 11, 12, 13, 16, 17, 18, 19, 22, 23, 28, 29, 30, 31$ (cf. Table 4.1.).

Table 4.1

The Uniform Design for Prime p and $p - 1$

p \ s	2	3	4	5	6	7	8	9	10	11	12	13	14	15	16	17	18	19	20	21	22	23	24	25	26	27	28	29	30
5	2	2	2																										
7	3	3	3	3	3																								
11	7	7	7	7	7	7	7	7																					
13	5	4	6	6	6	6	6	6	6	6	6																		
17	10	10	10	10	10	10	10	10	10	10	10	10	10	10	10														
19	8	8	14	14	14	14	14	14	14	14	14	14	14	14	14	14	14												
23	7	17	17	17	17	17	15	15	15	15	15	7	7	17	17	17	17	17	7	7	7								
29	12	9	16	16	16	16	8	8	8	8	8	14	14	14	8	8	8	8	8	8	8	8	18	18	18	18			
31	12	22	22	12	12	12	12	12	12	12	12	22	22	22	22	22	22	22	12	12	12	12	12	12	12	12	12	12	12

2. The uniform designs of $q = 8, 14, 20, 24, 26$ may be obtained by the uniform designs of $q = 9, 15, 21, 25, 27$ by omitting the corresponding last coordinates. For the case $q = 9, 15$ and 21, we may use (2.1). For the case $q = 25$ or 27, we may

use (2.2), since there exists primitive roots mod q although q is not a prime (cf. Table 4.2).

Table 4.2

The Uniform Design for q and $q-1$

q \ s	2	3	4	5	6	7	8	9	10	11
9	4	4, 7	4, 7, 2	4,7,2,5	4,7,2,5,8					
15	11	4, 7	4,7.13	4,7,13,2	2,4,7,11,14	2,4,7,11,14,13	2,4,7,11,14,13,8			
21	13	4. 10	4,10,13	4,10,16,19	4.10,13,16,19	4,10,13,16,19,20	4,5,8,10,11,17,19	2,4,5,8,10,11,17,19	2,4,5,8,10,11,16,17,19	2,4,5,8,10,11,13,16,17,19
25	11	11	11	11	4	9	8	8	8	8
27	8	8	20	20	20	16	16	16	20	5

q \ s	12	13	14	15	16	17	18	19	20
9									
15									
21	2,4,5,8,10,11,13,16,17,19,20								
25	8	8	8	8	8	8	8	8	8
27	5	20	20	20	5	5	5		

Table 4.3

$U_5(5^4)$

N_0 \ Factors	1	2	3	4
1	1	2	4	3
2	2	4	3	1
3	3	1	2	4
4	4	3	1	2
5	5	5	5	5

3. Example. For $s=4$ and $q=5$, we have uniform designs by Table 4.1

$$(k,\ 2k,\ 2^2k,\ 2^3k) \qquad (\mathrm{mod}\ 5), \qquad 1 \leqslant k \leqslant 5$$

or the form of Table 4.3 and we use the notation $U_5(5^4)$ similar to $L_{25}(5^6)$ in the orthogonal array.

Data in the uniform design may be treated by regression or step wise regression.

REFERENCES

[1]　中国科学院数学研究所统计组，方差分析，科学出版社，(1977).
[2]　Коробов Н. М., *ДАН СССР*, **6**(1959), 124, 1207—1210.
[3]　Hlawka, E., *Mon. Math.*, **2**(1962), 66, 140—151.
[4]　Коробов Н. М., *Вес. МГУ*, **4**(1959), 19—25.
[5]　华罗庚，王元，数论在近似分析中的应用，科学出版社，(1978).

第 25 卷 第 2 期　　　　　数 学 学 报　　　　Vol. 25, No. 2
1982 年 3 月　　　ACTA MATHEMATICA SINICA　　Mar., 1982

On Diophantine Approximation and Approximate Analysis (I)*

Wang Yuan (王元)

(Institute of Mathematics, Academia Sinica)

§1. Introduction

We use $\boldsymbol{\gamma} = (\gamma_{11}, \cdots, \gamma_{t1}, \cdots, \gamma_{1s}, \cdots, \gamma_{ts})$ to denote a point in st-dimensional Euclidean space R_{st}. We use the notations $\boldsymbol{\gamma}_i = (\gamma_{1i}, \cdots, \gamma_{ti})(1 \leqslant i \leqslant s)$ and $\boldsymbol{\gamma}_j = (\gamma_{j1}, \cdots, \gamma_{js})(1 \leqslant j \leqslant t)$. $\boldsymbol{q} = (q_1, \cdots, q_t)$, $\boldsymbol{k} = (k_1, \cdots, k_s)$ and $\boldsymbol{m} = (m_1, \cdots, m_s)$ will be vectors with integral components and $(\boldsymbol{x}, \boldsymbol{y}) = \sum_{i=1}^{s} x_i y_i$ the scalar product of $\boldsymbol{x}$ and $\boldsymbol{y}$. We use also ε to denote any pre-assigned positive number, $c(f, g, \cdots)$ a positive constant depending on f, g, $\cdots$ only, but not always with the same value, and α, c, a, b, $\cdots$ absolute constants.

Let G_s be the s-dimensional unit cube, i.e. the set of $\boldsymbol{x} = (x_1, \cdots, x_s)$, where $0 \leqslant x_i \leqslant 1$, $1 \leqslant i \leqslant s$. Let $P_n(k) = (x_n^{(1)}(k), \cdots, x_n^{(s)}(k))(1 \leqslant k \leqslant n)$ be a set of points in G_s. For any $\boldsymbol{a} = (\alpha_1, \cdots, \alpha_s) \in G_s$, let $N_n(\boldsymbol{a})$ be the number of points $P_n(k)(1 \leqslant k \leqslant n)$ satisfying

$$0 \leqslant x_n^{(i)}(k) < \alpha_i, \ 1 \leqslant i \leqslant s.$$

Then

$$D(n) = \sup_{a \in G_s} \left| \frac{N_n(\boldsymbol{a})}{n} - |\boldsymbol{a}| \right|, \ |\boldsymbol{a}| = \alpha_1 \cdots \alpha_s$$

is called the discrepancy of $P_n(k) \ (1 \leqslant k \leqslant n)$.

For a real number x, we use $[x]$, $\{x\}$ and $\langle x \rangle$ to denote respectively the integer part of x, the fractional part of x and the distance from x to the nearest integer, i. e. $\langle x \rangle = \min(x - \{x\}, \ 1 + \{x\} - x)$. We also denote $M(\boldsymbol{m}) = \prod_{j=1}^{s} \overline{m}_j^a (\overline{\ln m_j})^b$, where $a \geqslant 1, b \geqslant 0$ and $\bar{x} = \max(1, |x|)$.

Theorem 1. Suppose that q is an integer $\geqslant 2$ and that

$$\prod_{i=1}^{t} \langle (\boldsymbol{r}_i, \boldsymbol{m}) \rangle M(\boldsymbol{m}) > c(\boldsymbol{r}, a, b) \tag{1}$$

holds for any $\boldsymbol{m} \neq \boldsymbol{0}$. Then the set $((\boldsymbol{\gamma}_1, \boldsymbol{q}), \cdots, (\boldsymbol{\gamma}_s, \boldsymbol{q}))(\text{mod} 1)(1 \leqslant q_i \leqslant q, 1 \leqslant i \leqslant t)$ has discrepancy

---

* Received January 23, 1980.

$$D(q^t) \leqslant c(\boldsymbol{r},\ a,\ b,\ s,\ t)q^{-t+2st(a-1)}(\ln q)^{bs+t+s\delta_{1,a}}$$

where δ_{ij} denotes the Kronecker symbol.

Theorem 2. Suppose that q and $\boldsymbol{r}$ satisfy the conditions of Theorem 1. Suppose further that n is an integer $\geqslant q^{\frac{t+1}{\rho+1}}$ and that

$$\left| r_{ij} - \frac{h_{ij}}{n} \right| \leqslant c(\boldsymbol{r})n^{-1-\rho}, \quad 1 \leqslant i \leqslant t, \quad 1 \leqslant j \leqslant s,$$

where $\rho > 0$ and $h_{ij}(1 \leqslant i \leqslant t,\ 1 \leqslant j \leqslant s)$ are integers. Then the set $\left(\dfrac{(\boldsymbol{h}_1, \boldsymbol{q})}{n},\ \cdots, \right.$

$\left. \dfrac{(\boldsymbol{h}_s, \boldsymbol{q})}{n} \right) \pmod 1 (1 \leqslant q_i \leqslant q,\ 1 \leqslant i \leqslant t)$ has discrepancy

$$D(q^t) \leqslant c(\boldsymbol{r},\ a,\ b,\ s,\ t)q^{-t+2st(a-1)}(\ln q)^{bs+t+s\delta_{1,a}}$$

where $\boldsymbol{h}_i = (h_{1i},\ \cdots,\ h_{ti}),\ 1 \leqslant i \leqslant s.$

Similar results were obtained by Hua and Wang [1,2] and Niederreiter [3] for the case $t = 1$. It may benefitcial in practical use to make suitable choises of $\boldsymbol{r}$ and t.

<h3 align="center">§ 2. The Proof of Theorem 1.</h3>

Suppose always that $\boldsymbol{r}$ and $c(\boldsymbol{r},\ a,\ b)$ satisfy (1). We use $P_{t,M}$ to denote a t-dimensional parallelepiped defined by

$$a_i \leqslant x_i \leqslant b_i, \quad 1 \leqslant i \leqslant t,$$

where $\prod\limits_{i=1}^{t} (b_i - a_i) \leqslant M$. Denote also $Q = [2^{(a+b)s}M(\boldsymbol{m})c(\boldsymbol{r},\ a,\ b)^{-1}]$.

Lemma 1. There is at most 1 point $((\boldsymbol{r}_1,\ \boldsymbol{k}),\ \cdots,\ (\boldsymbol{r}_t,\ \boldsymbol{k}))$ in any parallelepiped of the type $P_{t,Q}$, where $|k_i| \leqslant |m_i|,\ 1 \leqslant i \leqslant s.$

Proof. If there are two points $((\boldsymbol{r}_1,\ \boldsymbol{k}'),\ \cdots,\ (\boldsymbol{r}_t,\ \boldsymbol{k}'))$ and $((\boldsymbol{r}_1,\ \boldsymbol{k}''),\ \cdots,(\boldsymbol{r}_s,\ \boldsymbol{k}''))$ in a $P_{t,Q}$, where $\boldsymbol{k}' \neq \boldsymbol{k}''$ and $|k_i'| \leqslant |m_i|, |k_i''| \leqslant |m_i|(1 \leqslant i \leqslant s)$. Then

$$\prod_{i=1}^{t} \langle (\boldsymbol{r}_i,\ \boldsymbol{k}' - \boldsymbol{k}'') \rangle \leqslant Q.$$

On the other hand, it follows from (1) that

$$\prod_{i=1}^{t} \langle (\boldsymbol{r}_i,\ \boldsymbol{k}' - \boldsymbol{k}'') \rangle > c(\boldsymbol{r},\ a,\ b)M(\boldsymbol{k}' - \boldsymbol{k}'')^{-1}$$

$$\geqslant c(\boldsymbol{r},\ a,\ b)2^{-as}\|\boldsymbol{m}\|^{-a} \left(\prod_{i=1}^{s} \overline{\ln 2\overline{m}_i} \right)^{-b}$$

$$\geqslant c(\boldsymbol{r},\ a,\ b)2^{-(a+b)s}M(\boldsymbol{m})^{-1} \geqslant Q,$$

which leads to a contradiction. Hence the lemma follows.

Lemma 2. Suppose that l is an integer $\geqslant 1$ and that M satisfies $0 < M \leqslant \dfrac{1}{2}$. Then the s-dimensional domain

$$M \leqslant x_1 \cdots x_s \leqslant lM, \quad M \leqslant x_i \leqslant \frac{1}{2}, \ 1 \leqslant i \leqslant s \tag{2}$$

can be covered by at most $2^{s-1}l(\log_2 M^{-1})^{s-1}$ parallelepipeds of the type $P_{l,M}$.

Proof. Since the domain (2) is the interval $M \leqslant x_1 \leqslant \min\left(lM, \frac{1}{2}\right)$ for the case $s=1$, it can be covered by at most l intervals of the type $[c, c + M]$, where c is a real number. Hence the lemma holds for $s = 1$. Suppose that k is a positive integer and that the lemma holds for $s = k$. Now we proceed to prove the validity of the lemma for $s = k + 1$. Divide the domain

$$M \leqslant x_1 \cdots x_{k+1} \leqslant lM, \ M \leqslant x_i \leqslant \frac{1}{2}, \ 1 \leqslant i \leqslant k + 1 \tag{3}$$

into $\left[\log_2 \frac{1}{M}\right]$ subdomains

$$2^{-i-1} \leqslant x_{k+1} \leqslant 2^{-i}, \ 1 \leqslant i \leqslant \left[\log_2 \frac{1}{M}\right], \tag{4}$$

where the left hand side of (4) should be replaced by M for $i = \left[\log_2 \frac{1}{M}\right]$. Consider a subdomain

$$2^{-i-1} \leqslant x_{k+1} \leqslant 2^{-i}, \tag{5}$$

where $i < \left[\log_2 \frac{1}{M}\right]$. Since

$$2^i M \leqslant \frac{M}{x_{k+1}} \leqslant x_1 \cdots x_k \leqslant \frac{lM}{x_{k+1}} \leqslant l2^{i+1}M.$$

Hence it follows from the assumption of the induction that the above k-dimensional domain can be covered by at most $2^k l (\log_2 M^{-1})^{k-1} = L$ (say) parallelepipeds of the type $P_{k,M2^i}$. Using these $P_{k,M2^i}$ as bases and 2^{-i} as height, we obtain a set of $P_{k+1,M}$ and so (5) can be covered by at most L parallelepipeds of the type $P_{k+1,M}$. Now consider the subdomain of (4) with $i = \left[\log_2 \frac{1}{M}\right]$. Since

$$\frac{1}{2} \leqslant M 2^{\left[\log_2\frac{1}{M}\right]} \leqslant x_1 \cdots x_k \leqslant \frac{lM}{M} = 2l \cdot \frac{1}{2},$$

it can be covered by at most L parallelepipeds of the type $P_{k,1/2}$. Using these $P_{k,1/2}$ as bases and $2M$ as height, we obtain a set of $P_{k+1,M}$ and so the subdomain of (4) with $i = \left[\log_2 \frac{1}{M}\right]$ can be covered by at most L parallelepipeds of the type $P_{k+1,M}$. Consequently domain (3) can be covered by at most

$$L\left[\log_2 \frac{1}{M}\right] \leqslant 2^k l(\log_2 M^{-1})^k$$

parallelopipeds of the type $P_{k+1,M}$. The lemma follows by induction.

Lemma 3. Suppose that $\alpha \geqslant 1$. Then

$$\sideset{}{'}\sum_{\substack{|k_j|\leqslant|m_i| \\ 1\leqslant i\leqslant s}} \frac{1}{\prod\limits_{i=1}^{t} \langle(r_i,\ m)\rangle^{\alpha}} \leqslant c(\alpha,t)Q^{-\alpha}(\ln Q)^{t-1+\delta_{1,\alpha}},$$

where $\sum'$ denotes a sum with an exception $k \neq 0 = (0,\ \cdots,\ 0)$.

Proof. Let $T_{l,Q}$ be the number of points $(\langle(r_1,\ k)\rangle, \cdots, \langle(r_t,\ k)\rangle)(|k_i| \leqslant |m_i|,\ 1 \leqslant i \leqslant s)$ in the domain

$$x_1 \cdots x_t \leqslant lQ, \quad Q \leqslant x_i \leqslant \frac{1}{2},\ 1 \leqslant i \leqslant t.$$

Since

$$\prod_{i=1}^{t} \langle(r_i,\ k)\rangle \geqslant c(r,\ a,\ b) \prod_{j=1}^{s} \bar{k}_j^{-a}(\overline{\ln \bar{k}_j})^{-b} > Q$$

by (1), we have $\langle(r_i,\ k)\rangle > Q(1 \leqslant i \leqslant t)$ and $T_{1,Q} = 0$. Hence by Lemmas 1 and 2

$$T_{l,Q} \leqslant c(t)l(\ln Q^{-1})^{t-1}$$

and thus

$$\sideset{}{'}\sum_{\substack{|k_j|\leqslant|m_i| \\ 1\leqslant i\leqslant s}} \frac{1}{\prod\limits_{i=1}^{t} \langle(r_i,\ k)\rangle^{\alpha}} \leqslant \sum_{l=1}^{[Q^{-1}]} \frac{T_{l+1,Q} - T_{l,Q}}{(lQ)^{\alpha}}$$

$$\leqslant Q^{-\alpha} \sum_{l=1}^{[Q^{-1}]} T_{l+1,Q} \left(\frac{1}{l^{\alpha}} - \frac{1}{(l+1)^{\alpha}}\right) + \frac{T_{[Q^{-1}]+1,Q}}{(([Q^{-1}]+1)Q)^{\alpha}}$$

$$\leqslant c(\alpha,t)Q^{-Q}(\ln Q^{-1})^{t-1} \sum_{l=1}^{[Q^{-1}]} \frac{1}{l^{\alpha}} + c(t)Q^{-\alpha}(\ln Q^{-1})^{t-1}$$

$$\leqslant c(\alpha,\ t)Q^{-\alpha}(\ln Q)^{t-1+\delta_{1,\alpha}}.$$

The lemma is proved.

Lemma 4. Suppose that h is an integer $\geqslant 1$ and $\alpha \geqslant 1$ and that $g(m)$ is a non-negative function of m. Then

$$\sideset{}{'}\sum_{\substack{|m_i|\leqslant h \\ 1\leqslant i\leqslant s}} \frac{g(m)}{\|m\|}$$

$$\leqslant \sum_{l=0}^{s} \frac{\alpha^{s-l}}{h^{l\alpha}} \sum_{i} \sum_{m_{i_{l+1}}=1}^{h} \cdots \sum_{m_{i_s}=1}^{h} \frac{1}{(m_{i_{l+1}} \cdots m_{i_s})^{\alpha+1}}$$

$$\times \sum_{|k_{i_1}|\leqslant h} \cdots \sum_{|k_{i_l}|\leqslant h} \sum_{|k_{i_{l+1}}|\leqslant m_{i_{l+1}}} \cdots \sideset{}{'}\sum_{|k_{i_s}|\leqslant m_{i_s}} g(k),$$

where $\sum\limits_{i}$ denotes a sum in which $i = (i_1,\cdots,i_s)$ runs over all the permutations of $(1,\cdots, s)$.

Proof.

$$\sum_{|m|\leqslant h}\frac{g(m)}{\overline{m}^{\alpha}}=g(0)+\sum_{m=1}^{h}\frac{1}{m^{\alpha}}(g(m)+g(-m))$$

$$\leqslant g(0)+\sum_{m=1}^{h}\left(\frac{1}{m^{\alpha}}-\frac{1}{(m+1)^{\alpha}}\right)\sideset{}{'}\sum_{|k|\leqslant m}g(k)+\frac{1}{h^{\alpha}}\sideset{}{'}\sum_{|k|\leqslant h}g(k)$$

$$\leqslant \alpha\sum_{m=1}^{h}m^{-\alpha-1}\sum_{|k|\leqslant m}g(k)+\frac{1}{h^{\alpha}}\sum_{|k|\leqslant h}g(k).$$

Similarly

$$\sideset{}{'}\sum_{|m|\leqslant h}\frac{g(m)}{\overline{m}^{\alpha}}\leqslant \alpha\sum_{m=1}^{h}m^{-\alpha-1}\sideset{}{'}\sum_{|k|\leqslant m}g(k)+\frac{1}{h^{\alpha}}\sideset{}{'}\sum_{|k|\leqslant h}g(k).$$

Hence

$$\sideset{}{'}\sum_{\substack{|m_i|\leqslant h\\1\leqslant i\leqslant s}}\frac{g(\boldsymbol{m})}{\|\boldsymbol{m}\|^{\alpha}}\leqslant \sideset{}{'}\sum_{\substack{|m_i|\leqslant h\\1\leqslant i\leqslant s-1}}\frac{1}{(\overline{m}_1\cdots\overline{m}_{s-1})^{\alpha}}\left(\alpha\sum_{m_s=1}^{h}m_s^{-\alpha-1}\sum_{|k_s|\leqslant m_s}g(m_1,\cdots,m_{s-1},k_s)\right.$$

$$\left.+\frac{1}{h^{\alpha}}\sum_{|k_s|\leqslant h}g(m_1,\cdots,m_{s-1},k_s)\right)$$

$$+\sum_{\substack{|m_i|\leqslant h\\1\leqslant i\leqslant s-1}}\frac{1}{(\overline{m}_1\cdots\overline{m}_{s-1})^{\alpha}}\left(\alpha\sum_{m_s=1}^{h}m_s^{-\alpha-1}\sideset{}{'}\sum_{|k_s|\leqslant m_s}g(m_1,\cdots,m_{s-1},k_s)\right.$$

$$\left.+\frac{1}{h^{\alpha}}\sideset{}{'}\sum_{|k_s|\leqslant h}g(m_1,\cdots,m_{s-1},k_s)\right)\leqslant\cdots$$

$$\leqslant \sum_{l=0}^{s}\frac{\alpha^{s-l}}{h^{l\alpha}}\sum_{i}\sum_{m_{i_{l+1}}=1}^{h}\cdots\sum_{m_{i_s}=1}^{h}\frac{1}{(m_{i_{l+1}}\cdots m_{i_s})^{\alpha+1}}$$

$$\times\sum_{|k_{i_1}|\leqslant h}\cdots\sum_{|k_{i_l}|\leqslant h}\sum_{|k_{i_{l+1}}|\leqslant m_{i_{l+1}}}\cdots\sideset{}{'}\sum_{|k_{i_s}|\leqslant m_{i_s}}g(\boldsymbol{k}).$$

The lemma is proved.

Lemma 5. Suppose that $\alpha\geqslant 1$ and h is an integer $\geqslant 2$. Then

$$\sideset{}{'}\sum_{\substack{|m_i|\leqslant h\\1\leqslant i\leqslant s}}\frac{1}{\left(\|\boldsymbol{m}\|\prod_{i=1}^{t}\langle(\boldsymbol{r}_i,\ \boldsymbol{m})\rangle\right)^{\alpha}}\leqslant c(\boldsymbol{r},\alpha,a,b,s,t)h^{s(a-1)\alpha}(\ln h)^{bs\alpha+t-1+s\delta_{1,a}+\delta_{1,\alpha}}.$$

Proof. From Lemmas 3 and 4, we have

$$\sideset{}{'}\sum_{\substack{|m_i|\leqslant h\\1\leqslant i\leqslant s}}\frac{1}{\left(\|\boldsymbol{m}\|\prod_{i=1}^{t}\langle(\boldsymbol{r}_i,\boldsymbol{m})\rangle\right)^{\alpha}}$$

$$\ll \sum_{l=0}^{s}h^{-l\alpha}\sum_{i}\sum_{m_{i_{l+1}}=1}^{h}\cdots\sum_{m_{i_s}=1}^{h}\frac{1}{(m_{i_{l+1}}\cdots m_{i_s})^{\alpha+1}}$$

$$\times \sum_{|k_{i_1}|\leqslant h} \cdots \sum_{|k_{i_l}|\leqslant h} \sum_{|k_{i_{l+1}}|\leqslant m_{i_{l+1}}} \cdots {\sum_{|k_{i_s}|\leqslant m_{i_s}}}' \frac{1}{\prod_{i=1}^{t} \langle (r_i, m) \rangle}$$

$$\ll \sum_{l=0}^{s} h^{-l\alpha} \sum_{i} h^{l a \alpha} \sum_{m_{i_{l+1}}=1}^{h} \cdots \sum_{m_{i_s}=1}^{h} \frac{(\ln h)^{bs\alpha + t - 1 + \delta_{1,a}}}{(m_{i_{l+1}} \cdots m_{i_s})^{-(a-1)\alpha + 1}}$$

$$\ll h^{s(a-1)\alpha} (\ln h)^{bs\alpha + t - 1 + s\delta_{1,a} + \delta_{1,\alpha}}$$

where the constants implied by the symbol $\ll$ depend only on r, α, a, b, s, t. The lemma is proved.

Lemma 6. Suppose that x is a real number. Then

$$\left| \sum_{k=1}^{n} e^{2\pi i k x} \right| \leqslant \min\left(n, \frac{1}{2\langle x \rangle} \right).$$

Lemma 7. (Erdös-Turán-Koksma). Suppose that r and h are positive integers such that $r \geqslant 1$ and $h > r/\eta$, where η satisfies $0 < \eta < 1/6$. Then the discrepancy of the set $P_n(k)$ $(1 \leqslant k \leqslant n)$ of G_s satisfies

$$D(n) < {\sum_{\substack{|m_l|\leqslant h \\ 1\leqslant l\leqslant s}}}' \frac{1}{\|\pi m\|} \left| \frac{1}{n} \sum_{k=1}^{n} e^{2\pi i (m, P_n(k))} \right| + (5s + 6)\eta$$

$$+ \frac{s2^s r^{r-1}}{\pi^{s+r} \eta^r h^r} (\ln 64h)^{s-1}.$$

(Cf. Hua and Wang[2]).

The proof of Theorem 1. By Lemmas 5 and 6,

$${\sum_{\substack{|m_l|\leqslant h \\ 1\leqslant l\leqslant s}}}' \frac{1}{\|\pi m\|} \left| \frac{1}{q^t} \sum_{\substack{q_j=1 \\ 1\leqslant j\leqslant s}}^{q} e^{2\pi i \sum_{k=1}^{s} m_k (r_k, q)} \right|$$

$$\leqslant \frac{1}{q^t} {\sum_{\substack{|m_l|\leqslant h \\ 1\leqslant l\leqslant s}}}' \frac{1}{\|\pi m\|} \prod_{j=1}^{t} \left| \sum_{q_j=1}^{q} e^{2\pi i (r_j, m) q_j} \right|$$

$$\leqslant \frac{1}{(2q)^t} {\sum_{\substack{|m_l|\leqslant h \\ 1\leqslant l\leqslant s}}}' \frac{1}{\|\pi m\| \prod_{i=1}^{t} \langle (r_i, m) \rangle}$$

$$\leqslant c(r, a, b, s, t) q^{-t} h^{s(a-1)} (\ln h)^{bs + t + s\delta_{1,a}}.$$

Take $r = 1, \eta = (4q)^{-t}$ and $h = (4q)^{2t}$ in Lemma 7. The Theorem follows.

$$\S 3. \text{ The Consequence of Theorem 1.}$$

Lemma 8. For almost all $r \in R_{st}$, we have

$$\prod_{i=1}^{t} \langle (r_i, m) \rangle M(m) \geqslant c(r, \varepsilon), \tag{6}$$

where $a = 1$ and $b = t + \dfrac{\varepsilon}{s + t}$.

(Cf. Wang and Yu [5]).

From Theorem 1 and Lemma 8, we have

Theorem 3. Suppose that q is an integer $\geqslant 2$. Then for almost all $r \in R_{st}$, the set $((\gamma_1, q), \cdots, (\gamma_s, q))(\mathrm{mod}\,1)(1 \leqslant q_i \leqslant q, 1 \leqslant i \leqslant s)$ has discrepancy

$$D(q^t) \leqslant c(r, \varepsilon, s, t)q^{-t}(\ln q)^{st+s+t+\varepsilon} \tag{7}$$

Now we shall give another proof of Theorem 3 based on an idea of Littlewood (Cf. Schmidt [4]) which leads to an improvement of the logarithmic order of (7)

Lemma 9. Suppose that $m \not\equiv 0$ and $\delta > 0$. Then

$$I(m) = \int_{G_{st}} \Big(\prod_{i=1}^{t} \langle (r_i, m) \rangle | \ln \langle (r_i, m) \rangle |^{1+\delta} \Big)^{-1} dr = c(\delta).$$

Proof. Suppose that $s = 1$. Then $m_1 \not\equiv 0$ and

$$I(m_1) = \int_{G_t} \Big(\prod_{i=1}^{t} \langle r_{i1} m_1 \rangle | \ln \langle r_{i1} m_1 \rangle |^{1+\delta} \Big)^{-1} d\gamma_1.$$

let $\beta_1 = (\beta_{11}, \cdots, \beta_{t1})$, where $r_{i1}m_1 = \beta_{i1}$, $1 \leqslant i \leqslant t$. Then from the periodicity of $\langle x \rangle$, we have

$$I(m_1) = m_1^{-t} \int_0^{m_1} \cdots \int_0^{m_1} \Big(\prod_{i=1}^{t} \langle \beta_{i1} \rangle | \ln \langle \beta_{i1} \rangle |^{1+\delta} \Big)^{-1} d\beta_1$$

$$= \int_{G_t} \Big(\prod_{i=1}^{t} \langle \beta_{i1} \rangle | \ln \langle \beta_{i1} \rangle |^{1+\delta} \Big)^{-1} d\beta_1 = c(\delta).$$

Without loss of generality, we may suppose that $m_s \not\equiv 0$ for the case $s > 1$. Let $r_{i1}m_1 + \cdots + r_{i,s-1}m_{s-1} = \theta_i (1 \leqslant i \leqslant t)$. Then

$$I(m) = \int_{G_{(s-1)t}} d\gamma_1 \cdots d\gamma_{s-1} \int_{G_t} \Big(\prod_{i=1}^{t} \langle \theta_i + r_{is}m_s \rangle | \ln \langle \theta_i + r_{is}m_s \rangle |^{1+\delta} \Big)^{-1} d\gamma_s$$

$$= \int_{G_t} \Big(\prod_{i=1}^{t} \langle r_{is}m_s \rangle | \ln \langle r_{is}m_s \rangle |^{1+\delta} \Big)^{-1} d\gamma_s = c(\delta).$$

The lemma is proved.

Theorem 3′. The right hand side of (7) may be replaced by $c(r, \varepsilon, s, t)q^{-t} \times (\ln q)^{s+t+\varepsilon}$.

Proof. We know from Lemma 9 that the series

$$\sum{}' \Big(\prod_{i=1}^{s} \overline{m_i}(\overline{\ln \overline{m_i}})^{1+\delta} \Big)^{-1} I(m)$$

is convergent and the series

$$\sum{}' \Big(\prod_{i=1}^{s} \overline{m_i} \Big(\overline{\ln \overline{m_i}} \Big)^{1+\delta} \prod_{j=1}^{t} \langle (r_j, m) \rangle | \ln \langle (r_j, m) \rangle |^{1+\delta} \Big)^{-1} \tag{8}$$

is convergent for almost all $r \in R_{st}$. Take r such that the series (8) is convergent and the inequality (6) is satisfied. Then

$$\left| \ln \langle (r_i,\, m) \rangle \right| \leqslant \left| \ln c(r,\, \varepsilon) \right| + \ln \prod_{i=1}^{s} \overline{m_i}(\overline{\ln \overline{m_i}})^{t+\frac{\varepsilon}{s+t}}$$

$$\leqslant c(r,\, \varepsilon,\, s,\, t)\overline{\ln \|m\|}.$$

Put $\delta = \dfrac{\varepsilon}{s+t}$. Then

$$\sum_{\substack{|m_i| \leqslant h \\ 1 \leqslant i \leqslant s}}{}' \frac{1}{\|m\| \displaystyle\prod_{i=1}^{t} \langle (r_i,\, m) \rangle}$$

$$\leqslant \max_{\substack{|m_i| \leqslant h \\ m \neq 0}} \prod_{i=1}^{s} (\overline{\ln \overline{m_i}})^{1+\delta} \prod_{j=1}^{t} |\ln \langle (r_j,\, m) \rangle|^{1+\delta}$$

$$\times \Sigma' \left(\prod_{i=1}^{s} \overline{m_i}(\overline{\ln \overline{m_i}})^{1+\delta} \prod_{j=1}^{t} \langle (r_j,\, m) \rangle |\ln \langle (r_j,m) \rangle|^{1+\delta} \right)^{-1}$$

$$\leqslant c(r,\, \varepsilon,\, s,\, t)(\ln h)^{(s+t)(1+\delta)} \leqslant c(r,\, \varepsilon,\, s,\, t)(\ln h)^{s+t+\varepsilon}.$$

Take $r = 1$, $\eta = (4q)^{-t}$ and $h = (4q)^{2t}$ in Lemma 7. The theorem follows.

§ 4. THE PROOF OF THEOREM 2.

Lemma 10. Suppose that the sets $P_n(k) = (x_n^{(1)}(k),\, \cdots,\, x_n^{(s)}(k))\ (1 \leqslant k \leqslant n)$ and $Q_n(k) = (y_n^{(1)}(k),\, \cdots,\, y_n^{(s)}(k))\ (1 \leqslant k \leqslant n)$ have discrepancies $D(n)$ and $E(n)$ respectively and that

$$|x_i^{(n)}(k) - y_i^{(n)}(k)| \leqslant \delta,\ 1 \leqslant k \leqslant n,\ 1 \leqslant i \leqslant s. \tag{9}$$

Then

$$|D(n) - E(n)| \leqslant s\delta$$

Proof. For $a \in G_s$, let $a' = (\alpha_1',\, \cdots,\, \alpha_s')$ and $a'' = (\alpha_1'',\, \cdots,\, \alpha_s'')$, where

$$\alpha_i' = \begin{cases} \alpha_i - \delta, & \text{if } \alpha_i - \delta \geqslant 0, \\ 0, & \text{otherwise.} \end{cases}$$

and

$$\alpha_i'' = \begin{cases} \alpha_i + \delta, & \text{if } \alpha_i + \delta \leqslant 1, \\ 1, & \text{otherwise.} \end{cases}$$

Let $N_n(a)$ and $M_n(a)$ denote respectively the numbers of points $P_n(k)$ and $Q_n(k)$ $(1 \leqslant k \leqslant n)$ belonging to the domain

$$0 \leqslant x_1 < \alpha_1,\, \cdots,\, 0 \leqslant x_s < \alpha_s.$$

Then we have

$$M_n(a') \leqslant N_n(a) \leqslant M_n(a'')$$

by (9). Denote

$$\sigma_1 = \left| \frac{M_n(a')}{n} - |a| \right| \quad \text{and} \quad \sigma_2 = \left| \frac{M_n(a'')}{n} - |a| \right|.$$

Then

$$\left| \frac{N_n(\boldsymbol{\alpha})}{n} - |\boldsymbol{\alpha}| \right| \leqslant \max(\sigma_1, \ \sigma_2).$$

Since

$$0 \leqslant |\boldsymbol{\alpha}''| - |\boldsymbol{\alpha}| \leqslant \delta + \alpha_1 \left(\prod_{i=2}^{s} \alpha_i'' - \prod_{i=2}^{s} \alpha_i \right) \leqslant \cdots \leqslant s\delta,$$

we have

$$\sigma_2 \leqslant \left| \frac{M_n(\boldsymbol{\alpha}'')}{n} - |\boldsymbol{\alpha}''| \ \right| + |\boldsymbol{\alpha}''| - |\boldsymbol{\alpha}| \leqslant E(n) + s\delta.$$

σ_1 also satisfies the inequality. Hence

$$D(n) \leqslant E(n) + s\delta.$$

Similarly

$$E(n) \leqslant D(n) + s\delta.$$

The lemma follows.

The proof of Theorem 2. Since

$$\left| \frac{(\boldsymbol{h}_i, \ \boldsymbol{q})}{n} - (\boldsymbol{\gamma}_i, \ \boldsymbol{q}) \right| < c(\boldsymbol{r}, \ s) q n^{-1-\rho}, \ 1 \leqslant i \leqslant s,$$

it follows by Theorem 1 and Lemma 10 that the discrepancy of the set $\left(\dfrac{(\boldsymbol{h}_1, \ \boldsymbol{q})}{n}, \ \ldots, \ \dfrac{(\boldsymbol{h}_s, \ \boldsymbol{q})}{n} \right)$ $(\mathrm{mod}\,1)(1 \leqslant q_i \leqslant q, \ 1 \leqslant i \leqslant t)$ satisfies

$$D(q^t) \ll q^{-t+2st(a-1)}(\ln q)^{bs+t+s\delta_{1,a}} + q n^{-1-\rho}$$
$$\ll q^{-t+2st(a-1)}(\ln q)^{bs+t+s\delta_{1,a}},$$

where the constants implied by the symbol $\ll$ depend only on $\boldsymbol{r}$, a, b, s, t. The theorem is proved.

REFERENCES

[1] Hua Loo Keng and Wang Yuan, On uniform distribution and numerical analysis (Number-theoretic method) (I) *Sci. Sin*, **4**: 16(1973), 483—505; (II) *Sci. Sin*, **3**: 17(1974), 331—348; (III) *Sci. Sin*, **2**: 18(1975), 184—198.

[2] Hua Loo Keng and Wang Yuan, Applications of number theory to numerical analysis, Science Press, Beijing, 1978.

[3] H. Niederreiter, Methods for estimating discrepancy, Applications of Number Theory to numerical Analysis (S. K. Zaremba, ed.), Academic Press, New York (1972), 203—236.

[4] W. M. Schmidt, Metrical theorems on fractional parts of sequences, *Trans. Amer. Math. Soc.*, **110**(1964), 493—518.

[5] Wang Yuan and Yu Kun Rui, A note on some metrical theorems in diophantine approximation, IHES/M/79/297.

第 25 卷 第 3 期　　　　　数　学　学　报　　　　　Vol. 25, No. 3
1982 年 5 月　　　　ACTA MATHEMATICA SINICA　　　　May, 1982

ON DIOPHANTINE APPROXIMATION AND APPROXIMATE ANALYSIS (II)*

Wang Yuan (王　元)

(Institute of Mathematics, Academia Sinica)

§ 1. Introduction

Let α and C be two positive constants. Let $E_s^\alpha(C)$ be the class of functions of G_s:

$$f(x) \sim \sum C(m)e^{2\pi i(m, x)},$$

Where the Fourier coefficients satisfy

$$|C(m)| \leqslant \frac{C}{\|m\|^\alpha}.$$

The notations introduced in Wang[21] are also used in this paper. Let l be the least integer $\geqslant \alpha$ and $\mu_{n, l, k}$ be the integer defined by

$$\left(\sum_{k=-n}^{n} z^k \right)^l = \sum_{k=-nl}^{nl} \mu_{n, l, k} z^k.$$

Theorem 1. Suppose that n is an integer $\geqslant 2$ and $\alpha > 1$. Suppose further that the inequality

$$\prod_{i=1}^{t} \langle (r_i, m) \rangle M(m) \geqslant c(\gamma, a, b), \tag{1}$$

holds for any $m \neq \mathbf{0}$. Then

$$\sup_{f \in E_s^\alpha(C)} \left| \int_{G_s} f(x)dx - \frac{1}{(2n+1)^{lt}} \sum_{\substack{q_k=-nt \\ 1 \leqslant k \leqslant t}}^{nt} \prod_{j=1}^{t} \mu_{n, l, q_j} f((\gamma_1, q), \cdots, (\gamma_s, q)) \right|$$

$$\leqslant Cc(\gamma, \alpha, a, b, s, t)n^{-\alpha t + \frac{st(a-1)a^2}{\alpha - 1}} (\ln n)^{bsa+t-1+s\delta_{1, a}}.$$

Similar results were obtained by Bahvalov[1], Haselgrove[5] Wang[20], Hua and Wang[8,9], Korobov[13] and Niederreiter[15] for the case $t = 1$. It may beneficial in practical use to make suitable choices of γ and t.

§ 2. The proof of Theorem 1

Suppose that $f \in E_s^\alpha(C)$. Since $\alpha > 1$, f has the absolutely convergent Fourier expansion

$$f(x) = \sum C(m)e^{2\pi i(m, x)}, \quad |C(m)| \leqslant \frac{C}{\|m\|^\alpha}.$$

* Received January 23, 1981.

Correction: The first part of this paper, titled «On Diophantine Approximation and Approximate Analysis (I)» was also received on January 23, 1981.

Since

$$C(0) = \int_{G_s} f(x)dx$$

and

$$\frac{1}{(2n+1)^{lt}} \sum_{\substack{q_k=-nt \\ 1\leqslant k\leqslant t}}^{nt} \prod_{j=1}^{t} \mu_{n,l,q_j} f((\gamma_1, q), \cdots, (\gamma_s, q))$$

$$= \frac{1}{(2m+1)^{lt}} \sum C(m) \sum_{\substack{q_k=-nt \\ 1\leqslant k\leqslant t}}^{nt} \prod_{j=1}^{t} \mu_{n,l,q_j} e^{2\pi i \sum_{\nu=1}^{s} m_r(\gamma_\nu, q)}$$

$$= C(0) + \sum' C(m) \prod_{j=1}^{t} \left(\frac{1}{2n+1} \sum_{q_j=-n}^{n} e^{2\pi i (r_j, m) q_j} \right)^{l},$$

we have by Lemma 6 of [21],

$$\sup_{f\in E_s^\alpha(C)} \left| \int_{G_s} f(x)dx - \frac{1}{(2n+1)^{lt}} \sum_{\substack{q_k=-nt \\ 1\leqslant k\leqslant t}}^{nt} \prod_{j=1}^{t} \mu_{n,l,q_j} f((\gamma_1, q), \cdots, (\gamma_s, q)) \right|$$

$$\leqslant C \sum' \frac{1}{\|m\|^\alpha} \prod_{j=1}^{t} \left| \frac{1}{2n+1} \sum_{q_j=-n}^{n} e^{2\pi i (r_j, m) q_j} \right|^{l}$$

$$\leqslant C \sum' \frac{1}{\|m\|^\alpha} \prod_{j=1}^{t} \left| \frac{1}{2n+1} \sum_{q_j=-n}^{n} e^{2\pi i (r_j, m) q_j} \right|^{a}$$

$$\leqslant C(\Sigma_1 + \Sigma_2),$$

where

$$\Sigma_1 = \frac{1}{(2n+1)^{\alpha t}} \sum_{\substack{|m_k|\leqslant h \\ 1\leqslant k\leqslant s}}' \frac{1}{\|m\|^\alpha \prod_{j=1}^{t} \langle (r_j, m) \rangle^\alpha}$$

ana

$$\Sigma_2 = \sum_{k=1}^{s} \sum_{|m_k|>h} \frac{1}{|m_k|^\alpha} \sum \frac{1}{\|m\|^\alpha},$$

in which h is an integer $\geqslant 2$ that will be determined later.

By Lemma 5 of [21], we have

$$\Sigma_1 \leqslant c(\gamma, \alpha, a, b, s, t) n^{-at} h^{s(a-1)\alpha} (\ln h)^{bsa+s-1+s\delta_{1,\alpha}},$$

Evidently

$$\Sigma_2 \leqslant c(\alpha, s) h^{-\alpha+1}.$$

Take $h = [n^{\frac{at}{\alpha-1}}]$. The theorem follows.

From Theorem 1 and Lemma 8 of [21], we have

Theorem 2. Suppose that n is an integer $\geqslant 2$ and $\alpha > 1$. Then the estimation

$$\sup_{f \in E_s^\alpha(C)} \left| \int_{G_s} f(\boldsymbol{x})d\boldsymbol{x} - \frac{1}{(2n+1)^{lt}} \sum_{\substack{q_k=-nt \\ 1 \leqslant k \leqslant t}}^{nt} \prod_{j=1}^{t} \mu_{n,l,q_j} f((\boldsymbol{\gamma}_1, \boldsymbol{q}), \cdots, (\boldsymbol{\gamma}_s, \boldsymbol{q})) \right|$$

$$\leqslant Cc(\boldsymbol{\gamma}, \alpha, \varepsilon, s, t)n^{-\alpha t}(\ln n)^{\alpha st+s+t-1+\varepsilon}, \tag{2}$$

holds for almost all $\boldsymbol{\gamma} \in R_{st}$.

Remark. By the use of the inequality

$$\sideset{}{'}\sum_{\substack{|m_k| \leqslant h \\ 1 \leqslant k \leqslant s}} \frac{1}{\|\boldsymbol{m}\|^\alpha \prod_{j=1}^{t} \langle (\boldsymbol{\gamma}_j, \boldsymbol{m}) \rangle^\alpha} \leqslant \left(\sideset{}{'}\sum_{\substack{|m_k| \leqslant h \\ 1 \leqslant k \leqslant s}} \frac{1}{\|\boldsymbol{m}\| \prod_{j=1}^{t} \langle (\boldsymbol{r}_j, \boldsymbol{m}) \rangle} \right)^\alpha \quad (\alpha \geqslant 1)$$

and the argument of the proof of Theorem $3'$ of [21], it may be proved that the right hand side of (2) may be replaced by

$$Cc(\boldsymbol{\gamma}, \alpha, \varepsilon, s, t)n^{-\alpha t}(\ln n)^{\alpha(s+t)+\varepsilon}.$$

§ 3. The case $\alpha = 2$.

For the case $\alpha = 2$, the weight $\mu_{n,l,k}$ in Theorem 1 may be simplified.

Lemma 1. Suppose that x is a real number. Then

$$\left| \sum_{k=-n}^{n} (n-|k|)e^{2\pi ikx} \right| \leqslant \min\left(n^2, \frac{1}{4\langle x \rangle^2}\right).$$

Theorem 3. Under the assumption of Theorem 1, we have

$$\sup_{f \in E_s^2(C)} \left| \int_{G_s} f(\boldsymbol{x})d\boldsymbol{x} - \frac{1}{n^t} \sum_{\substack{q_k=-n \\ 1 \leqslant k \leqslant t}}^{n} \prod_{j=1}^{t} \left(1 - \frac{|q_j|}{n}\right) f((\boldsymbol{\gamma}_1, \boldsymbol{q}), \cdots, (\boldsymbol{\gamma}_s, \boldsymbol{q})) \right|$$

$$\leqslant Cc(\boldsymbol{\gamma}, a, b, s, t)n^{-2t+4st(a-1)}(\ln n)^{2bs+t-1+s\delta_{1,a}}.$$

Proof.

$$\frac{1}{n^t} \sum_{\substack{q_k=-n \\ 1 \leqslant k \leqslant t}}^{n} \prod_{j=1}^{t} \left(1 - \frac{|q_j|}{n}\right) f((\boldsymbol{\gamma}_1, \boldsymbol{q}), \cdots, (\boldsymbol{\gamma}_s, \boldsymbol{q}))$$

$$= n^{-2t} \sum_{\substack{q_k=-n \\ 1 \leqslant k \leqslant t}}^{n} \prod_{j=1}^{t} (n-|q_j|) \sum C(\boldsymbol{m})e^{2\pi i \sum_{\nu=1}^{s} m_\nu(\boldsymbol{\gamma}_\nu, \boldsymbol{q})}$$

$$= C(\boldsymbol{0}) + \sideset{}{'}\sum C(\boldsymbol{m})n^{-2t} \prod_{j=1}^{t} \left(\sum_{q_j=-n}^{n} (n-|q_j|)e^{2\pi i(r_j, \boldsymbol{m})q_j} \right),$$

Hence by Lemma 1, we have

$$\sup_{f \in E_s^2(C)} \left| \int_{G_s} f(\boldsymbol{x})d\boldsymbol{x} - \frac{1}{n^t} \sum_{\substack{q_k=-n \\ 1 \leqslant k \leqslant t}}^{n} \prod_{j=1}^{t} (n-|q_j|) f((\boldsymbol{\gamma}_1, \boldsymbol{q}), \cdots, (\boldsymbol{\gamma}_s, \boldsymbol{q})) \right|$$

$$\leqslant C n^{-2t}\sum{}' \frac{1}{\|\boldsymbol{m}\|^2} \min\left(\frac{1}{\prod\limits_{j=1}^{t} \langle(\boldsymbol{r}_j, \boldsymbol{m})\rangle^2}, \ n^{2t}\right).$$

The theorem may be proved easily by the argument of the proof of Theorem 1.

§ 4. THE NUMERICAL INTEGRATION OVER $Q_s^a(C)$.

Let

$$\mu(x) = \begin{cases} \left(\cos\left(\frac{\pi}{2}\log_2|x|\right)\right)^2, & \text{if } \frac{1}{2} \leqslant |x| \leqslant 2, \\ 0, & \text{otherwise,} \end{cases}$$

$$\mu_t(x) = \mu(2^{1-t}x)$$

and

$$\mu_0(x) = 1 - \sum_{t=1}^{\infty} \mu_t(x).$$

suppose that $f(\boldsymbol{x})$ is a periodic function of s variables of which each variable has period 1 and that $f(\boldsymbol{x})$ has the Fourier expansion

$$f(\boldsymbol{x}) \sim \sum C(\boldsymbol{m})e^{2\pi i(\boldsymbol{m}, \boldsymbol{x})}.$$

If the series

$$\sum C(\boldsymbol{m})\lambda(\boldsymbol{m})e^{2\pi i(\boldsymbol{m}, \boldsymbol{x})}$$

is convergent almost everywhere, then its sum is denoted by $f(\boldsymbol{x})\odot\lambda(\boldsymbol{m})$. Let

$$\varphi_t(\boldsymbol{x}) = f(\boldsymbol{x})\odot \prod_{k=1}^{s} \mu_{t_k}(m_k) = \sum C_t(\boldsymbol{m})e^{2\pi i(\boldsymbol{m}, \boldsymbol{x})},$$

where $\boldsymbol{t} = (t_1, \cdots, t_s)$ denotes the vector with non-negative integer components and

$$C_t(\boldsymbol{m}) = C(\boldsymbol{m})\mu_{t_1}(m_1)\cdots\mu_{t_s}(m_s).$$

Let $Q_s^a(C)(\alpha > 0, \ C > 0)$ denote the set of continuous functions of G_s satisfying

$$\sup_{\boldsymbol{x}\in G_s} |\varphi_t| \leqslant C2^{-\alpha t_0}, \quad t_0 = t_1 + \cdots + t_s.$$

Lemma 2. $Q_s^a(C) \subset E_s^a(C2^s)$.

Lemma 3. If $f(\boldsymbol{x}) \in Q_s^a(C)$, then

$$f(\boldsymbol{x}) = \sum{}'' \varphi_t(\boldsymbol{x}),$$

where $\sum''$ denotes a sum in which $\boldsymbol{t}$ runs over all integral vectors with non-negative integer components.

(Cf. Bahvalov[2,3], Hua and Wang[10]).

Theorem 4. Suppose that $0 < \alpha \leqslant 1$. Then under the assumption of Theorem 1, we have

$$\sup_{f\in Q_s^a(C)}\left|\int_{G_s}f(\boldsymbol{x})d\boldsymbol{x}-\frac{1}{n^t}\sum_{\substack{q_k=1\\1\leqslant k\leqslant t}}^{n}f((\boldsymbol{\gamma}_1,\boldsymbol{q}),\cdots,(\boldsymbol{\gamma}_s,\boldsymbol{q}))\right|$$

$$\leqslant Cc(\boldsymbol{\gamma},\alpha,a,b,s,t)n^{-c_1+st(a-1)}(\ln n)^{bs+t+s\delta_{1,a}}.$$

Proof. Suppose that $f\in Q_s^a(C)$. Let

$$S(f)=\int_{G_s}f(\boldsymbol{x})d\boldsymbol{x}-\frac{1}{n^t}\sum_{\substack{q_k=1\\1\leqslant k\leqslant t}}^{n}f((\boldsymbol{\gamma}_1,\boldsymbol{q}),\cdots,(\boldsymbol{\gamma}_s,\boldsymbol{q})).$$

Then from Lemma 3, we have

$$S(f)=\sum{}''S(\varphi_t),$$

where

$$S(\varphi_t)=\int_{G_s}\varphi_t(\boldsymbol{x})d\boldsymbol{x}-\frac{1}{n^t}\sum_{\substack{q_k=1\\1\leqslant k\leqslant t}}^{n}\varphi_t((\boldsymbol{\gamma}_1,\boldsymbol{q}),\cdots,(\boldsymbol{\gamma}_s,\boldsymbol{q})).$$

Hence

$$\sup_{f\in Q_s^a(C)}|S(f)|\leqslant\Sigma_1+\Sigma_2,$$

where

$$\Sigma_1=\sup_{f\in Q_s^a(C)}\sum_{t_0\geqslant\log_2 n^t}{}''|S(\varphi_t)|$$

and

$$\Sigma_2=\sup_{f\in Q_s^a(C)}\sum_{t_0<\log_2 n^t}{}''|S(\varphi_t)|.$$

Obviously

$$\Sigma_1\leqslant C\sum_{t_0\geqslant\log_2 n^t}2^{1-\alpha t_0}\leqslant Cc(s)\sum_{l=[\log_2 n^t]}^{\infty}2^{-\alpha l}l^{s-1}\leqslant Cc(\alpha,s)n^{-\alpha t}(\ln n)^{s-1}.$$

Since $C_t(\boldsymbol{m})=0$ for $\|\boldsymbol{m}\|\geqslant 2^{t_0}$, we have

$$\sum_{\substack{q_j=1\\1\leqslant j\leqslant t}}^{n}\sum_{\|\boldsymbol{m}\|<2^{t_0}}C_t(\boldsymbol{m})e^{2\pi i\sum_{k=1}^{t}m_k(\boldsymbol{r}_k,\boldsymbol{q})}$$

$$=n^tC_t(\boldsymbol{0})+\sum_{\|\boldsymbol{m}\|<2^{t_0}}{}'C_t(\boldsymbol{m})\prod_{j=1}^{t}\left(\sum_{q_j=1}^{n}e^{2\pi i(\boldsymbol{r}_j,\boldsymbol{m})q_j}\right)$$

and so

$$|S(\varphi_t)|\leqslant n^{-t}\sum_{\|\boldsymbol{m}\|<2^{t_0}}{}'|C_t(\boldsymbol{m})|\frac{1}{\prod_{j=1}^{t}\langle(\boldsymbol{r}_j,\boldsymbol{m})\rangle}.$$

Hence by Lemma 2 and Lemma 5 of [21], we have

$$\Sigma_2 \leqslant n^{-t} \sideset{}{'}\sum_{\|m\|<n^t} \frac{1}{\prod_{j=1}^{t} \langle (r_j, m) \rangle} \Sigma'' |C_t(m)|$$

$$\leqslant n^{-t} \sideset{}{'}\sum_{\|m\|<n^t} |C(m)| \frac{1}{\prod_{j=1}^{t} \langle (r_j, m) \rangle}$$

$$\leqslant Cn^{-t} \sideset{}{'}\sum_{\|m\|<n^t} \frac{1}{\|m\|^{\alpha} \prod_{j=1}^{t} \langle (r_j, m) \rangle}$$

$$= Cn^{-\alpha t} \sideset{}{'}\sum_{\|m\|<n^t} \frac{1}{\|m\| \prod_{j=1}^{t} \langle (r_j, m) \rangle}$$

$$\leqslant Cc(\gamma, \alpha, a, b, s, t)n^{-\alpha t+st(\alpha-1)}(\ln n)^{bs+t+s\delta_1, \epsilon}.$$

The theorem follows.

From Theorem 4 and Lemma 8 of [21], we have

Theorem 5. Suppose that $0 < \alpha \leqslant 1$ and n is an integer $\geqslant 2$. Then

$$\sup_{f \in Q_s^{\alpha}(C)} \left| \int_{G_s} f(x)dx - \frac{1}{n^t} \sum_{\substack{q_k=1 \\ 1 \leqslant k \leqslant t}}^{n} f((\gamma_1, q), \cdots, (\gamma_s, q)) \right|$$

$$\leqslant Cc(\gamma, \alpha, \epsilon, s, t)n^{-\alpha t}(\ln n)^{st+s+t+\epsilon}. \tag{3}$$

holds for almost all $\gamma \in R_{st}$.

Remark. The right hand side of (3) may be improved to

$$Cc(\gamma, \alpha, \epsilon, s, t)n^{-\alpha t}(\ln n)^{s+t+\epsilon}$$

by the method used in the proof of Theorem 3' of [21].

§ 5. The error term of quadrature formula

Let

$$S(n, \gamma, f) = \int_{G_s} f(x)dx - \frac{1}{n^t} \sum_{\substack{q_k=-n \\ 1 \leqslant k \leqslant n}}^{n} \left(1 - \frac{|q_k|}{n}\right) f((\gamma_1, q), \cdots, (\gamma_s, q)).$$

Theorem 6.

$$\sup_{f \in E_s^2(C)} |S(n, \gamma, f)| \leqslant CW(n, f),$$

where

$$W(n, f) = n^{-t} \sum_{\substack{q_j=-n \\ 1 \leqslant j \leqslant t}}^{n} \left(1 - \frac{|q_j|}{n}\right) \prod_{k=1}^{s} (1 + 2\pi^2 B_2(\{(\gamma_k, q)\})) - 1.$$

in which $B_2(x) = x^2 - x + \dfrac{1}{6}$ denotes the Bernoulli polynomial.

Proof. It follows from the argument of the proof of Theorem 3 that the $W(n, f)$ may be given by

$$W(n, f) = n^{-2t} \sum' \frac{1}{\|\boldsymbol{m}\|^2} \sum_{\substack{q_j=-n \\ 1\leqslant j\leqslant t}}^{n} \prod_{k=1}^{t} (n - |q_k|) e^{\,2\pi i \sum\limits_{\nu=1}^{s} m_\nu(\boldsymbol{\gamma}_\nu,\, \boldsymbol{q})}$$

$$= n^{-2t} \sum_{\substack{k_\nu=0 \\ 1\leqslant \nu\leqslant t}}^{n-1} \sum_{\substack{q_j=-k_\nu \\ 1\leqslant j\leqslant t}}^{k_\nu} \prod_{k=1}^{s} \left(\sum_{m_k=-\infty}^{\infty} \frac{e^{2\pi i(\boldsymbol{\gamma}_k,\,\boldsymbol{q})m_k}}{\overline{m}_k^2} \right) - 1.$$

Since

$$\sum_{m=-\infty}^{\infty} \frac{e^{2\pi i m x}}{\overline{m}^2} = 1 + 2\pi^2 B_2(\{x\}),$$

We have

$$W(n, f) = n^{-2t} \sum_{\substack{k_\nu=0 \\ 1\leqslant \nu\leqslant t}}^{n-1} \sum_{\substack{q_j=-k_\nu \\ 1\leqslant j\leqslant t}}^{k_\nu} \prod_{k=1}^{s} (1 + 2\pi^2 B_2(\{(\boldsymbol{\gamma}_k,\, \boldsymbol{q})\})) - 1$$

$$= n^{-t} \sum_{\substack{q_j=-n \\ 1\leqslant j\leqslant t}}^{n} \left(1 - \frac{|q_j|}{n}\right) \prod_{k=1}^{s} (1 + 2\pi^2 B_2(\{(\boldsymbol{\gamma}_k,\, \boldsymbol{q})\})) - 1.$$

The theorem is proved.

Remark. In practical use, it may be suggested to take $\gamma_{ij}(1 \leqslant i \leqslant t, 1 \leqslant j \leqslant s)$ from an integral basis of a cyclotomic field $Q\left(\cos\dfrac{2\pi}{m}\right)$, where $\varphi(m) > 2st$. For example, we may take $\gamma_{ij}'s$ to be $2\cos\dfrac{2\pi l}{17}\,(1 \leqslant l \leqslant 6)$ for $s = 3$, $t = 2$ and $2\cos\dfrac{2\pi l}{19}\,(1 \leqslant l \leqslant 8)$ for $s = 4$, $t = 2$.

§ 6. Several conjectures

Concerning the assumption (1), we propose two conjectures.

$1°$. Suppose that $s \geqslant t$ and $\gamma_{ij}\ (1 \leqslant i \leqslant t,\ 1 \leqslant j \leqslant s)$ are st real algebraic numbers such that the determinants of all $t \times t$ minor matrices of $\begin{pmatrix} \gamma_{11}, & \cdots, & \gamma_{1s} & 1 & & 0 \\ & \cdots\cdots & & & \ddots & \\ \gamma_{t1}, & \cdots, & \gamma_{ts} & 0 & & 1 \end{pmatrix}$ are linearly independent over Q. Then

$$\prod_{i=1}^{t} \langle (\boldsymbol{r}_i,\, \boldsymbol{m}) \rangle \|\boldsymbol{m}\|^{1+\varepsilon} > c(\boldsymbol{\gamma},\, \varepsilon), \quad \boldsymbol{m} \neq \boldsymbol{0}.$$

$2°$. Suppose that $s \geqslant t$ and $\gamma_{ij} = c^{r_{ij}}$ where $r_{ij}\,(1 \leqslant i \leqslant t, 1 \leqslant j \leqslant s)$ are st different non-zero rational numbers. Suppose further that the determinants of all $t \times t$ minor matrices of $\begin{pmatrix} \gamma_{11}, & \cdots, & \gamma_{1s} & 1 & & 0 \\ & \cdots\cdots & & & \ddots & \\ \gamma_{t1}, & \cdots, & \gamma_{ts} & 0 & & 1 \end{pmatrix}$ are linearly independent over Q. Then

$$\prod_{i=1}^{t} \langle (r_i, m) \rangle \|m\|^{1+\varepsilon} > c(\gamma, \varepsilon), \quad m \neq 0.$$

1° and 2° are Schmidt[19] theorem and Baker[4] theorem respectively for the case $t = 1$. Let n be an integer ≥ 2. Consider the systems of linear congruences

$$\sum_{j=1}^{s} a_{ij} x_j + x_{s+i} \equiv 0 \pmod{n}, \quad 1 \leq i \leq t \tag{4}$$

and

$$\sum_{i=1}^{t} a_{ij} y_i + y_{j+t} \equiv 0 \pmod{n}, \quad 1 \leq j \leq s \tag{5}$$

and their solutions satisfying $-\left[\dfrac{n-1}{2}\right] \leq x_i, y_j \leq \left[\dfrac{n}{2}\right] (1 \leq i, j \leq n)$. Evidently $x_1 = \cdots = x_{s+t} = 0$ and $y_1 = \cdots = y_{s+t} = 0$ are the solutions of (4) and (5) respectively and they are called the trivial solutions. The other solutions are called the non-trivial solutions. Let $q = q(a_{11}, \cdots, a_{ts})$ denote the minimum of the product $\bar{x}_1 \cdots \bar{x}_{s+t}$, where $x_1, \cdots, x_{s+t}$ runs over all non-trivial solutions of (4). Further let

$$q^* = q^*(n) = \max_{1 \leq a_{ij} \leq n} q(a_{11}, \cdots, a_{ts}).$$

For equation (5), we may define $Q = Q(a_{11}, \cdots, a_{ts})$ and $Q^* = Q^*(n)$ similarly. Concerning q, Q and q^*, we have the following two conjectures.

3°. Let n be a positive integer. Then $q(n)$ and $Q(n)$ satisfy the inequality

$$\frac{Q^{s+t-1}}{q} \leq c(s, t) n^{(s+t)(s-1)}. \tag{6}$$

4°.

$$q^*(n) \geq c(s, t) n^t. \tag{7}$$

Let p_l denote the l-th prime number. Gelfond proved the inequality (6) for the case $n = p_l, s \geq 1$ and $t = 1$ (Cf. Korobov[14]) and Wang, Wang and Ren[22] generalized the result to the case $n = p_l$ and $s, t \geq 1$. Concerning the inequality (7), it is well-known that

$$q^*(F_{l+2}) > cF_{l+2} \text{ for the case } s=t=1, \text{ where } F_{l+2} = \frac{1}{\sqrt{5}}\left(\left(\frac{1+\sqrt{5}}{2}\right)^{l+2} - \left(\frac{1-\sqrt{5}}{2}\right)^{l+2}\right)$$

$(l \geq 1)$ denotes the Fibonacci sequence. Bahvalov [1] proved $q^*(p_l) \geq c(s)p_l/(\ln p_l)^s$ for the case $s \geq 1, t = 1$ and Wang, Wang and Ren[22] established $q^*(p_l) \geq c(s, t)p_l^t/(\ln p_l)^{s+t-1}$ for $s \geq 1, t \geq 1$.

5°. There exist a set of integers (n_l) $(1 < n_1 < n_2 < \cdots)$ and integral vectors $\boldsymbol{a} = \boldsymbol{a}(n_l) = (a_1, \cdots, a_s)(l = 1, 2, \cdots)$ such that

$$W(n_l) = \sup_{f \in E_s^\alpha(C)} \left| \int_{G_s} f(\boldsymbol{x}) d\boldsymbol{x} - \frac{1}{n_l} \sum_{k=1}^{n_l} f\left(\frac{a_1 k}{n_l}, \cdots, \frac{a_s k}{n_l}\right) \right|$$

$$\leq Cc(\alpha, s)(\ln n_l)^{s-1}/n_l^\alpha (\alpha > 1). \tag{8}$$

Hua and Wang[7] and Bahvalov[1] proved (8) independently for $s = 2, n_l = F_{l+2}$ and

$\boldsymbol{a}=(1,F_{l+1})(l=1,2,\cdots)$. Korobov[12] established that there exists an $\boldsymbol{a}=\boldsymbol{a}(p_l)$ such that $W(p_l)\leqslant Cc(\alpha,s)\dfrac{(\ln p_l)^{\alpha s}}{p_l^\alpha}$ and it was improved later to $W(p_l)\leqslant Cc(\alpha,s)\dfrac{(\ln p_l)^{\alpha(s-1)}}{p_l}$ by Bahvalov[1]. Sarygin[17] proved that $W(n)\geqslant Cc(\alpha,s)\dfrac{(\ln n)^{s-1}}{n^\alpha}$ holds for any integer $n\geqslant 2$ and integral vector $\boldsymbol{a}$.

Conjecture 5° may be derived from conjecture 4° for the case $t=1$

6°. There exist a set of integers (n_l) and integral vectors $\boldsymbol{a}=\boldsymbol{a}(n)$ such that the set $\left(\left\{\dfrac{a_1 k}{n_l}\right\},\cdots,\left\{\dfrac{a_s k}{n_l}\right\}\right)(1\leqslant k\leqslant n_l)$ has discrepancy

$$D(n_l)\leqslant c(s)\frac{(\ln n_l)^{s-1}}{n_l}$$

Zaremba[23] proved $D(F_{l+2})<c\,\dfrac{\ln F_{l+2}}{F_{l+2}}$ for $s=2$ and $\boldsymbol{a}=(1,F_{l+1})$. Korobov[13] and Hlawka[6] established independently that there exists an $\boldsymbol{a}=\boldsymbol{a}(p_l)$ such that $D(p_l)<c(s)\dfrac{(\ln p_l)^s}{p_l}$ for $s\geqslant 2$. Niederreiter[16] proved that there exists an $\boldsymbol{a}=\boldsymbol{a}(n)$ such that the relation $D(n)<c(s)\dfrac{(\ln n)^s}{n}$ holds for $s\geqslant 2$ and all integers $n\geqslant 2$.

It seems that the set (n_l) in, 5°, 6° may be even proposed to be the set of all positive integers $\geqslant 2$.

7°. For almost all $\boldsymbol{\gamma}=(\gamma_{11},\cdots,\gamma_{ts})$, the set $((\boldsymbol{\gamma}_1,\boldsymbol{q}),\cdots,(\boldsymbol{\gamma}_s,\boldsymbol{q}))\,(\mathrm{mod}\,1)$ $(1\leqslant q_i\leqslant q,\ 1\leqslant i\leqslant s)$ has discrepancy

$$D(q^t)\leqslant c(\boldsymbol{\gamma},\varepsilon,s,t)q^{-t}(\ln q)^{s+t-1+\varepsilon}.\tag{9}$$

By the use of continued fraction, Khintchine[11] proved the conjecture for $s=t=1$. Schmidt[18] proved that $D(q)\leqslant c(\boldsymbol{\gamma},\varepsilon,s)q^{-1}(\ln q)^{s+1+\varepsilon}$ holds for almost all $\boldsymbol{\gamma}$ for the case $s\geqslant 1$, $t=1$ and later Wang[21] generalized his result to $D(q^t)\leqslant c(\boldsymbol{\gamma},\varepsilon,s,t)q^{-t}(\ln q)^{s+t+\varepsilon}$ for $s\geqslant 1$, $t\geqslant 1$.

REFERENCES

[1] Bahvalov N. S., Approximate computation of multiple integals, *Vestnik Moscow Univ. Ser. Mat. Meh. Astr. Fiz. Him.*, 4 (1959), 3—18.

[2] Bahvalov N. S., On embedding theorems for class of functions with bounded derivatives, *Vestnik Moscow Univ. Ser. Mat. Meh. Astr. Fiz. Him.*, 3 (1963), 7—16.

[3] Bahvalov N. S., Optimal convergence bounds for quadrature processes and integration methods of Monte Carlo type for classes of functions, *Z. Vycisl. Mat. i Mat. Fiz.*, 4 Suppl (1964), 5—63.

[4] Baker A., On some diophantine inequalities involving the exponential function, *Canad. J. Math.*, 17 (1965), 616—626.

[5] Haselgrove C. B. A method for numerical integration, *Math. Comp.*, 15(1961), 323—337.

[6] Hlawka E., Uniform distribution modulo 1 and numerical integation, *Comp. Math.*, 16 (1964), 95—105.

[7] Hua Loo Keng and Wang Yuan, Remarks concerning numerical integration, *Sci Rec. New Ser.*, 4 (1960), 8—11.

[8] Hua Loo Keng and Wang Yuan, Numerical integration and its applications, Science Press, Beijing,

1963.

[9] Hua Loo Keng and Wang Yuan, On uniform distribution and numerical analysis (Number theoretic method) (I); (II); (III), *Sci. Sin.*, 4 (1973), 483—505; 3 (1974), 331—348; 2 (1975), 184—198.

[10] Hua Loo Keng and Wang Yuan, Applications of number theory to numerical analysis, Science Press, Beijing, 1978.

[11] Khintchine A., Ein Satz über Kettenbrüche, mit arithmetischon Anwendungen, Math., Z. (1923), 289—306.

[12] Korobov N. M., The approximate computation of multiple integrals, *Dokl. Akad. Hauk SSSR*, 124 (1959), 1207—1210.

[13] Korobov N. M., Number-theoretic methods in approximate analysis Fizmatigiz, Moscow, 1963.

[14] Korobov N. M., Several problems in the theory of diophantine approximation, *Uspehi Mat. Nauk SSSR*, 3 (1967), 83—118.

[15] Niederreiter H., Application of diophantine approximations to numerical integration, Diophantine approximation and its applications (C. F. Osgood, ed.), Academic Press, New York, 1973, 129—199.

[16] Niederreiter H., Existence of good lattice points in the sence of Hlawka, *Monatsh. Math.*, 86 (1978), 203—219.

[17] Sarygin I. F., A lower estimation for the error of quadrature formulas for certain classes of functions, Z. *Vycisl. Mat. i Mat, Fiz.*, 3(1963), 370—376.

[18] Schmidt W. M., Metrical theorems on fractional Parts of seguences, *TAMS.*, 110(1964), 493—518.

[19] Schmidt W. M., Simultaneous approximation to algebraic numbers by rationals, *Acta Math.*, 125 (1970), 189—201.

[20] Wang Yuan, On numerical integration and its applications (Number-theoretic method), *Shuxue Jinzhan*, 5 (1962), 1—44.

[21] Wang Yuan, On diophantine approximation and approximate analysis (I), *Acta Mathematica Sinica*, 25: 2(1982), 248—256.

[22] Wang Yuan, Wang Ling Xiang and Ren Jian Hua, A note on a transference theorem of the systems of congruences (to appear).

[23] Zaremba S. K., Good lattice points, discrepancy and numerical integration, *Ann. Mat. Pure Appl.*, 73 (1966), 293—317.

Chin. Ann. of Math.
11B:1(1990), 51—65.

NUMBER THEORETIC METHOD IN
APPLIED STATISTICS***

WANG YUAN(王　元)*　FANG KAITAI(方开泰)**
(Dedicated to the Tenth Anniversary of CAM)

Abstract

This paper gives some applications of number-theoretic method (or quasi Monte Carlo
method) for numerical evaluation of probabilities and moments of a continuous multivariate
distribution over a special domain such as cube, ball, sphere, simplex, etc., where the
uniformly distributed sets of points in such domains, which are useful in experimental
design, simulation, geometry probability, etc., are suggested. Some applications of number-
theoretic method in optimization are discussed also.

§1. Introduction

The problem of numerical evaluation of probabilities and moments is really a
problem of numerical integration. The number-theoretic method (or quasi Monte
Carlo method) for numerical evaluation of multiple integrals and for optimization is
based on the theory of uniform distribution $(u.\ d.)$. Let $K = [a_1,\ b_1] \times \cdots \times [a_s,\ b_s]$ be
a rectangle of R^s, $\boldsymbol{b} = (b_1,\ \cdots,\ b_s)'$, $\boldsymbol{x} = (x_1,\ \cdots,\ x_s)'$, and $F(\boldsymbol{x})$ be a continuous monotone
distribution function on K, which satisfies $F(\boldsymbol{b}) = 1$ and $F(\boldsymbol{x}) = 0$ whenever at least
one of the x_i is a_i. Note that a_i' s and b_i's may be defined to be $-\infty$ and ∞ respectively.
We use $\boldsymbol{x} \leqslant \boldsymbol{b}$ to denote that $x_i \leqslant b_i (i = 1,\ \cdots,\ s)$. For a set of points $P = (\boldsymbol{x}_k,\ k = 1,\ \cdots,\ n)$ in K and a rectangle $G = [a_1,\ x_1] \times \cdots \times [a_s,\ x_s]$, where $\boldsymbol{x} \leqslant \boldsymbol{b}$, let $N(P,\ G)$ be the
number of P satisfying $\boldsymbol{x}_k \in G$, and let

$$\sup_{G} \left| \frac{N(P,\ G)}{n} - F(\boldsymbol{x}) \right| = D_F(n,\ P).$$

$D_F(n,\ P)$ is called the F-discrepancy of P with respect to $F(\boldsymbol{x})$. If $P_n = (\boldsymbol{x}_1^{(n)},\ \cdots,\ \boldsymbol{x}_{k_n}^{(n)})$ is a sequence in K such that $k_n \to \infty$ as $n \to \infty$. and if $D_F(k_n,\ P) = o(1)$ as $n \to \infty$,
then P_n is called an F-uniformly distributed sequence. If $K = I^s$, where $I = [0,\ 1]$
and $F(\boldsymbol{x})$ is uniform distribution on I^s, i. e., $F(\boldsymbol{x}) = x_1 \cdots x_s$, then we omit the F in
the above notations (cf. Weyl [12], Hlawka and Muck [5], and Niderreiter [9]).

Manuscript received March 15, 1989.
* Institute of Mathematics, Academia Sinica, Beijing, China.
** Institute of Applied Mathematics, Academia Sinica, Beijing, China.
*** This Work is Supported by the Chinese National Science Foundation and the Science Foundation of
the Chinese Academy of Sciences.

If $P = \{x_k, k = 1, \cdots, n\}$ is a set I^s with decrepancy $D(n, P)$ or $D(n)$ for simplify and $f(x)$ is a functien of bounded variation in the sense of Hardy and Krause with total variation $V(f)$, then it is known that

$$\left| \int_{I^s} f(x) dx - \frac{1}{n} \sum_{k=1}^n f(x_k) \right| \leqslant V(f) D(n) \tag{1.1}$$

(See Koksma [7], Hlawka [4], Hua and Wang [6]).

Let D be a domain (for example, ball, sphere, simplex, etc.) in R^s. In this paper we shall pay more attention to numerical evaluation of

$$I = \int_D f(x) dv, \tag{1.2}$$

where dv is the volume element of D and D has a parameter representation. First by using a transformation a quadrature formula over D can be transferred to a quadrature formula over I^t, where t is the dimension of D. Another approach to this problem is to use a u. d. sequence in D. We shall start from a u. d. sequence in I^t. and then derive the u. d. sequences with respect to certain distribution functions, in particular, to some uniform distribution functions in some special domains: ball, sphere, simplex, etc., which are often useful in simulation, geometry probability, experimental design and many problems in statistics. More details are given in our next paper with the same title.

Another application of the u. d. sequences in D is in optimization. Let $f(x)$ be a continuous function on D, we want to find its global maximum M in D. There are many gradient methods for this kind of optimization problems (of. Avriel [2]). Unfortunately, there appear only few cases that the global maximum can be reached, and we can obtain in usual a local maximum if the function f is not unimodal, and the dimension of D is large, for example, dimension of $D \geqslant 5$, because the solution, in general, depends on the choice of initial point. Therefore, we use the following algorithm to find an approximate value of M.

$$m_1 = f(x_1),$$

$$m_{k+1} = \begin{cases} m_k, & \text{if } f(x_{k+1}) \leqslant m_k, \\ f(x_{k+1}), & \text{if } f(x_{k+1}) \geqslant m_k, \end{cases}$$

where $(x_1, x_2, \cdots)$ is a u. d. sequene in D, i. e., $P_n = \{x_1, \cdots, x_n\}$ is a u. d. sequence. After a large number n of steps, we may reasonably expect that m_n is close to M, if $f(x)$ satisfies some regularity conditions. We often use the following quantity to measure the uniformity of distribution of these points

$$d(n, D) = \max_{x \in D} \min_{1 < k < n} d(x, x_k), \tag{1.3}$$

where $d(x, x_k)$ denotes the Euclidan distance of x and x_k. $d(n, D)$ is called the dispersion of the set $\{x_k, k = 1, \cdots, n\}$. However one can show that if $D = I^s$, then

$$\sqrt{s}\, n^{-1/s} \leqslant d(n, D) \leqslant 2\sqrt{s}\, (D(n))^{1/s} \tag{1.4}$$

(cf. Zielinski [13] and Niederreiter [10]). This means that it is true that m_n is closed to the globle maximum M if n is large. In Section 4, we shall generalize the above result to some kind of D's and give some applications in statistics.

§2. Numerical Integration

Let D be a bounded domain in R^s. We are required to calculate the integral (1.2). Assume that tha dimension of D is s, $dv=\prod_{i=1}^{s} dx_i = d\boldsymbol{x}$ and $D \subset I^s$. Then it may be simply suggested to use the following formula

$$I = \int_{I^s} f(\boldsymbol{x}) I_D(\boldsymbol{x}) d\boldsymbol{x},$$

where $I_D(\boldsymbol{x})$ is the index functisn of D(cf. Hua and Wang [6]). This will lead to a big error sometimes, since $f(\boldsymbol{x}) I_D(\boldsymbol{x})$ may be discontinuous on the boundary of D. However, the domain D is often very special in statistics, so it is possible to reduce the integral over D to an integral over $I^t (t \leqslant s)$. More precisely, suppose that D has a representation

$$x_j = x_j(\varphi_1, \cdots, \varphi_t) = x_j(\boldsymbol{\varphi}), \; j = 1, \cdots, s, \tag{2.1}$$

where $\boldsymbol{\varphi} \in I^t$, and that x_j, $j = 1, \cdots, s$, have continuous derivatives with respect to φ_i, $i = 1, \cdots, t$, over I. Let

$$\boldsymbol{T} = (\partial x_j / \partial \varphi_i), \; i = 1, \cdots, t, \; j = 1, \cdots, s,$$

and let

$$J(\boldsymbol{\varphi}) = \det(\boldsymbol{T}\boldsymbol{T}')^{1/2}.$$

When $t = s$, $J(\boldsymbol{\varphi})$ is just the Jacobian of transfomation from $\boldsymbol{x}$ to $\boldsymbol{\varphi}$. Then we have

$$I = \int_D f(\boldsymbol{x}) dv = \int_{I^t} f(\boldsymbol{x}(\boldsymbol{\varphi})) J(\boldsymbol{\varphi}) d\boldsymbol{\varphi}, \tag{2.2}$$

where $d\boldsymbol{\varphi} = \prod_{i=1}^{t} d\varphi_i$. Therefore a quadrature formula over I^t induces a quadrature formula over D. Denote by $v(D)$ the volume of D. Then

$$v(D) = \int_{I^t} J(\boldsymbol{\varphi}) d(\boldsymbol{\varphi}). \tag{2.3}$$

Suppose further that $\varphi_1, \cdots, \varphi_t$ are independent, and

$$v(D)^{-1} J(\boldsymbol{\varphi}) = \prod_{i=1}^{t} f_i(\varphi_i),$$

where $f_i(\varphi_i)$ is the density functions of φ_i, $i = 1, \cdots, t$, and the corresponding distribution functions

$$F_i(x_i) = \int_0^{x_i} f_i(\varphi_i) d\varphi_i, \; i = 1, \cdots, t$$

satisfying $F_i(0) = 0$ and $F_i(1) = 1$, $i = 1, \cdots, t$. Let $F_i(x_i) = y_i$ and let $F_i^{-1}(y_i)$ denote the inverse function of $F_i(x_i)$, $i = 1, \cdots, t$. Then

$$\int_{I^t} f(\boldsymbol{\varphi}) J(\boldsymbol{\varphi}) d\boldsymbol{\varphi} = v(D) \int_{I^t} f(\boldsymbol{x}(\boldsymbol{F}^{-1}(\boldsymbol{y})) d\boldsymbol{y}, \tag{2.4}$$

where $\boldsymbol{F}^{-1}(\boldsymbol{y}) = (F_1^{-1}(y_1), \cdots, F_t^{-1}(y_t))'$, and $d\boldsymbol{y} = \prod_{i=1}^{t} dy_i$.

For a given set $\{\boldsymbol{b}_k = (b_{k1}, \cdots, b_{kt})', k=1, \cdots, n\}$ of I^t with discrepancy $D(n)$, we have a set $\{\boldsymbol{c}_k = F^{-1}(\boldsymbol{b}_k), k=1, \cdots, n\}$ which has F-discrepancy $D_F(n, \{c_k\}) = D(n)$ too, where $F(\boldsymbol{x}) = \prod_{i=1}^{t} F_i(x_i)$. Hence by (1.1), (2.2) and (2.4) we have

$$\left| \int_D f(\boldsymbol{x}) dv - v(D) \frac{1}{n} \sum_{k=1}^{n} f(\boldsymbol{x}(\boldsymbol{F}^{-1}(\boldsymbol{b}_k))) \right|$$
$$\leqslant v(D) D_F(n, \{\boldsymbol{c}_k\}) V(f(\boldsymbol{F}^{-1}))$$
$$= v(D) D(n) V(f). \tag{2.5}$$

By (2.2), (2.5), and the number theoretic method we thus may have the following two formulas for numerical evaluation of the multiple integral (1.2)

$$\int_D f(x) dv \cong \frac{1}{n} \sum_{k=1}^{n} f(\boldsymbol{x}(\boldsymbol{b}_k)) J(\boldsymbol{b}_k) \tag{2.6}$$

and

$$\int_D f(\boldsymbol{x}) dv \cong \frac{1}{n} v(D) \sum_{k=1}^{n} f(\boldsymbol{x}(\boldsymbol{c}_k))$$
$$= \frac{1}{n} v(D) \sum_{k=1}^{n} f(\boldsymbol{x}(\boldsymbol{F}^{-1}(\boldsymbol{b}_k))). \tag{2.7}$$

Both formulas have the same order of accuracy.

Define a set of D by

$$P = \{\boldsymbol{x}_k = \boldsymbol{x}(\boldsymbol{c}_k), k=1, \cdots, n\}. \tag{2.8}$$

The volume $v(\boldsymbol{\varphi} \leqslant \boldsymbol{y})$ of the domain in D defined by $\boldsymbol{\varphi} \leqslant \boldsymbol{y}$ is equal to

$$v(\boldsymbol{\varphi} \leqslant \boldsymbol{y}) = \int_{\varphi < y} J(\boldsymbol{\varphi}) d\boldsymbol{\varphi} = v((D) \prod_{i=1}^{t} F_i(y_i),$$

so that

$$v(\boldsymbol{\varphi} \leqslant \boldsymbol{y})/v(D) = \prod_{i=1}^{t} F_i(y_i).$$

Therefore, if we want the set P to be scattered uniformly over D, or that the ratio between the number $N(\boldsymbol{\varphi} \leqslant \boldsymbol{y})$ of P lying in the domain defined by $\boldsymbol{\varphi} \leqslant \boldsymbol{y}$ and n is approximately equal to the ratio between $v(\boldsymbol{\varphi} \leqslant \boldsymbol{y})$ and $v(D)$, we should take the set $\{\boldsymbol{c}_k, k=1, \cdots, n\}$ with lower F-discrepancy in I^t. Since $\{\boldsymbol{c}_k, k=1, \cdots, n\}$ has F-discrepancy $D(n)$, we have

$$\sup_{y \in I^t} \left| \frac{N(\boldsymbol{\varphi} \leqslant \boldsymbol{y})}{n} - \frac{v(\boldsymbol{\varphi} \leqslant \boldsymbol{y})}{v(D)} \right| = \sup_{y \in I^t} \left| \frac{N(\boldsymbol{\varphi} \leqslant \boldsymbol{y})}{n} - F(\boldsymbol{y}) \right|$$
$$= D(n). \tag{2.9}$$

We thus suggest an algorithm for obtaining a set P of D that is scattered uniformly in D from a known set with lower discrepancy in I^t. Now we give some examples.

Example 1. The domain D is a simplex $A_s = \{\boldsymbol{x}: 0 \leqslant x_s \leqslant x_{s-1} \leqslant \cdots \leqslant x_1 \leqslant 1\}$. Then D has a representation

$$x_j = \varphi_1 \cdots \varphi_j, \ j = 1, \cdots, s,$$

where $\varphi \in I^s$. We have

$$J(\varphi) = \prod_{i=1}^{s-1} \varphi_i^{s-i},$$

and

$$v(A_s) = \int_{I^s} J(\varphi) d\varphi = \prod_{i=1}^{s-1} \frac{1}{s-i+1} = 1/s!.$$

Therefore

$$f_i(\varphi_i) = (s-i+1)\varphi_i^{s-i}, \ i = 1, \cdots, s$$

are density functions over I^s with corresponding distribution functions

$$F_i(x_i) = \int_0^{x_i} f_i(\varphi_i) d\varphi_i = x_i^{s-i+1}.$$

For a given set $\{b_k, \ k = 1, \cdots, n\}$ in I^s with discrepancy $D(n)$, we have a set

$$c_k = F^{-1}(b_k) = (b_{k1}^{1/s}, \ b_{k2}^{1/(s-1)}, \ \cdots, \ b_{k,s-1}^{1/2}, \ b_{ks}), \ k = 1, \cdots, n$$

in I^s with F-discrepancy $D(n)$ too, and a set P of A_s:

$$x_k = (x_{k1}, \cdots, x_{ks})', \ k = 1, \cdots, n, \tag{2.10}$$

where

$$x_{kj} = \prod_{i=1}^{j} b_{ki}^{1/(s-i+1)}, \ k = 1, \cdots, n, \ j = 1, \cdots, s. \tag{2.11}$$

The set P satisfies (2.9).

Example 2.   Let D be the s-dimensional unit ball

$$B_s = \{x: x_1^2 + \cdots + x_s^2 \leqslant 1\}$$

which has a representation

$$x_j = \varphi_1 S_2 \cdots S_j C_{j+1}, \quad j = 1, \cdots, s-1,$$

$$x_s = \varphi_1 S_2 \cdots S_{s-1} S_s,$$

where $S_k = \sin(\pi\varphi_k)$, $C_k = \cos(\pi\varphi_k)$, $k = 2, \cdots, s-1$, $S_s = \sin(2\pi\varphi_s)$ and $C_s = \cos(2\pi\varphi_s)$ in which $\varphi \in I^s$. Then we have

$$J(\varphi) = 2\pi^{s-1}\varphi_1^{s-1} \prod_{i=2}^{s} S_i^{s-i}$$

and

$$v(B_s) = \int_{I^s} J(\varphi) d\varphi$$

$$= 2\pi^{s-1} \int_0^1 \varphi_1^{s-1} d\varphi_1 \prod_{i=2}^{s} \int_0^1 S_i^{s-i} d\varphi_i$$

$$= \frac{2}{s} \prod_{i=2}^{s} B\left(\frac{1}{2}, \ \frac{s-i+1}{2}\right),$$

because for any integer $m > 0$,

$$\int_0^1 (\sin(\pi x))^m dx = \frac{1}{\pi} B\left(\frac{1}{2}, \ \frac{m+1}{2}\right),$$

Therefore

$$f_i(\varphi_i) = \begin{cases} s\varphi_1^{s-1}, & \text{if } i = 1, \\ \pi(\sin(\pi\varphi_i))^{s-i} \Big/ B\left(\frac{1}{2}, \ \frac{s-i+1}{2}\right), & \text{if } i = 2, \cdots, s \end{cases}$$

are density functions over I^s with corresponfing distribution functions

$$F_1(x_1) = s\int_0^{x_1} \varphi_i^{s-1}d\varphi_1 = x_1^s,$$

$$F_i(x_i) = \frac{\pi}{B(1/2,\ (s-i+1)/2)} \int_0^{x_i} (\sin\pi x)^{s-i}dx,\ i=2,\ \cdots,\ s.$$

For a given set $\{\boldsymbol{b}_k,\ k=1,\ \cdots,\ n\}$ in I^s with discrepancy $D(n)$, we have a set $\{\boldsymbol{c}_k,\ k=1,\ \cdots,\ n\}$, where

$$c_{k1} = b_{k1}^{1/s},$$

$$F_i(c_{ki}) = b_{ki},\ i=2,\ \cdots,\ s,\ k=1,\ \cdots,\ n,$$

in I^s with F-discrepancy $D(n)$ too, and finally a set P of B_s:

$$\boldsymbol{x}_k = (x_{k1},\ \cdots,\ x_{ks})',\ k=1,\ \cdots,\ n, \tag{2.12}$$

where

$$x_{kj} = b_{k1}^{1/s}\prod_{i=2}^{j} S_{ki}O_{k,j+1},\ j=1,\ \cdots,\ s-1,$$

$$x_{ks} = b_{k1}^{1/s}\prod_{i=2}^{s} S_{ki} \tag{2.13}$$

in which $S_{ki} = \sin(\pi c_{ki})$, $O_{ki} = \cos(\pi c_{ki})$, $i=1,\ \cdots,\ s-1$, $S_{ks} = \sin(2\pi c_{ks})$, $O_{ks} = \cos(2\pi c_{ks})$, $k=1,\ \cdots,\ n$.

Example 3. Let D be the $s-1$ dimensional unit sphere

$$S^{s-1} = \{\boldsymbol{x}:\ x_1^2 + \cdots + x_s^2 = 1\}$$

which has a representation

$$x_j = \prod_{i=1}^{j-1} S_iO_j,\ j=1,\ \cdots,\ s-1,$$

$$x_s = \prod_{i=1}^{s-1} S_i,$$

where $S_i = \sin(\pi\varphi_i)$, $O_i = \cos(\pi\varphi_i)$, $i=1,\ \cdots,\ s-2$, $S_{s-1} = \sin(2\pi\varphi_{s-1})$ and $O_{s-1} = \cos(2\pi\varphi_{s-1})$ in which $\varphi \in I^{s-1}$. Then

$$J(\boldsymbol{\varphi}) = 2\pi^{s-1}\prod_{i=1}^{s-2} S_i^{s-i-1}$$

and

$$v(S^{s-1}) = \int_{I^{s-1}} J(\boldsymbol{\varphi})d\boldsymbol{\varphi} = 2\pi \prod_{i=1}^{s-2} B\left(\frac{1}{2},\ \frac{s-i}{2}\right).$$

Therefore

$$f_i(\varphi_i) = \pi S_i^{s-i-1}/B(1/2,\ (s-i)/2),\ i=1,\ \cdots,\ s-1$$

are density functions over I with corresponding distribution functions

$$F_i(x_i) = \frac{\pi}{B(1/2,\ (s-i)/2)} \int_0^{x_i} (\sin\pi t)^{s-i-1}\ dt,\ 0<i<s.$$

For a given set $\{\boldsymbol{b}_k,\ k=1,\ \cdots,\ n\}$ in I^{s-1} with discrepancy $D(n)$, we have a set $\{\boldsymbol{c}_k,\ k=1,\ \cdots,\ n\}$, where

$$F_i(c_{ki}) = b_{ki},\ i=1,\ \cdots,\ s-1,\ k=1,\ \cdots,\ n,$$

in I^{s-1} with F-discrepancy $D(n)$ too, and finally a set P of S^{s-1}:

$$x_k = (x_{k1}, \cdots, x_{ks})', \ k = 1, \cdots, n, \qquad (2.14)$$

where

$$x_{kj} = \prod_{i=1}^{j-1} S_{ki} C_{kj}, \ j = 1, \cdots, s-1, \qquad (2.15)$$

$$x_{ks} = \prod_{i=1}^{s-1} S_{ki}$$

in wihch $S_{ki} = \sin(\pi c_{ki})$, $C_{ki} = \cos(\pi_k c_i)$, $i = 1, \cdots, s-2$, $S_{k, s-1} = \sin(2\pi c_{k, s-1})$ and $C_{k, s-1} = \cos(2\pi c_{k, s-1})$, $k = 1, \cdots, n$.

Example 4.   The domain D is a part of the boundary of s–dimensional unit simplex

$$T_{s-1} = \{x: \ x_1 + \cdots + x_s = 1, \ x_i \geqslant 0, \ i = 1, \cdots, s\}$$

which has a representation

$$x_i = (S_1 \cdots S_{i-1} C_i)^2, \ i = 1, \cdots, s-1,$$
$$x_s = (S_1 \cdots S_{s-2} S_{s-1})^2,$$

where $S_i = \sin(\pi \varphi_i / 2)$, $C_i = \cos(\pi \varphi_i / 2)$, $i = 1, \cdots, s-1$ and $\varphi \in I^{s-1}$. We have

$$\det(TT') = \left(\pi^{s-1} \prod_{i=1}^{s-1} S_i^{2(s-i)-1} C_i\right)^2 \det(SS'),$$

where

$$S = \begin{pmatrix} -1 & C_2^2 & S_2^2 C_3^2 & \cdots & S_2^2 & \cdots & S_{s-2}^2 & C_{s-1}^2 & S_2^2 & \cdots & S_{s-2}^2 & S_{s-1}^2 \\ 0 & -1 & C_3^2 & \cdots & S_3^2 & \cdots & S_{s-2}^2 & C_{s-1}^2 & S_3^2 & \cdots & S_{s-2}^2 & S_{s-1}^2 \\ \vdots & \vdots & \vdots & \cdots & & & \vdots & & & & \vdots \\ 0 & 0 & 0 & \cdots & & & -1 & & & & 1 \end{pmatrix}.$$

Note that $\det(SS')$ is invarint if S is replaced by AS, where A is an $(s-1) \times (s-1)$ matrix with $\det A = \pm 1$. We now prove that there exists an $(s-1) \times (s-1)$ matrix A with $\det A = \pm 1$ such that

$$AS = \begin{pmatrix} -1 & 1 & 0 & \cdots & 0 \\ 0 & -1 & 1 & \cdots & 0 \\ \vdots & \vdots & \vdots & \vdots\vdots\vdots & \vdots \\ 0 & 0 & 0 & \cdots & 1 \\ 0 & 0 & 0 & \cdots & -1 \end{pmatrix} = V_{s-1},$$

say. In fact, if $s = 2$, then $S = (-1, 1)$, and the assertion is true. Suppose now that $s > 2$ and the assertion holds for $s-1$. Then

$$\begin{pmatrix} 1 & -S_2^2 & \cdots & 0 \\ 0 & 1 & \cdots & 0 \\ \vdots & \vdots & \cdots & \vdots \\ 0 & 0 & \cdots & 1 \end{pmatrix} S = \begin{pmatrix} -1 & 1 & 0 & \cdots & 0 \\ 0 & -1 & C_3^2 & \cdots & S_3^2 & \cdots & S_{s-1}^2 \\ \vdots & \vdots & \vdots & \cdots & \vdots \\ 0 & 0 & 0 & \cdots & -1 & 1 \end{pmatrix}.$$

By induction hypothesis, there is an $(s-2) \times (s-2)$ matrix A_1 with $\det A_1 = \pm 1$ and

$$A_1\begin{pmatrix} -1 & C_3^2 & \cdots & S_3^2 & \cdots & S_{s-1}^2 \\ 0 & -1 & \cdots & & & \vdots \\ \cdots & \vdots & \cdots & & & \vdots \\ 0 & 0 & \cdots & & -1 & 1 \end{pmatrix} = V_{s-2}.$$

Therefore

$$\begin{pmatrix} 1 & \mathbf{0} \\ \mathbf{0} & A_1 \end{pmatrix}\begin{pmatrix} 1 & -S_2^2 & \mathbf{0} \\ 0 & 1 & \mathbf{0} \\ \mathbf{0} & \mathbf{0} & 1 \end{pmatrix} S = V_{s-1},$$

and the assertion follows. We have

$$\det(SS') = \det(V_{s-1}V'_{s-1}) = \det\begin{pmatrix} 2 & -1 & \cdots & 0 & 0 \\ -1 & 2 & \cdots & 0 & 0 \\ \vdots & \vdots & \cdots & \vdots & \vdots \\ 0 & 0 & \cdots & 2 & -1 \\ 0 & 0 & \cdots & -1 & 2 \end{pmatrix} = \Delta_{s-1},$$

say. Since $\Delta_1 = 2$ and $\Delta_t = 2\Delta_{t-1} - \Delta_{t-2}\,(t>2)$, we have $\Delta_{s-1} = s$. Hence

$$J(\boldsymbol{\varphi}) = \pi^{s-1}s^{1/2}\prod_{i=1}^{s-1} S_i^{2(s-i)-1}C_i,$$

and

$$v(T_{s-1}) = \int_{I^{s-1}} J(\boldsymbol{\varphi})d\boldsymbol{\varphi} = s^{1/2}/(s-1)!.$$

Therefore

$$f_i(\varphi_i) = (s-i)S_i^{2(s-i)-1}C_i, \quad i=1, \cdots, s-1$$

are density functions over I with corresponding distribution functions

$$F_i(x_i) = \int_0^{x_i} f_i(\varphi_i)d\varphi_i = (\sin(\pi x_i/2))^{2(s-i)}, \quad i=1, \cdots, s-1.$$

For a given set $\{\boldsymbol{b}_k, \; k=1, \cdots, n\}$ in I^{s-1} with discrepancy $D(n)$, we have a set $\{\boldsymbol{c}_k, \; k=1, \cdots, n\}$, where

$$c_{ki} = (2/\pi)\arcsin(b_{ki}^{1/(2s-2i)}), \quad i=1, \cdots, s-1, \; k=1, \cdots, n.$$

Finally, we have a set P of T_{s-1}:

$$\boldsymbol{x}_k = (x_{k1}, \cdots, x_{ks})', \quad k=1, \cdots, n, \tag{2.16}$$

where

$$\begin{cases} x_{kj} = \prod_{i=1}^{j-1} b_{ki}^{1/(s-i)}(1 - b_{kj}^{1/(s-j)}), \quad j=1, \cdots, s-1, \\[2mm] x_{ks} = \prod_{i=1}^{s-1} b_{ki}^{1/(s-i)}, \quad k=1, \cdots, n. \end{cases} \tag{2.17}$$

§3. Some Applications

In this section we shall pay our attention to applications of number theoretic method in numerical evaluation of probabilities and moments of a continuous multivariate distributions. The basic quadrature formulas are given by (2.6) and (2.7). There are a number of methods to produce sets of points $\{\boldsymbol{b}_x, \ k=1, \cdots, n\}$ in I^s (see Hua and Wang [6]). In view, of our experiences, we will recommend using the following algorithm: Let $(h_1, \cdots, h_s; n)$ be an integral vector, where $h_1=1, \ 0<h_i$ $<n$ and g. c. d. $(h_i, n)=1, \ i=1, \cdots, s$. Let

$$\boldsymbol{p}_n(k) = (kh_1, \cdots, kh_s) = (q_{k1}, \cdots, q_{ks}) \pmod{n}, \ k=1, \cdots, n,$$

where $0<q_{ki}\leqslant n$. Set

$$b_{ki} = (2q_{ki}-1)/2n. \ i=1, \cdots, s, \ k=1, \cdots, n.$$

Then $\{\boldsymbol{b}_k\}$ is a set of points in I^s with lower discrepancy if $(h_1, \cdots, h_s; n)$ are carefully selected. A table of $(h_1, \cdots, h_s; n)$ for $1<s<19$ was contained in [6] as an appendix.

Example 5.   This problem was come from alloy steel industry (see Fang and Wu [3] in details). Let $\boldsymbol{x}$ be an $s\times1$ vector which denotes the percentage of chemical elements in an alloy steel and let $(\boldsymbol{\mu x})=(\mu_1, \cdots, \mu_t)'$ be the corresponding vector which stands for the quality of the steel. The regression equation between $\mu(\boldsymbol{x})$ and $\boldsymbol{x}$ is

$$\hat{\boldsymbol{\mu}}(\boldsymbol{x}) = \boldsymbol{a} + \boldsymbol{Bx},$$

where $\boldsymbol{a}$ and $\boldsymbol{B}$ are $t\times1$ and $t\times s$ matrices of regression coefficients and $\boldsymbol{x}$ belongs to a rectangle $K = [a_1, \ b_1] \times \cdots \times [a_s, \ b_s]$. Suppose that for each $\boldsymbol{x}\in K$, we have $\mu(\boldsymbol{x}) \sim N_t(\boldsymbol{a}+\boldsymbol{Bx}, \ \boldsymbol{\Sigma})$, the multivariate normal distribution, where $\boldsymbol{a}$, $\boldsymbol{B}$, and $\boldsymbol{\Sigma}$ can be used by their least square estimators. Let $T_i, \ i=1, \cdots, \ t$, be the constants such that the steel is said to be qualified if $\mu_i>T_i, \ i=1, \cdots, \ t$. Thus the probability that the alloy steel corresponding to $\boldsymbol{x}$ is qualified is equal to

$$p(\boldsymbol{x}) = \int_{T_1}^{\infty}\cdots\int_{T_t}^{\infty} n_t \ (\boldsymbol{y}, \ \hat{\mu}(\boldsymbol{x}), \ \boldsymbol{\Sigma}) \ d\boldsymbol{y} \tag{3.1}$$

where $\boldsymbol{y} = (y_1, \cdots, \ y_t)$, and $n_t(\boldsymbol{y}, \mu \ (\boldsymbol{x}), \ \boldsymbol{\Sigma})$ is the density of $N_t \ (\mu, \ \boldsymbol{\Sigma})$. The integral (3.1) can be evaluated by applying (2.6) if we choose suitable numbers $A_i, \ i=1, \cdots, \ t$, such that

$$p(\boldsymbol{x}) \cong \int_{T_1}^{A_1}\cdots\int_{T_t}^{A_t} n_t(\boldsymbol{y}, \ (\hat{\mu}\boldsymbol{x}), \ \boldsymbol{\Sigma}) \ d\boldsymbol{y}$$

$$= \prod_{i=1}^{t}(A_i - T_i) \frac{1}{n} \sum_{k=1}^{n} n_t(\boldsymbol{z}_k, \ \hat{\mu} \ (\boldsymbol{x}), \ \boldsymbol{\Sigma}), \tag{3.2}$$

where

$$\boldsymbol{z}_k' = (z_{k1}, \cdots, z_{kt}) = (T_1+(A_1-T_1)b_{k1}, \cdots, T_t+(A_t-T_t) \ b_{kt}),$$

$k=1, \cdots, \ n$ and $\{\boldsymbol{b}_k\}$ is a uniformly distributed set of points in $\boldsymbol{I}^t$. To illustrate the computational accuracy, set $\boldsymbol{\Sigma} = \boldsymbol{I}_5$, the 5×5 identity matrix and $\mu \ (\boldsymbol{x}) = \boldsymbol{0}, \ T_i = -1$, $A_i = 1, \ i=1, \cdots, \ 5$ in (3.2). We have

$$p = \int_1^{1-}\cdots\int_{-1}^{1} n_5(\boldsymbol{y}, \ \boldsymbol{0}, \ \boldsymbol{I}_5) \ d\boldsymbol{y}.$$

Now we have by (3.2) the following:

Table 1 shows that 5–digit accuracy for numerical evaluating a 5–fold integral is

Table 1.

n	approximate values of p
1069	0.148299406
2129	0.148295351
5003	0.148291410
8191	0.148291358
∞	0.148291347

obtained by the use of 1069 points only.

Example 6. The moments of order statistics. Let $X_1, \cdots, X_s$ be a sample from the population with distribution function $F(x)$ and density $f(x)$. Let $Y_s = X_{(1)} \leqslant Y_{s-1} = X_{(2)} \leqslant \cdots Y_1 = X_{(s)}$ be their order statistics. It is known that the joint density of $Y_1, \cdots, Y_s$ is given by

$$s! \prod_{i=1}^{s} f(y_i), \quad y_s < y_{s-1} < \cdots < y_1.$$

Then the order $m_1, \cdots, m_s$ mixed moment of $X_{(i)}$, $i=1, \cdots, s$ is defined by

$$\mu(m_s, \cdots, m_1) = s! \int_{D^*} \prod_{j=1}^{s} y_j^{m_j} f(y_j) dv, \tag{3.3}$$

where $D^* = \{-\infty < y_s < y_{s-1} < \cdots < y_1 < \infty\}$. There exist a and b such that

$$P(a < y_s, \ y_1 < b) \cong 1.$$

Taking a transformation $z_i = (y_i - a)/(b-a)$, $i=1, \cdots, s$, we have

$$\mu(m_s, \cdots, m_1) = s! (b-a)^s \int_{D} \prod_{j=1}^{s} [(a+(b-a)z_j)^{m_j} f(a+(b-a)z_j)] dv$$

where D is defined in Example 1. By Example 1, (2.6) and (2.7) we suggest the following two formulas for the calculation of $\mu(m_s, \cdots, m_1)$:

$$\mu(m_s, \cdots, m_1) \cong s! (b-a)^s \frac{1}{n} \sum_{k=1}^{n} \prod_{j=1}^{s} \Big[t_{kj}^{(m_j+1)-1} f\Big(\prod_{i=1}^{j} t_{ki}\Big)\Big],$$

where $t_{kj} = a + (b-a) b_{kj}$ and $\{\boldsymbol{b}_k, \ k=1, \cdots, n\}$ is a uniformly distributed set of points in I^s, and

$$\mu(m_s, \cdots, m_1) \cong s! (b-a)^s \frac{1}{n} \sum_{k=1}^{n} \prod_{j=1}^{s} [a+(b-a)c_{kj}]^{m_j} f(a+(b-a)c_{kj})$$

$$= s! (b-a)^s \frac{1}{n} \sum_{k=1}^{n} \prod_{j=1}^{s} [a+(b-a)b_{kj}^{1/(s-j+1)}]^{m_j} f(a+(b-a)b_{kj}^{1/(s-j+1)})$$

where $\{\boldsymbol{c}_k\}$ and $\{\boldsymbol{b}_k\}$ are given in Example 1.

Since the mixed moments of order statiftics of uniform distribution $U(0, 1)$ on $I = [0, 1]$ can be formulated. We give an example in Table 2 which shows the (see Table 2) acc uracies.

Example 7. In his study of compositional data, Aitchison [1] introduced in 1986 a so-called additive [logistic normal distribution. Let $\boldsymbol{T}_n$ be a domain defined in Example 4 with $n = N - 1$. Any $\boldsymbol{x}$ in T_n is called a composition. For a given $\boldsymbol{x} \in T_n$, we denote by $\boldsymbol{x}_{-N}$ the n-dimensional vector formed by the first n components of $\boldsymbol{x}$.

Table 2. Mixed moments of order statistics of $U(0, 1), s=7$

n	$E(X_{(1)}X_{(3)}X_{(5)})$	$E(X_{(1)}X_{(2)}^2X_{(3)}^2)$
418	0.03887518	0.00326433
597	0.03888518	0.00339034
828	0.03888511	0.00332116
1010	0.03887286	0.00332084
1220	0.03889159	0.00325168
∞	0.03888889	0.00326340

Let

$$\boldsymbol{y} = \log(\boldsymbol{x}_{-N}/X_N) = (\log(X_1/X_N), \cdots, \log(X_n/X_N))'. \tag{3.4}$$

The equation (3.4) yields an one to one mapping from T_n to R^n. A random vector $\boldsymbol{x} \in T_n$ is said to have an additive logistic normal distribution $AN_n(\boldsymbol{\mu}, \boldsymbol{\Sigma})$ if its corresponding $\boldsymbol{y} \sim N_n(\boldsymbol{\mu}, \boldsymbol{\Sigma})$.

Aitchison gave the formulas for $E(\log(X_i/X_j))$, $E(X_i/X_j)$, $\mathrm{Cov}(\log(X_i/X_j), \log(X_k/X_l))$, $\mathrm{Cov}(X_i/X_j, X_k/X_l)$. It seems difficulty for him to calculate $E(X_i)$ and $\mathrm{Cov}(X_i, X_j)$, which are required in many practictical problems.

The density function of $AN_n(\boldsymbol{\mu}, \boldsymbol{\Sigma})$ is given by

$$(2\pi)^{-n/2}(\det\boldsymbol{\Sigma})^{-1/2}\Big(\prod_{i=1}^{N} X_i^{-1}\Big)\exp\Big\{-\frac{1}{2}\Big(\log\frac{\boldsymbol{x}_{-N}}{\boldsymbol{x}_N}-\boldsymbol{\mu}\Big),\ \boldsymbol{\Sigma}^{-1}\Big(\log\frac{\boldsymbol{x}_{-N}}{\boldsymbol{x}_N}-\boldsymbol{\mu}\Big)\Big\} \tag{3.5}$$

and the mixed moment of $\boldsymbol{x}$ is

$$E(X_1^{t_1}\cdots X_N^{t_N}) = (2\pi)^{-n/2}(\det\boldsymbol{\Sigma})^{-1/2}\int_{T_n}\prod_{i=1}^{N}x_i^{t_i-1}$$

$$\exp\Big\{-(1/2)\Big[\Big(\log\frac{\boldsymbol{x}_{-N}}{X_N}-\boldsymbol{\mu}\Big),\ \boldsymbol{\Sigma}^{-1}\Big(\log\frac{\boldsymbol{x}_{-N}}{X_N}-\boldsymbol{\mu}\Big)\Big]\Big\}\mathrm{dv},$$

where dv is the volume element of T_n. By Example 4, we have

$$E(X_1^{t_1}\cdots X_N^{t_N}) = C\int_{I^n}\prod_{i=1}^{n}(C_i^{2t_i-1}S_i^{2(t_{i+1}+\cdots+t_N)-3})Q(\boldsymbol{\varphi})d\boldsymbol{\varphi},$$

where $\boldsymbol{\varphi} \in I^n$, $d\boldsymbol{\varphi} = \prod_{i=1}^{n}d\varphi_i$, $C = (2/\pi)^{-n/2}(\det\boldsymbol{\Sigma})^{-1/2}N^{1/2}$ and

$$Q(\boldsymbol{\varphi}) = \exp\{-(1/2)(\boldsymbol{g}(\boldsymbol{\varphi})-\boldsymbol{\mu})'\boldsymbol{\Sigma}^{-1}(\boldsymbol{g}(\boldsymbol{\varphi})-\boldsymbol{\mu})\}$$

in which

$$\boldsymbol{g}(\boldsymbol{\varphi}) = (\log(C_1^2)-\log(S_1^2\cdots S_n^2), \cdots, \log(S_1^2\cdots S_{n-1}^2C_n^2)-\log(S_1^2\cdots S_n^2))',$$

$$= 2(\log C_1-\sum_{i=1}^{n}\log S_i,\ \log C_2-\sum_{i=2}^{n}\log S_i, \cdots,\ \log C_k-\sum_{i=n}^{n}\log S_i)'.$$

By (2.6) or (2.7) we may obtain approximate values of any mixed moments of $AN_n(\boldsymbol{\mu}, \boldsymbol{\Sigma})$.

Example 8. We meet the problem of directional data in which some statistical distributions are defined over S^{s-1} (see Example 3). The so-called Langevin distribution which is an extension of Von Mises and Fisher's distribution has a density function

$$C \exp\{k\boldsymbol{\mu}'\boldsymbol{x}\}, \boldsymbol{x} \in S^{s-1},$$

where $\boldsymbol{\mu} \in S^{s-1}$, $k>0$ and C is the normaling constant (cf. Mardia [8]). Another distribution called Scheidegges–Watson distribution has a sensity functor

$$C \exp\{k(\boldsymbol{\mu}, \boldsymbol{x})^2\}, \ \boldsymbol{x} \in S^{s-1},$$

where C, k, and $\boldsymbol{\mu}$ have the similar meaning as before (cf. Watson [11]).

We can apply Example 3, (2.6) and (2.7) to calculate probabilities and mixed moments for these two kinds of distributions.

§ 4. Optimization

In this section we shall generalize inequlity (1.4) to some domains which have been discussed in the past sections. Then we give an example to show that the algorithm mentioned in Section 1 is powerful.

We suppose that the set of singularities of the transformation $\boldsymbol{x}=\boldsymbol{x}(\boldsymbol{\varphi})$, i. e., the set of solutions of $J(\boldsymbol{\varphi})=0$, is a set of D with dimension$<t$, where $\boldsymbol{x}$ is an s-dimensional vector, rnd t is the dimension of D. Hence for any given $s>0$, there exist a domain $\mathbb{C}$ and two positive constants c_1, c_2 depending only on s such that $v(\mathbb{C}) <s$ and

$$c_1 < f_i(\varphi_i) < c_2, \ i=1, \cdots, t,$$

where $\boldsymbol{\varphi} \in I^t \backslash \mathbb{C}$. Then $dF_i^{-1}(\varphi_i)/d\varphi_i$, $i=1, \cdots, t$ are positive and bounded over $I^t \backslash \mathbb{C}$ too.

First we take a set $\boldsymbol{b}_k = (b_{k1}, \cdots, b_{kt})'$, $k=1, \cdots, n$ in $I^t \backslash \mathbb{C}$ with lower discrepancy $D(n)$. Then we have shown that

$$\boldsymbol{c}_k = \boldsymbol{F}^{-1}(\boldsymbol{b}_k) = (F_1^{-1}(b_{k1}), \cdots, F_t^{-1}(b_{kt}))', \ k=1, \cdots, n$$

is a set in I^t with F–discrepancy $D(n)$ too, and finally we have a set in D: $\boldsymbol{x}_k = \boldsymbol{x}(\boldsymbol{c}_k)$, $k=1, \cdots, n$.

Let $\boldsymbol{x}=\boldsymbol{x}(\boldsymbol{\varphi})$, $\boldsymbol{x}^*=\boldsymbol{x}(\boldsymbol{\varphi}^*)$, $\boldsymbol{\varphi}=F^{-1}(\boldsymbol{\psi})$, $\boldsymbol{\varphi}^*=\boldsymbol{F}^{-1}(\boldsymbol{\psi}^*)$, $d\boldsymbol{x}=(dx_1, \cdots, dx_s)'$, $d\boldsymbol{\varphi}=(d\varphi_1, \cdots, d\varphi_t)'$, $d\boldsymbol{\psi}=(d\psi_1, \cdots, d\psi_t)'$, and S be the diagonal matrix

$$S = \mathrm{diag}(d\varphi_1/d\psi_1, \cdots, d\varphi_t/d\psi_t).$$

Then

$$d\boldsymbol{x}'d\boldsymbol{x} = d\boldsymbol{\varphi}'\boldsymbol{TT} d\boldsymbol{\varphi} = d\boldsymbol{\psi}'\boldsymbol{STT}'S d\boldsymbol{\psi}.$$

Since the elements of S and T are bounded in $I^t \backslash \mathbb{C}$, we have

$$d(\boldsymbol{x}, \boldsymbol{x}^*) = \int_{\boldsymbol{x}(\boldsymbol{\varphi})}^{\boldsymbol{x}(\boldsymbol{\varphi}^*)} (d\boldsymbol{x}' d\boldsymbol{x})^{1/2}$$

$$= \int_{\boldsymbol{\psi}}^{\boldsymbol{\psi}^*} (d\boldsymbol{\psi}'\boldsymbol{STT}'S d\boldsymbol{\psi})^{1/2} < c(\varepsilon) \int_{\boldsymbol{\psi}}^{\boldsymbol{\psi}^*} (d\boldsymbol{\psi}' d\boldsymbol{\psi})^{1/2}$$

$$= c(\varepsilon) d_t(\boldsymbol{\psi}, \boldsymbol{\psi}^*),$$

where $d_t(\boldsymbol{y}, \boldsymbol{z})$ denotes the Euclidean distance in R^t. Hence by (1.4), we have

$$d(n, D) = \max_{\boldsymbol{x} \in D} \min_{1 \leqslant k \leqslant n} d(\boldsymbol{x}, \boldsymbol{x}_k) \leqslant 2t^{1/2} c(\varepsilon) D(n)^{1/t}, \tag{4.1}$$

Thus if $\boldsymbol{x}_k$, $k=1, \cdots, n$, are scattered uniformly in D, then the maximum value of a function on these points may be taken as an approximate value of the global maximum of the function on D.

Example 9.   Additive logistic elliptical distributions. The so-called additive logistic elliptical distributions defined on T_{s-1} (see Example 4) are extensions of the additive logistic normal distributions mentioned in Example 7 and have density functions of the form

$$f(\boldsymbol{x}) = (\det \boldsymbol{\Sigma})^{-1/2} \prod_{i=1}^{s} x_i^{-1} g(\boldsymbol{x}), \tag{4.2}$$

where

$$g(\boldsymbol{x}) = g\left(\left(\log \frac{\boldsymbol{x}_{-s}}{x_s} - \mu\right)' \boldsymbol{\Sigma}^{-1}\left(\log \frac{\boldsymbol{x}_{-s}}{x_s} - \mu\right)\right) \tag{4.3}$$

and $\boldsymbol{x}_{-s}$ is an $(s-1)$-dimensional vector formed by the first $s-1$ components of $\boldsymbol{x}$. The mode of $f(\boldsymbol{x})$ can not be analytically formulated so far. However we may use uniformly distributed sets on T_{s-1} to calculate the approximate values of the mode.

When the function g in (4.3) has the form

$$g(u) = C \ (1+u/m)^{-p}, \ p > s/2, \ m > 0, \tag{4.4}$$

where

$$C = (\pi m)^{-s/2} \Gamma(p)/\Gamma(p-s/2),$$

the corresponding distribution is called additive logistic elliptical Pearson Type VII distribution. In this case to find the mode of $f(\boldsymbol{x})$ is equivalent to obtain the maximum of

$$h(\boldsymbol{x}) = \prod_{i=1}^{s} x_i^{-1} \left[1+\left(\log \frac{\boldsymbol{x}_{-s}}{x_s} - \mu\right) \boldsymbol{\Sigma}^{-1}\left(\log \frac{\boldsymbol{x}_{-s}}{x_s} - \mu\right)/m\right]^{-p}$$

over T_{s-1}. The results in Table 3 show that the approximate values of the mode in the case of $s=3$, $p=9$, $m=5.5$ and

$$\mu = \begin{pmatrix} 0 \\ 0 \end{pmatrix}, \ \boldsymbol{\Sigma} = \begin{pmatrix} 1 & -0.7 \\ -0.7 & 1 \end{pmatrix}$$

are closed to those of the mode $(1/3, 1/3, 1/3)'$.

Table 3   Approximate values of the mode

n	M_n	x_1	x_2	x_3
233	26.55203	0.3371035	0.3306723	0.3422242
377	26.82553	0.3061467	0.3598099	0.3340434
610	26.48600	0.3604561	0.3150540	0.3244899
4181	26.89095	0.3256147	0.3368701	0.3375152
10946	26.99783	0.3334971	0.3337690	0.3827339
17711	26.98296	0.3292827	0.3388427	0.3318746

We note in Table 3 that when n is increasing, the corresponding M_n, in principle, is increasing also. But sometimes $M_n < M_{n'}$, where $n > n'$, because the sets $\{c_k\}$ for different n may be completely distinct. Hence we suggest using a sequential method to improve the above result.

The following program is designed for our presented problem,

Step 1. Choose a uniformly distributed set of points $\{x_k,\ k=1,\ \cdots,\ n_0\}$ in $D_0=T_{s-1}$ with suitable n_0. Find the maximum M_0 of the function among these points, and assume that it is attained at $x_0^*=(x_{01},\ \cdots,\ x_{0s})'$.

Step 2. Find a small domain D_1 of D_0 such that $D_1\subset D_0$ and $x_0^*\subset D_1$. For instance, D_1 is a domain with x_0^* located near the gravity of D_1. More precisely, we cdoose $a_i,\ i=1,\ \cdots,\ s,$ such that

$$0\leqslant a_i<x_{0i},\quad i=1,\ \cdots,\ s.$$

Set $a=a_1+\cdots+a_s$ and $b_i=a_i+1-a,\ i=1,\ \cdots,\ s.$ Then

$$1\geqslant b_i\geqslant a_i+\sum_{j=1}^{s}x_{0j}-\sum_{k=1}^{s}a_k=x_{0i}+\sum_{\substack{j=1\\i\neq j}}^{s}(x_{0j}-a_j)\geqslant x_{0i},$$

$i=1,\cdots,s.$ Denote

$$D_1=\{x=(x_1,\ \cdots,\ x_s)':a_i\leqslant x_i\leqslant b_i,\ i==1,\ \cdots,\ s,\ x\in D_0\}.$$

Let $z_k,\ k=1,\ \cdots,\ n_1,$ be an uniformly distributed set of points in T_{s-1}. Then we have a set $\{x_k,\ k=1,\ \cdots,\ n_1\}$, where

$$x_{ki}=a_i+(1-a)z_{ki},\quad i=1,\ \cdots,\ s,\ k=1,\ \cdots,\ n_1,$$

which is uniformly distributed over D_1. Denote by M_1 the maximum of the function on x_k' s which is attained at the point x_1^*.

Step 3. Suppose that in the jth step we have found the maximum M_j of the function and the correrponding point x_j^*. By a similar method we can reduce the domain D_j to D_{j+1}, and make a set of points on D_{j+1}, by which we can find another maximum M_{j+1} of the function and the corresponding point x_{j+1}^*.

Repeat Step 3 until the search domain is smaller. The last maximum M_{j+1} is expected to be closed to the global maximum M of the function.

Applying the above program to our problem, we set $n_0=n_1=\cdots=223$ for each step, and the results are given in Table 4, which improve those in Table 3.

Table 4.　The sequencial method for optimization

No	a_i	b_i	M_i	x_1^*	x_2^*	x_3^*
1	0.0000	1.0000	26.55203	0.3271035	0.3306723	0.3422242
2	0.3000	0.4000	26.97836	0.3304331	0.3343395	0.3352274
3	0.3300	0.3400	26.99543	0.3327104	0.3330673	0.3342224
4	0.3230	0.3340	26.99994	0.3332711	0.3333067	0.3334223
the	global	maximum	27.00000	0.3333333	0.3333333	0.33333333

We have done many examples which all show that the present sequential method is advantageous.

References

[1] Aitchison, J., The statistical Analysis of Compositional Data, Chaman and Hall, London/New York, (1986).

[2] Avriel, M., Nonlinear Programming, Analysis and Methods Prentice–Hall, Inc. Englewood Cliffs, New Jersey, 1976.

[3] Fang, K. T. & Wu, C. Y., The extrime value problem of some probability function, *Acta Math. Appl. Sinica*, 2 (1979), 132—148.

[4] Hlawka, E., Funktionen von beschrankter Variation in der Theorie Gleichverteilnug, *Ann. Mat. pure Appl.*, **54**, (1961), 325—333.

[5] Hlawka, E. & Muck, R., A transformations of equidistributed sequences, in "Applications of Number Theory to Numerical Analyssi" (Zaremba.S. K. ed). Acad. Press, New York, 1972, 371—388.

[6] Hua, L. K. & Wang. Y., Applications of Number Theory to Numerical Analysis, Springer–Verlag (Heidelberg) and Science Press (Beijing), 1981.

[7] Koksma, I, F., Een algemeene stelling uit de theorie der gelijkmatige Verdeeling modulo 1, *Math. B (Zutphen)*, **11**, (1942—1943), 7—11.

[8] Madia, K. V., Statistcs of Directional Data, Acad. Press, New York, (1972).

[9[Niederreiter, H., Metric theorems on the distribution of sequenes, *Proc. Symp. Pure Math.*, 24, *AMS Pro. R. I.*, (1973), 195—212.

[10] Niederreiter, H., A quasi-Monte Carlo method for the approximate computation of the extreme values of a function, *Studies in Pure Math, Birkhauser, Basel*, (1983), 523-529.

[11] Watson, G. S., Statistics on Sphere, Wiley, New York, (1983).

[12] Weyl, H., Uber die Gleichverteilurg der Zahlen mod Eins, *Math. Ann.*, **77**. (1916), 313—352.

[13] Zielinski, R., On the Monte Carlo evaluaton of the extreme values of a function, *Algorytncy*, 2 (1965), 7—13.

Chin. Ann. of Math.
11B:3(1990), 384—394.

NUMBER THEORETIC METHODS IN APPLIED STATISTICS (II)

WANG YUAN (王　元)[*]　　FANG KAITAI (方开泰)[**]

Abstract

In this paper, the authors give some applications of F-uniformly distributed sequences, which are suggested in their previous paper under the same title, in experimental design, experiments with mixtures, geometric probability and simulation.

§1. Introduction

In our previous paper[13] we proposed a method to produce sets of points which are uniformly distributed over a domain D of R^s and gave some applications in numerical evalution of probabilties and moments of a continuous multivariate distribution and optimization. In this paper we shall give the applications of this kind of uniformly distributed sets of points in experim ental designs for both independent factors and experiments with mixtures in Section 2 and Section 3, and in geometric probability in Section 4. We also give examples to show the comparison between the number-theoretic method and some other methods.

The definitions and notations given in our paper [13] are retained hereafter.

§2. Uniform Design

If there are s factors and each factor has n levels, then the number of all possible ex-periments is n^s. The orthogonal array is to choose $O(n^2)$ experiments among them by the theory of orthogonal Latin squares and group theory. However the number of experiments in orthogonal array is still large if n is comparatively large. The number of experiments may be decreased by BIB (balanced incomplete blocks) method for the case of $s=2$ only. Hence it requires to find a method for decreasing the number of experiments.

Wang and Fang (1981) proposed a kind of experimental designs, the uniform

Manuscript received March 15, 1989.

* Institute of Mathematics, Academia Sinica, Beijing. China.

** Institute of Applied Mathematics, Academia Sinica, Beijing. China.

*** This work was supported by The Chinesse National Science Fund and Academia Sinica.

designs, by the number theoretic method which has been applied satisfactorily in designs of new products in textile industry, metallurgical industry, engineering industry and agriculture in China.

We offer a set of tables $U_n(b^c)$ of uniform designs, where n denotes the number of experiments, b the number of levels and c the maximum number of factors. For example, if there are 3 factors A, B, C and each has 11 levels A_i, B_i, C_k in a design, then a possible choice is to use the table U_{11} (11^6) listed in Table 1.

Table 1, $U_{11}(11^6)$

numbers\columns	1	2	3	4	5	6
1	1	2	3	5	7	10
2	2	4	6	10	3	9
3	3	6	9	4	10	8
4	4	8	1	9	6	7
5	5	10	4	3	2	6
6	6	1	7	8	9	5
7	7	3	10	2	5	4
8	8	5	2	7	1	3
9	9	7	5	1	8	2
10	10	9	8	6	4	1
11	11	11	11	11	11	11

There is a table attached to each $U_n(b^c)$ which indicates how to select columns for the s factors. For $U_{11}(11^6)$, the attached table is Table 2.

Table 2.   Table attached to U_{11} (11^6)

number of factors	recommended		columns			
2			1	5		
3		1		4	5	
4	1		2	4	5	5
5	1	2		3	4	
6	1	2	3	4	5	6

For our problem, the columns 1, 4, 5 are recommended. Finally we list the design of exreriments in Table 3. Therefore only 11 experiments are deigncd for 3 factors and each has 11 levels.

Tables of uniform design are obtained by an integer vector $(h_1, \cdots, h_s; n)$ in which $h_1 = 1$, $h_1 < h_2 < \cdots < h_s$ and g. c. d. $(h_i, n) = 1$, $i = 1, \cdots, s$. Let

$$P_n(k) = (kh_1, \cdots, kh_s) \equiv (q_{k1}, \cdots, q_{ks}) \pmod{n}, \tag{2.1}$$

where $0 < q_{kj} \leqslant n$, $k = 1, \cdots, n$, $j = 1, \cdots, s$. Table U_n (n^s) is formed by (q_{kj}). When $n = 11$, $s = 6$, $h_1 = 1$, $h_2 = 2$, $h_3 = 3$, $h_4 = 5$, $h_5 = 7$ and $h_6 = 10$, the corresponding table of uniform design is just Table 1.

Table 3. The design of experiments

No.	A	B	C
1	A_1	B_5	C_7
2	A_2	B_{10}	C_3
3	A_3	B_4	C_{10}
4	A_4	B_9	C_6
5	A_5	B_3	C_2
6	A_6	B_8	C_9
7	A_7	B_2	C_5
8	A_8	B_7	C_1
9	A_9	B_1	C_8
10	A_{10}	B_6	C_4
11	A_{11}	B_{11}	C_{11}

Since $1 \leqslant h_i < n$ and g. c. d. $(h_i, n) = 1$, $i = 1, \cdots, s$, the number of possible h_i is given by the Euler function $\phi(n)$

$$\phi(n) = n \prod_{p|n}\left(1 - \frac{1}{p}\right), \tag{2.2}$$

where p runs over the prime factors of n. Since $-h = n - h \pmod{n}$, the rank of matrix (q_{kj}) is at most $1 + \phi(n)/2$ if $n > 2$, i. e., the number of factors must be $\leqslant \phi(u)/2 + 1$. There are at most $\binom{\phi(n)/2}{s-1}$ possible choices of $\boldsymbol{h} = (h_1, \cdots, h_s)'$ since $h_1 = 1$. We want to obtain the "best" $\boldsymbol{h}$ among $\boldsymbol{h}$'s. Wang and Fang (1981) noted that a best $\boldsymbol{h}$ is the minimum of the function

$$D(\boldsymbol{h}) = \frac{1}{n} \sum_{k=1}^{n} \sum_{v=1}^{s}\left(1 - \frac{2}{\pi}\log\left(2\sin\left\{\frac{\pi k h_v}{n+1}\right\}\right)\right) \tag{2.3}$$

with respect to $\boldsymbol{h}$, where $\{x\}$ denote the fractional part of x, and its corresponding set of points is (2.1) which is called a uniform design.

When n is large it will cost much time for finding the best $\boldsymbol{h}$. We suggest to use a set of points of the type

$$Q_n(k) = (k, kb, \cdots, kb^{s-1}) \pmod{n}, \quad 1 \leqslant k \leqslant n \tag{2.4}$$

instead of $P_n(k)$, where b is an integer satisfying $1 < b \leqslant n/2$ and $b^i \neq b^j \pmod{n}$, $1 \leqslant i < j \leqslant s-1$. Integer b is usually chosen a primitive root mod n, if n is a prime number. We call also a set of points (2.4) with minimum $D(b)$ among all those $\boldsymbol{b} = (1, b, \cdots, b_{s-1})'$ a uniform design. Most of the tables of uniform design are produced by (2.4).

A table of uniform design for the case of even n can be obtained by omitting the last row of a table for $n+1$. For instance, the table $U_{10}(10^6)$ can be obtained by omitting the last row of $U_{11}(11^6)$ (Cf. Table 1).

Data in the uniform design can not be analysed by the usual analysis of variance, because the number of experiments is too small compared with the numbers of factors and levels. But we can treat the data by regression or stepwise regression.

Example 1. To design a Vinylon product we shall consider the following factors:

A: Temperature (C),

B: Time (m),

C: Concentration of methanol (g/l),

D: Concentration of sulphuric acid (g/l),

E: Concentration of mirabilite (g/l).

Each factor has 7 levels listed in Table 4.

Table 4. Factors and levels

factors\levels	1	2	3	4	5	6	7
A	64	66	68	70	72	74	76
B	14	16	18	20	22	24	26
C	18	20	22	24	26	28	30
D	206	212	218	224	230	236	242
E	70	70	85	85	85	100	100

If we use Latin square design, then we require 49 experiments which lead to the following linear regression equation

$$\hat{y} = -42.37 + 0.55x_1 + 0.38x_2 + 0.26x_3 + 0.10x_4 - 0.04x_5 \qquad (2.5)$$

with a multiple correlation coefficient $R = 0.97$ and a standard deviation $\hat{\sigma} = 0.83$. Here $\hat{y}$ stands for the quality of Vinylon. Now we use uniform design and choose $U_{14}(14^5)$, which designs for 14 levels. We repeat the original levels twice by the quasi-level method, and obtain the regression equation as follows

$$\hat{y} = -57.97 + 0.37x_1 + 0.46x_2 + 0.38x_3 + 0.17x_4 + 0.04x_5 \qquad (2.6)$$

with $R = 0.96$ and $\sigma = 1.13$. Equation (2.6) is close to (2.5). The result is not too bad because only 14 experiments are arranged.

§ 3. Experiments with Mixtures

If s factors $X_1, \cdots, X_s$ are non-negative and satisfy $X_1 + \cdots X_s = 1$, then the experiments are called the experiments with mixtures, which offen appear in designing the chemical and metallurgical products. In the last two decades, a lot of works appeared in the statistical literature have proposed some kinds of designs. Scheffe (1958) introduced the simplex-lattice designs and the polynomial models. He (Scheffe (1963)) suggested an alternative design, the simple-centroid design, to the general simplex-lattice. Draper and Lawrence (1965) proposed to use the designs which minimized the bias in the fitted model as well as the variance through minimizing the mean square error of the estimate of response over the experimental region. Thompson and Myers (1968) considered an elliptical region inside the simplex factor space by rotatable design. Snee (1973) suggested techniques for the analysis of mixture data. Cornell (1975) gave a suggestion of axial design and he ((1973), (1981)) gave a thorough review of this subject.

In this section we give a different approach to the experiments with mixtures by a uniformly distributed set of points over the simplex. This kind of designs has the same ad-vantages as the uniform design mentioned in Section 2. We pay attention to the best formulation of ingredients, and call this design the uniform design fo r experiments with mixtures (UDEM). Let

$$T_{s-1} = \{(x_1, \cdots, x_s): x_i \geqslant 0, \ i=1, \cdots, s, \ x_1 + \cdots + x_s = 1\}$$

be a part of the surface of the unit simplex in R^s. The idea of uniform design for experiments with mixtures is to design n experiments which are uniformly distributed on T_{s-1}.

Let $\{q_{ki}, \ i=1, \cdots, s-1, \ k=1, \cdots, n\}$ be a uniform design defined by (2.1). Then $\{b_{ki}\}$ with

$$b_{ki} = \frac{2q_{ki}-1}{2n}, \ k=1, \cdots, n, \ i=1, \cdots, s-1 \tag{3.1}$$

is a set of uniformly distributed points in I^{s-1}. In Section 2 of [13] we suggest a method to produce a uniformly distributed set of points $\{P_k, \ k=1, \cdots, n\}$ on T_{s-1} with $P_k = (x_{k1}, \cdots, x_{ks})'$ and

$$\begin{cases} x_{kj} = \prod_{i=1}^{j-1} b_{ki}^{1/(s-i)}(1-b_{kj}^{1/(s-j)}), \ j=1, \cdots, s-1 \\ x_{ks} = \prod_{i=1}^{s-1} b_{ki}^{1/(s-i)}, \ k=1, \cdots, n. \end{cases} \tag{3.2}$$

Figure 1 shows the uniformity of the set (3.2) when $s=3$ and $n=31$.

Example 2. Consider a regression model

$$Y = e + \sum_{i=1}^{s} e_i X_i + \sum_{i,j=1}^{s} e_{ij} X_i X_j + \epsilon,$$

where ϵ stands for a random error. Since $X_1 + \cdots + X_s = 1$, it can be reduced to the form

$$Y = e + \sum_{i=1}^{s-1} e_i X_i + \sum_{i,j=1}^{s-1} e_{ij} X_i X_j + \epsilon. \tag{3.3}$$

Consider the following special model

$$Y = X_1 + X_2 - 3X_1^2 - 3X_2^2 + X_1 X_2 + \epsilon, \tag{3.4}$$

where $\epsilon \sim N(0, \sigma^2)$. When σ is small (for example, $\sigma = 0.005$), we get the following data for $n=17$ and $s=3$ by the use of UDEM (3.2).

The corresponding regression model is

Table 5.   DATA

No	X_1	X_2	Y
1	.829	.076	-1.100
2	.703	.253	$-$ .541
3	.617	.102	$-$ .391
4	.546	.307	$-$ .157
5	.486	.045	$-$ .160
6	.431	.284	.038
7	.382	.564	$-$ .230
8	.336	.215	.146
9	.293	.520	$-$ .103
10	.252	.110	.163
11	.214	.439	.031
12	.178	.798	$-$ .889
13	.143	.328	.134
14	.109	.708	$-$ .644
15	.076	.190	.155
16	.045	.590	$-$ .388
17	.015	.029	.000

$$\hat{Y} = -0.0376 + 1.1162X_1 + 1.1197X_2 - 3.0842X_1^2$$
$$-3.0880X_2^2 + .8336X_1X_2 \tag{3.5}$$

which is close to the model (3.4). The multiple correlation coefficient of the equation (3.5) is $R = 0.9999$ and the estimate of standard deviation is $\delta = 0.0054$ which is close to $\sigma = 0.005$ too.

When σ is getting large, we can not get so excellent results. For example, consider the model

Table   6.

No	X_1	X_2	Y
1	0.817	0.055	8.508
2	0.684	0.179	9.464
3	0.592	0.340	9.935
4	0.517	0.048	9.400
5	0.452	0.201	10.680
6	0.394	0.384	9.748
7	0.342	0.592	9.698
8	0.293	0.118	10.238
9	0.247	0.326	9.809
10	0.204	0.557	9.732
11	0.163	0.809	8.933
12	0.124	0.204	9.971
13	0.087	0.456	9.881
14	0.051	0.727	8.892
15	0.017	0.033	10.139

$$Y = 10 + X_1 - 3X_1^2 - 3X_2^2 + X_1X_2 + \epsilon \tag{3.6}$$

where $\epsilon \sim N(0,\ \sigma^2)$ with $\sigma = 0.30$, and 15 experiments by (3.2). We obtain the data (Table 6) by simulation.

The correspondind regression equation now becomes

$$\hat{Y} = 10.0908 + 0.7972X_1 - 3.4542X_1^2 - 2.6733X_2^2 + 0.8884X_1X_2 \tag{3.7}$$

with $R = 0.9003$ and $\hat{\sigma} = 0.2891$. Note that this regression equation deviates from the model (3.6), because there are high correlations between X_1 and X_1^2. In the original model the response Y reaches its maximum 10.0857 at $X_1 = 0.171$ and $X_2 = 0.0286$. From the regression equation (3.7) it is easy to show that Y reaches its maximum at $X_1 = 0.105$ and $X_2 = 0.0196$ which $\hat{Y}$ in (3.6) is 10.0728 that is close to 10.0857. Hence, from the point of view of the best formulation of ingredients, the result of this example seems nice.

§4. Geometric Probability and Simulation

In this section, we use two "case study" methods to illustrate the applications of uniformly ditsributed sets of points over D to the problems in geometric probability and simulation. The readers can understand the general principle from these two examples without essential difficulties. The questions come from practical problems, and have no satisfactory solution for a comparatively long time. Now we propose algorithms for finding their feasible solutions by the use of number theoretic method.

A. The area of the intersection between a fixed circle and the union of a set of random circles. Given a unit circle K with centre at the origin. There are m random circles $O_1, \cdots, O_m$ with centres $P_1,\ \cdots,\ P_m$ and radii $R_1,\ \cdots,\ R_m$, respectively. Assume that

$$P_i \sim N_2(\mathbf{0},\ \sigma_i^2 \boldsymbol{I}_2),$$

where $\sigma_i > 0$ and $\boldsymbol{I}_2$ denotes the 2×2 identity matrix. Let S be the overlap area between K and the union of all random circles, i. e.,

$$S = K \cap (O_1 \cup \cdots \cup O_m).$$

It is required to know the distribution of S. It is easy to find the distribution of S for the case of $m = 1$, since the overlap area of two circles can be expressed explicitly in terms of the distance between their centres. When $m > 1$, it seems difficult to find a feasible method for finding the distribution of S. This is a problem of geometrical probability. A. natural way is to use simulation. The classical method is the so-called lattice points method. Let ABCD be the circumscribed square of the unit circle K as shown by Figure 3. Divide the square ABCD into n^2 equal subsquares of side $2/(n-1)$. We have n^2 lattice points

$$\left(-1+\frac{2i}{n-1}, \quad -1+\frac{2j}{n-1}\right), 0 \leqslant i, \ j \leqslant n-1$$

in ABCD. Suppose that there are N points lying in K. We now use comupter to produce m reandom circles with centres $P_i \sim N_2(\mathbf{0}, \ \sigma_i^2 I^2)$ and radii $R_i (i=1, \ \cdots, \ m)$. Suppose that M points among the N lattice points are covered by these m random circles. Then we get an obvservation $\pi M/N$ for the distribution of S. We then generate other m random circler and obtain another observation. Continuing this process, we have an empirical distribution of S. This method is called the method I. Its convergent rate is slow. The more serious thing is that its accuracy is low even if we take N large, because there are $O(\sqrt{N})$ points among the N lattice points located nearly the boundary of K.

However we may use a set of points (3.1) of $s=2$ with a linear transformation instead of the above n^2 lattic e points and do simulation as before (cf. Figure 3). We call this method the method II which gives faster convergent rate and higher accuracy than the method I. For example, we take $m=1$ and compare the results of S given by these two methods with the exact value of S. It takes more than 180 minutes by the computer IBMPC/XT and the method I to get a sample of size 1, 000 with an error 0.15, but it needs only 4 minutes by the same computer and the method II to obtain a sample of size 1, 500 with an error 0.02. In general, the method II is faster than the method I about $100 \sim 1000$ times.

Note that the set of points (3.1) of $s=2$ is defined in ABCD, but not in K. Inspiring by the numerical interation over K stated in [13], it is possible to define a set of points that is uniformly distributed over K by means of the set (3.1). Let

$$\begin{cases} x = r\cos(2\pi\theta), \\ y = r\sin(2\pi\theta), & 0 \leqslant \theta < 2\pi, \ 0 \leqslant r \leqslant 1, \end{cases} \tag{4.1}$$

and let $(\theta_i, \ r)$ $(1 \leqslant i \leqslant N)$ be a set of points (4.1). If each point of the set

$$Q_n(i) = (r_i \cos 2\pi\theta_i, \ r_i \sin 2\pi\theta_i), \quad 1 \leqslant i \leqslant N \tag{4.2}$$

has a suitable "weight" w_i, (4.2) may be considered as a uniformly distributed set in K (Cf. Figure 4). Since the transformation (4.1) has the Jacobian $2\pi r$, we define the weight for $Q_n(i)$ to be $2\pi r_i$. So if there are M points $Q_n(i_j) (j=1, \ \cdots, \ M)$ covered by the m random circles, we use

$$2\pi \sum_{j=1}^{M} r_{i_j}$$

to approximate the areas covered by these m random circles. We call this method the method III. In order to compare the methods II and III, we also take $m=1$ and consider the intersection area of two unit circles K and O with distance d between their centers (cf. Figure 5). We use a set of 1069 points of the type (3.1) in which

844 points are lying in K for the method II, and a set of 828 points of (3.1) is used for the method III. The result is given by Table 7.

Table 7

errors\d	0.1	0.75	0.8	1.3
method II	-0.55%	0.07%	0.19%	0.013%
method III	0.00%	0.04%	0.12%	0.08%

From Example 2 of [13] we may obtain a uniformly distributed set of points denoted by $\{x_k,\ k=1,\ \cdots,n\}$ over K, by which we may have another method——method IV. Now the simulation is based on $\{x_k,\ k=1,\ \cdots,\ n\}$ with the same manner as method II. Our simulation results show that the methods III and IV almost have the same accuracy.

B. The problem of coveringthe sphere by random belts with a fixed width.

This problem comes from the steel rolling [1]. People wish to increase the life of the roller by using a random rotary ball roller instead of a fixed roller. Its mathematical model may be stated as follows: Let S be the unit sphere $x_1^2+x_2^2+x_3^2=1$ and δ a constant satisfying $0<\delta<0.3$. Let R be a great circle which is uniformly distributed on S and $G_\delta(R)$ be the belt on S with width δ and with R as the equidistant partition curve. Let $G_{\delta_1},\ \cdots,\ G_{\delta_n},\ \cdots$ be a sequential sample of the population $G_\delta(R)$. For any $x\in S$, we denote by $D_N(x)$ the number of belts which cover x in the first N random belts. If there is a point in S that is covered by m belts where m is a given positive integer, we say that the roller is useless. For a given integer m, let T_m be the minimum of N such that $D_N(x)\geqslant m$ for some $x\in S$, i. e.,

$$T_m=\min\{N:\ D_N(x)\geqslant m,\ \text{for some}\ x\in S\}. \tag{4.3}$$

The T_m stands for the life of the roller. We wish to obtain the distribution of T_m and to find some way to increase the life of the roller.

It seems difficult to give a formula for the distribution function of T_m, which leads us to do the problem by simulation. In our simulation the following facts are used:

From Exmple 3 of [13] we can obtain a set $\{x_k=(x_{k1},\ x_{k2},\ x_{k3})',\ k=1,\ \cdots,\ n\}$ which is uniformly distributed on S. More precisely,

$$\begin{cases} x_{k1}=\cos(\pi c_{k1}), \\ x_{k2}=\sin(\pi c_{k1})\cos(2\pi c_{k2}), \quad k=1,\ \cdots,\ n, \\ x_{k3}=\sin(\pi c_{k1})\sin(2\pi c_{k2}), \end{cases} \tag{4.4}$$

where $F_i(c_{ki})=b_{ki},\ i=1,\ 2,\ k=1,\ \cdots,\ n,\ \{b_k=(b_{k1},\ b_{k2})',\ k=1,\ \cdots,\ n\}$ is a uniformly distributed set on I^2, and

$$F_i(x) = \frac{\pi}{B\left(\frac{1}{2}, \frac{3-i}{2}\right)} \int_0^x (\sin \pi t)^{(3-i-1)} dt, \quad i=1,\ 2,$$

i. e.,
$$F_1(x) = \frac{1}{2}(1 - \cos(\pi x))$$

and
$$F_2(x) = x.$$

Therefore
$$b_{k1} = \frac{1}{2}(1 - \cos(\pi c_{k1}))$$

$$b_{k2} = c_{k2}$$

and
$$\begin{cases} x_{k1} = 1 - 2b_{k1}, \\ x_{k2} = 2\sqrt{b_{k1} - b_{k1}^2} \cos(2\pi b_{k2}), \\ x_{k3} = 2\sqrt{b_{k1} - b_{k1}^2}\sin(2\pi b_{k2}), \quad k = 1,\ 2,\ \cdots,\ n. \end{cases} \tag{4.5}$$

Let
$$T_m^* = \min\{N : D_N(\boldsymbol{x}_k) \geqslant m, \text{ for some } k,\ 1 \leqslant k \leqslant n\}.$$

When n is sufficiently large, T_m^* is close to T_m and the distribution of T_m^* is close to that of T_m. Our simulation is based on T_m^*.

Given a point $\boldsymbol{v}$ on S, it corresponds a great circle R such that the normal direction of the plane including R is the direction of $\overrightarrow{ov}$. If we identify the points $\boldsymbol{v}$ and $-\boldsymbol{v}$, then it has a one–one correspondence between the points on S and the great circles and consequently the belts with width δ on S. Thus, to generate a random belt $G_\delta(R)$ which is uniformly distributed on S is equival ent to generate a point $\boldsymbol{v} \in S$ which is uniformly distributed on S. We also use $G_\delta(\boldsymbol{v})$ to denote the belt corresponding to $\boldsymbol{v}$. Our problem of simulation includes the following steps:

Step 1.   Give m and δ, for instance, $m=20$ and $\delta=0.2$.

Step 2.   Choose a suitable n (for example, $n=1069$) and produce a set of n points $\{\boldsymbol{x}_k,\ k=1,\ \cdots,\ n\}$ which are uniformly distributed on S.

Step 3.   By a standard technique of the simulation, generate sequentially the points $\boldsymbol{v}_1,\ \boldsymbol{v}_2,\ \cdots$, which are independent and uniformly distributed on S and consequently we have the corresponding belts $G_\delta(\boldsymbol{v}_1),\ G_\delta(\boldsymbol{v}_2),\ \cdots$.

Step 4.   If there are $N(N=1,\ 2,\ \cdots)$ random belts to be generated in the current step, account the number of belts covered $\boldsymbol{x}_k$ and denote by $D_N(\boldsymbol{x}_k)$. If $D_N(\boldsymbol{x}_k) = m$ for some k, go to Step 5, otherwise go back to Step 3 and generate the $(N+1)th$ random belt.

Step 5.   Account the number of random belts generated already. This number is an obcervation of T_m^*.

Repeat the above process n_0 times and obtain a sample of size n_0 of T_m^*. We take $n_0=5000$, the respective sample mean and the sample standard deviation are
$$\overline{T}_m^* = 99.7 \text{ and } \sigma(T_m^*) = 9.8.$$

Furthermore, the corresponding empirical distribution is close to the normal distribution.

By the same way, we obtain 20 samples of size $n_0=5000$ (total 100,000 observations) and find that the results are very close to each other.

Since the T_m (or T_m^*) stands for the life of roller, we note that sometimes T_m^* can be reached at 125 in 100.000 obcervations by the above simulation. We denote the corresponding normal directions by v_1^*, v_2^*, $\cdots$, v_{125}^*. This means that if $v_i=v_i^*$, $i=1$, 2, $\cdots$, are fixed, we always have $T_m^*=125$ in the case of $\delta=0.2$ and $m=20$, which is better than the above random choices of $\{v_i\}$. Is it possible to find another set of x_1^{**}, v_2^{**}, $\cdots$ to beat the above $\{v_i^*, i=1, \cdots, 125\}$? We may use the uniformly distributed sets of points on I^2 to produce the sets of $\{v_k^{**}, k=1, \cdots, n\}$ on S. First we choose $n=126$ and find the corresponding $T_m^*=126$. Then we increase n one by one untill the T_m^* can not be increased any more. Finally we find a set $\{v_k^{**}=(v_{k1}, v_{k2}, v_{k3})'$, $k=1, 2, \cdots, 155\}$ by which $T_m^*=155!$ The v_k^{**} is given by

$$\begin{cases} v_{k1}=1-2b_{k1}, \\ v_{k2}=2\sqrt{b_{k1}-b_{k1}^2}\cos(2\pi b_{k2}), \\ v_{k3}=2\sqrt{b_{k1}-b_{k2}^2}\sin(2\pi b_{k2}), \end{cases}$$

where the set $\{b_k=(b_{k1}, b_{k2})', k=1,\cdots, 155\}$ is produced by $(h_1, h_2; n)=(1, 20; 155)$ (See [13], Section 3 for details).

This example indicates that the number theoretic method wins the champion of 100, 000 experiments by Monte carlo method in our simulation.

Acknowledgement: We wish to thank Mr. Wei Gang for his excellent computer works.

References

[1] Cheng, P., An open problem in steel rolling, *Mathematics in Practice and Theory*, 2 (1983), 79—79.

[2] Cornell, J. A., Experiments with mixtures: A review. *Technometrics*, **15** (1973), 437—455.

[3] Cornell, J. A., Some comments on designs for Cox's mixture polynomial, *Technometrics*, **17** (1975), 25—35.

[4] Cornell, J. A., Experiments with Mixtures, designs, models, and the analysis of mixture data, 1981, Wiley, New York.

[5] Cox, D. R., A note on polynomial response functions for mixtures, *Biometrika*, **58** (1971), 155—159.

[6] Draper, N. R., & Lawrence, W. E., Mixture designs for three factors, *J. Royal Statist. Soc. B*, **27** (1965), 450—465.

[7] Hua, L. G., & Wang, Y., *Applications of Number Theory to Numerial Analysis*, Springer-Verlag and Science Press, 1981, Berlin/Beijing.

[8] Scheffe, H., Experiments with mixtures, *J. Rñyal Statist, Soc. B*, **20** (1958), 344—360.

[9] Scheffe, H., The simplex-centroid design for experiments with mixtures, *J. Royal Statist. Soc. B*, **25** (1963), 235—263.

[10] Snee, R. D., Techniques for the analysis of mixture data, *Technometrics*, **15** (1973), 517—528.

[11] Thompson, W. O. & Myers, R. H., Response surface design for experiments with mix-tures, *Technometrics*, **10** (1968), 739—756.

[12] Wang, Y. & Fang, K. T., A note on uniform distribution and experimental design, *Kexue Tongba*, **26** (1981), 485—489.

[13] Wang, Y. & Fang, K. T., Number theoretic methods in applied statistics, *Chin. Ann of Math.*, **11B**: 1(1990), 41—55.

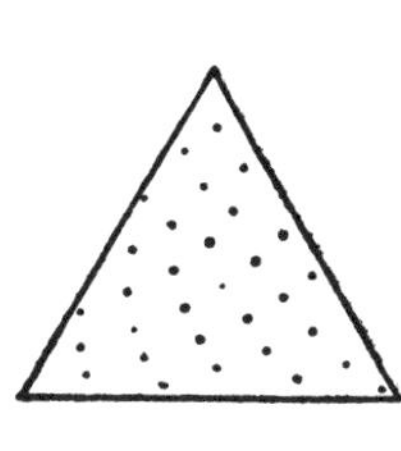

Fig. 1

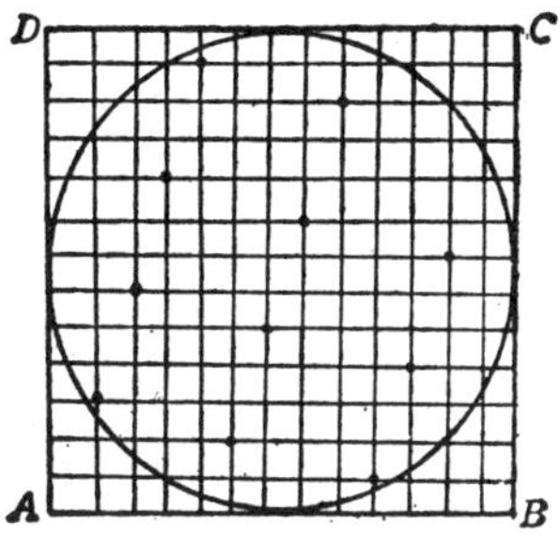

Fig. 2

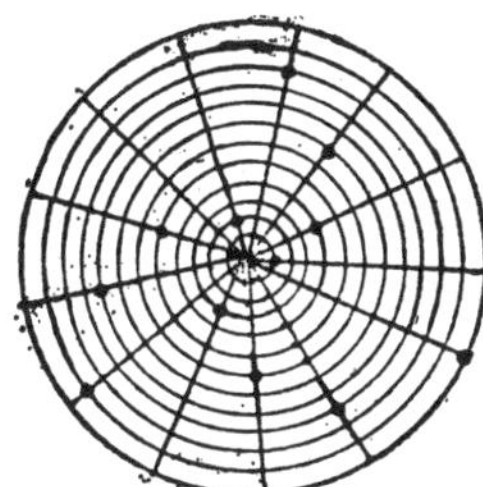

Fig. 3

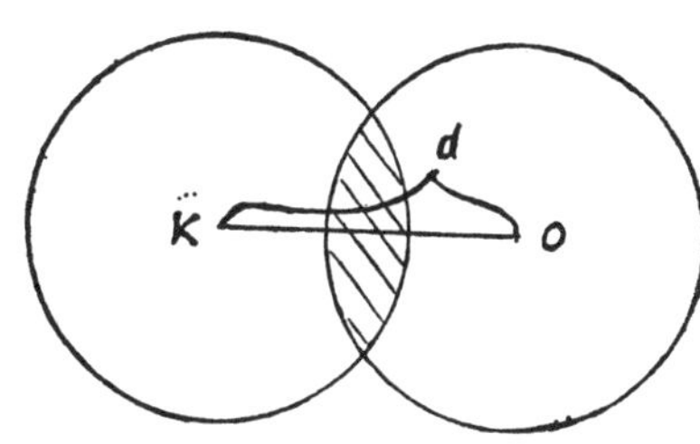

Fig. 4

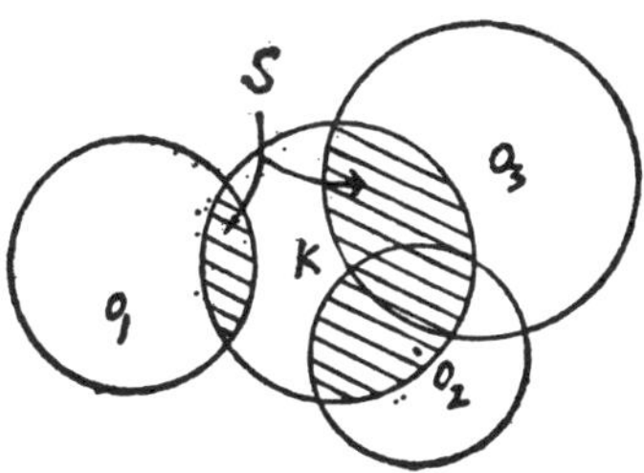

Fig. 5

Vol. 39 No. 3　　　　　SCIENCE IN CHINA (Series A)　　　　　March 1996

Uniform design of experiments with mixtures*

WANG Yuan (王　元)

(Institute of Mathematics, Chinese Academy of Sciences, Beijing 100080, China)

and FANG Kaitai (方开泰)

(Hong Kong Baptist University; Institute of Applied Mathematics, Chinese Academy of. Sciences, Beijing 100080, China)

Received July 3, 1995

Abstract　　　Consider a design of experiments with mixtures: $0 \leqslant a_i < x_i < b_i \leqslant 1$, $1 \leqslant i \leqslant s$, $x_1 + \cdots + x_s = 1$, where a_i, b_i, $1 \leqslant i \leqslant s$ are given constants. A method is proposed to treat this model by the theory of uniform distribution in number theory.

Keywords: experimental design, uniform design, experimental design with mixtures, number-theoretic net (NT-net), discrepancy.

As an application of the theory of uniform distribution in number theory to the experimental design, we propose a method, the so-called uniform design[1]. Suppose there are s factors $x_1, \cdots, x_s$. We may assume without loss of generality that the experimental domain is the unit cube $C^s = [0, 1]^s$ and each point in C^s corresponds to an experiment. The idea of uniform design is to find out a set of n points

$$\mathscr{P} = \{c_k = (c_{k1}, \cdots, c_{ks}), \ 1 \leqslant k \leqslant n\}$$

in C^s which has a lower discrepancy $D(P)$ in the sense of Weyl[2]. We call the set $\mathscr{P}$ a number-theoretic net or NT-net if $D(\mathscr{P}) = o(n^{-1/2})$ as $n \to \infty$.

There are a number of methods which can produce the NT-nets with different s and n to be compiled in tables[3, 4]. Then we may arrange n experiments on $\mathscr{P}$ and obtain a point c which has the best experimental result among these n results on $\mathscr{P}$. However, in practice, especially in chemical experiments and chemical engineering, some constraints should be added to the factors, for instance,

$$0 \leqslant a_i < x_i < b_i \leqslant 1, \ 1 \leqslant i \leqslant s, \ \sum_{i=1}^{s} x_i = 1, \tag{1}$$

where a_i, b_i are given constants. The experimental design with constraints is called the experimental design with mixtures. Cornell[5] gave a comprehensive exposition on the experimental design with mixtures, for example, he changed domain (1) by an ellipsoid and then transferred the original problem by an $(s-1)$-dimensional problem. The aim of the

---

* Project partially supported by the National Natural Science Foundation of China and Institute of Mathematics, Taipei.

present paper is to apply the theory of uniform distribution to the experimental design with constraints (1). The simplest model in (1) is that $a_i=0$ and $b_i=1$, $1\leqslant i\leqslant s$. This problem is equivalent to finding out a set of points $\{x_k,\ 1\leqslant k\leqslant n\}$ which are uniformly scattered on the simplex

$$S=\left\{x=(x_1,\cdots,x_s):x_i\geqslant 0,\ 1\leqslant i\leqslant s,\ \sum_{i=1}^{s}x_i=1\right\}.$$

We have suggested a method for finding out a uniformly scattered set of n points on S which is induced by a given NT-net of n points on C^{s-1} [4, 6, 7]. In this paper we shall generalize our method to the more difficult general case (1), i.e. we shall give a method for finding out some uniformly scattered sets on domain (1).

1 Inverse transform method

Let D be a bounded domain in the s-dimensional Euclidean space R^s and $x=(x_1,\cdots,\ x_s)$ be a random vector with a c.d.f. $F(x)$ defined on D which has a stochastic representation

$$x=x(\boldsymbol{\varphi}),\ \boldsymbol{\varphi}\in C^t,$$

where $t\leqslant s$ and $\boldsymbol{\varphi}=(\varphi_1,\cdots,\ \varphi_t)\in C^t$ is a random vector with independent components. Suppose $\dfrac{\partial x_j}{\partial\varphi_i}$, $1\leqslant i\leqslant t$, $1\leqslant j\leqslant s$, $1\leqslant j\leqslant s$ are continuous on C^t. Denote

$$T=\left(\frac{\partial x_j}{\partial\varphi_i}\right),\ 1\leqslant i\leqslant t,\ 1\leqslant j\leqslant s$$

and

$$J(\boldsymbol{\varphi})=\det(TT')^{1/2}.$$

$J(\boldsymbol{\varphi})$ is the volume element of D with respect to x. When $t=s$, $J(\boldsymbol{\varphi})$ is the usual Jacobian. Then it follows by the independence of $\varphi_i's$ that

$$\frac{1}{V(D)}\ J(\boldsymbol{\varphi})=\prod_{i=1}^{t}p_i(\varphi_i),$$

where $V(D)$ denotes the volume of D and $p_i(\varphi)$ the probability density function (p.d.f.) of φ_i, $1\leqslant i\leqslant t$. Therefore,

$$\frac{1}{V(D)}\int_{\boldsymbol{\varphi}\leqslant r}J(\boldsymbol{\varphi})\mathrm{d}\boldsymbol{\varphi}=\prod_{i=1}^{t}F_i(r_i),\tag{2}$$

where $\mathrm{d}\boldsymbol{\varphi}=\prod_{i=1}^{t}\mathrm{d}\varphi_i$, $\boldsymbol{\varphi}\leqslant r=(r_1,\cdots,\ r_t)$ means $\varphi_i\leqslant r_i$, $1\leqslant i\leqslant t$ and $F_i(r_i)$ denotes the cumulative distribution function (c.d.f.) of φ_i, $1\leqslant i\leqslant t$.

Let $\mathscr{P}=\{x_k,\ 1\leqslant k\leqslant n\}$ be a set of points on D and $x_i=x(\boldsymbol{\varphi}_i)$, $1\leqslant i\leqslant n$. Let

$$N(\mathscr{P},r)=\sum_{i=1}^{n} I(\varphi_i \leqslant r),$$

where $I(A)$ denotes the indicator function of A, i.e.

$$I(A)=\begin{cases} 1, & \text{if } A \text{ occurs,} \\ 0, & \text{otherwise.} \end{cases}$$

Then the empirical function of $x_1, \cdots, x_n$ may be defined by

$$F_n(r)=\frac{1}{n}\sum_{i=1}^{n} I(\varphi_i \leqslant r).$$

Then

$$\sup_{r\in C^t} |F_n(r)-F(r)| \tag{3}$$

is the Kolmogorov-Smirnov distance for the goodness-of-fit test of $F(r)$. If $F(r)$ is the uniform distribution of C^t, then by (2),

$$F(r)=\frac{V(t\leqslant r)}{V(D)} =\prod_{i=1}^{t} F_i(r_i),$$

where

$$V(t\leqslant r)=\int_{r\leqslant r} J(\varphi)\mathrm{d}\varphi.$$

Now we denote (3) by

$$D_F(\mathscr{P})=\sup_{r\in C^t}\left| F_n(r) - \frac{V(t\leqslant r)}{V(D)} \right|.$$

This is a measure for uniformity of the set $\mathscr{P}$ on D and it is called the F-discrepancy of $\mathscr{P}$. When $s=t$, $D=C^t$ and $x_i=\varphi_i$, $1\leqslant i\leqslant s$, $D_F(\mathscr{P})$ is just the discrepancy of $\mathscr{P}$ in the sense of Weyl[2] and is denoted by $D(\mathscr{P})$.

Let

$$\mathscr{P} =\{c_k=(c_{k1},\cdots,c_{kt}),1\leqslant k\leqslant n\}$$

be a sample from the uniform distribution on C^t with discrepancy $D(\mathscr{P})$. Denote by $F_i^{-1}(r)$ the inverse function of $F_i(r)$, $1\leqslant i\leqslant t$, and

$$x_k=x(F^{-1}(c_k)),$$

where

$$F^{-1}(c_k)=(F_1^{-1}(c_{k1}),\cdots,F_t^{-1}(c_{kt})), \ 1\leqslant k\leqslant n.$$

Now we proceed to find the F-discrepancy of the set $\mathscr{P}^*=\{x_k=x(F^{-1}(c_k)), 1\leqslant k\leqslant n\}$. Since

$$F_n(r)=\frac{1}{n}\sum_{i=1}^{n} I(F^{-1}(c_k)\leqslant r)=\frac{1}{n}\sum_{i=1}^{n} I(c_k\leqslant F(r)),$$

we have

$$D_F(\mathscr{P}^*)=\sup_{r\in C^t}\left|\frac{1}{n}\sum_{i=1}^{n}I(c_k\leqslant F(r))-\prod_{i=1}^{t}F_i(r_i)\right|=\sup_{u\in C^t}\left|\frac{1}{n}\sum_{i=1}^{n}I(c_k\leqslant u)-\prod_{i=1}^{t}u_i\right|=D(\mathscr{P}).$$

This means that the F-discrepancy of $\mathscr{P}^*$ is just the $D(\mathscr{P})$. Therefore, starting from a set $\mathscr{P}$ on C^t with discrepancy d, we can induce a set $\mathscr{P}^*$ on D with F-discrepancy d. This method is called the inverse transform method and we often call the uniformly scattered set of n points with F-discrepancy $o(n^{-1/2})$ the NT-net. We can construct the NT-nets with F-discrepancy $O(n^{-1}(\log n)^{s-1})$[3, 4].

2 The domain $S(a, b)$

Suppose $a=(a_1,\cdots, a_s)$ and $b=(b_1,\cdots, b_s)$ are two given vectors satisfying

$$0\leqslant a_i<b_i\leqslant 1, \ 1\leqslant i\leqslant s$$

and

$$a=\sum_{i=1}^{s} a_i<1 \ \text{ and } \ b=\sum_{i=1}^{s} b_i>1. \tag{4}$$

Denote by $S(a, b)$ the domain

$$S(a,b)=\left\{x=(x_1,\cdots,x_s): a_i<x_i<b_i, 1\leqslant i\leqslant s, \sum_{i=1}^{s} x_i=1\right\}.$$

Note that (4) is a necessary and sufficient condition that $S(a, b)$ is not empty. In fact if (4) holds, then $S(a, b)$ is not empty clearly. On the other hand, if $a\geqslant 1$ or $b\leqslant 1$, then $\sum_{i=1}^{s}x_i>1$ or $\sum_{i=1}^{s}x_i<1$ for all x with $a_i<x_i<b_i, \ 1\leqslant i\leqslant s$. Let

$$y_i=\frac{x_i-a_i}{b_i-a_i} \ , \ 1\leqslant i\leqslant s. \tag{5}$$

Then

$$\sum_{i=1}^{s} (b_i-a_i)y_i=\sum_{i=1}^{s} x_i-\sum_{i=1}^{s} a_i=1-a.$$

Write

$$d_i=\frac{b_i-a_i}{1-a}, \ 1\leqslant i\leqslant s.$$

We have

$$\sum_{i=1}^{s}d_i y_i=1, \ d_i>0, \ 1\leqslant i\leqslant s.$$

Therefore $S(a, b)$ and the condition (4) become

$$S(d)=\left\{y=(y_1,\cdots,y_s): y\in C^s, \sum_{i=1}^{s} d_i y_i=1, d_i>0, 1\leqslant i\leqslant s\right\}$$

268   SCIENCE IN CHINA (Series A)   Vol. 39

and

$$\sum_{i=1}^{s} d_i > 1, \tag{6}$$

respectively, where $d = (d_1, \cdots, d_s)$. Let

$$S^*(d) = \left\{ y = (y_1, \cdots, y_s): y_i > 0, \ 1 \leqslant i \leqslant s, \ \sum_{i=1}^{s} d_i y_i = 1, \ d_i > 0, \ 1 \leqslant i \leqslant s \right\}.$$

First we give an algorithm for finding an NT-net on $S^*(d)$. The transformation

$$\begin{cases} d_j y_j = z_1 \cdots z_{j-1}(1 - z_j), \ 1 \leqslant j \leqslant s - 1 \\ d_s y_s = z_1 \cdots z_{s-1}, \ z = (z_1, \cdots, z_{s-1}) \in C^{s-1} \end{cases}$$

is a one-to-one correspondence between C^{s-1} and $S^*(d)$. Let

$$\Delta = (\det HH')^{1/2} \ \text{ and } \ H = \left(\frac{\partial y_j}{\partial z_i} \right), \ 1 \leqslant i \leqslant s-1, \ 1 \leqslant j \leqslant s. \tag{7}$$

Then

$$\Delta = c(d) z_1^{s-2} \cdots z_{s-2}, \tag{8}$$

where

$$c(d) = (d_1 \cdots d_s)^{-1} \left(\sum_{i=1}^{s} d_i^2 \right)^{1/2}.$$

The proof will be given in sec. 4. Hence the volume of $S^*(d)$ equals

$$V(S^*(d)) = \int_{C^{s-1}} \Delta \mathrm{d}z = c(d) \int_{C^{s-1}} z_1^{s-2} \cdots z_{s-2} \mathrm{d}z_1 \cdots \mathrm{d}z_{s-1} = c(d)/(s-1)!.$$

We conclude that $Z_1, \cdots, Z_{s-1}$ are mutually independent and the p.d.f. and c.d.f. of Z_j are

$$p_j = (s - j) z^{s-j-1}$$

and

$$G_j = \int_0^z p_j(t) \mathrm{d}t = z^{s-j}, \ z \in C^1, \ 1 \leqslant j \leqslant s - 1$$

respectively. We can use the inverse transform method to get an NT-net on $S^*(d)$ as follows: start from an NT-net

$$\{ c_k = (c_{k1}, \cdots, c_{k,s-1}), \ 1 \leqslant k \leqslant n \}$$

on C^{s-1} [4]. Since the inverse function of $G_j(t)$ is

$$G_j^{-1}(z) = z^{\frac{1}{s-j}}, \ 1 \leqslant j \leqslant s - 1,$$

we have an NT-net on $S^*(d)$,

$$\{ y_k = (y_{k1}, \cdots, y_{ks}), \ 1 \leqslant k \leqslant n \}, \tag{9}$$

where

$$\begin{cases} y_{kj} = d_j^{-1} c_{k1}^{\frac{1}{s-1}} \cdots c_{k,j-1}^{\frac{1}{s-j+1}} \left(1 - c_{kj}^{\frac{1}{s-j}}\right), \\[2mm] y_{ks} = d_s^{-1} c_{k1}^{\frac{1}{s-1}} \cdots c_{k,s-2}^{\frac{1}{2}} c_{k,s-1}, \quad 1 \leqslant j \leqslant s-1, 1 \leqslant k \leqslant n. \end{cases}$$

Substituting them into (5), we have an NT-net $\{x_k = (x_{k1}, \cdots, x_{ks}), \ 1 \leqslant k \leqslant n\}$ on the domain

$$S(a) = \left\{ x = (x_1, \cdots, x_s) : 1 > x_i > a_i, 1 \leqslant i \leqslant s, \sum_{i=1}^{s} x_i = 1 \right\},$$

where

$$\begin{cases} x_{kj} = (1-a) c_{k1}^{\frac{1}{s-1}} \cdots c_{k,j-1}^{\frac{1}{s-j+1}} \left(1 - c_{kj}^{\frac{1}{s-j}}\right) + a_j \\[2mm] x_{ks} = (1-a) c_{k1}^{\frac{1}{s-1}} \cdots c_{k,s-2}^{\frac{1}{2}} c_{k,s-1} + a_s, \quad 1 \leqslant j \leqslant s-1, 1 \leqslant k \leqslant n. \end{cases}$$

Since the Jacobian of transformation (5) is a constant $\prod_{i=1}^{s}(b_i - a_i)$, the F-discrepancies of $\{x_k: 1 \leqslant k \leqslant n\}$ and $\{y_k: 1 \leqslant k \leqslant n\}$ are of the same order. So we study the $S(d)$ only in the latter.

3 Volumes of $S^*(d)$ and $S(d)$

Our purpose is to get an NT-net of n points on $S(d)$. The algorithm is that we choose (9) an NT-net $\mathscr{P}$ of m points on $S^*(d)$ of which there are nearly n points falling on $S(d)$. Since $\mathscr{P}$ is uniformly scattered on $S^*(d)$, the ratio $\dfrac{m}{n}$ should be approximately equal to the ratio of the volumes of $S^*(d)$ and $S(d)$, i.e.

$$r \equiv \frac{m}{n} \approx \frac{V(S^*(d))}{V(S(d))}.$$

Hence the estimation of the number m is reduced to the estimations $V(S^*(d))$ and $V(S(d))$. It follows from (7) that

$$V(S^*(d)) = c(d) \int_{z \in C^{s-1}} z_1^{s-2} \cdots z_{s-2} \, dz_1 \cdots dz_{s-1} = c(d)/(s-1)! . \tag{10}$$

Now we proceed to evaluate $V(S(d))$,

$$V(S(d)) = c(d) N(d), \tag{11}$$

where

$$N(d) = \int_{\substack{z \in C^{s-1} \\ y \in C^s}} z_1^{s-2} \cdots z_{s-2} \, dz_1 \cdots dz_{s-1}.$$

First we find out a recurrent formula which reduces the $(s-1)$-dimensional integral $N(d)$ to an $(s-2)$-dimensional integral. Secondly, we give the expression of $N(d)$ for the case $s=2$ and finally we get the expression of $N(d)$ for $s=3, 4, \cdots$ by successive iterations.

3.1 Reduction

We may suppose without loss of generality that

$$d_1 \geqslant d_2 \geqslant \cdots d_s. \tag{12}$$

Otherwise, we may change the order of variables such that (12) holds.

(i) If $d_s \geqslant 1$, then $S(d) = S^*(d)$, and

$$N(d) = \frac{1}{(s-1)!}. \tag{13}$$

(ii) If $d_1 \geqslant 1$ and $d_s < 1$, then there is no restriction on z_1, i.e. z_1 may take any value of C^1. Consider the domain for given $z_1 \in C^1$, $z_1 \neq 0$,

$$S^*\left(\frac{d_2}{z_1}, \cdots, \frac{d_s}{z_1}\right): \begin{cases} \dfrac{d_i}{z_1}\, y_i = z_2 \cdots z_{i-1}(1-z_i), \\ \dfrac{d_s}{z_1}\, y_s = z_2 \cdots z_{s-1}, 2 \leqslant i \leqslant s-1, \end{cases}$$

where $(z_2, \cdots, z_{s-1}) \in C^{s-2}$ and $(y_2, \cdots, y_s) \in C^{s-1}$. When $z_1 \in (0, d_s)$, we have $\dfrac{d_s}{z_1} \geqslant 1$ and (13) with s replaced by $s-1$.

$$N(d) = \int_0^{d_s} z_1^{s-2} dz_1 \int_{S^*\left(\frac{d_2}{z_1}, \cdots, \frac{d_s}{z_1}\right)} z_2^{s-3} \cdots z_{s-2} dz_2 \cdots dz_{s-1}$$

$$+ \int_{d_s}^1 z_1^{s-2} dz_1 \int_{S^*\left(\frac{d_2}{z_1}, \cdots, \frac{d_s}{z_1}\right)} z_2^{s-3} \cdots z_{s-2} dz_2 \cdots dz_{s-1}$$

$$= \frac{1}{(s-2)!} \int_0^{d_s} z_1^{s-2} dz_1 + \int_{d_s}^1 z_1^{s-2} dz_1 \int_{S^*\left(\frac{d_2}{z_1}, \cdots, \frac{d_s}{z_1}\right)} z_2^{s-3} \cdots z_{s-2} dz_2 \cdots dz_{s-1}.$$

It yields by (4) or (6) that z_1 must satisfy

$$\sum_{i=2}^s d_i \geqslant z_1$$

in the latter integral; otherwise, this integral is equal to zero. Therefore we obtain a recurrent formula

$$N(d) = \frac{d_s^{s-1}}{(s-1)!} + \int_{d_s}^{\min\left(1, \sum_{i=2}^s d_i\right)} z_1^{s-2} N\left(\frac{d_2}{z_1}, \cdots, \frac{d_s}{z_1}\right) dz_1. \tag{14}$$

(iii) If $d_1 < 1$, z_1 must satisfy $1 - z_1 \leqslant d_1$ since $d_1 y_1 = 1 - z_1$, $y_1 \in C^1$. Hence we get a recurrent formula

$$N(\boldsymbol{d}) = \int_{1-d_1}^{\min\left(1, \sum_{i=2}^{s} d_i\right)} z_1^{s-2} N\left(\frac{d_2}{z_1}, \cdots, \frac{d_s}{z_1}\right) dz_1. \tag{15}$$

3.2 $s=2$

$N(\boldsymbol{d})$ is equal to the measure of $z_1 \in C^1$ which satisfies

$$\begin{cases} d_1 y_1 = 1 - z_1, \\ d_2 y_2 = z_1. \end{cases}$$

We obtain the following

$$N(\boldsymbol{d}) = d_2 \;\; (d_1 \geqslant 1, d_2 < 1), \tag{16}$$

$$N(\boldsymbol{d}) = |1 - d_1 < z_1 < d_2| = d_1 + d_2 - 1 \;\; (d_1 < 1). \tag{17}$$

3.3 $s=3$

If $d_2 \geqslant 1$, $d_3 < 1$, then from (14) and (16)

$$N(\boldsymbol{d}) = \frac{d_3^2}{2} + \int_{d_3}^{\min(1, d_2+d_3)} z_1 N\left(\frac{d_2}{z_1}, \frac{d_3}{z_1}\right) dz_1 = \frac{d_3^2}{2} + \int_{d_3}^{1} z_1 \frac{d_3}{z_1} dz_1 = d_3 - \frac{d_3^2}{2}. \tag{18}$$

Similarly we have by (13)—(17) the following

$$N(\boldsymbol{d}) = (d_2+d_3)\min(1, d_2+d_3) - \frac{\min(1, d_2+d_3)^2}{2} - \frac{d_2^2+d_3^2}{2} \;\; (d_1 \geqslant 1, d_2 < 1), \tag{19}$$

$$N(\boldsymbol{d}) = d_1 - \frac{1}{2}(d_1^2+d_2^2+d_3^2+1) + (d_2+d_3)\min(1, d_2+d_3) - \frac{\min(1, d_2+d_3)^2}{2} \;\; (d_1 < 1, 1-d_1 < d_3), \tag{20}$$

$$N(\boldsymbol{d}) = d_3(d_1+d_2-1+\frac{1}{2}d_3) \;\; (d_1 < 1, d_3 \leqslant 1-d_1 < d_2), \tag{21}$$

$$N(\boldsymbol{d}) = \frac{1}{2}(d_1+d_2+d_3-1)^2 \;\; (d_1 < 1, d_2 \leqslant 1-d_1). \tag{22}$$

We can find complicated analytic expressions for the case $s \geqslant 4$, but we prefer to suggest the use of statistical simulation for an estimation of the ratio r: take an NT-net on C^{s-1},

$$\mathscr{P} = \{c_k = (c_{k1}, \cdots, c_{k,s-1}), 1 \leqslant k \leqslant m\},$$

where m is sufficiently large. If $\mathscr{P}$ has discrepancy d, then the induced set $\mathscr{P}^* = \{y_k = (y_{k1}, \cdots, y_{ks}), 1 \leqslant k \leqslant m\}$ on $S^*(\boldsymbol{d})$ has F-discrepancy d (see (9)). It yields from the uniformity of $\mathscr{P}^*$ on $S^*(\boldsymbol{d})$ that the number of $\mathscr{P}^*$ falling on $S(\boldsymbol{d})$ is asymptotically equal to

$$\left[m\,\frac{V(S(d))}{V(S^*(d))} \right] = n,$$

where $[x]$ denotes the integral part of $[x]$, i.e. $\dfrac{m}{n}$ may be regarded as an approximate value of the ratio

$$\frac{V(S^*(d))}{V(S(d))}.$$

4 The evaluation of Δ

Recall (7) and (8): $\Delta = (\det HH')^{\frac{1}{2}}$, where

$$H = \left(\frac{\partial x_j}{\partial y_i} \right)$$

$$= \begin{pmatrix}
-d_1^{-1}, & d_2^{-1}(1-z_2), & \cdots, & d_{s-1}^{-1}\dfrac{z_1\cdots z_{s-2}}{z_1}(1-z_{s-1}), & d_s^{-1}\dfrac{z_1\cdots z_{s-1}}{z_1} \\[2ex]
0, & -d_2^{-1}z_1, & \cdots, & d_{s-1}^{-1}\dfrac{z_1\cdots z_{s-2}}{z_2}(1-z_{s-1}), & d_s^{-1}\dfrac{z_1\cdots z_{s-1}}{z_2} \\[1ex]
\vdots & \vdots & \ddots & \vdots & \vdots \\[1ex]
0, & 0, & \cdots, & -d_{s-1}^{-1}\dfrac{z_1\cdots z_{s-1}}{z_{s-1}}, & d_s^{-1}\dfrac{z_1\cdots z_{s-1}}{z_{s-1}}
\end{pmatrix}.$$

First we shall prove the following lemma on determinant.

Lemma. *Let $a_i > 0$, $1 \leqslant i \leqslant s$, and*

$$\Delta_s = \begin{vmatrix}
a_1+a_2, & -a_2, & & & 0 \\
-a_2, & a_2+a_3, & \cdots & & \\
 & \ddots & \ddots & & \\
 & & \cdots & a_{s-2}+a_{s-1} & -a_{s-1} \\
0 & & & -a_{s-1} & a_{s-1}+a_s
\end{vmatrix}.$$

Then

$$\Delta_s = a_1\cdots a_s \sum_{i=1}^{s} a_i^{-1}.$$

Proof. The conclusion is obvious if $s=2$. Now by the mathematical induction we assume that $s \geqslant 3$ and that the conclusion is true for Δ_t, where $2 \leqslant t \leqslant s-1$. Then

$$\Delta_s = (a_1+a_2) \begin{vmatrix}
a_2+a_3, & -a_3, & & & 0 \\
-a_3 & a_3+a_4, & \cdots & & \\
 & \ddots & \ddots & & \\
 & & \cdots & a_{s-2}+a_{s-1} & -a_{s-1} \\
0 & & & -a_{s-1} & a_{s-1}+a_s
\end{vmatrix}$$

$$-a_2^2 \begin{vmatrix} a_3+a_4, & -a_4, & & & 0 \\ -a_4, & a_4+a_5, & \cdots & \cdots & \\ & & \cdots & & \\ & & \cdots & a_{s-2}+a_{s-1} & -a_{s-1} \\ 0 & & & -a_{s-1} & a_{s-1}+a_s \end{vmatrix}$$

$$=(a_1+a_2)a_2\cdots a_s\sum_{i=2}^{s} a_i^{-1} - a_2^2 a_3\cdots a_s\sum_{i=3}^{s} a_i^{-1}$$

$$= a_1\cdots a_s\sum_{i=2}^{s} a_i^{-1} + a_2^2 a_3\cdots a_s\sum_{i=2}^{s} a_i^{-1} - a_2^2 a_3\cdots a_s\sum_{i=3}^{s} a_i^{-1} = a_1\cdots a_s\sum_{i=1}^{s} a_i^{-1}.$$

The lemma is proved.

The evaluation of Δ: let A be an $(s-1)\times(s-1)$ matrix with $\det A=\pm 1$. Then Δ is invariant if H is changed by AH. Let $A_{ij}=(a_{uv})$, $1\leqslant u$, $v\leqslant s-1$, where $a_{uu}=1$, $1\leqslant u\leqslant s-1$, $a_{ij}=g$ and $a_{uv}=0$ for the remaining u, v. Then the difference between H and AH lies only on the fact that the i-th row of $A_{ij}H$ is equal to the i-th row of H plusing g multiple of the j-th row the H. Using these operations, we can find out a matrix A such that

$$AH=\begin{pmatrix} -d_1^{-1} & d_2^{-1} & & & \\ & -d_2^{-1}z_1 & d_3^{-1}z_1 & & 0 \\ & & \ddots & \ddots & \\ 0 & & & -d_{s-1}^{-1}z_1\cdots z_{s-2} & d_s^{-1}z_1\cdots z_{s-2} \end{pmatrix}.$$

Hence

$$\Delta^2=\det(HH')=\det(AHH'A')=(z_1^{s-2}\cdots z_{s-2})^2\det G,$$

where

$$G=\begin{pmatrix} d_1^{-2}+d_2^{-2} & -d_2^{-2} & & & 0 \\ -d_2^{-2} & d_2^{-2}+d_3^{-2} & \cdots & & \\ & \cdots & \cdots & \cdots & -d_{s-1}^{-2} \\ 0 & & \cdots & -d_{s-1}^{-2} & d_{s-1}^{-2}+d_s^{-2} \end{pmatrix}.$$

By the lemma it follows that

$$\det G=(d_1\cdots d_s)^{-2}\sum_{i=1}^{s} d_i^2=c(\boldsymbol{d})^2,$$

and therefore

$$\Delta=c(\boldsymbol{d})z_1^{s-2}\cdots z_{s-2}.$$

5   Examples

Example 1.   Suppose $\boldsymbol{a}=(0.6,\ 0.15,\ 0.05)$ and $\boldsymbol{b}=(0.8,\ 0.25,\ 0.15)$. Then $a=0.6+0.15$

$+0.05=0.8$, $b=1.2$ and $d=(1,\ 0.5,\ 0.5)$. By (10), (11) and (19), we have

$$N(d)=1-\frac{1}{2}-\frac{0.5}{2}=0.25$$

and

$$\frac{V(S^*(d))}{V(S(d))}=\frac{0.5}{0.25}=2.$$

Hence we may get an NT-net of nearly $[n/2]$ points on $S(d)$ (or $S(a,\ b)$) which is induced by an NT-net of n points on C^2. For example, if an NT-net of 17 points on C^2 is generated by the good lattice method, we can obtain 9 of them in $S(a,\ b)$ as follows:

$$x_1=(0.765\ 7,\ 0.175\ 2,\ 0.059\ 1),\quad x_2=(0.740\ 6,\ 0.176\ 2,\ 0.083\ 2),$$

$$x_3=(0.723\ 3,\ 0.161\ 3,\ 0.115\ 4),\quad x_4=(0.709\ 3,\ 0.227\ 4,\ 0.063\ 3),$$

$$x_5=(0.697\ 1,\ 0.207\ 5,\ 0.095\ 4),\quad x_6=(0.686\ 2,\ 0.180\ 1,\ 0.133\ 6),$$

$$x_7=(0.667\ 2,\ 0.239\ 9,\ 0.093\ 0),\quad x_8=(0.658\ 6,\ 0.204\ 1,\ 0.137\ 3),$$

$$x_9=(0.635\ 5,\ 0.232\ 2,\ 0.132\ 2).$$

Example 2. Suppose $a=(0.3,\ 0.2,\ 0.1,\ 0.05)$ and $b=(0.6,\ 0.5,\ 0.4,\ 0.2)$. Then $a=0.65$ and $d=\left(\dfrac{6}{7},\ \dfrac{6}{7},\ \dfrac{6}{7},\ \dfrac{3}{7}\right)$. By (14) we have

$$N(d)=\int_{\frac{1}{7}}^{1}z_1^2 N\left(\frac{6}{7z_1},\ \frac{6}{7z_1},\ \frac{3}{7z_1}\right)dz_1=\int_{\frac{1}{7}}^{\frac{3}{7}}z_1^2 N\left(\frac{6}{7z_1},\ \frac{6}{7z_1},\ \frac{3}{7z_1}\right)dz_1$$

$$+\int_{\frac{3}{7}}^{\frac{6}{7}}z_1^2 N\left(\frac{6}{7z_1},\ \frac{6}{7z_1},\ \frac{3}{7z_1}\right)dz_1+\int_{\frac{6}{7}}^{1}z_1^2 N\left(\frac{6}{7z_1},\ \frac{6}{7z_1},\ \frac{3}{7z_1}\right)dz_1=I_1+I_2+I_3,\ \text{say.}$$

For I_1, we have $\dfrac{3}{7z_1}\geqslant 1$, and so $N\left(\dfrac{6}{7z_1},\ \dfrac{6}{7z_1},\ \dfrac{3}{7z_1}\right)=\dfrac{1}{2}$ by (13). Therefore

$$I_1=\frac{1}{2}\int_{\frac{1}{7}}^{\frac{3}{7}}z_1^2 dz_1=\frac{26}{6\times 7^3}.$$

For I_2, we have $\dfrac{6}{7z_1}\geqslant 1$ and $\dfrac{3}{7z_1}\leqslant 1$, and so by (18),

$$I_2=\int_{3/7}^{6/7}z_1^2\left(\frac{3}{7z_1}-\frac{1}{2}\left(\frac{3}{7z_1}\right)^2\right)dz_1=\frac{27}{7^3}.$$

For I_3, we have $\dfrac{6}{7z_1}\leqslant 1$, $1-\dfrac{6}{7z_1}<\dfrac{3}{7z_1}$, $\dfrac{6}{7z_1}+\dfrac{3}{7z_1}\geqslant 1$. Therefore we deduce by (19),

$$I_3 = \int_{\frac{6}{7}}^{1} \left(-1 + \frac{15}{7z_1} - \frac{81}{2 \times 7^2 z_1^2} \right) z_1^2 dz_1 = \frac{88}{6 \times 7^3}.$$

Hence

$$N(d) = I_1 + I_2 + I_3 = \frac{276}{6 \times 7^3} = 0.134\,110\,787\cdots,$$

and by (10) and (11),

$$\frac{V(S^*(d))}{V(S(d))} = \frac{6^{-1}}{276/6 \times 7^3} = \frac{343}{276} = 1.242\,753\,662\,3\cdots.$$

This means that we may obtain an NT-net of nearly n points on $S(d)$ which is induced by an NT-net of $\left[\dfrac{343}{276} n \right]$ points on C^3.

References

1 Wang, Y., Fang, K. T., A note on uniform distribution and experimental design, *Kexue Tongbao*, 1981, 26: 485.

2 Weyl, H., Uber die Gleichverteilung der Zahlen mod Eins, *Math. Ann.*, 1916, 77: 313.

3 Hua, L. K., Wang, Y., *Application of Number Theory to Numerical Analysis*, Heidelberg and Beijing: Springer-Verlag and Science Press, 1981.

4 Fang, K. T., Wang, Y., *Number-theoretic Methods in Statistics*, London: Chapman and Hall, 1993.

5 Cornell, J. A., *Experiments with Mixtures, Design, Models, and the Analysis of Mixture Data*, New York: Wiley, 1990.

6 Wang, Y., Fang, K. T., Number-theoretic methods in applied statistics, *Chinese Ann. Math.* (Series B), 1990, 11: 51.

7 Wang, Y., Fang, K. T., Number-theoretic methods in applied statistics (II), *Chinese Ann. Math. Ser. B*, 1990, 11: 384.

LIST OF PUBLICATIONS BY WANG YUAN

I. Articles

1. On the representation of large even integer as a sum of a product of at most 3 primes and a product of at most 4 primes, *Acta Math. Sin.* **6**:3 (1956) 500–513, in Chinese.

2. On the representation of large even integer as a sum of a prime and a product of at most 4 primes (Conditional result), *Acta Math. Sin.* **6**:4 (1956) 565–582, in Chinese.

3. (with *A. Schinzel*) A note on some properties of the functions $\varphi(n), \sigma(n)$ and $\theta(n)$, *Bull. de L'Acad. Pol. Des Sci.* **4** (1956) 207–209.

4. On some properties of integral valued polynomials, *Shu Xue Jing Zhan* **3** (1957) 416–423, in Chinese.

5. On sieve methods and some of the related problems, *Sci. Rec.* (*New Series*) **1**:1 (1957) 9–12.

6. On sieve methods and some of their applications, *Sci. Rec.* (*New Series*) **1**:3 (1957) 1–5.

7. On the representation of large even number as a sum of two almost primes, *Sci. Rec.* (*New Series*) **1**:5 (1957) 15–19.

8. (with *A. Schinzel*) A note on some properties of the functions $\varphi(n), \sigma(n)$ and $\theta(n)$, *Ann. Pol. Math.* **4** (1958) 201–213; Corrigendum: **19** (1967).

9. A note on some properties of the arithmetic functions $\varphi(n), \sigma(n)$ and $d(n)$, *Acta Math. Sin.* **8**:1 (1958) 1–11, in Chinese.

10. (with *Ren Jian Hua*) A note on some properties of the Euler function $\varphi(n)$, *Proc. Northwestern Univ.* (1958) 15–23, in Chinese.

11. On sieve methods and some of their applications, *Sci. Sin.* **8**:4 (1959) 357–381; see also: *Acta Math. Sin.* **3** (1958) 413–429.

12. On sieve methods and some of their applications (II), *Sci. Sin.* **11**:12 (1962) 1607–1624; see also: *Acta Math. Sin.* **2** (1959) 87–100.

13. A note on the least primitive root of a prime, *Sci. Rec.* (*New Series*) **3**:5 (1959) 174–179.

14. On the least primitive root of a prime, *Sci. Sin.* **10**:1 (1961) 1–14; see also: *Acta Math. Sin.* **4** (1959) 432–441.

15. (with *Hua Loo Keng*) Remarks concerning numerical integration, *Sci. Rec.* (*New Series*) **4**:1 (1960) 8–11.

16. On the representation of large integer as a sum of a prime and an almost prime, *Sci. Sin.* **11**:8 (1962) 1033–1054; see also: *Acta Math. Sin.* **2** (1960) 168–181.

17. A note on interpolation of a certain class of functions, *Sci. Sin.* **10**:6 (1960) 632–636.

18. (with *Hua Loo Keng*) On the calculation of mineral reserves and hillside areas on contour maps, *Acta Math. Sin.* **11**:1 (1961) 29–40.

19. On numerical integration and its applications (Number-theoretic methods), *Shu Xue Jiang Zhan* **1** (1962) 1–44, in Chinese.

20. (with *Hua Loo Keng*) Finiteness and infinity, discrete and continuity, *Kexue Tongbao* (1963) 4–21, in Chinese.

21. On the estimation of character sum and its applications, *Shu Xue Jiang Zhan* **1** (1964) 78–83, in Chinese.

22. A note on the maximal number of pairwise orthogonal Latin squares of a given order, *Sci. Sin.* **13**:5 (1964) 841–843.

23. (with *Hua Loo Keng*) On diophantine approximations and numerical integrations I, *Sci. Sin.* **13**:6 (1964) 1007–1008.

24. (with *Hua Loo Keng*) On diophantine approximations and numerical integrations II, *Sci. Sin.* **13**:6 (1964) 1009–1010.

25. Remarks on the interpolations of a certain class of functions, *Sci. Sin.* **14**:4 (1965) 429–431.

26. (with *Shieh Shen Kang and Yu Kun Rui*) Remarks on the difference of consecutive primes, *Sci. Sin.* **14**:5 (1965) 786–788.

27. (with *Shieh Shen Kang and Yu Kun Rui*) Two results concerning the distribution of primes, *Proc. Univ. Sci. Tech. China* **1** (1965) 32–38, in Chinese.

28. (with *Zhu Yao Cheng and Jian Yun Cui*) Remarks on the number theoretic methods in numerical analysis, *Proc. Univ. Sci. Tech. China* **2** (1965) 213–218, in Chinese.

29. (with *Hua Loo Keng*) On numerical integration of periodic functions of several variables, *Sci. Sin.* **14**:7 (1965) 964–978; see also: *Proc. Univ. Sci. Tech. China* (1966) 1–12.

30. On interpolation of a certain class of a functions, *Kexue Tongbao* **9** (1966) 387–389, in Chinese.

31. On the maximal number of pairwise orthogonal Latin squares of order s (application of sieve methods), *Acta Math. Sin.* **16**:3 (1966) 400–410, in Chinese.

32. (with *Hua Loo Keng*) On uniform distribution and numerical analysis (I) (number theoretic methods), *Kexue Tongbao* **3** (1973) 112–114, in Chinese.

33. (with *Hua Loo Keng*) On uniform distribution and numerical analysis (II) (number theoretic methods), *Kexue Tongbao* **4** (1973) 165–166, in Chinese.

34. (with *Hua Loo Keng*) On uniform distribution and numerical analysis (I) (number theoretic methods), *Sci. Sin.* **16**:4 (1973) 483–505.

35. (with *Hua Loo Keng*) On uniform distribution and numerical analysis (II) (number theoretic methods), *Sci. Sin.* **17**:3 (1974) 331–348.

36. (with *Hua Loo Keng*) On uniform distribution and its applications for multi-dimensional functions of bounded variation, *Proc. Univ. Sci. Tech. China* **1** (1974) 39–67, in Chinese.

37. (with *Hua Loo Keng*) On uniform distribution and numerical analysis (III)— Number theoretic methods, *Kexue Tongbao* **12** (1974) 559–560, in Chinese.

38. (with *Hua Loo Keng*) On uniform distribution and numerical analysis (III) (number theoretic methods), *Sci. Sin.* **18**:2 (1975) 184–198.

39. (with *Pan Cheng Dong and Ding Xia Xi*) On representation of large even integer as a sum of a prime and an almost prime, *Kexue Tongbao* **8** (1975) 358–360, in Chinese.

40. (with *Pan Cheng Dong and Ding Xia Xi*) On the representation of every large even integer as a sum of a prime and an almost prime, *Sci. Sin.* **18**:5 (1975) 599–610; see also: *Proc. Shan Dong Univ.* **2** (1975) 15–26.

41. Remarks on a theorem of Davenport, *Acta Math. Sin.* **18**:4 (1975) 286–289, in Chinese.

42. (with *Hua Loo Keng*) A note on simultaneous diophantine approximations to algebraic integers, *Sci. Sin.* **20**:5 (1977) 563–567.

43. On Linnik's method concerning the Goldbach number, *Sci. Sin.* **20**:1 (1977) 16–30.

44. (with *Xu Guang Shan and Zhang Rong Xiao*) On number theoretic method in numerical integration of multi-dimensional space I, *Acta Appl. Math. Sin.* **1**:2 (1978) 106–114, in Chinese.

45. (with *Hua Loo Keng and Pei Ding Yi*) On a set of independent units of cyclofomic field, *Ziran Zazhi* **5** (1978) 6, in Chinese.

46. (with *Yu Kun Rui and Zhu Yao Cheng*) Remarks concerning a transference theorem of linear forms, *Acta Math. Sin. Act.* **22**:2 (1979) 237–240, in Chinese.

47. (with *W. M. Schmidt*) A note on a transference theorem of linear forms, *Sci. Sin.* **22**:3 (1979) 276–280.

48. (with *Wang Lian Xiang and Ren Jian Hua*) A note on a transference theorem of the systems of linear congruences, *Acta Math. Sin.* **24**:2 (1981) 303–307; see also: *Proc. Northwestern Univ.* **2** (1979) 12–23.

49. (with *Yu Kun Rui*) A note on some metrical theorems in diophantine approximations, *IHES/M* **297** (1979); see also: *Chin. Ann. Math.* **2** (1981) 1–12.

50. (with *Hua Loo Keng*) Applications of number theory to numerical analysis, in *Recent Progress in Analytic Number Theory*, eds. H. Halberstam and C. Hooley, Vol. 2 (Acad. Press, 1981), pp. 111–118.

51. (with *Fang Kai Tai*) A note on uniform distribution and experimental design, *Kexue Tongbao* **6** (1981) 485–489.

52. On diopantine approximation and approximate analysis, *Kexue Tongbao* **5** (1982) 468–472.

53. (with *Xu Guang Shan and Zhang Rong Xiao*) On number-theoretic method in numerical integration of multi-dimensional space II, *Acta Appl. Math. Sin.* **4** (1982) 414–417.

54. On diophantine approximation and approximate analysis I, *Acta Math. Sin.* **25**:2 (1982) 248–256.

55. On diophantine approximation and approximate analysis II, *Acta Math. Sin.* **25**:3 (1982) 323–332.

56. A note on the approximate solution of the Cauchy problem by number theoretic nets, *Chin. Ann. Math.* **3** (1982) 451–456.

57. On additive equations in an algebraic number field, *Kexue Tongbao* **5** (1985) 583–587.

58. (with *Shan Zun*) A conditional result on Goldbach problem, *Acta Math. Sin. (New Series)* **1** (1985) 72–78.

59. On a system of diophantine inequalities, *Kexue Tongbao* **17** (1987) 1162–1165.

60. Bounds for solutions of additive equations in an algebraic number field I, *Acta Ari.* **48** (1987) 117–144.

61. Bounds for solutions of additive equations in an algebraic number field II, *Acta Ari.* **48** (1987) 307–323.

62. Number theoretic method in numerical analysis, *Cont. Math. AMS* **77** (1988) 63–82.

63. Diophantine equations and dipohantine inequalities in algebraic number fields, *Cont. Math. AMS* **77** (1988) 83–94.

64. Diophantine inequalities for forms in an algebraic number field, *J. Number Theory* **29**:3 (1988) 324–344.

65. On homogeneous additive congruences, *Sci. Sin. (A)* **32**:5 (1989) 524–536.

66. On small zeros of quadratic forms over finite fields, *J. Number Theory* **31**:3 (1989) 272–284.

67. (with *Fang Kai Tai*) Number theoretic methods in applied statistics, *Chin. Ann. Math.* **11B** (1990) 51–65.

68. (with *Fang Kai Tai*) Number theoretic methods in applied statistics (II), *Chin. Ann. Math.* **11B** (1990) 484–494.

69. (with *Fang Kai Tai*) A sequential algorithm for optimization and its applications to regression analysis, in *Lecture Notes in Contemporary Mathematics A*, eds. Wang Yuan *et al.* (1990), pp. 17–28.

70. (with *Fang Kai Tai*) Applications of quasirandom sequence in statistics, *Proc. Asian Math. Conf.* (1990) 135–139.

71. Hua Loo Keng: A brief outline of his life and works, in *Number Theory in Honor of Hua Loo Keng*, eds. Gong Sheng *et al.* (Springer-Verlag and Science Press, 1991), pp. 1–14.

72. (with *M. V. Subbarao*) On a generalized Waring's problem in algebraic number fields, in *Number Theory in Honor of Hua Loo Keng*, eds. Gong Sheng *et al.* (Springer-Verlag and Science Press, 1991), pp. 265–277.

73. (with *Fang Kai Tai*) A sequential number theoretic methods for optimization and its applications in statistics, in *The Development of Statistics: Recent Contribution from China* (Longman, 1991), pp. 139–156.

74. (with *Fang Kai Tai*) A sequential algorithm for solving a system of nonlinear equations, *J. Comp. Math.* **9**:1 (1991) 9–16.

75. (with *Fang Kai Tai and H. L. Wong*) A new method for generating the uniform distribution on the unit sphere, *Tech. Rep. Hong Kong Baptist College* **15** (1992).

76. Small solutions of congruences, *J. Number Theory* **45**:3 (1993) 261–280.

77. On small zeros of quadratic forms over finite fields II, *Acta Math. Sin. (New Series)* **4** (1993) 382–389.

78. (with *Fang Kai Tai and P. M. Bentler*) Some applications of number theoretic methods in statistics, *Stat. Sci.* **9**:3 (1994), 416–428.

79. (with *Fang Kai Tai*) Uniform design of experiments with mixtures, *Sci. Sin. (A)* **39**:3 (1996) 264–275.

80. (with *Pan Cheng Dong*) Chen Jingrun: A brief outline of his life and works, *Acta Math. Sin. (New Series)* **12**:3 (1996) 225–233.

81. Pan Cheng Dong: A brief outline of his life and works, *Acta Math. Sin.* **3** (1998) 449–454.

II. Books and Monographs

1. (with *Hua Loo Keng*) *The Numerical Evaluation of Integrals* (Science Press, 1961), in Chinese.

2. (with *Hua Loo Keng*) *Numerical Integration and Its Applications* (Science Press, 1963), in Chinese.

3. *Prime Numbers* (Shanghai Education Press, 1978); (GuangDong Sci. Tech. Press, 1996), in Chinese.

4. (with *Hua Loo Keng*) *Applications of Number Theory to Numerical Analysis* (Springer-Verlag and Science Press, 1981); (Science Press, 1978), in Chinese.

5. *Goldbach Conjecture* (World Scientific, 1984), 2nd edition (2002); (Hei Long Jiang Education Press, 1987), in Chinese.

6. (eds. *Pan Cheng Biao and C. C. Yang*) *Number Theory and Its Applications in China*, Cont. Math. AMS, **77** (1988).

7. (with *Hua Loo Keng*) *Popularizing Mathematical Methods in the People's Republic of China* (Birkhäuser 1989); *Some Topics on Mathematical Modeling* (Hu Nan Normal Publisher, 1991), in Chinese.

8. (eds. *Zhang Gong Qing et al.*) *Lecture Notes in Contemporary Mathematics* (Science Press, 1990).

9. *Diophantine Equations and Inequalities in Algebraic Number Fields* (Springer-Verlag, 1991).

10. (eds. *Gong Sheng, Lu Qi Keng and Yang Lo*) *International Symposium in Memory of Hua Loo Keng*, Vol. I. *Number Theory*, Vol. II, *Analysis* (Springer-Verlag and Science Press, 1991).

11. (with *Fang Kai Tai*) *Number Theoretic Method in Statistics* (Chapman and Hall, 1994); (Science Press, 1996), in Chinese.

12. Hua Loo Keng (Kai Ming Press, 1995); (Jiu Zhang Press, 1995); (Jiang Xi Normal Publisher, 1999), in Chinese (Springer, 1999).

13. (with *Fong Yuen*) *Calculus* (Springer-Verlag, 1997).

14. *Wang Yuan, Selected Papers* (Hunan Normal Publisher, 1999), in Chinese.

15. Wang Yuan's Exposition on Goldlach Conjecture (Shandong Normal Publisher, 1999), in Chinese.

16. (with *Yang De Zhuang*) *Mathematical Career of Hua Loo* Keng (Science Press, 2000), in Chinese.